Concrete Recycling

Concrete Recycling
Research and Practice

Edited by
François de Larrard and
Horacio Colina

CRC Press
Taylor & Francis Group
Boca Raton London New York

CRC Press is an imprint of the
Taylor & Francis Group, an **informa** business

CRC Press
Taylor & Francis Group
6000 Broken Sound Parkway NW, Suite 300
Boca Raton, FL 33487-2742

Printed on acid-free paper

International Standard Book Number-13: 978-1-138-72472-3 (Hardback)

Visit the Taylor & Francis Web site at
http://www.taylorandfrancis.com

and the CRC Press Web site at
http://www.crcpress.com

MIX
Paper from responsible sources
FSC® C013056
FSC www.fsc.org

Printed and bound in Great Britain by
TJ International Ltd, Padstow, Cornwall

Contents

SECTION III
Recycled concrete from production to hardening stages
E. GARCIA-DIAZ

SECTION IV
Properties of hardened recycled concrete
P. ROUGEAU

SECTION V
Mix design of recycled concrete
T. SEDRAN

Committee

This book was written by participants to the French national project RECYBÉTON, managed by the IREX (Institut pour la Recherche et l'Expérimentation en Génie Civil) association from 2012 to 2018. The coordination was performed by the following list of fellows:

- Dr. François de Larrard, LafargeHolcim (editor)
- Dr. Horacio Colina, ATILH (coeditor)
- Mrs. Sophie Decreuse, CEMEX (coordinator of Section I)
- Dr. Laurent Izoret, ATILH (coordinator of Section II)
- Pr. Eric Garcia-Diaz, IMT Mines Alès (coordinator of Section III)
- Dr. Patrick Rougeau, CERIB (coordinator of Section IV)
- Dr. Thierry Sedran, IFSTTAR (coordinator of Section V, in charge of the manuscript preparation)
- Mr. Bernard Fouré, consulting engineer, retired (coordinator of Section VI)
- Mr. Patrick Dantec, consulting engineer (coordinator of Section VII)
- Dr. Adélaïde Feraille, Ecole des Ponts ParisTech (coordinator of Section VIII)
- Dr. Wilfried Pillard, EGF-BTP (coordinator of Section IX)
- Pr. Elhem Ghorbel, University of Cergy-Pontoise (representative from Ecoreb ANR project)

Foreword

This book, *Concrete Recycling—Research and Practice*, summarizes one of the most transversal and systematic teamworks that has been done so far around the world concerning the use of wastes in concrete production. I am referring to the French national project RECYBÉTON, which ran from 2012 to 2018, and benefited from the efforts of a large team of very competent researchers of the various scientific areas fundamental to understanding the behavior of concrete with incorporation of recycled materials and boosting its use by the construction industry.

The authors of this book made an effort to summarize the results of the project, published in around 50 reports and many papers in international conferences and scientific journals, comprising thousands of pages. The subjects are treated in a self-contained way, meaning that each chapter and subchapter of this book can be read and understood independently from the rest of this book but, at the same time, gives references to the other chapters for further reading and holistic understanding.

This book is written in a very pedagogical style, where the sequence of presentation of the various subjects helps the readers in understanding the barriers that need to be overcome in order to implement this new and more environmentally solution of making concrete, by far the most important structural material in the 20th and 21st centuries.

This book starts with the definition of the sources, processing and characterization techniques of the wastes, and recycled aggregates resulting from them.

Then, the various uses of the recovered materials are identified: as raw materials for cement production, as recycled aggregates, and as supplementary cementitious materials, and the consequences of their use in concrete mix design are analyzed.

This alternative material poses new problems in terms of batching, fresh-state behavior, and early-age performance, and these are analyzed next.

The performance of recycled concrete in the hardened state can only be understood by analyzing its microstructure. This allows explaining the mechanical and durability-related properties and also how this concrete behaves when subjected to elevated temperatures.

The mix design and specifications on concrete are then discussed, not only to identify those that are already adapted to the use of recycled concrete but also to present proposals on how to do that, considering the specificities of this new material and its components.

Reinforced concrete's performance depends on the interaction between a semi-homogeneous composite material, concrete, and discreet elements of steel. Naturally, the different behaviors of recycled concrete relative to conventional concrete affects the performance of reinforced concrete, and several issues concerning that are analyzed: reinforcement bond, compression, and flexural and shear behavior.

Then, the various possible applications of recycled concrete are analyzed through real application case studies, in order to call attention upon the specific new problems that this use raises in each situation. Quality control on site is naturally a major issue.

It is frequent to see claims that certain materials are environmentally friendly, but it is less common that these claims are actually demonstrated in a scientific way. This book properly addresses this issue by discussing aspects such as availability and scatter of properties of the raw materials, life cycle analysis, leaching, multi-recycling, and industrial upscaling.

This book ends by discussing the issue of upscaling the use of recycled concrete, by making sure that barriers are removed from the point of view of the various actors in the concrete industry, from the authorities to the end users.

In my opinion, this book will be a definite reference to all participants in the process of designing, manufacturing, applying, and benefiting from the wonderful material that concrete is. Quite often criticized as an unbearable load on the environment and a stagnated solution, concrete has proven over and over that not only it is in constant mutation but also it is by far the best alternative as structural material, bearing in mind the balance between costs, technical performance, and environmental impacts.

I wish to congratulate the editors of this book, and also all the participants in the various chapters and in the RECYBÉTON program, for their contribution to the enhancement of the knowledge on recycled aggregate and especially for their very practical common-sense view on how to go about research.

Jorge de Brito
Full Professor, IST/ULisbon, Portugal
Coauthor of
Recycled Aggregate in Concrete: Use of Industrial, Construction and Demolition Waste, Springer, Berlin, 2013.
Coeditor of
Handbook of Recycled Concrete and Demolition Waste, Woodhead Publishing Limited, 2013.
Lisbon, July 2018

Acknowledgments

The editors gratefully thank

- The Ministère de la Transition Ecologique et Solidaire/Direction de la Recherche et de l'Innovation for its financial support to the RECYBÉTON national project.
- The Agence Nationale de la Recherche (ANR) for the funding of the ECOREB project, in the frame of which a number of RECYBÉTON-related works were carried out.
- The "Institut pour la Recherche Appliquée et l'Expérimentation en Génie Civil (IREX)" for its administrative and financial management of RECYBÉTON (with a special mention to Brice Delaporte).
- All the participants of RECYBÉTON, with a particular thought for the owners of our six experimental construction sites, who dared to build a part of their projects with recycled concrete, for the advancement of construction sustainability practices.
- The members of the editorial committee and all the coauthors of this book. Thanks to their high competencies, good will, and hard work, we could achieve an important collective effort to summarize what we have learnt during a 6-year project about recycling concrete into concrete.
- Our external reviewers (François Toutlemonde, from ISTTAR, Fekri Meftah, from INSA de Rennes and Jacques Roudier, from IREX), who helped us in improving the quality of some of our chapters.

Editors

François de Larrard is scientific director of the R&D center of the LafargeHolcim group and the French national project RECYBÉTON. He formerly spent almost 30 years at IFSTTAR (formerly LCPC). He has been granted both the Robert l'Hermite medal and the G.H. Tattersall award by RILEM (Réunion Internationale des Laboratoires d'Etudes et de recherche sur les Matériaux), and is author of two books, including *Concrete Mixture Proportioning* (Taylor & Francis 1999).

Horacio Colina is the Research's managing director of ATILH (Association Technique de l'Industrie des Liants Hydrauliques), Paris, France, and director of the French national project RECYBÉTON. He is at ATILH from 2005 and formerly he was a researcher at ENPC (Ecole Nationale des Ponts et Chaussées), Paris, France, from 1999 to 2005, and at LAEGO (Laboratoire Environnement, Géomécanique et Structures), Nancy, France, from 1997 to 1999. He has also been Professor at National University of Salta, Salta, Argentina, from 1994 to 1997. He is author or co-author of 41 papers on peer-reviewed publications.

Editors

Contributors

F. Al-Mahmoud
Institut Jean Lamour
Université de Lorraine
Villers-lès-Nancy, Meurthe-et-Moselle, France

O. Amiri
GeM
Université de Nantes
Nantes, Loire-Atlantique, France

K. Apedo
Laboratoire ICube
Université de Strasbourg/INSA
Strasbourg, Bas-Rhin, France

A. L. Beaucour
L2MGC
Université de Cergy Pontoise
Pontoise, Val d'Oise, France

A. Ben Fraj
Laboratoire Eco-Matériaux Cerema
Sourdun, Seine-et-Marne, France

F. Benboudjema
LMT
ENS Cachan
Cachan, Val-de-Marne, France

A. Z. Bendimerad
GeM
Nantes, Loire-Atlantique, France

D. Blanc
INSA de Lyon - Laboratoire DEEP
Villeurbanne, Rhône, France

R. Bodet
UNPG
Paris, Seine, France

R. Boissière
Institut Jean Lamour
Université de Lorraine
Villers-lès-Nancy, Meurthe-et-Moselle, France

S. Braymand
ICube Laboratory
Mechanical department
University of Strasbourg
Strasbourg, Bas-Rhin, France

D. Bulteel
IMT Lille Douai
University of Lille
Lille, Nord, France

F. Buyle-Bodin
University of Lille
Lille, Nord, France

F. Cassagnabère
LMDC
Université de Toulouse
Toulouse, Haute-Garonne, France

B. Cazacliu
IFSTTAR
Bouguenais, Loire-Atlantique, France

X. Chomiki
Conseil départemental de Seine & Marne
Melun, Seine-et-Marne, France

H. Colina
ATILH
Paris-La Défense, Hauts-de-Seine, France

D. Collonge
LafargeHolcim
Paris, Seine, France

N. Comte
Syndicat des Recycleurs du BTP (SRBTP)
Paris, Seine, France

A. Cudeville
CLAMENS
Villeparisis, Seine-et-Marne, France

M. Cyr
Université de Toulouse
Toulouse, Haute-Garonne, France

P. Dantec
Consultant
Cournon d'Auvergne, Puy-de-Dôme, France

T. Dao
LafargeHolcim R&D
St Quentin-Fallavier, Isère, France

F. de Larrard
LCR (LafargeHolcim Research Centre)
Saint-Quentin Fallavier, Isère, France

C. De Sa
LMT
ENS Cachan
Cachan, Val-de-Marne, France

R. Deborre
Nacarat (groupe Rabot Dutilleul)
Euralille
Lille, France

S. Decreuse
Cemex
Rungis, Val-de-Marne, France

P. Devillers
Ecole Nationale Supérieure d'Architecture de Montpellier
Montpellier, Hérault, France

C. Diliberto
Institut Jean Lamour
Université de Lorraine
Villers-lès-Nancy, Meurthe-et-Moselle, France

A. Djerbi
IFSTTAR
Marne-la-Vallée, Seine-et-Marne, France

S. Favre
Léon Grosse
Paris, Seine, France

A. Feraille
Ecole des Ponts ParisTech
Marne-la-Vallée, Seine-et-Marne, France

S. Fonteny
UNPG
Paris, Seine, France

B. Fouré
Consulting engineer, retired Bougival, Yvelines, France

P. Francisco
CERIB
Epernon, Eure-et-Loir, France

E. Garcia
Lafarge Bétons France (groupe LafargeHolcim)
Clamart, Hauts-de-Seine, France

E. Garcia-Diaz
Centre des Matériaux des Mines d'Alès, IMT Mines Ales, France

E. Ghorbel
Université de Cergy Pontoise
Cergy-Pontoise, Val d'Oise, France

L. Gonzalez
INSAVALOR - Plateforme PROVADEMSE
Villeurbanne, France

F. Grondin
GeM
Ecole Centrale de Nantes
Nantes, Loire-Atlantique, France

M. Guo
GeM
Ecole Centrale de Nantes
Nantes, Loire-Atlantique, France

A. Hamaidia
Laboratoire des essais non-Destructifs par Ultrasons
Université de Jijel, Algeria

E. Hamard
IFSTTAR
Bouguenais, Loire-Atlantique, France

R. Idir
Laboratoire Eco-Matériaux Cerema, Ile de France Department
Sourdun, Seine-et-Marne, France

L. Izoret
ATILH
Paris-La Défense, Hauts-de-Seine, France

F. Jezequel
SIGMA BETON
L'Isle d'Abeau, Isère, France

E.-H. Kadri
L2MGC
Université de Cergy Pontoise
Cergy-Pontoise, Val d'Oise, France

R. Lavaud
Technodes S.A.S. (HEIDELBERGCEMENT Group)
Guerville, Yvelines, France

L. Le Guen
IFSTTAR
Bouguenais, Loire-Atlantique, France

G. Le Saout
Centre des Matériaux des Mines Alès, IMT Mines Ales, Herault, France

A. Lecomte
Institut Jean Lamour
Université de Lorraine
Villers-lès-Nancy, Meurthe-et-Moselle, France

N. Leklou
GeM
Université de Nantes
Nantes, Loire-Atlantique, France

T. Lenormand
LERM
Arles, Bouches-du-Rhône, France

A. Loukili
GeM
Ecole Centrale de Nantes
Nantes, Loire-Atlantique, France

J. Mai-Nhu
CERIB
Epernon, Eure-et-Loir, France

J.M. Mechling
Institut Jean Lamour
Université de Lorraine
Villers-lès-Nancy, Meurthe-et-Moselle, France

H. Mercado-Mendoza
Laboratoire ICube
Université de Strasbourg/INSA
Strasbourg, Bas-Rhin, France

L. Mongeard
ENS
Lyon, Rhône, France

I. Moulin
Lerm Setec
Arles, Bouches-du-Rhône, France

M. Mouret
LMDC
Université de Toulouse
Toulouse, Haute-Garonne, France

P. Natin
Vicat Technical Centre
Isle d'Abeau, Isère, France

P. Nicot
LMDC
Université de Toulouse
Toulouse, Hautes-Garonne, France

G. Noworyta
CTG—Calcia
Paris, Seine, France

E. Perin
Lerm Setec
Arles, Bouches-du-Rhône, France

W. Pillard
EGF.BTP
Paris, Seine, France

J.-M. Potier
SNBPE
Paris, Seine, France

C. Raillon
Syndicat National des Entreprises de Démolition (SNED)
Paris, Seine, France

S. Rémond
LGCgE, IMT Lille Douai
Université de Lille
Lille, Nord, France

F. Robert
CERIB
Epernon, Eure-et-Loir, France

D. Rogat
Sigma Béton
L'Isle d'Abeau

P. Rougeau
CERIB
Epernon, Eure-et-Loir, France

S. Roux
IJL UMR 7198
Villers-lès-Nancy, Meurthe-et-Moselle, France

E. Rozière
GeM
Ecole Centrale de Nantes
Nantes, Loire-Atlantique, France

M. Saillio
IFSTTAR
Marne-la-Vallée, Seine-et-Marne, France

M. Salgues
Centre des Matériaux des Mines Alès, IMT Mines Ales, France

L. Schmitt
CERIB
Epernon, Eure-et-Loir, France

T. Sedran
Ifsttar
Bouguenais, Loire-Atlantique, France

N. Serres
ICube Laboratory
Mechanical department
INSA Strasbourg, France

O. Servan
Setec TPI
Paris, Seine, France

J.-C. Souche
Centre des Matériaux des Mines Alès, IMT Mines Ales, France

Z.A. Tahar
L2MGC
Université de Cergy Pontoise
Pontoise, Val d'Oise, France

R. Trauchessec
Institut Jean Lamour
Université de Lorraine
Villers-lès-Nancy, Meurthe-et-Moselle, France

M. Verbauwhede
Bouygues TP
Paris, Seine, France

E. Vernus
INSAVALOR - Plateforme PROVADEMSE
Villeurbanne, Rhône, France

P. Vuillemin
EQIOM
Paris, France

G. Wardeh
Université de Cergy Pontoise
Cergy-Pontoise, Val d'Oise, France

P. Wolff
Laboratoire ICube
Université de Strasbourg/INSA
Strasbourg, Bas-Rhin, France

Introduction

F. de Larrard
LafargeHolcim R&D

H. Colina
ATILH

Concrete is the most important construction material in terms of volume and turnover. Its success comes, among other reasons, from its amazing versatility and its availability in most parts of this planet, being essentially a local material. However, this 21st century sees humanity facing its most difficult challenge since its appearance on this planet: how to keep it livable for the next generations, with a population approaching 10 billions of inhabitants, and a general shortage of nonrenewable resources? Most informed people agree nowadays with the necessity of turning from linear to circular economy. Concrete cannot miss this major paradigm change.

Concrete recycling into concrete must be developed, for at least four main reasons:

i. Although the mineral stock which can be used to produce concrete is almost unlimited—limestone, clay, and hard rocks—opening new quarries at accessible distances from cities becomes difficult in many countries. As a matter of fact, in the usage competition of land (residential, agricultural, industrial), the third category often loses the battle.
ii. Landfilling construction and demolition materials (C&DM) is more and more banned by public policies, as in Europe since the beginning of the millennium (1999/31/CE and 2003/33/CE European directives).
iii. There is a social pressure to shorten material transportation distances from production to use sites, in order to limit CO_2 emissions and impact of truck traffic on the populations. Hence, demolition materials are mostly generated in cities, where new construction sites need to be fed. Using C&DM in new construction is a double win, tackling transport generated by both material export and import at the urban zone scale.
iv. Finally, concrete must catch up with regard to others materials (steel, asphalt, etc. where recycling processes and streams have been established for years. It is not the place to perform a comprehensive life cycle analysis of concrete as a construction material compared to other solutions. Let us just note that, although concrete solutions have many merits, their ability to incorporate their own waste is an important aspect according to most environmental assessment methods.

What are the most significant data to assess the potential of concrete recycling? At the European scale, 196 Mt of recycled concrete aggregates (RAs) were produced in 2015[1] (mostly from C&DM). About 40% of the total aggregate productions were devoted to either ready-mix or precast concrete, which constituted a market of nearly 1 Bt. Therefore, according to this simple calculations, there would be a potential to substitute 20% of the aggregate phase by RA in the whole concrete production. We will see later that this figure is an underestimation of the potential, if we account for all inert materials coming from

[1] www.uepg.eu.

demolition (including natural rocks or granular materials from road subbase courses). In addition, to reach 20% as a mean percentage of recycling rate in concrete means to be able to reach locally much higher values, depending on the resource geography. Although general statistics for Europe were not found regarding the proportion of recycled aggregate in the aggregate consumption of concrete industry, it was found that in UK, this amount was about 5% in 2008. This country being the most advanced in this respect (followed by the Netherlands and Belgium),[1] the figure for the whole Europe is probably not higher that some 1%–2%, to be compared with the 20% potential. So it can be concluded that the European concrete industry could recycle substantially, but is not doing so significantly at the moment.

A large proportion of current RA streams is consumed in road subbases and embankments. Is this situation satisfactory? Should it be considered as fully sustainable in the present and near future? We don't think so for the following reasons:

- The RA stream will probably increase in the next years, because a large stock of residential buildings built during the 1950s and 1960s is coming to its end of life (especially in Europe). The need to cut the global energy expense in buildings leads not only to vast retrofitting plans but also to demolition/reconstruction given the difficulty to provide thermal insulation to old buildings;
- In the same countries, there are less and less new roads to be built, but mainly maintenance works to perform. As a matter of fact, these works deal with the road top layers and only require a low amount of good-quality natural aggregates. Therefore, the stream is growing, and the well is shrinking;
- Even if RA are excellent materials to build subbase layers, less noble materials (mixed C&DM, low-grade natural aggregates, municipal solid waste incineration bottom ash, nontoxic industrial wastes, etc.) can be used for the same purpose. Therefore, reserving RA for new concrete is a way to make a better use of the whole waste materials resource by the civil and construction sectors.

So there is a need to reorient the RA stream towards concrete, whatever the production process. Why haven't we seen yet such a trend to appear spontaneously? The academic community has long been paying attention to concrete recycling, through, for example, RILEM (TC 37-DRC working group started his works as early as in 1976). Numerous conferences have been devoted to this topic since this date. However, the concrete practitioners have been reluctant to turn to recycling. Indeed, the construction sector is still one of the most conservative ones, with many small enterprises and a rather low level of qualification, impairing a quick dissemination of innovative practices. Moreover, used (or second hand) goods generally bring an image of low quality and poor reliability, while the life expectancy of any construction ranges from 50 to 100 years or more. Fortunately, from time to time the construction sector also displays the ability to accelerate progress through collective, multi-partner R&D projects. In France, such *Projets Nationaux* (National Projects, a process sponsored by the Ministry in charge of Public Works) have taken place since the mid-1980s. Within the French National Projects devoted to concrete, after having dealt with high-performance concrete (VNB, BHP 2000), fiber-reinforced concrete (BEFIM), new placing processes (CALIBE), and self-compacting concrete (BAP), it was decided in 2010 to prepare a project promoting concrete recycling.

After 2 years of preparation, the RECYBÉTON *("RECYclage complet des BETONs")* National Project was launched in 2012. It encompassed 47 partners from both academic and practitioner sectors, including all professional unions of the aggregate, cement, admixtures, ready-mix concrete, precast concrete, building, and public works industries.

The main objective was to study and to promote the use of all materials generated by the demolition of concrete structures, either as RA used in substitution of natural aggregate or as a raw material in the cement production. A comprehensive R&D program was carried out through four internal calls within participants and covered all aspects from material processing to environmental analysis. Around 40 scientific and technical reports were produced. Furthermore, the feasibility of concrete recycling into concrete was demonstrated through a series of five experimental construction sites, covering a variety of areas (from building to civil engineering constructions). The project was partly financed by a subsidy from the French Ministry in charge of Ecology, by the member subscriptions, by donations from professional unions, and by self-support of active participants, for a total of 4.7 M€. A part of the R&D program was also financed by ANR (Agence Nationale de la Recherche) through the ECOREB *("ECOconstruction par le REcyclage du Béton")* project.

This book aims at presenting to the international public a summary of the work done. However, care was taken to place RECYBÉTON's output in the light of the general state-of-the-art, so that the reader can access to the most updated knowledge regarding the art of recycling concrete into concrete:

- Section I deals with RAs: how they are produced, and what are their key characteristics.
- Section II explains how ground, old concrete can be used in the cement process, either as an alternative raw material (to produce clinker) or as a supplementary cementitious material (to produce blended cements or to be added into the concrete mixer).
- Section III provides information on how to produce, cast, and cure recycled aggregate concrete (RAC).
- Section IV covers all important properties of hardened RAC, from microstructure to durability-related properties and behavior under fire.
- Section V is devoted to the RAC mix design: how to adjust an existing mix or to design a new one while matching specifications.
- Section VI addresses the combination of RAC with steel reinforcement. Classical design of reinforced concrete is revisited in the light of recycling.
- Section VII presents the various experimental construction sites carried out within RECYBÉTON, together with a collection of RAC utilizations in the precast industry;
- Section VIII raises the question of sustainability of concrete recycling: what is the available resource, should it be used in concrete, and at which conditions, could it be repeated in the future, etc.
- Section IX closes the topic by summarizing standards and national practices.

Each section is divided into several chapters having the format of a journal paper. Chapters start by an abstract followed by a state-of-the-art and then by the specific outputs produced by RECYBÉTON. Efforts were made to help the reader in readily finding relevant answers to his questions in a straightforward and concise way.

Most often this book is not providing suggestions for standards and code evolutions (this task being performed elsewhere by the RECYBÉTON project), although it is expected that a higher level and a better mastery of recycling will be permitted in the next editions of the various key documents. This book is complemented by a guide being currently written for all practitioner communities (as owners, material producers, consulting, and site engineers). It is the authors' hope that the corpus of experimental results and scientific analyses collected in this book will help the construction community to adopt more sustainable practices in its resource management. It can also inspire researchers to further study and optimize all aspects of concrete recycling.

Section I

Recycled aggregates

S. Decreuse
CEMEX

In order to foster the construction and demolition waste recycling process, it is necessary to have good quality aggregates. Furthermore, to allow the increase of recycling rates and the use of recycled aggregates in new concrete, it is necessary to have good quality sources of sorted wastes.

This section is subdivided in three chapters:

- Chapter 1 is a short review of demolition techniques and their impact on the quality of wastes and aggregates. No specific work has been done in the RECYBÉTON project on demolition techniques, so this chapter deals only with the state of the art of techniques used for demolition techniques. To complete this state of the art, a point is also made within the framework of RECYBÉTON on the current quality of aggregates on different recycling platforms in France to assess the efficiency of sorting on those platforms. More data dealing with the characteristics of these materials are provided in Chapter 29.
- Chapter 2 deals with the processing techniques of recycled concrete aggregates. The techniques of air sifting for light contaminants or dedusting, wet or dry density stratification, sensor-based sorting with different sensors' technologies, and multistage crushing and other mechanical, physical, or chemical methods to improve the release of the original natural aggregate.
- Chapter 3 concerns the characterization of recycled aggregates (RAs). The main physical and chemical characteristics of RA were measured in the RECYBÉTON project on coarse and fine recycled aggregates using experimental methods devoted to natural aggregates, in order to assess if they can be applied to this kind of product without any modification and if the variability of RA affects their accuracy.

Chapter 1

Demolition techniques and obtained materials

S. Decreuse
Cemex

N. Comte
Syndicat des Recycleurs du BTP (SRBTP)

C. Raillon
Syndicat National des Entreprises de Démolition (SNED)

CONTENTS

Abstract

There is actually an increasing effort to foster Construction and Demolition Wastes (CDW) recycling in many countries. To allow the increasing of recycling rates it is necessary to have good quality sources of wastes. Those wastes need to have been sorted and a deconstruction is preferable to a demolition even if for some operations it can be more expensive. Wastes Management plans have a high impact on the sorting but demolition techniques are also very important. The present chapter makes of a short review on existing demolition techniques and gives indications on their impact on recycling. The chapter also makes a point on the study of variability made in the RECYBÉTON project in order to check if sorting is actually efficient enough to allow recycling of demolition concrete into concrete.

1.1 INTRODUCTION

Recycled aggregates should have a high quality in order to be incorporated in concrete. For example, they should have a high content of concrete and rocky materials and not be contaminated by plaster, plastics, and wood. In order to get such characteristics, it is necessary to have an adequate demolition process that sorts out the different kinds of demolition materials. In this perspective, a process of deconstruction is better than a demolition process without presorting.

RECYBÉTON project has not addressed this subject through research on the different kinds of demolition processes and their impact on the characteristics of the aggregates but made a state of the art of the regulations and types of demolition techniques. A study has been conducted on the variability characteristics of the current production of aggregates in France (Jezequel 2013): geographic variability (16 recycling platforms) and time variability (2 recycling platforms during 24 months).

1.2 STATE OF THE ART

Numerous studies have proven that selective demolition is possible even technically and economically (RECYC-QUEBEC 1999).

In case of complete demolition, inert wastes represent 94% of the wastes generated by deconstruction, but it is truly only 60% as some non-inert wastes cannot be separated from the inert ones (e.g., remains of plaster on the concrete, wood and plastics from finishing works that have not been removed from the walls) (Laby 2007).

Currently, if recycling and recovery of pure concrete achieves a rate of more than 75%, the rate for mixed inert wastes only reaches 12% (CGDD 2010).

This situation should be enhanced in France through construction and demolition wastes (CDW) prevention and management plans that recommend to make a diagnosis before the demolition process (BIO IS 2015) and also through the European minimum target of 70% of reuse, recycling, or recovery of CDW wastes by 2020 established on the Waste Framework Directive (Directive 2006/12/EC revised by Directive 2008/98/EC).

1.2.1 Regulations and management of construction and demolition wastes in France and other countries

1.2.1.1 Different kinds of construction and demolition wastes

Construction and demolition (C&D) wastes can be classified as follows:

- Hazardous wastes
- Nonhazardous wastes (which can be further divided into inert wastes and nonhazardous non-inert wastes).

Inert waste mostly consists of wastes of mineral origin such as concrete, tiles and bricks, glass, and some reclaimed asphalt. These wastes are sometimes mixed together or slightly contaminated by plaster, plastics, and wood. These wastes are mainly found in the building structure and will be generated during the structure demolition.

To be used as recycled aggregates, demolition wastes have to maximize the amount of concrete and rocky materials.

The hazardous wastes are not concerned by the studies carried by RECYBÉTON. When some hazardous wastes are present in the CDW, they need to be separated from inert wastes. Some examples of these hazardous wastes can be bituminous mixtures containing coal tar, construction materials containing asbestos, and contaminated concrete from nuclear power plants.

1.2.1.2 Regulations applicable to C&D wastes

According to Tam (2013), there has been an overwhelming promotion of waste management tools and sustainable development activities in recent years, in many countries including China, Japan, the United States, Brazil, and most of the European countries.

Since 2011, the European Commission (EC) has financially supported the studies for the development of future European policies in the area of sustainable resource management. C&D wastes have been identified by the EC as a priority stream because of the large amounts of wastes that are generated and the high potential for reuse and recycling embodied in these materials. For this reason, the Waste Framework Directive 2 requires Member States to take any necessary measures to achieve a minimum target of 70% (by weight) of C&D waste by 2020 for the preparation for reuse, recycling, and other material recovery.

The main French incentives that could have an impact on the increase of concrete recycling and on the sorting of demolition wastes are as follows (BIO IS 2015):

- Articles 46 of Law 2009-967 of 3 August 2009 (known as Grenelle 1 law) and 190 of Law 2010-788 of 12 July 2010 (known as Grenelle 2 law) make preaudits compulsory on demolition sites. These preaudits aim to characterize all materials present on-site and to plan the CDW management.
- Decree no. 2011-610 of 31 May 2011 states this obligation, named "diagnosis on waste arising from demolition works", from March 2012 for certain categories of buildings (>1,000 m^2) (articles R. 111-43-R. 111-48 of the French Construction and Housing Code). This decree specifies the concerned buildings, the obligations of the developers regarding the diagnosis content and methodology, and the obligation to disclose the document and to fill a tracking form after demolition work completion.
- The National Framework for Waste Prevention (MEDDE 2012), drawn up in April 2012 by the "Prevention" working group of the Waste National Council 14, provided the basis for the National Waste Prevention Program 2014–2020.
- In order to reach this target, the National Waste Prevention Program 2014–2020 plans to launch several key actions including the review/refinement of the regulation related to demolition diagnosis. Such measures are also described in the 2015 Energy Transition Law (LOI no. 2015-992).

1.2.2 Demolition techniques

Various factors that influence the choice of the demolition technique or combination of demolition techniques used on a demolition site are as follows:

- The amount of workplace available
- The existence of local or national regulations leading to sorting of wastes
- The existence of environmental requirements (established by the owner)
- The location of the demolition site
- The volume of demolition
- The time allocated for completion of the demolition process

All the techniques described in the sections that follow need to be implemented by qualified companies, and security measures for workers and neighborhood need to be implemented.

1.2.2.1 Demolition techniques using handheld tools

This technique is mostly used for demolishing small volumes of structures and also as preparatory work before the demolition of larger structures (Da Costa 2009; Coelho and

De Brito 2013a; Brokk 2000). When used as a preparatory work, pollutants will be removed (asbestos, lead, etc.). Different tools such as mallets, picks, and hammers that can be electrical, pneumatic, or hydraulic are used. This kind of demolition is slow and expensive, but it allows a good sorting if well done.

1.2.2.2 Demolition techniques using rig-mounted tools

Compared to demolition techniques using handheld tools, this technique is much more powerful and much more effective (Coelho and De Brito 2013a; Brokk 2000). Different tools such as pliers, clippers, breakers, crushers, or hammers are attached to demolition excavators. The selection of the tool depends of the building's type (size, materials of the structure, etc.). The cost of use of such tools (in currency per ton of demolition material) is lower than that of handheld tools. When the building is too high for the biggest demolition excavators, mini-excavators can be placed at the top of the building to cut down a few levels before continuing with adapted demolition excavators (Figure 1.1).

1.2.2.3 Demolition by blasting

Blasting is a very effective demolition process which weakens the structure of the building or even makes it collapse (Coelho and De Brito 2013a; Brokk 2000). To collapse a building, different techniques of blasting can be implemented based on the building type. It can also be used in different parts of the building to have a demolition in a controlled manner. Before blasting, the building is cleared and materials are sorted out in order to obtain a good quality of inert wastes to be recycled. This stage is essential, because if the sorting has not been made before this demolition process, all the wastes will be mixed. These techniques also require expertise and a good structural knowledge of the building. It is generally used for buildings where classic demolition techniques cannot be implemented because of the building's size, the riskiness, or the lack of efficiency of classic techniques (Figure 1.2).

1.2.2.4 Other kinds of demolition process

Other kinds of demolition processes include old techniques such as sewing and drilling or pulling, as well as more recent techniques, such as thermal process, where materials are

Figure 1.1 Demolition using a breaker. (From Yves Soulabaille—Genier Deforge.)

Figure 1.2 Demolition by blasting. (From CARDEM.)

fused to be separated from the others, electrical process, or chemical process (Coelho and De Brito 2013a; Brokk 2000).

1.2.3 Demolition/deconstruction and costs

Deconstruction is the process of selective demolition with a good sorting of the different materials. As such, this process has a higher cost than conventional demolition. This over-cost is directly linked to the increase of duration of the demolition process, the increase of operations to be made in order to make a sorting of different materials, and also the increase of workforce needed.

In most cases, this increase in cost is not a constraint for environmental reasons; however, for demolition of buildings of more than 1000 m^2, an inventory of the different kinds of wastes has to be made and local recycling facilities for the different materials have to be identified. Finally, the processing of sorted wastes is less costly than the processing of mixed materials. Existence of such local processing facilities for different kinds of wastes is a good incentive for better sorting and deconstruction (Laby 2007) (Figure 1.3).

Figure 1.3 Sorting during selective deconstruction. (From BRUNEL.)

It means that if the structuring of the sector of recycling applied to concrete begins to work, this will be a good incentive for more deconstruction and more sorting, and then it will increase the quality of the recycled aggregates.

1.2.4 The obtained materials

In France, in 2014, the sources of C&D wastes break down as shown in Table 1.1 (SOEs 2017b).

The demolition wastes mostly consist of stony materials, such as concrete, mortars, and bricks and tiles. Other construction products can also be present in the obtained materials in small amounts but mostly as a pollution, for example, glass and bituminous mixtures.

Table 1.1 Volume of the different kinds of CDW

Waste nature	*Total quantity of waste generated* ($\times 10^6$ t)
Concrete	19.1
Bricks, tiles, ceramic, and slate	4.2
Glass	0.2
Bituminous mixtures containing no tar	11.2
Unpolluted soil and stones	114.8
Other materials from roadway demolition	37.5
Nonpolluted track ballast	2.2
Nonpolluted dredging spoil	2.8
Other inert wastes	1.1
Mixed inert wastes	18.1
Total inert waste	211.2

Note: Focus on 2017 inert CDW figures (data from the original 2014 comprehensive survey on CDW performed by SOEs 2017b).

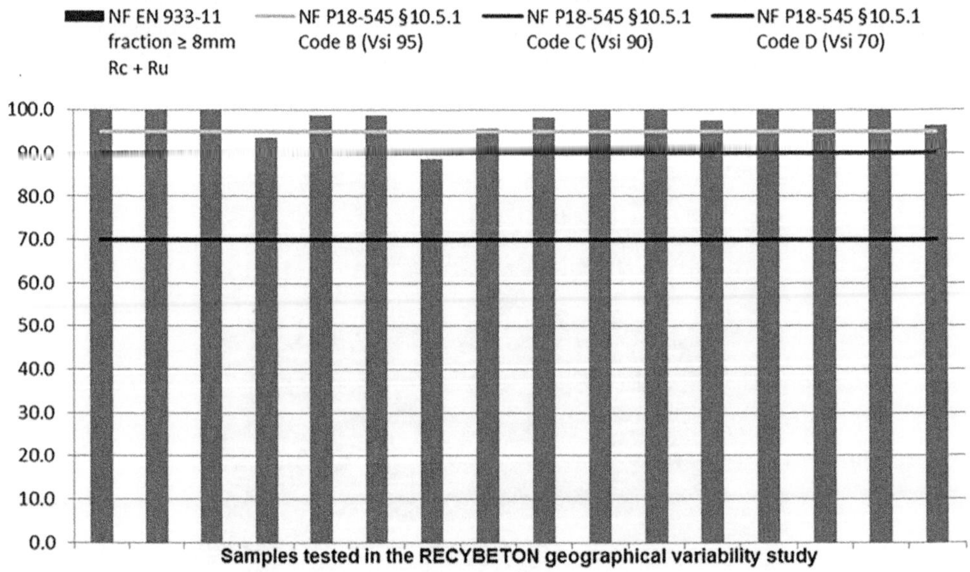

Figure 1.4 Concrete (Rc) and stony material (Ru) (in %) fractions in recycled aggregates submitted to the classification test (EN 933-11) in the geographic variability study.

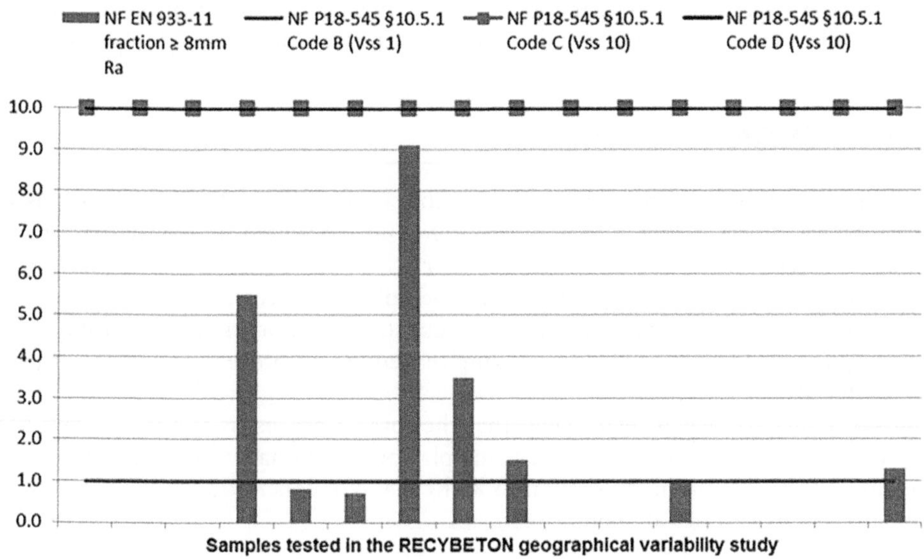

Figure 1.5 Tile and brick (Rb) fractions in recycled aggregates submitted to the classification test (EN 933-11) in the geographic variability study.

The study on 16 recycling facilities in France (Jezequel 2013) has shown that 14 of them were producing recycled aggregates that contained more than 95% of concrete and stony materials. The sorting of the sources then seems to be quite efficient (Figures 1.4 and 1.5).

1.3 RESEARCH NEEDS

The recycling rate of CDW will increase through the design of construction works that considers the future deconstruction process and through the sorting (PDR 2000; Bradley 2004). It will also be through the same levers that the use of recycled aggregates in concrete will increase.

Some other projects have started to consider the design for deconstruction in order to prevent the production of wastes and to maximize the recyclability of the wastes obtained. The project DEMODULOR (Château 2015) has worked on different processes, mainly on the shell of buildings. Some of the conclusions of DEMODULOR are good suggestions to increase recycling of CDW:

- Introduce a high recycling ratio of CDW for each operation.
- Introduce in the calls for tenders a clause for higher recycling rate bidder.
- Reinforce controls on construction sites.
- Ensure that waste management plans are effectively developed at the design of construction works step.

It would also be interesting to develop some automatic control devices in order to identify the constituents of the demolition materials and to ease and secure the receipt of demolition materials on recycling platforms or quarries. HISER project works currently on those kinds of devices, for example, in-line quality assessment system, based on laser-induced breakdown spectroscopy. These techniques have been applied to concrete aggregates (Zerbi et al. 2017), and preliminary results indicate that coupling the hyperspectral information and the robotic technology allows to achieve high-quality sorting of CDW (higher than 90% of purity).

1.4 CONCLUSIONS

From the information collected during the bibliographic review, the non-hazardous CDW constituents are mainly mineral components sometimes polluted by non-desired elements, for example, plastic, sulfates, and wood. The increase of recycling rates of CDW will necessarily pass through a purification step through demolition process.

The French regulations are providing strong constraints in order to structure the market and to fix quantitative targets for maximizing recycling and preventing the use of primary natural materials.

From variability studies, it seems clear that the sorting process is rather efficient, regardless of the used technology on the 16 recycling platforms that have been studied, 14 of them produce an output containing more than 95% (by mass) of concrete and stony materials, the absolute minimum of mineral content being 85%.

Chapter 2

Processing of recycled concrete aggregates

B. Cazacliu, L. Le Guen, and E. Hamard
IFSTTAR

S. Roux
IJL UMR 7198, Villers les Nancy

S. Braymand
ICube UMR 7357, Université de Strasbourg, Strasbourg

CONTENTS

Abstract

There are various patterns of construction and demolition waste (CDW) recycling plants combining different stages of the process like extraction of metallic elements and light elements, crushing, screening and clay flocculation. While most recycled plants have very simple processes and are hard to combine all the previous equipments, processes that are more complex could be proposed in practice and / or in the scientific literature. This sub-chapter briefly presents air-sifting for light contaminants or de-dusting, wet or dry density stratification, sensor-based sorting with different technologies, multi-stages of crushing to improve the old aggregate liberation and other mechanical, physical or chemical methods of liberation.

2.1 INTRODUCTION

Industrial recycled aggregates (RAs) are generally derived from a number of demolished concrete structures, contributing to large composition variability in the original concrete properties and large proportions of foreign materials (e.g., clay brick, asphalt, glass, gypsum, etc.). It is relatively easy to separate out constituents such as plastic, wood, metal, paper, and roofing materials. This is not the case for ceramics, asphalts, concrete, and stones. These components are almost always presented in different combinations of mixtures, reflecting that in practice these components are sparsely separated. Also in most cases, the RA particles are composed of two or more mineral components (stone, mortar, brick, gypsum, etc.—middling particles). To make the feed RA suitable for use in recycled materials, it is necessary to improve their uniformity and to decrease the level of the contaminants. However, the requested level of uniformity and purity highly depends on the different categories of use. Indeed, the recycled aggregates can be used as viable substitutes for natural aggregates in pavement subbase materials composed of a mixture of recycled concrete and brick rubble, even in cases with a significant amount of brick rubble (Arulrajah et al. 2012). On the contrary, for the concrete produced with recycled aggregates, the presence of cement mortar attached to natural aggregates (de Juan and Gutiérrez 2009), the presence of clay brick particles (Yang et al. 2011), and the presence of contaminants as waste glass cullet or wood chips (Poon et Chan 2007) have been identified as key factors that not only lower the quality of recycled concrete aggregate (RCA) but also increase the variability. As a consequence, the coarse recycled concrete aggregate composition admitted for concrete production is limited by standards to mixtures of at least 90% or 95% by weight of concrete or natural stone and has a total contaminant level typically lower than 1% of the bulk mass. In order to exclude any expansion phenomena and the damage of the microstructure in secondary concretes (Nixon 1978), the contaminant limit is generally stricter on allowable percentages of gypsum. Finally, most applications are however limited to coarse RA, while the required characteristics are difficult to ensure for recycled sand.

Nowadays, the industrial practice is mostly exclusively based on the selective demolition and, in some cases, on manual sorting to avoid the mixing of different materials. This results in poor quality RA as reflected in the marginal use of RA in structural concrete. Given the limitations on composition and the difficulty in extracting certain elements, the implementation of effective separation systems becomes key in RA processing. These separation systems include automated sorting systems and liberation of the components of middling particles. Synergy between mineral processing and recycling of construction and demolition waste (CDW) is obvious. The potential of technological transfer points the way towards more rational use of RA in more friendly environmental conditions.

2.2 STATE OF THE ART

There are various patterns of CDW recycling plants. Current sorting lines can combine different stages of the process (CEREMA 2015):

- Extraction of metallic elements using magnetic separators
- Extraction of undesirable light elements (plaster, wood, plastic, etc.) by sorting equipment (manual, air or hydraulic sifters, etc.)
- Crushing/screening or scalping to produce an elaborate material compatible with the wished-for end uses
- Clay flocculation, for instance, by froth flotation

However, as presented by Ademe (2011a) for France, most recycling plants have very simple processes and are far to combine all the previous equipment. Indeed, a number of papers in literature (see Bovea and Powell (2016) for a list of authors) consider simple processes as a basis for economical or environmental analysis. In this context, a typical process of a current well-equipped plant recycling inert debris to produce recycled concrete aggregate is explained as follows: The input waste materials are visually inspected during the reception, classified using excavators, and then stored in separate stockpiles. Recycled concrete is produced by reducing the size of concrete debris, obtained by breaking accessories such as hydraulic hammer and cutters. The concrete blocks are scalped, for instance, using a grizzly feeder and directed prior to a crushing stage. Ferrous scraps are segregated using overhead magnetic separators. The obtained material is sorted by fractional sieving in order to obtain the final concrete aggregates. Some other papers present more simplified processes.

In some cases, processes presented in the literature are significantly more complex, including complementary operations to improve the effectiveness of the process:

- Up to three stages of crushing (Kim 2014)
- Air shifting for contaminants such as paper, plastic, and wood (Coelho and de Brito 2013b; Lotfi et al. 2014)
- Rotary screens (trommels) for large gravel or other large contaminants (Montero et al. 2010; Tsoumani et al. 2015)
- Foucault current systems for nonferrous metals (Xing and Hendriks 2006)
- Air blowing for dedusting (Tam 2008)
- Washing plants including the treatment of the resulted sludge (Petavratzi et al. 2007)
- Spirals for heavy metal separation from fine RA (Coelho and de Brito 2013b)
- Air jig (Cazacliu et al. 2014), water jig (Weimann et al. 2003), or dry sand fluidized bed (Mulder et al. 2007) for coarse contaminants such as brick or gypsum
- Sensor-based sorting (Lotfi et al. 2014; Paranhos et al. 2016)
- Systems to separate mortar from RA (Koji and Sakai 2010)
- Magnetic separation of mortar and brick (Xing et al. 2002; Ulsen et al. 2013)
- Manual picking belts that enable separation of large disturbing substances between the first two crushing stages (Silva et al. 2014c)

The choice of the different automatic operations depends on the quantity of waste to be sorted and its composition but also on the technical and economic objectives of the company. In France, the Syndicat de Recycleurs du Bâtiment et des Travaux Publics (SRBTP 2014) suggests to not oppose systematically automatic sorting and simple or manual sorting which could be often complementary on a recycling platform.

2.2.1 Crushing

Depending on the size, the crushed concrete particles can present middling, which contains particles of natural aggregates mixed with mortars. Belin et al. (2014) showed that the amount of residual cement paste in RA mainly depends on the paste content of the initial concrete, the porosity of the initial cement paste, and the mechanical quality of the interface between the initial natural aggregates and the cement paste. Haase and Dahms (1998) stated that for the same particle size, aggregates produced from weaker parent concrete have less mortar content. They explained that the weaker interfacial transition zone (ITZ) favors the separation of the aggregates from the mortar during crushing and leaves more clean rock particles. However, the negative impact of a higher level of mortar in RA obtained from higher quality concrete is partially offset by the increase in strength and density of the mortar and the better old ITZ. This is demonstrated by the Los Angeles test values obtained for RA with different densities as obtained in the tests presented in Chapter 3.

A number of previous studies have also reported that, in general, the amount of mortar present in coarse RA decreases with an increase in the number of the crushing stages (Fleischer and Rubby 1999; de Juan and Gutierrez 2009; Florea and Brouwers 2013; Akbarnezhad et al. 2013b). Effects of the second crushing stage in reducing the mortar content of RA seemed to be more significant for the larger size fractions than for the 4–8 mm size fraction. Akbarnezhad et al. (2013b) observed that regardless of the number of crushing stages, the mortar content of RA generally increased with an increase in the parent concrete compressive strength. The water absorption could decrease significantly and linearly with the number of crushing stages. Noguchi et al. (2011) show that the water absorption of RA between 5 and 10 mm evolves from 8% to 4% and that of RA larger than 10 mm from 6% to 3% for zero (no) to three crushing stages. Nagataki et al. (2004) used a combination of jaw crusher and impact crusher followed by two mechanical grinding equipment to minimize the attached mortar of three concrete recipes, with 28-day compressive strengths of 28, 49, and 61 MPa. After the two crushing stages, the water absorption of the 5–20 mm corresponding RAs was 6.3%, 5.6%, and 4.9%. After the additional two stage mechanical grinding the water absorption decreased to 3.8%, 3.2%, and 3.1%, respectively. When comparing the behavior of secondary concrete produced with two RA fractions with that of a concrete designed with natural aggregate, Pedro et al. (2014) concluded that this caused more significant durability-related performance loss than mechanical performance loss for both recycled aggregate concretes. However, RA produced by a two-stage crushing—a jaw crusher followed by a hammer mill—gives better results than the RA produced by a single-stage jaw crusher, as the attached mortar is reduced by the two-stage crushing.

It should be noted that mortar is not always the primary parameter used for determining the quality of the recycled coarse aggregate (Nagataki et al. 2004). Sandstone coarse aggregate in the parent concrete had defects such as cracks, pores, and voids, diminished by the two crushing stages. The aggregate properties were consequently enhanced. This was confirmed by Ogawa and Nawa (2012), who observed that the jaw crusher slightly diminished the fraction of defects in the aggregate, while the change was more significant when processed with a ball mill or a granulator. In addition, the geometric characteristics changed only slightly after repeated crushing in the jaw crusher, but became round after being crushed in the ball mill and granulator.

Akbarnezhad et al. (2013b) found that the mortar content of RCA depended on the size of the natural aggregates in the parent concrete. They suggested that crushing the concrete to a maximum size close to that of its original aggregate might result in slightly less mortar content. Belin et al. (2014) observed that the parent concrete with rounded aggregate produced RA with lower paste content than that with crushed aggregate. This was explained by the weaker ITZ of the parent concrete designed with smooth natural rounded aggregate (de Larrard 1999a).

If further processing of the recycled coarse aggregate diminishes the attached mortar content, this also drastically diminishes the recovery of coarse aggregate and thus increases the proportion of fine recycled aggregate obtained from the processing. Noguchi et al. (2011) illustrate the evolution of the coarse particle proportion in the final product from 50% to 20% from zero to three crushing stages. In the same time, the resulting powder proportion increases from 5% to about 12%.

More fundamental studies of the crushing of concrete are discussed in the literature. This concerns the experiments on model concrete balls (Tomas et al. 1999) or disks (Neveu 2016) studied one by one under impact. Also, numerical studies using the discrete element method (DEM) were presented by Takahashi and Ando (2009).

2.2.2 Sieving

Less information is given in the literature on peculiarities related to the sieving of recycled aggregates. It can be mainly noted that the elimination of finer or coarser fractions could improve the concentration of concrete particles of the remaining fraction. Indeed, the grading of the different components of a recycled aggregate feed could be different. An industrial example was given by Asakura et al. (2010). They analyze the 0–5.6 mm fraction of mixed CDW rubble, after several treatment stages in five recycling plants in Japan. In this material, wood is coarser than the dense inert particles, while gypsum and organic matter are finer. Since gypsum is mainly distributed in the fine fraction, the amount of gypsum can be reduced by separating the fine fraction from mixed CDW (Montero et al. 2010).

RAs are more sensitive to fragmentation than natural aggregates (see also Chapter 3). In particular, sieving of RA could produce fragmentation. Schouenborg et al. (2004) illustrated the increase in the amount of material passing through the 8 mm sieve with the energy of sieving RA. They compared manual sieving and mechanical sieving during 2, 5, and 10 min. The increase in the amount of material passing was around 3%, 6.5%, and 8.5%, respectively. The increase was negligible (0.2%) for a natural aggregate.

2.2.3 Dedusting

The reuse and recycling of RA obtained by crushing/screening operations can be improved if the fine elements are be removed beforehand (Huang et al. 2002). The composition of this fraction is difficult to control and, among other induced artifacts, contains a higher percentage of contaminations (Asakura et al. 2010). The most common way of dedusting is the wet processing. However, washing generates water waste and forms sludge, which is difficult to recycle further. The alternative could be the use of a dry process after crushing, when the RA moisture is low.

The technical constraints of the dry recycled aggregate dedusting are close to those of crushed natural rock. The finest fractions from alluvial sand generally have a discontinuity in the grading curve, around 20–60 μm. This helps in fixing a convenient cut size to separate the fine particles (less than 0.1 mm). This is not the case for sand obtained from crushing where the granular distribution has some continuity in this range (Champeau and Potin 2003). Moreover, several additional difficulties must be faced with dry technology compared to wet technology: the cut size is higher and the efficiency is less, and the agglomeration of fine particles has an undesirable effect. Nevertheless, many pieces of literature show information on the possibility of carrying out the dedusting (Cazacliu and Huchet 2016). The considered technologies are either mechanical screening (Pettingell 2008; Meinel 2010) or air-shifter classification, which can work by injecting the material stream to be sorted into a chamber containing a column of rising air (Johansson 2014).

It should be noted, however, that the case of recycled sands is not completely similar to that of crushed natural rock sands. Indeed, recycled particles have a mixed mineralogy (concrete, rock, brick, etc.) and a heterogeneous structure (cement paste and rock elements for crushed concrete particles, for example). In some systems, these differences could improve the efficiency of the dedusting process, but in other systems, this efficiency could be reduced.

One of the rare well-documented equipment specifically designed for recycled aggregates is presented by Lotfi et al. (2014). This equipment combines an attrition milling and a wind-blowing system in order to separate the moist recycled aggregate into fine and coarse fraction (called advanced dry recovery technology). In the presented example, the recovery of fractions smaller than 4 mm from the extracted 4–16 mm fraction is less than 20%.

2.2.4 Sorting by density

As discussed above, air-shifter technologies are most commonly proposed for the separation of light materials such as wood, plastic, paper, and cardboard compared to the heavier ones. However, a larger spectrum of technologies can be proposed.

Gravity concentration is defined as the process in which particles of different sizes, shapes, and densities are separated from each other by gravity or by centrifugal force. This is called gravity concentration because the separation is mainly based on the density (specific gravity). Gravity concentration processes have high mass throughput and require low investment and operational costs. Moreover, there is no limit on the maximum particle size, and the process can be used for particles with varying size ranges.

Jigging is a separation process which involves repeated expansion (dilatation) and contraction (compression) of a particle bed, using a medium, usually water or air. The result is the stratification of the bed with increasing densities of the particle from the top to the base (Sampaio and Tavares 2005).

Hendriks and Xing (2004) used wet jigging in laboratory to separate concrete from brick coarse particles, concluding that this technology works well, mainly for particles >19 mm. Finer particles (2–5 mm) require careful control of operating parameters to obtain good results. Müller and Wienke (2004) did not obtain good results for the same process; however, they obtained more promising results on the gypsum/concrete separation. In subsequent research, Schnellert and Mueller (2011), using an industrial-scale equipment, showed a four- to fivefold reduction of the gypsum content in the output material compared to the initial content.

Air jigging was also investigated on brick/gypsum/concrete particle mixture. Concentration of concrete higher than 90% and gypsum content significantly lower than 1% were possible to be reached. Indeed, a 25-fold reduction of the gypsum content was seen in concrete concentrates (Sampaio et al. 2016). These results fit with the findings of Mulder et al. (2007), who observed that dry sand fluidized bed shows the highest product purities for 20 mm oversize and generates products with qualities that are similar to wet jigging.

Ambrós et al. (2017) tested air jig equipment for multicomponent separation of CDW. Tests results have indicated the technical feasibility of the use of batch jigging for the removal of undesired contaminants such as wood, paper, gypsum, and brick particles in only one single stage. Furthermore, the initial content of contaminants seems to have a negligible effect on the separation performance. For all cases tested, the product obtained showed contamination levels lesser than 1% in mass and approximately 90% in mass of pure concrete.

Elutriation separates particles based on their size, shape, and density, using a constant flow of gas or liquid in a direction usually opposite to the direction of sedimentation. The method was applied to narrow sieve fractions of recycled fine aggregates (Ulsen et al. 2013). It was observed that the cement paste and residual red ceramic particles decrease two times in heavy products compared to light ones. The method seemed effective since the natural aggregates had achieved a proper degree of liberation.

2.2.5 Sensor-based sorting

Automatic optical sorting was performed in the mining industry to process different ores, for materials having significant color or shape differences. Several other sensors could be used to discriminate between particles of different physical or chemical composition: dual energy X-ray transmission, laser-induced fluorescence, laser-induced breakdown spectroscopy (LIBS), X-ray fluorescence, near-infrared spectroscopy (NIR), light detection and ranging, Raman spectroscopy, and so on.

Color sorting can well be used to increase product quality and avoid the presence of unwanted materials in the stream to be recycled, for instance, for red or yellow brick (Xing et al. 2002). Total wood recovery reached 83% in tests conducted by Mulder et al. (2007). Also, removal of gypsum forms the heavy fraction showing a recovery of 94% and a glass recovery of 96%.

Using dual energy X-ray transmission, an effective split can be made between organic and inorganic materials, as well as an effective metal recognition in the same pass (Mulder et al. 2007).

NIR technology achieves good efficiency when sorting gypsum or lightweight autoclaved aerated concrete, but not metal, wood, and plastics (Vegas et al. 2015). In a differently configured NIR equipment, Palmieri et al. (2014) distinguished recycled aggregates from bricks, gypsum, plastics, wood, and foam.

With LIBS sensors, this single-shot method constitutes the most promising real-time methodology, the success of which depends on the quality of the model initially fitted based on a training dataset (Xia and Bakker 2014).

2.2.6 Liberation

- As the presence of this primary mortar is responsible for the behavior differences of recycled and natural aggregates (Braymand et al. 2015; Sri Ravindrarajah and Tam 1987), methods used to separate primary aggregates and mortars are discussed in the literature (Figure 2.1) (Torgal 2013; De Brito 2013). These methods using mechanical (wear, fragmentation, etc.), chemical (acid attack), or physical (heat treatment, microwave) principles are based on the differences of properties between the aggregates and the cement paste (Torgal 2013; Thomas and Jennings). Alternatively, the degradation of the ITZ, for example, by freeze-thaw cycles, allows the liberation of primary aggregates.

As a result of a literature review, Table 2.1 presents the main studied treatments.

Some studies propose single- or multicriteria comparative analyses of these methods (Torgal 2013); however, their efficiency is rarely discussed (Braymand et al. 2016).

With regard to efficiency, the first issue is the determination of the grain size considered as the maximum size constituting the primary mortar. The size-selective sieve for the final separation influences the mass lost value: the higher the sieve size, the more the mass loss (Braymand et al. 2016; Deodonne 2015). This parameter then influenced the

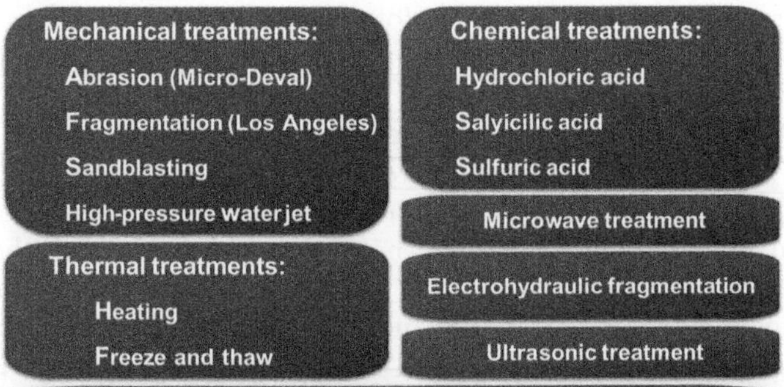

Figure 2.1 Separation methods and different processes.

Table 2.1 Separation methods: processes of literature

Treatment	*Industrialization*	*Energy consumption*	*Max mass loss (%)*	*Difficulties*	*References*
Mechanical	Easy	Los Angeles: 2 kWh/ton	60	Aggregate damage	Torgal (2013), NF EN 1097-1 (2011), Shima et al. (2005), Linß and Mueller (2004)
High temperature	Easy	250 kWh/ton	20	Aggregate damage	Bazant and Kaplan (1996), Rønning (2001), Yang et al. (2006)
Ultrasound	Specific equipment required	12 kWh/ton	70	health	Akbarnezhad and Ong (2013)
Low temperature		—	—	Obtaining a very low temperature	Menard et al. (2013), Tasong et al. (1999)
Microwave		—	71	Health	Thomas and Jennings, Akbarnezhad et al. (2011), Zhao et al. (2013a), Tam et al. (2007)
Chemical	Not possible	No	100	Health, waste produced from treatment	Akbarnezhad et al. (2013b), Momber (2004)
Electrohydraulic fragmentation	Very difficult, specific equipment, and high voltage required	2 kWh/ton	Recovery rate: 58.5	Low amount of treated material, electric installation needed	Menard et al. (2013), Bru et al. (2017)
Thermal-mechanical	Easy (hot)	Close to thermal	45	Aggregate damage	Bazant and Kaplan (1996)

determination of efficiency of the process studied and is one of the explanations of the high dispersion of average mortar content of RCA reported in the literature (Braymand et al. 2016).

The mass loss can also be influenced by primary aggregate damage during the treatment. The degradation part of the primary aggregates has been confirmed by Yoda et al. (2003) for mechanical process, Homand-Etienne (Homand-Etienne and Houpert 1989), and Zhao (2014) for high-temperature thermal treatment (higher than 600°C) and for acid treatment (Hansen 1986; Momber 2004).

Thus, the sole mass loss criterion cannot be considered for estimating the technical efficiency of the studied processes: other properties also have to be checked after treatment. Indeed, a high mass loss is not an indicator of the cleanliness of aggregates. If the density of treated aggregates is lower than that of natural aggregates, the recycled aggregates still contain mortars. In the case of a lack of knowledge about the initial properties of primary aggregates, technical efficiency evaluation requires assumptions about its composition. According to this analysis and observations reported by Akbarnezhad and Ong (2013), in the case of thermal treatment (Dao 2012), the results obtained highlighted that the mass loss is not correlated to the bulk density or to the 24 h water absorption (Braymand et al. 2016).

Although experimental procedures used to determine the contents and their limits are not always detailed in the literature (Akbarnezhad and Ong 2013), Dao (2012) proposes the following model to estimate the mortar content when primary concrete composition is known:

$$\text{Anat}_{\text{RA}} = \frac{(\rho_{\text{RCA}} - \rho_{\text{parentC}}) + (\rho_{\text{parentA}} - \rho_{\text{RA}}) \times \text{Anat}_{\text{parentC}}}{\rho_{\text{parentA}} - \rho_{\text{parentC}}} \qquad (2.1)$$

where Anat_{RA} is the volumetric proportion of natural aggregate in the RAC; $\text{Anat}_{\text{parentC}}$ is the volumetric proportion of natural aggregate from the composition of parent concrete; ρ_{RA} is the RA density; ρ_{parentC} is the parent concrete density; and ρ_{parentA} is the parent aggregate density.

This simple mix law is based on the hypothesis that the employed methods do not damage the primary aggregates.

From the perspective of an industrial process, global efficiency evaluation cannot be limited to technical efficiency measurements: energy consumption, noise generation, health risk, and production of second-generation waste have to be evaluated too. That is why it is difficult to compare the processes objectively.

2.3 RECYBÉTON'S OUTPUTS

2.3.1 Crushing

Crushing of RA was studied using a 10/20 mm RA fraction crushed by cone, jaw, and impact laboratory crushers. Three maximum sizes of the crushed aggregates for jaw and impact crushers and only a single maximum size for the cone crusher were studied. Results are presented in Figure 2.2 (Hamard and Cazacliu 2014a).

The aim of aggregate crushing is to reduce the granular fraction size. The production of aggregate fractions with narrow size range eases the control of the product. Figure 2.2a depicts the size range of crushed RA according to the maximum size of the crushed aggregate for cone, jaw, and impact laboratory crushers. With the same maximum size, the cone

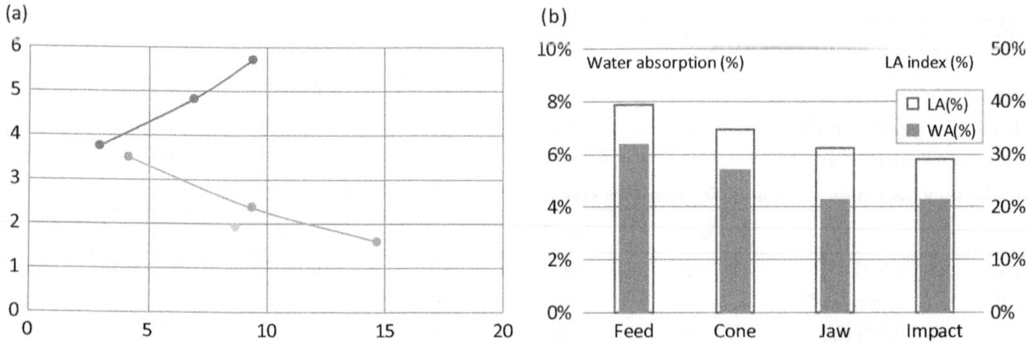

Figure 2.2 (a) Size ranges ((D95-D5)/D50) of crushed RA for cone, jaw, and impact crushers, according to the 95% passing diameter (D95) of crushed RA. Particle size distribution was determined according to EN 933-1. (b) Water absorption after immersion for 24 h (WA24) determined according to EN 1097-6 and resistance to fragmentation (LA) determined according to EN 1097-2 for RA source material and crushed RA thanks to cone, jaw, and impact crushers (Hamard and Cazacliu 2014a).

and jaw laboratory crushers produced the narrower RA fraction. Jaw crusher produced even narrower RA fraction for lower maximum size of the crushed aggregate. On the other hand, laboratory impact crusher produced narrower RA fraction for higher maximum size of the crushed aggregate.

Attached mortar of RA had a negative effect on mechanical resistance and water absorption. Crushing reduced the attached mortar content and thus enhanced the properties of RA. Water absorption and resistance to fragmentation of RA produced by the three laboratory crushers are presented in Figure 2.2b. Laboratory impact and jaw crushers produced the RA with the lowest water absorption, whereas the laboratory cone crusher produced RA with a water absorption capacity which has intermediate values between material source and other RA. The best resistance to fragmentation was obtained by the RA produced by the laboratory impact crusher, and the worst by the laboratory cone crusher. In this study, with regard to RA quality, the best laboratory crusher is the impact one and the worst is the cone one. Based on the available experience, it is difficult to extend these findings to industrial crushers.

2.3.2 Water Jigging

Water jigging, an equipment presented in Section 2.2.4, is a gravity separation process conditioned by the difference of density between the separated materials. While water jigs are commonly used for the beneficiation of ore in the mining industry, their use in the CDW sorting is a technological and process innovation. The ability of a laboratory water jig to recover concrete particles was investigated using the following tests. The experimental approach was to apply the jigging sorting to the different samples composed of building materials (Le Guen 2015). The tested samples were mixtures of particles of crushed concrete (RA), gypsum, and brick. The RA content was varied between 90%, 80%, and 60%. Table 2.2 presents the different samples used for the campaign.

The device used was a laboratory water jig ("stratificator"), made up of four main items (Figure 2.3):

- The container dedicated for materials, made of Plexiglas, is composed of 11 slides. It can be therefore easier to compose the sample studied in respect of the defined ratios.

Table 2.2 Composition of material in for each test using the water jig

(a) Binary samples with a 10/20 granular size

Materials		*Test 1*	*Test 2*	*Test 3*	*Test 4*	*Test 5*	*Test 6*
Size	RA (%)	90	80	60	90	80	60
10/20	Gypsum (%)	10	20	40	0	0	0
	Brick (%)	0	0	0	10	20	40

(b) Binary samples with a 4/20 granular size

		Test 7	*Test 8*	*Test 9*	*Test 10*	*Test 11*	*Test 12*
Size	RA (%)	90	80	60	90	80	60
4/20	Gypsum (%)	10	20	40	0	0	0
	Brick (%)	0	0	0	10	20	40

(c) Ternary samples with a 10/20 granular size

		Test 13	*Test 14*
Size	RA (%)	60	80
10/20	Gypsum (%)	20	10
	Brick (%)	20	10

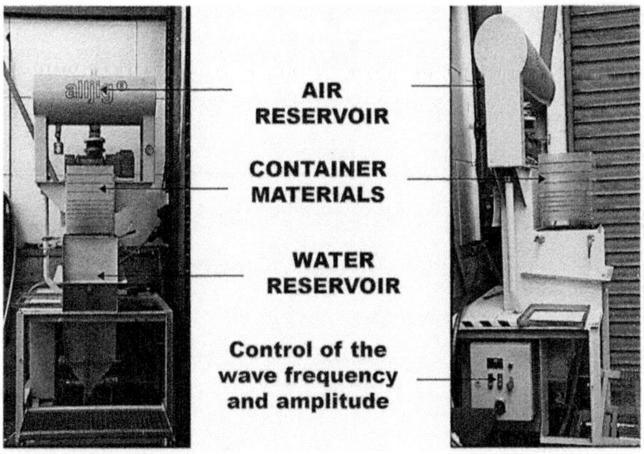

Figure 2.3 View of the Alljig water jig without materials in the container.

- The water reservoir is placed between the container and the air reservoir.
- The air reservoir is connected to a piston.
- The electrical table sets the operating parameters.

To begin the test, the materials were set up layer by layer in the water jig container, with the lightest material at the bottom of the container. After 20 min of jigging, the density of the particles in each one of the ten layers containing the mixture was measured according to EN 1097-6. Figure 2.4 presents the density profile for each tested sample (Table 2.2).

Each density profile presented a sharp discontinuity or decreased linearly along the height of the container. For the three tests carried out with a 10/20 mixture of RA and gypsum particles (tests 1–3), one can notice the cut of density varying with the ratio of gypsum. When gypsum is replaced by brick (tests 4–6), the density decreased more progressively with the height of the layer. The 4/20 mm mixtures followed the same trends. For this size, the concentration of concrete in the bottom layers seemed to be higher. Finally, for the tests carried out with a ternary mixture (RA, brick, and gypsum), the concentration of concrete in the lower layers remained high, but some mixing was observed between the brick and gypsum particles in the top layers (tests 13 and 14).

Finally, good concentration and recovery of concrete were obtained for the concrete particles in the bottom layers. The granular size did not change these trends significantly. In all cases, the stratification of ternary mixtures was less good than that of binary mixtures.

2.3.3 Sensor-based sorting

2.3.3.1 Laboratory tests

Optical and NIR sensor-based sorting are tested with liberated particles of CDW on a laboratory scale (Le Guen 2015). For these tests, the pictures of the sample are captured by a camera, a Nikon D200 SLR, and loaded into a sorting program called PACT developed by the sorting machine manufacturer *Tomra*. The analyzer device for the NIR test works is the Perkin Elmer FTIR Spotlight 400 spectrometer.

The tested particles contain concrete, mortar, brick, ceramic, asphalt, gypsum, glass, and metal, with a grain size between 10 and 30 mm.

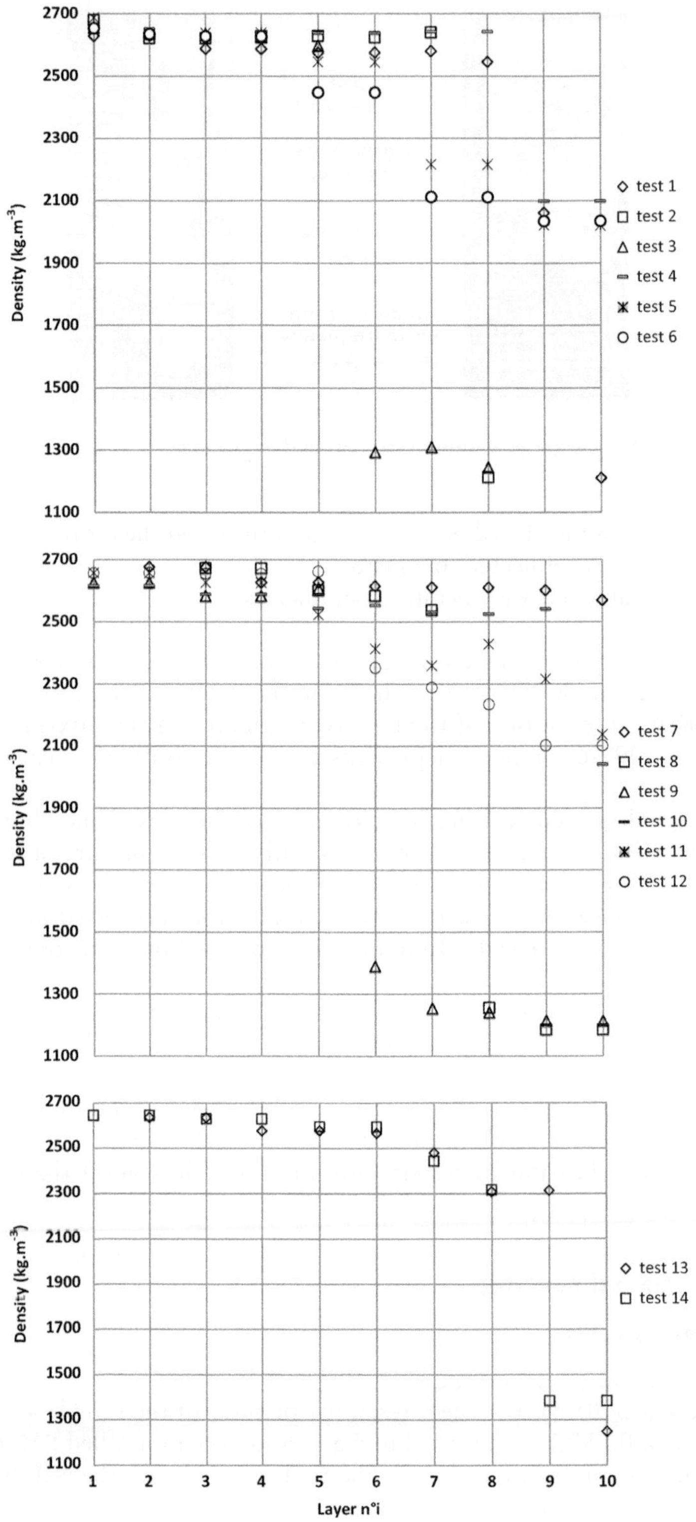

Figure 2.4 Density profiles. The layers are numbered from the bottom to the top.

The testing procedure is described as follows: forty-five particles are picked from the sample containing all eight fractions. For optical tests, the stones are photographed (wet and dry) and afterward processed by an optical separation program. A NIR spectrometer analyzes the same particles (dry; in this first stage only). The NIR spectra of the different fractions are compared between the wavelengths from 1,000 to 2,500 nm, which is common for NIR industrial sorters.

The optical sensor clearly differentiates the concrete particles from asphalt, brick, gypsum, metal, and glass. While differentiating concrete from ceramic and mortar particles, better results are obtained in wet condition.

NIR measurements show good results in the separation concrete from bitumen, glass, metal, and gypsum. Brick, ceramic, and mortar are found to be difficult to separate concrete from them by this technique.

2.3.3.2 *Industrial-scale tests*

Industrial-scale tests aiming the sorting of plaster from a "conventional" CDW are performed using an optical sorting technology. The technology was developed by Pellenc Selective Technologies. The detection of material and color is carried out by the combined NIR and X-ray sensors.

Regarding the results, the efficiency range defined by the ratio between the recovered materials and the materials in input is estimated to be 82%, with a mass loss of aggregates equal to 5% and a complete recovery of wood and gypsum. Data obtained with the industrial-scale tests demonstrate the efficiency of the optical sorting process for the application dedicated to CDW sorting with a high capacity of beneficiation.

2.3.4 Liberation

As shown in Figure 2.5, different processes are tested: mechanical wear, sandblasting, microwaves, and acid and thermal treatments (Braymand et al. 2016). These processes are tested with recycled aggregates collected from different origins.

Some criteria have to be considered to choose the relevant process: parent natural aggregate cleaning efficiency, non-damage of parent aggregate during the treatment, possible valorization of the second-generation waste, transferability to a semi-industrial scale, limited environmental and health impacts, and limited costs.

This primary stage indicates any single technology (low-energy microwave or sandblasting), which is inefficient. The chemical acid attack processes are efficient regarding technical objectives when parent aggregates do not react with the chosen acid. These methods could be considered as reference ones (Zhao et al. 2013a); however, due to their health and environmental impact, they can hardly be adapted to an industrial plan.

Any individual treatment is efficient enough to remove all the mortars without damaging primary aggregates: bulk density is lower than the value expected for a natural aggregate and water absorption is higher.

Combining the advantages of each process, described above, leads to the thermal–mechanical treatments proposed below (Shima et al. 2005). The selected relevant solution is to combine several processes and then their respective advantages. Figures 2.6 and 2.7 present the two chosen thermal–mechanical (hot and cold) processes, their parameters, and their variation ranges. The global process includes a pretreatment phase (water immersion or saturation), a thermal treatment, a mechanical posttreatment (wear and/or shocks), a separation step by sieving, and a quality check.

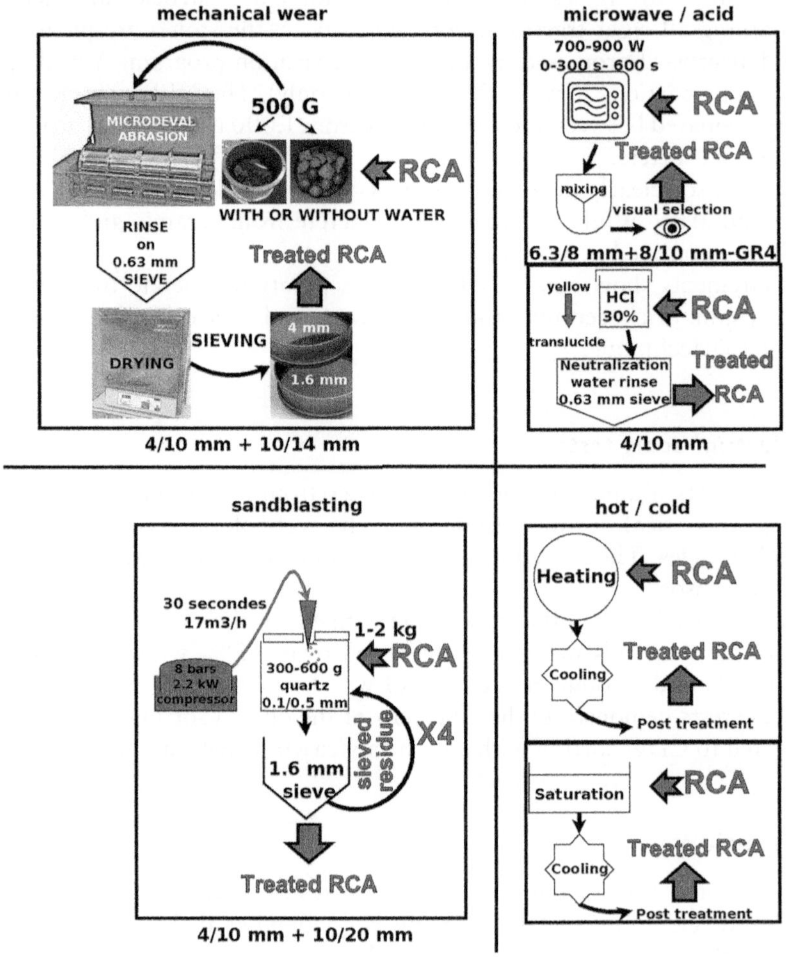

Figure 2.5 Simple separation processes tested in laboratory.

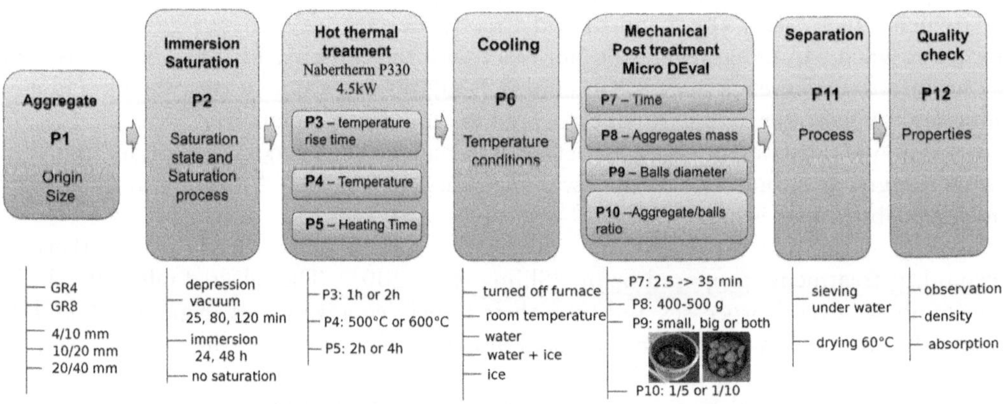

Figure 2.6 High-temperature thermal–mechanical processes and their parameter combinations (69 tests).

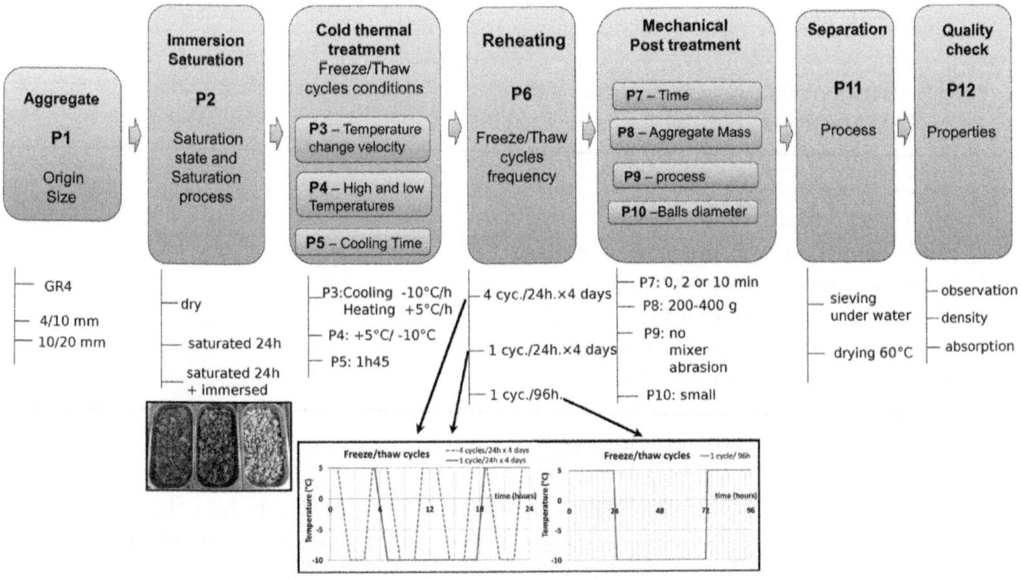

Figure 2.7 Low-temperature thermal–mechanical processes and their parameter combinations (54 tests).

Hot thermal processes lead to the damaging of the mortar by differential expansion and mineral phase (cement hydrates) destruction. A temperature higher than 500°C is required to obtain satisfying results, especially the portlandite degradation (de Juan and Gutiérrez 2009; Fares et al. 2010). Homand-Etienne and Houper (1989) confirm that a temperature of 600°C is needed and allows improving the mortar degradation in the case of parent siliceous aggregates thanks to quartz allotropic transformation. The efficiency of a complete process including presaturation or rapid cooling after heating (as presented by Bazant and Kaplan (1996) and confirmed by de Juan and Gutiérrez (2009)) is analyzed with the aim of identifying the relevant parameters.

Damage to primary mortar by freeze-thaw cycles is mainly due to thermal stresses, and the expansion of water during ice transformation and sample high saturation state can improve efficiency (Yang et al. 2006).

Damage to primary mortar during the hot thermal process is mainly due to the degradation of aggregates/mortar links if the temperature is more than 500°C. Next, a wear posttreatment for their separation is carried out by mechanical processes (as micro-deval equipment (MDE)). Mechanical posttreatment is efficient with an application time of at least 30 min, which is also confirmed by Yoda (Yoda et al. 2003). Enough processing duration and high-energy consumption are necessary to achieve a minimal efficiency of abrasion treatment by shocks (20 min for Los Angeles test).

Concerning the cleaning efficiency, most authors define this efficiency by the mass loss. First experimentations showed that this definition is insufficient. It should be observed that the density of treated clean aggregates is the same as that of parent aggregates. However, after thermomechanical tests, there is no correlation between sample mass loss and density of the treated material (Braymand et al. 2017a). For the following thermomechanical experiments, it was decided to use multicriteria selection as an efficient criterion: the ability to clean aggregate, their undamaging, the possibility of recovering the mortar residue with a view to its valorization, the adaptability to a semi-industrial scale, the limited environmental and health impacts, and the costs.

In order to illustrate these results, Table 2.3 presents the results obtained using the two optimized methods, the parameters of which are as follows:

- Hot thermomechanical treatment
  - No saturation
  - 1 h temperature increase up to 600°C
  - 2 h thermal treatment
  - Room temperature cooling
  - 10 min wear treatment with big balls
- Cold thermomechanical treatment
  - Samples saturated and immersed
  - Four cycles per day for 4 days
  - Temperature from 5°C down to −10°C
  - Temperature decreasing rate −10°C/h and temperature increasing rate +15°C/h
  - 10 min wear treatment with small balls

The mean density of the treated material before sorting (2.21) is less than that of the untreated material (2.24) because the treated material contains not only clean aggregates but also aggregates with undetached mortars and damaged aggregates. These results highlight the need to sort the material after treatment and sieving. At this time, visual checks and sorting are still the most appropriate methods for analyzing the cleanness of aggregates although observation can be subjective and damaged aggregate can be undetectable.

The mass percentage of cleaned aggregates that can be reused compared to treated recycled aggregates is less (19%). However, the aggregates obtained are of appropriate quality. Moreover, 24% of treated aggregates are cleaned but broken, 18% are composed of mortar, and 27% are partially cleaned. The rest is composed of fine particles. The density, porosity, and absorption measured with each type of aggregate are presented in Table 2.4.

2.4 RESEARCH NEEDS

The development of sorting equipment designed for CDW is incipient. Presently, the existing technologies are generally too expensive, and their use is not justified by a sufficient efficiency. Research effort is needed to adapt the existing technologies to the particularities of the CDW and to reduce the economic risk of the investment in new technologies. It should be underlined that the development of new equipment should consider the interaction between the different operation units, and research should focus on new integrated

Table 2.3 Thermomechanical—optimized process results

10/20 mm concrete aggregate	*Cold treatment*	*Hot treatment*
Mass loss after thermal treatment (%)	4.38	4.59
Mass loss after mechanical posttreatment (%)	11.35	25.59
MDE balls		
Total mass loss (%)	16.47	28.86
Fine particles (<1.6 mm) (%)	1.95	9.77
Sample initial density (%)	2.24	2.24
Treated sample density (%)	2.29	2.21

Table 2.4 Density, porosity, and absorption measured for each type of aggregate

Aggregate	*Density (g/cm³)*	*Porosity (%)*	*Absorption (%)*	*Picture*
Clean and undamaged 4–10 mm	2.72	2.70	0.99	
Clean and damaged	2.35	2.41	5.67	
Partially cleaned	2.39	8.21	3.43	
Mortar	2.09	17.67	8.46	
Untreated aggregate	2.25	12	6	
Initial 4–20 mm concrete aggregate	2.55	1	0.75	

processes. Of course, continuous sorting method should be used, as batch systems are not adapted to the process in a recycling plant.

2.5 CONCLUSIONS

The experiments carried out in the project brought some elements of better understanding for several operation units of the recycled aggregate processing, which could potentially improve the current practice.

Concerning crushing, no specific strategy allows to produce RA with both narrow size range and good quality. It was concluded that several strategies are possible, ranging from the highest liberation of the natural aggregate of the RA to the will of maximizing the reuse of CDW. Moreover, scale-one experiments are needed to confirm these laboratory findings.

The sorting by density was also investigated. The tests demonstrated the sorting capability of the water jig applied to concrete, brick, and plaster materials. The efficiency of this sorting process was validated for both binary and ternary mixtures. It appeared that the gypsum material was sorted of the mixtures where it was present. This phenomenon was noticeable for the mixture composed of two or three materials. This was due to a large difference in density between plaster and the other two materials (brick and crushed concrete). For the brick, the sorting efficiency was less important. It is more difficult to extract this material from a mixture with RA. This was due to a small difference in density between brick and RA in particular. It was however possible to improve the sorting with a large part of brick. Globally, the water jig process seemed to be a relevant process for the sorting of concrete, brick, and plaster materials. The sorting efficiency appeared to be correct up to relatively low levels of "undesirable" materials (about 10%). It seems pertinent to integrate this process with other processes, such as the sensor-based sorting, to fine-tune sorting at the end of the chain, especially to ensure the quality of the sorting between the concrete and the brick.

The test of the sensor-based sorting had brought promising results. First, the optical sensor could clearly differentiate red brick, white gypsum, dark asphalt, transparent glass, and dark gray metal from the concrete. It was more difficult to sort the mortar and the ceramic from concrete. Indeed, the ceramic appeared in many different colors, so some particles had similar color scheme compared to the concrete particles. The mortar had generally color scheme similar to the concrete. Nevertheless, under wet conditions, the existing small differences could be increased enough to potentially produce a clean concrete particle mixture. More generally, the optical sorting of wet particles had achieved better results than the dry sorting.

Next, NIR measurements also showed the capacity to distinguish the brick, gypsum, asphalt, glass, metal mortar, and ceramic from concrete. For the mortar and ceramic fractions, the results were more promising than for the optical sorting.

Finally, the liberation of original aggregates from recycled concrete was investigated. Few available technologies are likely to be transposed to an industrial scale at reasonable environmental and economic costs. High- or low-temperature treatment favors the damage to the mortar that can then be liberated from parent aggregates thanks to mechanical posttreatment.

Some complementary conclusions were also highlighted as follows:

- A high mass loss was not representative of an efficient treatment.
- High- or low-temperature thermal treatments favored the damage to the mortar which can be detached from the parent aggregate by mechanical posttreatment.
- For high-temperature thermal treatment, a temperature up to 600°C was necessary to degrade the constituents such as CSH and portlandite. However, this temperature led to the damage to calcareous and calcareous-silica aggregates.
- For low-temperature thermal treatment, industrial conditions had limited the value of the temperature to be applied and the duration of the treatment, which limited its efficiency. Immersion and saturation were necessary before cooling to ensure liberation of the mortar.
- To obtain a cleaned aggregate, a visual sorting would be necessary to complete sorting by sieving.
- The efficiency of treatments remained low since, for the studied treatments, the effective efficiency, characterized by the proportion of material cleaned and undamaged relative to the initial quantity, did not exceed 26%.

Chapter 3

Characterization of recycled concrete aggregates

S. Rémond
LGCgE, IMT Lille Douai

J.M. Mechling and R. Trauchessec
Institut Jean Lamour, Université de Lorraine

E. Garcia-Diaz
Centre des Matériaux des Mines d'Alès

R. Lavaud
Technodes S.A.S. (HEIDELBERGCEMENT Group)

B. Cazacliu
IFSTTAR

CONTENTS

Abstract

Recycled (concrete) Aggregates (RA) are mainly composed of an intimate mix between Natural Aggregates (NA) and adherent cement paste (ACP). Properties of RA strongly depend on the properties of each of these two phases and on their proportions. ACP possesses in particular a much larger porosity than NA which changes significantly RA characteristics in comparison to those of NA. RA exhibit larger water absorption and lower particle density than NA. Moreover, their chemical composition is also affected by that of ACP and NA which might play a role in some pathology like alkali silica reaction. Main physical and chemical characteristics of RA have been measured in the RECYBÉTON project on coarse and fine recycled aggregates using experimental methods devoted to NA. Results show that traditional characterization procedures used for NA can be applied very satisfactorily to coarse RA. On the contrary, the measurement of properties of fines RA is more difficult. Indeed, crushing of old concrete concentrates ACP in the finer granular fraction, resulting in important heterogeneities in the composition of recycled sand comparatively to coarse RA. Fine RA contain more ACP than coarse RA leading to larger absorption and lower particle density for the smaller fraction. In order to improve the characterization methods of fine RA, chemical and thermal methods have been developed in order to assess the ACP content of recycled sand. From theoretical relations between ACP content and physical properties of RA, important properties like water absorption coefficient can be measured more accurately. Moreover, because of their characteristics, it is difficult to replace completely fine NA by fine RA in concrete. One solution is to mix fine RA with fine NA. Thus, methods have been developed to precisely control the proportion of the two constituents in mixtures containing a low proportion of RA (less than 10%). Mechanical properties of RA like resistance to shocks and abrasion and application of models allowing the prediction of compressive strengths of mortars and concretes made with RA have also been investigated. The results show that RA exhibit a larger "ceiling effect" than traditional NA, however, they lead to good bond properties between RA and new cement paste.

3.1 INTRODUCTION

Recycled concrete aggregates (RAs) are composed of an intimate mix of natural aggregates (NAs) and adherent cement paste (ACP). They also contain impurities such as gypsum, asphalt, metals, plastic, soil, wood ... coming from other construction and demolition wastes (Silva et al. 2014c). RAs are therefore more heterogeneous than NAs. Also, due to their particular composition, they possess properties that differ significantly from those of NA. In particular, ACP gives RA higher porosity, higher water absorption (WA), and lower particle density than NA. The high WA of RA can have a particular influence on the properties of recycled concrete. A wrong determination of WA leads to large differences between the target efficient water-to-cement ratio (W_{eff}/C) and that prevailing in fresh concrete, inducing improper fresh and hardened properties for the in-place material. Unfortunately, the standard methods used for the characterization of NA are not always appropriate to RA, and some new procedures have to be developed. ACP and impurities such as gypsum affect the durability of recycled concrete. Production process of RAs, especially crushing, also influences their properties; different crushing processes lead to different shapes or different particle size distributions. In this chapter, the composition and chemical properties, the physical properties, and finally the mechanical properties of RA are successively studied. In each case, the peculiarities of RA in comparison with NA are pointed out and various specific experimental procedures developed for their measurement are presented. Moreover, a procedure for assessing the RA content in a mixture of RA and NA is proposed. Based on three main properties (absorption, soluble sulfate, and bituminous impurities) and different sources of recycled and natural sands, dosage reliability of the two constituents can be evaluated.

3.2 COMPOSITION AND CHEMICAL PROPERTIES OF RA

3.2.1 State of the art

3.2.1.1 Measurement of ACP content or adherent mortar content

When concrete is crushed, some cement paste remains attached to original aggregates. ACP is more porous than NAs generally used for the manufacture of concrete. Therefore, the quality and quantity of ACP are at the origins of the poorer properties of RA compared to NAs: lower particle density, higher WA, higher Los Angeles (LA) coefficient, and higher sulfate content. Knowing the ACP content of RA would be of great interest for the assessment of RA quality. Several methods have therefore been developed for its quantification. However, the measurement of ACP content can be difficult, and several experimental methods have been developed in order to determine adherent mortar (AM) content. AM is easier to quantify experimentally than ACP; however, its definition is more ambiguous than the latter. First, the definition of AM closely depends on the maximum particle size chosen for the separation between mortar and concrete (generally between 4 and 5 mm). Second, AM might be composed not only of the mortar of parent concrete but also of coarse NAs crushed during the process. AM therefore depends a lot on the experimental procedure used for its determination.

Several methods have been proposed in the literature for the measurement of ACP or AM. They are based either on thermomechanical, chemical, or geometrical methods, or even on combinations of these.

Thermomechanical methods attempt to separate AM from coarse aggregate by favoring the development of cracks between these two phases. De Juan and Gutiérrez (2009) used in particular a thermal method based first on the soaking of RA in water for 2 h in order to quasi-saturate, second on heating at 500°C for another 2 h, and third on a new soaking in cold water. Heating of saturated mortar at 500°C generates water vapor inside AM; moreover, sudden cooling of the heated aggregates induces stresses and cracks at the interface between the mortar and the aggregate. These two steps allow an easy separation between both materials. Some mortar generally remains attached to aggregates after these steps and has to be removed with a rubber hammer. The sample is finally screened on 4 mm sieve to separate the coarse aggregate from which all the mortar has been removed. AM content is calculated from the mass loss between the original recycled aggregates and the obtained coarse aggregates.

Chemical methods are based on the selective dissolution of the ACP. Solutions of hydrochloride acid allow for an efficient dissolution of ACP and have been used in several studies (Yagishita et al. 1994; Nagataki et al. 2004). This method permits to measure the ACP content (and not only the AM content), but it is not suitable for RA coming from concretes containing calcareous aggregates, which are also partly dissolved in hydrochloric acid.

Cement content in ACP can also be assessed using tracers of cement such as SiO_2, Al_2O_3, and Fe_2O_3 (Belin et al. 2014). RA has first to be ground down to 315 μm, and the resulting powder is diluted in (1/50) nitric acid solution. After filtration, the solution is analyzed and

ACP content can be determined from the ratio between the mass of SiO_2 per gram of RA and the mass fraction of SiO_2 contained in cement. This method can, however, only be used if the composition of original cement is known.

Abbas et al. (2007) proposed a combined chemical and thermomechanical method, based on freeze–thaw cycles applied on RAs previously immersed in a sodium sulfate solution. RAs are first dried at 105°C, weighted, and immersed 24 h in a 26% (by weight) sodium sulfate solution. They are then subjected to five freezing and thawing cycles (16 h at −17°C and 8 h in an oven at 80°C). The sodium sulfate solution is drained from the sample by washing aggregates with tap water over a 4.75 mm sieve, and RAs are then dried at 105°C and weighted. The residual AM content is obtained from the difference of mass before and after freeze–thaw cycles on the dry samples.

Several authors (Hansen and Narud 1983; Abbas et al. 2009) proposed to determine the AM content on coarse RA by observing a polished section of a new concrete made with coarse RA and new colored cement. Colored cement allows an easy identification of particles and an easy separation between NA and AM. The quantification of mortar surface area can then be carried out either by a linear traverse method (similar to that described in ASTM C457 (2012) for the measurement of air void content in hardened concrete) or by image analysis. The AM content can later on eventually be determined on a mass basis by analytical equations (Abbas et al. 2007).

De Juan and Gutiérrez (2009) compared the AM contents obtained in the literature with different RAs and different experimental methods (acid dissolution, production of new concrete with colored cement, or thermal treatment). Obtained values vary in a wide range and differ depending on the studied RA and used methods. They vary between 25% and 70% when measured with acid dissolution, between 25% and 65% with colored concrete, and between 40% and 55% with thermal methods.

Two alternative methods (microwaves and high-voltage pulses) used for the separation of cement paste and NA have also been developed in the frame of the COFRAGE project (Bru et al. 2014; Touzé et al. 2017). The goal here was to separate as much as possible the ACP from the NA in order to allow for an upcycling of RA.

The first method relies on the use of microwaves which induce a differential thermal expansion between ACP and NA, because of the different dielectric and thermal properties of these two phases. This treatment allows for weakening of the bond between ACP and NA prior to crushing. Laboratory experiments showed that the liberation degree of coarse aggregates submitted to impact crushing increased from 71.6% up to 90.1% if crushing was preceded by a microwave treatment. Liberation degree is defined here as the ratio between the mass content of liberated aggregates (particles without ACP, sorted visually) and the mass content of all aggregates included in a given size fraction. Results also showed that the liberation efficiency of NA decreased with the particle size.

The second method is based on the application of high-voltage pulses through blocks immersed in water. The high electric field creates local plasma at the interface between materials of different electrical properties resulting in thermal expansion accompanied by a radial shock wave. In contrast to microwaves, which are used as a pretreatment prior to crushing, electrical fragmentation can be used alone. Authors showed that the method allows for good liberation degrees (between 53% and 83%), the efficiency increasing with the size of particles.

3.2.1.2 *Parameters influencing ACP or AM contents*

Both ACP content and properties depend on the composition of the parent concrete. A larger proportion of cement and water in the parent concrete mix will result in larger ACP content in the crushed concrete. However, ACP is not homogeneously distributed in the different granular fractions of crushed aggregates. It has been shown in several studies that the AM content was lower for larger size RA (Etxeberria et al. 2007a; de Juan and Gutiérrez 2009; Topçu and Sengel 2004). Moreover, the distribution of ACP among granular classes of RA is largely influenced by the crushing process. Nagataki et al. (2004) studied the influence of the crushing method on the AM content in coarse RA. They showed that using only jaw and impact crushers, the AM contents of the RA (fraction 5/20 mm) obtained from high-, medium-, and low-quality concrete were 52.3%, 55.0%, and 52.3%, respectively. Using several grinding equipment after first crushing, the AM contents were reduced to 30.2%, 32.4%, and 32.3%, respectively. Several successive crushing processes can therefore decrease the AM content of coarse RA, whereas they increase the quantity of fine RA.

3.2.1.3 Mineralogical and chemical characteristics of RA

Little publication gives precise and exhaustive data concerning chemical and mineralogical characteristics of RA. Some studies were focused on the role of the residual AM on the RA's physical and mechanical general properties. Compared to NA (Tam et al. 2007; Al Bayati et al. 2016; Le et al. 2016), AM provokes higher porosity associated with lower density, an easier formation of tiny cracks during the concrete crushing process, and the weakest adhesion along the interfacial transition zone (ITZ). From a chemical point of view, AM also presents non-negligible contents in alkali elements and sulfate (Sánchez de Juan and Alaejos Gutierrez 2004b). Some authors Mahmoud (2005) and Sánchez de Juan and Alaejos Gutierrez (2004b) established correlations between the chloride and sulfate solubilities and AM contents in RA.

Most of the studies are dedicated to the properties of concrete incorporating recycled materials and provide characterizations of RA in accordance with the main classifications. For example, RILEM (1994) and BRE Digest (BRE 1998), respectively, define three types of RA considering the proportions of several materials such as masonry rubble, natural and recycled aggregates, brickwork, and concrete rubble. But the more common descriptions are now derived from the European Standard EN 933-11 (NF EN 933-11 2009) that defined five categories of materials in the margin of the (principal) concrete phase. It permits to describe and estimate the content of each contaminant and thus the RA's purity. The impact of the contaminant categories on fresh and hardened concretes has been otherwise evaluated by various studies (Martin-Morales et al. 2013).

On the other hand, various standards and guidelines concerning chemical requirements for the reuse of RA in concrete (Martin-Morales et al. 2013) deal more specifically with characteristics centered on total and soluble (in water and/or in acid) chloride and sulfate contents, or presence of organic matter. The French project DREAM (MRF 2014) defined a technical method to quickly determine the soluble sulfate amount produced by the leaching of demolition waste materials as RA, in order to define a potential reuse.

The knowledge of the detailed mineralogical and chemical characteristics of RA is nevertheless important to prevent risks as alkali–silica reaction (ASR) (detection of unstable siliceous phases with a microscope) and for the understanding of pathology mechanisms (see Sections 12.2.3 and 12.2.4). Calvo Pérez et al. (2002) conducted a detailed study on different RAs by observing thin sections under a petrographic microscope. Another way of study was chosen by Limbachiya et al. (2007) who characterized three coarse RAs (4–16 mm size fraction) arising from the London area. X-ray fluorescence (XRF) results show that the origin of the RAs has a negligible effect on the major elements. The mineralogy of the RA (X-ray diffraction [XRD]) indicates the presence of calcite and quartz associated to minerals directly proportional to the NA introduced in the original concretes. Experimental

RAs obtained by crushing experimental concretes after 28 days (Evangelista et al. 2015) show similar results with the predominance of quartz and calcite on XRD patterns. NAs correspond to small picks linked to their most abundant minerals (K–Na Feldspars, etc.). The hydration of the cement leads to the presence of gypsum and residual ettringite and portlandite. This study, like others (Hansen 1992; Sánchez et al. 2004), also concludes that smaller size fractions present high AM content with scarce presence of NA. The RA studied in the RECYBÉTON French National Project has been fully characterized, including chemical and mineralogical investigations.

3.2.2 RECYBÉTON's outputs

3.2.2.1 *Materials and sampling*

In the beginning of the RECYBÉTON project, stocks of homogeneous natural and recycled aggregates have been provisioned. Laboratories have then been supplied with representative samples throughout the duration of the project. The main characteristics of these materials are recalled in Appendix of this book.

3.2.2.2 *Measurement of AM or ACP content*

Braymand et al. (2016) tested different experimental methods proposed in the literature for the separation of attached mortar from RAs (mechanical wear, sandblasting, chemical attack, microwave, hot or cold treatment, or different combinations of these). They showed that none of the tested methods was able to remove completely the AM from RAs. Among the tested methods, combined thermal–mechanical processes were considered to be the most efficient in separating the AM from NAs. Authors also pointed out that assessing the separation efficiency was tricky, because it depended on the choice of the sieve used to separate mortar from NAs and because separation methods also resulted in damage to the NAs.

The literature has shown that several methods existed for the measurement of AM content of coarse RA, but very few of them were eventually suitable for the measurement of ACP content. It has also been shown that crushing of concrete induced a concentration of ACP in the smaller granular classes. Therefore, experimental methods are needed for the assessment of ACP in fine RA. Determining exactly the ACP content might be difficult. However, in some cases, the measurement of a quantity proportional to ACP gives sufficient information for the characterization of RA.

Among the existing methods, the chemical method, such as dissolution in hydrochloric acid solution, can be applied either to fine or to coarse RA. However, hydrochloric acid is not suitable for RA containing calcareous aggregates. Zhao et al. (2013b) have therefore developed a method based on the dissolution of ACP in the solution of salicylic acid in methanol. Salicylic acid allows selective dissolution of most of the phases composing the cement paste excluding the dissolution of major phases of NAs used for the manufacture of concretes. The developed procedure is given as follows:

- A representative sample of RA is dried at 105°C and ground until passing 0.2 mm.
- 0.5 g of dried sample is then immersed in a solution of 14 g of salicylic acid in 80 mL of methanol and stirred during 1 h.
- The solid fraction is filtered on a glass filter (pores: 10–16 μm) and washed four times using methanol (2–3 mm high on top of the filter).
- The solid residue is dried in an oven at 70°C for 30 min, and the ACP is then estimated from the soluble fraction in salicylic acid (SFSA).

Accuracy of the estimate of ACP by SFSA mostly depends on the amount of soluble and insoluble phases contained in the cement paste and in the NAs of RA. Calcium silicates and their corresponding hydrates are dissolved in salicylic acid, whereas calcium aluminates and their corresponding hydrates (except ettringite) are insoluble. Moreover, quartz, dolomite, and calcite are not dissolved in this acid. SFSA should therefore depend on the chemical composition of cement. In order to assess the impact of the cement type on SFSA, experiments were performed on two pure cement pastes (*W/C* ratio of 0.5) manufactured with either CEM II (gray cement containing calcite as a filler) and CEM I (white cement with no C_4AF). Two NAs, a siliceous and a calcareous one, were also tested.

Table 3.1 presents the obtained results. Dissolution of white cement paste is almost complete, showing that, in the case of white cement, SFSA is quasi-equivalent to ACP content. For the gray cement paste however, SFSA is much lower than cement paste content.

Table 3.1 also shows that both tested NAs present a very low SFSA. The SFSA of calcareous aggregate is somewhat larger than that of siliceous one, but in both cases, it can be considered that NAs are not affected by the used dissolution protocol. According to these results, SFSA could be used to determine a quantity proportional to ACP content in RA.

SFSA has been measured according to the proposed protocol on six fine RAs (Zhao et al. 2015a). Three RAs were obtained from the crushing of laboratory concretes cured more than 90 days (RA-OC1, RA-OC2, and RA-OC3) in a mini jaw crusher. The RAs selected for the RECYBÉTON National Project (RA-NP) and other two industrial RAs obtained from other recycling platforms (RA-I1 and RA-I2) were also tested. Each RA was divided into four granular classes (0/0.63, 0.63/1.25, 1.25/2.5, and 2.5/5 mm), and three dissolution tests were carried out on each fraction. Figure 3.1 presents the variation of SFSA as a function of mean size of granular classes. It shows that SFSA varies in a very large range, depending on the nature and granular class of RA. As seen previously, it is not possible to compare the absolute values of SFSA from one RA to another, in particular for industrial RA, because they are sourced from different parent concretes probably containing different types of cement. However, for a given RA, SFSA clearly increases when particle size decreases. Therefore, ACP increases when the size of particles decreases. This conclusion had already been drawn in the literature for coarse RA. It is shown here that it can be extended to fine RA.

Measurement of SFSA might be tricky, because it has to be carried out on very small samples, unless using large amounts of methanol. A strict sampling is therefore necessary. An alternative method used to study larger samples, based on the mass loss of RA between 105°C and 475°C, has therefore been proposed by Zhao (2014) and developed by Le et al. (2016). This method differs from the thermal methods used in the literature for the measurement of AM content. Here, mass loss is used to assess the quantity of bound water contained in the ACP. The first heating at 105°C allows removing free water from RA; the second one at 475°C allows dehydrating major part of hydrates, without affecting the NA. Once again,

Table 3.1 SFSA (% by weight) obtained on two cement pastes and two NAs

	Test 1	*Test 2*	*Test 3*	*Average*
White cement paste	95.5	96.4	94.9	95.6
Gray cement paste	62.6	63.1	63.3	63.0
Siliceous sand	0.8	0.9	0.9	0.9
Calcareous sand	3.4	3.0	3.2	3.2

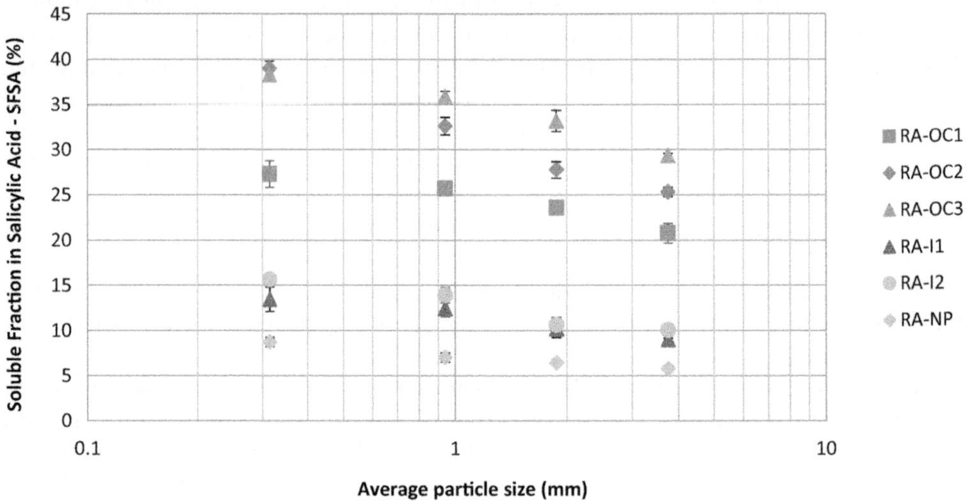

Figure 3.1 Variation of SFSA with the average particle size for different fine RAs.

the mass loss between 105°C and 475°C does not give the ACP content, but a quantity that is proportional to it. The experimental procedure is given as follows:

- A dry representative sample of RA is ground down to a particle size lower than 0.2 mm and put at 105°C for 24 h.
- 10 g (or more) of dried sample is then heated at 475°C for 24 h.
- The amount of bound water is calculated from the mass loss between 105°C and 475°C.

This method was carried out by Le et al. (2016) on a recycled sand and gave results similar to those obtained by Zhao et al. (2015a) with SFSA concerning the variation of ACP as a function of particle size.

3.2.2.3 Mineralogical and chemical characteristics of RA

This study was applied to nine RAs used in the RECYBÉTON project, four fine aggregates (size 0–4 or 0–6 mm), and five coarse aggregates (size 4–20, 4–10, or 10–20 mm), conditioned (crushing and sorted) from four recycling plants (identified 1–4). They were chosen to be representative of major French urban and industrial basins distributed throughout the country as well as to be representative of different French geological contexts (Table 3.2).

Table 3.2 Main geological contexts of the industrial basins where RAs have been produced

Platform	*Main geological contexts*
1	Periphery of the sedimentary basin of Paris—proximity of an ancient metamorphic basal complex
2	Alluvial basin of Rhône and Saône French rivers—proximity of the Alps Mountains
3	Center of the sedimentary basin of Paris
4	Alluvial basin of Rhine River—proximity of Vosges and Black Forest mountains (France and Germany)

The coarse RAs have been analyzed in accordance with the European Standard EN 933-11 (NF EN 933-11 2009). The results (Table 3.3) give an overview of the contaminants (typology and respective contents) present in the different RA coarse productions (by extension, the results for the fine fractions could be considered as similar). The purity of the products is then expressed by the Rc and Ru categories that correspond to fragments from the initial concretes. Except materials from the platform no. 1 that contain a significant part of bricks and bituminous rubbles (Rb+Ra≈20%), the other two platforms present coarse (and fine) RA rich in concrete fragments (>95%). The coarse RA from platform no. 2 could not be characterized, but few contaminants were visibly present.

Petrography and mineralogy of RA have been established with the complementary techniques of visual (macroscopic and microscopic) observations and XRD. Indeed, a macroscopic description of the coarser grains can easily and quickly reveal the main types of rocks (Table 3.4) used in the original concrete mixes. For the four platforms studied, the results are in conformity with the geological environment of their drainage basin despite two parameters that clearly appear: (a) Most of the concretes incorporate alluvial quartz sand, and (b) limestones are constantly present even in the production basin crossed by rivers, such as Rhine or Rhône, flowing from crystalline geological provinces (the Alps Mountains).

XRD analyses were conducted on each fine and coarse RA. Different materials (different particle size distribution) produced on the same platform with the same deconstruction wastes show very similar XRD patterns (Figure 3.2). In accordance with the previous remark concerning quartz and limestone predominance, quartz and calcite are the two minerals that systematically appear with a very strong intensity whatever the RA provenance is. The high content of quartz mainly results from the use of alluvial siliceous sand in most of the concrete mixes, eventually reinforced by the use of coarse aggregates containing quartz. The high content of calcite is explained by its two origins. The presence of calcareous aggregates and the carbonation of portlandite initially released during the cement hydration. Indeed, there was no detectable trace of portlandite on the different patterns.

Table 3.3 Constitution of the coarse RA according to the European standard EN 933-11

Platform	*Particle size*	*X*	*Rc*	*Ru*	*Rb*	*Ra*	*Rg*
1	4–20	0.02	73.1	4.0	17.3	2.5	0.30
3	4–10	0.02	65.5	32.5	0.6	0.8	0.03
3	10–20	0.01	94.9	0.03	3.8	1.5	0.02
4	4–20	0.02	85.0	13.1	1.8	0.3	0

X: Cohesive, metals, woods, plastics, plaster.
Rc: Concrete, concrete products, mortar, masonry concrete element.
Ru: Single aggregate, natural stone, aggregates treated with hydraulic binders.
Rb: Masonry elements made of clay, non-floating aerated concrete.
Ra: Bituminous materials.
Rg: Glass.

Table 3.4 Visual identification of the main rocks present in the RA

Platform	*Main types of rocks identified excluding quartz sand*
1	Limestones and metamorphic rocks (shales and quartzite)+artificial pollution (tiles and bricks)
2	Limestones, sandstones, various magmatic (granite, etc.) and metamorphic (micashist, etc.) rocks
3	Limestones rich in cherts
4	Limestones, sandstones, various magmatic (granite, etc.) and metamorphic (shales, Quartzite, etc.) rocks

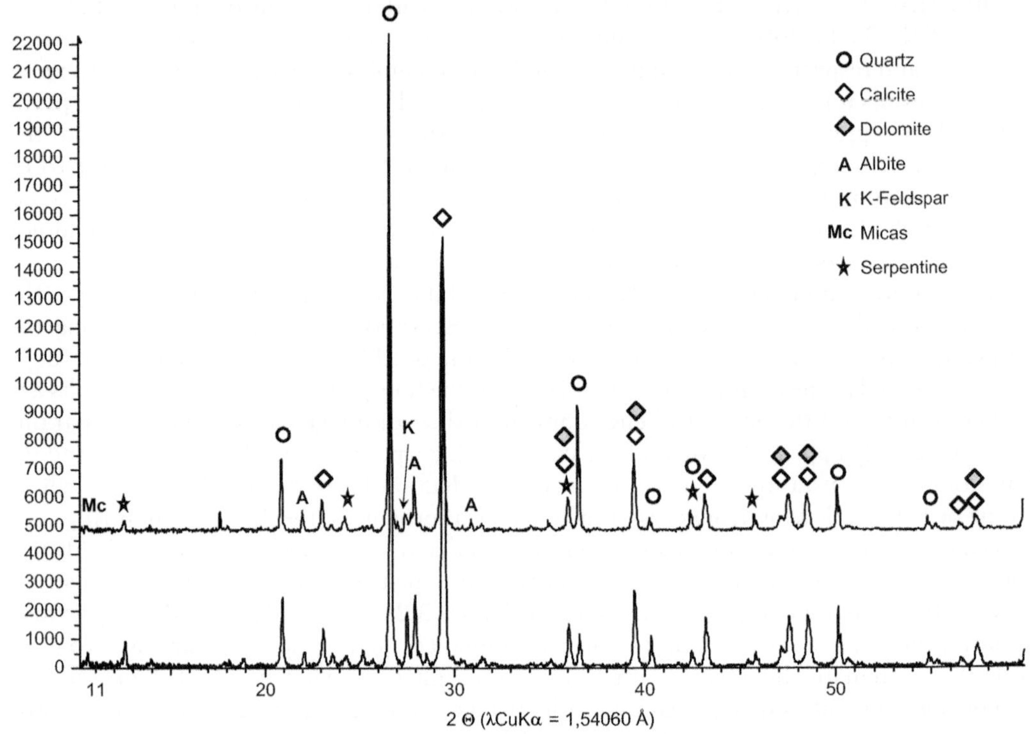

Figure 3.2 XRD patterns obtained on two RAs from platform no. 2: 0–4 mm (top) and 10–20 mm size fractions (bottom). 2Θ scale is from 8° to 69° (λCuK_α = 1.54060Å).

Only the major minerals that characterized the different rocks (NA) present small peaks that can rather be detected in coarser than in the finest granular fractions. The main crystals detected (XRD) for the different platforms are presented in Table 3.5. Silicates principally evolve between K–Na feldspars, plagioclases, and white micas. Dolomite sometimes weakly appears in addition to the calcite.

Other techniques have to be carried out to reveal the presence of minor phases such as unstable siliceous minerals or some altered phases suitable for soluble silica or alkali elements.

Observations of thin sections under a petrographic microscope have been carried out in order to round off the previous results. Six thin sections were made for each platform. The finest materials were sieved to prepare three samples, respectively, sized 0.16–1, 1–2, and 4–6.3 mm, whereas representative grains of the main rocks were chosen among the RA coarser aggregate. These different samples were then casted into a mold using thermosetting resin. The usual manufacturing process was then applied to each casting. Petrographic

Table 3.5 Main minerals (excluding quartz and calcite strongly present) detected with XRD

Platform	*Main minerals identified on XRD patterns, excluding quartz and calcite*
1	Dolomite and K–Na feldspars (microcline/albite)
2	Micas (muscovite) and K–Na feldspars (microcline/albite), chlorite
3	Dolomite and K-feldspars, very low peaks
4	Dolomite and K–Na feldspars (microcline/albite)

investigations confirm the XRD results and determine punctual minerals such as chlorite in a very small quantity. They also give information about the grain structures (Figure 3.3) and show that the finest grains are essentially composed of single crystals more or less surrounded by ACP or only formed with ACP. Quartz seems to be the predominant variety. When the size increases, the grains can contain several crystals linked by ACP or they can correspond to small pieces of rocks (that can be partially or totally surrounded by ACP). The coarser RAs (gravels) are characterized by a large diversity of rock fragments which is typical of the geological context of the area where aggregates and concretes were produced. In this last case, ACP represents the less abundant fraction of the RA volume, and it does not surround all the grains.

From the finest to the coarser grains, the ACP content slightly diminishes, whereas the mineralogical diversity increases in terms of ratios. ACP tends to form small particles during the crushing process. Hansen (1992) and the further analyses (XRF) attest the concentration of the ACP in the fine fraction (sand and filler particles). On the other hand, the mineralogical diversity of the coarser grains also increases the possibilities for the presence of unstable silica. Estimation of ASR risks induced by mineralogy and petrographic microstructures requires the same petrographic techniques for both NA and RA products.

The petrographic investigations carried out on different RAs show the presence of unstable silica phases (Figure 3.4), principally on the coarse RA. Observations of thin sections can be, for example, conducted in accordance with a standard such as the French one XP P18-543 (2015). Chalcedony, quartz with undulose extinction, altered silicates, secondary/micro quartz, volcanic glass, and so on were punctually observed and correspond to a feeble

Figure 3.3 Various aspects of the aggregates (platform no. 4) function of their sizes. Field: 3.2 × 2.6 mm (cross-polarized light).

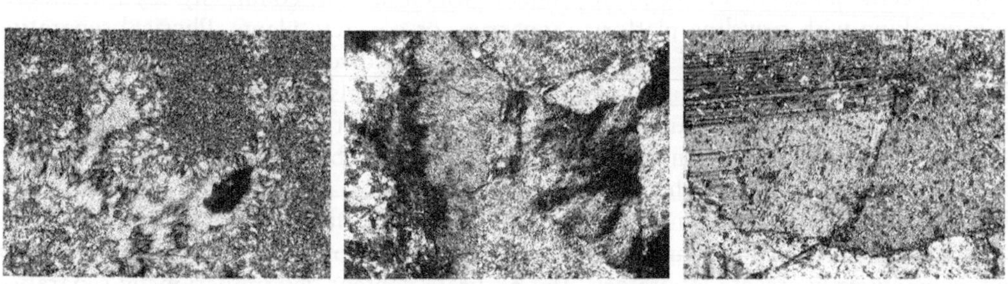

Figure 3.4 Unstable silica: (a) fibrous silica (chalcedony) in a chert; (b) altered feldspar. Field: 3.2×2.6 mm (cross-polarized light); (c) quartz with undulose extinction.

quantity. These investigations also confirmed that an important part of the silica contained in the RA from the platform no. 3 is constituted by chert.

Chemical compositions of each fine and coarse RA have been determined by XRF after some losses on ignition were done. Despite the presence of diversified siliceous rocks, the XRF results (Table 3.6) are consistent and do not present extreme variations of the main elements. The presence of large amounts of quartz and calcite could smooth these results. The respective content values evolve in relatively similar ranges. The standard deviations present some relatively low amplitudes (<10%) for the major elements, and they are less pronounced (average) for the fine aggregates than for the coarser ones.

ACP contains a non-negligible part of sulfate from the initial cement. The fine RAs are enriched in ACP as shown by petrographic investigations but also by the loss on ignition values (2% higher for fine RAs compared to coarser ones). Sulfate content is higher in the fine RAs compared to the coarser ones (0.46% in average to 0.30%).

Alkali elements present an average total content close to 0.8% for K_2O and 0.4% for Na_2O, for both fine and coarse RAs. There exist, however, some significant content variations from one RA to another, and relative standard deviations are therefore slightly higher for the alkalis than for the major elements, especially for Na_2O (relative standard deviations are, respectively, 65% and 42% for fine and coarse RAs).

These nine RAs produced on four platforms show the existence of great chemical and mineralogical similarities between the materials (fine and coarse RAs) of the same origin (platform). Quartz and calcite (i.e., limestones) are the most predominant minerals of the RAs independently from their geographic and geological contexts. Other minerals (and rocks) are nevertheless linked to the local geological situation. The more common phases are K–Na feldspars and plagioclases, white micas for the silicates, and dolomite for carbonates. Fine RA contains more ACP than coarser RA. RA chemical compositions of the major elements (Si, Ca, Al, Fe) evolve in the same range with a low relative standard deviation. Differences are relatively more pronounced for sulfates and alkali elements that are furthermore especially concentrated into the fine RA.

3.2.2.4 Assessment of the amount of RAs in a mixture between recycled and natural sands

Commercialization of recycled and natural sand mixtures for concrete production is a promising way for recycled sand valorization. However, quality control requires the possibility of assessing RA proportion in the mixtures for producers and users.

As RA contains ACP, recycled sands have lower density than natural sands so their separation could be done using heavy liquid, cyclone, and so on. However, these techniques require specific materials or chemical products which are not commonly used in laboratories working with recycling platforms, quarries, or concrete plants. Physical separation could also be based on color or shape differences, but the diversity, spatial, and temporal of NAs and RAs leads to a wide range of properties. RAs are also not homogeneous; there are particles with different color and density because of impurities and heterogeneous cement paste repartition among the sand grains. Moreover, it might be necessary to work on different granular fractions due to the difficulties in separating, based on density, particles with different volumes. Similarly, based on color or shape, it is difficult to sort the finest particles (<1 mm). Therefore, recycled and natural sand separation would not be convenient in common concrete or aggregate laboratory, and accurate content determination by separation would require the knowledge of NA and RA particle size distributions.

Properties of aggregates (granular distribution, absorption, etc.) commercialized for concrete production are regularly controlled. Knowing one of these properties (P) for (a) the

Table 3.6 Loss on ignition and chemical composition of the main elements for the nine tested RAs

	F1	*F2*	*F3*	*F4*				*C1*	*C2*	*C3*		*C4*			
Grading	*0/6*	*0/6*	*0/4*	*0/4*	*Av.*	*Std. dev.*	*Rel. std. dev.*	*4/20*	*4/20*	*4/10*	*10/20*	*4/20*	*Av.*	*Std. dev.*	*Rel. std. dev.*
LOI	21.55	20.56	18.06	17.09	19.32	2.09	*10.8*	19.17	20.99	16.64	15.71	15.96	17.69	2.30	*13.0*
SiO_2	46.61	47.42	58.45	57.04	52.38	6.23	*11.9*	45.64	51.51	57.75	58.82	54.11	53.57	5.30	*9.9*
Al_2O_3	4.54	5.58	2.47	4.76	4.34	1.32	*30.5*	5.05	4.41	2.55	2.39	5.35	3.95	1.39	*35.3*
Fe_2O_3	2.85	2.12	1.61	2.06	2.16	0.51	*23.8*	1.92	1.36	0.85	0.83	1.62	1.32	0.48	*36.3*
CaO	20.66	20.77	17.12	15.51	18.52	2.62	*14.2*	25.88	19.22	21.99	22.29	20.61	22.00	2.49	*11.3*
MgO	1.18	0.81	0.62	0.93	0.89	0.23	*26.5*	1.07	0.89	0.64	0.45	0.97	0.80	0.25	*31.6*
SO_3	0.62	0.39	0.47	0.35	0.46	0.12	*26.1*	0.58	0.29	0.12	0.14	0.39	0.30	0.19	*62.5*
K_2O	0.78	1.05	0.47	0.99	0.82	0.26	*31.8*	0.94	0.88	0.56	0.56	1.11	0.81	0.24	*30.0*
Na_2O	0.30	0.81	0.15	0.50	0.44	0.29	*64.8*	0.33	0.51	0.28	0.26	0.65	0.41	0.17	*41.5*
$Na_2O_{eq.}$	0.81	1.50	0.46	1.15	0.98	0.45	–	0.94	1.09	0.64	0.63	1.38	0.94	0.32	–

Source: Average and standard deviation are both calculated for the fine (F) and coarse (C) RAs. Standard deviations in italics are expressed relatively to the average. All values are expressed in percentages.

LOI, loss on ignition; Av., average; Std. dev., standard deviation.

mixture, (b) the recycled sand, and (c) the natural sand, RA content can be determined assuming additivity properties (equation 3.1).

$$\text{RA content} = \frac{P_{\text{mixture}} - P_{\text{NA}}}{P_{\text{RA}} - P_{\text{NA}}} \tag{3.1}$$

For this kind of dosage, precision depends on the recycled sand proportion, the contrast between the properties of the natural and recycled sands as well as the precision of the test. For example, error calculation shows that in blends containing a low proportion of recycled sand (10%), for properties measured with a relative error of 5% (e.g., $WA_{24,RA}=10.0\%\pm0.5\%$), the dosage error (equation 3.2) would be lower than 2% only if there is a contrast of 11 between the properties of recycled and natural sands (e.g., $WA_{24,RA}=11\times WA_{24,NA}$). For a property measured with a relative error of 10%, dosage with less than 2% of error is impossible (even with high properties contrast).

$$\text{Dosage error (\%)} = \left|\text{Measured RA content (\%)} - \text{Real RA content (\%)}\right| \tag{3.2}$$

Generally, differences between natural and recycled aggregates are important in terms of absorption, presence of impurities as well as soluble sulfate in water. Table 3.7 indicates values measured on recycled sand (three recycled plants) and natural sand typically used for concrete production (four quarries). Absorption and soluble sulfate in water were measured following the European standards NF EN 1097-6 (2014) and NF EN 1744-1+A1 (2014), whereas the number of bituminous particles was optically determined on three granular fractions (0.5/1, 1/2.5, and 2.5/4) after sieving under water. Using these seven sands and measuring these three properties for blends containing a low quantity (2.5%–10%) of recycled sand, 24 dosages (Table 3.8) show that, on average, these properties can be used for the assessment of RA amount in the mixture. However, dosage with an error less than 2% (equation 3.2) can only be achieved with soluble sulfate determination in repeatability condition. If absorption or impurities are used, dosage error can be higher than 6%. Reliability can be improved if the tests are repeated for the sands and mixture.

Table 3.7 Properties of recycled and natural sand

		Recycled sand	*Natural sand*
Absorption (WA, %)		6.3–8.6	0.5–2.0
Number of bituminous particles	In 100 g of fraction 2.5/4	136–160	0
	In 33.2 g of fraction 1/2.5	118–140	0
	In 9.8 g of fraction 2.5/4	92–145	0
Soluble sulfate in water (%)		0.09–0.22	<0.003–0.03

Table 3.8 Error (%) for dosage based on soluble sulfate, bituminous impurities and WA

	Error (%) for 24 dosages		
	Soluble sulfate	*Bituminous impurities*	*WA*
Average	1.0	2.6	2.8
Standard deviation	0.4	2.7	1.6
Minimum	0.1	0.1	0.8
Maximum	1.9	9.5	6.6

In conclusion, the presence of only 10% of recycled sand modifies significantly the natural sand characteristics such as absorption, presence of impurities, and soluble sulfate content. In the mix, recycled sand can be detected by measuring these properties, but accurate content determination is tricky. For blends with low recycled aggregate proportions, only soluble sulfate measurements in repeatability conditions allow dosage with an error less than 2%. Accurate determination of recycled sand proportion in reproducibility condition is therefore compromised. Producer and user controls or agreements would be more easily monitored if made directly on the blends' properties (absorption limits, etc.) and not on RA percentage.

3.3 PHYSICAL PROPERTIES OF RA

3.3.1 State of the art

3.3.1.1 *Particle density and WA of RA*

Physical properties such as particle density and WA coefficient are very important engineering properties for the reuse of RA in concrete. Particle density corresponds to the ratio between the oven-dried mass of an aggregate sample and the volume it occupies in water, including the volume of any internal sealed voids and the volume of any water-accessible voids. WA corresponds to the ratio between the mass of water absorbed by an aggregate sample after immersion in water and its oven-dried mass. Generally, RAs have higher WA and lower particle density than NAs because of their ACP content. WA is of particular importance because it determines the efficient water (W_{eff}) quantity in concrete, efficient water referring to the water present in the fresh cement paste, excluding that absorbed by aggregates.

Up to now, there is no particular standard for the measurement of WA and particle density of RA. Therefore, the standards used for NAs such as NF EN 1097-6 (2014), ASTM C127 (2015) (for coarse aggregates), and ASTM C128 (2015) (for fine aggregates) are also used for the characterization of RA. These methods are based on the same principle: Aggregates are first immersed for 24 h in water, and then WA coefficient (WA_{24}) is determined from the water content at saturated surface dry (SSD) state. In that state, accessible porosity of aggregates is supposed to be saturated by water, particle's surface being dry except a thin film of adsorbed water. The SSD state for coarse aggregates is reached directly by drying particles using dry cloths until no water is visible at their surface. The SSD state for fine aggregates is reached by progressively drying sand particles under warm air and is identified using a slump test, considering that when particle's surface is dried, the cohesion due to capillary forces vanishes.

Physical properties of RA vary in a wide range depending on the source and size (fine or coarse) of RA. Silva et al. (2014c) carried out a comprehensive study on RAs in more than 230 papers of the literature. They found that the particle density and WA_{24} of RA followed normal distributions. The mean particle densities of all the collected results were 2327 kg/m^3 for coarse RAs and 2,065 kg/m^3 for fine RAs. The WA_{24} values were 4.7% for coarse aggregates and 9.5% for fine aggregates.

Procedures used for NA might be less appropriate to RA. Tam et al. (2008) reported that the soaking of recycled aggregates in water might detach some ACP. Also the drying at 105°C to obtain the oven-dried mass could remove some bound water contained in the hydrates of ACP. Therefore, they proposed a new method named real-time assessment of WA (RAWA) for RA. In RAWA, the oven-dried mass is obtained after a heating at 75°C±5°C for 24 h. Then aggregates are placed into a pycnometer fully filled with distilled water, and water absorbed is recorded at different time intervals. This method not only allows the overcoming of the previous mentioned drawbacks, but also gives the kinetics of absorption of RA. Another method, based on hydrostatic weighing, was also proposed by Djerbi Tegguer (2012) in order to assess the kinetics of WA of aggregates. This method is based on the study of mass variation of a sample immersed in water. Both studies showed that, in contrast to the tested NA, WA was not necessarily completed within 24 h of immersion of RA into water. In some cases, up to 120 h is needed for a complete saturation. Moreover, they showed that the initial rates of absorption were very high and decreased progressively with time. Tam et al. (2008) reported, for example, that 80% of the absorption of RA was completed after 5 h of immersion. However, these methods do not allow for an accurate determination of absorption at the very beginning of the experiment, because air voids surrounding aggregates first have to be removed, which is particularly difficult for fine RA.

3.3.1.2 Influence of carbonation on the properties of RA

Depending on the duration and conditions of storage of RA after crushing, the ACP could be more or less carbonated. Indeed, crushing leads to a large increase in the specific surface area of the crushed concrete and could therefore accelerate the carbonation process. Carbonation of portlandite and C–S–H leads to the formation of larger volume products inducing a clogging of the porosity of ACP that could change the physical properties of RA.

Thiery et al. (2013) showed that the rate of CO_2 absorption increased sharply for particle sizes less than 2 mm with a liquid–water saturation degree below 0.4. Carbonation could therefore affect significantly the properties of fine RA. Zhang et al. (2015) observed that carbonation improved significantly the physical characteristics of fine RA, leading to a decrease in the WA coefficient, an increase in particle density, and a slight decrease in the crushing value. Crushing value was defined here as the mass loss percentage of a given granular class after loading at 25 kN (JGJ 52 2006). Zhao (2014) also observed a decrease in the WA coefficient and an increase in density for laboratory made RA submitted to accelerated carbonation. He also showed that carbonation could change significantly the measurement of ACP content. Carbonation indeed largely reduced the soluble fraction in salicylic acid and led to an increase in the mass loss of RA between 105°C and 600°C. This latter result was due to the fact that carbonated phases of the cement paste can start decomposing at about 500°C (Villain et al. 2007).

3.3.2 RECYBÉTON's outputs

3.3.2.1 Coarse RA

The particle density has been measured on the RA selected for the RECYBÉTON National Project for three different granular classes: 4/10, 10/20, and 4/20 (Laneyrie et al. 2014). Densities have first been determined three times on a single sample of each granular class (Table 3.9); then, densities of each class have been measured on five different samples

Table 3.9 Particle densities measured with standard method EN 1097-6 on a single sample of RA (Laneyrie et al. 2014)

Particle size	*No. of tests*	*Particle density (g/L)*	*Mean. dev. (g/L)*	*Relative mean dev. (%)*
4/10	3	2.27	0.003	0.13
10/20	3	2.24	0.006	0.27
4/20	3	2.22	0.004	0.18

(Table 3.10). These results show that a standard method can be applied efficiently to coarse RA for particle density measurements.

Jezequel (2014) carried out a study on the measurement uncertainties and pertinence of the standard procedure (NF EN 1097-6 2014) for the determination of particle density and WA of recycled coarse aggregates. Three technicians performed five tests on representative samples sourced from the same material. For each sample, particle density and WA_{24} of fraction 10/20 mm were determined. Results were analyzed according to NF ISO 5725-2 (1994) in order to check the applicability of the procedure to coarse RA. This study showed that repeatability and reproducibility values for particle density of coarse RA were less than standard requirements (Table 3.11).

WA coefficient of coarse RA was also determined according to the standard method EN 1097-6. Table 3.12 shows the results obtained from the three granular classes 4/10, 10/20, and 4/20 by Laneyrie et al. (2014).

Jezequel (2014) studied the repeatability and reproducibility of WA measurement on RA (Table 3.13). Repeatability and reproducibility values for WA of coarse RA were similar to standard requirements. The author, however, pointed out that the SSD state can be difficult to identify for RA. Indeed, surface roughness of RA makes it difficult to identify the

Table 3.10 Particle densities measured with standard method EN 1097-6 on five different samples of RA (Laneyrie et al. 2014)

Particle size	*No. of tests*	*Particle density (g/L)*	*Std. dev. (g/L)*	*Relative std. dev. (%)*
4/10	5	2.26	0.03	1.33
10/20	5	2.24	0.01	0.45
4/20	5	2.24	0.05	2.25

Table 3.11 Repeatability and reproducibility values for particle density of coarse RA (Jezequel 2014)

		Repeatability	*Reproducibility*
Particle density (Mg/m^3)	RECYBÉTON study	0.01	0.01
	Standard values[a]	0.031	0.042

[a] Annex I §I.1-Precision, NF EN 1097-6, 2014.

Table 3.12 WA coefficients measured with standard method EN 1097-6 on a single sample of RA (Laneyrie et al. 2014)

Particle size	*No. of tests*	*WA_{24} (%)*	*Mean. dev. (g/L)*	*Relative mean. dev. (%)*
4/10	3	6.16	0.26	4.22
10/20	3	5.83	0.10	1.72
4/20	3	6.69	0.12	1.79

Table 3.13 Repeatability and reproducibility values for WA of coarse RA (Jezequel 2014)

		Repeatability	*Reproducibility*
WA (%)	RECYBÉTON study	0.32	0.32
	Standard values[a]	0.3	0.4

[a] Annex I §I.1-Precision, NF EN 1097-6, 2014.

transition between wet and dry surfaces, which can easily be identified with NAs by the transition from glossy to mat of their surface.

RAs used in the RECYBÉTON project are classified "code C" with respect to WA, according to the standard NF P18-545 2011 §10, which prevents their use for the manufacture of concrete in some aggressive environments or in certain particular constructions (Aït Alaiwa et al. 2014).

The kinetics of absorption of RA has been measured with different experimental procedures in the RECYBÉTON project. Laneyrie et al. (2014) used the methodology of Tam et al. (2008), which uses a single sample for kinetics measurement and allows reducing the variability. Figure 3.5 shows the results obtained for the 4/10 mm fraction. Ninety-two percent of the 24 h absorption is reached after 10 min of immersion. After 5 min, that is to say, about the time needed for mixing of concrete, only 85% of absorption is reached. Therefore, adding the water quantity corresponding to WA_{24} during mixing could lead to a larger workability than expected.

Bendimerad et al. (2014) studied the kinetics of absorption of coarse natural and recycled aggregates having two saturation states (oven dried or after storage in a chamber at 20°C and 50% relative humidity). They used simultaneously two different methods for the measurement of absorption kinetics: the RAWA method presented by Tam et al. (2008) and the hydrostatic weighting used by Djerbi Tegguer (2012). They showed that both tests allowed for the measurement of absorption kinetics of coarse aggregates for 5 min after immersion. Both tests gave similar absorption rates and final values. In both cases, WA between immersion time and the first measurement was significant and had to be considered. Bendimerad et al. (2014) proposed to determine it *a posteriori* by measuring the final absorption using the standard method EN 1097-6. Water absorbed by RA after 24 h of immersion represented about 92% of the final absorption, indicating that 24 h immersion gave a good estimate of the long-term absorption. However, WA of RA between 1 and 24 h was significant, and the authors propose either to use the 1 h WA value to design RA mixtures or to fully saturate RA before mixing.

WA, porosity, and particle density were also determined using the methods proposed in the standard EN 1097-6 and NF P 18-459 (2010) (saturation after immersion under vacuum) (Omary et al. 2015). Table 3.14 presents the obtained results.

Table 3.14 shows that long-term WA of coarse RA can be estimated correctly from 24 h measurements, as recommended in EN 1097-6. Porosities measured on RA roughly correspond to those of traditional concrete. WA values measured with the two methods of EN 1097-6 are very close to each other and slightly less than those obtained under vacuum, which also influences particle density and porosity.

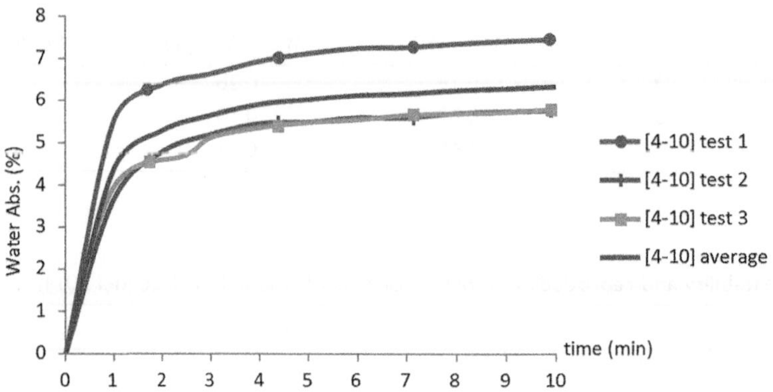

Figure 3.5 Kinetics of absorption of 4/10 RA during the first 10 min of immersion (Laneyrie et al. 2014).

Table 3.14 WA, particle density, and porosity measured with different standard protocols (Omary et al. 2015)

Method		*Coarse RA 4/10*			*Coarse RA 10/20*		
Time of immersion	*Measure*	*24 h*		*48 h*	*24 h*		*48 h*
Pycnometer (NF EN 1097-6 2014)	WA (%)	5.6		5.7	5.7		5.8
	ρ_{rd} (mg/m³)		2.29			2.26	
	Porosity (%)		13.0			12.7	
Under vacuum (NF P 18-459 2010)	WA (%)	6.1		6.4	6.2		6.2
	ρ_{rd} (mg/m³)		2.23			2.22	
	Porosity (%)		13.4			13.7	
Hydrostatic weighing (NF EN 1097-6 2014)	WA (%)	5.4		5.7	5.6		5.6

3.3.2.2 Fine RA

In the frame of RECYBÉTON, WA coefficient WA_{24} of the same fine RA has been determined by several laboratories, using different experimental procedures. According to the standard procedure (NF EN 1097-6 2014), WA_{24} values vary between 6.8% and 9.8% for fraction 0/4 mm and between 6.9% and 7.8% for fraction 0.063/4 mm (Sedran 2017; Jezequel 2013; Aït Alaiwa et al. 2014; Laneyrie et al. 2014; Le 2015; Cyr et al. 2014). The differences obtained for fraction 0/4 mm are very large (3% between the largest and lowest obtained values). Differences seem to be less for fraction 0.063/4 mm, but it has to be noticed in this case that the number of results (3) is smaller than for fraction 0/4 mm.

Jezequel (2014) carried out a study on the pertinence of the standard procedure (NF EN 1097-6 2014) for the measurement of particle density and WA of recycled sand (same fine aggregate as in the previous studies). Three technicians performed five tests on representative samples sourced from the same material. For each sample, particle density and WA_{24} of fractions 0/4 and 0.063/4 mm were determined. Results are analyzed according to NF ISO 5725-2 (1994) in order to check the applicability of the procedure to fine RA. They show that particle density varies between 2,080 and 2,260 kg/m³, these variations complying with the requirements of NF ISO 5725-2 (1994). However, WA_{24} values range between 6.1% and 10.0% for fraction 0.063/4 mm and between 5.8% and 9.7% for fraction 0/4 mm, showing that the standard procedure does not allow for an acceptable determination of WA.

Given these results, other experimental procedures were applied for the assessment of WA of fine RA.

Laneyrie et al. (2014) proposed a method based on the vacuum filtration of saturated sand in order to reach the SSD state of fine RA. This method allows avoiding the loss of very fine particles that could occur when the fine RA is dried under a current of warm air. WA_{24} values obtained were 11.2% for 0.063/4 mm and 12.8% for 0/4 mm (against 7.8% obtained with EN 1097-6 method), and the discrepancy of results was lower than that obtained with the standard procedure. The authors also studied the kinetics of absorption of water with the same fine RA. They observed that absorption was almost complete after 10 min of immersion into water.

Le (2015) applied the "evaporometry method" developed by Mechling et al. (2003) for the WA_{24} measurement of fine RA. This method is based on the change of the kinetics of drying of a granular bed submitted to a given temperature, which occurs when absorbed water starts to evaporate. The WA_{24} values obtained were 10.8% for 0/4 mm and 10.0% for 0.063/4 mm. The former value was close to that obtained with IFSTTAR method n°78 (Iffstar 2011), which was 10.6% for 0/4 mm. This method has been developed for the WA measurement of crushed sands. In that case, the SSD state is obtained by drying the saturated sand on successive sheets of absorbing paper, until no trace of humidity can be seen on these papers.

Zhao et al. (2013b) studied the WA of fine RAs as a function of their particle size with two different experimental methods: standard EN 1097-6 and IFSTTAR method No. 78. They showed that for particle sizes smaller than 0.63 mm, EN 1097-6 tended to underestimate WA, and IFSTTAR method led to overestimated values. However, for larger particle sizes, both methods gave similar results, suggesting that for coarser particles, these two protocols were able to identify accurately the SSD state. Knowing the soluble fraction in salicylic acid, they proposed a new method for the measurement of WA of fine RAs. Assuming that the physical properties of the ACP and NAs are the same in all the granular classes of RA, WA of RA can be written according to the following equation for every granular class composing the RA:

$$WA_{RA} = WA_{ACP} \times ACP + WA_{NA} \times (1 - ACP) \qquad (3.3)$$

where WA_{RA}, WA_{ACP}, and WA_{NA} are, respectively, the WA coefficients of the given granular class of RA, ACP, and NA, and ACP is the adherent cement paste content in the given granular class of RA. As seen in Section 3.2.2.2, the absolute value of ACP content is generally unknown. However, equation 3.3 shows that the variation of WA_{RA} should be linear as a function of every quantity that would be proportional to ACP content. Therefore, a linear relation should be obtained between the WA coefficient and either the SFSA or the mass loss between 105°C and 475°C ($ML_{105-475}$).

Le et al. (2016) applied this method for the measurement of WA_{24} of the RECYBÉTON's fine RA. They divided the sand into five granular classes: 0/0.5, 0.5/0.8, 0.8/1.6, 1.6/2.5, and 2.5/4 mm. Then, $ML_{105-475}$ was measured for each granular class, and the WA_{24} values of the three coarser fractions (0.8/1.6, 1.6/2.5, and 2.5/4 mm) were measured according to EN 1097-6 and IFSTTAR No. 78 methods. A linear equation has then been identified between WA_{24} and $ML_{105-475}$, which allows identifying the WA_{24} values of the two finer fractions from their $ML_{105-475}$ values (Figure 3.6). Knowing the mass proportion of each granular class in the 0/4 fine RA, the overall WA coefficient can be recalculated. These authors found

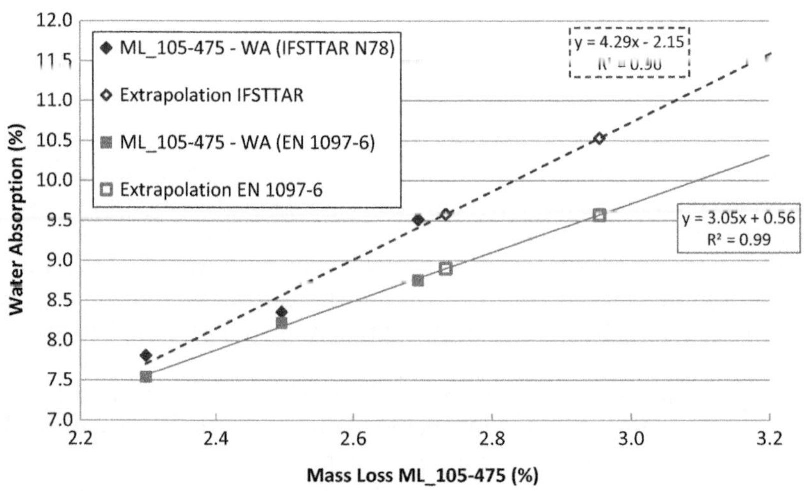

Figure 3.6 Variation of WA_{24} measured with EN 1097-6 and IFSTTAR methods as a function of $ML_{105-475}$ and extrapolation for WA_{24} identification of finer fractions.

WA_{24} of 8.7% and 9.1%, respectively, by extrapolating the WA_{24} values obtained by EN 1097-6 or IFSTTAR method.

Among all these methods, the one proposed by Zhao et al. (2013b) and Le et al. (2016) seems to be more accurate. Indeed, dividing recycled sand into four or five granular classes and measuring, for each class, the mass loss between 105°C and 475°C allow taking into account the variation of ACP content as a function of particle size. Moreover, with this method, the WA coefficient is determined only for the coarser granular fractions, for which WA can be measured accurately.

3.4 MECHANICAL PROPERTIES OF RA

3.4.1 State of the art

3.4.1.1 Resistance to chocks and abrasion

The LA method is most commonly applied to determine the resistance to chocks and abrasion of the recycled aggregate coarse particles. Indeed, scientific papers indicating LA value in relation to recycled aggregate characterization are about three times more than those associating the aggregate crushing value or crushing index (ACV), four times for the aggregate impact value (AIV) and six times for the micro-Deval (MDE) test. The proportions are very similar when considering RA in relation to these tests.

The recommendation for coarse recycled aggregate resistance to fragmentation categories for the use of concrete in the European standard NF EN 206 (2014) is based on the LA coefficient as specified in EN 1097-2 (2010) for natural coarse aggregate. As general trend, the abrasion measured for recycled building materials in the LA test ranges from 20% to 40%. It is thus slightly greater than the reference values for natural mineral aggregates. It is the mortar content of the RA which induces higher LA value for the corresponding original NA (Zega et al. 2010). It is also observed that both types of aggregates, NA and RA, show an almost identical increase in the generated abrasion in line with decreasing bulk density, as well as almost the same degree of variation (Mueller 2015).

In the case of the recycled aggregates, the LA coefficient increases with the percentage of masonry (Barbudo et al. 2012; Cameron et al. 2012), with the proportion of mortar particles (Barbudo et al. 2012), and with the mortar content of the crushed concrete particles (de Juan and Gutiérrez 2009). The quality of the crushed concrete has also a significant impact on the LA test result. De Juan and Gutiérrez (2009) also reported a correlation between the strength of the parent concrete and the LA coefficient of the subsequent RA, while the proportional relationship between the parent concrete strength and the mortar content of the RA particles was not clear.

Thus, the relation between the LA and the mortar content of the recycled concrete particles could be led by adverse effects. The greatest LA in RA occurs in concretes with a weaker matrix (Zega et al. 2010; Tavakoli and Soroushian 1996a). Nevertheless, for a similar crushing procedure and size of the original NA, the RA produced from the stronger concrete contained a relatively higher amount of mortar (Belin et al. 2014) but with better adherence of mortar to NA (Akbarnezhad et al. 2013a).

In general, the LA test cannot be regarded as an accurate description for the sample's resistance to fragmentation, as it includes fine particles generated by rounding and abrasion, and thus inevitably overestimates the degree of fragmentation. When testing NAs, Bach (2013) observed that for some rock types, the fragmentation process appears to be still under way even at the end of the test, whereas for other rock types, the process comes to an end after only relatively few revolutions of the test drum. Erichsen et al. (2011) demonstrate that during the tumbling time, the aggregate passes through two stages of fragmentation. The first stage occurs during the initial phase and is characterized by degradation into intermediate fractions (bellow the initial graduation but higher than the LA test limit sieve). Production of fine material takes place at the same time at a constant rate. The second stage only occurs in aggregates with weak mechanical strengths and is initiated if the intermediate fraction passes an optimum level. Further degradation is dominated by production of fine material. It can be considered that these different mechanisms are expected to occur in the testing of RA. To study this phenomenon, Erichsen et al. (2011) propose to measure both LA value and residual LA value (aggregate still in the initial grading), and to follow the evolution of the granular size for different number of rotations of the drum.

Even if the different abrasion tests involve different fracture mechanisms, the abrasion classification of RA by different tests is generally rather well correlated—see, for instance, Butler et al. (2012) for the correlation between ACV and MDE or Dhir and Paine (2007) for the correlation between LA and MDE. However, the correlation between ACV or LA and AIV is not clear. As a consequence, the tendencies presented above for the LA value could also be observed for the MDE or ACV values. For instance, the crushing index is greater for recycled aggregate than for the NA (Wang et al. 2011) and, for the recycled aggregates, the AIV increases with the reduction of the parent concrete strength (Padmini et al. 2009; Liu et al. 2016a; Wang et al. 2011).

3.4.1.2 Models for concrete compressive strength prediction

Usual models for compressive strength consider that concrete could be described as a two-phase material: a stiff inorganic inclusion (the aggregate) dispersed in a matrix (cement paste) considered as homogeneous at the mesoscale. The cement paste is generally the weakest part of the composite, and classic laws, such as Féret's law, consider that the compressive strength is the product of a constant which depends on the nature of aggregate and a function of cement paste composition:

$$\mathrm{fc}_{28} = K_g \mathrm{fcm}_{28} \frac{1}{\left[\frac{\rho_c}{\rho_w}\left(\frac{W + \rho_w V_a}{C}\right)\right]^2} \tag{3.4}$$

where:

K_g is the Féret aggregate constant: $4.5 < K_g < 5.5$ for NA
fc_{28} is the compressive strength of concrete at 28 days
fcm_{28} is the ISO strength of cement at 28 days
ρ_c and ρ_w is the specific gravity of cement and water
W and C is the mass of water and cement for 1 m^3 of concrete
V_a is the volume of air for 1 m^3 of concrete

de Larrard (1999a) proposes a modification of Féret's law to consider the influence of an increase of the maximal paste thickness (MPT) of the concrete (expressed in millimeters) on the reduction of the compressive strength:

$$\mathrm{fc}_{28} = K_g \mathrm{fcm}_{28} \frac{1}{\left[\frac{\rho_c}{\rho_w}\left(\frac{W + \rho_w V_a}{C}\right)\right]^2} (\mathrm{MPT})^{-0.13} \tag{3.5}$$

The maximal paste thickness is given by the following equation:

$$MPT = D_{max}3\sqrt{\left(\frac{g^*}{g}-1\right)} \tag{3.6}$$

where:

D_{max} is the maximal diameter of the coarse aggregate
g is the volume fraction of fine and coarse aggregates in concrete
g^* is the packing density of fine and coarse aggregate skeleton for a compaction index of 9

Previous laws are not sufficient to describe the ceiling effect of some aggregates such as limestone aggregates for which the strength of the concrete is not strictly proportional to the strength of the matrix in the high strength of paste area.

de Larrard (1999a) proposed an empirical hyperbolic equation to model this nonlinearity:

$$fc_{28} = \frac{p\,fcm_{28}}{(q\,fcm_{28}+1)} \tag{3.7}$$

fc_{28} is the compressive strength of concrete at 28 days (MPa)
fcm_{28} is the Compressive strength of matrix at 28 days (MPa)
p is an empirical constant
q is an empirical constant (MPa^{-1})

The compressive strength of the matrix can be calculated according to the following equation:

$$fcm_{28} = 13.4Rc_{28}\left[\frac{V_c}{V_c+V_w+V_a}\right]^{2.85}(MPT)^{-0.13} \tag{3.8}$$

where:

V_c, V_w, and V_a are the volumes of cement, water, and air in the matrix, respectively
Rc_{28} is the cement strength

For low matrix strength values, equation 3.7 can be approximated by equation 3.9. Parameter p appears as a parameter which describes the quality of the bond between the aggregate and the cement paste:

$$fc_{28} = p\,fcm_{28} \tag{3.9}$$

For very high matrix strength values, the strength of the concrete tends to an asymptotic value, which is the ratio between p and q parameters:

$$\lim_{fcp28\to\infty} fc_{28} \to \frac{p}{q} \tag{3.10}$$

Table 3.15 summarizes p and q values for aggregates from several sources of rocks (de Larrard 1999a). High p values near 1 are characteristic of limestone aggregates known to provide an excellent bond with the cement paste. Aggregates with low p values do not have

Table 3.15 Mechanical characteristics of NAs

Type	*Origin*	*p*	*p/q (MPa)*	*fc (MPa)*[a]	*Debonded aggregate (%)*[b]
Crushed hard limestone	Boulonnais	1.11	337	160	0
Crushed semihard limestone	Arlaut	0.96	241	111	0
Crushed basalt	Raon l'Etape	0.65	∞	250	14
Quartzite	Cherbourg	1.15	244	–	0
Rounded flint	Crotoy	0.58	∞	285	13

[a] Compressive strength of the rock.
[b] Percentage of debonded aggregate after 28 days of splitting test.

the ceiling effect and concrete made with these aggregates present debonded grains after 28 days of splitting tests.

Simplified (equations 3.4–3.6) and complete approaches (equations 3.7–3.10) have been applied with success to recycled concrete by Dao (2012). The classes of strength are, respectively, 35 and 65 MPa at 28 days of hardening. Fine and coarse recycled aggregates are used. Recycled aggregates are made by the crushing of natural concretes whose compositions are known. For the simplified approach, the average error of the model is 3.5 MPa if K_g is extrapolated by the experimental way and 5 MPa if K_g is derived from relationship between K_g and resistance to abrasion of the aggregate evaluated with MDE test (see Section 3.4.2.1 and equation 3.11).

$$K_g = -0.0952\text{MDE} + 8.3927 \tag{3.11}$$

where:
MDE is the micro-Deval coefficient (%)

Complete approach has a better precision with an average error of 2.1 MPa. Table 3.16 provides p and q values for recycled aggregates of different origins (Dao 2012).

Recycled aggregates have high p values (>1) characteristic of aggregates which develop excellent bond with the cement paste as natural limestone aggregates (see Table 3.15). Recycled aggregates are characterized by lower p/q ratios than NAs (see Table 3.15) and are also impacted by a stronger "ceiling effect" which could limit the strength of recycled concrete based on high-performance paste with low water-to-cement ratio. For NAs, p/q ratio is a decreasing function of the compressive strength of the original rock (de Larrard 1999a), and no clear relationship has been found between p/q ratios and compressive strength of original concrete for recycled aggregates (Dao 2012).

Table 3.16 Mechanical characteristics of recycled aggregates according to Dao (2012)

		Fine aggregate		*Mix fine/coarse aggregate*	
Class of original concrete	*Original aggregate*	*p*	*p/q (MPa)*	*p*	*p/q (MPa)*
C35	Rounded siliceous	1.08	135	1.64	109
C65	Rounded siliceous	1.06	177	1.88	125
C35	Crushed semihard limestone	1.07	178	1.13	226
C65	Crushed semihard limestone	1.36	151	1.83	141

3.4.2 RECYBÉTON's outputs

3.4.2.1 Resistance to chocks and abrasion

The LA equipment was used to test different sizes of RA (4/6, 6/10, 10/14, and 14/20) and, as a reference aggregate, a siliceous crushed NA (10/14) (Figure 3.7). It may be recalled that the number of standard drum turns in the LA test is 500. In this study, we are intending to evaluate the linearity of the behavior with the number of turns. So, in these tests, the number of drum turns was varied from 20 to 6,000.

For the NA, the linearity of the creation rate of fine particles (passing 1.6 mm sieve) with the number of turns was maintained up to 2,000 revolutions. The behavior was different for all RA fractions. Indeed, the creation rate of fine particles was not linear. At low number of turns (under 500 or 750), this rate was significantly higher than for the NA. At higher number of turns (above 1,000), the rate is comparable for the NA. As a consequence, one can consider that the attached mortar was the first to be fragmented by the metallic ball impacts. Once the content of attached mortar was decreased, the behavior of the RA was similar to that of the original NA.

This conclusion was better analyzed by measures of WA of RA which had undergone different number of turns in the LA equipment (Figure 3.8). The tests were performed on particles which stayed after the test in the initial size of the RA. The results indicated that the capacity to absorb water drastically decreased at the early number of turns. The WA tended to stabilize at a level higher but close to the level of a common NA. In fact, the results suggest that not all the attached mortar, but mainly the more brittle, is separated by the early number of turns in the LA equipment. One can then suppose that the hard mortar still attached to the original aggregate of the RA at about 500 revolutions still induced a higher level of WA but had a similar strength with respect to the LA test stresses. It should be pointed out that the evolution of the WA capacity is very sensitive to the size of the RA, basically to their initial attached mortar content. This was much fewer the case for the behavior to fragmentation. The fine creation is higher for smaller particles, but the differences between the different sizes were on the second order compared with the difference of the RA and the NA.

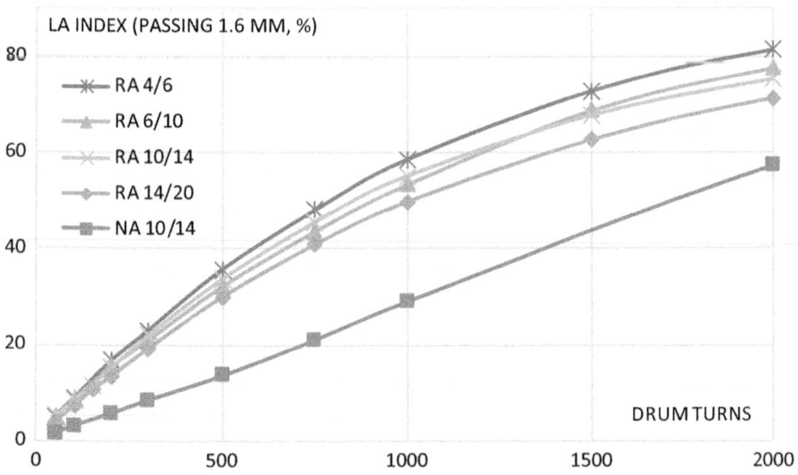

Figure 3.7 Mass fraction of particles lower than 1.6 mm sieve for different number of drum turns in the LA equipment for NA and RA of different sizes.

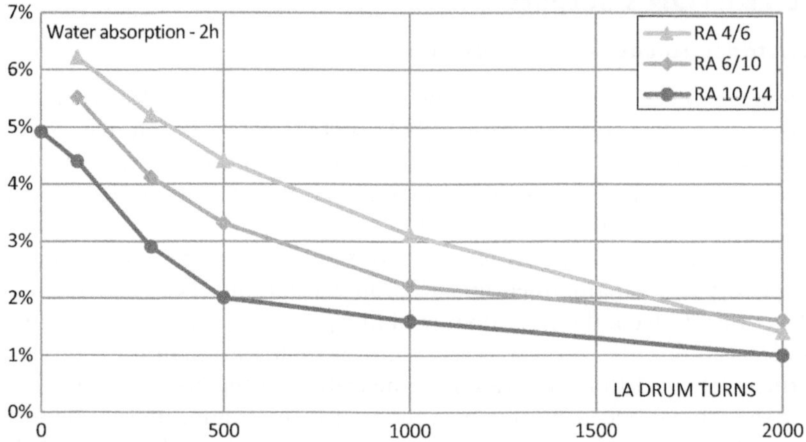

Figure 3.8 The WA measured after 2 h of immersion in water for aggregates after test in LA equipment with different number of drum turns.

As demonstrated by Artoni et al. (2017) when analyzing the same data, the LA test gives information on the cleavage mechanisms. In order to estimate the attrition behavior, tests were performed on the MDE equipment on the three sizes of RA. The number of standard drum turns in the MDE test is 10,000, but, as in the LA test, different numbers of turns of the drum were tested here (Figure 3.9). In these tests, each experimental point was obtained on a different sample. Duplication of tests showed a very good repeatability of the same magnitude as the difference between the 6/10 and 10/14 RA. So, the observed behavior of the sizes 6/10 and 10/14 was very similar, probably identical. This was not the case for the smaller fraction 4/6. One can assume that the MDE metallic balls act as an abrasion mechanism for particles having similar size as the balls (10 mm) and as a cleavage mechanism for significantly smaller particles as is the case of the fraction 4/6.

By crossing the information between the LA and MDE tests, one can conclude that the fracture of the RA particles in the surface (attrition) is independent of the grain size, while

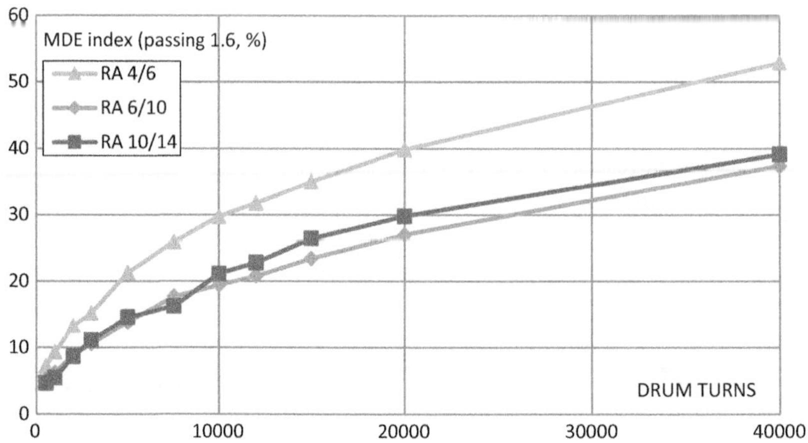

Figure 3.9 Mass fraction of particles lower than 1.6 mm sieve for different number of drum turns in the MDE equipment for RA of three different sizes.

the breakage of coarser fragments from the RA particles (cleavage) is more pronounced for the smaller fractions.

3.4.2.2 *Application of models to estimate RA mechanical properties*

The *p* and *p/q* parameters have been identified for the recycled fine aggregate of RECYBÉTON project. The identified values are, respectively, 1.3 for *p* parameter and 100 for *p/q* parameter (Le et al. 2017). These results are in accordance with previous studies (see Table 3.16). Microstructural analysis (see Chapter 9) confirms the excellent continuity between old and new cement pastes at 28 days of hardening for mortars made of dried fine recycled aggregates (Le et al. 2017). Finally, high *p* and low *p/q* values explain why substitution of NA with RA can increase the strength at low strength level and decrease it at high strength level.

3.5 RESEARCH NEEDS

The results of RECYBÉTON project presented here have confirmed that RAs possess high WA coefficient, which could lead to a wrong estimation of the efficient water-to-cement ratio in concrete. It has also been shown that standard methods could be used with satisfying accuracy for the characterization of coarse RA, and some modified methods have been proposed for better evaluation of fine RA properties. These methods should allow for a good estimation of the WA coefficient after 24 h of immersion of aggregates. However, the measurement of WA in the laboratory after an immersion in water during 24 h differs largely from that occurring during mixing of fresh concrete. In real application cases, RAs partly saturated are immersed in the fresh cement paste. The kinetics and amount of absorbed water could be significantly different from those estimated with standard procedures and could depend on both the initial saturation state of RA and the characteristics of the fresh cement paste. This difference could be negligible for NAs possessing low absorption coefficients, but it might be much more pronounced with RA given their large WA. A better understanding of the kinetics of absorption of water into RA and of the respective influences of the initial state of saturation and of the medium in which absorption is carried out is still needed.

Regarding the recycled sand proportion assessment in a mixture with natural sand, improving soluble sulfate measurement method would allow a more accurate and faster measurement. Monitoring of the soluble sulfate content in recycled aggregates and their mixture would also allow a better control of the RA quality.

3.6 CONCLUSIONS

The main chemical, physical, and mechanical properties of RA have been measured with various experimental methods on the frame of the RECYBÉTON project.

Results show that the ACP content of RA depends on the properties of the parent concrete, the crushing process, and the granular class of particles: the larger the particles, the lower the ACP content. New methods for the estimation of ACP content have been proposed, which allow measuring quantities that are proportional to the cement paste content (soluble fraction in salicylic acid or mass loss between 105°C and 475°C).

Physical properties such as particle density and WA have been measured on both coarse and fine RAs using standard and nonstandard experimental procedures. The accuracy of used methods depends a lot on the size of particles. For coarse particles, repeatability and reproducibility studies show that standard methods allow for a very satisfying measurement of particle density and WA coefficient of RA. The kinetics of absorption has been studied for these materials, and the results show that most of the absorption (92%) is achieved after 24h of immersion, suggesting that WA_{24} could give a good estimate of the total amount of absorbed water in the long term. However, using WA_{24} for the calculation of absorbed water in concrete could lead to an overestimation of absorbed water in the mix. For fine RA, standard procedures allow for a satisfying measurement of particle density. However, WA cannot be determined accurately. A new method based on the correlation between ACP content and WA has been developed, allowing for a better characterization of fine RA absorption.

Mechanical properties of coarse and fine RAs have been assessed in order to apply models of the literature for the prediction of compressive strengths of mortars and concretes. These studies show that RAs present an important ceiling effect, which could limit the compressive strength of mortars and concrete when high strength cement pastes are used. On the contrary, RAs allow for a very good bond with new cement paste.

LA tests performed on coarse RA show that, in contrast to NA, the rate of creation of fine particles as a function of the number of turns is not linear. The attached mortar is the first to be fragmented, which results in a drastic decrease in the WA coefficient at the early number of turns. The WA after 2000 turns tends toward that of common NA but remains slightly higher. The comparison between LA and MDE results shows that the fracture of RA particles in surface (attrition) does not depend on the grain size, while the breakage of coarser fragments (cleavage) is more pronounced for smaller fractions.

Aggregate tests devoted to assessing the risk of ASR are presented in Chapter 12.

Section II

Binder incorporating recycled concrete fraction

L. Izoret
ATILH

Sections III and IV show that fine (sand) fraction obtained from concrete crushing (which represents about 50% of the total mass of processed material, depending on the type and use of crushers), is easier to use in the new concrete mixes compared to coarse fraction.

Therefore, as the fine fraction is a "fatal product" from the crushing process, it may be an alternative solution for using it "as it is" (as sand fraction) or for further reducing the grain size in order to transform it into a cement component or a concrete addition.

The above-mentioned alternatives represent the three viable solutions in order to recycle the sand fraction:

a. As an alternative raw material for Portland raw meal
b. As a main Portland cement constituent other than clinker
c. As concrete addition

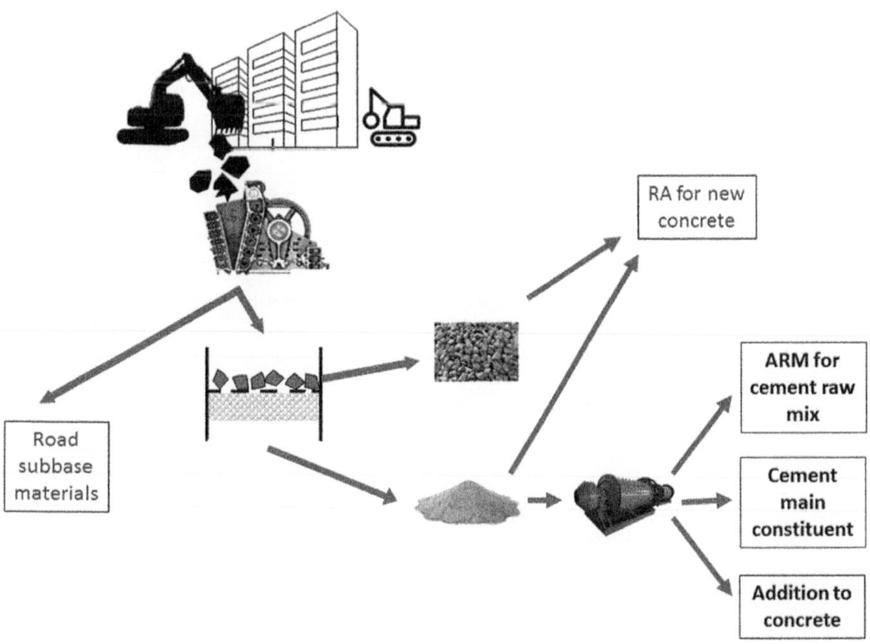

The first one will use the sand fraction without granlometric reduction ("as it is"), and the last two will use the sand fraction after a further grain size reduction (finely ground). These three solutions are schematically represented as in the following figure (see bottom-left options):

This section includes two chapters. The aim of Chapters 4 (option a) and 5 (options b and c) is to estimate the feasibility of each of these solutions, even if option a had already been largely investigated (Galbenis and Tsimas 2006; De Schepper et al. 2013; Schoon et al. 2015; Marrocoli et al. 2016) but remained to be demonstrated in France at a full industrial scale.

Chapter 4

Recycled concrete sand as alternative raw material for Portland clinker production

L. Izoret
ATILH

C. Diliberto, J.M. Mechling, and A. Lecomte
Institut Jean Lamour-Université de Lorraine

P. Natin†
Vicat Technical Centre

CONTENTS

Abstract

As explained in the general introduction, considering the fine fraction of crushed demolition concrete as alternative raw materials for Portland clinker production is one of the three options for recycling concrete. This way of recycling sands from crushed concretes

† Deceased.

is based on the rough assumption that the chemical and mineralogical composition of a given concrete is mainly governed by the nature of the coarse and fine aggregates, which can be assumed in first approximation as siliceous-calcareous (even if in some geographical regions, pure limestone or pure siliceous aggregates can be found), which brings into the chemical system the desired main elements: calcium and silica. The potential value of the use of an alternative raw material for Portland clinker production is a classic problem which can be solved by applying a semi-experimental approach. In this approach, a set of a few experiments can determine, from the chemical composition of each material used, the quantitative proportions of the raw components through the calculation of the potential composition of the future clinker (Bogue 1929) and the burnability indices. This approach has been implemented using a calculation sheet under MS-Excel run under Solver in order to maximize the replacement rate of raw materials by Recycled Concrete Sands (RCS) in optimizing the potential composition of the clinker and the burnability indices. In this chapter, we present the technological evaluation of the opportunity to use the fine fraction extracted from crushed demolition concretes as Alternative Raw Materials (ARM) for Portland Clinker production. The above methodology has been successfully applied to recycled concretes from four geographical areas in France (with four distinct geological contexts), followed by a full-scale industrial Portland clinker production.

4.1 RAW MIX CALCULATIONS: PRINCIPLES AND METHODS

4.1.1 Principles

Cement raw mix is a polymineral and polydisperse mix (Chatterjee 1991) whose composition can vary within a wide range due to the variability in the used raw material. Toward this variability, the composition of the raw mix obeys some fundamental principles established with regard to the very first of Vicat's works (Vicat 1817). These principles stipulate that the proportions of the main components are not given by chance as, after their thermal dissociation during the pyroprocessing step, they determine the ratio of the relevant major oxides (SiO_2, Al_2O_3, Fe_2O_3, and CaO). Hence, under proper burning conditions, new mineralogical phases with hydraulic properties are formed similar to clinker-forming minerals, such as C_3S, C_2S, C_3A, C_4AF, and free lime (fCaO); the quantity of free lime, sometimes considered as the "fifth phase" of clinker Harrisson (2010), is an indicator of burning quality (Fundal 1996).

These principles are still the same but have been refined and developed in quantitative terms. The purpose of calculating the raw mix composition is to determine the quantitative proportions of the raw components in order to give the clinker the desired chemical and mineralogical compositions. The desired composition is given by initial assumptions on C_3S and C_3A contents as well as liquid-phase formation (Taylor 1997). The methodology is based on a two-step semi-empirical approach: (i) chemico-mineralogical characterization of the candidate material and proposed raw mix and (ii) laboratory burnability testing based on raw mix calculations (Chatterjee 1991; Fundal 1996).

4.1.2 Method

The raw meal composition of an industrial Portland cement with the adequate proportions of traditional raw materials (TRM), such as limestone, marl, clay, bauxite, and iron oxide, has been taken as a reference. From this real composition, different mix proportions were derived by calculation in introducing recycled concrete sands (RCS) as alternative raw materials (ARM).

The mix proportions of the various components were adjusted to keep the values of the lime saturation factor (LSF), silica modulus (SM), alumina modulus (AM), and also the liquid content at 1,400°C and 1,450°C in the range of those of the reference clinker. These values are important for raw compositions because they generally determine the quality of cement clinkers (Christensen and Johansen 1979; Oliveira 1992; Lea 2003). The equations for calculating LSF, SM, AM, and the percentage of liquid at 1,400°C and 1,450°C are shown as follows:

$$\text{LSF}: 100\times CaO/2.8\times SiO_2+1.18\times Al_2O_3+0.65\times Fe_2O_3 \quad (4.1)$$

$$\text{SM} = SiO_2/Al_2O_3+Fe_2O_3 \quad (4.2)$$

$$\text{AM} = Al_2O_3/Fe_2O_3 \quad (4.3)$$

$$\%\text{ of liquid at }1{,}450°C: 3Al_2O_3+2.25\times Fe_2O_3+MgO+K_2O+Na_2O \quad (4.4)$$

$$\%\text{ of liquid at }1{,}400°C: 2.95Al_2O_3+2.20\times Fe_2O_3+MgO+K_2O+Na_2O \quad (4.5)$$

The quantitative phase composition was estimated using the procedure of Bogue calculation (Bogue 1929). The calculation is based on the fact that the Fe_2O_3 occurs as C_4AF, the remaining Al_2O_3 occurs as C_3A, and the C_3S and C_2S contents are obtained after solving two simultaneous equations considering the CaO content attributable to C_4AF, C_3A, and free lime content (fCaO). The equations are as follows:

$$\%C_3S = 4.071\times(\%CaO - CaO_f) - (7.6\times\%SiO_2+6.718\times\%Al_2O_3+1.43\times\%Fe_2O_3) \quad (4.6)$$

$$\%C_2S = 8.6\times\%SiO_2+5.07\times\%Al_2O_3+1.08\times\%Fe_2O_3-3.071\times(\%CaO-CaO_f) \quad (4.7)$$

$$\%C_3A = 2.65\times\%Al_2O_3-1.69\times\%Fe_2O_3 \quad (4.8)$$

$$\%C_4AF = 3.04\times\%Fe_2O_3 \quad (4.9)$$

The various materials were crushed together in a laboratory ring mill and homogenized in a "Turbula" shaker mixer for 24 h. The powder was then pressed into the form of pellet; then placed in a platinum crucible; burned in a high-temperature electrical furnace at a heating rate of 10°C/min; and maintained at 1,200°C for 20 min, then at 1,300°C for 20 min, at 1,400°C for 20 min, and finally at 1,450°C for 20 min. At the end of each level of temperature, the sample is quickly cooled to room temperature and ground into fine powder.

4.2 PRELIMINARY LABORATORY STUDIES

4.2.1 Chemical and mineralogical characterizations of industrial RCS

Four RCSs of 0/4 or 0/6 mm fraction size reclaimed from industrial production sites were used for the cement raw meals. They were collected from crushing units, established in four French areas that use different geological natural aggregates: two are mainly siliceous aggregates (denoted S)—granite from Vosges Mountain "S#1" and chert gravels from the Seine River Valley "S#2"—and the other two are silico-calcareous aggregates (denoted SC) from

Table 4.1 Chemical composition of industrial RCS

	Siliceous aggregates		*Silico-calcareous aggregates*	
	S#1 (wt%)	*S#2 (wt%)*	*S#1 (wt%)*	*S#2 (wt%)*
LOI	17.09	18.06	21.55	20.56
SIO_2	57.04	58.45	46.61	47.42
Al_2O_3	4.76	2.47	4.54	5.58
Fe_2O_3	2.06	1.61	2.85	2.12
CaO	15.51	17.12	20.66	20.77
MgO	0.93	0.62	1.18	0.81
MnO	0.09	0.07	0.12	0.07
SO_3	0.35	0.47	0.62	0.39
K_2O	0.99	0.47	0.78	1.05
Na_2O	0.5	0.15	0.3	0.81
TiO_2	0.12	0.08	0.19	0.17
P_2O_5	0.06	0.05	0.14	0.11
SrO	0.12	0.1	0.12	0.14
Total	99.64	99.72	99.66	100

Siliceous aggregates
S#1 Granite from Vosges Mountain.
S#2 Chert from Seine River Valley.
Silico-Calcareous aggregates
SC#1 From Lille.
SC#1 From Lyon (Rhone River Valley).

Lille "SC#1" and from the Rhone River Valley "SC#2". The chemical compositions of the ground and homogenized sands are given in Table 4.1.

The chemistry of these sands is characterized by high values of CaO and SiO_2, and the minor components are Al_2O_3 and Fe_2O_3. As expected, the contents of these oxides vary according to the geological origin of the aggregates used in the production of concrete: the calcium oxide content is higher for the silico-calcareous aggregates than for the siliceous ones. Sands are mainly composed of quartz and calcite (Figure 4.1). Albite and microcline are other minerals that can be identified by X-ray diffraction (XRD) and confirmed by optical microscopy examination, which appear in lower quantities than those of quartz and calcite (Figure 4.2).

The ubiquitous presence of quartz in these ARMs is definitely a question that must be addressed as this mineral is well known to "poison" the raw meal reactivity (Christensen 1979; Fundal 1996) by the formation of belite clusters from quartz grain (Maki et al. 1995). A special focus will be made through the grindability study.

A Thermogravimetric/differential thermal analysis TGA/DTA of the samples completes the characterization of these materials (Figure 4.3): a weight loss is observed for all the samples at 100°C, which corresponds to the loss of free water and to the dehydration of cement paste hydrates (CSH, Afm...). At 450°C, a small endothermic peak is observed for some samples (essentially silico-calcareous sands), which corresponds to the dehydration of portlandite. The endothermic signal intensity is low, which indicates that the portlandite is present in very small quantities and that it is strongly carbonated over time. At 573°C, an endothermic peak corresponding to the transformation of the alpha quartz to beta quartz is observed. Finally, an endothermic peak is observed near 800°C, which corresponds to the decarbonation of calcite, and a significant cumulative weight loss is observed: about 10% for the siliceous sands and 13% for the silico-calcareous one corresponding to

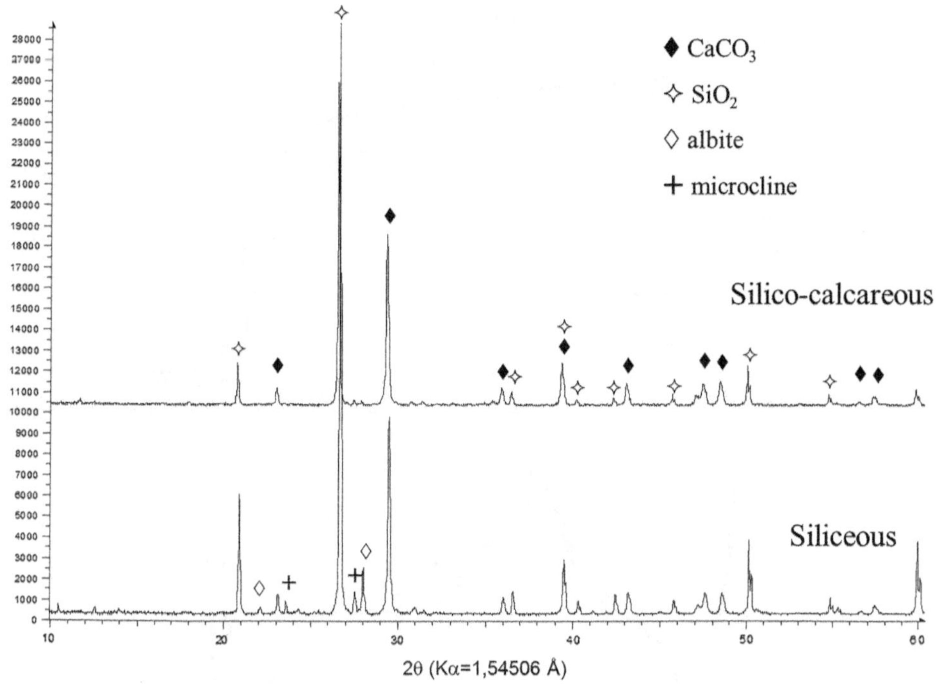

Figure 4.1 XRD analysis of the RCS.

Figure 4.2 Mineralogical analysis by optical microscopy (1, limestone; 2, Twinned microcline (K-feldspars e.g. $KAl[Si_3O_8]$).

a calcium carbonate content of 20%–25% for the siliceous sands and 30%–34% for the silico-calcareous sands.

4.2.2 Clinker raw meal preparation and grindability

The grindability of materials is an important parameter for the preparation of cement raw meals. Several guidelines on threshold values have been given on the maximum size of quartz and limestone particles (Reggad 1993; Fundal 1996) in order to minimize the departure from clinker quality target based on the potential composition. An identical sample

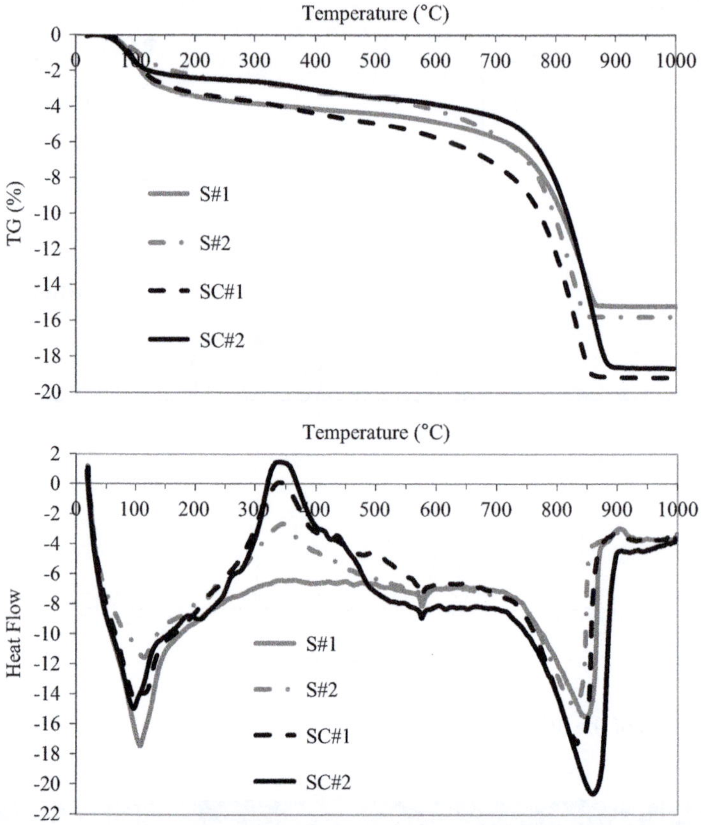

Figure 4.3 TGA (top) and DTA (bottom) of the sands.

mass of 0/4 particle size was ground in a laboratory bond ball mill during the same duration and with the same frequency, until a zero residue to 250-µm sieve. The residue to 250 µm was weighed after each grinding sequence and meticulously placed back in the jar mill, with passers-through, so the sample always has the same starting mass. The percentage of residue on the sieve of 250 µm according to the grinding time is shown in Figure 4.4.

The RCSs, with the exception of the siliceous #1, have a quite similar behavior, and height passages are enough to get almost no residue on the sieve of 250 µm and for a cumulative time of 105 s.

In a second step, each RCS sample with an identical mass was ground simultaneously at the same frequency, and the percentages of passers-through were determined on the sieve of 0.315, 0.25, and 0.2 µm (Table 4.2). This experiment confirms that the siliceous sand S#1 is more difficult to grind; the petrographic nature of the material (granite) and, in particular, the presence of quartz may explain this difference by comparison with siliceous sand S#2 mainly composed of microquartz (chert).

4.2.3 Burnability study

RCSs have a chemical composition similar to that of clay. A first experimental study (Mix 1) was based on a total replacement of clay by the RCS samples, which leads to the substitution rates close to 11% for the siliceous sands and to 14% for the silico-calcareous ones.

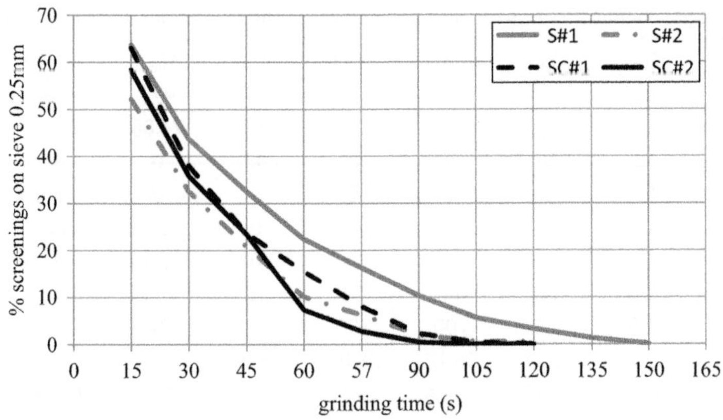

Figure 4.4 Evolution of the sieve (%) 250 μm according to the milling time.

Table 4.2 Grading profile of ground RCSs

	S#1	*S#2*	*SC#1*	*SC#2*
Passers-by 0.315 mm (%)	99.7	100	100	100
Passers-by 0.25 mm (%)	95.6	99.55	99.8	99.9
Passers-by 0.2 mm (%)	84.8	95.6	96.7	99.3

A second series of experiment (Mix 2) was conducted by substituting clay and marl by RCS sands, which leads to the substitution rates reaching 19% for the siliceous sands and 24% for the silico-calcareous ones. Bauxite and iron oxide were also added to the meal to compensate for the lack of alumina. The optimized mix proportions (constitutions), chemical compositions, potential clinker compositions, and relevant indexes of the tested raw mixes are presented in Table 4.3.

TGA and DTA were performed in order to follow the dehydration and decarbonation of the cement raw meal (Figure 4.5). The mass of the cement raw meals is lost due to the dehydration of hydration products and clays (near 150°C–250°C and 450°C–500°C, respectively) and due to decarbonation at 800°C. The weight loss at this temperature is considered the most important. The onset of the liquid phase (C_3A and C_4AF) was observed through the endothermic peak at about 1,350°C. The total mass loss of 35%–36% is the expected one and equal to that of the reference raw meal. The substitution by the RCSs does not change the temperature of the onset of the liquid phase.

The reactivity of the cement raw meals was evaluated on the basis of free lime (fCaO) contents (chemical titration with Ethylen-Diamine-Tetraacetic Acid; EDTA) after sintering at 1,200°C, 1,300°C, 1,400°C, and 1,450°C (Figure 4.6).

The free lime content decreases significantly with the temperature ranging between 1,200°C and 1,300°C (the formation of C_3S by the reaction between C_2S and CaO). The free lime contents are relatively high at 1,450°C for the cement raw meals with siliceous sands. The presence of quartz in the sands can explain the difficulty in burning the cement raw meal. The free lime contents of clinker synthesized with silico-calcareous sands and of the reference clinker are similar to each other at 1,450°C.

The mineralogical analysis of the phases by XRD shows the presence of the four main clinker's phases: C_3S, C_2S, C_3A, and C_4AF (Figure 4.7). The free lime content (by mean of chemical titration) remains the same at 1,450°C for the tested Mixes 1.

Table 4.3 Proportions, chemical compositions, and relevant indexes for different raw mixes

	Ref. Mix	*Mix1 S#1*	*Mix1 S#2*	*Mix1 SC#1*	*Mix1 SC#2*	*Mix2 S#1*	*Mix2 S#2*	*Mix2 SC#1*	*Mix2 SC#2*
Limestone (wt%)	38	40	40	38	38	74.85	75.2	70.6	71.2
Marl (wt%)	48	48	48	47	47	0	0	0	0
Clay (wt%)	14	0	0	0	0	0	0	0	0
Bauxite (wt%)	0	1	1	1	0	5.5	8.5	4.2	4.6
Iron Ore (wt%)	0	0	0	0	0	0.8	0	1.0	0.8
RCS (wt%)	0	11	11	14	15	18.8	16.3	24.2	23.5
Sum	100	100	100	100	100	100	100	100	100
LoI (975°C) (wt%)	35.6	36.07	36.18	36.01	35.94	35.9	36.1	35.9	35.8
SiO_2 (wt%)	13.12	13.51	13.66	13.60	13.81	13.82	13.74	13.81	13.82
Al_2O_3 (wt%)	3.91	3.09	2.84	3.15	2.98	3.32	3.92	3.02	3.36
Fe_2O_3 (wt%)	2.63	2.30	2.25	2.42	2.25	2.49	1.85	2.79	2.45
CaO (wt%)	42.83	43.24	43.41	42.93	43.15	43.29	43.35	43.09	43.25
MgO (wt%)	0.96	0.93	0.89	0.96	0.92	0.63	0.57	0.72	0.63
SO_3 (wt%)	0.64	0.45	0.47	0.49	0.46	0.23	0.25	0.31	0.25
K_2O (wt%)	0.57	0.38	0.32	0.37	0.42	0.19	0.08	0.19	0.25
Na_2O (wt%)	0.07	0.11	0.07	0.09	0.17	0.13	0.06	0.11	0.22
TiO_2 (wt%)	0.19	0.13	0.13	0.14	0.12	0.14	0.19	0.14	0.14
P_2O_5 (wt%)	0.08	0.07	0.07	0.08	0.08	0.05	0.04	0.07	0.06
Sum	100.6	100.3	100.3	100.2	100.3	100.2	100.1	100.1	100.2
LSF	97.98	99.14	99.31	97.51	97.40	96.39	96.38	96.39	96.26
SR	2.00	2.51	2.68	2.44	2.64	2.38	2.38	2.38	2.38
A/F	1.49	1.34	1.26	1.30	1.33	1.34	2.12	1.08	1.37
Liq.Ph 1400°C (%)	29.4	24.4	22.9	25.1	23.8	25.1	25.5	24.8	25.2
Liq.Ph 1450°C (%)	29.9	24.8	23.3	25.5	24.2	25.5	26.0	25.3	25.7

The proportions of the phases (calculated using Bogue's equations and determined by Rietveld's analysis) are reported in Table 4.4. The sum of the C_3S and C_2S amounts of the substituted clinker is close to that obtained in a Portland clinker (even if the C_3S/C_2S ratio is sometimes different between the Bogue calculation and the Rietveld analysis). The highest LSF value for the first series results in a higher C_3S content. The higher value of free lime content of clinker manufactured with siliceous sands is also confirmed by XRD.

Scanning Electron Microscope (SEM) observations on polished sections of the clinkers show large and angular grains of C_3S and rounded grains of C_2S (Figure 4.8). C_3A and C_4AF are the interstitial phases. The mineralogical phases (shape, size...) are the same as those of the reference clinker; the presence of quartz grains does not generate belite clusters.

4.2.4 Conclusions of the laboratory tests

Laboratory clinkerization tests show that the substitution of one or two traditional components of the cement raw meal by RCS leads to a viable clinker if the chemical composition of the cement raw meal is well balanced (phases of the same composition, same morphology, and same range of amounts of the various phases compared to the reference clinker). The geological origin of the sands (in particular, the silica content) has an influence on the

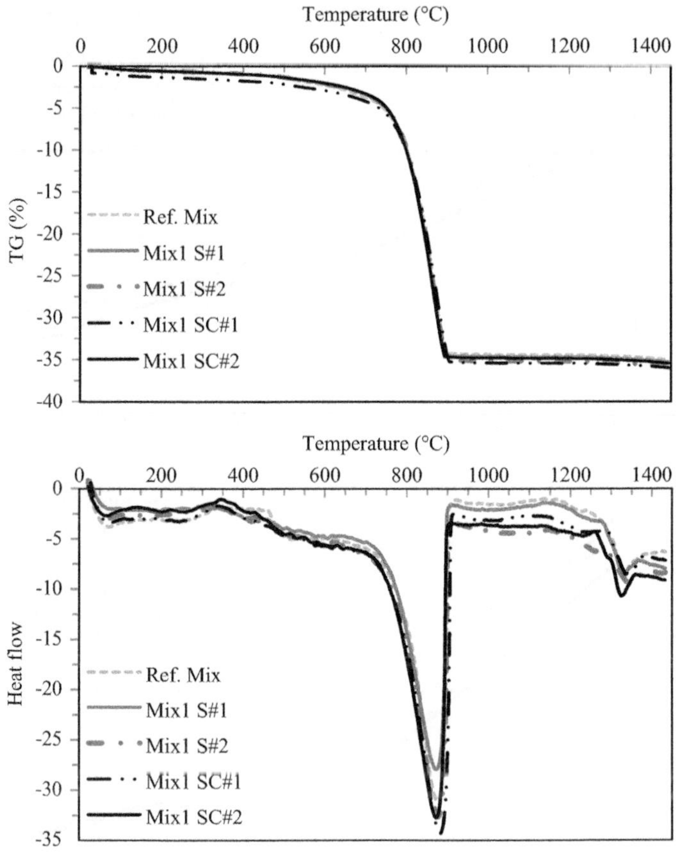

Figure 4.5 TGA (top) and DTA (bottom) curves of the cement raw meals.

clinkerization stage, making the substitution rate to vary between 11% and 25%. These results were satisfactorily enough to decide to move forward in the experimentation by means of producing a "recycled clinker" at the full industrial scale.

4.3 INDUSTRIAL PORTLAND CLINKER PRODUCTION WITH RCSs AS ARM

4.3.1 Introduction

Previous results of burnability tests applied to synthetic raw meals led to the conclusion that a significant rate of substitution of TRM by RCS is achievable, up to 25% by mass, without severe loss of clinker quality. These positive results allowed us to take the decision to run a full-scale industrial trial. The main interest of any industrial trial is to escape from scale effects regarding both milling and burning processes, once the burnability trials have been carried out with positive results and taken as feasibility tests. This methodology is currently used in the French cement industry when a manufacturer desires to accept a new material as ARM. In our case, the aim of such a trial is to produce a "modified clinker", denoted here a "recycled clinker". This is based on an alternative raw meal integrating a significant

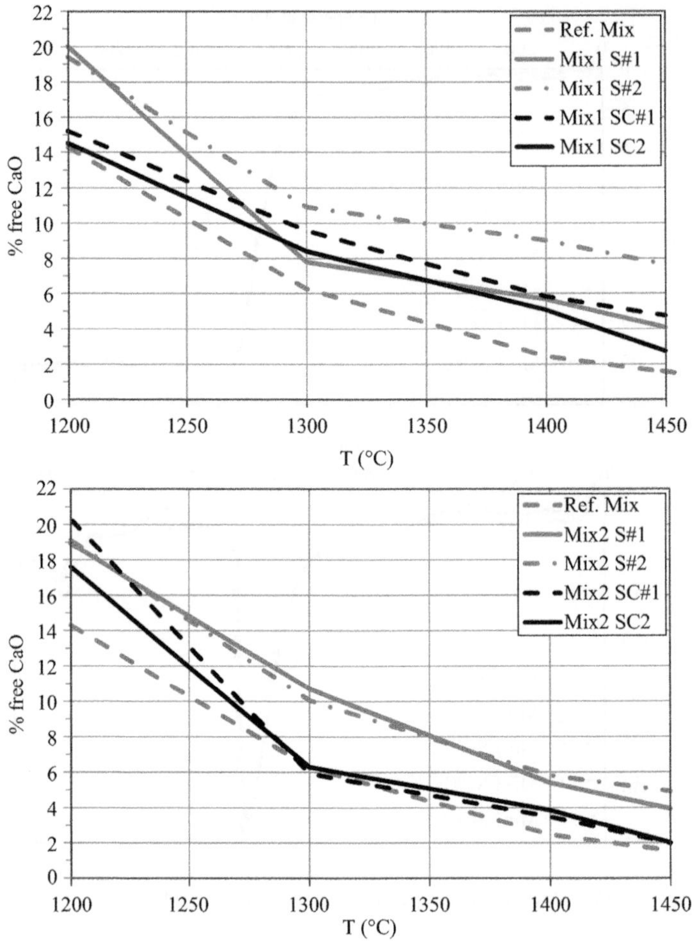

Figure 4.6 Free lime content according to temperature for the tested mixes (1 and 2).

proportion of the ARM of interest, RCS for instance, and comparing its characteristics and properties with the standard industrial production.

4.3.2 Materials and methods

The trial was carried out at Créchy cement plant, belonging to VICAT Group and located in the central part of France, not far from Vichy in the Auvergne region. This plant was chosen for its general simple layout and equipment, especially the linear and compartmented clinker storage hall and linear pre-homogenization hall allowing such tests with a limited amount of material and opportunities to isolate the test production. This plant has a medium-rated capacity of 425 kt of clinker/year equipped with a discontinuous pre-homogenization hall, a raw ball mill, (Polysius birotator) of 120 t/h output, feeding homogenization silos in order to feed a Polysius rotary kiln (75 m length, 4.40 m diameter) dry process equipped with a Dopol four-stage preheater without pre-calciner. This kiln is fired with 60%–65% of alternative fuels (waste wood, oil, humid and dry slurries, and water-based wastes G2000).

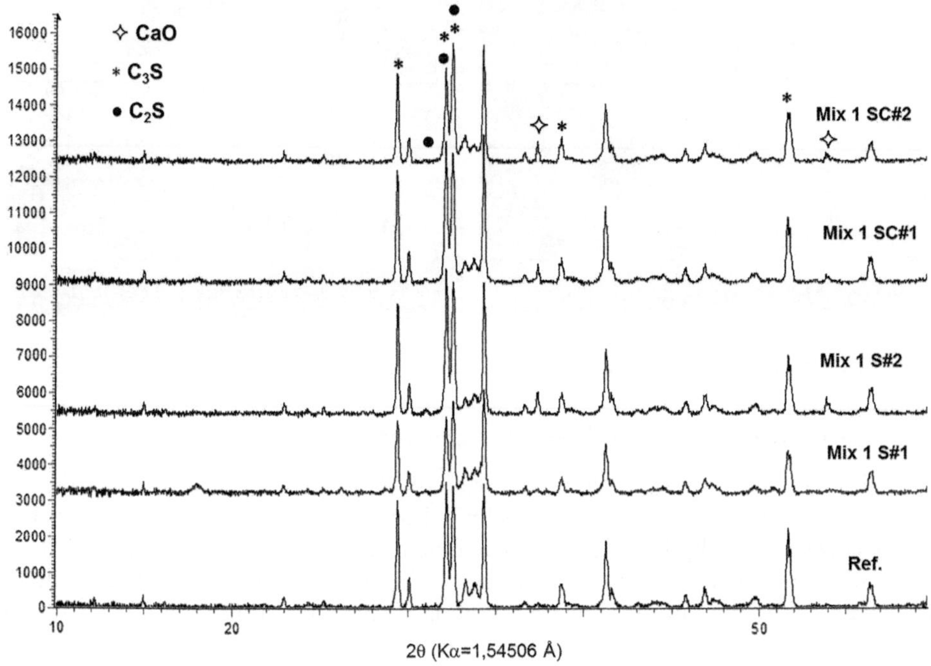

Figure 4.7 XRD patterns of the substituted clinkers.

Table 4.4 Bogue calculation and Rietveld analyses of clinkers

	Mix1 S#1		*Mix1 S#2*		*Mix S#1*		*Mix1 SC#2*		*Mix2 S#1*		*Mix2 S#2*		*Mix2 S#1*		*Mix2 SC#2*	
Mineral phase	*Bogue*	*Rietveld*	*Bogue*	*Rietveld*	*Bogue*	*Rietveld*	*Bogue*	*Rietveld*	*Bogue*	*Rietveld*	*Bogue*	*Rietveld*	*Bogue*	*Rietveld*	*Bogue*	*Rietveld*
C_3S (%)	67.6	68.7	69.4	67.5	67.4	67.8	68	66.3	66.9	60.6	54.8	64.9	67.9	56.4	67.1	69.7
C_2S (%)	9.7	9.2	9.1	14.8	10.1	13.4	10.6	13.7	11.4	14.4	20.4	9.7	10.6	20.9	11.1	6.6
C_3A (%)	6.7	8.1	5.8	2.6	6.6	5.1	6.4	6.9	7.2	8.3	11.4	15.7	5.1	7.5	7.4	11.7
C_4AF (%)	10.9	11.5	10.7	10.8	11.5	10.5	10.7	10.1	11.8	14.1	8.8	6.5	13.2	14.2	11.6	11.2
CaO (%)		2.3		2.4		1.4		1.5		0.9		3.1		1		0.7

In these conditions, the current daily production is 1,350 t of clinker/day. Cement production is assured with a Polysius cement mill single chamber equipped with a third-generation separator, with a capacity of 120 t/h of CEM I at 3,950 ± 220 cm^2/g.

4.3.2.1 Traditional raw materials

The TRMs are mainly local materials in the surroundings of the plant; the range of order of proportions of the major components (98 wt%) is limestone (51 wt%), marl (36 wt%), foundry sands (6 wt%), and excavated earth from civil works (5 wt%). The minor components, external but traditional materials, are bauxite, iron ore, and gypsum, representing 2 wt% in total.

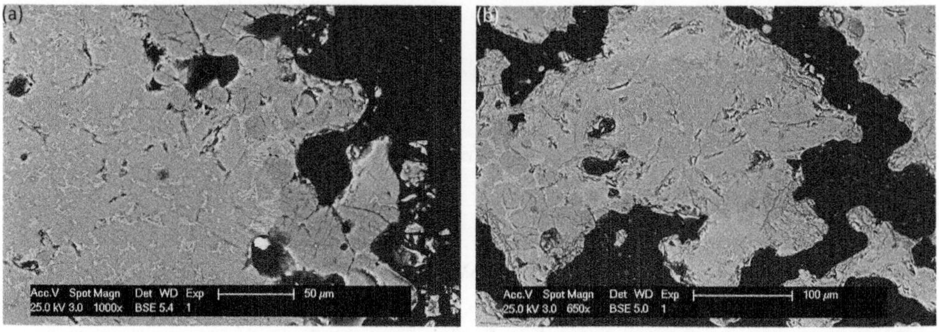

Figure 4.8 SEM micrographs of substituted clinkers.

4.3.2.2 *Alternative raw material/recycled concrete sand*

After preliminary calculations, 900 t (metric) of RCS was reclaimed from the crushing plant, transported to Créchy, and stockpiled at the cement plant in order to build a pre-homogenization stockpile of 5,160 metric tons from the 21st to 22nd April 2016.

Given the mineralogical composition of the RCS, in particular the quartz content, the main strategy of the raw mix design was dominated by substituting the foundry sands by RCS in order to remain with the acceptable quartz content.

4.3.2.3 *Chemical and mineralogical analyses*

Chemical analysis was performed by Bruker S4 X-ray fluorescence (XRF) spectrometer combining results on pressed powder pellets for alkalis, sulfur, and chlorine and on fused bead for the major and minor elements.

Mineral phases were identified by XRD on pressed pellets by Bruker D4 Endeavour X-ray diffractometer using Cu Kα line under 40 kV and 40 mA. Mineral phases were quantified by Rietveld's analysis using Topaz software (Taylor et al. 2000; Schmidt and Kern 2001; Le Saoût et al. 2011). The comparison of mineralogical compositions between analytical methods (XRD versus Bogue) was made and led to the conclusion that if the correlation is strong enough, the XRD/Rietveld analysis gives higher values compared to Bogue/XRF-derived values, as shown in Figure 4.9.

Microstructural analysis of clinker granules was performed on the polished sections obtained after immersion in epoxy resin followed by cutting and scrubbing with abrasive disks P600 (30 μm) and P1200 (15 μm). Finishing polish is obtained by felt disks impregnated by diamond pastes of granulometries between 6 and 1 μm. Microscopic observations have been made by a metallographic Zeiss Axioscope microscope.

4.3.2.4 *Cement grinding, physical and mechanical performance of the cement*

Cement grinding was performed at pilot scale by a cement ball mill of the following characteristics: 3 m long, 0.9 m internal diameter, two chambers of 2 and 1 m long, respectively, equipped with one O-SEPA aeraulic separator, third generation. In these conditions, the output flow rate is 500 kg/h.

The physical characteristics and the mechanical performance of the cements were determined according to the relevant European standard, that is, NF EN 196-1 for mechanical strength, NF EN 196-3 for setting time, and NF EN 196-6 for fineness.

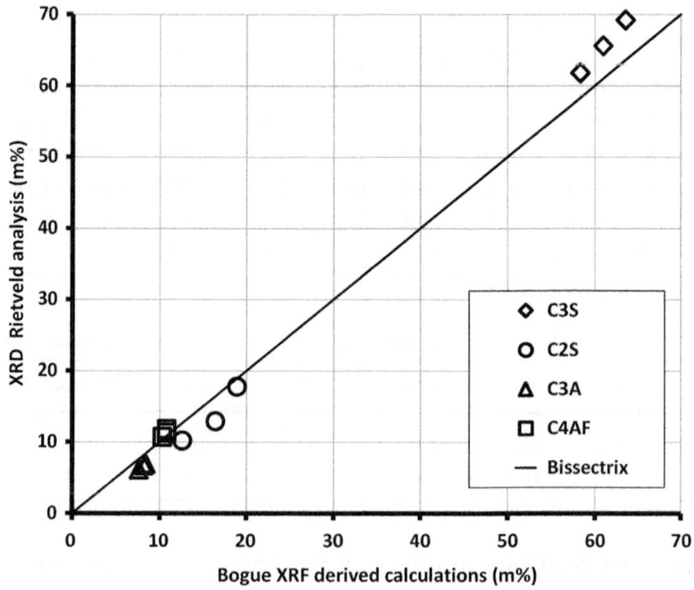

Figure 4.9 Mineralogical composition of recycled clinker: XRD/Rietveld analysis versus traditional Bogue/XRF-derived calculations.

4.3.3 Results and discussion

4.3.3.1 Chemical and mineralogical analyses of industrial RCS

Of the 900 metric tons of RCS, three samples of 5 kg each were taken from the stockpile during the build-up phase. Chemical and mineralogical contents of these samples were analyzed for data collection in order to feed the calculation tool for raw mix optimization.

The results of the chemical and mineralogical characterizations are given in Table 4.5, showing the chemical homogeneity of the stockpile.

4.3.3.2 Raw mix design: pre-homogenization stockpile

A stockpile of 5,160 metric tons of raw materials, including RCS, was built up for the purpose of the trial. In order to refine the mix proportions of the constituents according to the classic rules for pre-homogenization build-up, a composite sample corresponding to 1 h of production was taken at the sampling tower and analyzed every hour.

Considering the original design of the current industrial raw mix and given the quartz content in the RCS (average 43% by mass), the best strategy to get a similar raw mix design was to replace foundry sands by RCS.

The targets for the ratios and indices of the raw mix to be constituted at the pre-homogenization stockpile were the following: LSF: 98/MS: 2.6/A/F: 1.7/SO3: 0.35%, leading to the final raw mix design in comparison with the current values (Table 4.6). From Table 4.6, it is noticeable that the incorporation of the RCS was made possible by replacing the foundry sands with a slight increase in limestone content and minor materials but with a significant reduction in marly component (from 36% to 25%). The final raw mix composition of the trial stockpile was found very similar to the reference stockpile. Concerning the quartz content, the trial value was measured at 9.8 wt%, very similar to the reference value, between 8 and 9 wt%.

Table 4.5 Chemical and mineralogical compositions of three representative samples of RCS

	H_2O (wt%)	LoI (wt%)	SiO_2 (wt%)	Al_2O_3 (wt%)	Fe_2O_3 (wt%)	CaO (wt%)	MgO (wt%)	SO_3 (wt%)	K_2O (wt%)	Na_2O (wt%)	Cl (ppm)	Quartz (wt%)
#1	8.14	16.9	43.2	6.70	1.84	26.5	0.94	0.47	1.46	1.45	0	43.3
#2	7.51	17.6	42.0	6.13	2.02	27.1	1.13	0.37	1.35	1.70	250	41.2
#3	7.57	16.3	43.4	6.95	1.99	24.9	1.12	0.38	1.43	1.78	0	43.6
Average	7.74	16.9	42.9	6.59	1.95	26.2	1.06	0.41	1.41	1.64	83	42.70
Std dev.		0.65	0.76	0.42	0.10	1.14	0.11	0.06	0.06	0.17		1.29

Table 4.6 Raw mix constitution for traditional and recycled raw mixes and quartz content

	Limestone (wt%)	Marl (wt%)	Foundery sand (wt%)	Concrete sand	Excavated earth (wt%)	Minor comp'ts (wt%)	Total (wt%)	Quartz (wt%)
Traditional raw mix	51,0	36,0	6,0	0,0	5,0	2,0	100	8,6
Recycled raw mix	54,5	25,6	0,0	14,6	2,3	3,0	100	9,8

4.3.3.3 Raw meal composition: building up the homogenization silo

At Créchy cement works, the agreed strategy for the homogenization process is by batch; the objective of this is that the cumulative chemistry of control samples from raw mill outlet will reach the target at the end of silo filling operation. At the very beginning of the stockpile consumption, a control sample was taken every hour at the raw mill outlet in order to adjust as fast as possible the chemistry of the raw meal already in the silo. The obtained raw meal is then homogenized by aeraulic recirculation for 2.5 h. In these conditions, four homogenization silos corresponding to 5,760 metric tons of ground raw meal were filled. In the running conditions of the raw milling operation with the modified raw meal, no significant difference was noticed, in comparison with the reference conditions, as it can be seen in Table 4.7.

4.3.3.3.1 *Raw meal fineness at raw mill outlet*

As shown in Table 4.8, the fineness of the "recycled" raw meal is very close to the reference values of the current raw meals for the normal clinker and the SR3 clinker currently produced at Créchy cement works, but a bit lower at 90-μm sieve.

4.3.3.3.2 *Special focus on quartz content*

Given the current practice to use foundry sands as a quartz-bearing constituent in the raw mix, a specific awareness has been developed by the plant with regard to quartz content of

Table 4.7 Chemical modulus during raw meal homogenization process

Target	102		2.56		1.75	
Homo	LSF		MS		A/F	
	Average	*St. dev.*	*Average*	*St. dev*	*Average*	*St. dev.*
3306	101.55	4.43	2.53	0.06	1.75	0.033
3307	101.84	3.11	2.57	0.06	1.77	0.026
3308	101.63	3.23	2.58	0.07	1.81	0.009
3309	104.16	6.46	2.68	0.07	1.75	0.029

Table 4.8 Fineness characteristics of recycled raw meal versus reference values for current raw meals

	Current raw mix[a]	*Raw mix RECYBÉTON*	*Raw mix SR3*[a]
% > 200 μm	1.4	1.2	1.1
% > 90 μm	15.8	14.5	14.9
% > 45 μm	35.5	33.8	35.2

[a] Reference values, June 2016.

Table 4.9 Quartz content value of RCS raw meal versus reference values for current raw meals

	Current raw mix[a]	*Raw mix RECYBÉTON*	*Raw mix SR3*[a]
Total quartz	7.20	8.17	11.1
% quartz > 45 μm	14.23	13.92	19.6
% Qz > 45 μm (raw)	5.1	4.71	7.2
% Qz > 45 μm vs total Qz	71.9	57.9	59.9

[a] Reference values, June 2016.

the raw meal. Considering the quartz content of RCS, a special focus has been dedicated to this important feature. The mineralogical analysis was carried out by XRD on sieved samples at 45 μm after calibration, using the "dosed aliquot" methodology. According to the values in Table 4.9, the quartz content due to RCS is fully comparable to the reference values relevant for the two current clinkers, indicating the appropriateness of the chosen strategy and calculations. The values shown in Table 4.9 are mean values of several samples taken during the raising of the homogenization silo level.

4.3.3.4 Burning the mix and clinker production

4.3.3.4.1 Operational conditions

Timing: Feeding the kiln with the alternative raw feed started on Saturday April 30th, 2016, at 9:00 am and ended on Monday May 2nd, 2016, at 10:00 pm (61 h). During this period of time, 3,000 metric tons of clinker was produced; of them, 2,700 was stockpiled separately from the current clinker in a dedicated area.

From the operator's point of view, RECYBÉTON clinker is a bit more difficult to burn than the current one with some thermal instabilities and coating/decoating formation. However, the easiness to burn is not so different compared with sulfate-resisting clinker (SR), and the transition periods (in and out) were run smoothly.

4.3.3.4.2 Kiln feed characteristics

Samples were taken once per shift (composite sample every 8 h) in which the following characterization analyses were performed at the works laboratory: loss on ignition (LoI); complete chemical analysis by XRF; mineralogical analysis by XRD; fineness through sieve analysis (refusals at 200, 90, and 45 μm); and quartz content analysis on full raw feed and on fraction greater than 45 μm.

Except for a slightly high A/F ratio, targeted criteria (LSF and MS) are meeting expectations, very close to the current values of the normal raw feed. Table 4.10 gives the average values of useful ratios for alternative and reference kiln feed.

Table 4.10 Average values for LSF, MS, and A/F ratios for alternative and reference kiln feeds

KILN FEED	*Fineness*			*LoI 950°C*	*LSF*		*MS*		*A/F*		*Quartz content*			
Unit	*(%) >200 µm*	*(%) >90 µm*	*(%) >45 µm*		*Average*	*S.Dev*	*Average*	*S.Dev*	*Average*	*S.Dev*	*(%) Tot. Qz*	*(%) Qz > 45 µm*	*(%) Qz > 45 µm*	*(%) Qz > 45 µm*
RECYBÉTON	1.1	15.3	34.3	35,6	101.9	1.4	2.59	0.02	1.75	0.05	7.4	5.3	72.1	15.5
Std raw meal June 2016	1.2	14.8	33.6	35,96	102.3	1.8	2.59	0.06	1.65	0.08	6.5	5.3	81.6	15.8

4.3.3.4.3 Clinker characteristics

4.3.3.4.3.1 CONTROL OF PRODUCTION

During the production period, one sample of clinker was taken twice per shift (every 4 h), corresponding to 3,000 metric tons of clinker produced. Of them, the first 300 t was classified as low-quality clinker, corresponding to 10% of the production, versus 11% of the typical production in 2016. From these 14 samples representing the clinker production, silicate and free lime contents were determined from XRF and XRD analyses. The results, given in Table 4.11, show that the C_3S content is a bit lower (63%) compared to the reference (64%) with a slightly better standard deviation; accordingly, the C_2S content shows the reverse trend for the average, with a better scattering value. Regarding the free lime content, which is considered an indicator of quality for burnability, the quantitative values are almost identical for both reference and trial clinkers, with the same scattering data. With such values, the demonstration of the feasibility of producing Portland clinker with a raw meal based on a substantial proportion (15%) of RCS is definitely given, from the industrial point of view.

4.3.3.4.4 Laboratory characterizations

4.3.3.4.4.1 CHEMISTRY

The chemical compositions of the two recycled clinkers (denoted "RECYBÉTON 1 and 2") compared to the reference clinker show moderate LSF, for example, 95 compared to reference at 96, associated with a medium SM, for example, 2.4–2.5, indicating good clinker-forming conditions with, however, a trend to form crust. The calculated proportion of liquid phase confirms the combination ability satisfactorily.

In terms of degree of sulfatization, the DS indices show comparable values (54–58), indicating excess of alkalis relative to sulfur, which is the normal chemical context of Créchy plant.

Table 4.11 Silicates and free lime (fCaO) content of recycled clinker versus reference values (Créchy 2016)

Target	*>55%*		*—*		*<2.5%*	
Clinker	*C_3S*		*C_2S*		*CaO f*	
	Average	*Std dev.*	*Average*	*Std dev.*	*Average*	*Std dev.*
RECYBÉTON clinker	62.8	4.1	15.8	3.5	1.33	0.9
Average Créchy 2016	64.2	5.5	14.7	5.1	1.37	0.9

4.3.3.4.4.2 MINERALOGY

Rietveld refinements of the three clinkers are very consistent, showing slightly lower total C_3S content and subsequently slightly higher C_2S content associated with similar C_3A and C_4AF contents.

4.3.3.4.4.3 MICROSTRUCTURE

Based on structural observations made on polish sections under the optical microscope (Figure 4.10), the clinker granules are found with a typical zonation between core and crown; the core is denser and harder. This structural feature is comparable to the one found

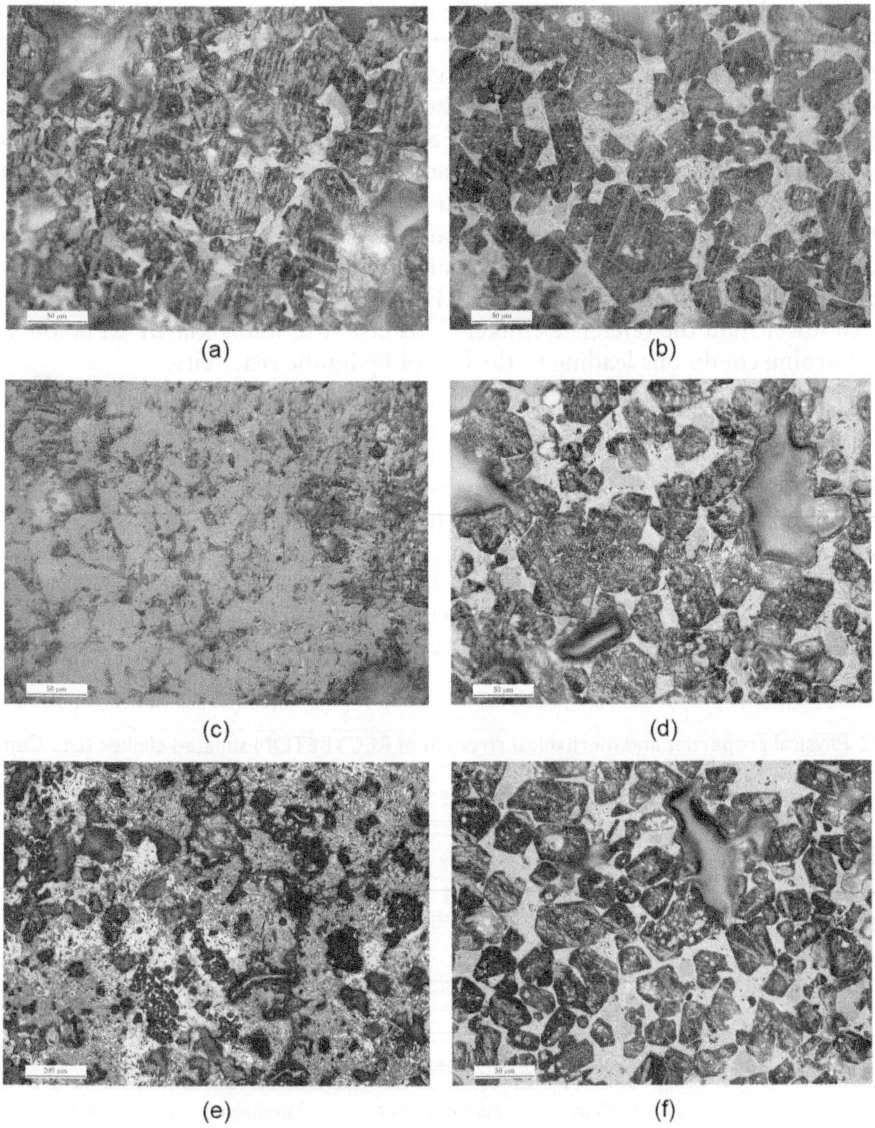

Figure 4.10 Microscopic microstructure of clinker's crown versus core (794: reference clinker before the trial; 795: RECYBÉTON clinker; 796: reference clinker just after the trial).

in the reference clinker. In textural terms, crystal size distribution of the RECYBÉTON clinker is very similar to that of the reference clinker and relatively homogeneous with an average size of alite crystals around 30 µm and a moderate proportion of free lime clusters. Considering the amount of quartz brought by the RCS, a special attention was paid to belite clusters, which are known to bring losses of mechanical strength. In this respect, it has been found that these belite clusters are relatively rare as shown in Figure 4.10, giving the appearance of the clinker's microstructure and texture.

4.3.3.4.4.4 MECHANICAL PERFORMANCES (1): SULFATED CLINKER LABORATORY SCALE

An average sample representative of the production period has been made from 14 grab samples, taken twice per shift, reaching approximately 30 kg of material. After homogenization, 1 kg has been extracted using a blade divider for the determination of the chemical and mineralogical compositions (see above). The rest has been used to produce a "sulfated clinker" ground at 4,000 cm^2/g (by ball mill grinding at the laboratory scale) with gypsum addition (5% by mass). This sample was used to measure mechanical performance, for example, compressive strengths on mortar, according to EN 196-1, water demand, and fineness. Table 4.12 gives the physical characterization and the mechanical performance by comparison with reference value by means of a "merit factor", decreasing in function of time ranging from 94% to 91% between 1 and 28 days, respectively. These performances are in accordance with the mineralogical characteristics of the clinker (Table 4.13), giving a lower C_3S content by three points for the RECYBÉTON clinker, as well as a rate of cubic C_3A much lower than the reference clinker. This last criterion is known to be the result of difficult burning conditions leading to the loss of hydraulic reactivity.

4.3.3.4.4.5 MECHANICAL PERFORMANCES (2): SEMI-INDUSTRIAL GRINDING

Table 4.14 gives the physical characterization and the mechanical properties of the cement ground at pilot ball mill showing that the mechanical strengths of the RECYBÉTON clinker ground as a CEM I 52.5N cement type are very similar and slightly better at early times compared to the same cement type under industrial grinding (2016 average); surprisingly, the value at 28 days is lower for RECYBÉTON compared to the reference cement, taking into account that the physical parameters are quite identical, that is, d50 and ground

Table 4.12 Physical properties and mechanical strength of RECYBÉTON sulfated clinker (Lab Cement)

	d50 (µm)	*SSB (cm^2/g)*	w/c	*ST(i) (min)*	*SO3 (%)*	*Cs1d (Mpa)*	*Cs2d (Mpa)*	*Cs7d (Mpa)*	*Cs28d (Mpa)*
RECYBÉTON sulfated CK						19,1	29,2	46	56,6
Avg Créchy 2016	12,7	4060	0,304	238		20,3	31,5	49,9	62,4
Diff (RECYBÉTON -ref)						−1,2	−2,3	−3,9	−5,8
Merit factor						94	93	92	91

Table 4.13 Mineralogical composition of RECYBÉTON clinker

Clinker	*Sum of alites*	*Sum of belites*	*Alite/belite ratio*	*% C_3A cub/total C_3A*
RECYBÉTON clinker	62.7	16.2	3.87	33
Average 2016	65.5	14.1	4.89	50

Table 4.14 Physical characterization and mechanical performance of CEM I "RECYBÉTON" ground at pilot scale

	d50 (µm)	*SSB (cm^2/g)*	w/c	*ST(i) (min.)*	*SO3 (%)*	*Cs1d (Mpa)*	*Cs2d (Mpa)*	*Cs7d (Mpa)*	*Cs28d (Mpa)*
RECYBÉTON pilot grinding	10,8	3992	0,349	272		20,2	34,6	51,9	59
Avg Créchy 2016	12,7	4060	0,304	238		20,3	31,5	49,9	62,4
Diff (RECYBÉTON -ref)						–0,1	3,1	2	–3,4
Merit factor						100	110	104	95

d50 median diameter; SSB, blaine specific surface; *w/c*, water-to-cement ratio; ST(i), setting time initial; Cs, compressive strength.

fineness; this loss of performance can be due to the differences in the mineralogical composition between the two clinkers and the difference in the water demand.

4.4 CONCLUSIONS

In this chapter, we have demonstrated that the mineralogical composition of RCS is mainly dominated by quartz and calcite with some minor amount of feldspars and micas.

Through burnability tests at the laboratory scale, we have shown that despite their quartz content, RCSs are suitable as ARM with a maximum substitution rate of value of 25% by mass. These conclusions have been satisfactorily enough to decide to go for a full-scale trial of Portland clinker production in a cement plant.

The raw mix design optimization incorporating RCS has been easily performed leading to a raw meal with a significant proportion of substitution: 15% by mass. This proportion has been achieved by the substitution of foundry sands, existing in the current raw mix design. The clinker production of this "recycled raw meal" showed few differences in the burning conditions compared to the TRM, in operation terms.

The clinker characterization has shown that the chemical and mineralogical characteristics are very similar to those of the reference clinker and led to very acceptable mechanical performance by means of a pilot grinding, giving good performance on the working site.

This demonstration shows that it is possible to recycle demolition concrete sand in the Portland cement process with good performance and open the way to industrial optimization.

This is a confirmation that the current practices of the French cement industry in terms of raw material substitution are also valid for recycling fine particles from demolition concretes.

Chapter 5

Recycled concrete as cement main constituent (CMC) or supplementary cementitious materials (SCM)

M. Cyr
LMDC, Université de Toulouse

C. Diliberto and A. Lecomte
Institut Jean Lamour-Université de Lorraine

L. Izoret
ATILH

CONTENTS

Abstract

Among the three ways of recycling fine recycled aggregate (FRA) in the cement and concrete process, the first one is the use it as components of clinker raw mix raw does not raise any question in terms of standardization as long as, through burning process, the result of it will give, by design, a Portland clinker with the expected mineralogical composition and the subsequent usage properties (Chapter 4). The second and third ways are, respectively, the incorporation of FRA as cement main constituents other than clinker and considering FRA as concrete addition or supplementary cementitious material (SCM) after grinding; they are evaluated in this chapter. The first part of this chapter is devoted to the evaluation of the physicochemical properties of composed Portland cements, produced at a semi-pilot scale obtained by replacement of limestone, considering their mineral composition. The FRA can be considered as a serious potential candidate to be recognized as cement main constituent. In the second part of the chapter, the evaluation of the opportunity to use the fine fraction extracted from crushed demolition concretes is presented considering FRA as mineral additions in binder-based mortars or concrete mixes. Two types of fines particles are described and were studied: fines particles from dust collector, obtained during the process of recycled concrete crushing for the production of fine and coarse aggregates and fines particles obtained from the grinding of fine recycled aggregate. The work includes the characterization of these materials, the study of their hydraulic and pozzolanic activity, and their effect on the properties of fresh and hardened mortars when used in replacement of a fraction of Portland cement. The FRA can be considered as type I addition (filler) and even as type II addition (pozzolanic), since it was observed that they seem to have a long-term binding activity.

5.1 INTRODUCTION

In Chapter 4, a CEM I 52.5N cement type from a clinker based on a raw mix incorporating a significant amount (15% by mass) of fine recycled aggregate (FRA) in substitution of traditional raw materials was studied and evaluated.

In terms of performance on International Organization for Standardization (ISO) mortar compared to actual industrial production, the "recycled" CEM I 52.5N exhibits a quite identical mechanical strength between 1 and 7 days, and at 28 days, there is a loss of performance by 6%, which can be explained by a slightly higher water demand and differences in the mineralogical composition of the clinker.

The use of FRA as components of clinker raw mix does not raise any question in terms of standardization as long as, through burning process, the result of it will give, by design, a Portland clinker with the expected mineralogical composition and the subsequent usage properties.

The second way to recycle FRA is to consider this material as a main constituent of cement other than clinker. The experiments presented in this chapter will explore the possibilities of production of a CEM II cement type, based on a normal Portland clinker and traditional limestone, by progressively substituting limestone by FRA over a range of composition from a CEM II/A-M (LL-FRA) to a CEM II/B-M (FRA-LL), and the end-members of this series are a CEM II/A-LL (15) 42,5R and a CEMII/B-FRA (25).

The third way to recycle FRA is to consider it as supplementary cementitious material (SCM) to be used as concrete addition. Two types of fine particles are described and were studied: fine particles from dust collector obtained during the process of recycled concrete crushing for the production of fine and coarse aggregates and fine particles obtained from the grinding of FRA. The work includes the characterization of these materials, the study of their hydraulic and pozzolanic activities, and their effect on the properties of fresh and hardened mortars when used in replacement of a fraction of Portland cement. The FRA can be considered as type I addition (filler) and even as type II addition (pozzolanic), since it was observed that they seem to have a long-term binding activity.

5.2 STATE OF THE ART

5.2.1 Cement composition: principles and methods

Cement composition is governed by rules described in the European Standard EN 197-1 and its French national version NF EN 197-1. This fundamental standard defines the main (more than 6% by mass) and the secondary constituents (less than 5% by mass) of the

Table 5.1 Cumulative heat of hydration (J/g) of the cements at 12, 24, 48, 51, and 120h

	CEM II/A-LL	*CEM II/A-M (LL-FRA)*	*CEM II/B-M (LL-FRA)*	*CEM II/B-FRA*
12h	217.5	210.3	181.8	156.2
24h	287.5	276.6	247.9	233.8
48h	309.5	296.2	268.5	254.4
51h	309.5	297.1	268.9	255.9
120h	316.7	302.7	278.2	268.6

cement. Among the main constituents, we find (a) Portland clinker and (b) main constituents other than clinker. These last constituents are basically limestone (L, LL), ground granulated blast furnace slag (S), fly ash from coal combustion (V, W), pozzolanic materials (P, Q), calcined clays (T), and silica fumes (D). Among the secondary constituents, we find any material from the cement manufacturing process, such as limestone, under burnt clinker or sometimes called burning fines. In addition, calcium sulfate is necessarily added to the other (main + secondary) constituents in order to optimize the setting time.

Depending on mass proportions of each constituent, a given cement composition will fall in the field of CEM I (clinker + calcium sulfate), CEM II/A-LL (clinker + limestone + calcium sulfate), or CEM III/A (clinker + Ground Granulated Blast Furnace Sag (GGBFS) + calcium sulfate), for example. The agreed compositions of the standard EN 197-1 are given in Table 5.1 in detail.

5.2.2 Standardization perspective

As mentioned earlier, the essence of EN 197-1 is to well define which materials can be used as cement constituent. Apart from the agreed list lying in the standard, there is not a single opportunity to self-declare a new material as cement constituent. In other words, the cement composition is based on "well-known and proven constituents" principle, in order to ensure the constancy of performance and stability of cements, even for a long period of time.

This logic is, of course, conservative for the security of construction reasons, but since a recent past, it could have been considered by some as a barrier to innovation, leading to a limitation of opportunities to develop cements with low CO_2 impact. However, it is critical to remind here the essential requirement no. 1 of from the European Construction Product Regulation (CPR), which is "stability and mechanical strength" for protection of goods and people.

For this reason, a new standard has been developed by CEN/TC51 as a technical report in order to give guidelines for a technical dossier in order to support the European standardization of new cement compositions. This standard (CEN/TR 16912, guidelines for a procedure to support European standardization of cements, published in 2016) makes a distinction between:

- a new cement based on Portland clinker and "well-known and proven" main constituents other than clinker but with undescribed (new) proportions according to Table 5.1 of EN 197-1,
- a cement based on Portland cement clinker mixed with one or several new main components, and
- a new cement based on chemical system differing substantially from Portland clinker.

Our current situation with FRAs is clearly belonging to the second option of the standard, for which FRAs will be considered as a new main constituent of cement other than clinker.

Regarding the concrete standard NF EN 206, the situation is rather similar where the suitability of a constituent can be recognized through a specific product standard, for example, EN 12620 for limestone fillers or EN 15167-1 for ground granulated blast furnace slag. Where there is no European Standard for a particular constituent, which refers specifically to the use of this constituent in concrete conforming to this standard, or where there is an existing European Standard that does not cover the particular product, or where the constituent deviates significantly from the European Standard, the establishment of suitability may result from the following:

- A European technical assessment (ETA) that refers specifically to the use of the constituent in concrete conforming to this standard and
- Provisions valid in the place of use of the concrete, which refers specifically to the use of the constituent in concrete conforming to this standard.

This standardization perspective will guide us in order to prepare the technical dossier, based on the following RECYBÉTON outputs, whose clear objective is to give to the standardization committee a comprehensive overview of the new material and its properties when mixed with cement, in order to ensure a proper and relevant usage in all circumstances.

5.3 RECYBÉTON OUTPUTS

5.3.1 Semi-pilot production of composed Portland cement: CEM II/A-M and CEM II/B-M: limestone substituted by FRA

5.3.1.1 Cement compositions and working methodology

Given the different types of materials accepted as the main cement constituent other than clinker, the first task was to determine which type of substitution can be relevant. Considering the mineralogical composition of FRA (see Chapter 4)—quartz and calcite without hydraulic or pozzolanic properties—it became obvious that FRA could potentially substitute limestone in CEM II/A or B-LL cement types. We have indeed considered the progressive replacement of limestone by FRA, which subsequently led to the following cement compositions:

- CEM II/A-LL15 42.5R: clinker + 15% limestone + 6% gypsum
- CEM II/A-M (LL7-FRA8): clinker + 7% limestone + 8% FRA + 6% gypsum
- CEM II/B-M (LL9-FRA16): clinker + 9% limestone + 16% FRA + 6% gypsum
- CEM II/B-FRA25: clinker + 25% FRA + 6% gypsum

All these cements were produced at semi-pilot scale by crushing and milling in a ball mill, followed by sieving operation and granular reconstitution in order to achieve a particle size distribution close to an industrial cement, CEM II/A-LL 42.5R from Port la Nouvelle Plant (Lafarge France) taken as a reference. This way of production was chosen, as under industrial conditions, the cement production is demanding a very high quantity of material (approximately 2,000 t); moreover, it requires a full cleaning and refurbishment of cement silo in order to prevent from any pollution from previous cement types.

5.3.1.2 *Evaluation of the physicochemical properties of the composed Portland cement*

The evolution of properties on cement and on mortars was observed in accordance with the cement standard NF EN 196 and its relevant parts.

5.3.1.2.1 *Characteristics of dry (unhydrated) powders*

The specific gravity of cements decreases with the addition of FRA due to their low value as compared to clinker. It varies from 3.065 g/cm^3 for the CEM II/A-LL to 2.966 g/cm^3 for the CEM II/B-FRA. The Blaine specific surface is similar for all the cements (4,500–4,760 cm^2/g). This result is also confirmed by laser granulometry.

The cements are mainly composed of calcium oxide and silicon oxide. A few increase in this last oxide content is observed with the addition of FRA and reaches 27% for the CEM II/B-FRA. The free lime content is similar for all the cements (close to 2%), but the insoluble residue increases linearly with the FRA content in cement and reaches 9.0% for the CEM II/B-FRA. This is due to the presence of quartz in FRA [confirmed by X-ray fluorescence and X-ray diffraction (XRD)].

The loss of ignition does not really decrease with the substitution of limestone fillers by the FRA due to the presence of carbonates, sulfates, CSH, etc in the FRA.

5.3.1.2.2 *Characteristics of cement pastes (hydrated)*

The incorporation of FRA in cement leads to a slight stiffening of the paste, which must be compensated by additional water to obtain the normal consistency. The water demand value increases by 8% between the CEM II/A-LL and CEM II/B-FRA. This can be explained by the presence of microporous hardened cement paste of the FRA (Hansen 1992; Evangelista et al. 2007, 2015).

The initial setting time increases with the rate of incorporation of FRA and decreases with the clinker rate. For the CEM II/A, the addition of FRA does not appear to affect the initial and final setting times. For the CEM II/B, the increase in the FRA content from 16% (with 9% of limestone filler) to 25% delays the initial setting time by 30 min but there is no effect on the final setting time.

5.3.1.2.3 *Characteristics of mortars*

The values of the heat of hydration, determined by Langavant's semi-adiabatic calorimeter, are reported in Table 5.1: cumulative heat of hydration (J/g) of the cements at 12, 24, 48, 51, and 120 h. The addition of FRA decreases the heat of hydration of the cements. The maximum temperature is observed later when FRA is added to the cements. This phenomenon was also observed for the initial setting time, which is higher for the CEM II/BFRA. The cements with FRA are therefore cements with a lower heat of hydration, an interesting quality for particular uses.

The values of compressive strength and the relative compressive strength compared to the CEM II/A-LL according to time are shown in Figure 5.1. The compressive strength of the cements that contain FRA decreases for each term, compared to the control value (CEM II/A-LL). This decrease is more important for the CEM II/A than for the CEM II/B. The decrease is also much visible for the short term (1 or 2 days) than for the long term (28 or 90 days). The relative compressive strengths are close to each other for 7 days.

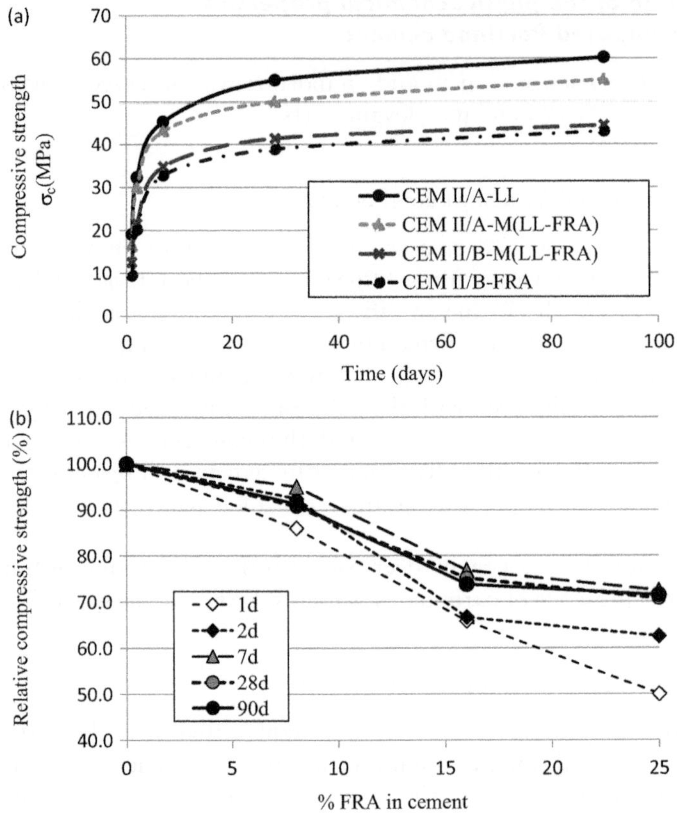

Figure 5.1 (a) Compressive strength and (b) relative compressive strength compared to the CEM II/A-LL one according to time.

The recycled concrete sands do not behave as inert and seem to have a long-term binding activity. One can note also that the lack of limestone fillers (between the cement with 16% FRA-9% LL and the cement with 25% FRA-0% LL) does not result in a real loss of performance.

FRA-containing cements show mechanical strength values that may be consistent with 42.5N and 32.5N, depending on the substitution rate from CEM II/A and CEM II/B, respectively, having sufficient properties for common uses. Obviously, cement strength class can only be given on a statistical basis and will need further studies from industrial production.

5.3.2 Cement combination concept and addition to concrete

5.3.2.1 Introduction

Fine particles coming from recycled concrete sand were evaluated as possible mineral additions for the replacement of a part of the cement in concrete applications. Additions are defined as finely divided mineral constituents used to improve or give specific properties to the concrete. In EN 206, they are separated in two categories: type I addition, which are almost inert, such as filler or pigment, and type II addition, that is, pozzolanic or latent hydraulic addition, such as fly ash, silica fume, or ground granulated blast furnace slag.

Recycled concrete not only contains hydrated cement paste and aggregates, but also contains anhydrous compounds such as clinker or additions that may not be completely inert when reused in new concrete (Braga et al. 2012, 2014). Moreover, siliceous minerals in the aggregates may still have certain reactivity. Thus, fine mineral particles from the recycled concrete could be seen as a type I addition, that is, similar to a filler due to the inert minerals present in the aggregates and hydrated paste, or a type II addition, which can present some pozzolanic or hydraulic activity, thanks to the remaining anhydrous cement, additions, or reactive natural aggregates (Evangelista and de Brito 2007a; Kim et al. 2012; Oksri-Nelfia et al. 2016; Topic et al. 2017).

The aim of this work was to evaluate the activity of fine mineral particles from the recycled concrete on pastes and mortars. The fine particles used as additions had two origins:

- Fine particles from dust collector (FRA-DC—fines recycled aggregate from dust collector), obtained during the crushing process of recycled concrete for the production of fine and coarse aggregates. As the industrial process is not producing high amounts of these fines (at least not enough to represent a possible source of addition for industrial applications), another source of addition has been tested.
- Fine particles obtained by grinding of FRA (G-FRA), which could be a source of addition, as the reuse of recycled concrete sand remains challenging due to its high water demand in concrete.

The work included the characterization of FRA-DC and G-FRA, the study of their hydraulic and pozzolanic activities, and their effect on the properties of fresh and hardened mortars when used in replacement of a fraction of Portland cement.

5.3.2.2 Characteristics of FRA-DC and G-FRA

FRA-DC samples were collected during the crushing process of recycled concrete in an industrial platform near Paris (Gonesse F-95). G-FRA samples were obtained from the grinding of four sands coming from different industrial platforms of concrete crushing with distinct main petrographic nature-based concretes: Paris (Pa) with flint aggregates, South of Lyon (Ly) with silico-calcareous aggregates, Strasbourg (St) with granitic aggregates, and Lille (Li) with silico-calcareous aggregates. The 5-kg batch grinding was carried out in a rod mill for 30 and 120 min, which allowed reaching specific surface areas (Blaine) around 4,500 and 7,500 cm^2/g, respectively. The d_{50} of the fines were between 9 and 20 μm. It could be noted that similar grinding times were necessary for all sand particles to reach the correct fineness.

Table 5.2, which gives the properties of the different fines (G-FRA and FRA-DC), shows that FRA-DC contained more residues of hydrated cement paste than G-FRA. The differences in the compositions of the four G-FRA were due to the nature of the initial aggregate in the concrete.

5.3.2.3 Hydraulic and pozzolanic activities of FRA-DC and G-FRA

5.3.2.3.1 Hydraulic activity

Small amounts of anhydrous compounds coming from unreacted cement were found in the fines. In order to evaluate whether these anhydrous phases could react in contact with water, samples of G-FRA and FRA-DC were mixed with deionized water (water/fine ratio of 0.40 and 0.55 for G-FRA and FRA-DC, respectively), and the hydration was followed by isothermal calorimetry at 20°C for several days.

Table 5.2 Chemical, mineralogical, and physical characteristics of G-FRA and FRA-DC

	G-FRA				
	Pa	*Ly*	*St*	*Li*	*FRA-DC*
Chemical composition (%)					
CaO	17.1	20.8	15.5	20.7	36.7
SiO_2	58.5	47.4	57.0	46.6	22.6
Al_2O_3	2.3	5.6	4.8	4.5	3.2
Fe_2O_3	1.6	2.1	2.1	2.9	1.4
SO_3	0.5	0.4	0.4	0.6	1.6
Na_2O	0.2	0.8	0.5	0.3	0.1
K_2O	0.5	1.1	1.0	0.8	0.4
MgO	0.6	0.8	0.9	1.2	1.2
LOI	18.1	20.6	17.1	21.6	32.3
Mineralogical composition					
Aggregates	Mainly quartz (<50%), calcite (20%–30%), dolomite (3%–5%), feldspars, micas				Quartz (<20%), calcite (45%), dolomite (6%)
Anhydrous phases	C_2S and C_3S				C_2S and C_3S
Hydrated phases	C-S-H (<18%), portlandite (traces), AFt (traces)				Portlandite (2%), ettringite, C-S-H (<60%)
Physical properties					
Specific gravity (kg/m^3)	2,440	2,490	2,480	2,450	2,180

After 72 h, the heat released by the mixtures was between 2 and 7 J/g, far from the typical values found for Portland cement (300–350 J/g). It should be noted that FRA-DC, which contained more cement paste (and thus probably more anhydrous cement), did not give higher values of heat release. By taking mean values of 5 and 325 J/g for FRA-DC/G-FRA and Portland cement, respectively, and by considering that the anhydrous cement in FRA-DC/G-FRA had completely reacted, it would mean that FRA-DC/G-FRA contained around 1.5% of anhydrous cement (5/325 × 100), which seems to be realistic and coherent with existence of hollow shell (Hadley) hydration grains (these are grains that contain a void within the original boundary of the anhydrous cement grain).

After the end of the test, the paste cylinders were demolded to check the consequence of the slight reactivity, but no setting or hardening was detected even after 28 days of curing in water.

5.3.2.3.2 Pozzolanic activity

Silica might be released from old paste or aggregates (see Section 12.5 dealing with alkali–silica reaction) or from remaining unreacted additions and thus leads to a pozzolanic reaction between FRA-DC and G-FRA. The reactivity of FRA-DC/G-FRA was assessed on pastes made up of hydrated lime (4 parts in mass), fines (1 part), and water (2.5 parts) followed by the consumption of lime up to 90 days (samples protected from carbonation) by thermogravimetric analysis (TGA) and XRD measurements. The fines used had a specific surface area of 7,500 cm^2/g to improve their reactivity.

Figure 5.2 shows the results obtained on G-FRA pozzolanic activity. It was clearly seen on the derivative of the TGA curves (DTG) that the hydrated lime content (portlandite—CH) decreased over time (mostly in the first 28 days), whereas new hydrated phases appeared, as shown by the increase in the chemically bound water at temperatures below

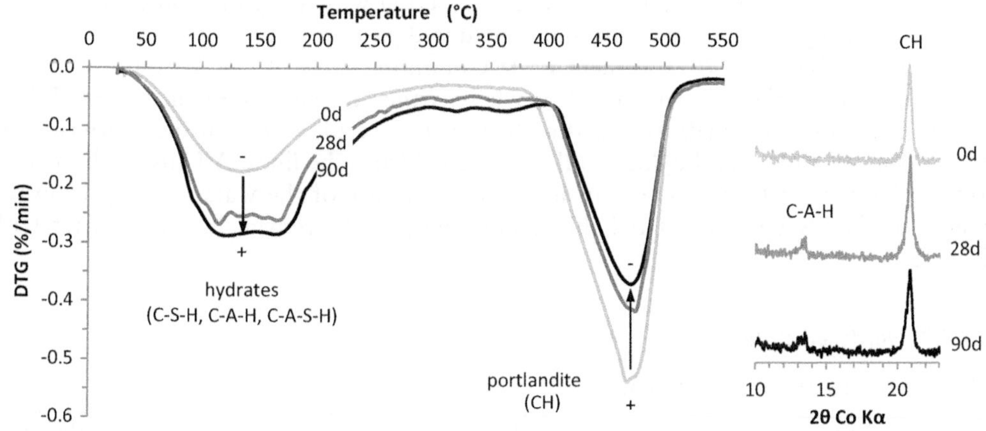

Figure 5.2 Pozzolanic reactivity of G-FRA Pa with hydrated lime, as followed by DTG and XRD.

300°C (some hydrates such as C-S-H were already present in G-FRA, at age 0). These new phases were the products of reaction between the hydrated lime and the Si and Al released from the fines. They were probably composed of not only C-S-H and C-A-S-H, but also hydrated calcium aluminate phases (C-A-H) such as C_4AH_x, as seen on the XRD diagram (right of Figure 5.2).

The quantification of the hydrated lime consumed gave values ranging from 25% (G-FRA Ly) to 60% (FRA-DC), with 40% for G-FRA Pa (example of Figure 5.2). Although it seems significant, the chemically bound water of the new hydrates was only between 1% and 2.5% of the total mass of the dry samples. Assuming only pozzolanic C-S-H (C/S of 1.1 and between 1.4 and 4 mol of water/molecule), then it would represent between 3% and 15% of C-S-H in the hydrated pastes, which is relatively low. Isothermal calorimetry on the pastes showed heat releases ranging between 5 and 10 J/g for 5 days. Metakaolin, recognized as an active pozzolan, gave 50 J/g in the same test (Martin Cyr, RECYBÉTON Report, Fig 33, section 4.4.2.4).

It means that G-FRA/FRA-DC is able to consume lime and produce slight amounts of hydrates, so they can be considered as pozzolanic materials. However, it does not mean that the effect on the properties of hardened materials will be significant.

5.3.2.4 Effect of FRA-DC/G-FRA on the properties of mortars

The common practice when evaluating additions consists in using them in mortars in replacement of a fraction of Portland cement. FRA-DC and G-FRA were combined to two types of cements: CEM I 52.5 R and CEM II/A-L 42.5, at the replacement rates (p) of 15%, 25%, and 35%. A quartz filler (4.400 cm²/g) was sometimes used to compare the activity of the fines with that of a chemically inert powder. The mortars, composed of three parts of sand (by mass), one part of binder (cement + fines), and a water/binder ratio of 0.5, were prepared according to NF EN 196-1.

5.3.2.4.1 Fresh state and setting

The results at fresh state were obtained from the flowing time by LCL device, according to NF P18-452. When used at 25%, no significant effect of G-FRA was seen just after mixing, but a small decrease in workability was observed after 60 min compared with the reference mortar without fines. FRA-DC gave a similar trend, but with a slightly poorer workability at all ages. It was probably due to the higher hydrated paste content of this fine.

According to NF EN 196-3, cement paste containing 25% of G-FRA did not change the setting time of the cement (i.e., 3 h), whereas the cement paste containing 25% of FRA-DC accelerated the hydration within 20 min. This result was confirmed with isothermal calorimetry and could be due to alkalis leached from the fines. Alkalis are known as good accelerators of Portland cement hydration. The use of the water coming from the leaching of the FRA-DC validated this assumption, as the cement hydrated faster with this water.

5.3.2.4.2 Hardened state

5.3.2.4.2.1 STRENGTH ACTIVITY INDEX

Compressive strengths were measured at 1, 28, and 90 days (conservation: 1 day in the mold and the rest in water) and are expressed as strength activity index (SAI), defined as the ratio (in percent) of the compressive strength of standard mortar bars prepared with (100 – p) % test cement plus p% of addition by mass, to the compressive strength of standard mortar bars prepared with 100% test cement, when tested at the same age. Figure 5.3 compares the different G-FRA and FRA-DC to chemically inert filler for a replacement rate of 25% of both cements. It can be seen that

- SAI was in the range of 0.59–0.74 for CEM I and 0.53–0.75 for CEM II/A-L,
- The effect of age was more significant for CEM II/A-L,
- General tendencies did not seem to emerge when comparing G-FRA and FRA-DC,
- SAI of G-FRA and FRA-DC was most of the time lower than SAI of filler having an equivalent fineness.

It should be noted that increasing the fineness of the fines had only a limited effect in the case of G-FRA Pa (results not presented here).

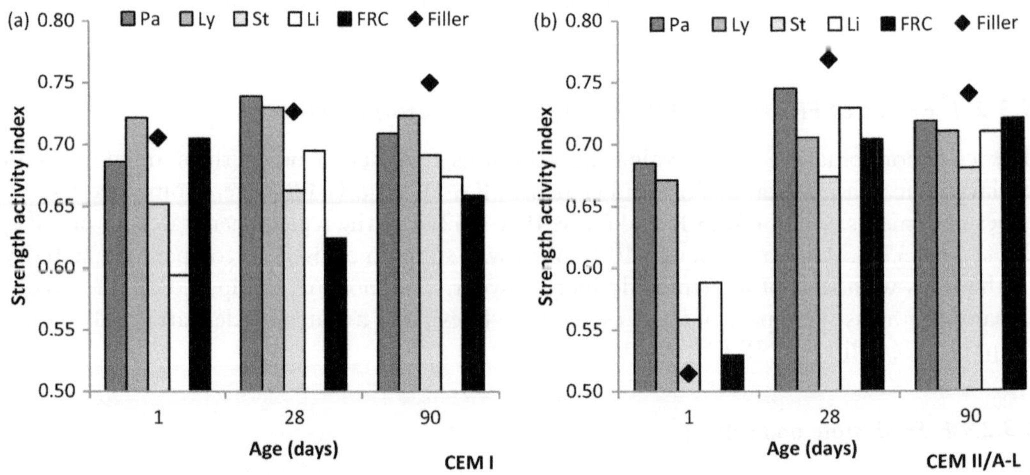

Figure 5.3 SAI of mortars containing 25% of fines (4,500 cm²/g) in replacement of (a) CEM I and (b) CEM II/A-L, at 1, 28, and 90 days.

5.3.2.4.2.2 QUANTIFICATION OF THE ACTIVITY

The replacement of cement by a powder could not only have positive effects such as pozzolanic reaction, but also lead to a negative effect on strength, which can be explained by a dilution effect related to the partial replacement of the cement by a nonhydraulic material when the water content is maintained at a constant value. To quantify the effect of G-FRA/FRA-DC on the strength by taking the dilution effect into account, an analysis based on the application of an empirical strength model was used. In the present case, the model proposed by Bolomey was used (equation 5.1):

$$S = K_B\left(\frac{C}{W+V} - 0.5\right) \qquad (5.1)$$

where S denotes the strength, C and W are the masses of cement and water, respectively, V represents the mass of water filling the air in the mixture, and K_B is the Bolomey coefficient, calculated from the reference mixtures without G-FRA/FRA-DC.

The analysis consisted in calculating the strength expected for mortars containing G-FRA/FRA-DC using the composition parameters of the mixtures. The calculation was based on the assumption that the strength only depended on cement and water contents; the G-FRA and FRA-DC are inert fillers that only participated in the reduction in the cement proportion. Thus, this calculation allowed the real amounts of cement (dilution effect of the G-FRA/FRA-DC) in mortars to be considered. It also permitted the assumption that the G-FRA/FRA-DC had no chemical effect on the calculated strengths. So, the differences between the measured and calculated strengths could be imputed to the chemical effect of the G-FRA/FRA-DC.

Figure 5.4 shows the results of the calculation in comparison with the measured values. At 1 day, the experimental values were around the dilution curve, showing that G-FRA/FRA-DC had no significant physical or chemical effects on mortar strength. At later ages, some of the values were slightly above the dilution curve, without being over the data of the chemically inert filler. Thus, although G-FRA and FRA-DC showed a small pozzolanic activity, the results presented here tend to prove that it did not help in improving significantly the long-term strength of the mortars.

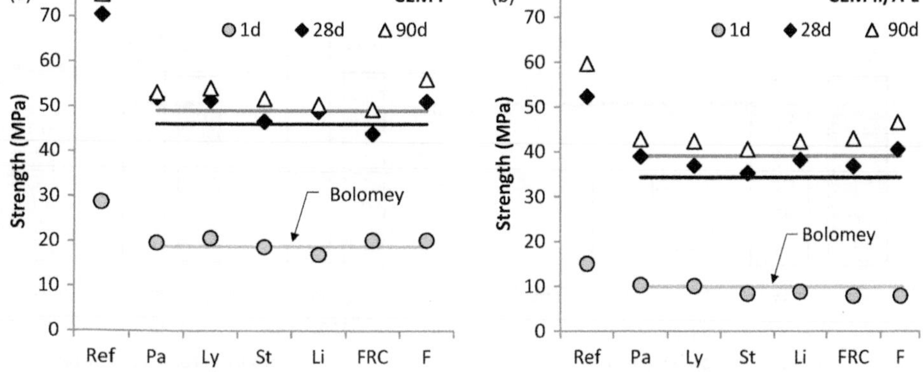

Figure 5.4 Comparison of the measured strength of mortars containing 25% of fines, with the dilution curves calculated with Bolomey's model, for (a) CEM I and (b) CEM II/A-L.

5.3.2.4.2.3 *k*-VALUE CONCEPT

The *k*-value concept describes the cement-equivalent contribution of a mineral addition (by means of the efficiency factor *k*) to concrete compressive strength. *k* reaches the values between 0 (no contribution) and 1 (Portland cement-equivalent contribution) or even higher. According to EN 206, the *k*-value is defined in regard to durability performance of concretes with type II additions, although it is calculated on the basis of strength measurements. If appropriate, a proxy criterion for durability is the comparison of the compressive strength.

Among the different ways of calculating the efficiency factor *k*, the one used here was based on equation 5.1, by replacing the term *C* by *C* + *kA*, where *A* was the amount of addition in the mixture (equation 5.2).

$$S = K_B\left(\frac{C + kA}{W + V} - 0.5\right) \quad (5.2)$$

Figure 5.5 presents the results of the calculation of the efficiency factors *k* for the G-FRA and FRA-DC, used in the replacement of (a) 25% of CEM I (different origins of G-FRA/FRA-DC at 4,500 cm²/g) and (b) 15%, 25%, and 35% of CEM I (G-FRA from Paris, at 4,500 and 7,500 cm²/g), after hydration time of 28 days. It can be seen from Figure 5.5a that

- For FRA-DC, the *k*-value was less than 0 at 28 days (around 0 at 90 days), meaning that this fine had no overall physical or chemical contribution to the strength of the mortar. It could be due to the high hydrated paste content with C-S-H not being under seeds form, which probably included compounds having a negative effect on the long-time hydration of the cement;
- For G-FRA, the results were quite different depending on the origin of the fine. G-FRA Pa and G-FRA Ly were slightly higher than the filler; G-FRA Li and especially G-FRA St were strongly lower and had only a limited activity on the strength performance of mortars.

Figure 5.5b, which gives the effect of the fine content and the increased fineness of G-FRA Pa on the *k*-value, shows that increasing the fineness led to higher values of *k*, meaning that

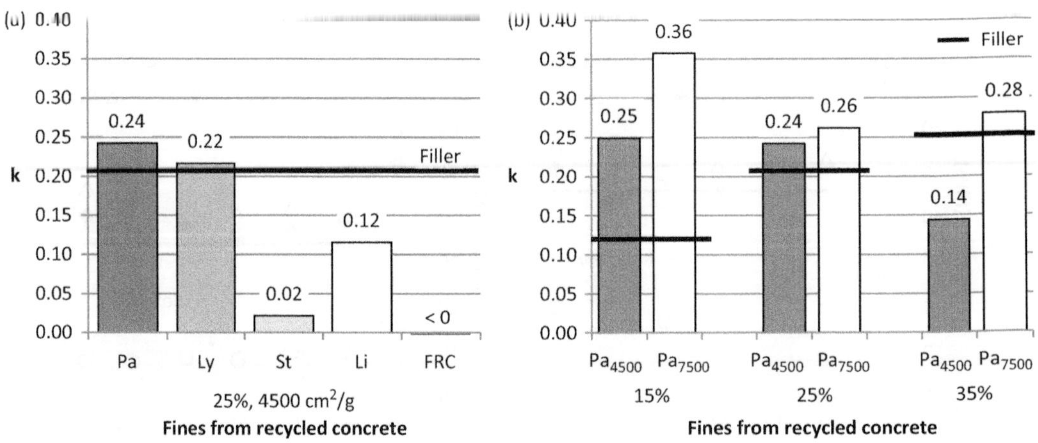

Figure 5.5 *k*-Value concept at 28 days applied to G-FRA and FRA-DC, based on Bolomey's law, and (a) 25% of fines at 4,500 cm²/g and (b) 15%, 25%, and 35% of G-FRA from Paris at 4,500.

grinding more the G-FRA could logically improve its activity. The *k*-values of G-FRA Pa were higher than the ones of the filler, for fine contents up to 25%, and it could be the same for 35%, as long as the fineness of the G-FRA is increased.

It could also be seen that higher amounts of G-FRA tend to reduce the *k*-values. It means that, as for FRA-DC but to a lower extent (probably due to the lower hydrated paste content of G-FRA), G-FRA probably contained compounds that had a negative impact on the strength development of mortars. The limitation of the fine content could thus be an option to improve the efficiency of G-FRA, as using 15% of G-FRA Pa increased *k* compared to the filler used at the same content.

It could be noted that the filler presented *k*-values increasing with the replacement rate. This surprising result was subjected to an analysis of sensitivity, which shows that a variation of 1 MPa on the 15% or 35% mixtures could lead to almost equivalent *k*-values. This result is an illustration of the high sensitivity of the *k* coefficient regarding experimental errors (Cyr et al. 2000).

5.4 RESEARCH NEEDS

In order to confirm the properties of FRA-based cement compositions, it would be necessary to switch to industrial-scale trials in order to confirm from a statistical point of view the achievable strength classes depending on the natural variations in a given source of FRA. In addition, a source of demolition concrete from a limestone region has also to be tested and could give better results. Durability tests will have also to be delivered.

Considering the ubiquitous presence of quartz, it is also necessary to characterize the particle size distribution (PSD) of crystalline silica under industrial grinding conditions to assure that there no harmful effect on human health.

When FRA is used as a concrete addition, the test made showed that the best that can be expected is a filler-type behavior with slight or no pozzolanic effect. In the current technological situation, grinding remains an option to recycle FRA as filler in concrete. However, considering other ways of recycling FRA (Portland clinker raw mix or main constituent in cement), the cost study will probably help to make the final decision.

5.5 CONCLUSION

When FRA is used as a cement constituent, the experiments made showed clearly that, considering the mineralogical composition of FRA from different geological regions (siliceous, granitic, and silico-calcareous natures), introduction of FRA in the substitution of limestone is realistic to obtain current cements with acceptable usage properties and mechanical strength.

The tests made on cement pastes and mortars containing fines from dust collector (FRA-DC—fines from recycled concrete) and fines obtained from the grinding of recycled concrete sand (G-FRA—grinded recycled concrete) showed that these materials could at best behave as fillers, although they had a slight pozzolanic activity.

On the one hand, FRA-DC seemed to be less efficient in terms of contribution to the strength of the mortar. On the other hand, G-FRA must be ground before being used, but they showed a better efficiency to be used as type I addition. Moreover, G-FRA could represent a way of reusing recycled FA in the case where this fraction cannot be used easily in new concretes.

Section III

Recycled concrete from production to hardening stages

E. Garcia-Diaz

Centre des Matériaux des Mines d'Alès, IMT Mines Ales

Recycled concrete aggregates (RAs) have specific characteristics relative to the presence of the porous old attached cement paste or mortar but also to the presence of some impurities as plaster and gypsum for example. In general, compared to natural aggregates, RAs

- Are more lightweight and more angular, and have lower packing densities.
- Are more porous and more brittle and have bigger water absorption.
- Are likely to reduce the efficiency of superplasticizers because of potential sulfate and impurity interactions.

These specific characteristics must be taken into account to design concrete with stable properties at the early age from batching to setting. What are according to the type of Recycled Aggregate Concrete (RAC) (building and civil engineering applications) the critical levels for fine and coarse RA substitution beyond which these properties are significantly impacted? As it is advisable for lightweight aggregates, should we or not pre-saturate fine and coarse RA before the mixing?

This section is subdivided in three chapters:

- Chapter 6 concerns the batching of RAC with a focus on attrition mechanisms for different modes of mixing with dried and also saturated RAs.
- Chapter 7 deals with workability (slump follow-up) and rheological stability (yield stress and plastic viscosity follow-ups) of RAC during the two first hours after the mixing: the influence of RA content and water saturation degree and also of Portland cement-superplasticizer couples is investigated.
- Chapter 8 concerns bleeding, shrinkage, and cracking of concrete at the early age. Behaviors of Normal Aggregate Concrete (NAC) and RAC for building and civil engineering applications are compared. Influence of the way of drying (with or without wind), RA content and water saturation degree is also studied.

Chapter 6

Batching recycled concrete

J.-M. Potier
SNBPE

B. Cazacliu
IFSTTAR

CONTENTS

Abstract

There are obstacles to the use of Recycled Aggregates in Ready-Mix Concrete (RMC), one of them being the need of additional silos, often impossible for plants in urban areas, while it is in these areas that recycled aggregates are mostly available. Another main issue is the control of moisture in the aggregates and the impact on fresh concrete rheology. It is generally admitted that the RA high level of water absorption generates difficulties in the correct water proportioning. Also, the water absorbed between end of mixing and casting can change the consistency of the concrete during transport. RECYBÉTON project studies influence of the pre-saturation level on the RAC rheology (see Chapter 7 for details). There is no significant adverse effect of the pre-humidification of RA on the rheological properties of fresh concrete at the end of mixing. The mixing time seems sufficient for under-saturated RA to absorb a large part of the water necessary to their saturation. Another topic was abrasion during mixing, in order to better understand this evolution and improve the recycled aggregate concrete mix-design, the friability of recycled concrete aggregate was studied under different mixing configurations. The reduction of the size of the RA during mixing seems to be caused by a combination of attrition and fragmentation phenomena which are dependent of the MDE of the RA and the content on attached cement pastes and mortars.

The fresh cement paste protects the RA during the mixing and limits these phenomena. However, for mixing time inferior to 1 min, which is frequently the case in industrial batching, the attrition mechanism is not very intense and does not seem impact the quality of the concrete production. All these results are in good agreement with usual method of batching used with success for the experimental construction sites of RECYBÉTON project (see Chapter 7).

6.1 INTRODUCTION

Concrete is widely used as a construction material in buildings and civil infrastructure, but, presently, recycled concrete aggregate (RA) is primarily used in road construction where it replaces natural aggregates (NAs) in subbases. But this is not enough to fully participate to circular economy; it is necessary to reuse "old" concrete in "new" one.

Issues of concrete recycling are of first importance. France produces 18 million tons of deconstructed concrete on the 240 million tons of inert waste from the construction industry.

And, even if France has significant and quality geological resources, inequality in the distribution of the resource on the territory and the growing difficulty to obtain authorizations to open new quarries make the choice of using this new type of resource unavoidable, especially in urban areas.

6.2 STATE OF THE ART

6.2.1 General

The use of aggregates from demolition concrete recycling in new concrete is not recent; the first state of the art on the subject was published by Réunion Internationale des Laboratoires et Experts des Matériaux[International Union of Laboratories and Experts in Construction Materials] (RILEM) in 1978 (Nixon 1978).

The first extensive and well-documented use of materials from the demolition of buildings as aggregate in fresh concrete was during and just after the Second World War. At this time, the rubble left after the bombardment of cities, especially after aerial bombing, was used in concrete for rebuilding (Buck 1972).

In many countries, concrete made with recycled aggregates is no longer merely a research field; it is already a practical reality (Grübl and Nealen 1998; Koulouris et al. 2004; Poon and Chan 2007).

6.2.2 Professional specifications

In 1994, *Specifications for Concrete with Recycled Aggregates, RILEM Recommendations for demolition and reuse of concrete and masonry*, RILEM (1994), were published, being the first international recommendations for the application of recycled aggregates in concrete production. In this publication, different categories of recycled coarse aggregates were defined.

In this publication, it was also recommended that recycled aggregates should mainly come from crushed concrete, known as RA, due to its higher quality. In addition, the recommendation was given to use only the coarse fraction of recycled aggregates, so as not to affect the final properties of concrete.

In 2009, in Portugal, experimental research to assess the practicality of using fine RAs to produce new concrete had been realized (Evangelista and de Brito 2010).

The conclusion is that, for durability reasons, the total replacement of the natural fines by recycled ones in a concrete mix may present some serious difficulties. However, for smaller replacement ratios (e.g., 30%), the use of recycled fines for structural concrete production is feasible.

The European concrete standard EN 206 and its French Annex (NF EN 206/CN) contain recommendation on the use of coarse RA in concrete (Annex E and Annex NA E for the French version). No specific information is given in this standard on the influence on concrete production itself (see also Chapter 14).

In fact, little information exists in standards on the specific problems linked to fabrication and production of concrete using RA in ready mixed concrete (RMC).

In October 2014, the Architectural Institute of Japan established and published the Recommendation for Mix Design, Production and Construction Practice of Concrete with Recycled Concrete Aggregate (Noguchi 2015). This recommendation gives information on the recommended mix design methods, production methods, transportation methods, construction methods, and quality control methods for concrete with recycled aggregate. Specifically for production, the requirements were as follows:

- The stockyard for accepting recycled aggregates should be furnished with partitions to avoid blending with other aggregate and should be regularly cleaned and well maintained.
- The plant should have several type of aggregates silos.
- The plant should have pre-wetting equipment to stabilize the surface moisture of recycled aggregates before use.

6.2.3 The fabrication of concrete using RA in RMC

6.2.3.1 *Moisture content*

Another main issue is the control of moisture in the aggregates and the impact on consistency and rheology of the concrete. Basically, it is difficult to measure online the water content of the recycled aggregates, whereas the quantity of water they bring into the mixture is very significant. If it is generally difficult to measure the moisture of coarse aggregates, it is even more complicated with RA because RA is also affected by its much higher water absorption capacity. Indeed, in the microwave moisture measurement sensors, the absorbed water gives different response than free water not absorbed by the aggregate. This is exemplified in Figure 6.1 showing the response of a sensor for common natural sand containing lot of fine particles and for 10/14 RA (Cazacliu 2013). The prototype sensor, developed by HydroStop, is a noncontact microwave technology, composed from a transmitter and a receiver. The microwaves pass through the sample of aggregates, and the phase shift angle between the emitted and received waves is measured. According to its design, the sensor is expected to measure the moisture of coarse aggregate with similar precision as for sand.

One can observe a mostly linear evolution of the phase shift angle with the aggregate moisture, for the two materials, on a large range of high moisture levels. The behavior of the sensor became poorly correlated with the moisture for the RAs when the level of moisture is fewer than 3.5%; this limit value was close, slightly lower than the water absorption capacity, which was determined at around 4.2%. A similar nonlinearity could be observed for the measurement of the natural sand moisture when it is inferior to the water absorption (which is 0.6% for this sand). But in the second case, the range of water contents misevaluated is lower and the consequences on concrete regularity are smaller.

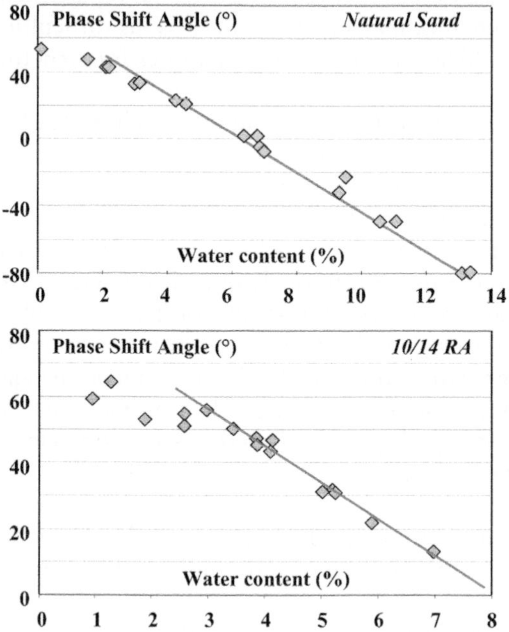

Figure 6.1 Phase-shift angle evolution of a microwave sensor with the moisture content of a natural sand containing high level of fines and a 10/14 RA.

6.2.3.2 *Pre-saturation of RA*

It is generally admitted that the RA high level of water absorption (Sagoe-Crentsil 2001; Quattrone et al. 2016) and heterogeneity (Joseph et al. 2015) generate difficulties in the correct water proportioning from one batch to another. Also, the water absorbed between end of mixing and casting can change the consistency of the concrete during transport. Belin et al. (2014) studied RA for which up to 5% of water was absorbed between 100 s and 1 h. They expect that workability loss due to water absorption should be different from one RA to another. They also discussed that the water absorption kinetics of RA could be different in pure water or when they are immersed in a cement paste. Indeed, the work of Bello (Bello 2014) shows that the water absorption kinetics of lightweight aggregates immersed in a cement paste strongly depended on the water-to-powder ratio of the paste.

Therefore, questions arose about the influence of the initial moisture of the RA on the RAC mixing result. In order to avoid problems on the consistency of the RAC, González et al. (2013) proposed immersing the aggregates in water for 10 min, with eventually a brief air drying and subsequent elimination of surface water from the aggregate. The complete saturation in order to stop any water transfer into the aggregates had already been proposed by some authors (Hansen 1992).

Barra and Vazquez (1996) and Poon et al. (2004) analyzed the effect of oven drying, air drying, or completely saturating surface-dry RA before mixing, by making constant the total water content into the mixture. They conclude that air-dried RA gave improved workability and compressive strength of the concrete, as compared to oven drying or saturated surface-dry conditions. The authors suggested that saturation point should not be reached because of the risk of bleeding. However, for recycled concrete fine aggregates, Ji et al. (2013) obtained that the cracking sensitivity of the concrete increases with the initial moisture of the aggregates, from oven-dried to saturated surface-dried.

In evidence, the use of differently moistened RA by keeping the added water constant (so with different total water content) results in an increase in workability and a decrease in the compressive strength for the most pre-wetted aggregates (Mefteh et al. 2013; Sánchez Roldán et al. 2016). Mefteh et al. (2013) also studied the slump loss for RAs in oven-dried, saturated surface-dried, and intermediary pre-wetted conditions, maintaining the added water constant. The slump loss did not seem to be influenced by the initial moisture of the aggregates (or consequently by the initial slump value). In fact, the slump loss was similar to that of the concrete produced with NAs. This result could be interpreted as the effect of a fast absorption of the RA from this study, which probably mostly occurs during the mixing time. Indeed, Salgues et al. (2016) confirmed that a partially saturated RA could have fast absorption kinetics, implying a complete saturation during the mixing time. However, more recently, Khoury et al. (2017) demonstrated that RA pre-wetted several weeks induces a larger amount of total water absorbed after the mixing than RA pre-wetted at a same initial level just before the mixing.

For the practical applications, Etxeberria et al. (2007a) suggested using sprinklers on aggregate piles in the plant. A deeper analysis of a pre-saturation procedure, defined to correspond to a practical and executable procedure, which could be applied when producing concrete in a plant was proposed by Ferreira et al. (2011).

6.2.3.3 *Mixing procedure—effect of the loading sequence*

The sand enveloped with cement (SEC) concrete mixing procedure for concrete with NA was proposed in the early 1980s as a method to improve the compressive strength of the normal concrete (using NAs) (Higuchie 1980; Hayakawa and Itoh 1982). The method yielded up to 25% of increase in compressive strength as compared to normal mixing method (one-stage mixing) for the same total water content. It was demonstrated that the benefit resulted from an improvement in the cement paste–aggregate interface (Tamimi 1994).

In the late 1990s, it was observed that the process for RA to absorb a part of mixing water before mixing or during mixing process is effective to achieve higher strength of concrete (Kurowa et al. 1999). Consequently, the SEC method was adapted to the specificity of RA, by increasing the amount of the first part of water (50% of the total water) and by delaying the introduction of the cement after the loading of this first amount of water (Ryu 2002a; Otsuki et al. 2003). This double-mixing method improved the mechanical behavior (compressive strength by about 15% and tensile strength by about 20%–25%) and the durability (chloride penetration and carbonation resistances) of RAC compared with normal single-stage mixing. This beneficial effect is explained by an improvement in the properties of the new interfacial transition zone in the RAC when the two-stage mixing method is used (Otsuki et al. 2003; Li et al. 2012). This mixing method, termed two-stage mixing approach (TSMA) (Tam et al. 2005), was more deeply studied in the recent years. Tam et al. (2007) showed an increasing beneficial effect with the RA replacement level. Poon and Chan (2007) used the procedure for concrete using fine RA with a reduction in the adverse effects such as strength reduction. Babu et al. (2014) confirmed that this mixing method improved compressive strength and durability properties (water absorption, sorptivity, chloride ion penetration, drying shrinkage, and abrasion resistance) for high-strength concrete. The mixing approach used had no effect on the frost resistance of air-entrained RAC (Liu et al. 2016a).

Brand et al. (2015) have shown that TSMA is more effective in increasing the mechanical properties of RAC if the RAs are partially saturated, as opposed to their use in oven-dried or saturated conditions.

Other mixing procedures were studied. The cement paste encapsulating aggregate method (first mixing the cement paste and then incorporating coarse aggregates and sand) had an intermediate mechanical effect on the RAC between the normal mixing and the TSMA (Liu et al. 2016b). The sand-enveloped mixing approach (SEMA) gave better results than the TSMA in terms of the compressive strength of the RAC. The SEMA mixes sand, cement, and three-fourths of the total water before the addition of RA, allowing the sand particles to mix more readily with the cement and water, and thus, less water will be absorbed by the RA (Liang et al. 2015).

6.2.3.4 Coating of aggregates

The specific mixing procedure such as TSMA allows the cementitious slurry to coat the surface of RA conducting to higher compressive strength of the RAC. It was observed that the soakage of RA in a pozzolanic liquid increases the strength of RAC but has adverse effects on the workability of fresh concrete (Kurowa et al. 1999). Several researches focus on combining the use of mineral addition and mixing procedures (Ryu 2002a). For instance, the effect of the addition (fly ash or slag) on the compressive strength and chloride ion penetration resistance of the RAC was shown to be improved by a triple-stage mixing method compared to that by a double-stage mixing (Kong et al. 2010). Also, Tam and Tam (2008) showed that combining the additions of silica fume and the TSMA can develop a stronger interfacial layer around aggregates and hence a higher strength of the concrete.

More drastically, Liang et al. (2015) proposed surface pretreatments for the formation of a new layer of cement paste coating on the RA applied 7 days before casting of concrete samples.

6.2.3.5 Mixing method

Mixing method could also improve the mechanical properties or the workability of the RAC. For instance, Teramoto et al. (1998) proposed to grout soft cement paste into microcracks of RA under partial vacuum. They observed that the use of high water-reducing agent makes the effect of mixing under decreased pressure on compressive strength higher.

An alternative method is the mixing method using vibrators. Heng et al. (2005) demonstrated an improvement in the compressive strength by more than 20% at 91 days compared to the concrete mixed without vibration. However, the effect was observed only for low-quality RA, not for high-quality ones.

6.2.3.6 Use of premixed aggregates

The Japanese recommendation Noguchi (2015) emphasizes some obstacles to the use of recycled aggregates in RMC, one of them being the need of additional silos, often impossible for plants in urban areas, while it is in these areas that recycled aggregates are mostly available. To solve this problem, one of the solutions is the use of premixed aggregates, where a part of NAs are substituted by recycled aggregates directly in the quarry.

6.3 RECYBÉTON'S OUTPUTS

6.3.1 Influence of the pre-saturation level on the RAC rheology

The study carried out in the RECYBÉTON program dealt with the fresh-state behavior of concretes containing recycled fines or coarse aggregates (see Chapter 7). The main goal was to study the evolution of the rheology of concrete (C25/30 and S4 consistency) between

0 and 90 min, as a function of the initial water content of RA. The mixtures studied included a concrete with 100% of recycled coarse aggregates used at three saturation levels (absorption +1%, 0.75, 0.3Ab) and a concrete with 30% by weight of fine RA used at five saturation levels (1.2, 1.0, 0.87, 0.5, and 0.33Ab). The total water content remained constant for each design (100% recycled coarse aggregates or 30% recycled sand). It was decided to keep constant the initial consistency, corresponding to the high limit of the S4 class (slump value of 21 cm). Accordingly, the content of superplasticizer was adjusted only for 0S-100G composition with RA moisture state of 1.2Ab (compared to the theoretical design, 7% increase). The main results are the following:

- The mixing time seems sufficient for undersaturated RA to absorb a large part of the water necessary to their saturation. This result is in good agreement with usual method of batching (no pre-saturation of RA, added water for a given effective water adjusted to the water content of RA) used with success for the experimental construction sites (Chapter 7).
- If devices allowing pre-saturation are present, the results suggest that the moisture content of RA should be ranged from 0.8 to 1.0 Ab as a compromise between the yield value and the viscosity to minimize the adverse effect of time on the flow properties of concrete.

6.3.2 Abrasion during mixing

6.3.2.1 Program

During mixing, constituents of concrete undergo stresses that will alter their size and that may produce additional fine elements. In order to better understand this evolution and improve the RAC mix design, the friability of RA (99% of crushed industrial concrete) was studied under different mixing configurations.

- *Planetary* 30-L mixing system.
- Intensive vertical 5-L Erich mixer and eccentric—*cocurrent* or *countercurrent* configurations (agitator turning in the sense, respectively, the opposite sense of the inclined rotating vessel); the agitator speed varied from 150 rpm (slow speed) to 300 rpm (nominal laboratory speed) and 500 rpm (very fast speed).
- 40-L *rotating drum* concrete mixer.

Three different aspects were evaluated: the coarse aggregate mass loss (mass of fraction inferior to 2.5 mm generated), the grading, and the angularity at different mixing time.

First, three types of an initially 10- to 14-mm aggregate were tested in a typical concrete mixture: two qualities of RA, with Micro Deval en presence d'Eau [Micro Deval with Water] (MDE) values of 21 and 27, respectively, and a reference NA, with MDE value of 6. In some configurations, the RA was loaded wet or dry. The mixing procedure was as follows: 35 s of dry mixing followed by up to 300 s of wet mixing after the water loading.

Complementary tests were performed on the MDE-21 10/14 RA and a 14/20 RA of same origin and having the same MDE value, mixed alone, that is, without the other dry components. The mixing was achieved under dry conditions in the *planetary* 30-L mixing system and under wet conditions in the drum mixer.

6.3.2.2 Results

The degradation of RA drastically increased when the mixing speed was raised up to 500 rpm (Figure 6.2). This was valid for both cocurrent and countercurrent configurations

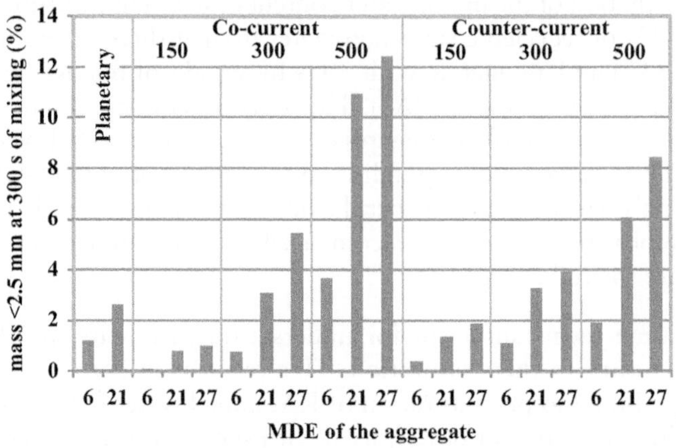

Figure 6.2 The mass fraction inferior to 2.5 mm after 300 s of wet mixing in the three mixing configurations for the three types initially 10/14 aggregate data from the Moreno et al. (2015).

and affected all types of aggregates: MDE values of 6, 21, and 27. However, the cocurrent mixing was by far more damaging than the countercurrent was. At the usual mixing speed (300 rpm), the mass loss was similar in cocurrent and countercurrent mixing, and similar with the result of the usual planetary mixing as well. Finally, when the mixing speed was further decreased, the cocurrent mixing became slightly less damaging than the countercurrent did. Globally, the mass loss increased faster than the mixing speed in the cocurrent method, whereas the nonlinearity seemed much reduced in the countercurrent method.

For the MDE-21 RA, the mass loss increased faster with the mixing time, at short mixing than at longer mixing (Figure 6.3). This can easily be explained by the composite composition of this aggregate: mortar and natural stone. One can suppose that the

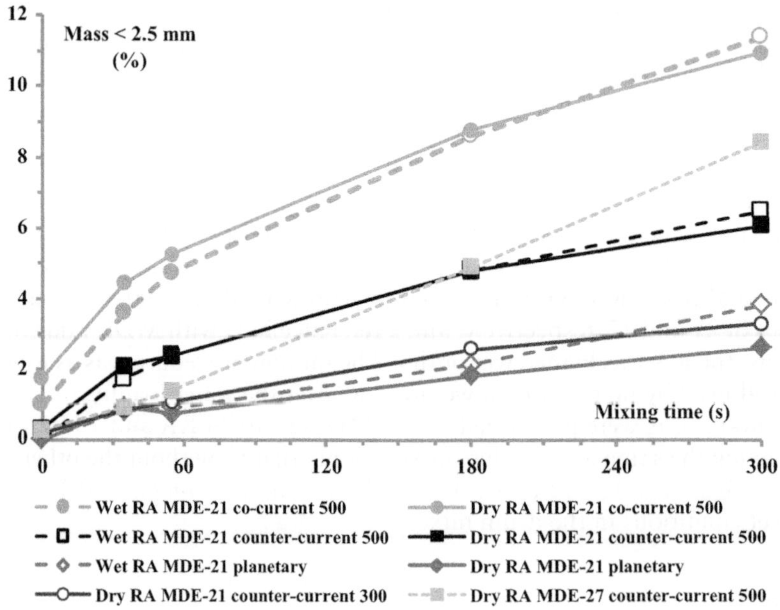

Figure 6.3 Evolution of the fraction inferior to 2.5 mm of an initially 10/14 RA; data from Moreno et al. (2015).

peripheral mortar was first fractured and then the aggregate strengthened, while more and more of its borders became constituted from stone or higher quality mortar. The mortar content of the MDE-27 RA was higher, and this could explain the constant mass loss rate during mixing of this aggregate. It should be noted that the MDE-27 RA was produced from the MDE-21 RA by density separation using a water jig (see Chapter 2 for a description of the method).

In normal laboratory setting of the mixers' configuration and usual mixing times, the mass loss for NA was less than 1% of the coarse aggregate. This percentage reaches 3% for good-quality RA (MDE value of 21) and 5% for lower quality RA (MDE value of 27) (Figure 6.3). This last value could potentially be higher given the method by which the MDE-27 RA was produced (see explanation above).

The initial moisture of the RA, oven-dried—"dry", or saturated—"wet", did not impact the fracturing behavior during mixing (Figure 6.3).

By analyzing the grading evolution during mixing, it was shown that both cleavage (creation of intermediate size particles) and attrition (creation of small particles) mechanisms influenced the aggregate degradation (Moreno et al. 2016). The angularity evolution showed that RA surface becomes smoother and the edges become more rounded after mixing (Moreno et al. 2015; Moreno et al. 2016). This suggests that the attrition mechanism is globally more influent than the cleavage for the tested mixing methods. However, the configuration of mixing significantly influenced the proportion of attrition and cleavage mechanisms.

For the RA mixed without other dry components such as cement, the grading of the RA evolved similarly as during the mixing in a concrete mixture in the vertical planetary mixer (Hamard and Cazacliu 2014b). At opposite, mixing in a rotating drum produced only a slight attrition of the wet RA. Indeed, the mass of particles created by mixing between 1.6 and 6.3 mm was almost the same as the mass of particles less than 1.6 mm in the planetary mixer, whereas in the rotating drum, the mass loss is only composed of fine particles (Figure 6.4). The behavior of coarser RA is similar, with very similar attrition effect, regardless of the initial size of the RA. Of course, the cleavage in the planetary mixer produces larger intermediate size particles from initially larger RA. However, a more detailed analysis of the grading evolution with the mixing time has shown (Hamard and Cazacliu 2014b) that for both RA sizes, the cleavage produces a majority of particles of around 4 mm. This could be explained by the separation of primary sand particles from the parent concrete under the shearing imposed by the planetary mixing.

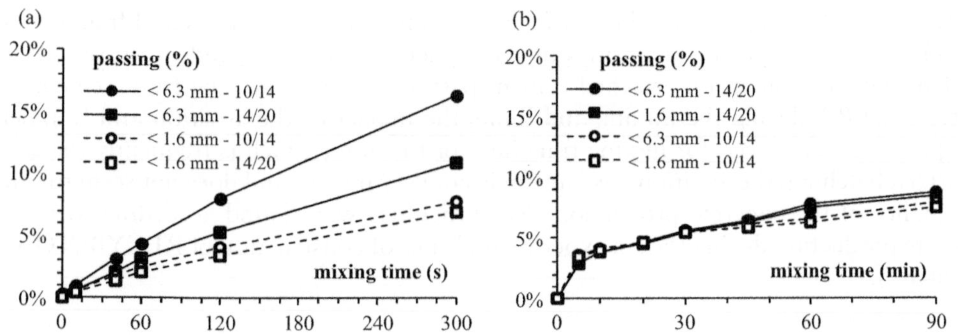

Figure 6.4 Evolution of the mass fraction inferior to 1.6 and 6.3 mm in (a) vertical planetary mixer (dry RA) and (b) rotating drum mixer (wet RA); the RA in these tests had a MDE value of 21 (both 10/14 and 14/20 fractions) (analysis provided by E. Hamard).

Water absorption tests were conducted on different coarse granular fractions after mixing, in order to analyze the evolution of the mortar content with the mixing time. As a general remark, the mortar content was higher for smaller coarser particles and the amount of mortar decreases with the mixing time (Hamard and Cazacliu 2014b). The behavior in the rotating drum was similar but the evolutions are strongly reduced.

6.3.2.3 *Conclusion*

It can be concluded that mixing in intensive mixer can produce adverse effects on the RA, by reducing its size and producing supplementary powder, mainly when the mixing time is increased, as suggested by many authors for RAC. One can suggest that the mixture design should be done in the real mixing conditions and the mixing time should be kept constant between different batches.

6.4 RESEARCH NEEDS

One of the ways to have a better control of the rheology of the concrete may be to develop a real-time tracking of the water content adapted for RA. To improve quality of the aggregates, it will also be interesting to have a better tracking of sulfate content. Some studies on the impact of introduction of recycled concrete fines on abrasion of the aggregates and their influence on mixing (saturation, evolution on rheology during transport) may also be carried out.

6.5 CONCLUSION

Even if some questions are still pending, the experiences of the past and the experimental test works show that using RA in concrete is possible. Nevertheless, to encourage it, the use of premix aggregate should be encouraged; however, this process presents a number of drawbacks that must be addressed (need to keep virgin aggregate stocks for certain markets such as architectonic concrete, difficulty to quantify the amount of recycled materials in a fine aggregate fraction, and an increase in transportation distance). There is no significant adverse effect of the pre-humidification of RA on the rheological properties of fresh concrete at the end of mixing. The mixing time seems sufficient for undersaturated RA to absorb a large part of the water necessary to their saturation. The reduction in the size of the RA during mixing seems to be caused by a combination of attrition and fragmentation phenomena. The respective intensities of these phenomena are dependent on the MDE of the RA and the content on attached cement pastes and mortars. The fresh cement paste protects the RA during the mixing and limits the impact of the attrition and fragmentation processes. However, for mixing time inferior to 1 min, which is frequently the case in industrial batching, the attrition mechanism is not very intense and does not seem to impact the quality of the concrete production. No difficulties were found according to recycled concrete production during the six experimental sites of construction of RECYBÉTON (see Chapter 7).

Chapter 7

Effect of recycled aggregate on fresh concrete properties

E.-H. Kadri and Z.A. Tahar
L2MGC

M. Mouret
LMDC

D. Rogat
Sigma Béton

CONTENTS

Abstract

The objective of this research is to study the effect of initial moisture of RA and also the cement-admixture combination on the behavior of RAC with different substitution percentages of coarse and/or fine NA by RA. Regarding the rheological properties, the measured parameters are: torque and yield stress, torque/ rotation rate ratio at steady state flow and plastic viscosity, workability, air content and density. Furthermore, in order to follow the evolution of these quantities versus time the tests were carried out from the end of mixing to 90 min of age. Water demand and superplasticizer contents are fixed to have the same initial S4 class of workability (a slump of 200±20 mm is targeted). There is no significant adverse effect of the pre-humidification of RA on the rheological properties of fresh concrete at the end of mixing. The mixing time seems sufficient for under saturated RA to absorb a large part of the water necessary to their saturation. The moisture content of RA should be ranged from 0.8 to 1.0 Ab as a compromise between the yield value and the viscosity to minimize the adverse effect of time on the flow properties of concrete. The properties of recycled aggregates concrete depend on the couple cement/admixture and the level of substitution on recycled aggregates. This means that there is indeed an influence of recycled aggregate on the compatibility (physico-chemical equilibrium) of the couple cement/admixture. This influence is more noticeable on the fine aggregate than on coarse aggregate. The results also show that for concrete based on recycled aggregate and in our conditions of test, the higher the percentage of substitution, the higher the rheological loss. We can distinguish also concretes made with recycled fine aggregate for which beyond a substitution rate of 30% a degradation is observed and concretes made with recycled coarse aggregates for which an only few degradation is observed for 100% of substitution.

7.1 INTRODUCTION

Among the studies to develop the concrete with recycled aggregates (RAs), some have focused on the effect of coarse aggregate on recycled concrete properties in hardened state. Some studies on the recycled fine aggregate (RFA) mortars have also been conducted; the effect of the latter on spreading and mortar resistance has been studied.

The literature also lacks complete works carried out on the effect of RFA and/or recycled coarse aggregate (RCA) on recycled concrete properties in the fresh state. Furthermore, few studies had the objective to compare the RFA and RCA on the parameters to fresh concrete.

Based on this lack of findings, research work needs to be conducted. This would complement the literature and aim for a better use of RAs in concrete mixer industry. Indeed, an understanding of the rheological and mechanical behaviors of recycled concrete paves the way for optimization and therefore the use of these concretes by companies. These objectives will be achieved all the more that this work will consider the use of RAs outgoing directly from crushing plant (without having to go through a thorough sorting process, costly for businesses).

7.2 STATE OF THE ART

The rheology of fresh concrete is a relatively young science. There have been many attempts to characterize the consistency of fresh concrete by a variety of technological tests, but few researchers have applied continuous media mechanics to the rheological behavior of fresh concrete (Bartos 1992; Ferraris 1996). Rheological properties of the concrete mixture are characterized by the parameters of Bingham material model (i.e., the yield stress and the plastic viscosity). Their value depends mostly on constituents' properties and mixture composition, time passed from the moment of mixing the constituents, and temperature.

Once the stress passes the yield value, the mixture flow occurs with velocity proportional to the plastic viscosity. The rheological properties of fresh concrete are determined by the so-called rheometers, which measure the shear stress at varying shear rates.

Unfortunately, the inherent properties of concrete make it impossible to use the rheometers designed for neat fluids without any solid particles.

Recent studies recall that the behavior of fresh concrete with natural aggregates (NAs) depends on several parameters including the nature of the admixture, its dose, and the nature of the aggregates and their dimensions (Banfill 2011; Ngo et al. 2010; Vázquez et al. 2014; Wallevik and Wallevik 2011).

In previous studies, the analysis on properties of concretes in the fresh state was not given importance. A recent study (Tahar et al. 2016) showed that the rheology of RA concrete (RAC) is negatively influenced by the substitution in fine RAs (FRAs) beyond 30%.

Use of RA in high-performance concrete is not a common practice for yet, simply due to the reductions in mechanical properties as well as durability properties. Further, it has been found that cement paste in RA contributes to a lowered relative density and higher water absorption than virgin aggregates. There is a significant increase in the porosity of the concrete when NA is replaced by RAC (Gómez-Soberón 2002a).

Zhao et al. (2015b) investigated the influence of fine RA (FRA) on the properties of mortars and concluded that slump of mortars containing dried FRA is always larger than that of mortars containing saturated FRA. In this context, the objective of this research is to study the effect of initial moisture state of RA and also the cement–admixture combination on the behavior of RAC with different substitution percentages (15%, 30%, 70%, and 100%) of coarse and/or fine NA by RA. The variation in the rheological properties of fresh concrete over time is also studied from the end of mixing (T_0) to 90 min of age (T_0+90′).

7.3 RECYBÉTON'S OUTPUTS

7.3.1 Effect of initial moisture state of RA on the rheology of fresh concrete over time

The high water absorption capacity of RAs influences the properties of concrete in the fresh state. That is why several studies focused on the effect of the moisture content of RA on the fresh concrete workability (see, for instance, Mefteh et al. (2013)). Most often, the results obtained, which show how the pre-saturation state of RA is significant with respect to the water demand or the evolution of workability over time, are issued from "single-point" tests such as slump test of slump-flow test. Nevertheless, a slump or a spread value cannot discriminate correctly the flowability of different concrete designs (de Larrard 1999b). Therefore, "multipoint" measurements like those taken by means of a rheometer are relevant to assess the influence of the moisture state of RA on the flow properties of fresh concrete, from immediately after mixing till the delivery to the building site.

7.3.1.1 *Experimental program*

7.3.1.1.1 Raw materials and concrete designs

Among the concrete designs developed for the National Project RECYBÉTON (see Appendix), two of them were selected in the C25/30 strength class:

- 0S-100G for which 100% by weight of the natural coarse aggregate (NCA) is replaced by the RCA and
- 30S-0G for which 30% by weight of the natural fine aggregate (NFA) is replaced by the recycled one.

Regarding the literature, the replacement rates studied are extreme; in particular, the dose of the RFA corresponds to the high limit, above which the mechanical and durability properties of concrete can be impaired (see, for instance, (Evangelista and de Brito 2007a)).

Irrespective of the tested hydric state (Table 7.1), it is important to note that the total water (added water+water absorbed by aggregates) was kept constant for each concrete design.

Table 7.1 Moisture conditions tested (fraction of Ab)

(%)	*0S-100G*	*30S-0G*
NA	1.0	1.0
RA	0.30, 0.75, 1.20	0.33, 0.50, 0.87, 1.0, 1.2

7.3.1.1.2 Pre-humidification of the aggregates

Table 7.2 gathers together the density and 24-h water absorption coefficients (Ab) of NA and RA. Los Angeles (LA) and Micro-Deval (MDE) abrasion coefficients are also shown in Table 1.1 for the coarsest fractions.

Even though it has been shown in the literature that the measurement procedure described in NF EN 1097-6 may underestimate Ab in the case of RFA by excessive drying to eliminate the menisci between angular and small particles (Le et al. 2016), Ab values were used as a reference state to define comparatively the differently tested hydric states (Table 7.1). Hence, below (beyond) the Ab coefficient, aggregates are considered as undersaturated (sur-saturated); at the Ab values, aggregates are considered to be in a saturated surface-dried (SSD) state.

From Table 7.1, the following observations can be predicted:

- Whatever the concrete design, SSD NA was always used to highlight only the influence of the moisture content of RA on the flow of concrete.
- A detailed declination of moisture content values of the RFA was done to assess precisely the evolution of flow properties when only the RFA is incorporated.
- Drying of RA at 80°C was necessary each time the initial water content was greater than the targeted hydric states (0.3 Ab for RCAs and 0.33 Ab for RFA).

The pre-humidification procedure was always as follows for a batch:

- Sampling homogenization by quartering,
- Water content measurement,
- Sample placed in a non-water absorbent container,
- Water amount adjustment to achieve the targeted moisture content (24h prior to mixing),
- Water content homogenization by rotating the container up to the mixing.

7.3.1.1.3 Mixing and simulated transport

For a given concrete composition and moisture content of RA, two 25-L batches were carried out in an open-pan mixer with vertical axis (Figure 7.1). Two blades exert a planetary motion on concrete, and a rotating edge scraper pushes the concrete toward the blades so that dead zones are avoided near the wall of the pan.

The mixing sequence was as follows:

- Introduction of coarse aggregate, half of fine aggregate, cement, and limestone filler, and the other half of fine aggregate and mixing for 1 min,

Table 7.2 Density, water absorption coefficient, LA, and MDE coefficients of NA and RA

		Density (kg/m³)	*24-h Ab (wt%)*[a]	*LA (wt%)*	*MDE (wt%)*
NA	0/4 mm	2,580	0.8	–	–
	4/10 mm	2,710	0.5	–	–
	6.3/20 mm	2,710	0.5	16	17
RA	0/4 mm	2,180	9.2	–	–
	4/10 mm	2,290	5.2	–	–
	10/20 mm	2,260	5.1	37	23

[a] Measurement according to the French Standard NF EN 1097-6.

- Introduction of added water with the chemical admixtures and mixing for 0.5 min,
- Continuous mixing for 3.5 min.

The first batch was dedicated to the characterization at the end of mixing (T_0), whereas the second batch was used for the characterization over time (T_0+45′ and T_0+90′). Then, in order to simulate the transport conditions by mixer truck, the second batch was agitated continuously from T_0 and up to T_0+45′ (or T_0+90′) in a concrete mixer with inclined axis and equipped with a frequency variator (Figure 7.2). Agitation was done first at low rotational speed (3 rpm), and the delivery time was simulated by applying a higher speed (12 rpm) for 2 min before the characterization time.

Figure 7.1 Mixer used in the study (ColloMatic XM2-650-Collomix).

Figure 7.2 Concrete mixer equipped with frequency variator to simulate the agitation during transport and delivery to the site.

7.3.1.1.4 Characterization tests

Two tests were performed in parallel at the different tested terms: the Abrams cone slump test according to the French Standard NF EN 12350-2 and rheological test by means of a concrete rheometer. At this step, it was decided to keep constant the initial consistency, corresponding to the high limit of the S4 class (slump value of 21 cm). Accordingly, the content of superplasticizer was adjusted only for 0S–100G composition with RA moisture state of 1.2 Ab (compared to the theoretical design, 7% increase). No action was required for the other compositions that have been tested.

Rheological measurements were taken with a device developed by CAD instrumentation, which operates at the controlled speed and measures the torque necessary to shear a 5-L concrete sample at a defined number of revolutions. The shearing tool was a four-bladed impeller (Figure 7.3).

Each sample was placed in a single layer by rapid vibration (3 s) in the cylindrical vessel. As soon as the shearing tool was rotated, measurements were taken under vibration using a vibrating table (50 Hz, 1 g) (Figure 7.3). The shear history was composed of a fast increase in rotational speed to 30 rpm to remove the time-dependent properties (Legrand 1972), and a stepwise decrease in speed (30, 20, 10, and 5 rpm) (Figure 7.4). The stepwise change in speed was made as soon as the variation in torque was less than 5 N/cm, assuming a steady-state flow.

7.3.1.2 Results

7.3.1.2.1 Slump

Starting from an almost constant value at T_0 (Figure 7.5), likely explained by the systematic incorporation of the superplasticizer that levels the variation in the intergrain water amount in all cases as a function of the moisture content of RA, a slump decrease was observed over time in all cases. The irreversible process of hydration is then preponderant but is aggravated when the hydric state of RA (coarse aggregates or fine aggregate) is beyond or close to the saturation condition.

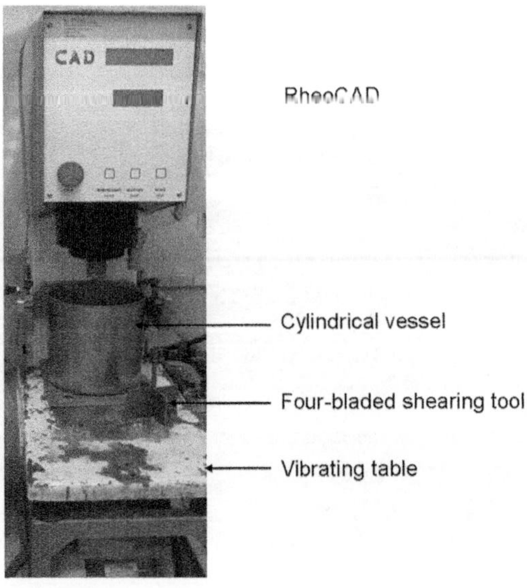

Figure 7.3 Device for rheological tests.

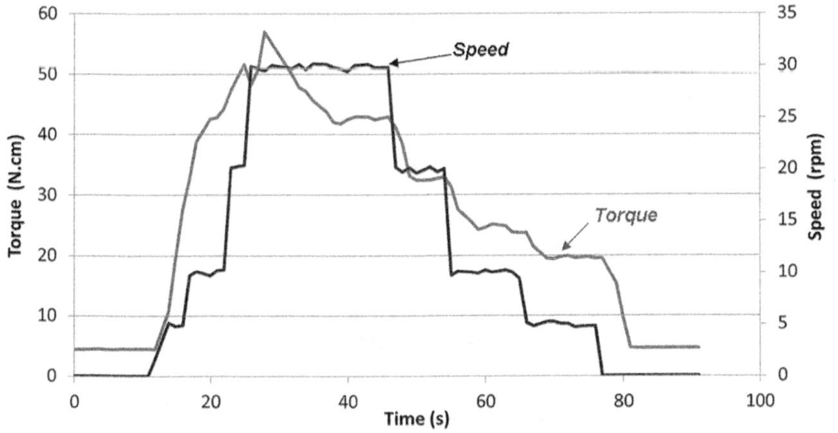

Figure 7.4 Typical shear history applied during rheological measurements.

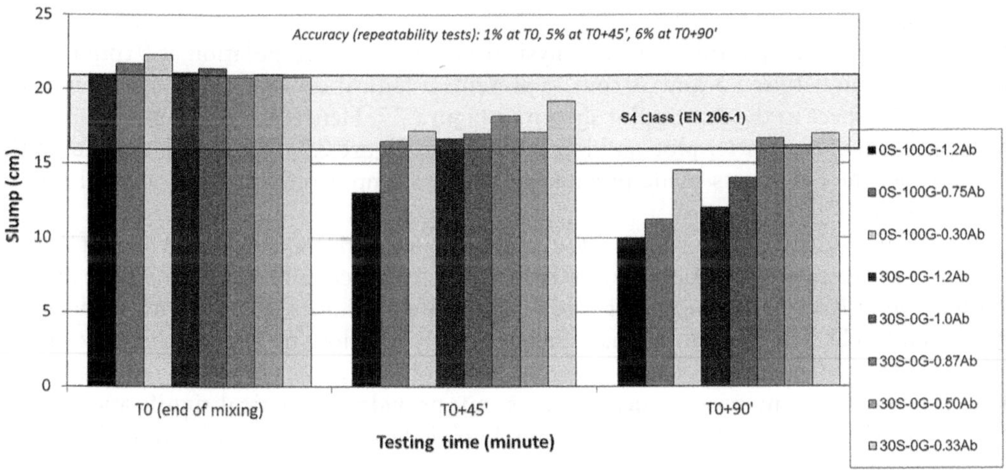

Figure 7.5 Time evolution of slump as a function of the concrete design and the RA moisture content.

Because of different experimental conditions (concrete designs did neither incorporate chemical admixtures nor incorporate mineral ones, and no agitation was applied between testing times), some studies reported different results indicating that totally dried RA led to a significant slump loss in comparison with pre-wetted or SSD RA (Poon et al. 2004; Mefteh et al. 2013).

7.3.1.2.2 Rheological parameters

The typical evolution between the torque C and the rotational speed ω is shown in Figure 7.6. Two points can be studied from the Figure 7.6.

- A linear relationship between C and ω is always observed, where C_0 and K are the parameters fitted by the least-squares method:

$$C = C_0 + K\omega \tag{7.1}$$

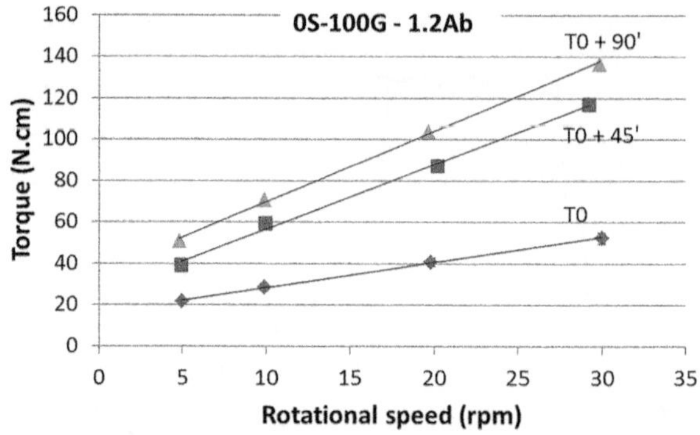

Figure 7.6 Typical torque–rotational speed relationship over time at steady-state flow.

- Due to the material aging (structuration during the hydration process although continuous agitation exists after mixing), the flow is always altered over time, whatever the moisture content of RA may be.

Assuming that the C_0 parameter has a physical meaning as extrapolation at 0 rpm of measurements taken between 5 and 30 rpm, it presents a logical evolution as a function of the slump with respect to the scatterplot shown in Figure 7.7. Hence, in the limits of the maximal dispersion of the scatter plot (solid lines shown in Figure 7.7) that do not depend on the testing time, the comments made previously for the slump evolution apply as well for the yield torque C_0.

Once the flow is initiated (yield torque exceeded), the K parameter gives an idea of viscous properties of the concretes. As shown in Figure 7.8, irrespective of the testing time and the grading size of RA (fine aggregate or coarse aggregate), K values, known with an accuracy of 5%, tend to be lower when the moisture state of RA is close to but less than the saturation state. This information is inconsistent with that obtained from the slump values defined as reference data in any project specifications, but provides a practical significance not yet recorded in the literature to the best of our knowledge: in the case of vibrated concrete

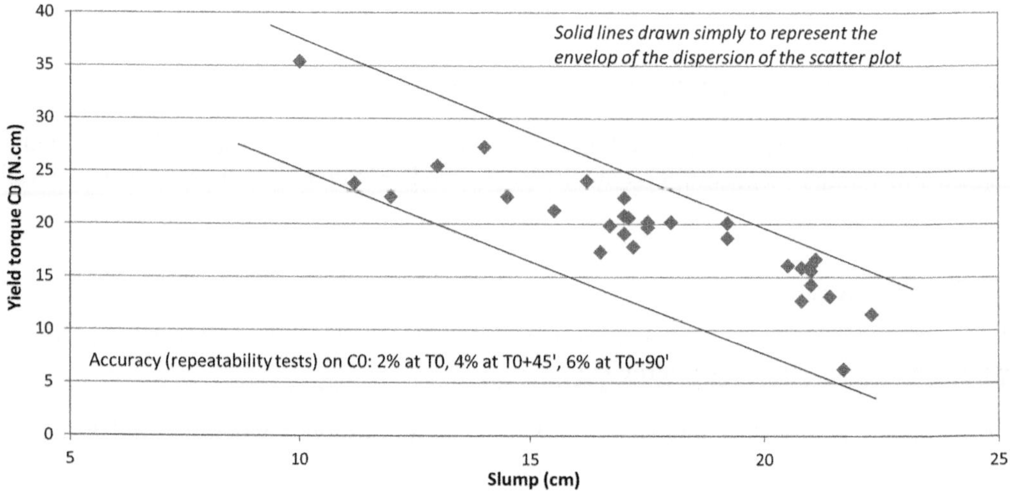

Figure 7.7 Dynamic yield torque C_0 as a function of the slump.

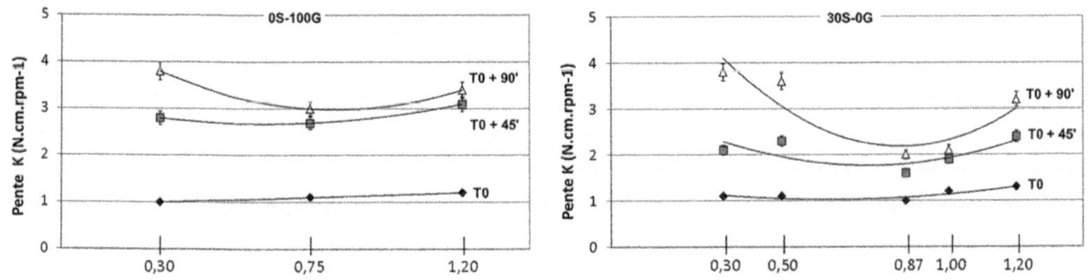

Figure 7.8 Evolution of the viscous properties of concrete as a function of the moisture content of RCA (left) and RFA (right).

incorporating chemical admixtures (superplasticizer and/or retarders) during placing, once vibration initiates the flow, concretes incorporating almost saturated RA show the best flowability.

7.3.2 Characterization of the rheological behavior of recycled concrete

Five types of cement are used: CEM I 52.5R CE CP2 (A), CEM I 52.5R CE (B), CEM I 52.5N CE CP2 (C), CEM I 52.5N SR3 CE PM CP2 (D), and CEM I 52.5N CE ES CP2 (E). Cement B has a higher C_3S content, and Cement A and Cement D have a higher C_3A content than the other cements. Coarse aggregates are two NCA fractions of limestone from crushed rock (4/10 and 10/20 mm) and two RA fractions (4/10 and 10/20 mm). Fine aggregates correspond to NFA of limestone from crushed rock (0/4 mm) and RFA (0/4 mm) (see Appendix).

The grain size distribution of natural and RAs used is shown in Figure 7.9. The RFA is coarser than NFA. The fineness modulus of RFA (i.e., 3.27) was significantly higher than that of NFA (i.e., 2.25).

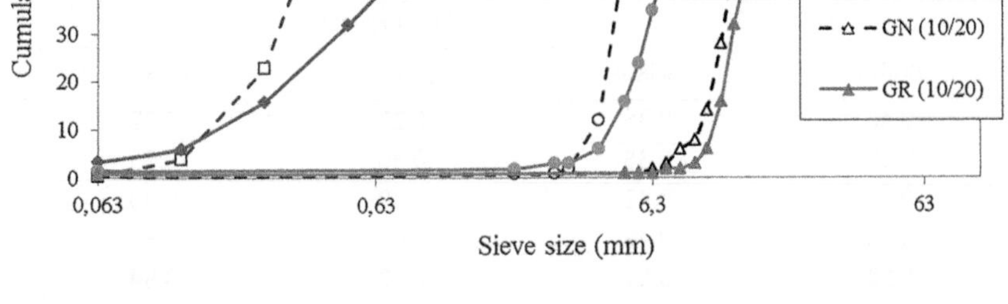

Figure 7.9 Grading of natural and RFA.

The water absorption of NFA is 0.9%, while that of RFA, NCA, and RA is 10%, 0.4%, and 5.5%, respectively. It is clearly evident that RA has relatively higher water absorption capacity compared to NA (see Chapter 3). The higher water absorption capacity of RA is due to the mortar gangue. Visual observations show that RA presents a rough cracked surface compared to a smoother surface of NA, which confirms the high porosity of RA (de Juan and Gutiérrez 2009; Tahar et al. 2017).

Four admixtures (A1, A2, A3, and A4) were used: water reducing (A1 and A4) and high water reducing (A2 and A3) admixtures; moreover, there are polycarboxylate (A1 and A4) and ether polycarboxylic (A2 and A3) admixtures.

Keeping the same granular skeleton and the same quantity of cement, two concrete families (with couple cement/admixture A/A1, B/A2, C/A3, D/A4, and E/A4) and 40 different concrete mix proportions have been tested. These couples have been selected according to the feedback from the French company SIGMA-BETON. Either natural fine or coarse aggregates were partially replaced (15%, 30%, 50%, 70% and 100%) with RAs. RAs are pre-saturated to the saturation value +1%.

7.3.2.1 Influence of substitution of RAs

In order to limit the number of mixes and to be able to compare them on a common basis, a constant initial slump of 200±20mm for the different percentages of substitution (0%, 15%, 30%, 50%, 70%, and 100%) has been imposed and the water content and an amount of admixture varied in consequence. The evolution of the admixtures represents as a function of the percentages of substitution of fine aggregate and coarse aggregates, which are summarized in Tables 7.3 and 7.4, respectively. Because of a possible absorption of a part

Table 7.3 Admixture dosage for BSR (%)

	Admixture (%)				
(%)	*BSR-A*	*BSR-B*	*BSR-C*	*BSR-D*	*BSR-E*
0	0.40	0.35	0.25	0.31	0.30
15	0.40	0.35	0.25	0.31	0.30
30	0.40	0.40	0.35	0.39	0.36
50	0.50	0.53	0.51	0.49	0.48
70	0.60	0.61	0.60	0.59	0.57
100	0.78	0.75	0.73	0.75	0.73

Table 7.4 Admixture dosage for BGR (%)

	Admixture (%)				
(%)	*BGR-A*	*BGR-B*	*BGR-C*	*BGR-D*	*BGR-E*
0	0.40	0.35	0.25	0.30	0.30
15	0.40	0.35	0.30	0.35	0.32
30	0.45	0.42	0.39	0.40	0.36
50	0.48	0.46	0.45	0.48	0.44
70	0.52	0.51	0.50	0.54	0.49
100	0.55	0.54	0.52	0.55	0.51

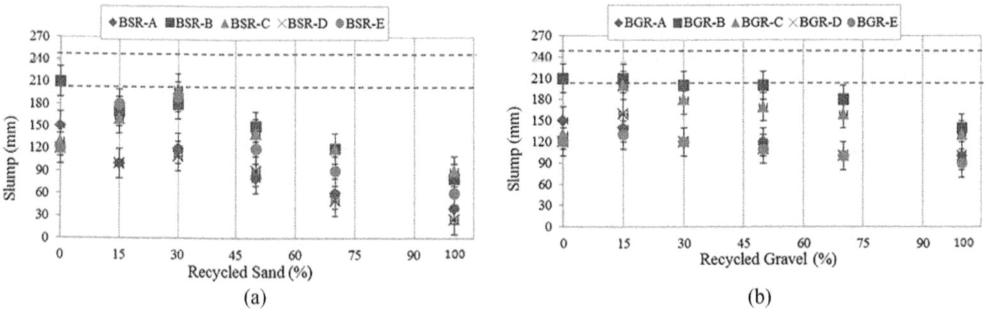

Figure 7.10 Variation in slump with percentage of substitution at T90 in (a) RFA and (b) RA.

of the admixture by the porous RAs, the contents shown in these tables do not necessarily match the efficient contents especially for the high substitution rates.

7.3.2.1.1 The fresh density

The variation in the fresh density of concrete with respect to the percentage of substitution in RA is shown in Figure 7.10. The fresh density decreases when the substitution rate of fine or coarse NAs by, respectively, RFA or RCA increases because of the lower densities of RAs. RFA has a lower density than RCA: 2.1 kg/L for RFA and 2.3 kg/L for RCA (see Appendix). So, the decrease in density is stronger with the substitution of RFA, which implies that we could use the RA where lightweight concrete is needed and where the heavy dead weight is a problem such as in bridge constructions.

7.3.2.1.2 The rheological behavior

The principle is to rotate a vane at different speeds in the form of cross in a cylindrical sample of fresh concrete, and the torques exerted to maintain the rotation were determined.

A rheological test is performed by imposing a decreasing rotational speed to the vane interrupted by a stabilization stage to take the measurements. Every second, the control software of the stirrer saves the torque (M) corresponding to the imposed rotation speed (Ω). Moreover, the same type of rheometer should be used to measure and compare the rheological parameters of concrete (torque and torque rotation rate ratio at the steady state). Indeed, each rheometer gives results that differ significantly depending on its configuration.

7.3.2.1.3 Description of the apparatus

The version of the rheometer is composed of three main parts (Figure 7.11): an agitator with speed electronic control for recording the torques via an RFA232, a container, and a steel vane. The agitator is the key element for the tests. It is attached to an arm, via a jaw, which is itself fixed on a tripod. The agitator is driven by a computer using software (watch & control). The container is a cylinder of 30 cm in diameter and 25 cm high. To optimize the adhesion of concrete and to prevent the sliding of concrete on the outer wall of the container, steel rods were welded. The vane in the chuck of the agitator and a blade in form of a perpendicular double U are fixed. The blade has a height of 15 cm and a diameter of 10.5 cm.

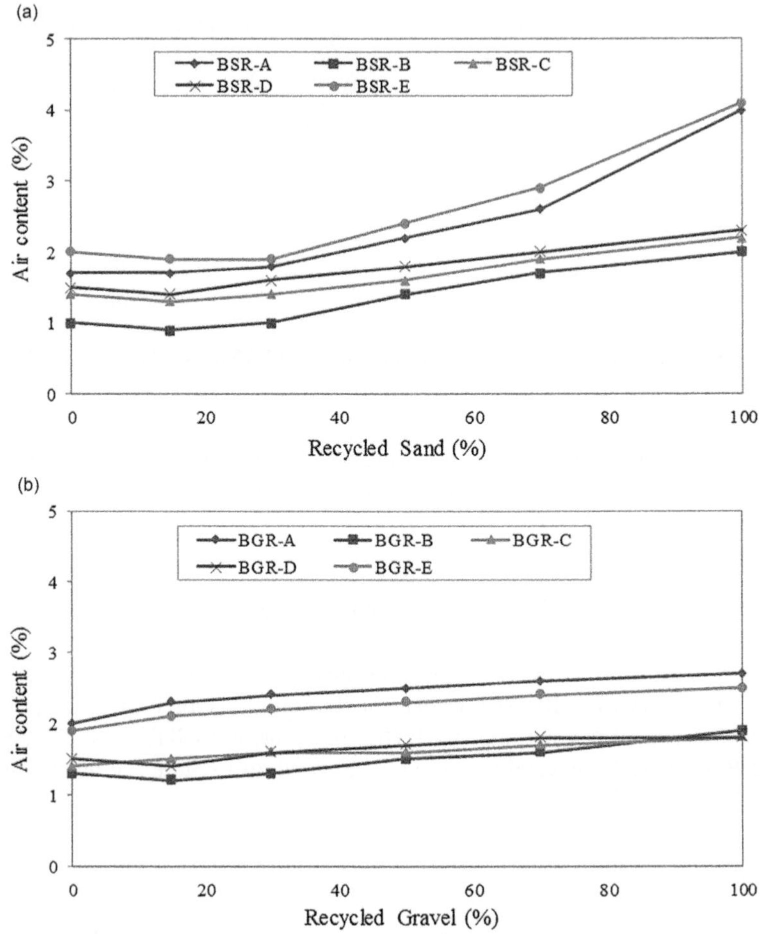

Figure 7.11 Variation of air content of concrete with percentage of RA: (a) RFA and (b) RA.

7.3.2.1.4 Final test procedure

The measurements are taken with the velocity profile of Soualhi et al. (2016).

- Step 1: Filling concrete into the container. It is done by three layers. Each layer receives 25 strokes of stitching using a rod.
- Step 2: Positioning of the blade. The blade is centered and thrust into the concrete until the surface of concrete and leaving a layer of concrete of 10 cm thick below the blade.
- Step 3: Measuring torques (M). They correspond to torques to be imposed to maintain speeds when the container is filled with concrete.
- Step 4: Using the measurements and calculation of rheological parameters.

Rheological parameters (yield stress and plastic viscosity)[1] at the initial time (T_0) are shown in Figures 7.12 and 7.13. From these figures, it is noted that

[1] Note from the editors: although these parameters are expressed with fundamental units, they cannot be considered as intrinsic physical parameters of the material, but rather as an empirical assessment of its fresh rheological behavior.

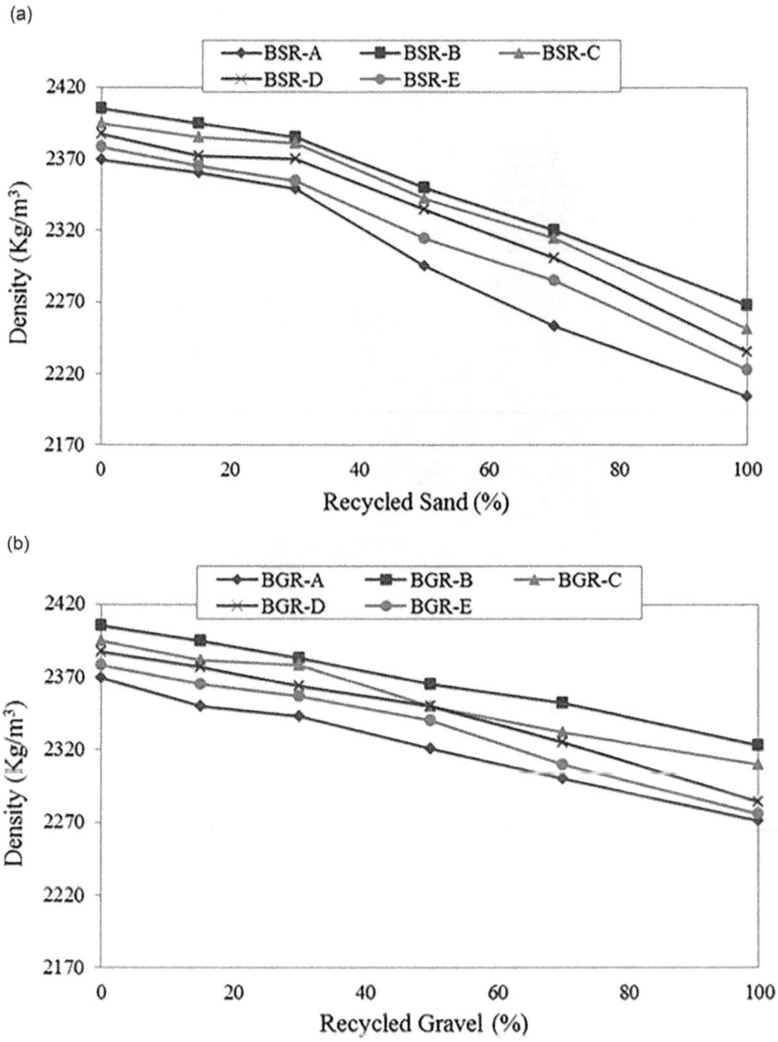

Figure 7.12 Variation in density of concrete with percentage of RA: (a) RFA and (b): RA.

- For concretes based on RCA yield stress and plastic viscosity, rheological parameters are rather constant whatever the substitution rate may be. These results agree with the objective to maintain the initial slump close to 200 mm thanks to an appropriate admixture Hu and Wang (2011), Wallevik (2006), Murata and Kukokawa (1992) obtained similar results.
- For concrete based on RFA, rheological parameters increase for a substitution rate over 30%. In this range of substitution, we observe a degradation of rheological properties even for a significant increase in admixture (see Table 7.3).

The substitution of NA by RA generally declines the compactness of the granular mixture because of a more angular form and lower packing densities of RAs (Chakradhara et al. 2010; Lin et al. 2004; Djerbi Tegguer 2012). To compensate this loss of compactness and

Figure 7.13 The rheometer used (Soualhi et al. 2016).

to maintain rheological properties, we have to increase paste fluidity or paste volume (de Larrard 2000; Sedran and de Larrard 1994; Soualhi et al. 2015). An increase in the fluidity of the fresh cement paste thanks to an increase in admixture content (see Table 7.4) allowed us to maintain rheological properties for concretes based on RA. This way to do is inefficient for concrete based on RFA for a substitution rate over 50%. In this range of substitution,

- The loss of compactness is too high to be compensated by an increase in paste fluidity.
- Some physical–chemical interactions between impurities or particles of old cement paste brought by the RFA and the admixture–cement couple of the fresh cement paste could occur and cause a stiffening of the concrete.

7.3.2.1.5 Air content

The variations in air content with the percentage of substitution in RA are shown in Figure 7.14. These variations are in good accordance with the rheological behaviors described in the previous paragraph:

- As rheological properties, air contents are rather constant whatever the rate of RCA substitution.
- Air content increases for a substitution rate of RFA over 30%. These increases are the consequences of the loss of rheological properties for these concretes (Figure 7.12). Indeed, RFA has a greater porosity than NFA; the shape and roughness of the RFA prevent air bubbles to extract concrete during vibration (Topçu and Şengel 2004; Raeis et al. 2015). This causes an increase in the air content for a high percentage of RFA.

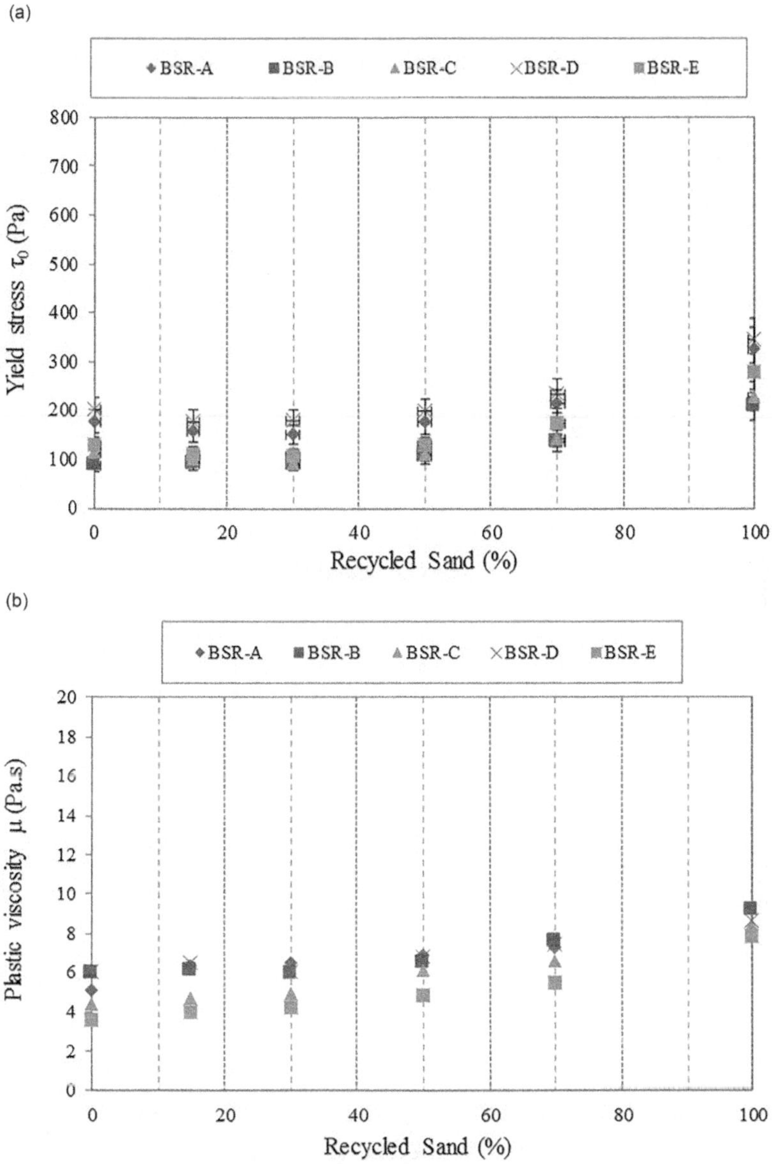

Figure 7.14 Variations in rheological parameters of concrete with percentage of RFA at τ_0: (a) yield stress and (b) plastic viscosity.

7.3.2.2 Interaction between the cement–admixture couple and recycling

The value of slump at 90 min after the output of mixer (named: T90) of the various mixes of concrete with RFA is presented in Figure 7.15.

It is possible to classify the cement–admixture couples into two groups:

- Couples B and C based on high water reducing admixture (ether polycarboxylique A2 and A3) for those highest slump values are targeted for a given content of RCA or RFA. For RCA concretes, the initial class of slump (class S4) is maintained at T90 until 70% of substitution. For 100% of RCA substitution, an S3 class value (150 mm) close

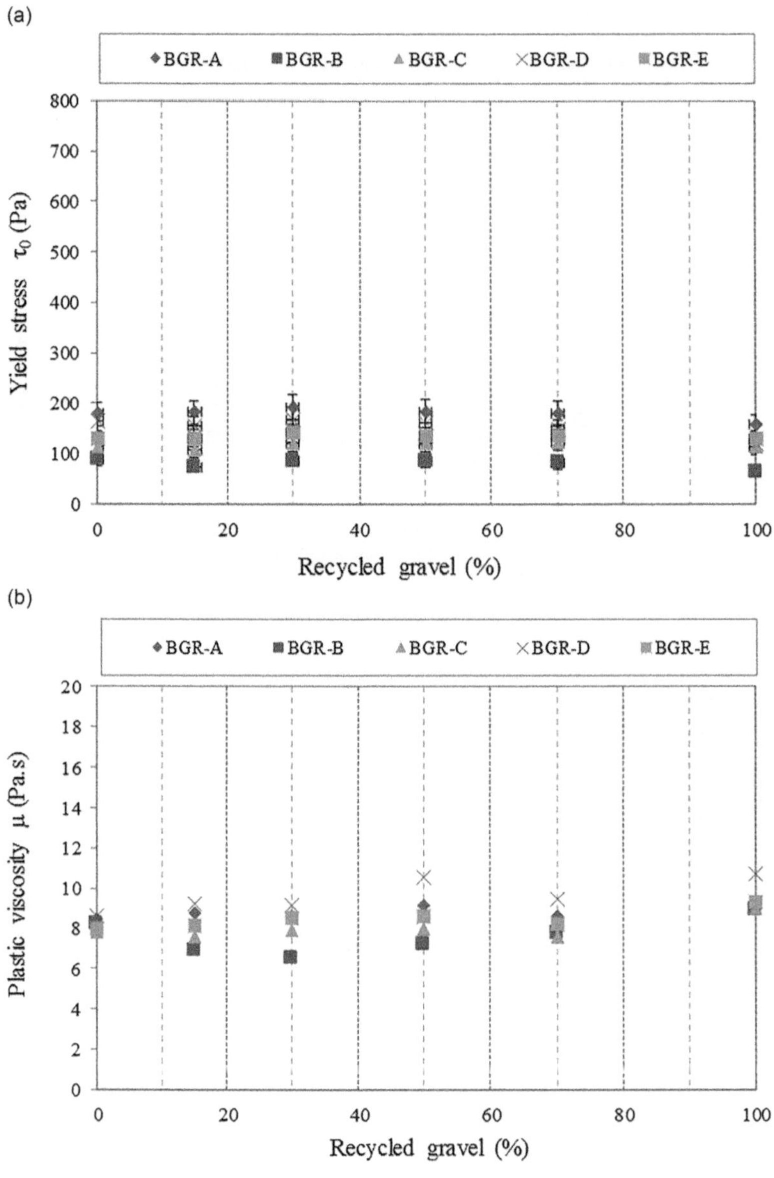

Figure 7.15 Variations in rheological parameters of concrete with percentage of RA at τ_0: (a) yield stress and (b) plastic viscosity.

to the S4 class value is reached. For FRA concretes, the threshold value is close to 30% of substitution and had been identified at the initial state from the rheological characteristics. RCA concretes with couple B seem to give better slump than RCA concretes with couple A. This behavior could be due to the lower amount of C_3A in Cement B, which delays the degree of hydration of the cement and increases the workability of the mixture. This result is similar to that obtained in (Kadri et al. 2009).

- Couples A and D are based on water reducing admixture (polycarboxylate A1 and A4). The threshold values according to the RA substitution are similar to those of the previous group and close to 30% for fine aggregate and close to 100% for coarse aggregate. This time we lost one class of workability (S4 to S3) for substitution ratios below the threshold values and even for the natural concretes. These couples are less efficient than couples B and C to maintain the initial slump value.

The couple E (based on Cement E and admixture A4) is more difficult to classify. For RCA concretes, this couple has a behavior close to the less efficient couples A and D. For RFA concretes, its behavior seems to be between the less efficient (couples A and D) and the most efficient (couples B and C).

Finally, it is possible thanks to an appropriate admixing and combining an appropriate cement–admixture couples (high water reducing admixture or superplasticizers arerecommended) and an appropriate admixture content to batch stable fresh recycled concrete with a good workability (S4 class) and recycling until 100% of coarse aggregate or 30% of fine aggregate.

7.4 RESEARCH NEEDS

The potential effects of accelerated stiffening by fines or impurities in the fine aggregate should be further investigated. The porosity of fine and coarse aggregates that can absorb a part of the admixture that is no longer available to fluidify the cement paste can be an explanation. Finally, the non-monotonic effect of the pre-saturation rate, especially for RFA, on the rheological behavior should be better understood.

7.5 CONCLUSION

There is no significant adverse effect of the pre-humidification of RA on the rheological properties of fresh concrete at the end of mixing. The mixing time seems sufficient for undersaturated RA to absorb a large part of the water necessary for their saturation. If a pre-saturation process is integrated to the concrete batching process, the targeted moisture content of RA should be ranged from 0.8 Ab to 1.0 Ab to minimize the adverse effect of time on the flow properties of concrete.

Threshold values for RA recycling to obtain a stable fresh concrete have been identified: 30% of substitution for fine aggregate is possible, whereas 100% of substitution of coarse aggregate is possible. However, a satisfactory behavior can only be obtained after a careful choice and dose of admixture in the fresh mix.

Chapter 8

Recycled concrete at early age

E. Rozière, A.Z. Bendimerad, and A. Loukili
GeM

J.-C. Souche, M. Salgues, and E. Garcia-Diaz
C2MA

P. Devillers
Ecole Nationale Supérieure d'Architecture de Montpellier

CONTENTS

Abstract

This chapter presents the methodology and results of a comprehensive experimental study about early age behavior of concrete with recycled concrete aggregates (RCA). This study includes the determination of RCA total absorption and absorption rate, the behavior of fresh concrete under standard and severe drying conditions, and the assessment of properties of hardening concrete. The RCA used in this study (provided by RECYBÉTON Project) showed high absorption rate. They reached 90% of the nominal absorption within the first 2 h, and their absorption was even faster in fresh cement paste. The monitoring of RCA water saturation in cement paste also showed a release of absorbed water when exposed to severe drying. When shifting from standard to severe drying, the increase of evaporation rate did not result in increased shrinkage to mass-loss ratios. However higher shrinkage magnitudes were reached because plastic shrinkage development was accelerated in severe drying conditions thus it could develop before setting. The air entry pressure concept was used to analyze the cracking sensitivity of fresh concrete. It was assessed on various concrete mixtures with and without RCA. A strong correlation was found between air entry pressure and negative capillary pressure at cracking time. The Free water to Binder (W_f/B) ratio was found to influence directly the air entry pressure. The (W_f/B) ratio accounts for the effects of initial water saturation of RCA. The evolution of plastic shrinkage, Young's modulus, and tensile strength were continuously monitored from fresh state to 24 h. The initial water saturation of RCA had limited influence on these properties. Restrained shrinkage tests would be necessary to assess the influence of RCA proportion on cracking sensitivity and estimate relaxation.

8.1 INTRODUCTION

Shrinkage-induced cracking affects the durability of concrete and concrete structures. This actually favors the ingress of chemically aggressive agents and the corrosion of steel reinforcement. Several phenomena are likely to promote cracking of concrete at an early age. The first one is due to the build-up of negative capillary pressure in fresh cement paste (Radocea 1994; Slowik et al. 2008). Negative capillary pressure can be caused by evaporation when concrete is no longer covered by bleeding and/or curing water. As soon as air entry pressure is reached, cracks are likely to appear. Plastic shrinkage develops when the evaporation rate exceeds the bleeding rate (Wittmann 1976). Thus, the accelerated drying of fresh concrete is likely to increase plastic shrinkage (Turcry and Loukili 2006; Mbemba 2010). The second phenomenon is restrained shrinkage on hardening concrete (see Figure 8.3). Plastic shrinkage cannot freely develop in most of the concrete structures. Moreover, the elastic modulus of concrete starts increasing from very early age, which is likely to induce significant stresses, whereas tensile strength is still relatively low. Therefore, the time period that covers the setting time and early hardening can be considered as critical for cracking (Hammer et al. 2007; Roziere et al. 2015). Recycled concrete aggregates (RCA) are reported to lead to higher shrinkage and creep, and lower strength and elastic modulus, for a given binder content. As a consequence, concrete mixtures are designed to achieve constant slump or strength. However, the studies dealing with early-age behavior are rather scarce. Moreover, RCA are porous aggregates with variable total absorptions and absorption rates; thus, their initial water saturation is likely to have a significant influence on effective water content of cement paste and thus on all concrete properties.

A comprehensive study has been designed in order to characterize RCA and fresh and hardening concrete, from mixing to 24 h. Two laboratories have been involved in this study: Center of materials Mines Alès (C2MA), at Ecole des Mines d'Alès, France, and Civil Engineering and Mechanics Research Institute (GeM), at Ecole Centrale de Nantes, France. Concretes were tested under severe and standard drying conditions, respectively. Severe drying actually promotes the firstly described phenomenon, that is, cracking of the fresh concrete. Two main parameters were investigated: the proportion of RCA aggregates and their initial water saturation.

Firstly, this chapter gives a description of phenomena causing cracking of concrete at the early age and related experimental results in the state-of-the-art part. Then, the two-part testing program is presented. The experimental results and discussions include evaporation rates, plastic shrinkage magnitudes, and cracking sensitivity. They first deal with the behavior of fresh concrete and then with hardening concrete.

8.2 STATE OF THE ART

The study presented in this chapter mainly deals with plastic shrinkage and cracking of concrete with RCA at the early age. The results available in the literature are rather scarce but existing knowledge can be used to design a new study on the influence of RCA. Two time periods are critical for cracking of fresh and hardening concrete: an early period ranging from placing to 1–4 h and a late period that covers the setting time and early hardening.

The phenomena that take place during the first critical time period are not specific of concrete mixtures. They can also characterize the drying of soils or other concentrated suspensions (Radocea 1994; Slowik et al. 2008). The water layer progressively disappears due to evaporation (Figure 8.1). Menisci are eventually formed between the particles at the sample surface. This results in the build-up of capillary pressure. As the curvature of menisci increases, the system of tensile forces between particles becomes unstable, and then, air starts to penetrate the pores. This finally causes local capillary pressure "breakthrough". The air entry concept is used to describe the phenomenon (Slowik et al. 2008). The air entry pressure is the capillary pressure at which the air is likely to penetrate the system. When the capillary pressure reaches a limit value called air entry value, air starts filling the largest pores and the capillary tension compresses the smaller pores and causes particle arrangement (Slowik and Ju 2011). From this time, the volume change is no longer equal to the loss of water, and the horizontal strains start to develop. Cracking was found to appear in the places where air entry pressure is reached; thus, air entry can be correlated with cracking (Souche et al. 2016).

Air entry depends on the composition of the material and its particle size distribution. In soil mechanics, the air entry value is usually determined from the volumetric water content versus suction curves. Other methods have been developed to determine the air entry for cement-based materials (Slowik et al. 2008; Souche et al. 2016).

RCA were found to influence the bleeding. A significant content of water is actually absorbed by such aggregates during mixing. Poon et al. reported that using RCA at the saturated surface-dried (SSD) state may result in bleeding during casting (Poon et al. 2004), at constant free water-to-cement (*W*/*C*) ratio of 0.57.

The second critical time period covers the setting time and early hardening. When plastic shrinkage is restrained, tensile stresses develop. Cracking was found to occur when tensile stresses exceed the tensile strength (Ravina and Shalon 1968; Hammer et al. 2007). Moreover, the evolution of the tensile strain capacity shows a minimum at this time period. The significant increase in the modulus actually appears earlier than the significant increase in the tensile strength; thus, the strain capacity firstly decreases, and then, it increases with the tensile strength (Roziere et al. 2015).

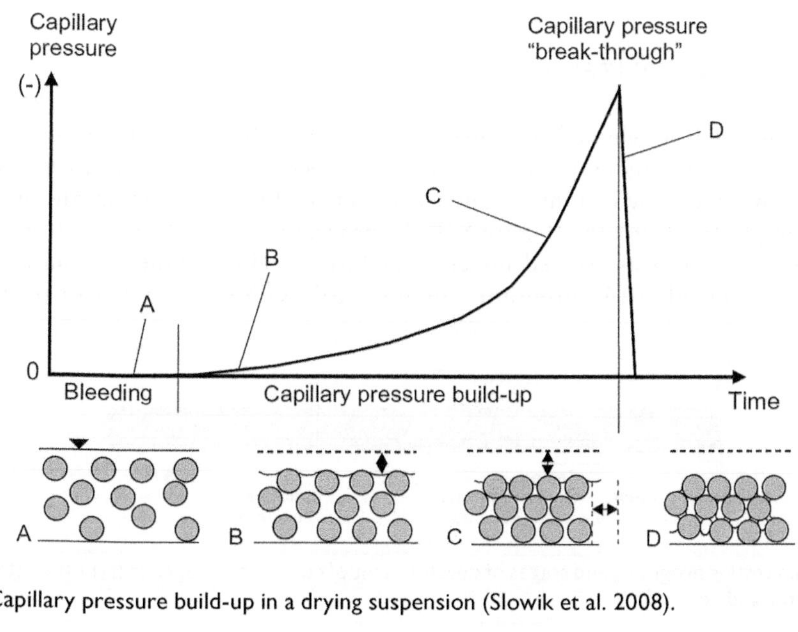

Figure 8.1 Capillary pressure build-up in a drying suspension (Slowik et al. 2008).

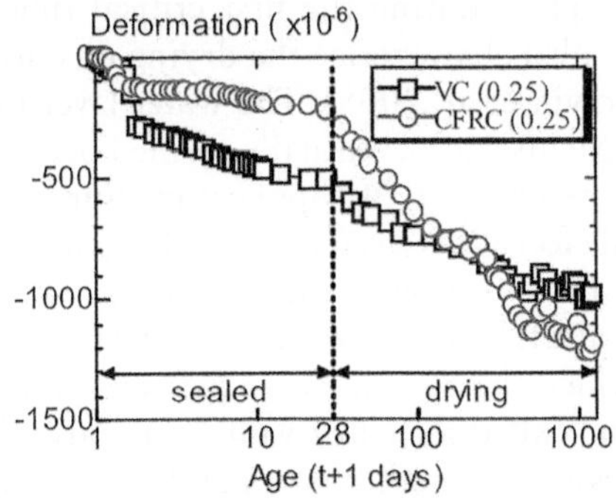

Figure 8.2 Deformation of high-strength concrete with virgin aggregate (VC) and high-strength concrete with recycled aggregate (CFRC), *W/C* = 0.25 (Maruyama and Sato 2005).

RCA influences all the parameters involved in restrained shrinkage-induced cracking. On the one hand, increasing RCA proportions results in higher drying shrinkage (Figure 8.2) and sometimes in lower strength. Equivalent or higher strength can be reached through mix design. On the other hand, Young's modulus is reduced and creep increases. As a consequence, it is not easy to predict the influence of RCA on cracking sensitivity. Moreover, the initial water saturation of porous aggregates is likely to influence the plastic shrinkage and other concrete properties (Cortas et al. 2014). As shown in Figure 8.2, shrinkage was lower in sealed conditions (Maruyama and Sato 2005). Saturated recycled aggregates could actually mitigate autogenous shrinkage by water desorption mechanism, in a self-curing process. The water absorbed at setting time is influenced by the absorption kinetics of aggregates. If the aggregates are initially dry or partially saturated, the water remaining in cement paste at setting time is higher than the theoretical effective water content.

For example, Alhozaimy (2009) has shown that dry limestone aggregates did not fully absorb the part of added water required to compensate the absorption of aggregates, and thus, lead to an increase in the actual *W/C* ratio of the mixes, increasing the initial slump and decreasing the compressive strength.

8.3 RECYBÉTON'S OUTPUTS

The water absorbed or released by porous aggregates is likely to influence the behavior of concrete at the early age, especially shrinkage-induced cracking. Thus, two experimental programs were designed to investigate the effects of RCA proportion and initial water saturation on plastic shrinkage (Figure 8.3). In program 1, the effect of external restraint on hardening concrete was studied under "standard" drying conditions characterized by low wind (Bendimerad 2016). Program 2 mainly dealt with the behavior of fresh concrete

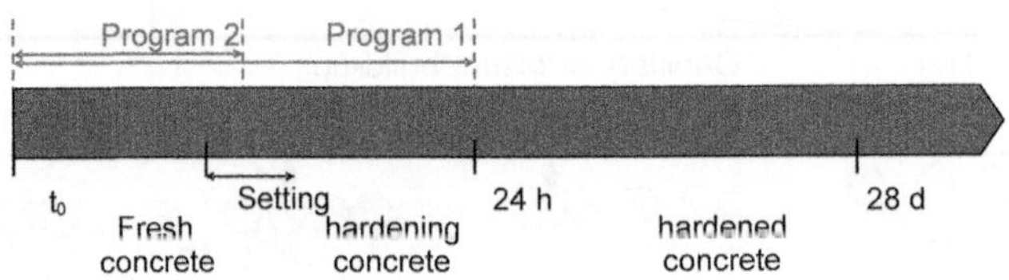

Figure 8.3 Main testing programs and stages of development of concrete properties (t_0 first contact between water and cement).

(Souche 2015). In order to promote the build-up of capillary pressure and cracking before setting, the wind speed was increased. This is referred to as "severe" drying conditions in the following sections.

8.3.1 Experimental program

8.3.1.1 Testing program

Two series of concrete mixtures were studied (Sedran 2017): one for building applications called B and another for civil engineering applications called GC. They were designed with different total water/binder ratios (*W/B*) in order to reach C 25/30 and C 35/45 strength class.

The water absorption of aggregates is a key parameter in concrete mix design. Moreover, aggregates can be partially or totally saturated when they are introduced in concrete mixer. This initial water saturation of aggregates influences the early-age as well as the long-term behavior of concrete, even if the total water is kept constant (Cortas et al. 2014). The use of recycled aggregates in concrete actually modifies the available water at the different stages. To clarify these exchanges in water into fresh concrete, some notions are required.

First, the pre-saturation water (W_{ps}) corresponds to the amount of water in the aggregates before mixing. The additional water (W_{ad}) is the complementary water introduced in the mixer. The absorption water (W_{abs}) is the water corresponding to saturated aggregates at nominal absorption, dried in surface (SSD state). The water remaining in the cement paste is called the free water (W_f). The total water (W_{tot}) introduced in the mix is given by the following equation:

$$W_{tot} = W_{abs}(A) + W_f = W_{ps} + W_{ad} \tag{8.1}$$

In the study presented in this chapter, considering that "A" is the initial water saturation of aggregates corresponding to the coefficient of water absorption measured at 24 h (EN 1097-6 2001), the 1.2A concretes have been made with aggregates at 120% SSD initial water saturation, which means a pre-saturation degree corresponding to the nominal absorption plus approximately 1% of dry mass of aggregates.

Two test programs have been conducted. The samples were tested under normal drying conditions, program 1 or under severe drying conditions, program 2, that is to say under wind (Table 8.1). In program 1, the free water content was kept constant. In program 2, added water was kept constant. For example, a building concrete (B) in program 1 with natural sand and recycled coarse aggregates pre-saturated at 1.2A is named: 1B0S100G-1.2A. All studied concretes are detailed in Table 8.2.

Table 8.1 Exposure conditions

	Program 1		*Program 2*	
	Constant free water		*Constant added water*	
	Standard drying conditions		*Severe drying conditions*	
Concrete family	B	GC	B	GC
RH (%)	50	50	45	45
Wind (m/s)	<0.3	<0.3	8	8

Table 8.2 Testing program

		Substitution rate			
		0S0G	*0S30G*	*30S0G*	*0S100G*
Initial water saturation of RCA	0.3A	–	–	–	1-B0S100G_0.3A
	0.5A	–	–	–	2-B0S100G_0.5A 2-GC0S100G_0.5A
	0.7A	–	–	–	1-B0S100G_0.7A
	1A	1-B0S0G_1A 1-GC0S0G_1A	1-B0S30G_1A	1-B30S0G_1A	1-B0S100G_1A 1-GC0S100G_1A
	1.2A	2-B0S0G_1.2A 2-GC0S0G_1.2A	–	–	1-B0S100G_1.2A 2-B0S100G_1.2A 2-GC0S100G_1.2A

8.3.1.2 Raw materials and concrete mixtures

The recycled aggregates used in this study had a relatively high coefficient of water absorption of 8.9% for sand and from 5.6% to 5.8% for 4/10 and 10/20 fractions, respectively (Sedran 2013).

All the concrete mixtures were designed to reach C 25/30 and C 35/45 strength classes for building concretes (B) and civil engineering concretes (GC), respectively, with different proportions of fine or coarse RCA. They were made up of the same batch of Portland cement CEM II/A-L 42.5 N proceeding from the same cement plant. The cement had an estimated Bogue composition of 61% C_3S, 0% C_2S, 7.9% C_3A, 12% C_4AF, and Na_2Oeq of 0.59%, a Blaine fineness of 370 m^2/kg, and a density of 3.09 kg/m^3.

Limestone filler was taken into account into binder content (B), defined as the sum of cement and limestone filler contents (in kg/m^3).

The tested concrete mixtures were designed following the RECYBÉTON formulas (Sedran 2017) (Tables 8.3 and 8.4).

Table 8.3 Concrete mixtures tested in program 1

	1-B				*1-GC*	
	0S0G	*30S0G*	*0S30G*	*0S100G*	*0S0G*	*0S100G*
NG 6,3/20 (kg/m^3)	820	829	462		810	
RG 10/20 (kg/m^3)			296	701		682
NG 4/10 (kg/m^3)	267	190	228		264	
RG 4/10 (kg/m^3)				163		158
NS 0/4 (kg/m^3)	780	549	813	806	771	782
RS 0/4 (kg/m^3)		235				
Cement, *C* (kg/m^3)	270	276	276	282	299	336
Limestone, *L* (kg/m^3)	45	31	31	31	58	53
Superplasticizer SP (kg/m^3)	0.747	0.798	0.861	0.798	1.26	1.31
W_f (kg/m^3)	180	185	185	189	175	185
W_{tot} (kg/m^3)	194.6	221.4	212.3	241	188.7	235.1
W_f/B (kg/m^3)	0.57	0.60	0.60	0.60	0.49	0.48
Volume of paste (L/m^3)	285	287	287	293	302	321
Packing density g^*	0.785	0.79	0.781	0.777	0.784	0.772
MPT (mm)	0.80	0.86	0.82	0.89	0.90	1.05

Table 8.4 Concrete mixtures tested in program 2

	2-B			2-GC		
	0S0G	*0S100G-1.2A*	*0S100G-0.5A*	*0S0G*	*0S100G-1.2A*	*0S100G-0.5A*
NG 6,3/20 (kg/m^3)	820			810		
RG 10/20 (kg/m^3)		701	701		682	682
NG 4/10 (kg/m^3)	266			264		
RG 4/10 (kg/m^3)		163	163		158	158
NS 0/4 (kg/m^3)	780	806	806	771	782	782
RS 0/4 (kg/m^3)						
Cement, C (kg/m^3)	270	282	282	299	336	336
Limestone, L (kg/m^3)	45	31	31	58	53	53
Superplasticizer SP (kg/m^3)	1.35	1.40	1.40	2.10	2.18	2.18
W_f (kg/m^3)	166	174	149.1	151	157	133
W_{tot} (kg/m^3)	180	232	207	165	213	189
W_f/B (kg/m^3)	0.53	0.56	0.48	0.42	0.40	0.34
Vol. of paste (L/m^3)	289	292	282	292	304	293
Packing density g*	0.785	0.777	0.777	0.784	0.772	0.772

8.3.1.3 Testing procedures

The testing procedures used in programs 1 and 2 are, respectively, described in GeM (Turcry 2004; Bendimerad 2016; Mbemba 2010; Souche 2015). The tests started 20 min after the first contact between water and cement. For all curves, the origin of time is the beginning of drying.

8.3.2 Behavior of fresh concrete before setting

8.3.2.1 Effect of environmental conditions on recycled concrete behavior

The plastic shrinkage evolution plotted as a function of weight loss (Figure 8.4) shows similar trends of samples in both drying conditions. The development of plastic shrinkage was slightly delayed for concretes of program 2. They actually showed higher bleeding (independently from drying conditions) than concretes tested in program 1. Then, the second phase showed a constant shrinkage rate.

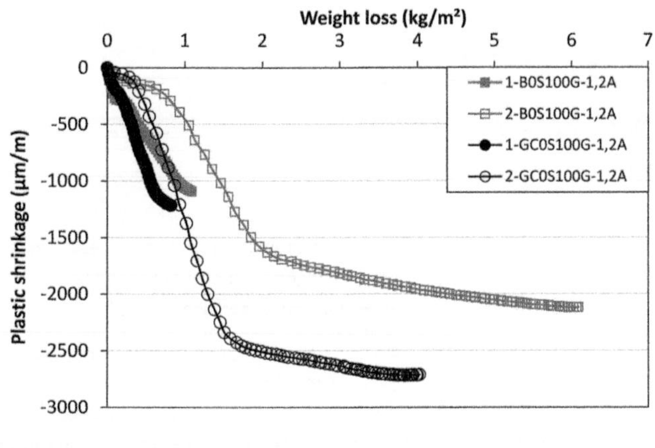

Figure 8.4 Plastic shrinkage versus weight loss.

The slope of the linear part of the curves was identical in both programs, for a given concrete mixture. Finally, during the third phase, the shrinkage rate gradually decreased until the end of the test, and the weight losses resulted in increasing air content in the pore system. The beginning of the third phase appeared at lower weight loss for concretes of program 1. In standard drying conditions, the decrease in plastic shrinkage rate actually occurred between the initial and final setting times, when shear modulus became significant (Bendimerad et al. 2016). In severe conditions, plastic shrinkage mostly developed before 2 h. At this age, the stiffness of concrete was still negligible; thus, the development of plastic shrinkage was not limited, and it reached higher values. The increase in drying rate had a stronger influence on the plastic shrinkage of civil engineering concretes, characterized by lower initial *W*/*B* ratio.

8.3.2.2 Effect of substitution rate and initial water saturation of RCA

The water content of recycled aggregates was measured at different ages under sealed and drying conditions. The purpose of these tests is to evaluate hydric transfers between recycled aggregates and cement paste over time with and without drying. The experimental protocol is inspired by Bello's work (Bello 2014) on lightweight aggregates and is described in Figure 8.5.

After mixing, concrete is stored under endogenous or desiccated conditions. Approximately every 10 min, a sample of fresh concrete is sieved through a 5-mm sieve in order to separate cement paste and coarse recycled aggregates. The wet aggregates are weighed and dried in an oven for 2 h. Once aggregates are dried, they are weighed again (Salgues et al. 2016).

Results are given in Figure 8.6 for 2-B0S100G-0.5A concrete. The purpose of these tests was to evaluate hydric transfers between recycled aggregates and cement paste over time with and without drying (Salgues et al. 2016). The behavior of recycled aggregates depends

Figure 8.5 Test protocol to measure water in coarse recycled aggregates during time (Salgues et al. 2016)

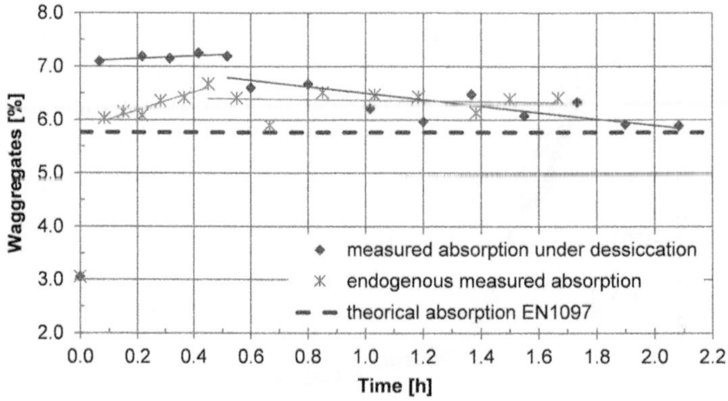

Figure 8.6 Water content of recycled aggregates (g of water/g of dry aggregates) in concrete 2-B0S100G-0.5A under drying and endogenous conditions (Salgues et al. 2016).

on their initial saturation state. Initially, unsaturated aggregates absorb water while being mixed to nearly reach its maximum absorption, as shown in Figure 8.6.

Slight water absorption can occur up to 30 minutes after pouring fresh concrete. Therefore, the water absorption of the recycled aggregates in the cement paste is very fast, and the result is in accordance with the absorption of recycled aggregates in the water (De Brito et al. 2011; Bendimerad et al. 2015). Nevertheless, after this first absorption phase, under severe wind conditions, aggregates from 2-B0S100G-0.5A release water due to the lack of water in concrete and paste. Endogenous conditions for B0S100G-0.5A concrete do not highlight this water transfer (Figure 8.6).

According to these first results, it is interesting to monitor the variation in the volume of water in the aggregates, cement paste, and concrete. The different phases are shown in Figure 8.7. In the mixer, the unsaturated recycled aggregates complete their absorption until 90% of the nominal value. Then, the absorption can continue after casting during the first half of an hour. After casting, as soon as the desiccation begins, the concrete homogeneously loses water. This water is provided by aggregates and/or cement paste.

Figure 8.8 shows the plastic shrinkage value at 7.2 h as a function of *W/B* ratio between total water and binder. Under standard conditions, the plastic shrinkage magnitude decreased with an increase in *W/B* ratio, whereas it increased under severe conditions.

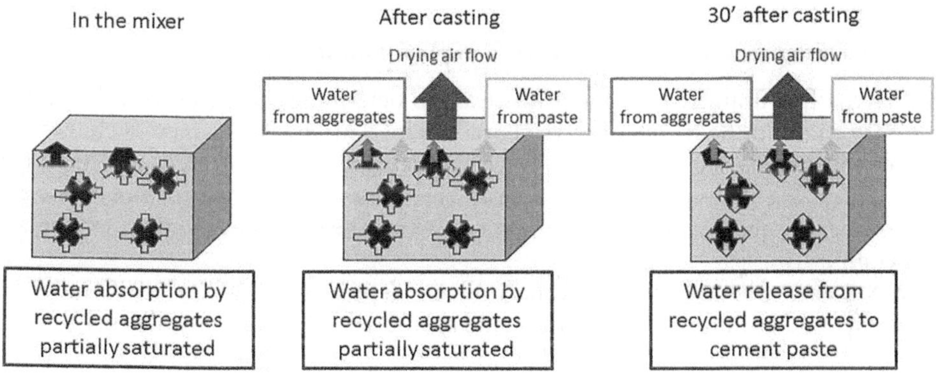

Figure 8.7 Hydric transfers in concrete, cement paste, and recycled aggregates over time for B0S100G-0.5A concrete (Salgues et al. 2016).

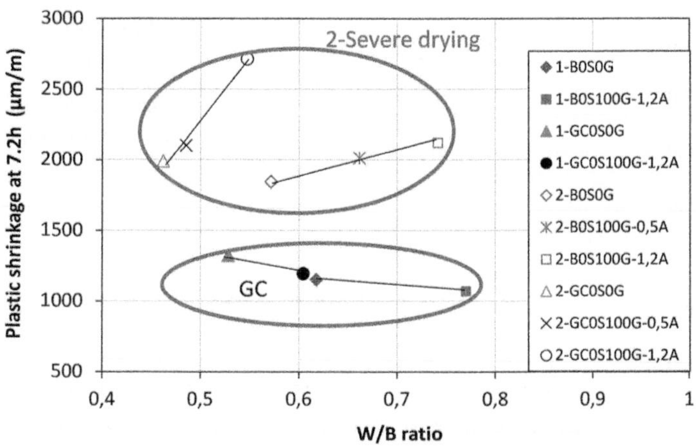

Figure 8.8 Plastic shrinkage at 7.2 h versus *W/B* ratios (total *W/B* ratio).

The increase in *W/B* ratio generally favors bleeding, which delays and mitigates the development of plastic shrinkage. Severe conditions did not result in higher plastic shrinkage-to-weight loss ratios, but they significantly influenced the magnitude of plastic shrinkage. At high evaporation rates, the development of plastic shrinkage was accelerated, and thus, it could develop before the initial setting time, which was not possible in standard drying conditions (Figure 8.4).

For building concretes under severe drying conditions, the shrinkage versus *W/B* ratio can be plotted as a linear function (Figure 8.8). For civil engineering concretes, another linear function was found, with a higher slope. These correlations could be explained by the influence of mix design and the behavior of RCA. RCA concrete mixtures actually had higher paste proportions (Table 8.4), and shrinkage magnitude increases linearly with paste volume (Hansen and Almudaiheem 1987). Moreover, severe drying is likely to cause drying of RCA (Maruyama and Sato 2005; Wyrzykowski et al. 2015), whereas they can provide internal curing in sealed conditions or standard drying conditions. Finally, increasing RCA proportion results in an increase in drying shrinkage of concrete because of their original paste content and lower modulus.

Finally, the influence of RCA on plastic shrinkage depends on drying conditions. In severe conditions, it resulted in an increase in shrinkage, whereas it mitigated shrinkage development in standard drying conditions.

8.3.2.3 Cracking sensitivity under severe drying conditions

In this part, a fine drained soil model was used to determine the air entry value. Firstly, it is necessary to build the fresh concrete characteristics curves. The concrete characteristics curves are defined as the relationship between water content (%) and capillary pressure (hPa) in a semi-log description. According to the method developed for fine soils, the air entry value can be approximated as the point at which the two lines intersect (Souche 2015; Souche et al. 2016).

Figure 8.9 shows that the air entry value is highly dependent on effective *W/B* ratio (W_f/B). Binder content takes into account all the cements and all the calcareous fillers. Coarse RCA

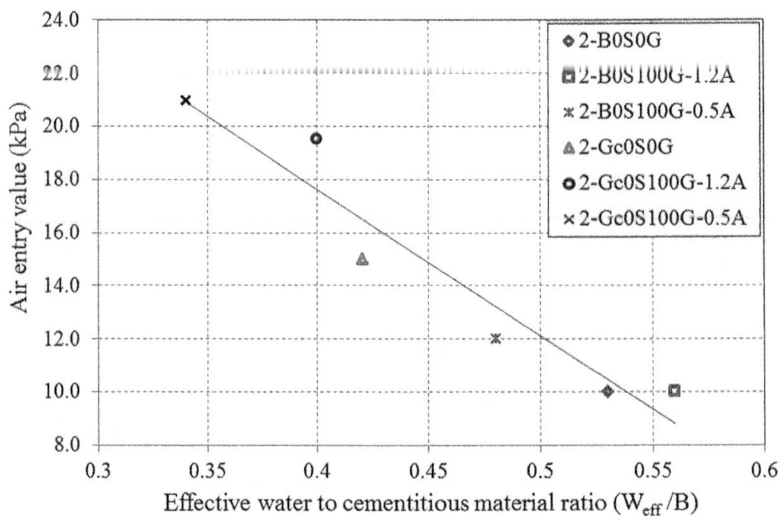

Figure 8.9 Influence of W_f/B ratio on air entry value (Souche et al. 2016).

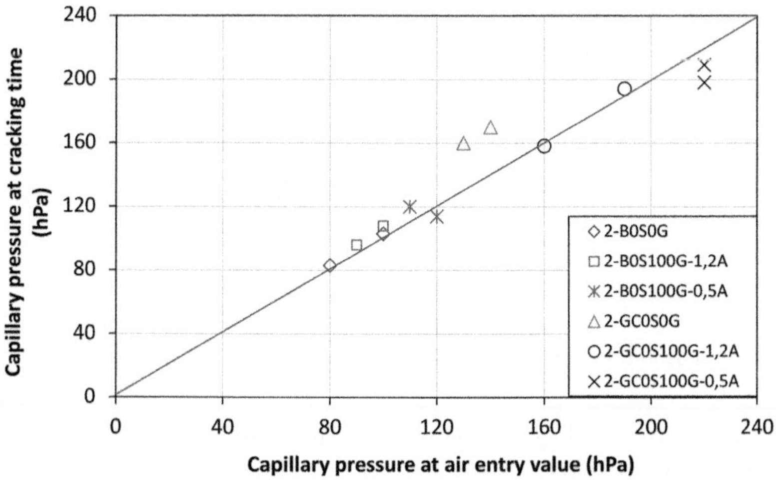

Figure 8.10 Capillary pressure at cracking time as a function of air entry values (Souche 2015).

do not seem to influence the air entry value as the air entry value is a physical characteristic of the paste. Then, the field of capillary depressions is disturbed by the air entry, and the identification of permeability coefficient starting from the capillary depression gradient is not possible anymore. After the air entry, fresh concrete becomes an unsaturated porous medium.

For each concrete, capillary pressure at cracking time can be compared to air entry value. Figure 8.10 represents the measured capillary pressure values at cracking time as a function of the calculated air entry value for all concretes. For conventional concrete (NC) and recycled concrete (RC), Figure 8.10 brings to light a strong linear correlation between air entry value and cracking pressure.

Cracking occurs around the air entry. For NC, it confirms the results from previous studies (Slowik et al. 2008; Slowik and Ju 2011): air entry is a critical period for cracking of fresh concrete. For RC, the same behavior as NC is observed. The local overstresses cause cracking. According to the pre-saturation degree, the unsaturated RCA complete their absorption modifying the effective water of the paste. On the opposite, the initially oversaturated RCA could provide a potential internal curing to the paste under desiccation. So according to Salgues et al. (2016), the water contained into RCA had a direct impact on concrete cracking time at the early age.

8.3.3 Behavior of hardening concrete

8.3.3.1 Effect of substitution rate

The graphs (Figure 8.11) showed the influence of the percentage of substitution of natural aggregates by coarse or fine RCA on 24-h plastic shrinkage magnitude (Turcry and Loukili 2006).

In this study, all the curves were initialized at the age of initial deformation defined in Bendimerad et al. (2016), when the temperature of fresh concrete first reaches the temperature of the mold (around 1 h). B0S30G and B0S100G concretes showed nearly the same plastic shrinkage magnitude, close to 1,100 µm/m. When they are compared to the NC, the difference is significant but lower than 10%. This can be explained by the maximal paste thickness (MPT) concept (de Larrard and Sedran 1994; de Larrard and Belloc 1997). Considering that the aggregates mix is filled with cement paste, the MPT actually represents

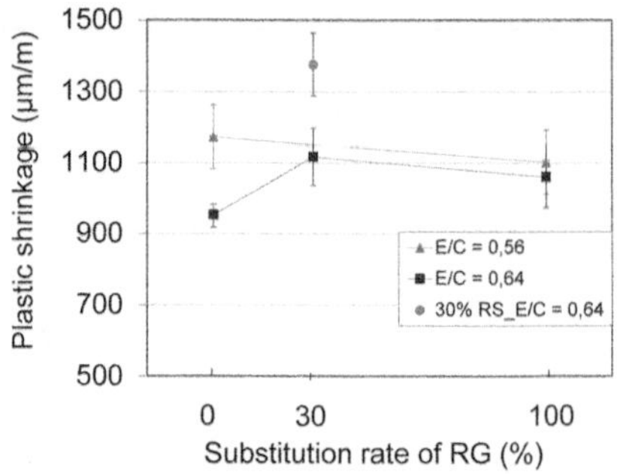

Figure 8.11 Effect of substitution rate on plastic shrinkage.

the paste that is in free shrinkage configuration. The values of the MPT of the three concrete mixtures were actually very close (Figure 8.12). The most significant plastic shrinkage concerned concrete with recycled sand. Because of the higher surface area of fine RCA, the bleeding water is lower than in the control mix, and the risk of plastic shrinkage related to insufficient curing increases.

This study aimed at providing data to understand the evolution of the shrinkage-induced cracking sensitivity of different conventional and RCs at the early age. The concrete tensile strength was experimentally assessed 24h after casting using the direct tensile test. The percentage of substitution of recycled gravel or sand significantly affected the mechanical properties, such as elastic modulus and tensile strength (Figure 8.13).

8.3.3.2 Effect of initial water saturation

The plastic shrinkage measurements on the B0S100G concrete with different initial water saturation of gravel showed minor differences. The magnitude at 24h was between 870 and 1,055µm/m (Figure 8.14a). The initial setting times were around 6h, and the final setting times were around 11h. From the final setting to 24h, shrinkage did not significantly increase.

The mixtures with initial saturation 1B0S100G-0.3A and 1B0S100G-1.2A showed the lowest values. The concrete with 0.3A had the highest added water (W_{add} = 218.4kg/m^3, $W_{aggr.}$ = 22.6kg/m^3), which implies a higher bleeding rate (Almusallam et al. 1998). This additional

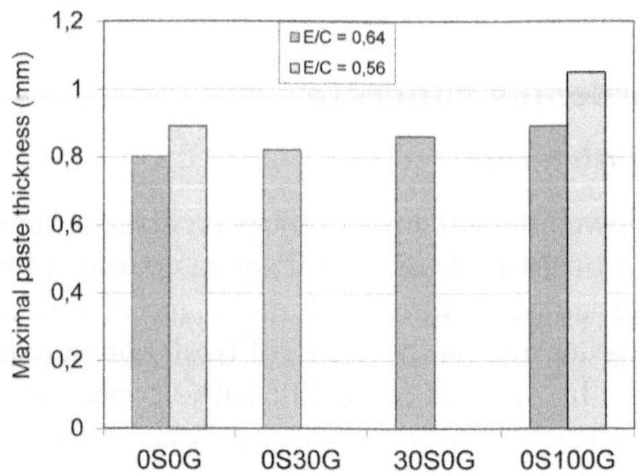

Figure 8.12 Effect of substitution rate on MPT.

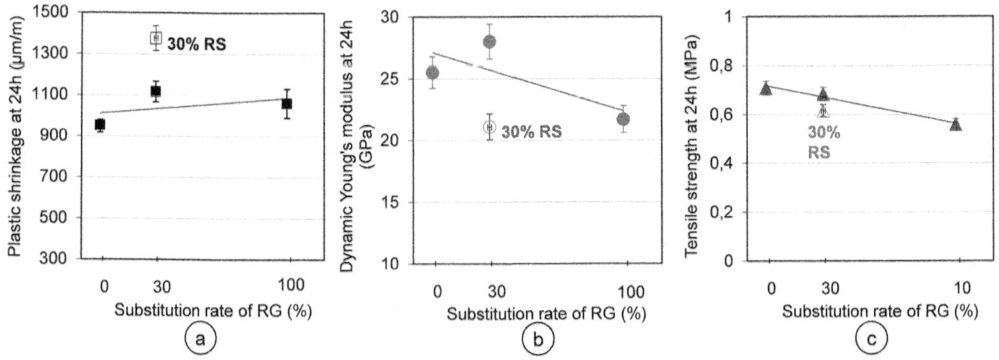

Figure 8.13 Influence of substitution rate of RCA on early-age properties: Plastic shrinkage (a), Young's modulus (b), Tensile strength (c).

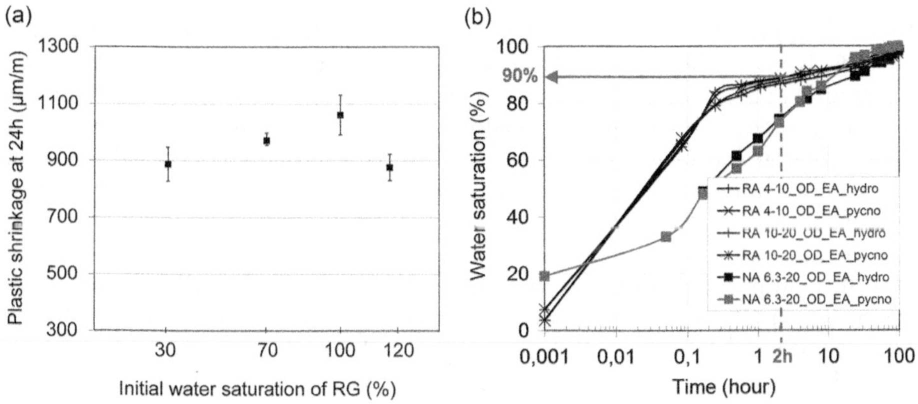

Figure 8.14 Correlation between plastic shrinkage and rate of water absorption. (a) Effect of water saturation of recycled aggregate concrete on plastic shrinkage; (b) influence of aggregate type on water absorption rate (Bendimerad et al. 2015).

bleeding water mitigated the plastic shrinkage of concrete at 0.3A. The concrete with oversaturated gravel (1.2A) had a relatively low shrinkage as well. This mixture 1.2A (W_{add} = 179.8 kg/m^3, $W_{aggr.}$ = 61.2 kg/m^3) was characterized by the same total water as 0.3A concrete. Oversaturated gravel actually releases water in cement paste and provides self-curing in concrete (Zhutovsky et al. 2002). This internal curing was also observed in the case of RC (Salgues et al. 2016).

Absorption measurements on different aggregates showed a relatively high water saturation rate of recycled aggregates during the first hour (Bendimerad et al. 2015). After 2 h, the water saturation actually reached 90% (Figure 8.14b). Therefore, when the development of plastic deformation begins, the major part of water saturation has already occurred, which explains the limited effect of initial water saturation.

The values of mechanical properties (elastic modulus and tensile strength) at 24 h show that the effect of initial water saturation of coarse aggregates was negligible (Figure 8.15).

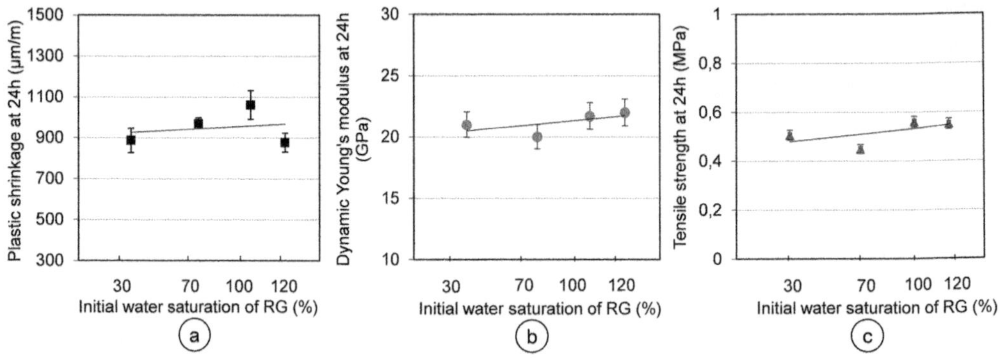

Figure 8.15 Influence of initial water saturation of RCA on early-age properties of concrete I-B0S100G (Bendimerad et al. 2016).

8.4 RESEARCH NEEDS

Few studies deal with the influence of RCA on early-age behavior of concrete. A comprehensive study involving two laboratories has been designed to understand the influence of initial water saturation and RCA proportion on water movements in fresh concrete and properties of hardening concrete. This study presented in this chapter mainly dealt with coarse aggregates. This material was characterized by relatively high purity, low original paste content, and high absorption rate. The results obtained with fine RCA (sand) were totally different, which underlines the need to take into account the properties of RCA. Further research would include experimental studies on the influence of RCA purity and fine RCA proportions. Although their proportion in RCA is low, the fines (powder produced by crushing original concrete) have a significant influence on plastic shrinkage because of high water absorption.

In terms of testing and analyses, the investigation on the cracking sensitivity lacks the assessment of tensile creep or relaxation at the early age. These properties cannot be easily determined by direct testing. However, restrained shrinkage tests allow an indirect estimation. For instance, temperature stress testing machine system provides the evolution of free and restrained deformations, from which creep and relaxation can be deduced (Bendimerad 2016).

RCA influence the early-age behavior of concrete at different levels, and the effects are often coupled. As a consequence, experimental studies do not always allow separating the effects due to RCA intrinsic properties and their behavior in cement paste. Thus, multiscale modeling appears as a promising approach to understand these complex phenomena.

8.5 CONCLUSIONS

Two experimental programs were connected to provide data on the influence of RCA proportion and initial water saturation on plastic shrinkage, cracking risk, and concrete properties at the early age. Several conclusions can be drawn from this comprehensive study.

- Due to the relatively high porosity, RCA absorb water that can be released in cement paste. In sealed conditions or under moderate drying conditions, this provides internal curing that mitigates plastic shrinkage. However, under severe conditions, high evaporation rates are likely to trigger the drying of aggregates, which results in an increase in shrinkage magnitude.

- Air entry concept can be used to investigate the influence of RCA on cracking risk of fresh concrete. Cracking time was assessed through specific testing (ASTM C1579), and experimental results actually showed a strong correlation with air entry determined from water content versus capillary pressure curves of the studied concretes.
- The coarse RCA used in this study had high absorption rate; thus, the water content remaining in the cement paste was close to the theoretical free water content. A relatively low influence of initial water saturation was actually observed when keeping the total water constant. At the variable total water contents, the air entry pressure linearly decreased with free *W/B* ratio (W_f/B), whatever the type of aggregates.
- The proportion of RCA had a significant influence on properties of concrete: plastic shrinkage, Young's modulus, and tensile strength. However, it is not possible to conclude on cracking sensitivity from these results, as the influence of RCA on relaxation has not been assessed in this study.

From the experimental results, careful curing of RCA concretes can be recommended. The studies actually show a strong coupling between initial water saturation, properties of fresh concrete, and drying exposure on plastic shrinkage and cracking risk. RCA proportion and initial saturation do not induce higher cracking sensitivity in themselves, but they are likely to promote cracking in severe drying conditions (initially dry RCA and high evaporation rate).

ACKNOWLEDGMENTS

Support from the Agence Nationale de la Recherche (National Research Agency, France) within ECOREB project and Project national (National Project, France) RECYBÉTON is gratefully acknowledged.

Section IV

Properties of hardened recycled concrete

P. Rougeau
CERIB

Previous sections have presented the peculiarities of recycled aggregate (RA), especially their microstructure and porosity. Their intrinsic characteristics are determined by the nature of natural aggregate (NA) and the amount and characteristics of the old cementitious paste. Recycled aggregate concrete (RAC) can have specific hardened properties depending on the nature and substitution rate of RA.

The questions investigated in this section for each hardened property are as follows:

- Does the RA play a specific role?
- What are the mechanisms involved?
- Are the usual tests and methodologies adapted to RAC?
- What are the required conditions to produce RAC with similar properties than NAC?

This section is subdivided into five chapters:

- The first one (Chapter 9) concerns the microstructure of RAC with a focus of the interfacial transition zone between the old and the new cement paste in RAC and the transport of water from the oversaturated RA to the cement paste.
- The second one (Chapter 10) describes the influence of the introduction of RA on the instantaneous properties of concrete. The relations between compressive and tensile strength and elastic modulus considering the substitution rate of RA are investigated.
- The third one (Chapter 11) concerns delayed deformations and fatigue properties, particularly shrinkage and creep deformations.
- The fourth one (Chapter 12) deals with durability-related properties: properties related to the corrosion risks of the reinforcement (carbonation, chloride migration, gas permeability, porosity), the resistance to freeze/thaw cycles, the risks linked to alkali/silica reactions and to the presence of sulfates (ettringite/thaumasite formation).
- Finally, the fifth one (Chapter 13) deals with the behavior under fire. The spalling of RAC and the fire resistance of two beams are studied.

Chapter 9

Microstructure of recycled concrete

E. Garcia-Diaz and G. Le Saout

Centre des Matériaux des Mines d'Alès, IMT Mines Ales

A. Djerbi

IFSTTAR, Marne-la-Vallée

CONTENTS

Abstract

Interfacial Transition Zone (ITZ) between old and new cement paste in recycled mortars and concretes is characterized. The microstructure is investigated during the first phase of hardening (2 and 28 days) for mortars and up to 1 year for concretes. Mortars and concretes made with over-saturated fine and coarse recycled aggregates develop more porous ITZ with lower anhydrous profiles. Transport of water from the over-saturated recycled aggregates to the cement paste, and water lens formation by micro-bleeding effect, are expected to explain this phenomenon. On the other hand, mortars made with dried fine recycled aggregates develop "denser" ITZ with anhydrous profile similar to that observed on mortars made with dried natural fine aggregates with the same targeted average *W/C* ratio. Transport of portlandite from the new cement paste to the old cement paste is observed during the first 28 days of hardening. This phenomenon is stronger for mortars made with over-saturated fine aggregates. The reduction of *W/C* ratio to obtain C25, C35 and C45 recycled concretes improves slightly the porosity profiles. But concretes based on 100% of recycled coarse aggregates have higher porous ITZ than concretes based on 30% of recycled fine and coarse aggregates. The low porosity of the ITZ of the C45 concrete based on 30% of fine and coarse recycled aggregates could be the consequence of a positive curing effect, with a contribution of the recycled aggregate water to cement hydration because of the partial desaturation during the hardening of the cement paste with a low initial water cement ratio (*W/C* close to 0.41). A treshold value for paste porosity which separates porous networks wealkly interconnected (low increase of permability with porosity) and porous network very interconnected (high increase of permeability with porosity) has been identified. The value is closed to 17%–18% of porosity.

9.1 INTRODUCTION

Recycled aggregates (RAs) are composed of a mixture of natural aggregates (NAs) roughly coated with hardened cement paste or mortar. The presence of this carbonated and porous old paste could modify mechanisms governing the formation of interfacial transition zone (ITZ) between aggregate and cement paste. The consequence could be a specific microstructure formation between the old and new cement paste. The objective of this chapter is the first to characterize by scanning electron microscopy (SEM) the interphase area between the old and new paste by semiquantitative measurement on mortars and concretes. Dried and oversaturated aggregates have been used to appreciate the effect of water transfer between RA and new cement paste on this ITZ. Then, the porosity of the concrete microstructures is correlated with intrinsic gas permeability to identify the area of weakly interconnected network improving durability.

9.2 STATE OF THE ART

The ITZ is a cement paste zone between the aggregate and the homogeneous bulk paste. Several authors (Barnes et al. 1978; Crumbie 1994; Kjellsen et al. 1998; Langton and Roy 1980; Monteiro et al. 1985; Olivier et al. 1995; Scrivener et al. 1988; Scrivener and Gartner 1987; Scrivener and Pratt 1996; Zimbelman 1985) have studied in detail the microstructure of the ITZ and its forming mechanisms. This zone is characterized by a microstructure gradient: the porosity increases in the ITZ from the bulk cement paste to the aggregate surface. The microstructure gradient is mainly the consequence of the "wall effect" exerted by the coarse aggregate on the fine cement particles. Because of this "wall effect", the initial cement content decreases and the water content increases from the bulk cement paste to the coarse aggregate surface. The range of order of the depth of the disturbed area by the "wall effect" is approximately of several tenths microns and corresponds to the initial depth of the ITZ. During the hardening and because of a filling of the porosity by the hydrated products, the depth of the ITZ generally decreases. At the same time, a mechanism of transport of the hydrated products (mainly the Portlandite) occurs from the cement-rich to the cement-lean areas. Because of this Portlandite transport, we could observe a formation of a "duplex film" rich in Portlandite of few microns at the aggregate surface. Other mechanisms could play a role in the porosity formation of the ITZ. We could note the microbleeding around the coarse aggregates during the concrete casting and vibration (Metha 1986; Crumbie 1994) and the unilateral growth of Portlandite in the duplex film (Garboczi and Bentz 1996). In the presence of porous aggregate with high absorption capacity, the mechanisms of ITZ formation are influenced by the water transport between the aggregate and the fresh cement paste. For lightweight aggregates with a macropore network (several hundred microns to several millimeters) as expanded clay or shale aggregates, a dry initial state and a water absorption during the concrete making lead to a denser ITZ and a bond zone of better quality (Wasserman and Bentur 1996; Elsharief et al. 2005; Lo et al. 2005). According to Nguyen et al. (2014) for a targeted net water/cement ratio (*W/C*), mortars based on oversaturated porous limestone aggregates have more porous ITZ than mortars based on the same dry aggregates. On the other hand, ITZ of mortars based on dry and oversaturated nonporous limestone aggregates has the same porosity. Zhao et al. (2013a) obtained similar results with mortars containing fine RA (FRA). On the contrary, Tam et al. (2005) showed that the quality of the new ITZ between the new and old cement paste is improved by using wet aggregates.

9.3 RECYBÉTON'S OUTPUTS

9.3.1 Mortars' microstructure

9.3.1.1 Materials and method

9.3.1.1.1 Mortars

The FRA was sieved and recomposed from six size fractions to obtain the same grain size distribution as the natural one. Fine NA (FNA) and FRA were used in the dry state or oversaturated state for mortar fabrication. Mix proportions are presented in Table 9.1.

The oversaturated state corresponds to a water amount of absorbed water plus 10% ($WA_{24\,h}$ + 10%). Water amount of absorbed water was close to 10% for FRA and 0.5% for FNA (see Chapter 3). An effective *W/C* ratio of 0.5 is targeted. The volume fraction of the fine aggregate in the mortars is constant and close to 60%. The cement is a CEM II/A-L 42.5N.

9.3.1.1.2 Microstructure characterization and nano-indentation tests

For the microstructural investigations, pieces of hydrated samples of $1 \times 2 \times 2\,cm^3$ were examined using a Quanta 200 FEG SEM from Field Electron and Ion Company (FEI) coupled to an Oxford INCA X-sight energy-dispersive X-ray spectroscopy analyzer. Backscattered electron (BSE) imaging has been used to study the ITZ in mortars following the lead of Scrivener and Gartner (1987).

For the nano-indentation investigations, the sample preparation is similar to the SEM but with a sample size of $1 \times 1.5 \times 1.5\,cm^3$. The instrument is a SEM type of HITACHI S-4300SE/N with a modified Berkovich-type used indenter. The Martens micro-hardness (H_M) value based on the maximum depth obtained by the micro-hardness test has been chosen to analyze the results, which is given by the following equation:

$$H_M(\text{GPa}) = \frac{P_{max}}{A_r} = \frac{P_{max}}{26.968 \times h_m^2} \tag{9.1}$$

P_{max} (mN) is the applied maximum load; h_m (µm) is the maximum depth of penetration measured; $A_r = 26.968 \times h_m^2$ corresponds to the contact area of the indenter for the maximum depth of penetration measured.

For more details on the methods used, we can refer to Le et al. (2017).

Table 9.1 Mix proportions of the mortars with different moisture states of fine aggregate

	Mortar with FNA		Mortar with FRA	
Moisture state of fine aggregate	*Dry*	*Oversaturated*	*Dry*	*Oversaturated*
Name	NM-Dry	NM-Sat	RM-Dry	RM-Sat
W_{eff}/C	0.5	0.5	0.5	0.5
Dry fine aggregate (g)	1350	1350	1155	1155
Water absorption (%)	0.5	0.5	10.0	10.0
Cement (g)	450	450	450	450
Effective water (g)	225	225	225	225

9.3.1.2 Results

Examples of microstructures at 2 and 28 days are shown in Figures 9.1 and 9.2, respectively. At 2 days, the contrast between the mesoporosity aspect of the young new paste and the microporosity aspect of the carbonated old paste allows us to define the border between the two pastes. A precipitation of Portlandite is observed close to this border and defines a duplex film. The thickness of this duplex film is larger for mortars based on oversaturated FRA and is yet discernible at 28 days for these mortars. On the other hand, for mortars made with dry FRA, an excellent continuity is observed at 28 days and it is not possible to distinguish the border between the old and the new paste at this age.

The anhydrous distributions at 2 days of hardening are shown in Figure 9.3. Mortars based on dry FNA (NM-Dry) and dry FRA (RM-Dry) have the same distributions. The pastes surrounding these aggregates have been impacted by a similar wall effect and have also a similar initial W/C ratio distribution in the ITZ. Recycled mortars based on oversaturated

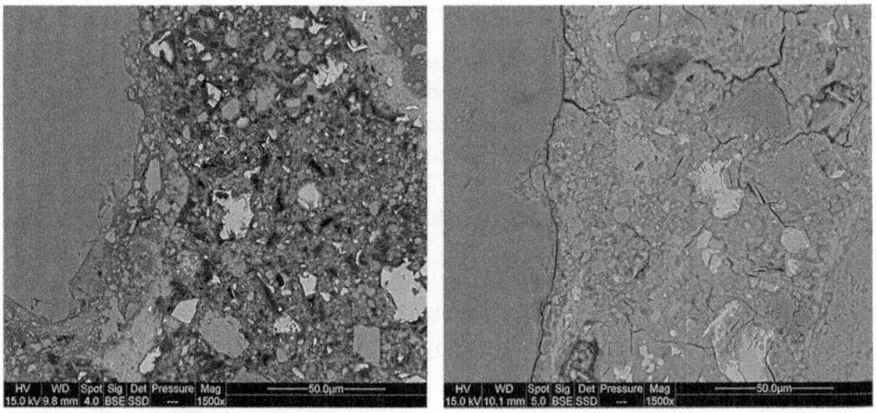

(a) (b)

Figure 9.1 Images of mortar based on dry FRA (a: 2 days; b: 28 days).

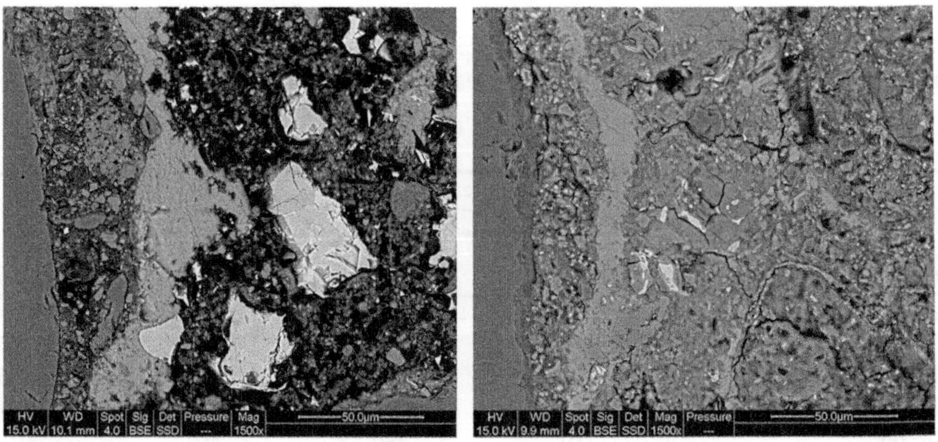

(a) (b)

Figure 9.2 Images of mortar based on saturated FRA (a: 2 days; b: 28 days).

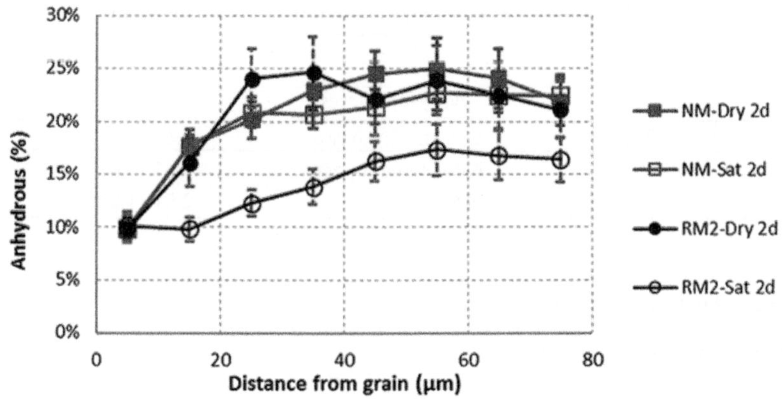

Figure 9.3 Anhydrous distributions in ITZ at 2 days.

FRA (RM-Sat) have a different anhydrous distribution characteristic of a cement paste with a higher W/C ratio. Similar results have been observed by Nguyen et al. (2014) for the oversaturated limestone aggregates. For this mortar, it was not possible to reach the targeted W/C ratio, and we probably underestimated the quantity of water given to the cement paste by the oversaturated FRA: a part of the absorbed water could migrate in the paste during this early period. Bello et al. (2017) showed that a pre-saturated expanded clay aggregate with an initial dry surface state (water content nearly 30%) can release a quantity of water corresponding to 5% of its dry mass in the fresh cement paste during the first 5 min of mixing. The mesoporosity distributions at 2 days are given in Figure 9.4. FRA mortars based on RA are denser in the first 10 µm because of the film duplex of Portlandite. The distribution of the dry FRA mortar is characteristic of a more porous surrounded paste, which could be induced by Portlandite transportation of the new to the old paste. The higher porosity of interphase of mortars based on oversaturated FRA could be the combination of higher W/C ratio and Portlandite transportation.

Micro-hardness values of new cement pastes are lower than those of carbonated old pastes (Table 9.2). There is no significant influence of the initial saturation degree of the FRA on the micro-hardness of the new cement paste. The main contrast stays, however, between the rigid original aggregate and the two more deformable cement pastes.

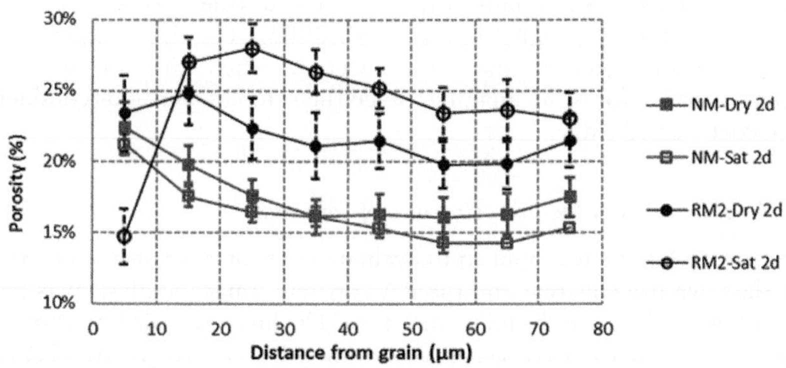

Figure 9.4 Porosity distributions in ITZ at 2 days.

Table 9.2 Results of nano-indentation tests of mortars based on the recycled sand

		H_M (MPa)	
Mortars	*Analyzed phases*	*Average*	*SD*
RM-Dry 28 days (RM-Sat 28 days)	Old cement paste	0.67 (0.64)	0.13 (0.3)
	New cement paste	0.41 (0.43)	0.18 (0.15)
	NA	2.51 (1.82)	1.13 (0.53)

9.3.2 Concrete microstructure

9.3.2.1 *Materials and method*

9.3.2.1.1 Mix compositions

A total of nine mixes have been studied:

- Three reference concretes with NA corresponding to three strength classes (C25/30, C35/45, and C45/55 according to EN 206 specifications), named NAC.
- Three concretes for which 30% of FNA were replaced by FRA and 30% of coarse NA (CNA) were replaced by coarse RA (CRA), named 30S30G.
- Three concretes for which 100% of CNA were substituted by CRA, named 0S100G.
- CEM II/A-L 42.5 Portland cement and limestone filler were used as binder and addition, respectively. Crushed limestone aggregates were used as the NA and RA sourced from construction and demolition wastes. Properties of aggregates and mix designs of concretes are reported in the appendix: mix design of recycled concrete. Due to the higher water absorption capacity of RA by comparison with NA, water pre-saturation of the RA is adopted before adding it to the mix. In this study, the RA were oversaturated at 1%, which means that the water content of RA will be equal to the water absorption plus 1%.

9.3.2.1.2 Specimens' casting and curing

For each composition, cylindrical specimens of 11 × 22 cm were cast in steel molds and compacted using a vibrating table. All specimens were unmolded after 24 h from casting and immersed in water. After 1 year of cure, small samples of size 35 mm × 35 mm × 10 mm were taken from these specimens. These samples were dried under vacuum with silica gel at 45°C for 14 days to remove the free water; drying at this temperature did not appear to cause cracking (Baroghel-Bouny et al. 2002). The dry samples were impregnated with epoxy resin, and then, the samples were polished with various steps to create a smooth, plane surface for SEM imaging. Since the samples were not conductive, a very thin metallic coat was added.

9.3.2.1.3 SEM observations of the microstructure of concretes

The BSE imaging technique was applied to evaluate the properties of the microstructure of the ITZ for the reference concrete and the RA concrete. Image analysis was performed to quantify the different phases of the microstructure. The image acquisition protocol was carried out near the aggregates; it consisted of obtaining BSE for the entire ITZ surrounding the selected aggregate. At the beginning, low magnification images were captured in order to locate the ITZ, and then, images of high magnification were taken to quantify porosity and anhydrous cement profiles. These were evaluated on strips of 10 µm wide starting at the aggregate surface (or old paste for RA) and up to 100 µm away from the aggregate. Strip segmentation and thresholding were performed by Olympus Stream software in order to quantify the different phases of the microstructure. The percentage area corresponding to porosity or unhydrated cement is related to the strip. In this study, the porosity appears red and the anhydrous are blue as illustrated in Figure 9.5. It took approximately ten images to cover the ITZ around aggregate on three samples of each type of concrete. To carry out a comparative study, the microstructure study of ITZ was done on the coarse fraction around 10–16 mm.

9.3.2.2 Results

9.3.2.2.1 ITZ of reference concretes

Figure 9.5 shows the microstructure of the ITZ in reference concretes (C25-0S0G, C35-0S0G, and C45-0S0G). The porosity and anhydrous profiles of reference concretes are given in Figure 9.6. It is well known that the microstructure properties of cementitious paste depend on the mix parameters such as the *W/C* ratio; the increase in the latter results

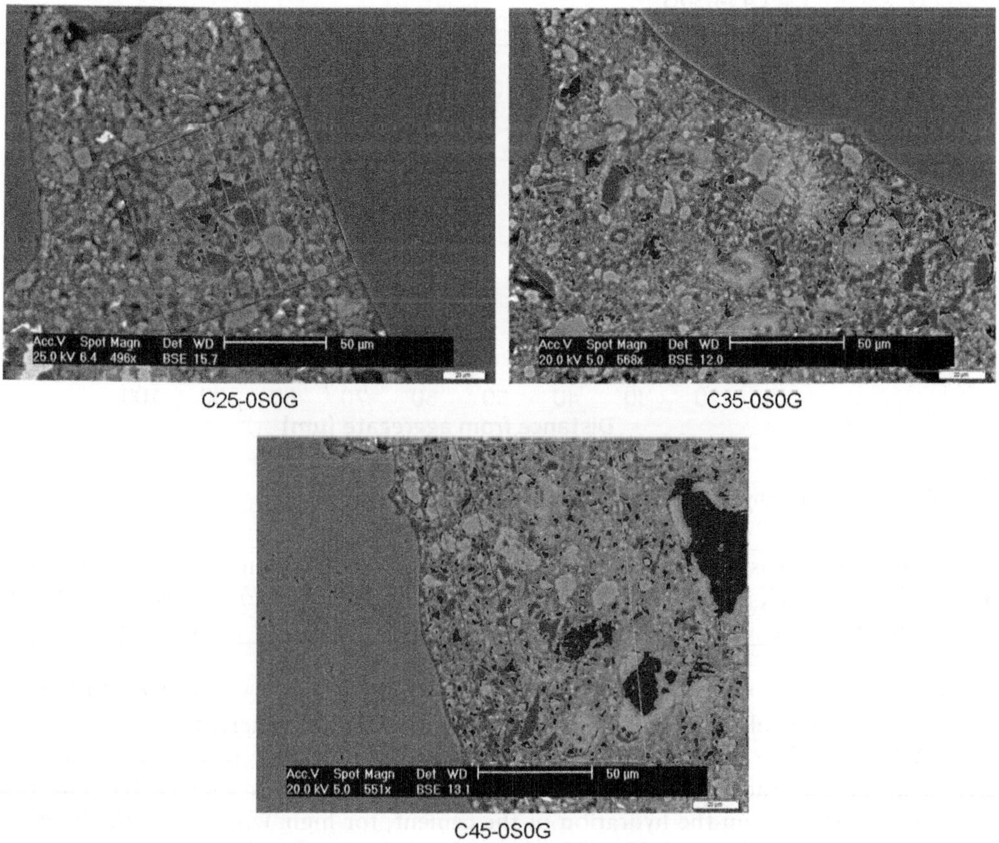

Figure 9.5 Image analysis of reference concretes.

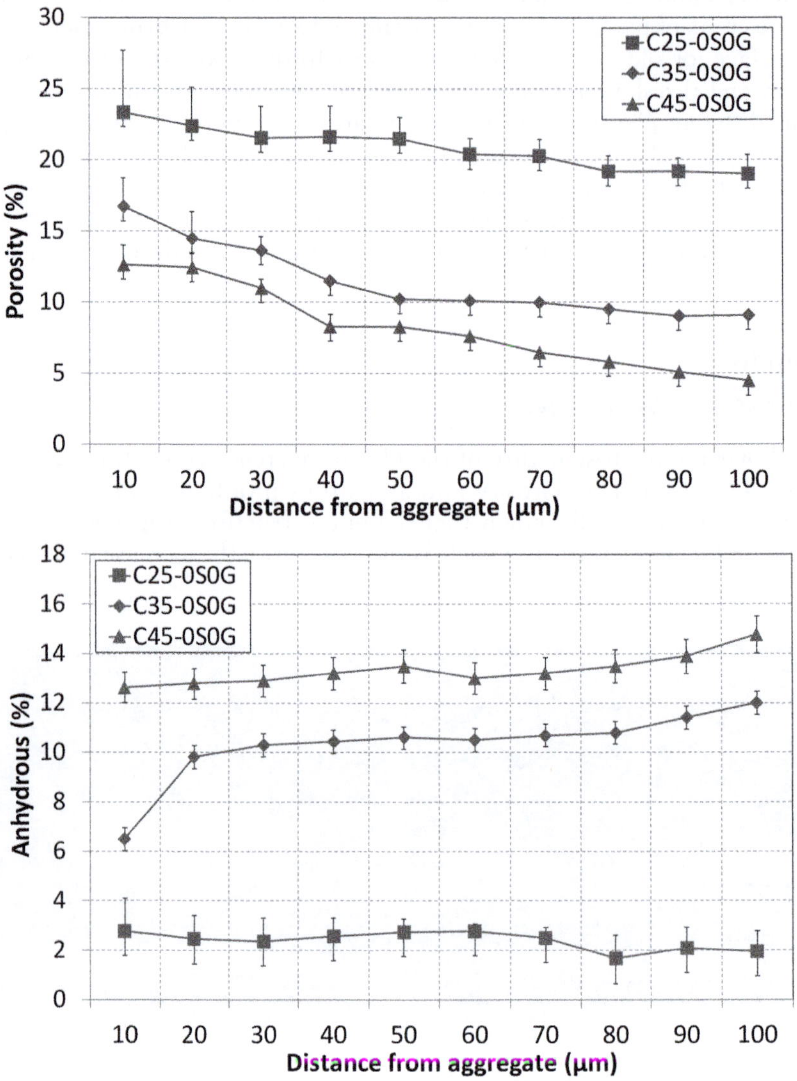

Figure 9.6 Porosity and anhydrous profiles of reference concretes.

in an increase in porosity. The porosity profile of C25-0S0G is higher than that of C35-0S0G and C45-0S0G, and it shows the same trend for the three concretes: an increase in the porosity at the aggregate interface and a reduction in porosity with distance from the aggregate.

This is explained by the wall effect (Scrivener 1999; Bentur and Alexander 2000), which generates more void volume at the interface of the aggregate and thereafter a local increase in the *W/C* ratio. The ratio of the interface porosity to the bulk paste increases with a decrease in the *W/C* ratio and can reach a factor of about 3 for the C45-0S0G. The anhydrous content depends on the hydration of the cement, for high *W/C* ratios; the hydration of C25-0S0G is well advanced. The C25-0S0G has a lower anhydrous content compared to the other concretes, that is, 2.8% to the aggregate interface, whereas the C35-0S0G and C45-0S0G have a anhydrous content of 6.7% and 12.7%, respectively.

9.3.2.2.2 ITZ of C25 RA concretes

Despite the age of one year, it is possible to distinguish the old and the new cement paste. The new paste is slightly porous and has anhydrous content, whereas the old paste appears more compact and cracked (Figure 9.7)—these cracks are probably generated by the crushing process. A greater porosity and micro-cracks are observed in the ITZ between the new and the old paste.

The profiles shown in Figure 9.8 are obtained from the image analysis of zooms evaluated in the Figure 9.7. The concretes studied have different *W/C* ratios: 0.67 for C25-0F100C, 0.63 for C25-30S30G, and 0.66 for C25-0S100G. Despite the reduction in the *W/C* ratio, the porosity until 100 µm from the RA/paste interface of the RA concretes is higher than the reference concretes (Figure 9.8), and the slope of porosity variation is higher for the RA concretes than for the NA concretes. The ratio of the interface porosity to the bulk paste is 2.3 for C25–30S30G and 1.6 for C25-0S100G, whereas it is 1.2 for C25-0S0G. This can be probably explained by the displacement of water from RA (the old mortar paste) to new mortar paste. Observation of porosity and anhydrous profiles can give the following interpretation:

- The RA can diffuse water, which then affects the local *W/C* ratio at the interface of the old paste, which will result in more porosity and less anhydrous content up to a distance of 60 µm from the aggregate (for C25-30S30G) compared to the reference concrete.

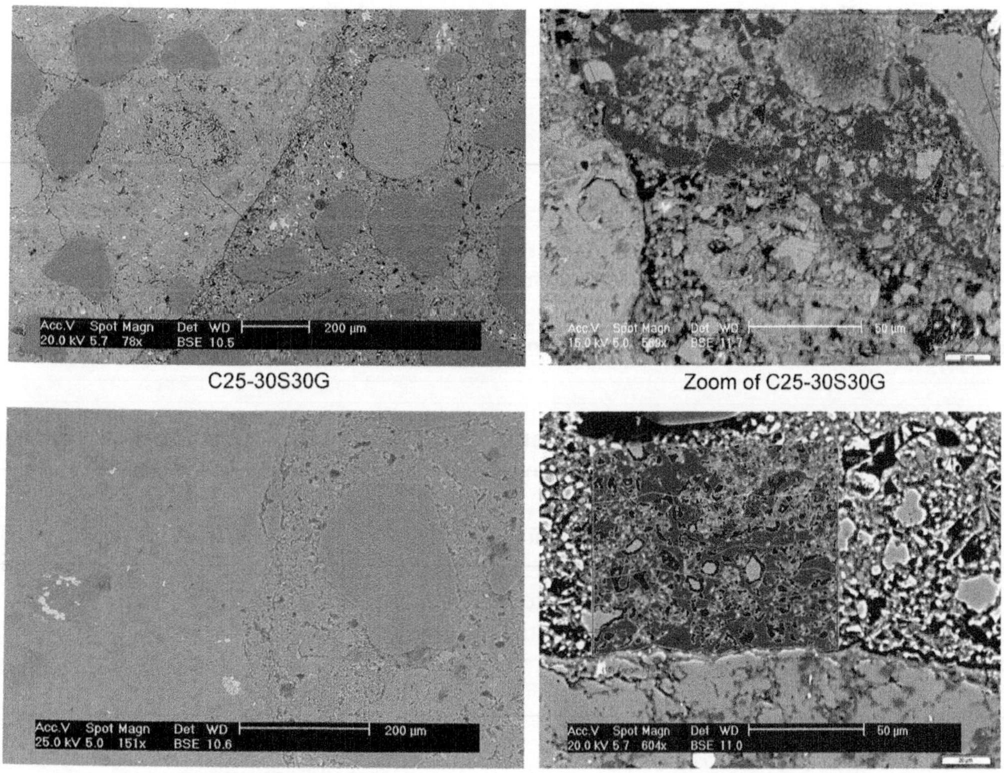

Figure 9.7 Image analysis of RA concretes, C25.

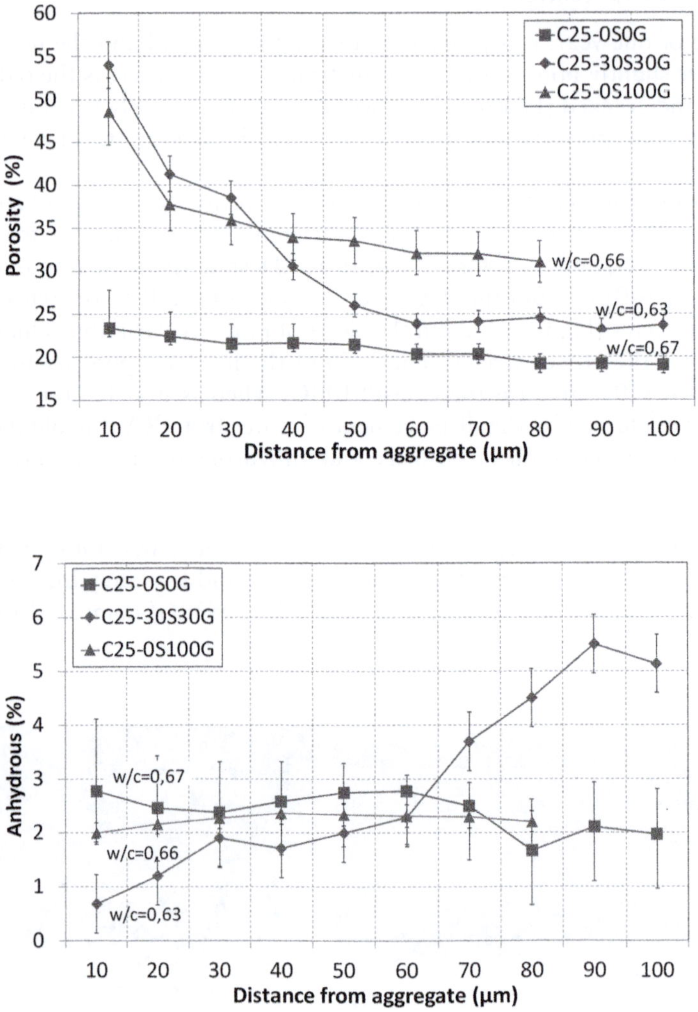

Figure 9.8 Porosity and anhydrous profiles of RA concretes, C25.

- In this study, the reduction in *W/C* ratio for the RA concretes does not decrease the porosity up to a distance of 100 µm from the interface of RA.

It is observed that the porosities of the concretes C25-30S30G and C25-0S100G are relatively close for a distance of 35 µm from the interface of the RA (the old paste). Beyond this distance, the porosity stabilizes for concrete C25-30S30G, whereas the porosity stabilizes beyond 60 µm for concrete C25-0S100G: these distances can represent the ITZ of these concretes.

9.3.2.2.3 ITZ of C35 RA concretes

The profiles shown in Figure 9.9 are obtained from the image analysis of zooms evaluated in Figure 9.10. An increase in porosity and a reduction in anhydrous content compared to

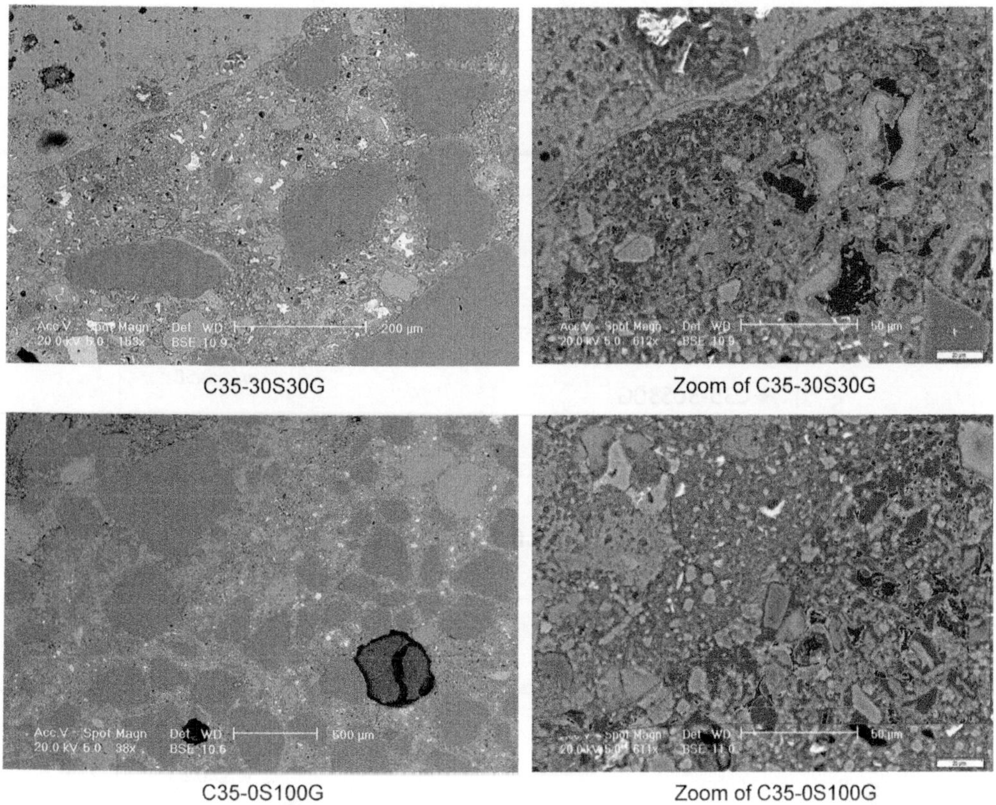

Figure 9.9 Porosity and anhydrous profiles of RA concretes, C35.

the reference concretes are observed, despite the reduction in the *W/C* ratio: 0.58 for C35-0S0G, 0.53 for C35-30S30G, and 0.54 for C35-0S100G. This result can be explained by the displacement of water from the RA, which creates more pores and less anhydrous content.

The comparison between the concretes shows that the porosity and the anhydrous content depend on the *W/C* ratio. The porosity of concrete C35-0S100G is higher than that of C35-30S30G0, and the anhydrous ratio of concrete C35-0S100G is lower than that of concrete C35-30S30G. The porosity for concrete C35-0S100G stabilizes beyond 60 μm from the interface of the old paste, whereas it continues to decrease for concrete C35-30S30G. A slight decrease in the anhydrous rate of RA concrete is observed beyond 60 μm.

9.3.2.2.4 ITZ of C45 RA concretes

The profiles shown in Figure 9.11 are obtained from the image analysis of zooms evaluated in Figure 9.12. The *W/C* ratio is close to 0.43 for C45-0S0G and C45-0S100G and 0.41 for C45-30S30G.

The porosity profile of C45-0S100G is higher than that of reference but with a lower anhydrous content. The porosity profile of C45-30S30G is lower than that of reference and C45-0S100G. The porosity of the RA concrete stabilizes beyond 60 μm (Figure 9.11).

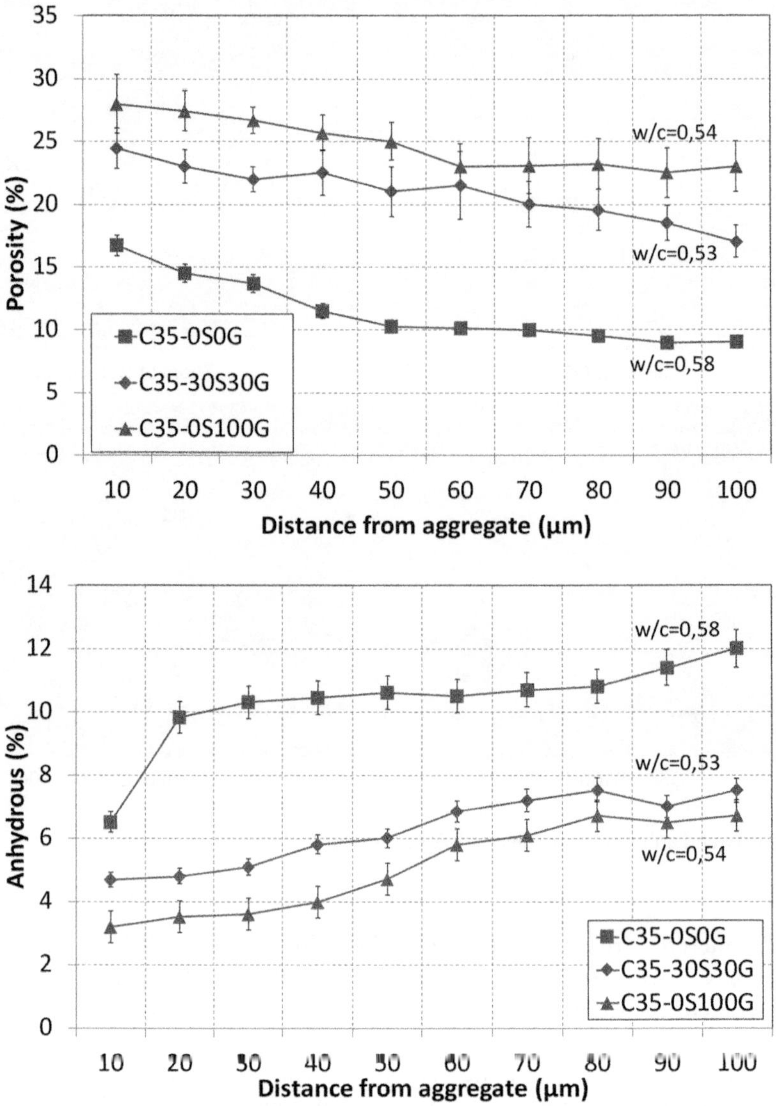

Figure 9.10 Image analysis of RA concretes, C35.

9.3.2.2.5 Comparative analysis of RA concretes

The effect of the *W/C* ratio on the porosity and anhydrous profiles of the RA concretes is shown in Figure 9.13. The reduction in the *W/C* ratio improves the porosity profile; the ratio of the interface porosity to the bulk paste is higher for the *W/C* ratios 0.43 and 0.41, which is of 2.55 and 2.7, respectively. The porosity profile stabilizes at about 50–60 µm. The anhydrous ratio depends on the *W/C* ratio of the new paste and the water release from the RA. The profiles of the anhydrous increase with a decrease in the concrete strength classes following a variable trend. The profile can be stable for a distance of 100 µm from the interface of RA for the concrete ratios 0.66 and 0.41, whereas it shows a sudden variation for the concrete ratios 0.43 and 0.63 for a distance of 40 and 60 µm from the interface of RA.

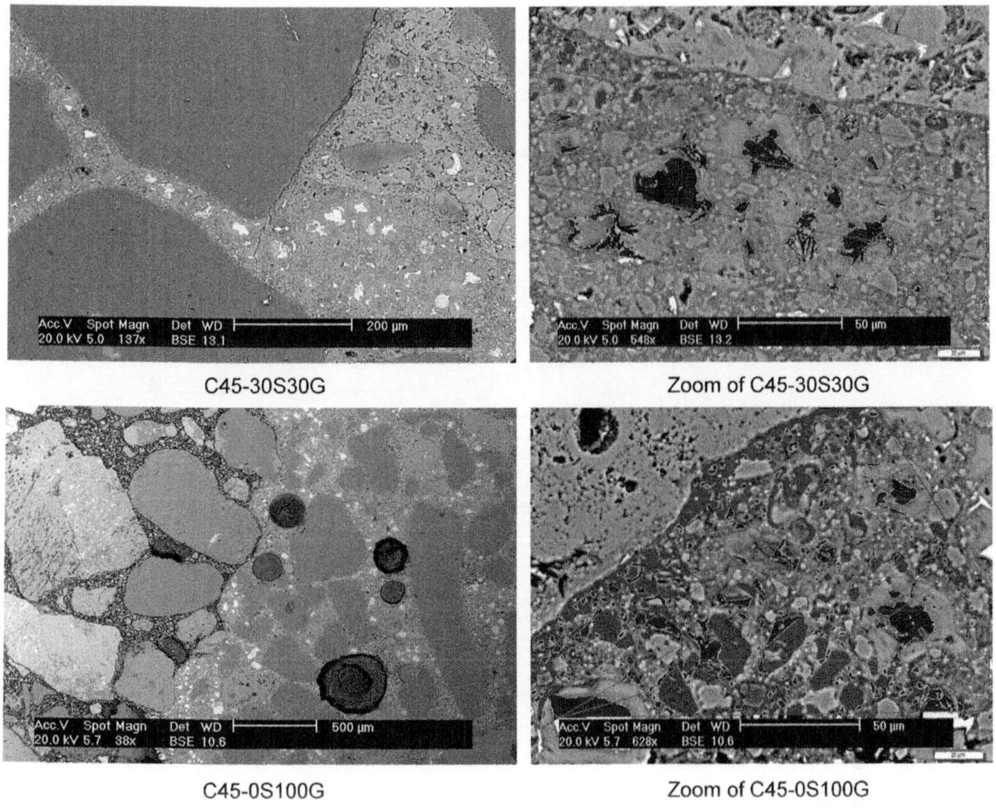

Figure 9.11 Porosity and anhydrous profiles of C45 RA concretes.

Figure 9.14 summarizes the gas permeability of the concretes (see Chapter 12) versus the porosity in the paste at a distance of 100 µm of the coarse aggregates. For C25 and C35 concretes, the substitution of FNA and CNA by FRA and CRA induces a significant increase in permeability and porosity, respectively. The substitution of 100% coarse aggregates is more disadvantageous than the substitution of 30% of fine and coarse aggregates for the transfer properties. As previously indicated, the higher porosities of the recycled concretes not only coud result in a transfer of a part of the absorbed water of the RA to the cement paste during the mixing or the casting and before the setting (concrete at the fresh state) but also could result in a transfer of Portlandite from the new to the old cement paste during the hardening. These phenomena would concern fine and coarse aggregates. For coarse aggregates, mainly another phenomenon could be the cause of an increase in porosity by the blocking during the bleeding of a part of mobile water, which could form a water lens at the bottom of the aggregate. This microbleeding effect has been observed by Crumbie (1994). For coarse RA, this effect could be increased because of the especially angular form of these particles. Figure 9.14 allows us to define a threshold value for porosity (close to 17%–18%), which separates porous networks weakly interconnected (low increase in permeability with porosity) from porous networks very interconnected (high increase in permeability with porosity). Benz and Garbockzi (1991) found a similar value for the porosity threshold by cement paste microstructure modeling. The substitution of NAs by RA for C25 class and to a lesser extent for C35 class induces the formation of a very interconnected porous network.

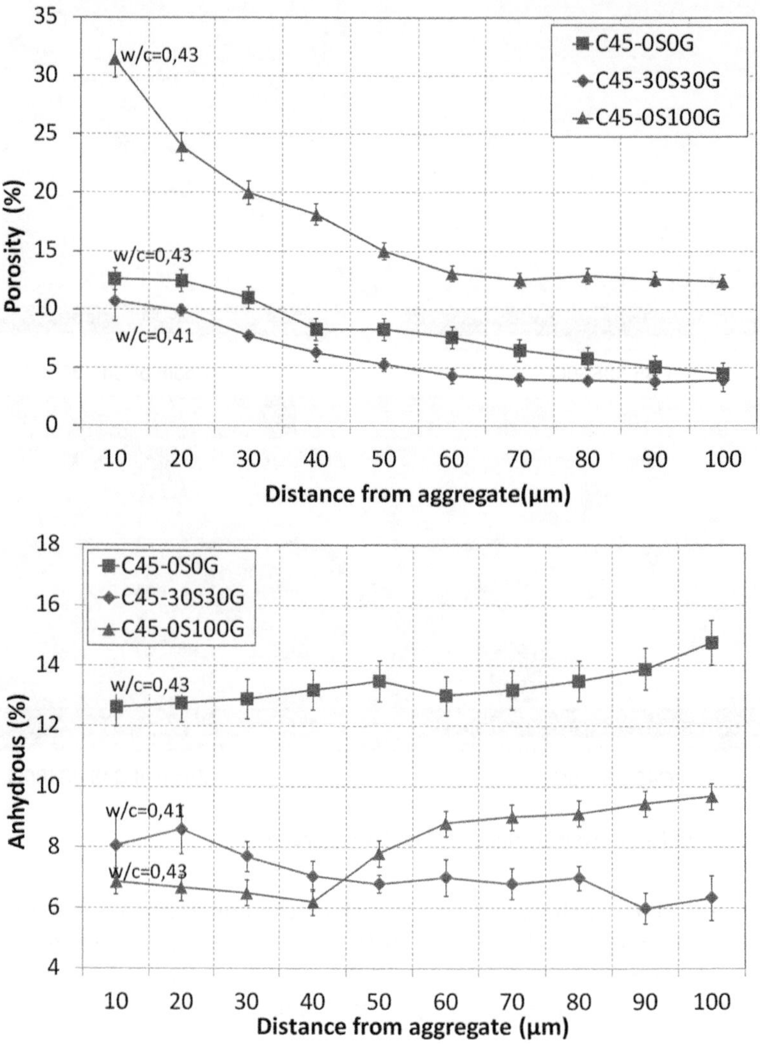

Figure 9.12 Image analysis of C45 RA concretes.

9.4 RESEARCH NEEDS

It could be interesting to evaluate the impact of the use of partially saturated coarse and fine aggregates on the quality of the ITZ of recycled concrete and by consequence on the transfer properties of these concretes. Further investigations are required to try to quantify the relative impact of water transfer and the microbleeding effect on the microstructure of the ITZ and transfer properties of recycled concretes especially for coarse aggregate recycling.

9.5 CONCLUSION

The use of oversaturated FRA and CRA generates higher porosity and lower anhydrous profiles in ITZ between the old and new cement paste in mortar and concrete of normal classes

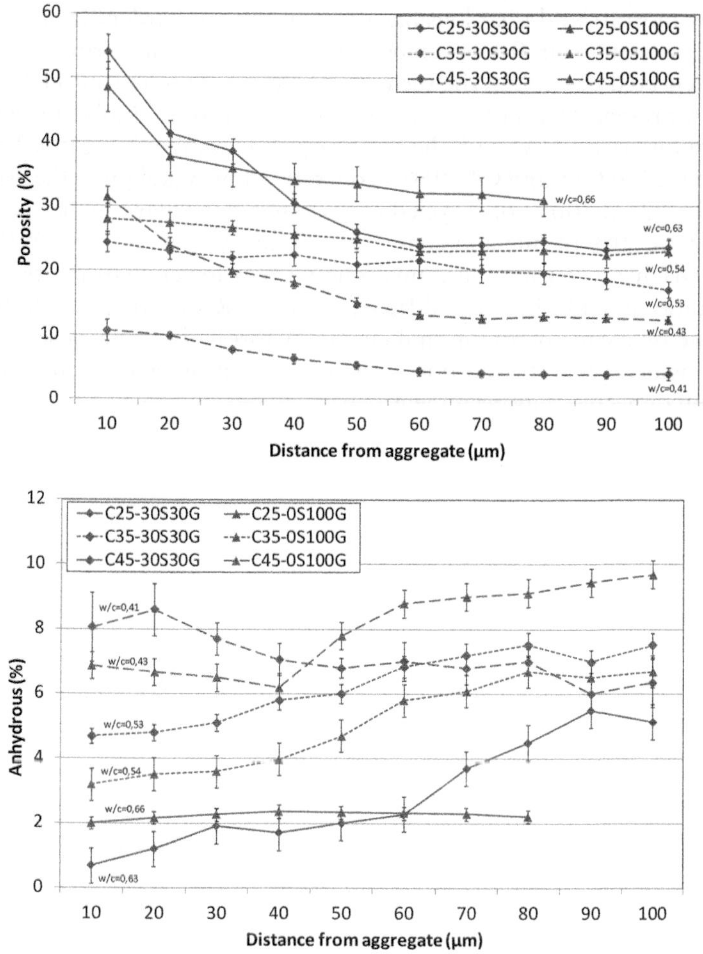

Figure 9.13 Effect of the *W/C* ratio on porosity and anhydrous profiles of RA concretes.

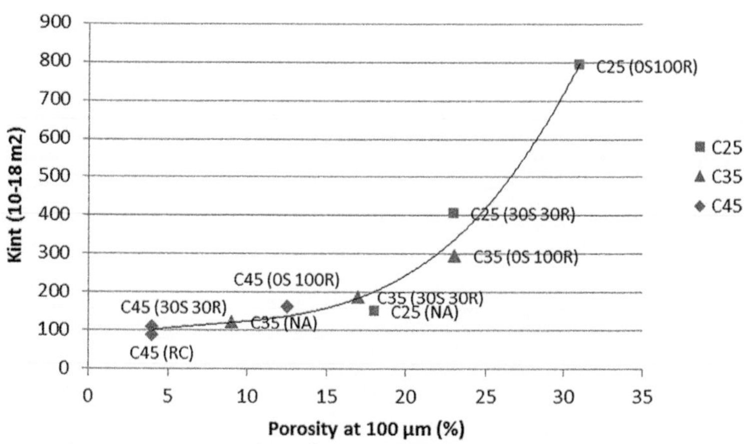

Figure 9.14 Gas permeability versus porosity at 100 µm of the coarse aggregate.

of strength (C25 and C35). The formation of this "porous" ITZ is identified from 2 days of hardening on mortars and is still visible after one year of hardening on concretes. Water release at the fresh state from RA to cement paste and water lens formation by microbleeding effect are expected to explain the formation of these more porous ITZ. Mortars made with dried FRA present "denser" ITZ with anhydrous profile similar to that observed on mortars made with dried FNAs with the same targeted average *W/C* ratio. The reduction in *W/C* ratio from C25 to C35 and C45 0F0G concretes improves slightly the porosity profiles. But concretes based on 100% of recycled coarse aggregates have higher porous ITZ than concretes based on 30% of fine and coarse aggregates. Concrete C45-30S30G with a low *W/C* ratio of 0.41 has porosity and anhydrous profiles lower than the reference concrete with the *W/C* ratio of 0.43: the water diffused from the RA participates in the hydration of the cement paste by means of an internal curing. A threshold value close to 17%–18% for paste porosity, which separates porous networks weakly interconnected and porous networks very interconnected, has been identified.

Chapter 10

Instantaneous mechanical properties

E. Ghorbel
Université de Cergy Pontoise

T. Sedran
Ifsttar

G. Wardeh
Université de Cergy Pontoise

CONTENTS

Abstract

This chapter describes the influence of the introduction of recycled aggregates (RA) on the instantaneous properties of concrete. In the first part, existing models linking compressive and tensile strength and elastic modulus of natural aggregates concretes (NAC) to their mix design are validated and adapted for recycled aggregates concrete (RAC). These models make it possible to consider properly and on a case-by-case basis the respective quality of NA and RA in the mix design process of RAC. In the second part, more global and statistical models describe the effect of the introduction of RA on compressive and tensile strength, elastic modulus, and peak and ultimate. This section provides elements of reflection for an extension of the design codes to the case of RAC. It was thus pointed out that design codes relationships dedicated to assess the mechanical properties and the stress–strain compressive curve of NAC are not adequate to predict the behavior of RAC. It is established that the elastic modulus E_{cm}, the tensile splitting strength $f_{ctm,sp}$, the peak strain ε_{c1} and the ultimate strain ε_{cu1} are related to the mean compressive strength f_{cm} and to a parameter taking into account the effect of the recycled aggregates replacement rate. Furthermore, an analytical stress–strain expression is proposed for concretes incorporating recycled aggregates.

10.1 NOTATION

For results analysis, the following ratio is defined:

$$\mathrm{VF}_i = \frac{V_i}{\sum_j V_j} \tag{10.1}$$

where VF_i is the volume fraction of the aggregate i and V_j the respective volume content for the different aggregate fractions in the concrete mix. Generally, aggregates are separated in fine, small, and coarse aggregate (CA) fractions, recycled or natural.

$$\Gamma_\mathrm{v} = \frac{\sum V_\mathrm{RA}}{\sum V_\mathrm{(NA+RA)}} \tag{10.2}$$

where Γ_v is the volume recycling coefficient describing the ratio between the volume of recycled aggregate (RA) (fine aggregates (FAs) and coarse aggregates (CAs)) and the overall volume of aggregate in a concrete mix.

$$\Gamma_\mathrm{m} = \frac{\sum M_\mathrm{RA}}{\sum M_\mathrm{(NA+RA)}} \tag{10.3}$$

where Γ_m is the mass recycling coefficient describing the ratio between the mass of RA (FAs and CAs) and the overall mass of aggregate in a concrete mix.

10.2 MECHANICAL PROPERTIES VERSUS MIX DESIGN

Because of their composite nature (see Chapter 3), RAs may display different (and often but not always lower) mechanical properties compared to natural aggregates (NAs). This influences the mechanical properties of concrete when NA is replaced by RA. This section proposes models predicting the mechanical properties of concrete from its mix design. This approach allows accounting quantitatively for the respective quality and dosage of NA and RA through a fine characterization of RA.

10.2.1 Compressive strength

10.2.1.1 State of the art

Numerous papers are available, describing the influence of the introduction of recycled concrete aggregates on concrete compressive strength. Yet, it is actually difficult to raise clear conclusions, for different reasons:

- Concrete made with RA is always compared to reference concrete made with NA, but the choice of the NA itself may affect the influence of RA. In other words, the same RA will have a different impact if the NA it replaces has excellent or medium mechanical properties.

- Different strategies are adopted to compare concrete with varying recycling rates: by considering the same total water-to-cement (W_{total}/C) ratio or the same free water-to-cement (W_{eff}/C) ratio, or else keeping the workability constant which lead to various W_{eff}/C ratios.
- Finally, the influence of RA, like for NA, depends on its shape, size, mechanical properties, etc.

An important review of 236 papers was recently published dealing with the factors affecting the different properties of RA sourced from construction and demolition waste, intended for concrete production (Silva et al. 2014b). Among those papers and more dedicated to compressive strength, a further analysis of 119 papers is detailed in the study of Silva et al. (2014a). Based on a statistical study, the authors first developed a classification for RA based on water absorption and oven-dried density (see Figure 10.1). Recycled concrete CAs are mainly classified as A and B, while recycled concrete FAs are mainly classified as B and C (Silva et al. 2014b) because of a higher remaining cement paste content (see Chapter 3). Then, they established a monograph describing the statistical influence of RA on compressive strength, for different recycling ratios and aggregate quality index (see Figure 10.2).

This graph is interesting for two reasons: first, it is useful to approximately evaluate the expected effect of RA on concrete compressive strength, when keeping the W_{eff}/C ratio constant; and secondly, it illustrates some well-known general trends:

- The increase in RA ratio generally leads to a decrease in compressive strength, at the same W_{eff}/C. The decrease may be amplified if concretes with the same cement content and workability are compared, as the introduction of RA may also lead to a higher water demand.
- The poorer mechanical quality of RA compared to NA mainly explains the decrease. In fact, NAs are produced from rocks with compressive strength generally higher than 50 MPa, whereas RAs are generally crushed from concrete with strength lower than 50 MPa. For a given recycling level, the decrease in compressive strength is lesser when RAs are sourced from a concrete with a higher compressive strength. For example, RA sourced from Ultra-High Performance Concretes (UHPC) can be recycled in UHPC without significant change compared to NA (Sedran et al. 2010).

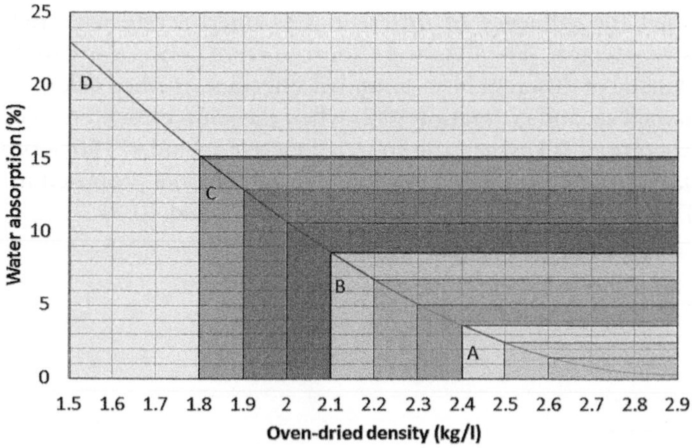

Figure 10.1 Mechanical classification of aggregates sourced from construction and demolition waste (concrete, masonry, or mixed). (Adapted from Silva et al. 2014b.)

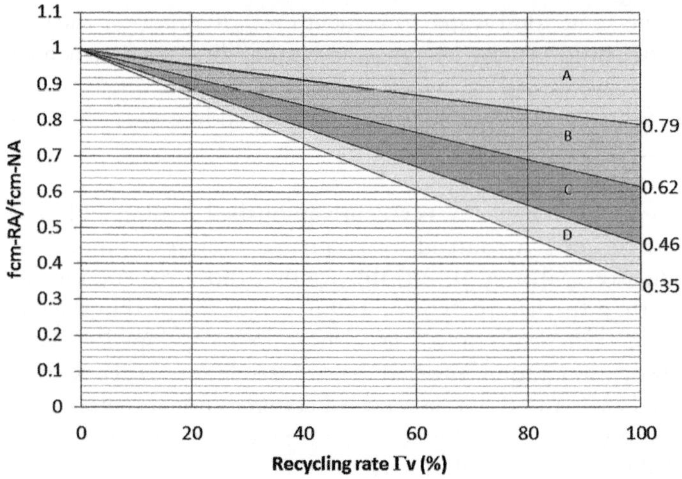

Figure 10.2 Statistical influence of the recycling on concrete with the same W_{eff}/C ratio (lower limit of the 95% confidence interval), $f_{cm,NA}$ mean compressive strength of reference concrete including only NAs, and $f_{cm,RA}$ mean compressive strength including different ratios of RAs. (Adapted from Silva et al. 2014a.)

- The recycled FAs are generally of poorer mechanical quality compared to recycled CAs, as explained above. So, they are expected to reduce the compressive strength more for the same volume of replacement.

Even if this approach is an important step to illustrate the global tendency for the role of RA on compressive strength, it suffers from two main drawbacks: first, it gives only a statistical rough evaluation of compressive strength, because concretes used to build Figure 10.2 are made of various NA and RA. Secondly, it is not a convenient tool to design a concrete with a given compressive strength.

10.2.1.2 RECYBÉTON outputs

Dao (2012) studied the influence of different RAs on concrete with recycled aggregate (RAC) compressive strength. The RAs were sourced from concretes with various compressive strength and made of different NAs (rounded siliceous, crushed calcareous, and mixed), The RAs were named second generation (G2) when the source concrete was made of 100% NA and third generation (G3) when the source concrete included 100% of G2 (Dao 2012; Dao et al. 2014). Dao used a modified version of Féret's model proposed in the study of de Larrard (1999a) to analyze his results (see equations 10.4 and 10.5)

$$f_{cm} = k_g \times Rc_{28}^{ce} \times \left(\frac{v_c}{v_c + v_e + 0.5 \cdot v_a} \right)^2 \times MPT^{-0.13} \qquad (10.4)$$

with

$$MPT = D_{max} 3\sqrt{\frac{g^*}{g} - 1} \qquad (10.5)$$

where

- k_g is a mechanical parameter accounting for the influence of aggregate, depending on its own strength and bonding quality with the cement paste.
- Rc_{28}^{ce} is the compressive strength class of cement at 28 days according to NF EN 197-1.
- v_c, v_e, and v_a are, respectively, the cement, water, and air volumes contained in the mixture.
- MPT is the maximum paste thickness (de Larrard 1999a).
- D_{max} is the diameter corresponding to 90% of passing of the granular skeleton.
- $g = \sum_j V_j$ is the volume fraction of the granular skeleton in the mixture.
- g^* is the packing density of the granular skeleton. It can be calculated by the way of the compressible packing model (de Larrard 1999a).

For analysis purposes, it is assumed that the factor k_g of the overall aggregate skeleton is calculated according to the following equation, where $k_{g,j}$ is the mechanical parameter for the aggregate fraction j:

$$k_g = \sum_j VF_j k_{g,j} \tag{10.6}$$

Around 100 concretes and mortars with different compressive strength were produced with 100% of RA, within this study. Knowing the composition of the concretes, compressive strength was measured on Ø16 × 32 cm cylinders, and using the previous equations, Dao calculated the $k_{g,s}$ and $k_{g,g}$ terms corresponding, respectively, to the FA and CA fractions for the different RA. He concluded that for RA, $k_{g,s}$ could be approximated by a constant value of 4.42 and $k_{g,g}$ deduced from the attrition coefficient MDE (standard NF EN 1097-1) according to the following equation:

$$k_{g,g} = -0.0952\text{MDE} + 8.3927 \tag{10.7}$$

Doing so, he was able to predict the experimental values of compressive strength with a mean error of 5 MPa for RAC made of 100% of recycled (G2) or multi-recycled (G3) concrete aggregates.

As a next step, the same approach was used in RECYBÉTON project to analyze different mixes with various sand and CA recycling rates (see Table 10.1), with the components described in Appendix. Equations 10.4–10.6 were then used to calibrate the $k_{g,j}$ terms of the different fractions (see Table 10.2) to have a good fitting between experimental and predicted compressive strength, which is shown in Figure 10.3. It is noteworthy that equation 10.7 seems to overestimate the $k_{g,j}$ of coarse RA which displays a MDE of 23 leading to $k_{g,j} = 6.2$.

The model is interesting because it separates the influence of aggregate from the other parameters (cement class and efficient water-to-cement ratio). Moreover, by fitting the $k_{g,i}$ terms for NA and RA fractions individually, it allows an intrinsic characterization of RA which does not depend on the reference concrete to which RAC is compared to. In the present case, the coarse RA is of lower quality compared to the NA CA as its $k_{g,j}$ is lower. In the same way, Table 10.2 confirms a general trend: for the same volume of replacement, the negative effect of fine RAs on compressive strength is stronger than that of coarse RA.

Table 10.1 RECYBÉTON concrete mixtures (Sedran 2017)

Mix (kg/m³)	*1*	*2*	*3*	*4*	*5*	*6*	*7*	*8*	*9*	*10*	*11*
Added water	190	186	185	313	305	297	245	236	236	227	220
Efficient water	180	176.2	175.6	206	199.9	194.1	189.2	181.9	183.3	183	178
Cement CEMII/A-L 42,5 N	270	309	351	307	352	403	285	310	353	270	363
Limestone filler	45	53	61	47	51	54	32	54	53	45	38
Natural sand 0/4	780	768	751	—	—	—	803	794	778	531	515
Natural gravel 4/10	266	262	257	—	—	—	—	—	—	171	165
Natural gravel 6.3/20	820	807	789	—	—	—	—	—	—	518	502
Recycled sand 0/4	—	—	—	823	810	798	—	—	—	228	221
Recycled gravel 4/10	—	—	—	151	149	146	163	161	157	144	140
Recycled gravel 10/20	—	—	—	447	440	433	700	692	678	151	146
Retarder	—	—	—	—	—	—	—	—	—	—	—
Superplasticizer	1.35	2	2.3	1	1.23	2.1	1.5	2	2.29	1	1.6
Nat. sand (%)[a]	43	43	43	0	0	0	45	45	45	29.5	29.5
Nat. coarse agg (%)[a]	57	57	57	0	0	0	0	0	0	36.3	36.3
Recy. sand (%)[a]	0	0	0	60	60	60	0	0	0	15.7	15.7
Recy. coarse agg (%)[a]	0	0	0	40	40	40	55	55	55	18.5	18.5
Γ_m (%)	0	0	0	100	100	100	51,8	51,8	51,8	30	30
Γ_v (%)	0	0	0	100	100	100	55	55	55	33,2	33,2
R_c (on Ø16 × 32 cm cylinder) at 28 days	33.9	42	49.9	25.8	31.5	38.3	28.5	37.7	43	27.7	45.2
R_{tb} (on Ø16 × 32 cm cylinder) at 28 days	3.3	3.8	4.2	2.2	2.2	3.0	2.8	3.2	3.6	2.7	4.0

[a] Volume fraction of the overall granular skeleton, VF_i see equation 10.1.

Table 10.2 Fitted values of $k_{g,j}$ assuming $k_{g,j}$ is the same for all NA fractions

Aggregate	$k_{g,j}$
NA (sand, 4/10 and 6.3/20)	5.767
RA (sand)	4.42
RA (4/10 and 10/20)	5.173

Assuming that reference concrete with NA and concrete with 100% of RA have almost the same MPT as shown in Figure 10.2, it is possible to extract statistical ranges of $k_{g,RA}/k_{g,NA}$ ratio, where the first term stands for completely recycled skeleton and the second one for a natural skeleton. Thus, we obtain the following ranges for RA classes C, B, and A, respectively: [0.46; 0.62], [0.62; 0.79], and [0.79; 1].

Note that, in cases where NA is of poor quality and RA of good quality, the RA may have a positive influence on compressive strength for the same W_{eff}/C as shown in the study of Dao (2012) and Dao et al. (2014).

10.2.2 Tensile splitting strength

In the literature, the splitting tensile strength f_{ctm} of concrete with NAs is often deduced from f_{cm} using the power law (see Section 10.3.3). For example, equation 10.8 was proposed (de Larrard 1999a). The parameter k_t describes the role of the overall granular skeleton and

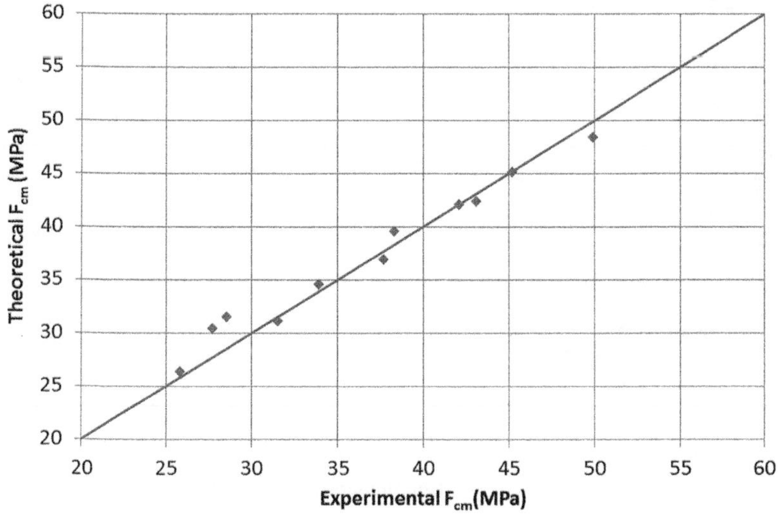

Figure 10.3 Comparison between experimental and theoretical values of compressive strength for mixes in Table 10.1, with $k_{g,j}$ in Table 10.2. Mean error = 1.07 MPa.

depends on the nature of this skeleton. For analysis purposes, it is assumed that this factor k_t is calculated according to following, where $k_{t,j}$ is the mechanical parameter for the aggregate fraction j.

$$F_{ctm,sp} = k_t \times F_{cm}^{0.57} \tag{10.8}$$

with

$$k_t = \sum_j VF_j k_{t,j} \tag{10.9}$$

The model was tested on the mixes presented in Table 10.1, containing various fine and coarse aggregate recycling ratios. Equations 10.8 and 10.9 were used to calibrate the $k_{t,j}$ terms of NA and RA (see Table 10.3) to fit the experimental tensile strength. The very good quality of the prediction is shown in Figure 10.4. A further validation is made in the study of Ajdukiewicz and Kliszczewicz (2007) (see Table 10.5 and Figure 10.5).

As for compressive strength, fitting the $k_{t,i}$ terms for NA and RA fractions individually allows an intrinsic characterization of RA, which does not depend on the reference concrete to which RAC are compared to. In the present case, the coarse RA is of lower quality compared to the NA CA as its $k_{t,j}$ is lower.

In the study of Silva et al. (2015b), the authors analyze 600 mixes from several papers and conclude that recycling rate has no effect on the relationship between f_{ctm} and f_{cm}.

Table 10.3 Fitted values of $k_{t,j}$ assuming that $k_{t,j}$ is a constant for all NA fractions and another one for the RA

Aggregate	$k_{t,j}$
NA (sand, 4/10 and 6.3/20)	0.453
RA (sand, 4/10 and 6.3/20)	0.364

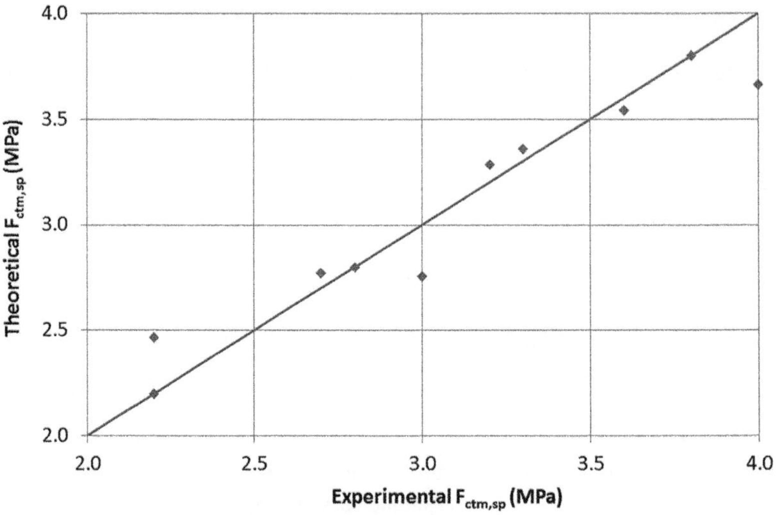

Figure 10.4 Comparison between experimental and calculated values of splitting tensile strength for mixes in Table 10.1, with $k_{t,j}$ in Table 10.3. Mean error = 0.12 MPa.

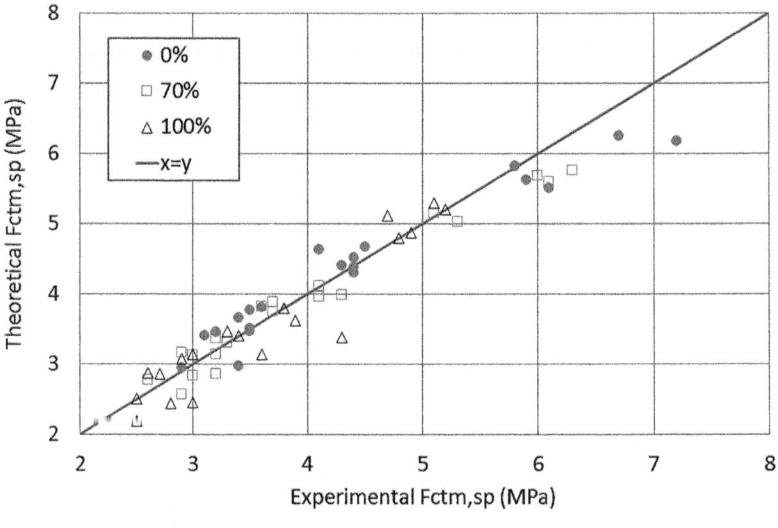

Figure 10.5 Comparison between experimental and calculated values of splitting tensile strength from Ajdukiewicz and Kliszczewicz (2007). Mean error = 0.23 MPa. In legend: Γ_v.

Khoshkenari et al. (2014) conclude in the same way when studying seven concretes at three different ages with 100% of recycled CA and up to 100% of recycled FA. The same conclusion is attained in the study of Sanchez de Juan and Aleajos (2004a) on 24 concretes with 0%, 20%, 50%, or 100% of recycled CA. As shown in Figure 10.6, other studies exhibit significant positive or negative influence of the tensile strength (i.e., on k_t). In fact, the influence of the RA depends on the relative quality of the NA which is compared to.

In many cases, RAs are used locally to avoid transportation cost, so they are obtained from concrete made with the local NA. So in Dao (2012) and Ajdukiewicz and Kliszczewicz (2007), the authors compare the influence on tensile strength of RA with that of NA they

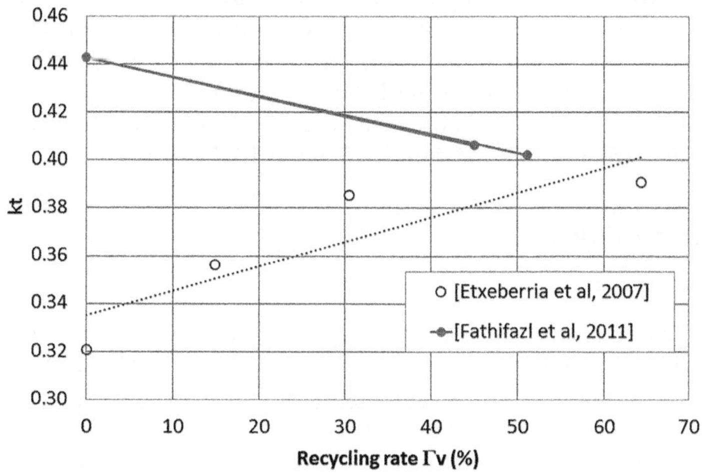

Figure 10.6 Values of overall k_t in concrete mixes with various recycling rate (only CAs are recycled). (From Etxeberria et al. 2007b and Fathifazl et al. 2011a.)

are made of. The k_t values calculated from these studies are summarized in Tables 10.4 and 10.5. The results show that, contrary to what we might have expected, RAs do not inherit k_t values close to that of the corresponding NA: the k_t values of RA are, respectively, 7% and 13% less than those of NA, and there is a poor correlation between them. The same conclusion can be observed from Figure 10.6, with data from the study of Fathifazl et al. 2011a, where RAs are also sourced from the NA they are compared to. Moreover, Dao was unable to relate the k_t values obtained for RA with other classical property of NA (LA, MDE…). The influence of the compressive strength of the concrete where the RAs are sourced from is also not clear (negative in the study of Dao (2012), positive in the study of Ajdukiewicz and Kliszczewicz (2007) for a given nature of NA).

Table 10.4 Traction coefficients from the study of Dao (2012), with RA1 RA sourced from a concrete with a compressive strength around 45 MPa and RA2 RA sourced from a concrete with a compressive strength around 75 MPa

Nature of NA	$k_{t\,NA}$	$k_{t\,RA1}$	$k_{t\,RA2}$	*Mean*, $k_{t\,RA}/k_{t\,NA}$
Siliceous FA and CAs	0.435	0.44	0.417	0.985
Calcareous FA and CAs	0.431	0.376	0.377	0.874
Sil. FA and calc. CAs	0.471	0.446	0.433	0.933
			Mean	0.931

Table 10.5 Traction coefficients from the study of Ajdukiewicz and Kliszczewicz (2007)

	k_t				f_{cm} *of source concrete (MPa)*	
Nature of NA	*NA*	*RA1*	*RA2*	*Mean*, $k_{t\,RA}/k_{t\,NA}$	*RA1*	*RA2*
Rounded quartzite	0.373	0.312	0.365	0.908	30	45
Crushed granite	0.423	0.371	0.389	0.898	60.5	73.1
Crushed basalt	0.445	0.345	0.382	0.817	72.4	110.1
			Mean	0.874		

It seems difficult to conclude on the real effect of recycling rate on the relationship between f_{cm} and $f_{ctm,sp}$, and the k_t values of RA and NA must be fitted on experimental data if precise mix design optimization is needed. Yet it can be noted that RAs generally exhibit a lower (up to 13%) value of k_t compared to the RA they are made of. Finally, a comprehensive examination of all available results more often displays a negative effect of recycling on the tensile–compressive strength relationship (see Section 10.3.3).

10.2.3 Elastic modulus

10.2.3.1 State of the art

RAs contain residual cementitious paste, which generally confers to them a lower elastic modulus compared to that of NA. A consequence is that RAC generally display a lower elastic modulus compared to concrete with natural aggregate (NAC), as presented in Section 10.3.2.

The tri-sphere model cited in the study of de Larrard (1999a) has proved to be efficient for the calculation of elastic modulus of concrete made with NA using the following equation:

$$E_{cm} = \left(1 + 2g \frac{E_g^2 - E_m^2}{\left(g^* - g\right)E_g^2 + 2\left(2 - g^*\right)E_g E_m + \left(g + g^*\right)E_m^2}\right)E_m \tag{10.10}$$

where

- $g = \sum_j V_j$, the volume of the granular skeleton in the mix.
- g^*, the packing density of the granular skeleton. It can be calculated using the compressible packing model (de Larrard 1999a).
- E_m, the elastic modulus of the cement paste. It is calculated from the concrete compressive strength: $E_m = 226R_c$.
- E_g, the elastic modulus of the granular skeleton. When different sources of aggregates are used in a concrete, the overall E_g is calculated with the following, where $E_{g,j}$ is the elastic modulus of the different fractions j:

$$E_g = \sum_j VF_j E_{g,j} \tag{10.11}$$

10.2.3.2 RECYBÉTON outputs

In the study described in Section 10.2.1.2, Dao (2012) used this model to calibrate the elastic modulus of recycled FAs and CAs from mortars and concrete made only with RA. The RAs were crushed from concrete with varying compressive strength and made with different aggregates (Dao 2012; Dao et al. 2014). The author found that the elastic modulus of recycled FAs and CAs was quite similar (less than 7% difference) and that the mean value $E_{g\,RA}$ can be calculated from the elastic modulus of the source concrete E_s and one of the source aggregates E_{gs} with the following equation:

$$E_{g\,RA} = 0.65E_s + 0.35E_{gs} \tag{10.12}$$

E_s and E_{gs} are generally not known precisely; nevertheless, local NA as well as the type of concrete used to produce RA is often identified. In that case, equation 10.12 can give a

Table 10.6 Elastic modulus of RECYBÉTON reference mixes. Compositions are given in Appendix

Mix	*C25/30-0S-0G*	*C25/30-30S-30G*	*C25/30-0S-100G*	*C25/30-100S-100G*	*C35/45-0S-0G*	*C35/45-30S-30G*	*C35/45-0S-100G*	*C35/45-100S-100G*
E_{cm} (GPa)	37	29.5	27	25	36	33	31.5	31
Γ_v (%)	0	34.4	55	100	0	34.4	55	100
Γ_m (%)	0	30	52	100	0	30	52	100

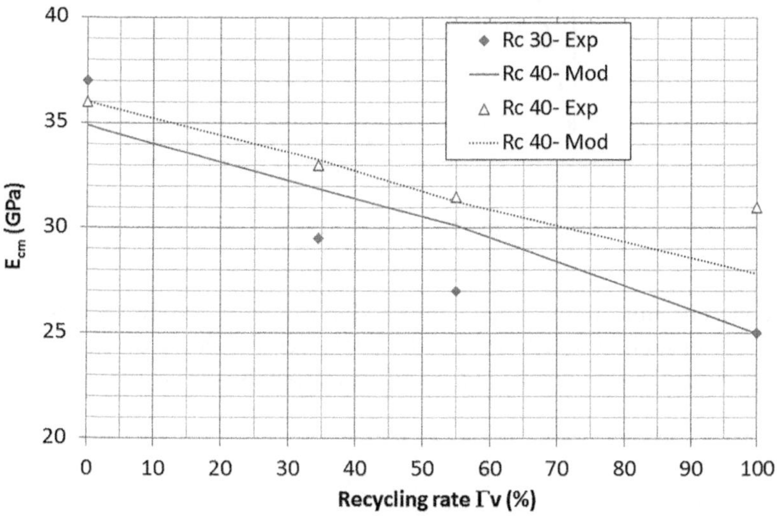

Figure 10.7 Comparison between experimental values of elastic modulus (see Table 10.6) and those calculated with equations 10.10 and 10.11. The elastic modulus of RA and NA is fitted to minimize the error. The optimization gives $E_g = 49.5$ GPa for RA, $E_g = 66.2$ for NA, and a mean error = 1.4 GPa.

rough estimation of $E_{g\,RA}$. Typical values of $E_{g\,RA}$ range between 35 and 60 GPa which are lower than those of NA and demonstrate that elastic modulus of RAC decreases when recycling rate increases. The other reason is that introduction of RAC may need to increase the paste volume in order to maintain the workability.

A further validation of the model was obtained on two concrete families with 30 and 40 MPa target compressive strength within RECYBÉTON project (Figure 10.9). For each family, four recycling rates were selected. The compositions of mixes are summarized in the Appendix, and the experimental results are given in Table 10.6. Figure 10.7 confirms that equations 10.10 and 10.11 can predict the evolution of the elastic modulus with the recycling rate with a mean error of 1.4 GPa, once the elastic modulus of NA and RA is calibrated.

10.3 MECHANICAL PROPERTIES VERSUS DESIGN CODES[1]

The analysis of reinforced concrete structures requires realistic stress–strain material models to reproduce the real behavior of the structure. It is commonly accepted that the monotonic

[1] The models proposed in this section do not commit the RECYBÉTON National Project, nor the French representatives of the Eurocode 2 commission, nor IFSTTAR in the positions they may take vis-à-vis the revision of Eurocode 2, to take into account the use of recycled concrete aggregates.

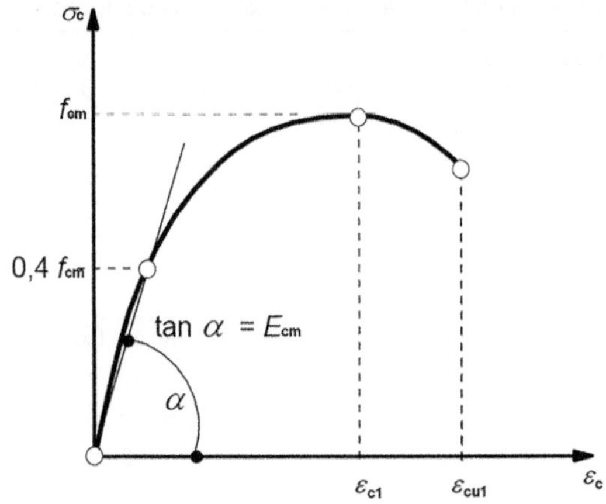

Figure 10.8 Stress–strain curve according to EC2.

behavior of concrete subjected to axial compression can be approximated by a stress–strain curve such as the given one by EN 1992-1-1 (2004) (see Figure 10.8). Moreover, the evolution of the mechanical characteristics, f_{cm} and f_{cmt}, depends on the age, t, of the concrete in accordance with the following equation:

$$\frac{P(t)}{P(28\text{ days})} = \beta_{cc}(t); \quad \beta_{cc}(t) = \exp\left\{s\left[1-\left(\frac{28}{t}\right)\right]^{1/2}\right\} \tag{10.13}$$

$P(t)/P(28\text{ days})$ is the normalized considered property, t is the age expressed in day, and s is the coefficient which depends on the type of cement. For CEM, II $s = 0.2$. For the secant modulus E_{cm}, EC2 recommends the relation $\frac{E_{cm}(t)}{E_{cm}} = \beta_{cc}^{0.3}(t)$.

Hence, we propose to check the validity of EC2 (EN 1992-1-1 2004, part 3.1.5) when RAs are incorporated into concretes and to propose, if required, modifications on the basis of the specific experimental results of this study and those of the literature. The adequacy of each model is estimated by calculating the following:

- The sum of squares of error $\text{SSE} = \sum_{i=1}^{n}\left(\hat{P}_i - P_i\right)^2$ that characterizes the deviations of experimental values P_i from their predicted ones, $\hat{P}_i$. The smaller SSE, the more reliable the predictions obtained from the model.
- The deviations of the experimental points from their mean $\overline{P} = P_{mean}$ named $\text{SST} = \sum_{i=1}^{n}\left(P_i - \overline{P}\right)^2$. This parameter quantifies how much the data points, P_i, vary around their mean value $\overline{P}$. This term is calculated to normalize SSE to obtain the coefficient $\rho = \frac{\text{SSE}}{\text{SST}}$. The closer this coefficient is to 0, the more the experimental points approach the model. On the contrary, the higher the coefficient, the more scattered the experimental points around the line describing the model.

- The ratio $\xi = \frac{\bar{\hat{P}}}{\bar{P}}$ is another parameter used to estimate the difference between the experimental mean value $\bar{P}$ and the predictive mean one $\bar{\hat{P}}$. When the parameters of the model are obtained by fitting the experimental values, SSE tends to 0 and therefore $\xi = 1$

with

$$E_{cm} = 22{,}000\left(\frac{f_{cm}}{10}\right)^{0.30}$$

$$\varepsilon_{c1} = 0.7\left(f_{cm}\right)^{0.31}$$

$\varepsilon_{cu1} = 3.5‰$ for $f_{ck} \le 50\,\text{MPa}$ otherwise $\varepsilon_{cu1}(‰) = 2.8 + 27\left[\left(98 - f_{cm}\right)/100\right]^4$
and

$$\frac{\sigma_c}{f_{cm}} = \frac{k\eta - \eta^2}{1 + (k-2)\eta} \text{ with} \begin{cases} k = 1.05\dfrac{E_{cm}\varepsilon_{c1}}{f_{cm}} \\ \eta = \dfrac{\varepsilon_c}{\varepsilon_{c1}} \end{cases}$$

10.3.1 Development of compressive strength with time

In a recent study (Omary et al. 2016), two series of mixes were studied (with NA only and with RA at various replacement rates, respectively). The evolution of the mean compressive strength with age is presented in Figure 10.9b with the corresponding EC2 prediction proposed initially for NA concrete and expressed in the form of equation 10.13 with $\frac{f_{cm}(t)}{f_{cm}(28\text{ days})} = \beta_{cc}(t)$ and $\beta_{cc}(t) = \exp\left\{s\left[1 - \left(\frac{28}{t}\right)\right]^{1/2}\right\}$.

The obtained results reveal that the development of the compressive strength is independent of RA replacement ratio.

10.3.2 Elastic modulus

The assessment of the average dynamic and static modulus of elasticity, named E_d and E_{cm}, respectively, was conducted at different ages. It can be observed (Figure 10.10b) that equation 10.13 with $\frac{P(t)}{P(28\text{ days})} = \frac{E_{cm}(t)}{E_{cm}(28\text{ days})}$ well describes this evolution regardless of the class of compressive strength with a determination factor $R^2 = 0.97$. The variation of elastic modulus presented in Figure 10.10a shows that it decreases with the increase in replacement ratio for the same class of compressive strength. Similar observations are reported in the literature, and the reduction in the elastic modulus can be attributed to the higher total cement paste volume in RAC (Xiao et al. 2006a; Cabral et al. 2010; Omary et al. 2016).

Several empirical relationships were developed in the literature in order to predict the elastic modulus from the compressive mean strength (Ravindrarajah and Tam 1985;

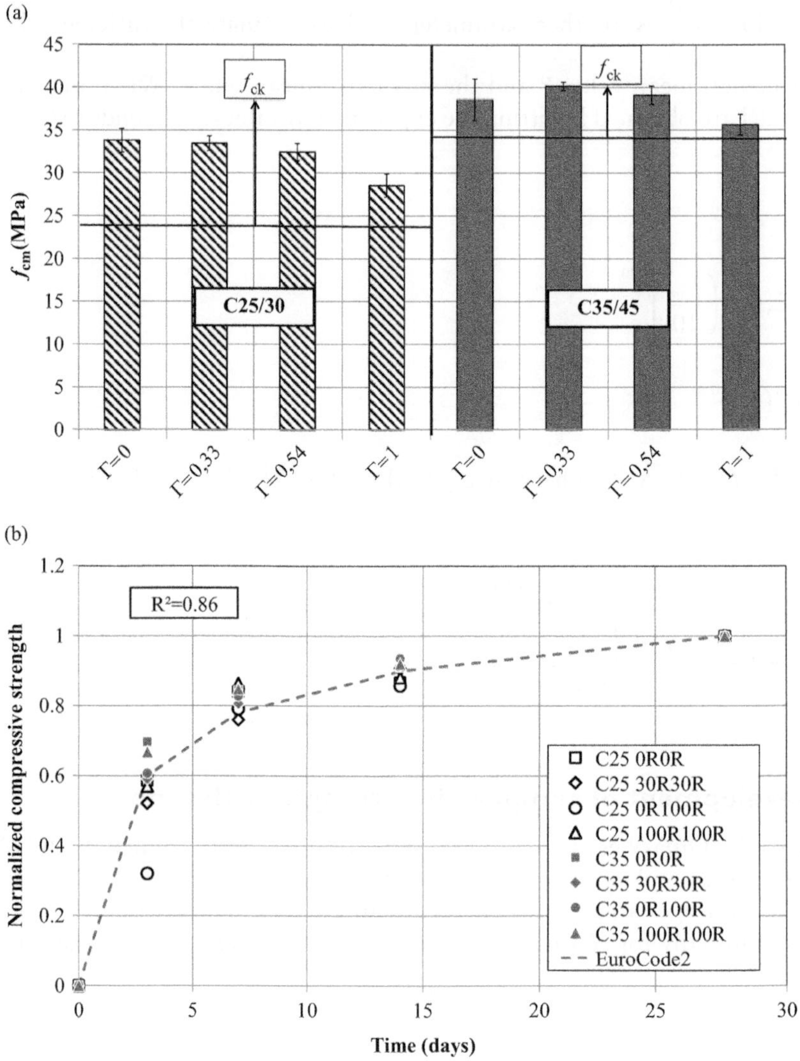

Figure 10.9 (a) Variation of the compressive strength for different replacement ratios Γ_m and (b) variation of compressive strength with age.

EN 1992-1-1 2004; Aslani and Nejadi 2012; Wardeh et al. 2015a). Most of them are developed for NAC and are expressed as follows:

- In EC2-1-1 (Section 3.1.5) (EN 1992-1-1 2004), the secant modulus E_{cm} is expressed as $E_{cm}(\text{MPa}) = 22{,}000\left(\frac{f_{cm}}{10}\right)^{0.30}$. However, it is recommended to reduce the obtained value by 10% and 30% for limestone and sandstone aggregates, respectively, and to increase it by 20% for basalt aggregates.
- The CEB/FIB model code (CEB-FIB 2010) provides the expression of the initial tangent modulus E_{ci} as $E_{ci}(\text{MPa}) = 21{,}500\alpha_E\left(\frac{f_{cm}}{10}\right)^{1/3}$ where α_E is a coefficient depending

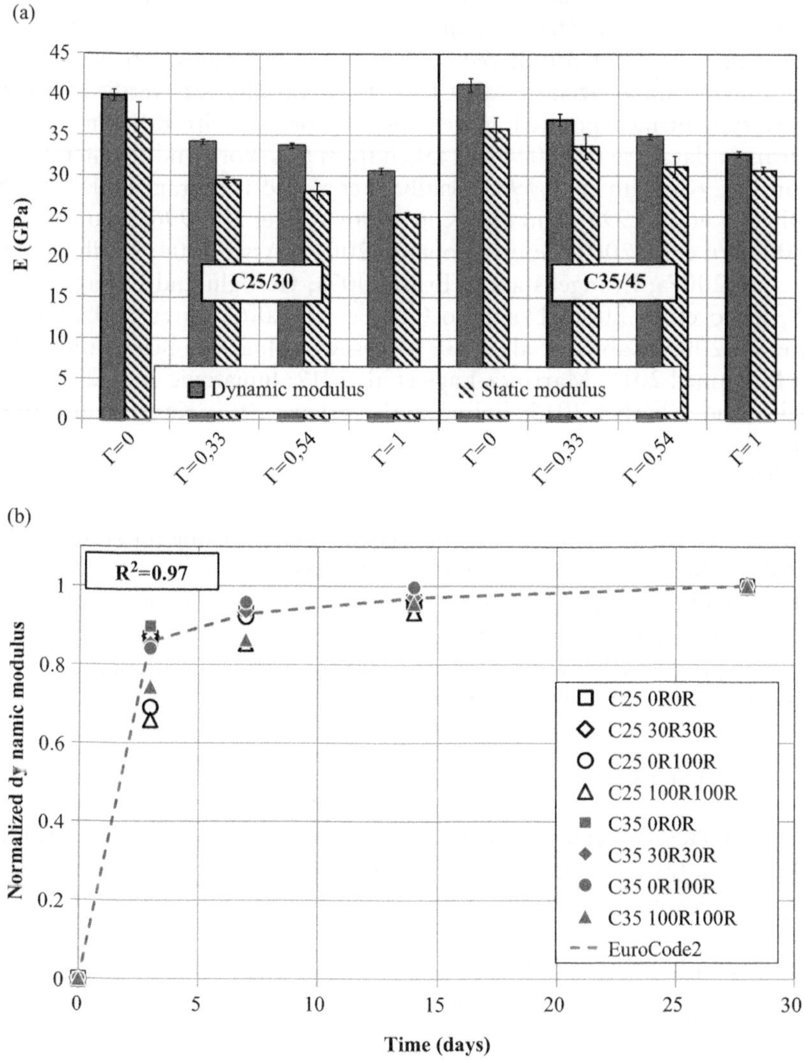

Figure 10.10 (a) variation of modulus of elasticity by class of compressive strength for different replacement ratios and (b) variation of elastic modulus with age.

on the type of aggregates ($\alpha_E = 0.7$ for sandstone and $\alpha_E = 1.2$ for basalt). The initial tangent modulus is also approximately equal to the dynamic modulus and conventionally applicable at very low stress levels.

In structural design, the most generally used modulus is the secant modulus, E_{cm}, or the tangent modulus E_c, with $E_c = 1.05E_{cm}$ and $E_c = 0.85E_{ci}$ conducting to the following expressions $E_c(\text{MPa}) = 18{,}275\alpha_E\left(\frac{f_{cm}}{10}\right)^{1/3}$ (CEB/FIB) or $E_c(\text{MPa}) = 23{,}100\left(\frac{f_{cm}}{10}\right)^{0.30}$ (EC2).

In the literature, the nature of aggregates is not necessarily mentioned. Hence, a mean value of the parameter α_E for CEB/FIB model is adopted, using the available database. It shows that $(\alpha_E)_{NA} = 1.126$ independently of the NA origin.

All the empirical models established to estimate the elastic modulus from the compressive strength of concretes are therefore in the form of $E_c = \xi f_{cm}^{\beta}$. For all these models, the effect of RA content was not considered explicitly. The most adequate equations proposed in the literature are summarized in Table 10.7. Their validity was verified for concretes by specifying the type of incorporated aggregates (NA or RA with different recycling ratios $0 \le \Gamma_v \le 1$) using a database of 360 values taken from this work and the literature including 81 values for RAC (Cedolin et al. 1983; Baalbaki et al. 1991; Bairagi et al. 1993; Shen et al. 1995; de-Oliveira et al. 1996; Kim and Kim 1996; Wee et al. 1996; Shannag 2000; Dong and Keru 2001;Wu et al. 2001; Gomez-Soberon 2002a; Assié 2004; Karihaloo et al. 2006; Etxeberria et al. 2007a; Evangelista and Brito 2007a; Gesoğlu and Ozbay 2007; Casuccio et al. 2008; Zhao et al. 2008; Domingo-Cabo et al. 2009; Fares 2009; Fathifazl et al. 2009;Shen et al. 2009; Cabral et al. 2010; Malesev et al. 2010; Belén et al. 2011; Kumar et al. 2011; Mohamed 2011; Martínez-Lage et al. 2012; Ignjatovic et al. 2013; Manzi et al. 2013; Wardeh and Ghorbel 2015; Folino and Xargay. 2014; Kang et al. 2014; Wardeh et al. 2015a).

It shows that

- For all models, the ability to predict the elastic modulus using the compressive strength diminishes when RAs are incorporated to the mixes.
- The EC2 model (equation 10.14) is the less predictive one among the three models especially when RAs are incorporated ($0 \le \Gamma_v \le 1$).
- The model proposed by Wardeh et al. (2015), equation 10.16, seems to be the most suitable one irrespective of the type of aggregates.

In CEB/FIB model (CEB-FIB, 1990) expressed as $E_c(\text{MPa}) = 18{,}275\alpha_E\left(\frac{f_{cm}}{10}\right)^{1/3}$, a mean value of the parameter α_E is calculated, using the available database, for $\Gamma_v = 1$. It shows that α_E diminishes when the concrete is elaborated with RA: $(\alpha_E)_{NA} = 1.126$, when only NA are used and $(\alpha_E)_{RA} = 0.998$, when 100% of RA is incorporated in concretes irrespective of the origin of RAs.

To improve the prediction of the elastic modulus of concretes incorporating RAs, we propose to modify the expressions as follows: $E_c = \zeta\left(1 - \alpha_{E_c}\Gamma_m\right)\left(\frac{f_{cm}}{10}\right)^{\beta}$. The value of α_{E_c} is obtained by minimizing the residues between experimental and predicted elastic modulus of concretes regardless of the origin of aggregates. Table 10.8 summarizes the obtained results which are given as follows:

Table 10.7 Elastic modulus models for concrete

Equation no.	*Equation*	*All type of aggregates, $0 \le \Gamma_v \le 1$*	*NAs*	*RAs, $0 \le \Gamma_v \le 1$*
(10.14), (EN 1992-1-1 2004)	$E_c = 23{,}100\left(\frac{f_{cm}}{10}\right)^{0.30}$	$\rho = 0.41, \zeta = 1.1$	$\rho = 0.41, \zeta = 1.1$	$\rho = 1.48, \zeta = 1.2$
(10.15), (CEB-FIB 2010)	$E_c(\text{MPa}) = 20{,}578\left(\frac{f_{cm}}{10}\right)^{1/3}$	$\rho = 0.29, \zeta = 1.0$	$\rho = 0.32, \zeta = 1.0$	$\rho = 0.75, \zeta = 1.1$
(10.16), (Wardeh et al. 2015a)	$E_c = 17{,}553\left(\frac{f_{cm}}{10}\right)^{0.42}$	$\rho = 0.23, \zeta = 1.0$	$\rho = 0.26, \zeta = 1.0$	$\rho = 0.43, \zeta = 1.2$

Table 10.8 Modified elastic modulus models for concrete

Equation no.	*Modified equations*	α_{Ec}	*All type of aggregates,* $0 < \Gamma_v < 1$	*RAs* $0 < \Gamma_v < 1$
10.17	$E_c = 23{,}100\left(1 - \alpha_{E_c}\Gamma_m\right)\left(\frac{f_{cm}}{10}\right)^{0.30}$	0.328	$\rho = 0.34, \zeta = 1.1$	$\rho = 0.55, \zeta = 1.0$
10.18	$E_c = 20{,}578\left(1 - \alpha_{E_c}\Gamma_m\right)\left(\frac{f_{cm}}{10}\right)^{\frac{1}{3}}$	0.179	$\rho = 0.26, \zeta = 1.0$	$\rho = 0.42, \zeta = 1.0$
10.19	$E_c = 17{,}553\left(1 - \alpha_{E_c}\Gamma_m\right)\left(\frac{f_{cm}}{10}\right)^{0.42}$	0.131	$\rho = 0.23, \zeta = 1.0$	$\rho = 0.37, \zeta = 1.0$

- The modification of EC2 (equation 10.17) enhances the prediction of the elastic modulus of concretes incorporating RA.
- The equation 10.18 describes the modified CEB/FIB model. It can be noticed that the modification well improves the predictions.
- The equation 10.19 established in the study of Omary et al. (2017) provides the best prediction of the elastic modulus of RAC.

Models of EC2 and Wardeh et al. given in Tables 10.7 and 10.8 are illustrated in Figure 10.11 for concretes elaborated regardless of the type of aggregates. It can be concluded that equation 10.19 can be applied for elastic modulus. The incorporation of RAs leads to its decrease.

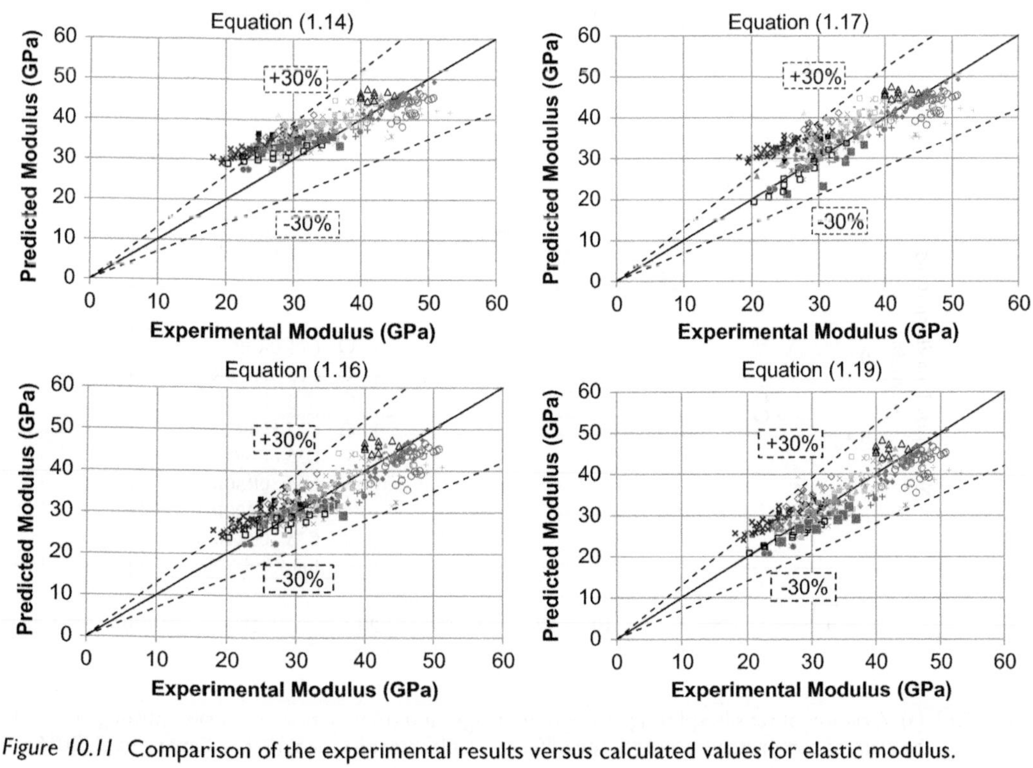

Figure 10.11 Comparison of the experimental results versus calculated values for elastic modulus.

10.3.3 Prediction of tensile splitting strength

The effect of recycling on tensile strength was described in Section 10.2.2. Let us examine the tensile strength development in a particular study. The average splitting tensile strength, $f_{ctm,sp}$, for each mix was measured in five $\phi 11 \times 22$ cm specimens at the age of 28 days (Omary et al. 2016). The evolution of the normalized tensile strength with age is well described by EC2 (Section 10.3.1). Figure 10.12a shows that equation 10.13 with $\frac{P(t)}{P(28\text{ days})} = \frac{f_{ctm,sp}(t)}{f_{ctm,sp}(28\text{ days})}$ fits well the experimental results with an acceptable correlation.

Several authors intended to correlate the tensile splitting strength to the compressive strength (EN 1992-1-1 2004; Aslani and Nejadi 2012; Kou and Poon 2013; de Larrard 1999a;

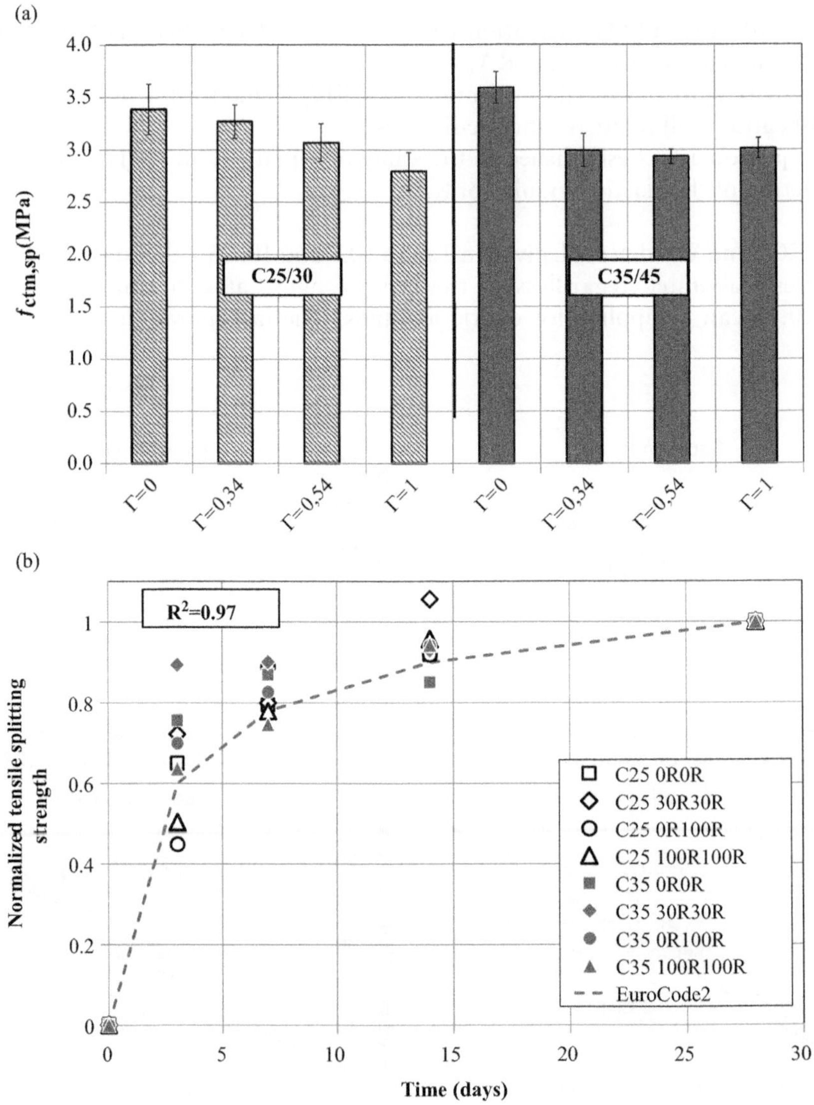

Figure 10.12 (a) Variation of tensile splitting strength with age, and (b) variation of tensile splitting strength by class of compressive strength for different replacement ratios. (From Omary et al. 2016.)

Omary et al. 2017). All the proposed models may be written in the form $f_{ctm,sp} = \eta_1 (f_{cm})^{\eta_2}$ except the expression of the EC2 (equation 10.20) which considers the characteristic compressive strength, f_{ck}, instead of the mean compressive strength, f_{cm}. The applicability of relationships recapitulated in Table 10.9 to all type of concrete was assessed using the materials of this work and 167 experimental values found in the literature and distributed in 67 NAC and 100 RAC (Casuccio et al. 2008; Evangelista and De Brito 2007a; Belén et al. 2011; Gomez-Soberon 2002a; Malesev et al. 2010; Kou and Poon 2013; Etxeberria et al. 2007a; Bairagi et al. 1993; Kameche et al. 2012; Kim and Kim 1996; Breccolotti et al. 2015; Butler et al. 2011).

The results demonstrate the following:

- All the proposed empirical models are unable to adequately predict the tensile splitting strength of concretes containing RA.
- Equation 10.22 allows the best predictions of tensile splitting strengths regardless of the type and nature of aggregates.

To consider the effects of incorporating RA, these expressions were modified by introducing the coefficient $(1-\alpha_{f_{ctm}}\Gamma_m)$. The parameter $\alpha_{f_{ctm}}$ is obtained by minimizing the residues between experimental and predicted tensile splitting strengths of concretes regardless of the type and origin of aggregates, while keeping the parameters η_1 and η_2 previously calculated. The parameters ρ and ζ are given in Table 10.10. It can be seen that

- This modification improves predicted tensile strengths values.
- The incorporation of RA leads to a diminution of the tensile splitting strength even if the mean compressive strength is constant. Moreover, this diminution seems to be significant for replacement ratios, Γ_v, higher than 33%.

Table 10.9 Tensile splitting strength models for concretes

		Concretes elaborated with		
Equation no.	*Equations*	*NA*	*RA, $(0 \le \Gamma_v \le 1)$*	*NA + RA, $(0 \le \Gamma_v \le 1)$*
10.20, (EN 1992-1-1 2004)	$f_{ctm,sp} = 0.33(f_{ck})^{2/3}$, $f_{ck}(\text{MPa}) = f_{cm} - 8$	$\rho = 0.37$, $\zeta = 1.0$	$\rho = 0.31$, $\zeta = 1.1$	$\rho = 0.60$, $\zeta = 1.1$
10.21, (de Larrard 1999a)	$f_{ctm,sp} = 0.453(f_{cm})^{0.57}$	$\rho = 0.41$, $\zeta = 1.1$	$\rho = 0.26$, $\zeta = 1.2$	$\rho = 0.76$, $\zeta = 1.2$
10.22, (Omary et al. 2017)	$f_{ctm,sp} = 0.364(f_{cm})^{0.608}$	$\rho = 0.26$, $\zeta = 1.0$	$\rho = 0.20$, $\zeta = 1.1$	$\rho = 0.43$, $\zeta = 1.1$

Table 10.10 Tensile strength models for RAC

Equation no.	*Modified equations*	$\alpha_{f_{ctm}}$	*RAs, $0 < \Gamma_v \le 1$*	*All type of aggregates, $0 \le \Gamma_v \le 1$*
10.23	$f_{ctm,sp} = 0.33(1-\alpha_{f_{ctm}}\Gamma_m)(f_{cm}-8)^{2/3}$	0.282	$\rho = 0.32$, $\zeta = 0.9$	$\rho = 0.47, \zeta = 1.0$
10.24	$f_{ctm,sp} = 0.453(1-\alpha_{f_{ctm}}\Gamma_m)(f_{cm})^{0.57}$	0.221	$\rho = 0.24$, $\zeta = 1.0$	$\rho = 0.31, \zeta = 1.1$
10.25	$f_{ctm,sp} = 0.364(1-\alpha_{f_{ctm}}\Gamma_m)(f_{cm})^{0.608}$	0.142	$\rho = 0.20$, $\zeta = 1.0$	$\rho = 0.28, \zeta = 1.0$

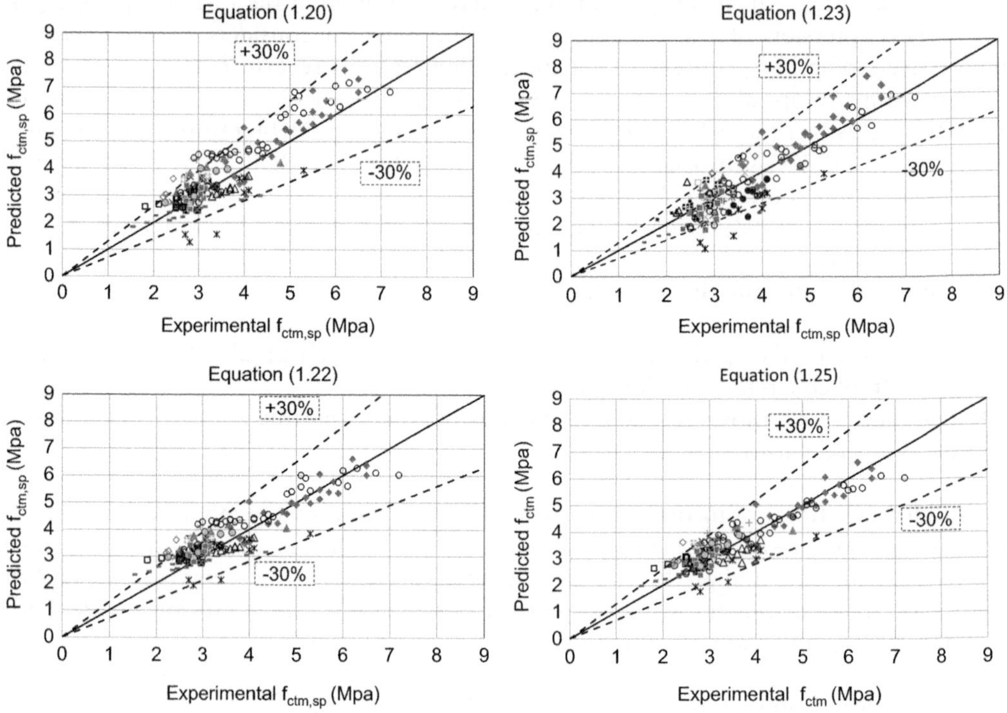

Figure 10.13 Predicted and experimented tensile splitting strength of concretes regardless of aggregates type.

- The equation 10.25 derived from equation 10.22 established in the work of Omary et al. (Omary et al. 2017) is the most suitable for the prediction of the tensile splitting strength of concretes incorporating RAs. The comparison between experimental data versus EC2 and models of Omary et al., summarized in Tables 10.9 and 10.10, is illustrated in Figure 10.13.

10.3.4 Relationships for peak and ultimate strains

The peak strain, ε_{c1}, is the strain corresponding to the maximum compressive strength. The results obtained for two series, illustrated in Figure 10.14a, show that when RA content increases, the peak strain increases too. These results are consistent with those found in the literature (Wardeh et al. 2015a; Xiao et al. 2006a; Belén et al. 2011; Breccolotti et al. 2015; Martínez-Lage et al. 2012; Folino and Xargay 2014; Wee et al. 1996; Kumar et al. 2011; Carreira and Chu 1985; Anis et al. 1990; Dhonde et al. 2007). The increase in the peak strain with the replacement ratio is due to the incorporation of RA independently of the volume of the paste, which increases in this study as the replacement ratio increases for a same strength class. Moreover, the higher the strength class, the higher the ε_{c1}. These results are also in accordance with the literature. As a matter of fact, Wee et al. (1996) showed that the peak strain increases when the class of compressive strength increases but decreases when the paste volume rises for the same class of compressive strength. Moreover, researchers who introduced RA by keeping the paste volume constant have noticed an increase in ε_{c1} (Belén et al. 2011; Breccolotti et al. 2015).

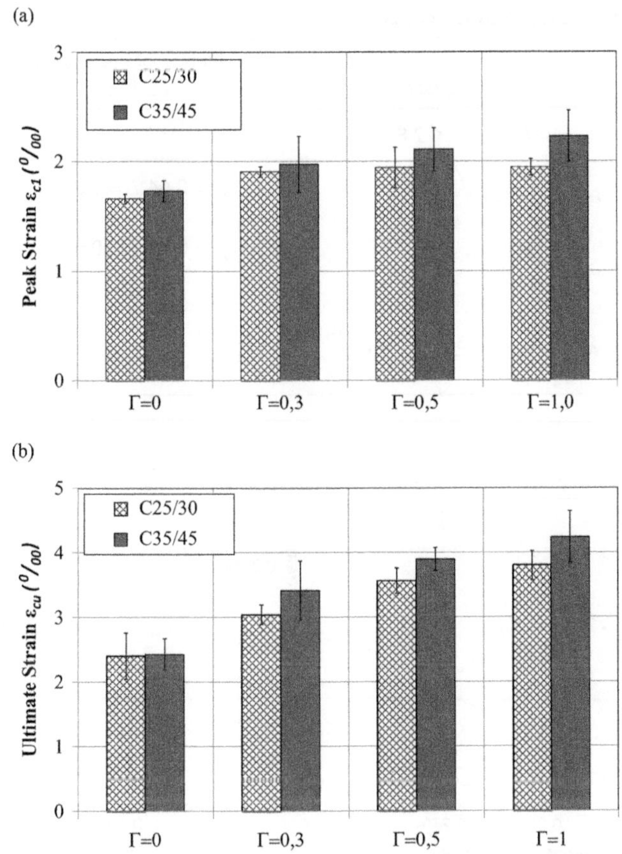

Figure 10.14 (a) Peak strain for different replacement ratios (Γ_m), and (b) ultimate strain for different replacement ratios (Γ_m). (From Omary et al. 2016.)

The ultimate strain, ε_{cu}, is evaluated in the descending branch and corresponds to the strain obtained for a stress equal to $0.6 \times f_{cm}$. The results presented in Figure 10.14b show that the value of this strain increases when the replacement ratio increases.

For the prediction of peak strain, relationships initially proposed for concrete elaborated with NA (EN 1992-1-1 2004; Wardeh et al. 2015a; Fouré 1996) have been used, and their reliability to estimate the experimental values is assessed (Table 10.11).

The expression proposed in the framework of EC2 is given by equation 10.26, whereas the one established in the study of Wardeh et al. (2015a) is illustrated by equation 10.27.

The expression given in the study of Fouré (1996) depends on the mean compressive strength f_{cm} and on the modulus E_{c_i} (see Chapter 20). In this chapter, it is stated that all the mechanical characteristics should depend only on f_{cm}. Therefore, the effects of E_{c_i} are introduced by means of the parameter $k_0 = \frac{E_{c_i}}{f_{cm}^{1/3}}$ calculated using the equation. Here, a mean value of $k_0 = 11{,}237$ is adopted, and the proposed model is given by equation 10.28.

The results show (see Table 10.11 and Figure 10.15) the following:

- EC2 expression (equation 10.26) cannot adequately predict the peak strain of concretes independently of the type of aggregates.

Table 10.11 Peak-strain relationships for concrete

Equation no.	*Equation*	*All type of aggregates,* $0 \le \Gamma_v \le 1$	*NAs*	*RA*
10.26, (EN 1992-1-1 2004)	$\varepsilon_{c1} = 0.7(f_{cm})^{0.31} \le 2.8$	$\rho = 1.26, \zeta = 1.1$	$\rho = 1.48,$ $\zeta = 1.1$	$\rho = 1.11,$ $\zeta = 1.1$
10.27, (Wardeh et al. 2015a)	$\varepsilon_{c1} = 1.1(f_{cm})^{0.175}$	$\rho = 0.64, \zeta = 1.0$	$\rho = 0.45,$ $\zeta = 1.0$	$\rho = 0.94,$ $\zeta = 1.0$
10.28, (Fouré 1996)	$\varepsilon_{c1} = \left(\frac{k}{k_0}\right) f_{cm}^{2/3};$ where $k_0 = 11{,}237, k = 1 + \frac{0.16k_0}{f_{cm}^2 + 800}$	$\rho = 0.99, \zeta = 0.9$	$\rho = 0.81,$ $\zeta = 0.9$	$\rho = 1.26,$ $\zeta = 0.9$

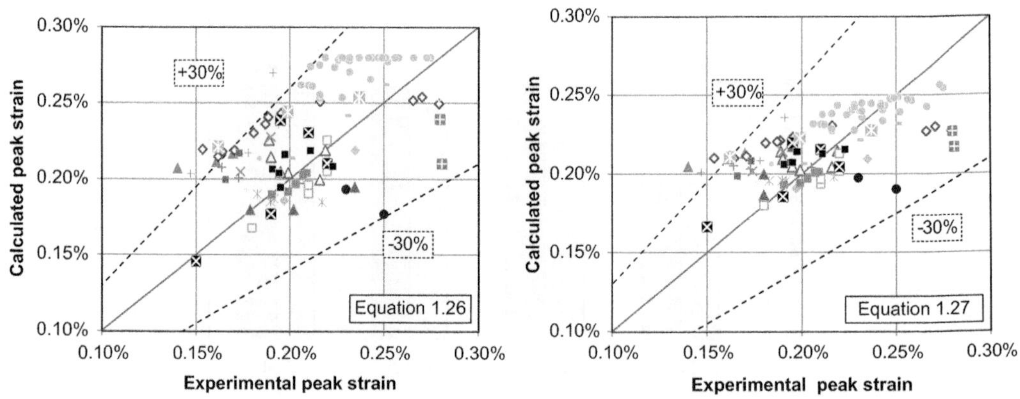

Figure 10.15 Peak strain for different replacement ratios calculated values versus experimental ones for concretes regardless of the aggregates type.

- The incorporation of RA increases the scattering between the estimated and experimental values irrespective of the model used.
- The introduction of the coefficient $(1 - \alpha_m \Gamma_m)$ did not lead to an improvement of peak-strain prediction for all studied models.
- The most adequate model estimating the peak strain is given by equation 10.27 (see Figure 10.15).

It has been established that the ultimate strain, ε_{cu}, evaluated in the descending branch as the strain obtained for a stress equal to $0.6 \times f_{cm}$, increases when the replacement ratio increases. An empirical expression is suggested for calculating the ultimate strain, which suitably fits the experimental results:

$$\varepsilon_{cu} = \left(0.00298 - 0.0625\left\langle\frac{50 - f_{cm}}{100}\right\rangle^4\right)(1 + 0.2\Gamma_m); \quad \text{for } f_{ck} \le 50(\text{MPa}) \qquad (10.29)$$

More data would be needed to confirm the validity of this relationship.

Table 10.12 Stress–strain models

Origin of the model	*Modified models*	*Parameters*	*Equation no.*
EC2 (EN 1992-1-1 2004)	$\frac{\sigma}{f_{cm}} = \frac{k\eta - \eta^2}{1 + (k-2)\eta}$	$k = \frac{E_c \varepsilon_{c1}}{f_{cm}}$ $\eta = \frac{\varepsilon}{\varepsilon_{c1}}$, with E_c equation 10.19 and ε_{c1} equation 10.27	10.30
Carreira and Shu (1985)	$\frac{\sigma}{f_{cm}} = \frac{\beta \frac{\varepsilon}{\varepsilon_{c1}}}{\beta - 1 + \left(\frac{\varepsilon}{\varepsilon_{c1}}\right)^{\beta}}$	$\beta = \frac{1}{1 - \frac{f_{cm}}{E_{ci}\varepsilon_{c1}}}$, $E_{ci} = \frac{E_c}{0.85}$, and E_c equation 10.19 and ε_{c1} equation 10.27	10.31
Popovics (Fouré 1996)	$y = \frac{kx + (k'-1)x^2}{1 + (k-2)x + k'x^2}$, Where $x = \frac{\varepsilon}{\varepsilon_{c1}}$ and $y = \frac{\sigma}{f_{cm}}$	$k' = 3.33 - 2.33\frac{2X-1}{X^2} - \frac{k}{X}$, where $X = 1 + \frac{20}{f_{cm}}$, $\varepsilon_{c1} = \left(\frac{k}{k_0}\right) f_{cm}^{2/3}$	10.32

10.3.5 Stress–strain relationship under uniaxial compression

Many authors and design standards have proposed a simple analytical shape for the full stress–strain curve for NAC.

In this study, the analytical expressions initially proposed by EC2 (EN 1992-1-1 2004; Carreira and Shu 1985), and Popovics in Fouré (1996) for NAC were modified and expanded to RAC. The expressions are given in Table 10.12.

In Figure 10.16, a comparison between the experimental and predicted compressive stress–strain curves using EC2 and Carreira and Shu models is shown for all concretes studied by Omary et al. (2016). It clearly shows that equations 10.30–10.32 predict quite well the compressive behavior of concretes up to the ultimate strain irrespective of the type of aggregates. Moreover, it appears that the model represented by equation 10.31 overestimates the strains in the descending branch and leads to high ductility, whereas equation 10.30 underestimates them. Equation 10.32 seems to be the most suitable to predict the overall compressive behavior up to failure for all types of concretes differing by the strength class and the type of aggregates. However, this is not surprising since this equation was fitted on these few data. More data would be necessary to establish a general equation that could be used for any recycled concrete at the design stage.

10.4 CONCLUSIONS AND RESEARCH NEEDS

In this chapter, experimental results for the mechanical properties of RAC are presented and discussed. Moreover, an extensive database was constructed gathering experimental results from literature and own research.

From these data, the first part of this chapter proposes three sets of equations to predict the mechanical properties of RAC from mix composition:

- Equations 10.4–10.6, for compressive strength;
- Equations 10.8 and 10.9, for tensile strength;
- Equations 10.10 and 10.11, for elastic modulus.

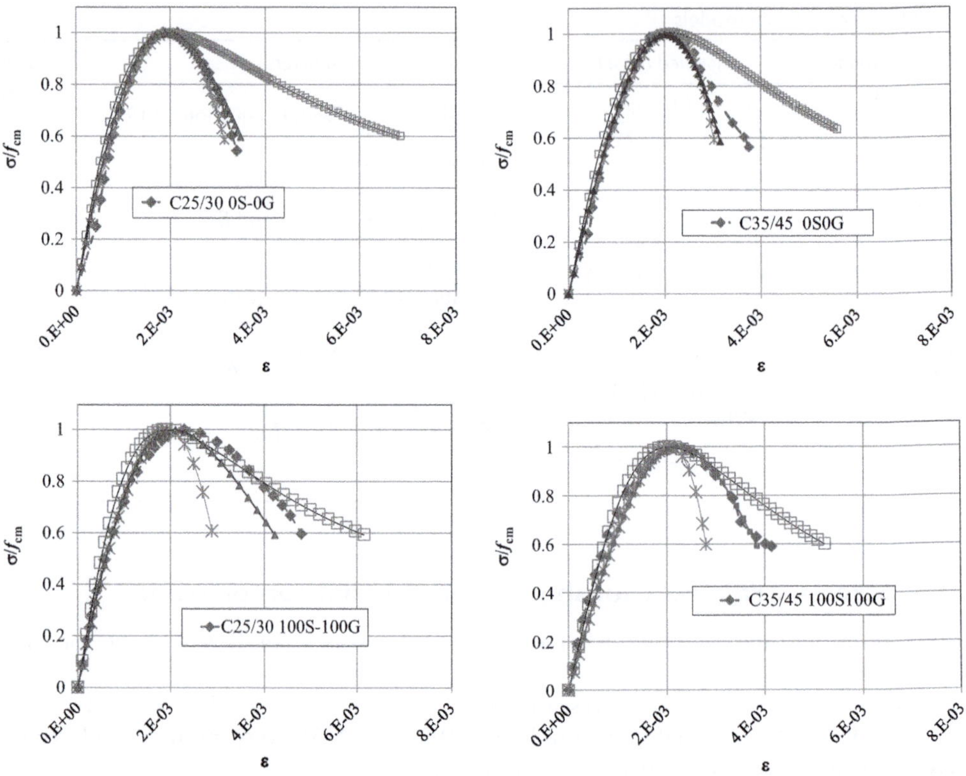

Figure 10.16 Comparison of the stress–strain models (equations 10.30, 10.31, 10.32) and the experimental results.

For that purpose, three parameters, $k_{g,j}$, $k_{t,j}$, and $E_{g,j}$ must be calibrated for FA and CA, by measuring the compressive strength, tensile splitting strength, and elastic modulus, respectively, on some mortars and concretes, including these RAs.

The main interest of this approach is that the determination of $k_{g,j}$, $k_{t,j}$, and $E_{g,j}$ gives an intrinsic characterization of RA, independent of the reference concrete the RAC is compared to, and of other mix design parameters, such as water-to-cement ratio. Then, it is possible to precisely evaluate the influence of RA on mechanical properties irrespective of the water-to-cement ratio, the nature of NA, and the recycling rate of FA or CA.

Yet even if this method is easy, it may be burdensome, so it would be interesting to find a direct evaluation of these parameters. For example, it could be interesting to confirm more widely the relevance equation 10.7, proposed in the study of Dao et al. (2014), to predict the $k_{g,j}$ from the abrasion resistance (MDE) of RA. Unfortunately, it was not possible to find papers in technical literature with all the data necessary to test the model.

In addition to the experimental results obtained in this study, data available in the literature were used in order to provide further elements for the development of equations that could be used at the design stage. The most important findings are given as follows:

- For the same class of compressive strength, the introduction of RA induces a decrease in elastic modulus. This is due to two phenomena that act simultaneously: the lower elastic modulus of RA in comparison with that of NA and the highest volume of paste in RAC needed to keep the slump and the strength class constants, compared to the reference.

- For the same class of compressive strength, the introduction of RA most often induces a decrease in the tensile strength. The magnitude of this decrease depends on the quality of the RA and can be the same than the one observed for concretes elaborated with poor-quality NA.
- The peak strain and the ultimate stain of RAC are higher than those of NAC. They increase with the increase in RA content.
- The slope of the post-peak branch increases for both series C25/30 and C35/45 when the replacement ratio increases.
- Code-type relationships dedicated to assess the mechanical properties of NAC are not satisfactory in predicting the behavior of RAC. All the proposed models devoted to the prediction of the tensile splitting strength $f_{ctm,sp}$, elastic modulus E_{cm}, and the ultimate strain ε_{cu1} are phenomenological and depend on the mean compressive strength, f_{cm}, as well as in a term $(1-\alpha\Gamma_m)$, where Γ_m is the mass ratio of RA (see equations 10.25, 10.19, 10.27, and 10.29, respectively). Moreover, the evolution of these mechanical characteristics depends on the age, t, of the concrete in accordance with EC2 model.
- The comparison between experimental stress–strain curves obtained in this work and some modified models shows that equations 10.31 and 10.32 describe satisfactory the descending branch up to failure. However, it must be noted that the model given by equation 10.31 overestimates the strains, whereas the one described by equation 10.32 underestimates the post-peak behavior.

ACKNOWLEDGMENTS

Technical support from the project ANR ECOREB is gratefully acknowledged for Section 10.3.

Chapter 11

Delayed mechanical properties

F. Grondin, A. Z. Bendimerad, M. Guo, A. Loukili, and E. Rozière
Institut de Recherche en Génie Civil et Mécanique UMR 6183,
Centrale Nantes - Université de Nantes - CNRS, Nantes, France

T. Sedran
Ifsttar, Materials and Structures Department, Bougenais, France

C. De Sa and F. Benboudjema
LMT-Cachan, ENS-Cachan, CNRS, Université Paris Saclay, Paris, France

F. Cassagnabère and P. Nicot
LMDC, Université de Toulouse, France

B. Fouré
Bougival

CONTENTS

Abstract

Delayed deformations of concrete are known to induce a risk of cracking in structures and excessive bending. Delayed deformations are mainly of two types: shrinkage and creep. The shrinkage tests were carried out for two substitution rates of natural aggregates with 30% and 100% recycled aggregates (RA). State-of-art and the experimental results show that shrinkage and creep generally increase when RA are incorporated in concrete, due to a higher amount of the cement paste. Differences in micro-cracking are observed very locally on the concrete's skin. Three-point bending creep tests show the same conclusions. It is difficult to distinguish natural concrete from recycled concrete at macroscopic scale. Two types of modelling have been used to describe the mechanisms associated with shrinkage and creep. The first, conducted at macroscopic scale on a homogeneous material, shows that damage seems more severe in the concrete's skin subjected to desiccation. The second, carried out at mesoscopic scale to evaluate the creep, allows considering the influence of the attached old mortar. This modelling makes it possible to better understand the influence of RA on the location of micro-cracks. Experimental fatigue tests have been conducted within RECYBÉTON. They show that the endurance of concrete slightly decreases when RA are introduced.

11.1 INTRODUCTION

During their service life, concrete structures are submitted to mechanical loads which imply delayed deformations. These deformations are mainly of two types: shrinkage and creep. These phenomena are known due to internal mechanisms into the microstructure of concrete: pore pressure, interaction between the cement paste matrix and aggregates, the viscoelasticity of the cement paste, etc. Several studies have shown the effect of the aggregate properties on these delayed deformations (Cortas et al. 2014; Cortas 2012). Results have shown the influence of the mineralogical nature of aggregates and the water content. So, the deformation ability of aggregate has a significant role on the delayed deformations. Only few studies have been performed on the effect of the recycled coarse aggregates and recycled sand (RS) on shrinkage and creep of concrete. Because their nature is different compared to natural aggregates and their microstructure can lead to variable water absorption, it is important to have a better understanding of their influence on the delayed deformations of concrete. A state of the art describes a synthesis of results obtained by researchers over the world. With the objective to give more explanations and offer to construction actors some recommendations, French laboratories and industries have worked together, and this chapter presents some results on shrinkage and flexural creep (realized at GeM Institute at Ecole Centrale de Nantes), compressive creep (realized at LMDC at the University of Toulouse), and fatigue strength (realized at Ifsttar Nantes). Shrinkage results gained by Sigma Béton during the RECYBÉTON first experimental construction site were also added.

11.2 STATE OF THE ART ON SHRINKAGE AND CREEP OF RECYCLED AGGREGATE CONCRETE

11.2.1 Influence of the substitution rate of recycled aggregates on shrinkage

The influence of the substitution rate of natural aggregates by recycled aggregate (RA) on long-term shrinkage has already been studied in numerous studies (Hansen and Boegh 1985; Tavakoli and Soroushian 1996b; Sagoe-Crentsil et al. 2001; Gómez-Soberón 2003; Katz 2003; Domingo-Cabo et al. 2009; Fathifazl et al. 2011b; Dao 2012; Manzi et al. 2013; Pedro et al. 2014). It has been shown that recycled aggregate concretes (RACs) exhibit a shrinkage increase of 15%–60% in proportion to the substitution rate. The most studies have considered only the coarse aggregate substitution.

This phenomenon is explained by the high water absorption rate of coarse RAs, which are very porous materials because of the attached old mortar to aggregate which is estimated to 31.5% and 18% for the fractions 4/8 and 8/20, respectively (Domingo-Cabo et al. 2009). At

28 days, for concrete with 20% of replacement rate, there is a very little difference compared to natural concrete, whereas after 6 months, there is a difference of 4% (Figure 11.1). On the other hand, for concrete of 50% and 100% substitution of natural coarse aggregate by RA, increase of 12% and 70% for shrinkage was observed, respectively.

The effect of the initial water saturation of RA on long-term shrinkage was little studied in the bibliography, whereas it is an important parameter because of its high water absorption rate (Tam and Tam 2008; Djerbi Tegguer 2012; Bendimerad et al. 2015). Brand et al. (2015) studied three saturation levels of RA: oven-dried (OD), completely saturated (CD) aggregates, and partially saturated (80% CD). The author also studied the effect of mixing procedures; for the first time, the mixing was made according to the ASTM standard (ASTM C192 2007) noted (NMP: normal mixing process), and secondly, according to the method developed by Tam (Tam and Tam 2008), noted (TSMA: two-stage mixing approach). During the first 24 h, no effect of the initial water saturation state of gravel and the mixing procedure (TSMA or NMP) on the shrinkage was observed. Nevertheless, at long term, it can be noted that all concretes according to the TSMA procedure make less shrinkage, more particularly concrete with 80% CD. On the other hand, concrete with OD aggregates (NMP) shows the higher value of shrinkage. The author (Brand et al. 2015) explains that concrete with OD RA may not sufficiently reach a fully saturated condition after mixing. The free drying shrinkage is greater for concrete with OD RA since the effective *w/c* ratio would be greater than concrete with fully or partially saturated RA. The hypothesis that the free drying shrinkage at 90 days could be statistically lower for concrete with fully or partially saturated RA relative to concrete with OD RA is verified.

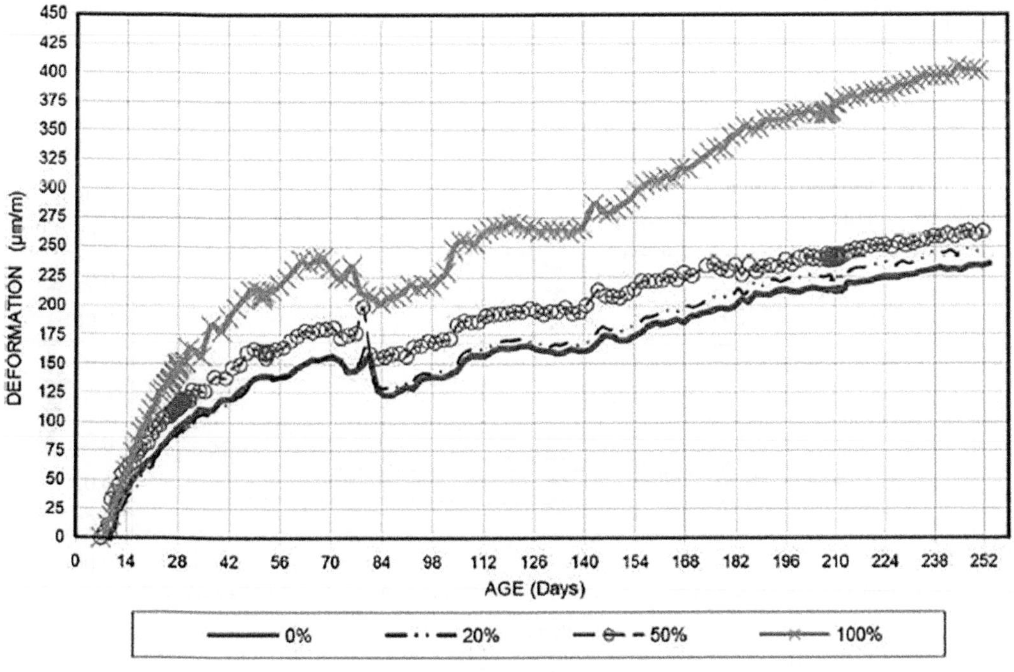

Figure 11.1 Shrinkage of RAC with different substitution rate of RA (Domingo-Cabo et al. 2009).

11.2.2 Influence of the substitution rate of RAs on creep

Similar observations were made in creep studies: creep increases with the use of RA (Hansen 1986, 1992; Gómez-Soberón 2003; Domingo-Cabo et al. 2009; Fathifazl et al. 2011b; WRAP 2007; Marinkovic and Ignjatovic 2013).

Figure 11.2 shows that for a constant compressive force equal to 40% of the strength applied to the specimen, the creep deformation of the recycled concrete for substitution levels of 20%–100% is higher than that of a natural aggregate concrete by 35%–51%, respectively. Viewed from another angle, creep leads to the relaxation of elastic stresses and subsequently the reduction in tensile stresses in case of restrained shrinkage (ACI-224R-01 2001). Fan et al. (2014) studied the effect of the attached old mortar on aggregate, on the creep characteristics of RAC. The content, elastic modulus, and characteristic of the attached old mortar can be responsible for this difference. Considering this influence, they adapted the Neville's model (Neville 1983) to predict the creep of recycled concrete.

According to the study of Gomez-Soberon (2002b), the characteristics and the composition of the RAC give an influence on the short- and long-term mechanical behavior. In particular, creep desiccation of RAC is significantly affected by comparison with that of natural aggregate concrete from 30% replacement rate of natural aggregate (NA) by RA (Figure 11.2). The study (WRAP 2007) deals with RA replacing the coarse aggregates only. The ratio of replacement with respect to the total mass of aggregates was 0%, 30%, or 60% (this last value notably higher than 50% suggests that the total mass in cause was that of the coarse aggregates, so the effective ratio may be only 15% or 30%). The compressive strength was in the classes C20 or C35. Four mixes were studied, with two shrinkage tests and two creep tests for each mix, with a duration of about 6 months. The results show that shrinkage of RAC was equal or slightly smaller than that of NAC. On the contrary, creep was notably higher for RAC than for NAC, at maximum nearly twice. From long-term extrapolation of the experimental curves, the ratio of RAC creep to NAC creep has approximate values given in Table 11.1.

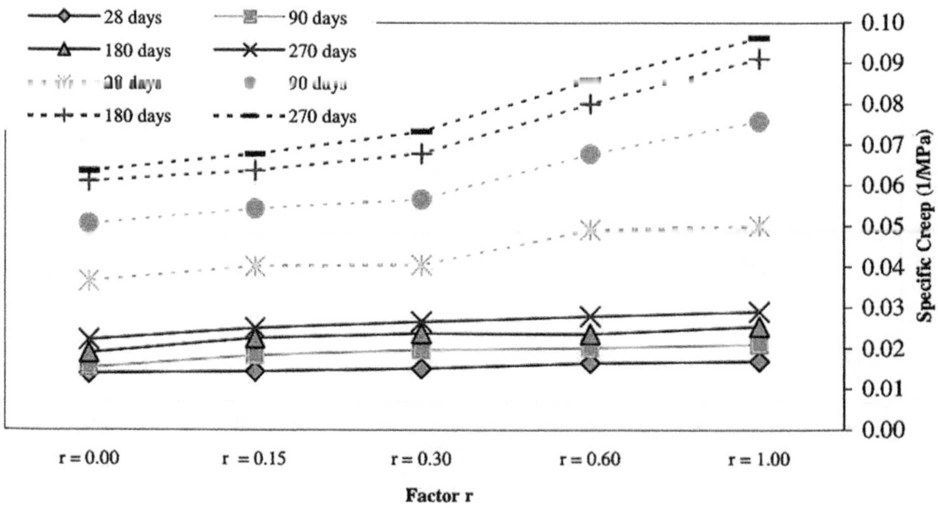

Figure 11.2 Basic creep (full lines) and desiccation creep (dot lines) of RAC for different substitution rates (Gomez-Soberon 2002a).

Table 11.1 Relative effect of recycling on creep (results from (WRAP 2007))

Mix	*C20*		*C35*	
RA ratio	30%	60%	30%	60%
Creep RAC/NAC	1.6	2.4	1.8	2.2

Extrapolation for a ratio of 100% of RA would give at maximum RAC creep about four times NAC creep (and even more if the effective RA ratio is less than 60%). From another source (Marinkovic and Ignjatovic 2013), a compilation of results shows the creep of RAC being 50% higher at a maximum than that of NAC.

11.2.3 Influence of RAs on fatigue in flexion of concrete

Pavements are a traditional tank for recycled materials, mainly in the lower layers whether they are bonded or unbonded. However, for several decades, pavement concrete has also been an attractive outlet for recycled concrete aggregate, with, for example, the double-lift recycling technology developed by the Austrians since 1990 Switzerland (FHWA 2007; Gaspar et al. 2015). For a rational design of a pavement containing a concrete layer, it is necessary to evaluate two material properties of the material: its elastic modulus and its flexural fatigue behavior (SETRA-LCPC 1997). We focused on the first in Chapter 10 and on the second in this chapter.

The fatigue behavior of concrete in compression, tension, or flexion is often described by the following Wholer's curve (Arora and Singh 2015):

$$\frac{\sigma_{\log N}}{\sigma_0} = \mathrm{A} + \mathrm{B} \log N \tag{11.1}$$

where σ_0 represents the static strength, $\sigma_{\log N}$ the cyclic flexural stress leading to concrete failure at N cycles, and $\sigma_{\log N}/\sigma_0$ the endurance of concrete at N cycles and A and B empirical parameters depending on the loading mode and concrete mixture. B is negative.

The French pavement design method focuses on flexural behavior of the material and assumes that A = 1 while B is fitted on fatigue tests with a value of σ close to σ_6. Since the tests are scattered, the standard deviation of log(N) is a very important value for pavement design (SETRA-LCPC 1997) and its determination requires numerous tests (see NF 98-233-1 French standard). Unfortunately, only very few data are available in the literature on flexural fatigue behavior of unreinforced concrete containing RA. For compression fatigue behavior, Xiao et al. (2013b) showed that the 2 million cycles endurance is higher with RA compared to that with ordinary concrete (0.68 vs. 0.58). However, Thomas et al. (2014a,b) showed that it decreases from 0.65 to 0.55 as the RA rate and compressive strength increase (see Figure 11.3). Although each value of endurance is extrapolated from ten fatigue tests at different stress level, no clear conclusion is given concerning the evolution of the scattering with the recycling rate. Some data are also available from four-point bending fatigue tests. Sobhan and Krizek (1999); Sobhan et al. (2016) claimed that a cement-treated base material with a complete RA skeleton behaves as conventional pavement base materials. But Xiao et al. (2013c) founded a decrease in the endurance for concrete made of 100% of RA. Arora and Singh (2015, 2016) have tested two concretes in four-point bending fatigue with 100% of RA. They have obtained a 2 million cycles endurance of 0.58 for ordinary concrete and confirmed a decrease for the RAC (0.5). The important numbers of experimental points make it possible to calculate the standard deviation SN according to NF 98-233-1 French

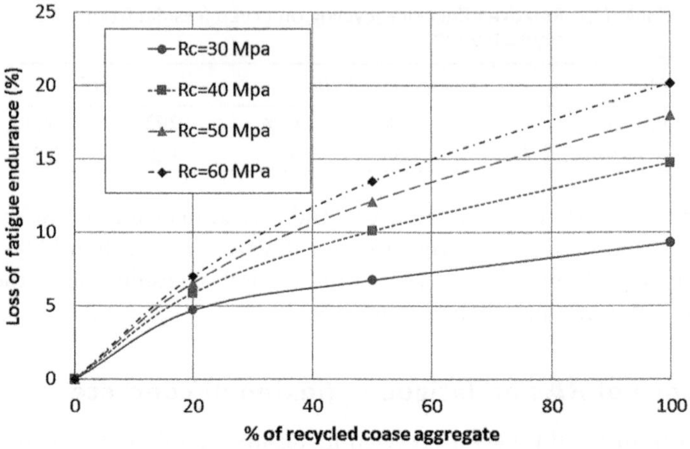

Figure 11.3 Loss of 2 million cycles endurance as a function of RA rate (Thomas et al. 2014b).

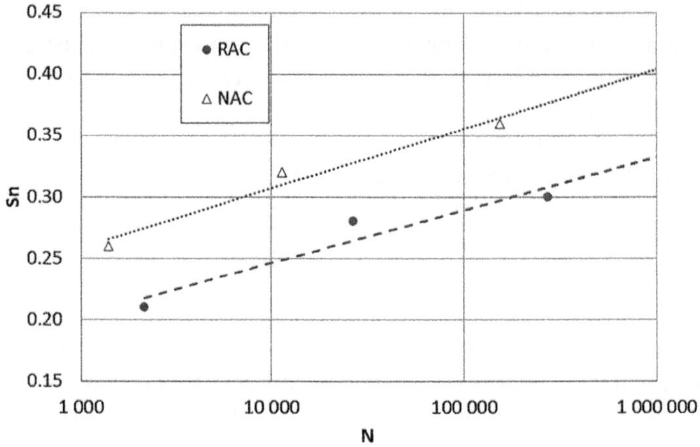

Figure 11.4 Evolution of standard deviation of logN (SN) as a function of stress level (Arora and Singh 2016).

standard, as a function of stress level. Figure 11.4 shows that the recycling rate leads to a moderate increase in SN.

In conclusion, the state-of-the-art review shows a trend for an endurance decrease when adding RA to concrete. Nevertheless, only few reliable data are available in the literature and in particular with a sufficient number of specimens to consider the scattering of fatigue life span that is of paramount importance for pavement design.

11.3 EFFECT OF RAs ON THE SHRINKAGE OF CONCRETE

11.3.1 First experimentation campaign

The studied mortar and concrete mixtures (see Appendix) come from the report produced under the theme 0 of the National Project RECYBÉTON "Development of the reference concrete formulas" written by Sedran (2017). Long-term shrinkage measurements are carried out on $2 \times 2 \times 16\,cm^3$ specimens for mortars and on $7 \times 7 \times 28\,cm^3$ specimens for concrete. At each end, the specimen is equipped with metal studs cast in the material. These allow the specimen to be placed vertically on the test bench, and its length is tracked with an LVDT sensor. Demolding takes place 24 h after water-to-cement contact. For each mixture, measurements are carried out on specimens in autogenous and drying conditions. A specimen is used to measure the mass loss under the same drying conditions, in order to correlate it with the desiccation shrinkage. The test takes place in a controlled temperature and humidity room (20°C, 50% RH), and the data acquisition is done automatically every hour. The first measurement takes place approximately 20 min after demolding. The following model proposed by Torben (Hansen and Mattock 1966) is calibrated on shrinkage measurements at 7 and 40 days for RS mortar:

$$\varepsilon(t) = \frac{t}{N_s + t}\varepsilon_\infty \tag{11.2}$$

where ε_∞ represents the long-term shrinkage and N_s the time needed to reach the half of the long-term shrinkage. We then obtain N_s (7j) = 2.45; N_s (40j) = 2.43; (7j) = 1,388 µm/m; (40j) = 1,348 µm/m. Given the linear dependence of time N_s on the square-drying radius, a measurement time of 7 days on 2 × 2 × 16 cm³ specimens corresponds to about 86 days on 7 × 7 × 28 cm³ specimens.

Tests on RAC were carried out to measure the autogenous deformations by the BTJADE device developed at IFSTTAR (Boulay 2007) (Figure 11.5).

It is observed that there is virtually no autogenous shrinkage. The most important deformation is for concrete with natural aggregates (about 25 µm/m). Recycled sand concrete or RAC has lower values. In the case of 100% RAC, there is a swelling from 24 h. RA appears to act as a reservoir by providing water to minimize self-desiccation. This mechanism was observed in the case of light aggregates (Kohno et al. 1999; Zhutovsky et al. 2002), as well as, for crushed aggregates of returned concrete and RA (Maruyama and Sato 2005; Kim and Bentz 2008). These authors demonstrate the effect of light aggregates with the decrease in the autogenous shrinkage by 40%–70% compared to natural aggregate concretes.

Now, neglect the autogenous part of shrinkage. Table 11.2 shows the effect of the proportion of RA on the amplitude and characteristic time of shrinkage. In comparison with 0S0G, the increase in the amplitude for 0S100G RAC is limited to 7%–11%, respectively, for $w/c = 0.53$

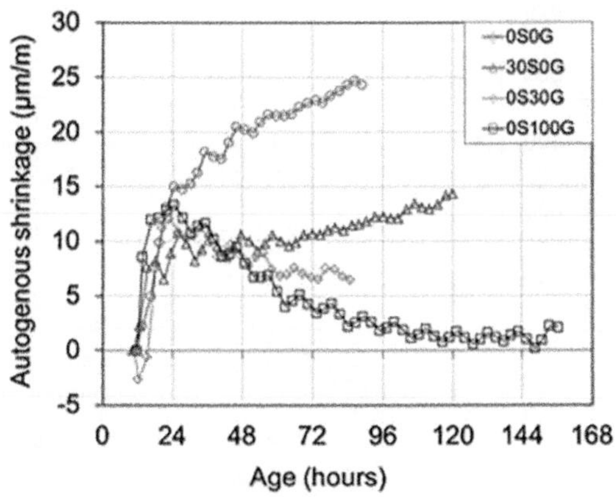

Figure 11.5 Autogenous shrinkage of RAC.

Table 11.2 Influence of coarse RA proportion on total shrinkage (autogenous + desiccation) and the characteristic time of shrinkage

	0S0G	*0S30G*	*0S100G*
w/c = 0.64			
ε_∞ (μm/m)	593	654	661
N_s (days)	15.4	17.6	23.7
w/w = 0.53			
ε_∞ (μm/m)	651	—	696
N_s (days)	15.5	—	19.7

and 0.64. There is a fairly clear tendency to increase the characteristic time of shrinkage with the proportion of RA. This evolution has already been demonstrated on mortars and can be related to the finer porosity of the cement paste of the original concrete. The most important part of the desiccation shrinkage is related to the mass loss which causes a decrease in the internal moisture and subsequently an increase in the capillary pressure which is the motor of the shrinkage. The behavior follows three steps. In the first phase, the mass loss causes almost no shrinkage. This relatively brief phase is associated with water evaporation from the largest pores at the periphery of the specimen (Khelidj and Loukili 1998). It increases with the percentage of aggregate substitution. This phase is more pronounced for concrete with high *w/c*. For concrete with a lower *w/c* (<0.35), this zone is virtually nonexistent, and the shrinkage–mass loss relation becomes linear (Bissonnette et al. 1999). The second phase seems always linear. It is characterized by the decrease in the internal moisture and the beginning of the desiccation shrinkage. However, in the last stage, the mass loss begins to stabilize with a continuous increase in shrinkage. In this third phase, we do not observe a mass recovery because of the possible carbonation of the material (measurement stop at 110 days). By translating mass loss into equivalent distance (Samouh et al. 2016), we obtain the results presented in Figure 11.6. A given drying depth can reflect two different configurations. In the case of RA, the paste and aggregates dry, whereas in the case of natural aggregates, the drying depth represents only drying of the cement paste. The values of the drying depths are close to the value of $D_{max}/2$ (where D_{max} represents the maximal aggregate diameter), which highlights the existence of a wall effect which affects concrete skin (Figure 11.7). Its rapid drying results in incomplete hydration or even microcracking.

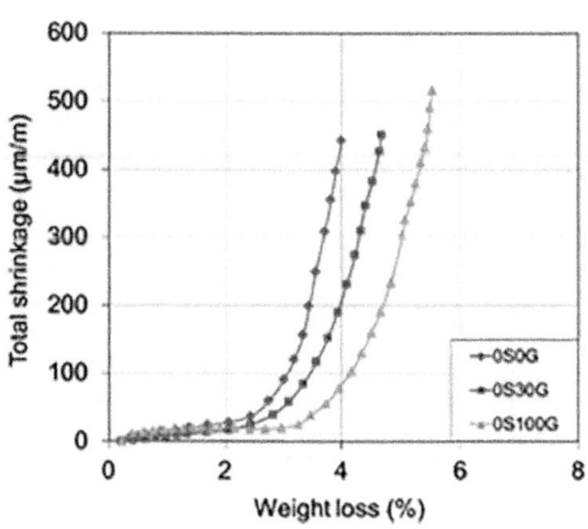

Figure 11.6 Influence of RA substitution on the total shrinkage–mass loss relation.

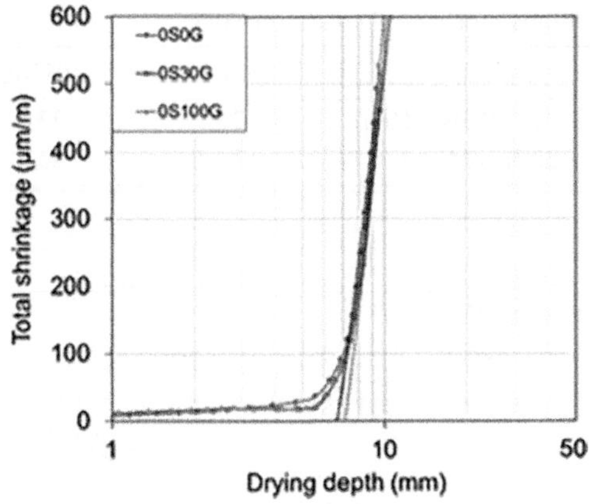

Figure 11.7 Influence of RA substitution on the total shrinkage–drying depth relation.

11.3.2 Second experimentation campaign

In this experimental program, three concretes were studied in order to assess the influence of RA incorporation on delayed strain properties.

Table 11.3 summarizes the designs of mixed concretes according to the specifications in the study of Sedran (2017). For three concretes, compressive strength of 35 MPa is targeted at 28 days of age. All physical properties and chemical composition of raw constituents are available in the study of Sedran (2017). Described concretes have been used for shrinkage study (present part) and for compressive creep.

After mixing according to Sedran (2017), concretes were casted in cylinders (Ø11 × 22) cm^3 to measure at 28 days of age compressive strength (NF EN 12390-3 2012), elasticity modulus (NF EN 12390-13 2014), and shrinkage (the conservation process is given in detail in [RILEM TC107-CSP]). Table 11.4 presents the results of F_{cm28d} and E_{28d} used in this part and in "Compressive Creep". The results and discussion are presented through a systematic comparison between 0S0G (reference concrete) and 30S30G and 0S100G (concretes with RAs). Figure 11.8 presents the variation of desiccation shrinkage with a duration from 28 to 146 days.

Table 11.3 Concrete designs (with dried aggregates)

Batch (kg/m³)	*0S0G*	*30S30G*	*0S100G*
Total water	185	220	238
Efficiency water	175	179	185
CEMII/A-L 42,5 N	299	321	336
Limestone filler	58	44	53
Natural sand 0/4	771	491	782
Natural coarse aggregates 4/10	264	168	
Natural coarse aggregates 6,3/20	810	542	
RS 0/4		214	
RAs 4/10		142	158
RAs 10/20		164	682
Set retardant		1.3	
Superplasticizer	2.1	1.64	2.18

Table 11.4 Instantaneous mechanical characteristics

Concrete	*0S0G*	*30S30G*	*0S100G*
Compressive strength at 28 days (MPa)	36.7	37.6	34.8
Elasticity modulus at 28 days (GPa)	36.7	33.4	30.0

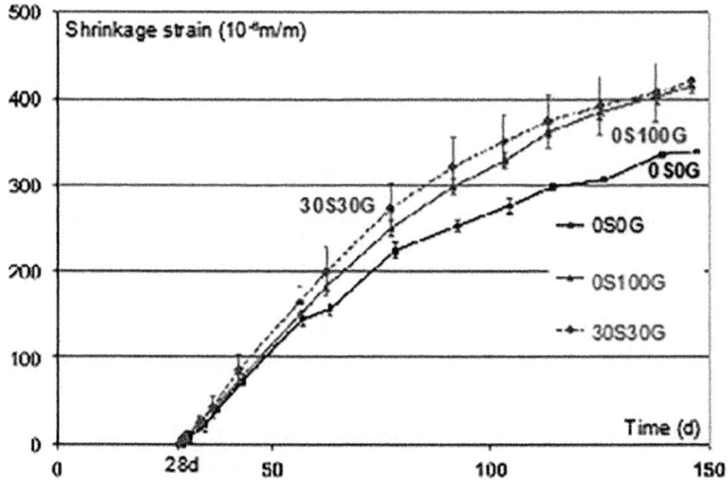

Figure 11.8 Desiccation shrinkage evolution on time.

Considering only the desiccation shrinkage, the experimental scatter and the experimental pattern (beginning of measurement at 28 days), both concretes with RAs (30S30G and 0S100G), exhibit strains that are similar in kinetics and amplitude. At the end of the tests (146 days), the measured values for desiccation shrinkage were about 340 and 420 µm/m, for 0S0G and RAC, respectively. So, comparatively to the reference concrete (only with natural aggregates, 0S0G), for both RACs, the addition of RAs increases the desiccation shrinkage values by around 24%. One can note that, after 28 days (beginning of test), no average autogenous shrinkage was detected for all concretes. It should be noted that the standard deviation for the results of endogenous shrinkage was included between 3 and 61 µm/m.

11.3.3 Third experimentation campaign

Chapter 22 presents the Chaponost experimental construction site carried out during the RECYBÉTON project. Six mixtures with the replacement rate ranging from 0% to 100% were used to build a parking lot. The mixes were sampled at the concrete plant, and conventional tests were performed, including drying shrinkage according to NFP 18-427. $7 \times 7 \times 28\,cm^3$ prisms were cast, demolded at one day, and cured at 20°C, 50% RH. The deformations were measured from 1 to 90 days. Mix proportions and test results are displayed in Figure 20.2. Figure 11.9 displays the relationship between recycling rate, calculated as the total mass of RA (irrespective of the fraction) divided by the total aggregate mass, against drying shrinkage at 90 days. Although some mixes contain only coarse RA whereas others contain only fine RA, it is interesting to note that all mixes are aligned on the same master curve. According to this curve, a total recycling essentially doubles the level of drying shrinkage.

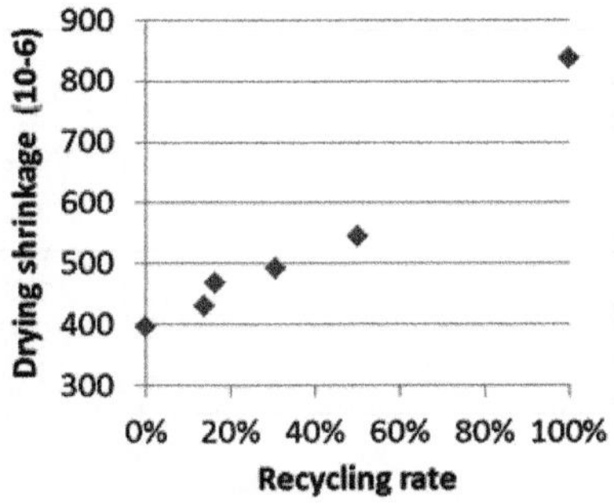

Figure 11.9 Drying shrinkage of Chaponost mixes, as a function of recycling rate.

11.3.4 Numerical study of shrinkage-induced cracking

The prediction of the thermomechanical behavior of RAC in structures needs the improvement of numerical models by considering the properties of RA. So, a numerical model is suggested in this section. The evolution of hydration (equation 11.3) and effects in terms of temperature rise and evolution of mechanical properties have to be considered so as to be able to correctly predict the development of stresses during the hydration and drying at long term. This model is commonly used in the literature (e.g., Briffaut et al. 2011): hydration is considered to be thermo-activated (through the use of Arrhenius law) and exothermic (latent heat) (equation 11.4). Mechanical properties are directly related to the hydration degree using power law and percolation threshold (equation 11.5).

$$\dot{\xi} = \tilde{A}(\xi)\exp\left(-\frac{E_a}{R \cdot T}\right) \tag{11.3}$$

$$C\dot{T} = \nabla(k\nabla T) + L\dot{\xi} \tag{11.4}$$

$$E(\xi) = E_\infty \bar{\xi}^{a1} \quad f_t(\xi) = f_{t\infty} \bar{\xi}^{a2} \tag{11.5}$$

where E_a [J/mol] represents the activation energy, R [J/K mol] the universal gas constant, T [K] the temperature, ξ the hydration degree, $\tilde{A}(\xi)$ [s^{-1}] the normalized chemical affinity, L [J/m^3] the hydration latent heat, k [W/m K] the thermal conductivity, C [J/m^3 K] the thermal capacity, $f_{t\infty}$ [MPa] the final tensile strength (when $\xi = \xi_\infty$, i.e., the final hydration degree), E_∞ [MPa] the final Young modulus, and a1 and a2 parameters governing the evolutions of the laws.

In case of thin structural elements (i.e., slab-on-grade), the latent part is negligible. An approximation of the hydration degree evolution is obtained by considering a linear evolution of hydration degree with the measured compression strength. In Section VII, this kind of approximation has been used to obtain the evolution of the hydration degree for the different volume fractions of recycled concrete (from 0% to 100% of RA (G) and RS (S)) (Figure 11.10). The evolution of the hydration degree progress is almost identical for all the formulations. The effect of RA seems not to be significant on the hydration evolution with time.

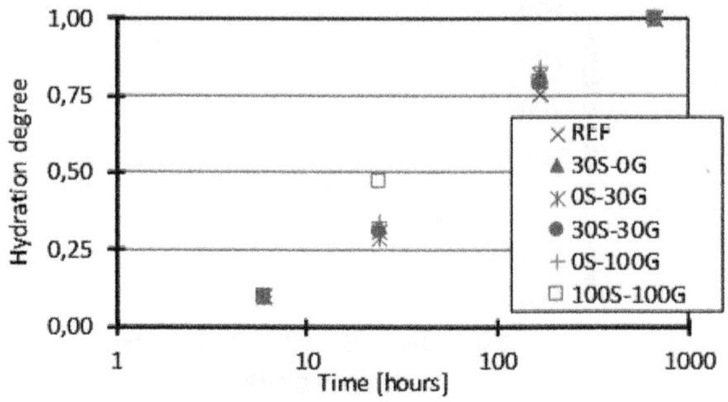

Figure 11.10 Comparison between different volume fractions of concrete for the evolution of hydration degree (XS-YG, X% of recycled fine aggregate, Y% of RA).

The autogenous shrinkage (ε_{au}) is directly related to the hydration degree evolution (Mounanga et al. 2006) and can be modeled by a linear relation with respect to the hydration degree (Briffaut et al. 2011):

$$\varepsilon_{au} = -\kappa(\xi)\bar{\xi}1 \text{ with } \bar{\xi} = \left\langle \frac{\xi - \xi_0}{\xi_\infty - \xi_0} \right\rangle \quad (11.6)$$

where $\kappa(\xi)$ represents an evolution function, $\langle \cdot \rangle_+$ the operator positive part, and ξ_0 the mechanical percolation threshold (Torrenti and Benboudjema 2005).

After casting, the recycled concrete structures are submitted to two types of transfers:

- Thermal transfers due to hydration and environmental temperature
- Water mass transfers due to the imbalance between internal and environmental relative humidity.

Shrinkage is also coupled to creep. For example, a basic representation can be used by combining one Kelvin–Voigt chain to a damper in series. The chain and damper coefficients depend on the hydration degree. To increase numerical simulations precision, a damper is added in series to predict in a more precise way the partially reversible character of creep strain and the fact that the basic creep has a logarithmic evolution at long term.

The experimental results show that we observe a proportionality between the drying creep strain in compression ε_{dc} (and in tension) and the drying shrinkage $\dot{\varepsilon}_{ds}$, linked by a material parameter λ_{dc}. The model proposed by Bazant and Chern (1985) for the prediction of drying creep strains is retained:

$$\dot{\varepsilon}_{dc} = \lambda_{dc}\dot{\varepsilon}_{ds}\tilde{\sigma} \quad (11.7)$$

Finally, the rheological model is extended to the states of multiaxial stresses by the use of a creep Poisson coefficient that can be taken equal to the elastic Poisson coefficient Benboudjema and Torrenti (2008). In case of drying in a room temperature, it is sufficient to consider only the mechanism of liquid water permeation which seems to be predominant during drying of concrete under these conditions. So, the use of the preservation of the water liquid body leads to the resolution of a nonlinear parabolic differential equation. The evolution of water content with relative humidity can be represented by the sorption isotherm (liquid water saturation degree vs. capillary pressure), using the van Genuchten model. By considering that drying shrinkage results from the mechanical strain of the solid skeleton under the capillary pressure, a poromechanical model for unsaturated media can be used (Sciumé et al. 2013). Numerical results of creep coefficients (ratio between strain and applied stress) for concrete, discussed in Chapter 22, are compared to the ones from Gomez-Soberon (2002b) in Figure 11.11. We can observe that the creep coefficient values are similar between the two studies. This allows validating the numerical prediction proposed for basic and drying creep. Moreover, 100% recycled concrete differs significantly from the other formulations with higher values of creep predicted. Referent concrete (0% of recycling) is close to the other formulations including substitution ratios inferior to 100% of recycling.

Use of cracking indicator due to shrinkage is not sufficient to predict where cracking will occur in the structure. Incorporating a damage model allows having access to crack localization. The one developed by Mazars (1984), which considers only principal extension strains for the prediction of material degradation, has been adopted and extended to take into account hydration.

Figure 11.12 presents the geometry and the mesh used to study the susceptibility to cracking of different mixtures of recycled concretes in ring tests. The final results for the 100% RAC after 200 h, in terms of relative humidity and orthoradial stresses (with a first elastic calculation), are shown in Figure 11.12. We observe that drying is very slow and not achieved. For stresses, drying gradients induces very high-tension stresses at the surface, exceeding the tensile strength of the material. A superficial microcracking is so expected. The fact that drying shrinkage is restrained by steel also induces tensile stresses in concrete, balanced by compression stresses in steel.

When considering damage (Figure 11.13), two different cracking patterns are obtained at the end of calculation. Only 100% RAC presents a through-cracking, whereas the other formulations just show a superficial cracking (on the concrete skin), which occurs classically because of drying gradient between surface and core of concrete. A great damage value at the interface between steel and concrete is also observed, which is representative of a debonding observed generally experimentally for these kind of ring tests Briffaut et al. (2011). Microscopic characterization should be done in order to validate the results obtained by this modelling approach.

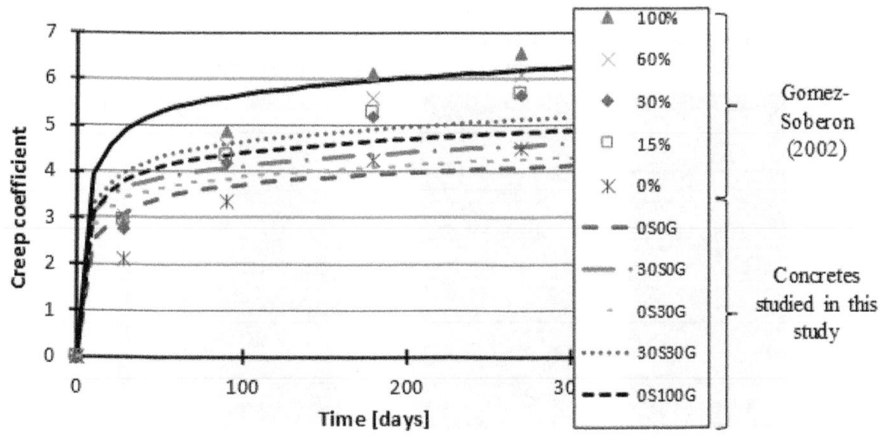

Figure 11.11 Evolution of total creep for different volume fractions—comparison between experimental (Gomez-Soberon 2002b) and numerical results.

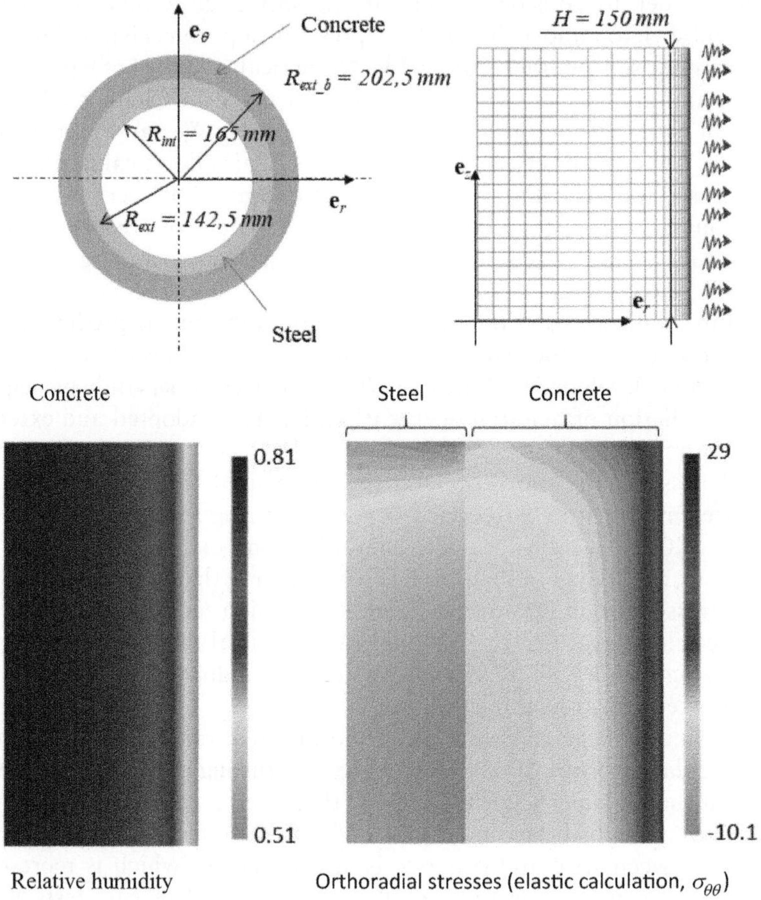

Figure 11.12 Geometry and mesh (axisymmetric calculus) for ring test numerical study on different mixtures of recycled concrete.

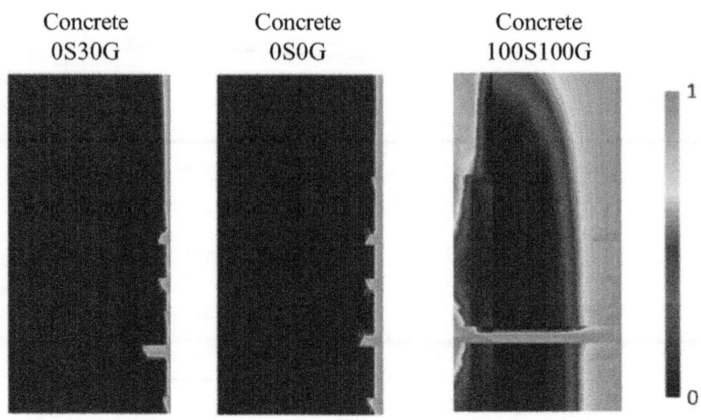

Figure 11.13 Damage into three different concrete mixtures: 0S30G, 0S0G, and 100S100G. Damage variable is going from 0 for sound material to 1 for fully cracked material.

11.4 EFFECT OF RAs ON THE CREEP OF CONCRETE

11.4.1 Compressive creep of recycled concrete

As for the shrinkage (Section 11.3.2), same concretes (0S0G, 0S100G, and 30S30G) were tested for creep behavior in either desiccation mode (hydric exchange with the surroundings) or autogenous mode (no hydric exchange with the surroundings). Two Ø11 × 22 cm^3 samples were used for each condition.

Before the first test (28 days of age), specimens were stored at 20°C in a plastic film (some of them were partially damaged during transport). The delayed strains were tracked for 146 days with the setup (LVDT sensor) shown in Figure 11.14. The loading rate was equal to 40% of the compressive strength measured at 28 days. The same loading rate was applied to 0S0G (14.7 MPa), 30S30G (15.0 MPa), and 0S100G (13.9 MPa). The total delayed strains can be considered as the sum of creep, instantaneous, creep, and shrinkage strains (equation 11.8). Hence, creep can be dissociated from the elastic and shrinkage strains although the different phenomena are interdependent (Cassagnabère et al. 2009).

$$\varepsilon_{dl}(\tau = 40\% f_{c28d}, t) = \varepsilon_{ins}(\tau = 40\% f_{c28d}) + \varepsilon_{cr}\ (\tau = 40\% f_{c28d}, t) + \varepsilon_{sh}(t) \quad (11.8)$$

where $\varepsilon_{dl}(\tau = 40\% f_{c28d}, t)$ represents the delayed strains at age t (loading rate of 40% of the strength at 28 days), $\varepsilon_{cr}(\tau = 40\% f_{c28d}, t)$ the creep strains at age t (loading rate of 40% of the strength at 28 days), $\varepsilon_{ins}(\tau = 40\% f_{c28d})$ the instantaneous elastic strains at age t (loading rate of 40% of the strength at 28 days), and $\varepsilon_{sh}(t)$ the shrinkage strains at age t.

The total and autogenous specific creep versus time (from 28 to 146 days) curves are plotted in Figure 11.15 (desiccation and autogenous modes). We must remember that, in the specific creep format, creep strain values are divided by loading values. These figures clearly show that the specific creep strains are significantly modified when NAs are partially replaced with RCA but with different tendencies depending on conservation mode. The increase in the creep strains can be linked to the higher amount of the cement paste. As for the desiccation mode, the gap between the curves becomes constant from 126 days on; at 146 days and comparatively to the reference concrete, a 8% decrease in strain is achieved when 100% coarse aggregates are replaced (79.8 µm/m for 0S0G and 73.4 µm/m for 0S100G) and a 13% increase is observed as for concrete with RS and RAs (90.0 µm/m for 30S30G). As for the endogenous mode, the variation is different with the desiccation

(a) (b)

Figure 11.14 Creep setup—(a) loading setup, (b) specimen with LVDT sensor.

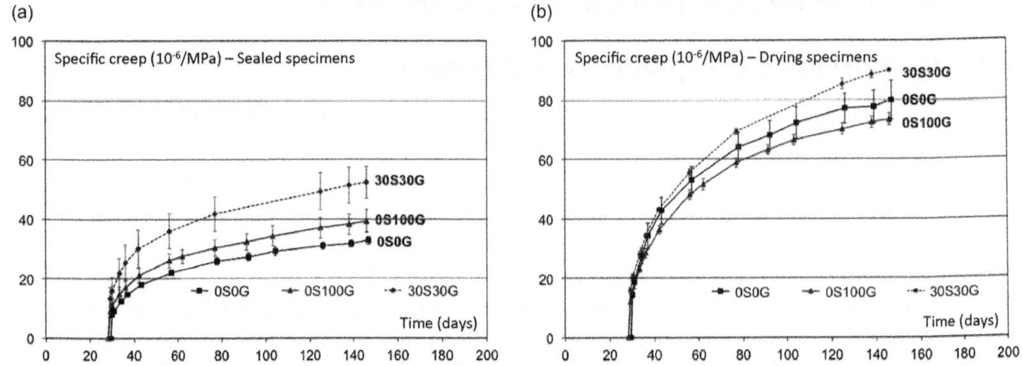

Figure 11.15 Specific creep strains evolution on time (a) autogenous mode, (b) desiccation mode.

mode; the difference between the curves remains constant from 77 days; at 146 days, and comparatively to the reference concrete, the increase is about +20% and +59% (for 0S100G and 30S30G, respectively).

These results show that the RCA incorporation in a concrete contributes to a delayed strain modification under loading and specifically for the creep in the desiccation and endogenous modes. It is noteworthy that the incorporation of 100% RAs seems to restrict the desiccation creep. However, this effect could apply only at short term, where the internal humidity of concrete is still high, due to the water reserve stored in RA.

The creep coefficient is the ratio between creep strain at t days (146 days) and elasticity strain at 28 days of age ($f_{c28d}(t_0)/E_{c28d}$) (Table 11.5):

$$C_{creep} = \frac{\varepsilon_{cr}\left(\tau = 40\% f_{c28d}, t\right) \times E_{c28d}}{f_{c28d}} \quad \text{with } t = 146 \text{ days} \tag{11.9}$$

With these coefficients calculated at 146 days, two noteworthy features can be observed. In comparison with the reference concrete (0S0G), the concrete incorporating RS and gravel (30S30G) present higher values in both modes. Quite the reverse, as for the concrete designed only with RAs (0S100G), the values are higher than the ones of the reference concrete that favors creep behaviors. To complete this study of the RCA incorporation impact in a concrete design on delayed strain, a comparison of experimental values obtained on shrinkage and creep tests with values obtained by prediction according to Eurocode 2 standard—"Calculus of concrete structures" (NF EN 1992-2 2006) in annex B "Creep and shrinkage strains"—is given. In EC2, various prediction models enable to define shrinkage and creep strains in endogenous and desiccation modes. To use the EC2 models, various inputs must be used. Table 11.6 summarizes inputs to calculate delayed strains.

Table 11.7 presents various results related to the experimental and the predicted delayed strain values (creep and shrinkage) for the endogenous and desiccation modes. From Table 11.7,

Table 11.5 Creep coefficient calculated at 146 days according to equation 11.9

Creep coefficient (μm/m)	*0S0G*	*30S30G*	*0S100G*
Autogenous	1.27	2.06	1.24
Desiccation	3.08	3.55	2.31

Table 11.6 Creep coefficient calculated at 146 days

Used inputs	*0S0G*	*30S30G*	*0S100G*
t_0 (days)	28	28	28
RH (%)	50	50	50
Specimens (cm)	Ø11 × h22	Ø11 × h22	Ø11 × h22
$f_{cm}(t = 28 \text{ days} = t_0)$ (MPa)	36.7	37.6	34.8
$f_{ck} = f_{cm}$–8 MPa (MPa)	28.7	29.6	26.8
$\sigma_{t0} = 40\%\ f_{cm}(t_0)$ (MPa)	14.7	15.0	13.9
$E_{cm}(t = 28)$ (MPa)	36,700	33,400	30,000
$E_c = 1.05\ E_{cm}(t = 28 \text{ days})$ (MPa)	38,535	35,070	31,500

Table 11.7 Results of delayed strains at 146 days (experimental values/EC2 predicted values)

Delayed strains (μm/m)	*0S0G*	*30S30G*	*0S100G*
Shrinkage strains			
Endogenous	-/22	-/25	-/17
Desiccation	339/517	420/508	415/538
Creep strains			
Endogenous	490/181	588/208	779/203
Desiccation	1,189/812	1,341/906	1,097/963

some comments can be made. As for the shrinkage, it seems that the predicted values according to EC2 models overvalue the experimentally measured values (from +21% to +53%) in spite of accuracy estimated to 30% for creep model according to the EC. As for the creep, the opposite would happen for the desiccation mode with an underestimation of the predicted values of creep comparatively of the experimental ones (from –12% to –74%).

To conclude on the impact of replacement of natural aggregate NA with RA produced from concrete wastes on creep behavior, the following results could be retained. A significant increase in the autogenous creep was observed for concrete incorporating RCA. On the other hand, it is noteworthy that for 100% of recycled gravel, the desiccation creep was reduced. Finally, the volume of paste, very different for the three concretes, demonstrate explain these modifications. EC2 models for creep and shrinkage prediction seem to be unsuitable when RCA was considered in concrete design.

11.4.2 Flexural creep of recycled concrete

The influence of RA on the flexural creep of concrete beams is presented in this section. For this purpose, three notched concrete beams (0S0G, 0S30G, and 0S100G) were submitted to a three-point bending creep test during three months. A constant load equal to 40% of the maximal strength (F_{max}) is applied. The bending creep tests are carried out on $10 \times 20 \times 80\ cm^3$ prismatic specimens using specific creep devices developed at GeM (Omar et al. 2009) placed in a temperature-controlled room at 20°C (±2°C) and in a hygrometry of 50% (±5%). The adjustment of the load applied to the beam is done by positioning a mass of 500 kg along the beam constituting the support. The total deflection is measured at the middle of the beam with a LVDT sensor. The deflection includes the instantaneous displacement due to load application and creep. Given the size of the studied specimens, the shrinkage component of the concrete beams was not measured. In this case, creep is defined as the difference between the total deflection under load and the instantaneous elastic deflection, measured after 2 min and corresponding to an equilibrium of the defection before the viscoelastic phenomenon takes place. Specific delayed displacements are calculated by subtracting the elastic part of the total deflection and dividing by the bending stress:

$$\sigma = \frac{3}{4}\frac{\rho g h l^2}{(h-a_0)^2} + \frac{3}{2}\frac{Fl}{b(h-a_0)^2} \tag{11.10}$$

where ρ represents the mass volume of concrete, l the length between the supports, F the applied load, h and b, respectively, the height and the depth of the beam, and a_0 the notch length.

Figure 11.16 shows the specific delayed displacements for the three concrete types. Note that creep kinetics are comparable for natural aggregate concrete and 30% RAC with an amplitude 1.2 times greater after 3 months of loading for the last. For concrete with 100% RA, the creep kinetic is faster and the amplitude is practically double compared to the natural aggregate concrete. The creep amplitude of the natural aggregate concrete after 3 months of loading is reached only after 4 weeks of loading by the 100% RAC. The analysis of the failure of these concretes (Guo et al. 2016) leads to give two assumptions: either RA creep or microcracks weaken the RAC under the constant load. To try to check these assumptions, we can analyze the creep evolution by plotting the ratio between the specific delayed displacement at time t on the final specific displacement measured at the end of the test as a function of time (Figure 11.17). The evolutions of the relative delayed displacement are comparable for the three concretes. This suggests that the creep mechanism is identical for recycled concrete and natural concrete. But, the final creep strains are different due to the higher amount of the cement paste and, if microcracks are present, due to a more brittle interface surrounding the RA made from the attached old mortar. In other words, viscoelasticity and/or microcracks are of the same type in the three types of concrete, but their amplitudes are different. The experimental tests performed show the influence of RA on shrinkage and creep of concrete. But these tests are time limited. So, to be able to prevent

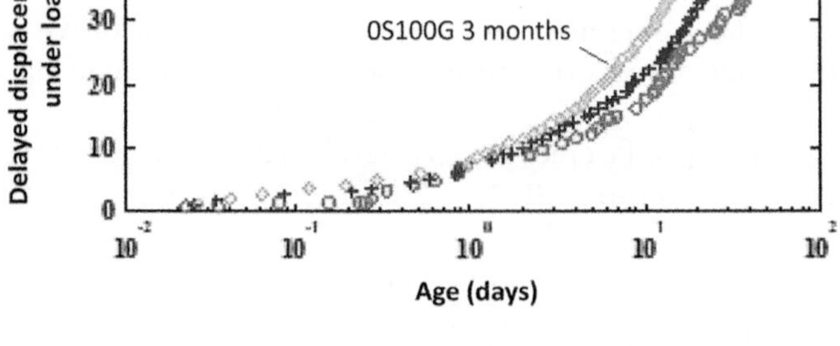

Figure 11.16 Delayed displacements of concrete beams under creep.

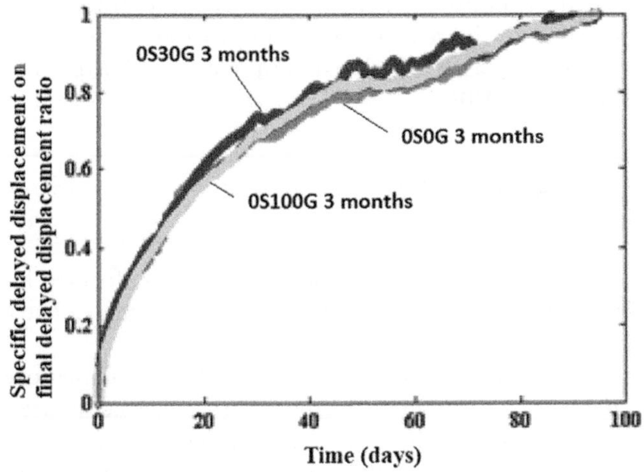

Figure 11.17 Relative delayed displacements of concrete beams under creep.

recycled concrete structures from cracking at early age and at long term or to quantify in a first time the degradation of their properties, it is necessary to build predictive modeling, including all the mechanisms involved.

11.4.3 Mesoscopic model for the creep of RAC

The simulation of the creep of concrete needs the knowledge of the creep compliance of its matrix (mortar) and aggregates. The creep of mortar needs the knowledge of the cement paste and sand. And the creep of the cement paste needs the knowledge of the hydrated and unhydrated phases. So, the creep of concrete depends mainly on the viscoelasticity of the latest components. In the case of RAC, at the concrete scale the attached old mortar on aggregates has to be considered too as an interfacial transition zone (Grondin and Matallah 2014) (Figure 11.18). Because it is formed as an ordinary concrete, its properties can be determined by the same method. The present model considers that the cement paste is formed by a matrix, constituted by C-S-H and pores, and inclusions formed by the other hydrated phases (V_{hyd} = [CH, ettringite, gypsum, C_3AH_6, FH_3]) and the residual cement clinker (C_3S, C_2S, C_3A, C_4AF) (Farah et al. 2013). With the

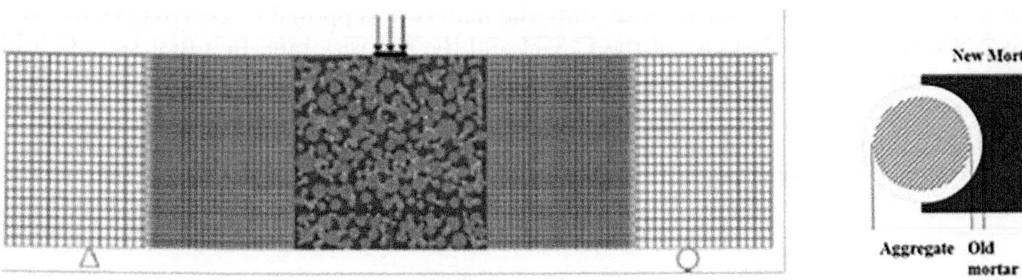

Figure 11.18 Mesh of RAC beams submitted to a three-point bending creep test (left) and representation of RA with the attached old mortar (right).

advancement of the hydration process, the clinker's volume fraction decreases, whereas the hydrate's volume fraction increases. According to the Arrhenius law, these volume fractions can be calculated based on the stoichiometric relations of cement (Bernard et al. 2003; Grondin et al. 2010):

$$V_k^P(t) = \sum_{l=1}^{n} V_0^i \frac{n_k^P \rho_c M_k}{n_l^R \rho_k M_l} \xi_l(t) \quad k = 1, m \tag{11.11}$$

where V_0^i represents the residual clinker volume, $V_k^P(t)$ the new formed hydrates volume, M the molar mass [g/mol], ρ the mass density [g/cm^3] and n the mole. The index k represents the products (clinker), l the reactants, and c the cement.

At the cement paste and mortar scales, the local problem is defined over a representative elementary volume defined by a volume V formed by two distinct phases: a matrix V_m and n inclusions V_i ($i = 1, n$). The matrix has a viscoelastic behavior, where its compliance (J) is defined by a generalized Kelvin model with four chains. A constant tensile load F is applied on the top surface Γ_1 of V according to the unit normal vector $\boldsymbol{n}$, and the bottom surface Γ_2 is fixed. These conditions imply local displacements fields $\boldsymbol{u}(\boldsymbol{y})$, local strain fields $\boldsymbol{\varepsilon}(\boldsymbol{y},t)$, and local stress fields $\boldsymbol{\sigma}(\boldsymbol{y},t)$ in each point $\boldsymbol{y}$ of V. The relation between local viscoelastic strains and local viscoelastic stresses is given by the following relation:

$$\varepsilon^v(y, t) = J(t) \otimes \sigma^v(y, t) \tag{11.12}$$

According to the Kelvin–Voigt model, $J(t)$ is given as follow:

$$J(t) = \frac{1}{E} + \sum_{i=1}^{i=3} \frac{1}{k_i}\left(1 - e^{\frac{-t}{\tau_i}}\right) \tag{11.13}$$

where k_i represents the stiffness coefficient, and τ_i the characteristic time.

The resolution of the local problem gives a relation between the average stresses $\langle\sigma\rangle_V$ and the average strains $\langle\varepsilon\rangle_V$ linked by the effective creep tensor J^{hom}:

$$\langle\varepsilon\rangle_V = J^{\text{hom}} \langle\sigma\rangle_V \tag{11.14}$$

At the concrete scale, the viscoelastic model is coupled to the damage model developed by (Saliba et al. 2013). The calculation of the creep of recycled concrete requires the determination of a large number of parameters. Indeed, the physical model presented here considers all the phases of the material and their elastic, viscoelastic, and fracture properties. On the scale of the cement paste, only the matrix is supposed to be viscoelastic, and this is due to the viscoelasticity of the C-S-H and the porosity rate. In a first step, C-S-H properties are determined from an inverse analysis from the concrete scale to C-S-H. The calculation at the concrete scale was calibrated on the experiment presented in Section 11.3.1, and "descending" on the CSH scale, the formula of (Ricaud and Masson 2009) is used to deduce the properties of CSH independent of the composition of the concrete and of the age:

$$k_i^{\text{CSH}} = k_i^{\text{CSH+pores}} \frac{3 \cdot f_p}{4 \cdot (1 - f_p)} \tag{11.15}$$

where f_p represents the volume fraction of pores in the matrix of the cement paste.

For the four chains of the creep model, the k_i values for the creep tests are given as follows: $k_1^{CSH} = 16.56$ GPa; $k_2^{CSH} = 5.96$ GPa; $k_3^{CSH} = 2.65$ GPa; $k_4^{CSH} = 0.53$ GPa. From these parameters it is then possible to calculate the creep properties of all cement pastes and therefore any mortar. For the old mortar attached, as its composition is unknown, assume that it is included among 16 formulations of ordinary mortar: four different *w/c* ratio (0.3, 0.4, 0.5, and 0.6) and four different sand-on-cement ratios (3, 3.46, 5, and 7). The hydration model makes it possible to estimate the average elastic properties of all the mortar formulas, and the creep model gives the average compliance. To limit errors, choose the lowest and highest values to limit the result on the concrete. Finally, the fracture parameters (f_t and the cracking energy G_f) of the old mortar attached are calculated according to the relationships of the damage model:

$$G_f = \frac{h}{2}\frac{(1+\nu)f_t^2}{2E} = \frac{h}{2}\frac{(1+\nu)E^2 \cdot \varepsilon_0^{\,2}}{2E} = \frac{h}{4}(1+\nu)E \cdot \varepsilon_0^{\,2} \tag{11.16}$$

where h represents the height of the finite element, and ε_0 the damage threshold. This allows the relationship between the old and the new mortar to be written as:

$$\frac{G_f^{\text{oldm}}}{G_f^{\text{newm}}} = \frac{E^{\text{oldm}}}{E^{\text{newm}}} \tag{11.17}$$

Also, the damaging elastic behavior of the mortar makes it possible to write $f_t^{\text{oldm}} = E^{\text{oldm}} \cdot \varepsilon_0$, with $\varepsilon_0 \in (5e^{-5}, 1e^{-4})$

Concrete beams are simulated by considering a mesoscopic mesh at the center of the beam and a homogeneous mesh at the extremities, whose properties are those resulting from the experimental measurements (Figure 11.16). Table 11.8 presents the set of parameters and the methods of determination (Lit., literature; Hyd., hydration model; Ass., assumption; Cal., calculated by the multiscale model; Calib., calibrated). The characteristic creep times of the four chains are given as follows: $\tau_1 = 0.1$ day, $\tau_2 = 1$ day, $\tau_3 = 10$ days, and $\tau_4 = 100$ days, respectively.

Specific numerical creep and experimental creep are shown in Figure 11.19. It is shown that for concrete with 100% RAs, the numerical displacement is less than that of the experiment. A recent study has shown that RAs are less resistant to shock and friction (Omary et al. 2015). Thus, the influence of the RA on the creep displacement of concrete should be considered in the simulation. For this purpose, the RA viscoelastic coefficient is calibrated to the value of 23 GPa for the four chains. This makes it possible to reproduce the creep of 100% RAC and 30% RAC without other calibration (Figure 11.19).

Table 11.8 Coefficients used in the creep-damage model

	E [GPa]	ν [-]	f_t [MPa]	G_f [N/m]	k_1 [GPa]	k_2 [GPa]	k_3 [GPa]	k_4 [GPa]
Natural aggregate	78 Lit.	0.24 Lit.	6 Lit.	80 Lit.	∞ Hyp.	∞ Hyp.	∞ Hyp.	∞ Hyp.
RA	54.2 Lit.	0.24 Lit.	6 Lit.	80 Lit.	∞ Hyp.	∞ Hyp.	∞ Hyp.	∞ Hyp.
Old mortar (min–max)	27.2–30.2 Hyd.	0.33–0.33 Hyd.	1.36–1.51 Mul. Cal.	139–154 Mul. Cal.	90–250 Mul. Cal.	45–76 Mul. Cal.	22–37 Mul. Cal.	4–7 Mul. Cal.
New mortar	27.45 Hyd.	0.33 Hyd.	1.6 Mul. Cal.	140 Calib.	892 Mul. Cal.	203 Mul. Cal.	98 Mul. Cal.	21 Mul. Cal.

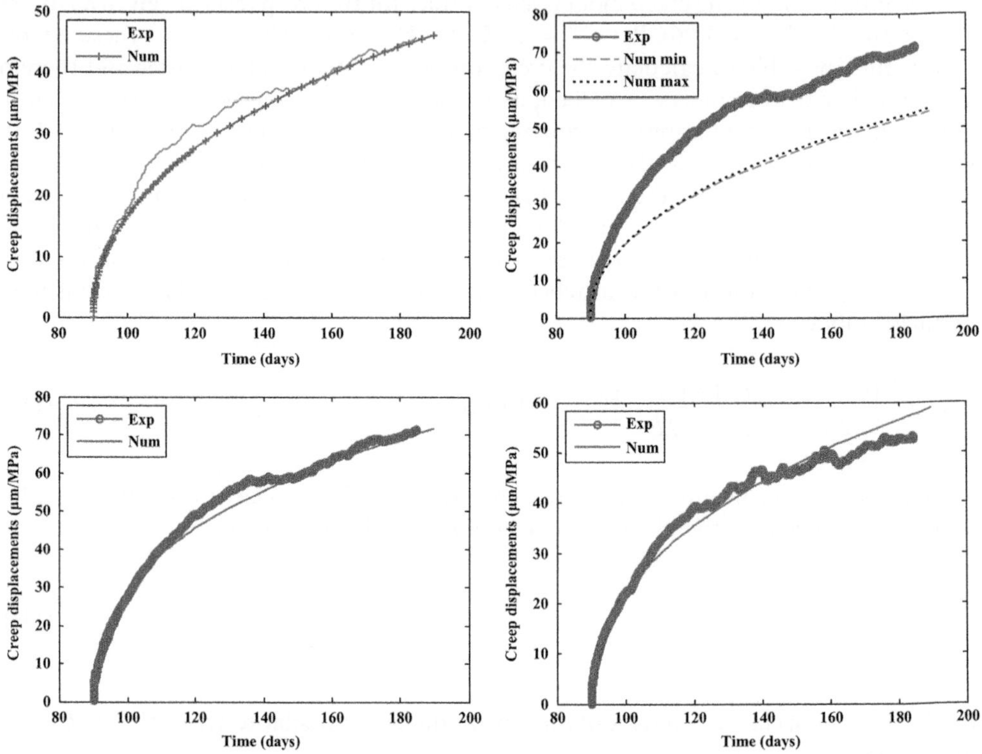

Figure 11.19 Comparison between measured and calculated creep displacements for 0S0G (top left), 0S100G without calibration (top right), 0S100G after calibration (bottom left), and 0S30G (bottom right).

The model allows to predict the creep of RAC by considering the influence of the attached old mortar. The determined parameters must make it possible to stop calibrating any coefficient. An example is given for several substitution rates (Figure 11.20).

11.5 FATIGUE OF RECYCLED CONCRETE

11.5.1 Experimental study and results

Three concretes with different recycling rates were designed with the components selected in RECYBÉTON project (see Appendix) in order to reach almost the same splitting tensile strength $R_{tb} = 3.3\,MPa$, classically used for concrete wearing course in France (Table 11.9). The fatigue behavior of these concrete was evaluated according to NF 98-233-1 French standard with two-point bending tests. All the details are given in the study of Sedran and Le Mouel (2016). Four batches of 90 L of each mix were produced to cast:

- Twelve trapezoidal samples for static bending test (NF P98-232-4 1994)
- Twenty-four trapezoidal samples tested in fatigue at a stress level close to σ_6 (NF P98-233-1 1994)
- Eight cylinders for compression tests (NF EN 12390-3 2012) and elastic modulus (NF EN 12390-13 2014)

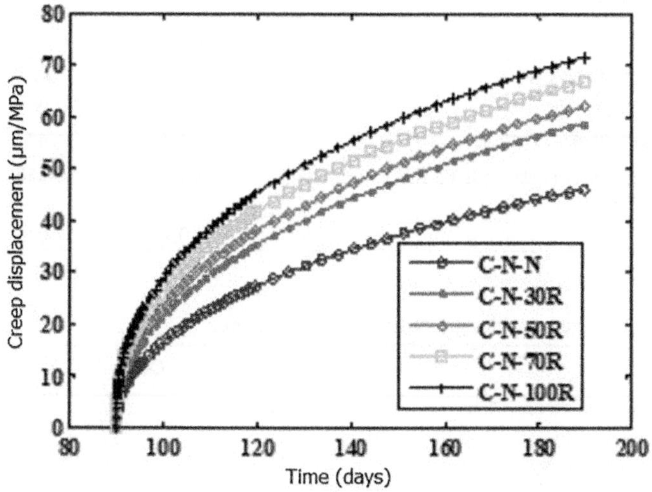

Figure 11.20 Creep displacements for different RA substitution rates.

Table 11.9 Mix design of tested concrete

Mix (kg/m³)	*0S0G*	*50S100G*	*100S100G*
Added water	179.5	245	271
Free water	169.1	167.5	176.7
CEMII/A-L 42,5 N Cement	308	385	443
Limestone filler	45	50	63
Natural sand 0/4	747	337	
Natural gravel 4/10	256		
Natural gravel 6.3/20	785		
RS 0/4		337	653
Recycled gravel 4/10		229	296
Recycled gravel 10/20		615	430
Retarder		2.7	3.6
Superplasticizer	0.7	2.3	3.5

All tests were carried out around 95 days to limit the influence of cement hydration during tests. Table 11.10 shows that the reference concrete displays the same order of fatigue parameters (SN close to 1 and endurance around 0.54) as tabulated in SETRA-LCPC (1997). It also confirms that there is a moderate decrease in endurance and an increase in the standard deviation when increasing the RA content.

In order to more quantitatively evaluate the impact of recycling, we have designed a typical structure corresponding to Sheet no 21 in the French pavements catalogue (SETRA-LCPC 1998) using the fatigue parameters of Table 11.10. This structure consists in a continuously reinforced concrete layer on a 5-cm class 3 bituminous-bounded well-graded aggregate layer laid on a 120 MPa platform. Table 11.11 summarizes the results of calculations accounting for a traffic of 3,000 trucks per day and direction. At first analysis, it shows that the economic and environmental values of the replacement of natural aggregates by RA are offset by a strong increase in cement consumption per m² of pavement. Yet a more detailed analysis is needed.

Table 11.10 Mechanical results

	0S0G	*50S100G*	*100S100G*
Compressive strength at 95 days (MPa)	44.6	55.7	57
E-modulus at 95 days (GPa)	37.2	29.5	28.8
σ_0 (splitting strength) at ~95 days (MPa)	4.7	4.5	4.0
σ_6/σ_0 at ~100 days (where σ_6 corresponds to the stress which generates failure at 10^6 cycles)	0.505	0.46	0.48
SN at ~100 days	0.7	0.9	1.2

Table 11.11 Concrete pavement design for the three concretes

	C-0S-0G	*C-50S-100G*	*C-100S-100G*
σ_{limit} (maximum allowable stress, MPa)	1.26	0.89	0.78
Minimal concrete layer thickness (cm)	22	27	29
Cement consumption (kg/m^2)	68	104	128
Increase in cement consumption compared to reference concrete (%)	0	53	90

The value σ_{limit} is directly related to SN and σ_6. Therefore, the endurance is an important parameter but σ_0 as well. The flexural strength σ_0 can be calculated from mix proportions with the following equations:

$$R_{tb} = k_t \; k_g \; f\left(\frac{W}{C}\right) \quad \text{((de Larrard 1999a), see also Chapter 10)} \tag{11.18}$$

$$\sigma_0 = 1{,}44 \; R_{tb} \; \text{(SETRA-LCPC 1997)} \tag{11.19}$$

where *f*() represents a decreasing function of *w*/*c* ratio, and k_g and k_t, granular parameters controlling the compressive strength and the splitting tensile strength, respectively. The literature concludes that RA and natural aggregates generally exhibit very similar k_t. Even when compared to natural aggregates they are made of, RAs seem to exhibit only a limited (between 7% and 13%) decrease in k_t. However, in the particular case of this study, RAs display a very low k_t and natural aggregates a high one, respectively, 0.364 and 0.453 (Sedran 2017). Thus, the selection made for the RECYBÉTON project has undoubtedly introduced a bias, because the natural aggregate and the RA do not come from the same region. This leads, for given slump and σ_{limit}, to a strong increase in the cement demand for concrete with RA, compared to the reference.

According to the scientific mix design approach presented in Chapter 15, it is possible to recalculate the theoretical composition and the elastic modulus of 50S100G and 100S100G for different k_t, with the same slump and σ_0 as the reference concrete. For example, if we adopt the same k_t for RA as for NA, as claimed by literature, and we keep the fatigue constants of Table 11.10, the theoretical overconsumption of cement drops to 22% (83 kg/m^2) for 50S100G and 39% (94 kg/m^2) for 100S100G. If we adopt a more conservative value $k_t = 0.42$, the theoretical overconsumption is, respectively, 30% and 52%. In that latter case, we can evaluate the theoretical overconsumption of cement with the following equation:

$$\Delta C = 0.44\,S + 0.08\,G \tag{11.20}$$

where S and G represent, respectively, the mass replacement ratios of fine and coarse NA by RA. This equation shows that the fine RAs have a higher influence compared to the coarse RA on cement overconsumption. Moreover, according to this model, the overconsumption will be limited to 2.4% for a concrete where only 30% of coarse NAs are replaced by RA, as classically done in several countries.

11.5.2 Discussion

Tests carried out in this study seem to confirm that high recycling rates (100% of the coarse aggregate, 50%–100% of the sand) lead to a slight deterioration in the fatigue behavior of pavement concretes. Thus, we can observe a decrease in endurance and an increase in SN standard deviation that are not completely balanced by the concomitant decrease in the elastic modulus, in the calculation of the pavement design. However, the evolutions are moderate and probably tainted with dispersion. Before the generalization of these results, it would be necessary to verify them further on other mixes.

The pavement design made in this study shows that high recycling rates lead to large overconsumption of cement per m^2 of pavement, which is detrimental to the economic and environmental benefits of recycling. However, it should be noted that the particular choice of RA made in the framework of RECYBÉTON project leads to values that are probably overestimated due to a particularly poor tensile behavior of the aggregates and not directly to a strong degradation of fatigue behavior itself (i.e., endurance σ_6/σ_0 and standard deviation SN). In addition, the technical community seems to be moving towards lower recycling rates (around 30% for coarse aggregate and no RS), as suggested in NF EN 206/CN. This makes it possible to limit greatly (around 2.5%) the increase in the cement demand for given slump and compressive strength. In this case, fatigue performance is likely to be close to that of natural aggregate concretes, since the deviations in the fatigue parameters observed in the present study are limited and represent maximum expected values. However, this point requires further investigation.

In conclusion, the fatigue behavior of RAC must be more widely verified but does not appear to be a blocking factor for the development of moderate recycling in pavement concretes.

11.6 CONCLUSION

A state-of-the-art and experimental tests on the delayed deformations and fatigue properties of RAC have been carried out.

The state-of-the-art and experimental tests carried out by French Project RECYBÉTON show that long-term shrinkage increases in RAC due to a higher amount of the cement paste and depending on their characteristics and substitution rate:

- Until 30% of coarse aggregate replacement rate, the increase is generally limited to 10%.
- Beyond 50%, the increase is greater. Some studies reported an increase equal to 70% for a 100% replacement rate. However, it is important to note that such an increase is not systematic, +24% has been obtained for 0S100G RAC tested in RECYBÉTON project. On the other hand, industrial mixes used on the Chaponost construction site displayed a double-shrinkage of the fully recycled concrete, as compared to the control mix.

The results also show that RA can play a positive role regarding autogenous shrinkage, due to high amount of its initial water content.

Concerning compressive creep deformation of RAC:

- A significant increase (i.e., higher than 35%) of total creep deformation can be observed from 20% replacement rate of NA by RA (that is to say an increase in creep about 175% for the total replacement of NA by RA, assuming a linear extrapolation).
- However, the previous tendency cannot be generalized to all RAC. Gomez-Soberon (2002b) point no significant increase in creep until 30% replacement rate. In fact, a large number of parameters are involved in creep phenomenon: characteristics of RA (volume proportion and compactness of the old concrete paste, nature of natural aggregate, initial water degree saturation, creep history of old concrete, etc.), paste volume proportion of RAC, *w/c* of RAC cementitious paste.

Experimental tests carried on flexural creep show an increase of about 45% with 100% of RA, but the difference is under 10% with 30% of RA. The difference with the compressive creep could be due to the tensile area at the bottom of the beam in this case which localizes stresses in the cement paste, which is more present in RAC. In the compressive creep tests, the effect of the amount of the cement paste is less significant.

A study was also realized to evaluate the effect of high recycling rates (100% of the coarse aggregate, 50%–100% of the sand) on flexural fatigue, for pavement applications. Tests carried out seem to confirm that even full recycling leads only to a slight deterioration in the fatigue behavior of pavement concrete due to a decrease in endurance (approx. –5%) and an increase in SN standard deviation (approx. +0.5 in absolute value). In conclusion, the fatigue behavior of RAC must be more widely verified but does not appear to be a blocking factor for the development of moderate recycling in pavement concretes (coarse aggregate recycling rate around 30% with no fine RA).

Chapter 12

Durability-related properties

P. Rougeau, L. Schmitt, and J. Mai-Nhu
CERIB

A. Djerbi and M. Saillio
IFSTTAR

E. Ghorbel
Université de Cergy Pontoise

J.M. Mechling, A. Lecomte, and R. Trauchessec
Institut Jean Lamour-Université de Lorraine

D. Bulteel
University of Lille

M. Cyr
Université de Toulouse

N. Leklou and O. Amiri
Université de Nantes

I. Moulin and T. Lenormand
LERM

CONTENTS

Abstract

This chapter presents the work carried out to study the influence of recycled concrete aggregates on concrete durability properties. Numerous previous studies have highlighted that recycled concrete aggregates are more porous than natural aggregates (NAs) and can lead to a decrease in the durability properties. The durability properties investigated by the French national project RECYBÉTON and the ANR project ECOREB are the properties related to the corrosion risks of the reinforcement (carbonation, chloride migration, gas permeability, and porosity), the resistance to freeze/thaw cycles, the risks linked to alkali–silica reactions (ASRs), and the presence of sulfates (ettringite/thaumasite formation). The results obtained show that the porosity accessible to water is a durability indicator not relevant in itself to sufficiently predict the risks of corrosion. It is more rigorous to consider at least the chloride diffusion coefficient and the resistance of concrete against carbonation penetration. The recycled aggregates (RAs) predictably decrease the performances of the concrete relating to the transfer properties with an intensity which depends on the intrinsic characteristics of RAs (porosity), the substitution rate, and the compactness of the cement matrix of the new concrete. By optimizing concrete mix (reduction of the *W/B* ratio in particular), it is easy to produce concretes that are as resistant as concretes only constituted by NAs. The frost resistance of RA depends on the characteristics of the original concrete. RA concrete (RAC) can be resistant to freeze/thaw cycles with or without deicing the salts depending on the frost resistance of RA and the application of the rules of formulation (binder content, *W/B* ratio, entrained air content and so forth). Concerning the risks of ASRs, RA can release significant amount of water-soluble alkalis in particular for the fine aggregates that contain an important fraction of adherent cement paste (ACP). RA can also contain unstable silica phases provided by some specific NA and possible pollutants (broken tiles). The studies also show that the actual recommendations can be used. Meanwhile, the main tests currently used for NA have to be adapted (higher water absorption of RA), and nevertheless, some of them (microbar test) are not adapted to RA. The studies on ettringite and thaumasite formation due to the presence of sulfates lead to recommendations in order to avoid the risk of disorders: 0.3% and 0.2% for the maximum water-soluble sulfate content, in RA and total aggregate, respectively.

12.1 INTRODUCTION

As a behavior at fresh state and of mechanical properties, the durability properties are the key points for the reuse of recycled aggregate (RA) in new concrete. The first of these properties are the durability transport properties (Section 12.3.1), which are determinant for most durability properties, especially those which control the reinforcement corrosion. The studies presented here focus on the durability properties, which can be used in the field of performance-based approach:

- Porosity and gas permeability, these properties often play the role of durability indicators
- Accelerated carbonation and chloride ions migration under electric field, generally used as performance tests.

Then the resistance to freeze/thaw cycles with or without deicing salts of concrete with RA is discussed (Section 12.3.2). Sections 12.3.3 and 12.3.4 deal with the internal reactions, alkali-silica reaction (ASR), and ettringite/thaumasite formation.

Previous chapters have presented the specificities of RA, their microstructure, and porosity. Their intrinsic characteristics are determined by the nature of natural aggregate (NA) and the amount and characteristics of the old paste. The questions investigated in this chapter for each durability properties are as follows:

- Does the RA play a specific role?
- What are the mechanisms involved?
- Are the usual durability tests and methodologies adapted to RA concrete (RAC)?
- What are the required conditions to produce RAC with similar durability than natural aggregate concrete (NAC)?

Existing knowledge about each durability-induced parameter is first reviewed in the "State-of-the-Art" section. Then, new results from either ANR ECOREB or RECYBÉTON projects are presented in the sections dealing with the various properties.

12.2 STATE OF THE ART

12.2.1 Durability transport properties

Carbonation is a well-known phenomenon which induces a risk of reinforcement corrosion with time. In the atmosphere, the percentage of CO_2 is 0.04% in average. When carbon dioxide penetrates into the cement matrix, the portlandite is consumed and calcite is formed. It induces a drop of pH in the interstitial solution from more than 13 to 8–9. As the pH of the cement matrix decreases with carbonation, the reinforcement is not protected anymore, depassivation occurs, and corrosion can begin if enough quantities of oxygen and water are present.

The carbonation of RAC is well documented in the literature (Otsuki et al. 2003; Cui et al. 2004; Evangelista and De Brito 2010; Sim and Park 2011; Dao 2012; Xiao et al. 2012; Xiao et al. 2013; Lotfi et al. 2015; Silva et al. 2015a).

The effects of RA incorporation in concrete on carbonation depth are linked to several factors:

- The substitution rate (Evangelista and De Brito 2010; Dhir et al. 1999)
- The cement content (Dao 2012)
- The characteristics of the old concrete from which RAs are coming (Xiao et al. 2012; Ryu 2002b)

- The type of grinding (Pedro et al. 2014)
- The nature and the quality of RA (bituminous materials, bricks, concrete, glass, and so forth) (Bravo et al. 2015)
- The curing of concrete (Amorim et al. 2012)
- The use of superplasticizer to reduce the *W/C* ratio (Buyle-Bodin and Hadjieva-Zaharieva 2002)

The large survey carried out by Silva et al. (2015a) found at the same time the heterogeneity of the results and the major conclusions.

- RA coming from crushed masonry, may include aerated and lightweight concrete blocks, and ceramic bricks, causes greater carbonation depths than recycled concrete aggregate.
- There is a greater probability that concrete mixes made with fine recycled concrete aggregate (FRA) exhibit greater carbonation depths than those of mixes with coarse recycled concrete aggregate (CRA).
- Generally, as the replacement of NA by RA level increases, the carbonation depth of RAC also increases.
- Depending on the parameters kept constant between NAC and RAC, the ratio of carbonation depths can vary: 1 to almost 2.5 in the case of CRA substitution, 1 to almost 8.7 in the case of FRA substitution.
- It is possible to produce RAC with similar carbonation resistance to those of conventional concrete mixes by optimizing water/binder ratio and the nature of binder.

Chloride penetration is, together with carbonation, the main reason responsible for the depassivation of the reinforcement. Similar tendency than those for carbonation is highlighted in literature (Otsuki et al. 2003; Evangelista and De Brito 2010; Sim and Park 2011; Xiao et al. 2013c; Xiao et al. 2014; Wil 2015; Lotfi et al. 2015; Bravo et al. 2015; Sucic and Lotfy 2016):

- The increment of the chloride diffusion coefficient varies linearly with replacement ratio of NA with RA.
- The replacement of FRA leads to chloride diffusion coefficient values higher than those obtained with CRA.
- A wide scatter of results can be found in literature due to the difference of methodology for the making of concrete mixes.
- Like conventional concrete, chloride migration can be improved by reducing water/binder ratio or using blast furnace slag, fly ash, and silica fume.
- An improvement of chloride migration resistance is observed when fine ceramic aggregate replaces fine NA, probably due to the pozzolanic nature of this material.

Some authors show that replacing coarse natural aggregate (CNA) by CRA increases the chloride migration (De Brito et al. 2010; Kou and Poon 2010). An increase in the chloride diffusion coefficient by 34% is measured by Evangelista and De Brito (2010) when Fine Natural Aggregate (FNA) is replaced in totality by FRA in concrete with a *W/C* ratio equal to 0.41 for NAC and 0.48 for RAC. The impregnation of these aggregates by a polyvinyl alcohol improves the resistance of concrete against chloride penetration. Other studies obtain similar chloride diffusion coefficient between NAC and RAC (Abbas and Fathifalz 2009; Sucic and Lotfy 2016).

Permeability of concrete is closely related to its microstructure. This transport property depends on the porosity, connectivity of the pores, cracks, and water saturation content of the concrete. Since the RAs are porous and, in some cases, cracked due to the grinding process, the incorporation of the RA in concrete affects its permeability. Most of the studies have clearly indicated that gas permeability of RAC made with total and partial substitution of NA by CRA is higher than that of NAC (Limbachiya et al. 2000; Olorunsogo and Padayachee 2002; Gonçalves et al. 2004; Kwan et al. 2012). However, no effect was observed for 30% replacement by some authors (Limbachiya et al. 2000).

It is observed that the difference of gas permeability between NAC and RAC decreased with the following parameters:

- The curing time (Kwan et al. 2012) suggests that the decrease in intrinsic permeability is due to the continuation of the hydration process in the cement system. For low *W/C* ratio, the hydration reactions induce the self-desiccation of the cement paste and, hence, cause the capillary spaces to become narrower.
- Curing condition (Buyle-Bodin and Hadjieva-Zaharieva 2002) reported that the air permeability for RAC was 6 times more than that of NAC after water curing and about 20 times after air curing.
- The increase in cement content in concrete was observed by Gonçalves et al. (2004) and using lower water-binder ratio (Lotfi et al. 2015).
- Mix process by two-stage mixing approach: first, coarse and fine aggregates are mixed with half of water; and second, cementitious material is added and mixed with rest of the water, which reduces the air permeability to about 41% in 182 days (Tam and Tam 2007).

12.2.2 Resistance to freeze/thaw cycles

Most of the studies report that the freeze-thaw resistance of RAC is lower than those of usual concretes (Buck 1977; Malhotra 1978; Coquillat 1982) and depends, in particular, on the degree of water saturation of the material (Zaharieva et al. 2004). Other researchers (Richardson et al. 2011) show that using an air-entraining agent or RA-based concretes can also prove to be as durable as concretes composed of NAs.

A study is carried by Yildrin et al. (2015) on the influence of RA water presaturation on freeze-thaw cycle resistance. The following specimen characteristics were investigated: *W/C* ratio (0.7, 0.6, 0.5) percentage of NA replaced by RA (0%, 50%, 100%) and degree of RA saturation (0%, 50%, 100%). The maximum particle size of RA was 8 mm, with a water absorption equal to 6.2%. It was found that saturation and semi-saturation of the 50% RCA enhanced the concrete resistance to freeze-thaw cycle. The possible effect of internal curing is mentioned by the authors.

In their overview, Xiao et al. (2013) reported several results as follows:

- Freezing and thawing resistance of RAC is similar to the corresponding NAC for a similar strength
- Some studies reported that the strength loss rate of RAC can be higher than that of NAC; the main reason is their higher water absorption
- As for NAC, it is necessary to use air entrainment.

Kaihua et al. (2016) show that the frost resistance of RAC is closely related to the properties of the old concrete of RA. The RA coming from a no air-entrained concrete exhibited poor frost resistance.

12.2.3 ASR and recycled concrete aggregates

ASR poses a durability problem to concrete. It can induce cracking and severe damage in concrete structures. The origin of ASR is a chemical reaction between four essential compounds included in concrete: reactive silica into aggregates, a sufficiently high concentration of alkali, the presence of portlandite, and a moisture level above 70%–80% relative humidity (RH). The chemical mechanisms were described in the literature (Stanton 1940; Dent Glasser and Kataoka 1981; Poole 1992; Wang and Gillott 1991). The consequences of the ASR due to gel formation are mechanical damages with different mechanisms that were described by Diamond (1989); Dron et al. (1998); Jones (1997); Prezzi et al. (1997); Dent Glasser (1979); and Chatterji (1989).

Recycled concrete aggregates (RA) are composed of two very different fractions: the adherent cement paste (ACP) representative from the compositions of the various cements used in the original mixes and the NAs (fine and coarse aggregate) also introduced in the mixes. The cementitious fraction contains a small amount of alkali elements while the aggregates recover many types of very different rocks (from carbonated to siliceous) that eventually include potential reactive phase (minerals and/or structures). Knowing that (local) aggregates represent about 70% of the concrete volume, the petrographic and chemical characteristics of the RA are logically linked to its industrial production basin and its geology. It is much marked for the coarser recycled grains because fine aggregates used into initial concretes are often alluvial resources where quartz is predominant. Then recycled concrete used as aggregates could contain significant quantities of soluble alkalis and potentially reactive (PR) silica, and the reuse of these RAs may present a risk for ASR (Etxeberria 2004; Shehata et al. 2010; Adams 2012).

In order to limit the risk of ASR in concrete made of NA, several ways exist regarding the formulation of the concrete itself. For instance, the use of nonreactive (NR) aggregates, the limitation of the alkali content in the formulation, or the use of supplementary cementing materials (SCM) is among the methods that have shown to be efficient to avoid or limit the expansion due to ASR (Scott 2006; Hong and Glasser 2002; Kawabata and Yamada 2015). In France, these methods have been included in a standard FD P18-456 2004 in order to limit the risk of ASR when using NAs. However, it remained to be proven if they could be applicable directly to the recycled concrete aggregates.

12.2.4 Risks due to ettringite and thaumasite formation

One of the problems in the use of RAs for the manufacture of new concrete is the presence of contaminants in the RAC (Nixon 1978). Khalaf and De Venny (2004) studied the recycling of demolished masonry rubble as coarse aggregate in a new concrete. They showed that mortar, gypsum, organic matter, chlorides, sulfates, and glass could be found in the RAs, decreasing their durability-induced properties. Other studies have highlighted the high concentrations of water-soluble sulfates of some RAC (Barbudo et al. 2012; Tovar-Rodríguez et al. 2013), which are therefore out of specification. For example, standard NF EN 12620 states that RAC must have a maximum rate of 0.2% water-soluble sulfates ($SS_{0.2}$). Large-scale analyses of an RAC stock representative of four large French production basins have sometimes shown high levels of water-soluble sulfates in some samples (PN RECYBÉTON thème 2 2014). In such cases, an internal sulfate attack (ISA) may happen, which consists of a delayed swelling of hydraulic mixtures due to the formation of a significant amount of secondary or delayed ettringite (Neville 2004). It appears in the presence of easily mobilizable sulfates and generally requires simultaneous specific conditions (moist environment, presence of aluminates and possibly alkalis in the cement paste, high temperature of the concrete at an early age, and so forth). In concretes incorporating RAC, sulfate inputs may have several origins, independent of those contained in cement. Soluble sulfates can not only be released from old concrete but also from accidental external pollution, such as plaster coatings (it should also be noted that in addition to ettringite, thaumasite can also be formed at low temperature).

12.3 RECYBÉTON'S OUTPUTS

12.3.1 Durability transport properties

12.3.1.1 Experimental procedures

12.3.1.1.1 Mix compositions

Three strength classes are studied (C25/30, C35/45, and C45/55 according to NF EN 206/CN 2014 specifications), and for each class, three different concretes are mixed. So, a total of nine mixtures are studied (Table 12.1). For C45/55, the lower compressive strength obtained is due to a lower cement content and a higher *W/C* ratio in comparison of those of corresponding NAC.

A cement CEM II/A-L 42.5 and a limestone filler are used as binders. The equivalent binder is calculated according to the NF EN 206/CN 2014 rules. The superplasticizer content is adjusted in order to keep constant the workability (slump between 19 and 20 cm which corresponds to an S4 class according to the standard NF EN 206/CN 2014. The slump measurement is performed following the standard NF EN 12350-2 2012. Crushed limestone aggregates are used as the NAs and the RAs come from construction demolition wastes. The main properties of the NAs and RAs are studied in RECYBÉTON project and are displayed in Table 12.2. Due to the presence of the old cementitious paste, the RAs present a higher porosity than NA which leads to a lower density and a higher water absorption.

Due to the higher water absorption capacity of RA by comparison with NA (Table 12.2), water presaturation of the RA is adopted before mixing. In this study, the RAs are

Table 12.1 Mix proportions of concretes, per cubic meter

	C25/30			*C35/45*			*C45/55*		
	NAC	*F30C30*	*C100*	*NAC*	*F30C30*	*C100*	*NAC*	*F30C30*	*C100*
Water (L/m^3)	184	177	176	178	158	172	168	148	150
CEM II/A-L 42,5 N (kg/m^3)	267	282	279	299	324	336	390	371	369
Filler (kg/m^3)	45	49	70	58	44	53	100	65	73
FNA (0/4) (kg/m^3)	772	492	794	769	495	782	732	483	775
FRA (0/4) (kg/m^3)	0	233	0	0	216	0	0	229	0
NCA (4/10) (kg/m^3)	264	167	0	264	169	0	250	164	0
RCA (4/10) (kg/m^3)	0	151	161	0	143	158	0	148	157
NCA (6,3/20) (kg/m^3)	811	539	0	808	546	0	769	529	0
RCA (10/20) (kg/m^3)	0	167	691	0	164	682	0	162	676
Superplasticizer (kg/m^3)	0.14	0.13	0.08	0.29	0.66	0.25	1.23	1.28	0.92
w/equivalent binder	0.57	0.54	0.53	0.49	0.46	0.47	0.34	0.35	0.36
Slump (cm)	19.5	20	19.5	20	20	20	19.5	19	19
Compressive strength at 28 days (MPa)	31.8	32.4	29.4	40.7	46.6	39.8	61.4	58.8	51.2

Table 12.2 Properties of the aggregates

	Relative density of particle, Dr (g/cm³)	*Water absorption, A (%)*
FRA (0/4)	2.08	8.9
CRA (4/10)	2.29	5.6
CRA (10/20)	2.26	5.8
FNA (0/4)	2.58	0.8
CNA (4/10)	2.71	0.51
CNA (6,3/20)	2.71	0.46

oversaturated at 1%, meaning that the water content of RAs for mixing will be equal to the water absorption plus 1%. The RA and the water quantity to add are introduced in a hermetic barrel. Rolling up the barrel for 10 min on the floor leads to increase the homogeneity of the water in the RA. Then the barrel is at rest 2 h before concrete mixing.

12.3.1.1.2 Specimens' casting and curing

For each composition, prismatic specimens (7 cm × 7 cm × 28 cm) and cubic specimens (15 cm × 15 cm × 15 cm) are used to determine the physical properties of concrete. All prismatic samples are casted in steel molds and all cubic samples in plastic molds. They are compacted using a vibrating table. Neither segregation nor bleeding is observed. All specimens are unmolded 24 h after casting. Two types of curing are studied:

- Wet curing: in water (20°C) until the day of testing (90 days in this case)
- Dry curing: 3 days in water (20°C) and 87 days in laboratory conditions (20°C ± 2, 50% ± 5 RH).

Then some cylindrical specimens of Ø5 × 10 cm are taken from the center of the cubic specimens in order to carry the chloride migration test. Before this test, the samples are vacuum saturated with alkaline solution (NaOH 0.1 M) for 72 h.

The prismatic specimens are used to perform the carbonation test.

12.3.1.1.3 Procedures for the characterization of the transport and durability properties

Porosity to water was measured by total saturated water method according to NF P18-459 2010 standard. The samples have been weighted both below water and in air before being oven dried at a temperature of 105°C until constant mass. The average of three measurements was used to evaluate the water porosity.

The carbonation test is performed according to the European standard project (prCEN/TS 12390-12 2010). After the curing period, the samples are placed in a room at 20°C ± 2 and 50% ± 5 RH for 15 days. After this preconditioning, the 27 specimens (three for each concrete) are placed in the CO_2 enclosure where the CO_2 concentration is 4% ± 0.5%. The RH for this test is 55% ± 5% and the temperature is 20°C ± 2°C. After 90 days in the enclosure, the specimen is split and pulverized with phenolphthalein. The depth reached by the pH drop front is measured.

Chloride migration test is performed according to XP P18-462 (2012) standard. Three cylindrical test specimens are tested for each composition. Each specimen is placed between two compartments of a cell where flat silicone circular seals ensure that the system is leak tight. The solutions are made with NaOH (0.025 mol/L) + KOH (0.083 mol/L) in the upstream and the downstream compartment. NaCl (0.513 mol/L) is added in the upstream solution. A tension (30 V) is applied between the two compartments in order to force the chloride ions movement from upstream to downstream. The test is carried out at 20°C ± 5°C. The test duration is about 24 h for these concrete mixes. At the end of the test, the specimen is split and pulverized with silver nitrate. The depth reached by the chloride in the specimen at a predetermined age is measured and then used to calculate the chloride diffusion coefficient.

An apparatus known as the CEMBUREAU permeameter was used for the determination of gas permeability according to XP P18-463 (2011) standard. This is a constant head permeameter, and oxygen gas is used as the permeating medium. A pressure difference up to 0.3 MPa is applied to the specimens in the pressure cells which are sealed by a tightly fitting polyurethane rubber pressing under high pressure (0.7 MPa) against the curved surface. To determine the intrinsic permeability, the equation proposed by Klinkenberg (1941) was used.

12.3.1.2 Results and interpretation

12.3.1.2.1 Carbonation and porosity

In cementitious materials, carbon dioxide transport is essentially linked to the following main mechanisms:

- The Portlandite proportion: By reacting with CO_2 to precipitate calcite, it slows down the CO_2 migration and contributes to a reduction of the porosity because of the highest molar volume of the calcite
- The concrete porosity: its proportion, size, and connectivity of pores
- The degree of water saturation: The migration coefficient of CO_2 through water is 10,000 time slower than in the air.

The results confirm the key role of *W/B* ratio (Figure 12.1): the higher is the ratio, the greater the carbonation depth is. The performances of C45/55 concretes (lower *W/B* ratio) with respect to the carbonation phenomenon are thus better than those of C35/45 concretes, which are themselves higher than those of C25/30. Concretes having a low *W/B* ratio have a lower porosity, so CO_2 penetrates easier into the cement matrix. The pores are also of smaller sizes which lead to a higher saturation rate, a factor unfavorable to the migration of CO_2. In these concretes, the clogging effect also has more intensity.

The results presented in Figure 12.1 show the significant increase in the porosity accessible to water of the concrete as a function of the rate of substitution of NAs by RAs. It is observed that the carbonation depths are not strictly correlated with the total porosity of the concretes. For example, the C25/30 NAC and C45/55 F30C30 concretes have porosities close to 14.8% and 14.2%, respectively. On the other hand, their carbonation depths differed significantly: respectively 24.9 and 7.9 mm for the dry cure and 15.7 and 5.7 mm for the wet cure. Therefore, the porosity, an indicator of general durability, does not allow to immediately anticipate the behavior of recycled concretes with regard to the risks of corrosion of the reinforcements by carbonation of the concrete cover.

The porosity of the RAs appears to play a role in carbonation, which is all the more important as the *W/B* ratio of the new concrete is high. The results presented in Figure 12.1 show that the carbonation depths of the C25/30 C100 concrete are much higher than those of the C25/30 NAC and C25/30 F30C30 concrete. In the case of wet curing, this trend is not observed for C35/45 and C45/55 concretes. These results are in the direction of a lower

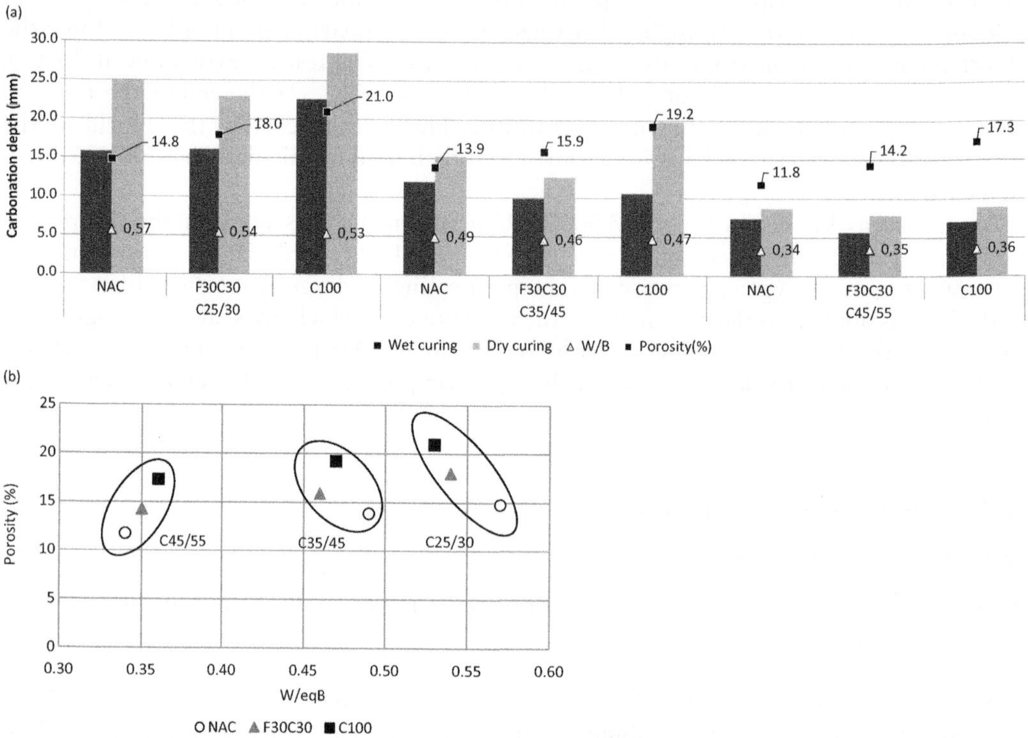

Figure 12.1 Influence of curing on the carbonation depths of RAC (a); porosity versus W/eqB (b).

impact of the RAs when the compactness of the cementitious matrix is high. In the case of C35/45 and C45/55, the cementitious matrix of the new concrete acts as a diffusion barrier and the RAs probably result from a less compact and therefore more permeable concrete.

It is possible that this diffusion barrier effect, particularly marked in the case of C45/55 concretes, is accentuated by the wall effects. During concrete pouring, some of the aggregates come near the walls of the mold. This local loosening of the granular stack near a wall, also called "wall effect", due to aggregates that can only move in substantially parallel directions, partially conditions the microstructure of the skin concrete. Indeed, during the vibration, the center of the large aggregates cannot approach closer to the walls of the mold than to their radius. In the first millimeters from the surface exposed to CO_2, the proportion of cement matrix/aggregate is therefore higher, which could accentuate the effect of the cement matrix as a diffusion barrier.

For all strength classes, a slightly lower carbonation is observed for the F30C30 concretes than for the NAC. In the case of C25/30 and C35/45, the main cause is probably a lower *W*/*B* ratio in F30C30 concretes compared to NAC, which leads to a more compact cementitious matrix and therefore a difficult CO_2 penetration through the paste. This assumption is not necessarily contradicted by the results obtained on C45/55 insofar as the diffusion barrier effect could be predominant for these concretes. Another hypothesis is that in the case of F30C30 concretes, the rate of pore saturation at the start of the accelerated carbonation test is higher. This would result from the high porosity of the RAs and the fact that they are presaturated before manufacture. The RAs would then act as a water reservoir, which would lead locally to an increase in the liquid water content and thus to a decrease in the carbonation depth (decrease in the CO_2 transfer properties). If this assumption is true, C100 concretes should also benefit from the same phenomenon and should have lower carbonation depths or the like, which is not the case. This mechanism may be more pronounced in the case of FRA than CRA.

At the end of the 90-day curing period, the amount of Portlandite is considered to be the same between moist cured and dry cured concretes, although it is likely that the concretes in the water for 90 days lead to a higher level of Portlandite due to favored hydration reactions. The porosity of the concretes cured in water and in laboratory conditions can also be considered very close. The main difference between the two series of concrete shown in Figure 12.1 is the water content of the concretes at the start of the accelerated carbonation test. Drying in the open air or in an oven has the effect of reducing the degree of water saturation of the concrete and allowing a better diffusion of the CO_2 within the concrete. A dry cure therefore accelerates the phenomenon of carbonation.

Thus, the difference observed in Figure 12.1 between the wet curing and the dry curing is in accordance with the expectations. The depth of carbonation increases for all the concretes by 39% in average when a dry cure is carried out. The difference is greater for a lower resistance class: 42% in average for C25/30 versus 28% in average for C45/55. The desaturation is faster in the less compact concretes. It appears that dry curing has a greater influence on the carbonation resistance of less compact concretes (*W/B* ratio higher). A more compact cementitious matrix limits the impact of other factors (cure, amount of RAs) on the concrete.

It is possible to reduce the *W/B* ratio to increase the compactness and compensate the negative effects of RA. Studies in the literature also go in this direction (Silva et al. 2015a). However, the authors draw attention to the need to adapt the superplasticizer content in order to maintain the workability of concrete.

In conclusion, it appears that:

- It is easy to produce concretes made of RA, even at high rates of substitution, resistant to carbonation
- When the new concrete is compact, it seems that the cementitious matrix acts as a diffusion barrier—the effect of which should be more marked as the difference in compactness between the old and new cement matrices is great.
- The positive effect associated with the presence of water in the RAs which would act as a reservoir and would maintain a higher saturation rate, thus preventing the advance of the carbonation front, is neither confirmed nor contradicted; if it exists, it appears as a second-order effect.

Finally, the coexistence of many mechanisms described above, without mentioning the conditions for carrying out the accelerated carbonation test from one study to another, explains the variability of the results presented in the literature (Figure 12.2).

12.3.1.2.2 Chloride migration

The results obtained show that the coefficient D_{app} increases with the ratio *W/B* ratio and the rate of substitution in RAs (Figure 12.3) which confirms the results of the literature. Chloride ions penetrate more easily into the coating concrete containing RAs. This is most likely due to the old cement matrix of RAs, which is more porous than the new cement matrix.

In contrast to the observations made during the carbonation tests, the increase in the chloride ion diffusion coefficient is higher in the C45/55 concretes than in the C25/30. Apparent diffusion coefficients are higher for a lower class of strength, but the increase in the apparent

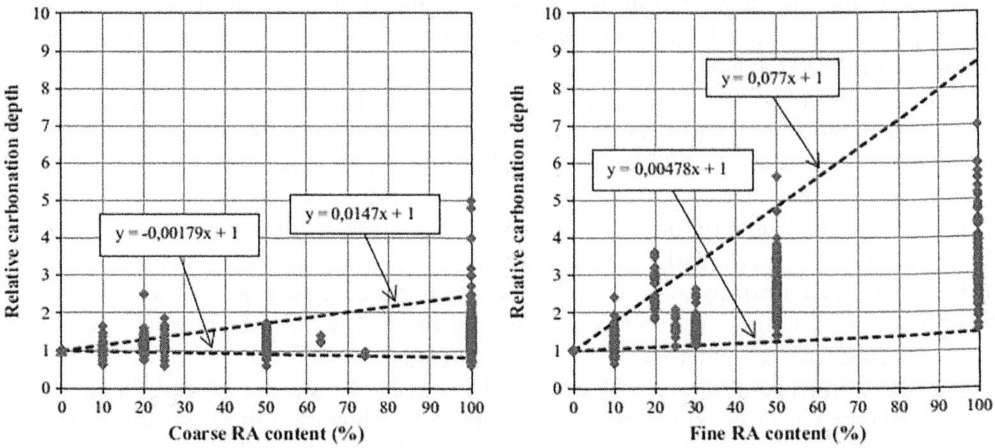

Figure 12.2 Relative carbonation depth versus replacement level for CRA and FRA (Silva et al. 2015a).

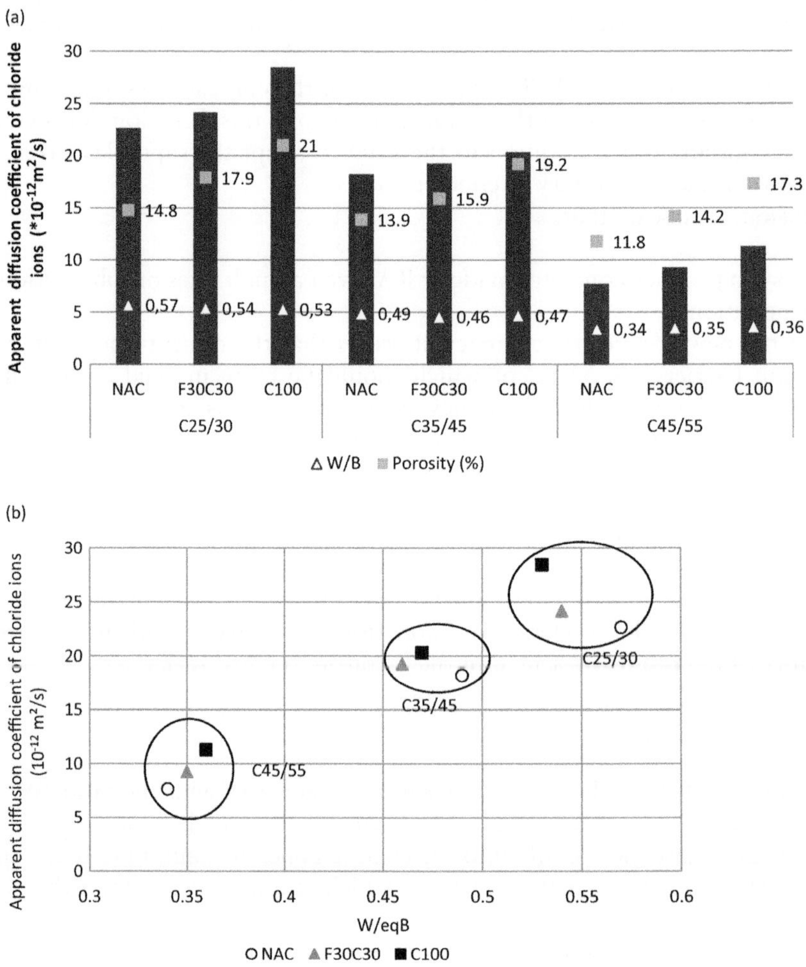

Figure 12.3 Chloride ion diffusion coefficient of RAC function to the resistance classes (a) or W/eqB (b).

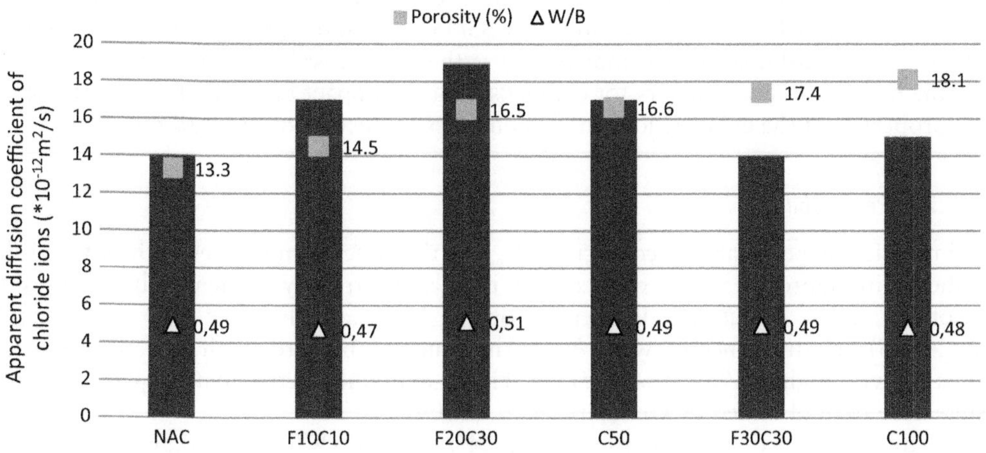

Figure 12.4 Chloride ion diffusion coefficient of C35/45 RAC.

diffusion coefficient within a class is greater when NAs are replaced by RAs in a concrete with low *W/B* ratio (Figure 12.3). By considering the C25/30 and C35/45 concrete ranges, the substitution of 30% NA with RA leads to a low increase in diffusion coefficient (less than 10%). A same substitution rate leads to an increase in the apparent diffusion coefficient of about 21% for C45/55 concretes. These trends should be approached with caution given the uncertainties of the test measurements.

Some tests have also been realized on a different set of C35/45 concretes among which some contain a lower quantity of RA (Figure 12.4). Same tendencies are observed. It appears that the coefficient of variation of the chloride apparent diffusion coefficient increase when RAs are added (6.9%–13.8% for RAC instead of 2.8% for NAC).

These results can be explained by the presence of the old cement paste around the RA, which probably corresponds more to the cementitious matrix of a concrete formulated with a *W/B* ratio close to a C25/30 concrete rather than a C45/55 one. Apparently, the diffusion barrier effect promoted for carbonation does not seem to play such a strong role for chloride. However, it is not possible to exclude this effect. Indeed, the chloride test is carried out on specimens with sawn faces, which excludes by definition the faces in contact with the molds where the wall effects could accentuate the diffusion barrier phenomena. On the other hand, since the test specimens are always saturated during the test, the possible water reservoir effect of the RAs cannot be mobilized (assuming it exists). In reality, the concrete in contact with chlorides is always a coffered or molded face, exposed in most cases to carbonation process which decreases the ability of the cement matrix to fix the chlorides (disadvantage) but which leads to a clogging effect of the cementitious matrix (advantage).

In conclusion, it appears that:

- The chloride ion migration test tends to overestimate the influence of RA
- The RAs, at least those used in this study, have relatively little influence if the substitution rate is less than 30% simultaneously for FRA and CRA; for a 100% CNA substitution rate, the scattering diffusion coefficient of a CX/Y strength class concrete is close to the lower grade class concrete: C45/55 C100≈C40/50 NAC; C35/45 C100≈C30/37 NAC; one can effectively observe in Figure 12.3 that the value of C45/55 C100 corresponds to the mean of C45/55 NAC and C35/45 NAC values which could correspond

to C40/50 NAC value if this concrete had been realized. The same observation can be done for C35/45 C100 (mean value of C35/45 NAC and C25/30 NAC values) as well.

- As for the carbonation phenomenon, it is therefore possible, if necessary, to play on the formulation parameters for compensating the impact of the RAs on the transport properties of the chloride ions in the concretes (reduction of the *W/B* ratio).

12.3.1.2.3 Permeability

The intrinsic gas permeability coefficient (k_{int}) of the concrete mixtures with NA and RA are shown in Figure 12.5. The results show that k_{int} is strongly influenced by the proportion of RA. These evolutions also depend on the *W/B* ratio. For low *W/B* ratio (C35/45 and C45/55), the difference between k_{int} of the concrete mixtures with NA and RA is less compared with higher *W/B* ratio (C25/30). This is due to the better matrix quality. Similar results have been obtained by Gonçalves et al. (2004).

One can note that despite a denser paste, C45/55 C100 (*W/B* ratio=0.36) has a similar permeability than C25/30 NAC (*W/B* ratio=0.57) and C45/55 F30C30 (*W/B*=0.35) has similar permeability than C35/45 NAC (*W/B* ratio=0.49). It means that it is necessary to significantly reduce the water/equivalent binder ratio when RA is added to obtain similar values of permeability. An amount of 25% RA requires more or less a *W/B* ratio decrease equal to 0.05 to achieve the same level of k_{int}.

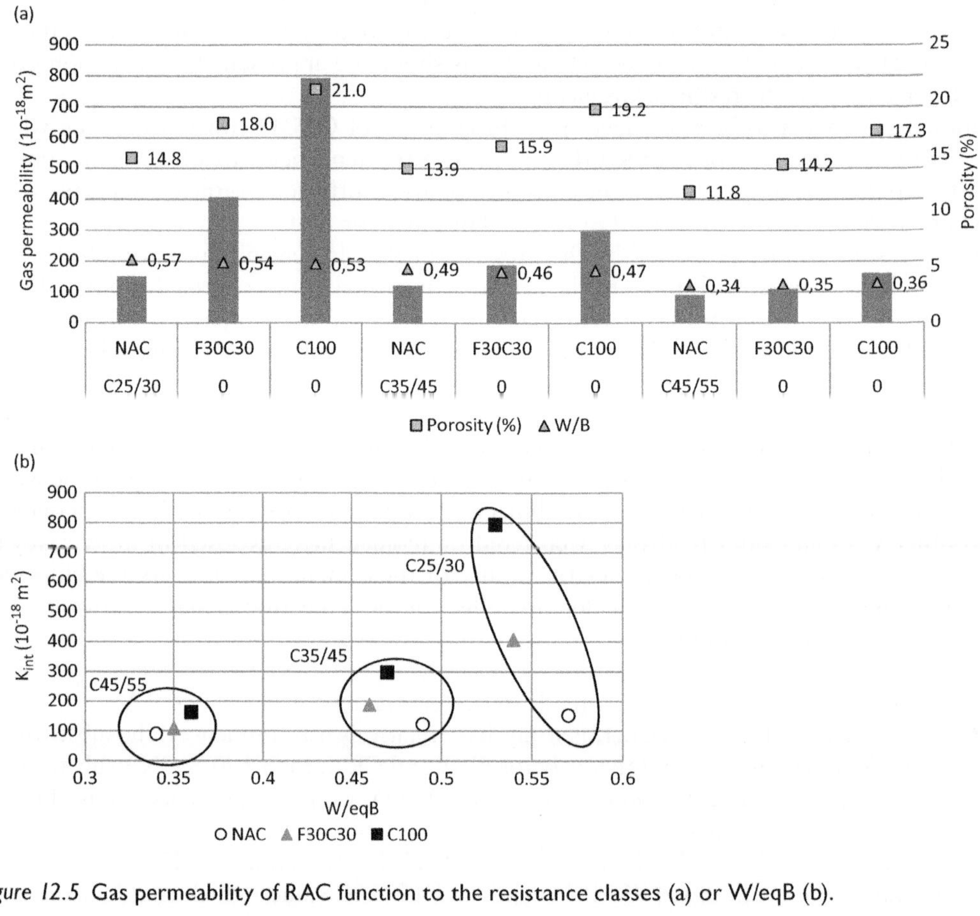

Figure 12.5 Gas permeability of RAC function to the resistance classes (a) or W/eqB (b).

12.3.1.3 Conclusion

The rate of substitution of the RAs plays an important role on the porosity accessible to the water of the concretes. The increase in the rate of incorporation of RAs is accompanied by an increase in the porosity of the concretes. This increase is attributed to the cement paste of the original concrete contained in the RAs.

The porosity accessible to water is an indicator of durability useful in first approach, notably in the context of a control of the regularity of concretes during the construction. However, it does not appear to be relevant in itself to sufficiently understand the risks of corrosion in RAC. When a more thorough analysis of corrosion risks is necessary, it is more rigorous to consider at least the chloride diffusion coefficient and the resistance of concrete against carbonation penetration. The use of these durability indicators is particularly relevant, on the one hand because they integrate more the characteristics of the concrete, including its porosity, and on the other hand because most of the operational durability models work with them.

It appears that the RAs predictably decrease the performances of the concrete relating to the transfer properties with an intensity which depends on the intrinsic characteristics of RAs (porosity), the substitution rate, and the compactness of the cement matrix of the new concrete. By optimizing concrete mix (reduction of the *W/B* ratio in particular), it is easy even with a high rate of substitution to produce concretes that are resistant to carbonation and to migration of chlorides as concretes consisted of NAs.

When the new concrete is compact, its cement matrix acts as a diffusion barrier, the effect of which should be all the more marked as the difference in compactness between the old and new cement matrices is great. In this study, the effect of this diffusion barrier is clear for carbonation but less in the case of chlorides, which may be due to the test procedure.

The positive effect linked to the presence of water in the RAs which would act as a reservoir and would allow to maintain a higher saturation rate, favoring the densification of the microstructure because of the new hydrates formed and thus opposing the advancement of carbonation or chlorides, is neither confirmed nor questioned. If it exists, it appears as a second-order effect.

12.3.2 Freezing/thawing resistance of air-entrained RACs

12.3.2.1 Introduction—concrete characteristics

In France, there are three standards for the measurement of freeze/thaw resistance of concretes: NF P 18-424 and NF P 18-425 for internal freezing and XP P 18-420 for the resistance to freeze/thaw cycles with deicing salts. Mix proportions of tested concretes are given in Table 12.3. Their characteristics regarding compressive strength class, cement content, air content, and *W/B* ratio comply to exposure class XF3 of NF EN 206/CN.

12.3.2.2 Results

The French standards NF P18-424 2008 and NF P18-425 2008 proposed two criteria to estimate the frost resistance of concretes: the first indicator is related to the longitudinal dimensional variation of prismatic concrete samples (10 cm × 10 cm × 40 cm), whereas the second one is linked to the transverse resonance frequency. A concrete is considered frost resistant after 300 freeze/thaw cycles, according to the French standards: $\Delta l/l_0 \leq 500$ µm/m and fundamental resonance frequency ratio $\geq 60\%$. As it can be seen on Figure 12.6, the dimensional variations are under the threshold value.

Table 12.3 Mix proportions of concretes, per cubic meter

	C30/37		
	NAC	*F20G30*	*F20G30*
Water	151	147	154
CEM II/A-L 42,5 N	342	334	350
Limestone filler	39	38	40
FNA (0/4)	739	628	535
FRA (0/4)	0	157	229
NCA (4/10)	254	134	157
RCA (4/10)	0	255	281
NCA (6,3/20)	777	461	497
Superplasticizer	0.63	0.61	0.81
Air entraining agent	0.11	0.12	0.09
w/equivalent binder	0.44	0.44	0.44
Slump (cm)	20	20	18
Air content (%)	5	7	4
Compressive strength at 28 days (MPa)	40.0	37.5	43.5

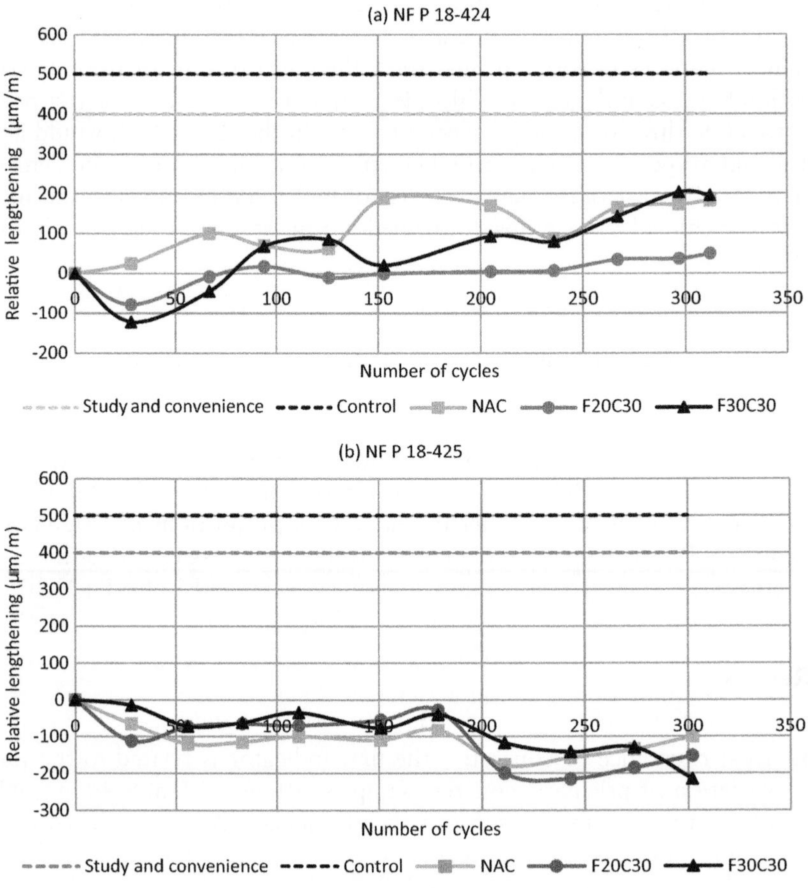

Figure 12.6 Resistance to 300 freeze/thaw cycles according NF P 18-424 (a) and NF P 18-425 (b) procedures.

A third method for internal freezing with another cycle has also been used (Figure 12.7). All results are given in the study by Omary (2017). This study leads to the same tendency.

For resistance to scaling according to the XP P 18-420 standard, the loss of mass was under 500 g/m². The three concretes also have a good resistance to freeze/thaw cycles in the presence of deicing salts (Figure 12.8).

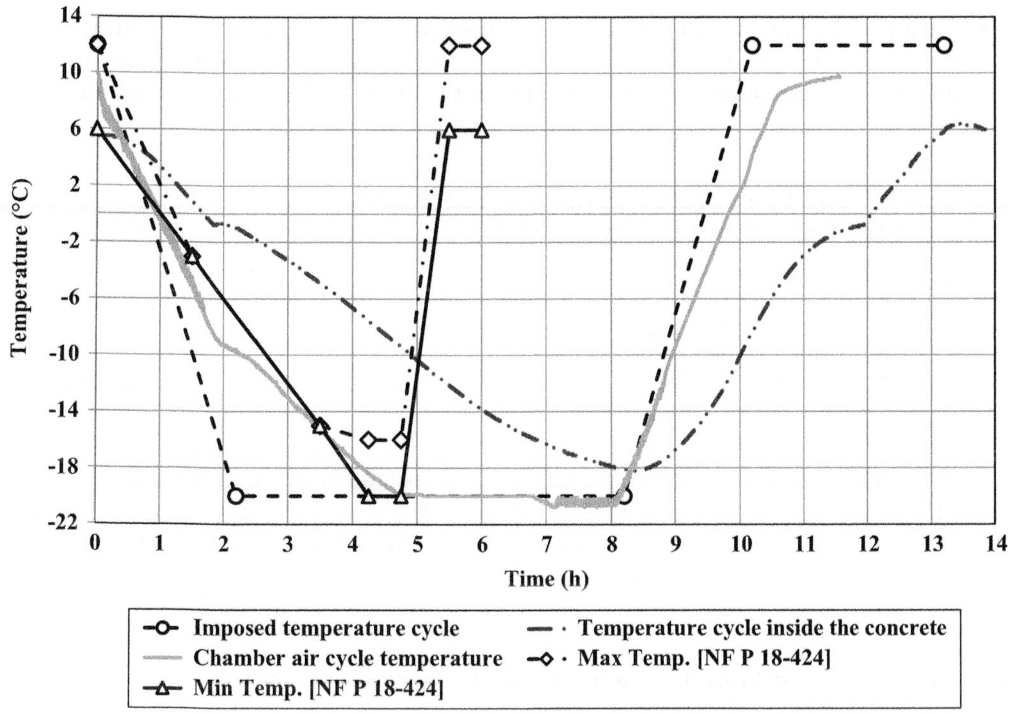

Figure 12.7 The imposed freezing/thawing cycle in the third method compared to the NF P 18-425 and 18-424.

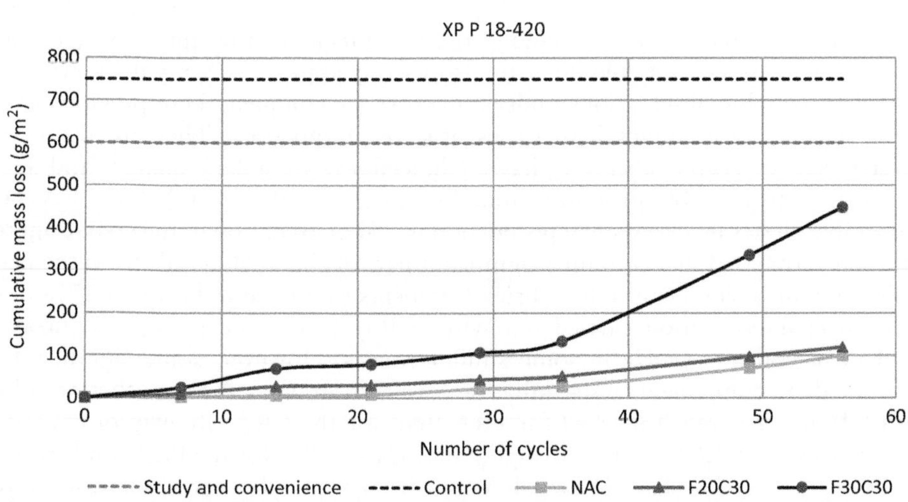

Figure 12.8 Resistance to 56 freeze/thaw cycles with deicing salt according XP P 18-420 procedure.

12.3.2.3 Conclusion

On the basis of these experimental results, the following conclusions and recommendations can be put forward:

- Results from the literature show that the frost resistance of RA is very dependent on the characteristics of the old concrete: frost resistance of NA, air content, and *W/B* ratio of the cementitious paste.
- PN RECYBÉTON results show that it is possible to produce concretes with RA resistant to freeze/thaw cycles with or without deicing salts. It is necessary to check the intrinsic frost resistance of RA and to respect the rules of formulation (binder content, *W/B* ratio, entrained air content, and so forth).

12.3.3 ASR due to recycled concrete aggregates—characterization and mitigation

12.3.3.1 Introduction

This study included the following:

- The quantification and localization of soluble alkalis and silica in recycled concrete aggregates
- The evaluation of the reactive potential of recycled concrete aggregates by the means of aggregate dissolution kinetics and mortar tests (autoclave and microbar)
- The verification on concrete incorporating recycled concrete aggregates of the main methods allowing the limitation of abnormal expansion due to ASR. This part focuses on the recycled concrete aggregates and alkali contents, as well as the use of metakaolin (MK) or ground-granulated blast-furnace slag (GGBS) as inhibitors of ASR.

12.3.3.2 Soluble alkalis and silica in recycled concrete aggregates

This part of the study was applied to the nine RAs (four fine and five coarse) selected for RECYBÉTON. Their main properties are given in Chapter 3. Those RAs are characterized by a relative mineralogical and chemical homogeneity despite their varied provenance. Mineralogical investigations showed that quartz and calcite are always the preponderant minerals. In accordance with the geological environment of the initial NA, K-feldspars and plagioclases, micas, and dolomite can also weakly appear on X-ray diffraction (XRD) patterns. Petrographic observations under a microscope complete these previous results in order to determine other minerals that present lower proportions. They are also necessary to detect unstable specific siliceous phases (chalcedony, secondary quartz, and so forth) or special siliceous unstable structures (undulose quartz, and so forth). Most RAs contain those unstable phases in a very small proportion (<1%) in exception to the coarse aggregates from the platform no. 3 that contain a significant part of chert. Materials from the platform no. 1 also contain a significant rate in brick fragments estimated to be about 17% of CRA.

Chemical characterization carried out with X-Ray Fluorescence analysis showed that major element contents evolve in some similar ranges. However, some significant variations of the alkali components (especially for Na_2O) from one RA to another are observed (Table 12.4) and it is much marked for FRA than for the CRA. By way of consequence, $Na_2O_{eq.}$ evolves from 4,600 to 15,000 mg/kg (0.46%–1.50%) for the FRA, and from 6,300 to 13,800 mg/kg (0.63%–1.38%) for CRA. But from a global point of view, the K_2O and Na_2O average contents are very close for both fine and CRA: respectively 8,100 and 4,200 mg/kg (≈0.8% and 0.4%).

Table 12.4 Alkali content for the nine tested RA (mg/kg)

	FRA1	*FRA2*	*FRA3*	*FRA4*				*CRA1*	*CRA2*	*CRA3*	*CRA3*	*CRA4*			
Grading	*0–6*	*0–6*	*0–4*	*0–4*	*Average*	*Std. dev.*	*Std. dev.*	*4–20*	*4–20*	*4–10*	*10–20*	*4–20*	*Average*	*Std. dev.*	*Std. dev.*
K_2O	7,800	10,500	4,700	9,900	8,200	2,600	31.8	9,400	8,800	5,600	5,600	11,100	8,100	2,400	30.0
Na_2O	3,000	8,100	1,500	5,000	4,400	2,900	64.8	3,300	5,100	2,800	2,600	6,500	4,100	1,700	41.5
$Na_2O_{eq.}$	8,100	15,000	4,600	11,500	9,800	4,500	—	9,400	10,900	6,400	6,300	13,800	9,400	3,200	—

Average and standard deviation (mg/kg) are both calculated for FRA and CRA. Standard deviations in italic are expressed in percentages relatively to the average.

Table 12.5 Soluble alkalis (K_2O, Na_2O and $Na_2O_{eq.}$) and dissolution rates of the nine tested RA

	FRA1	*FRA2*	*FRA3*	*FRA4*		*CRA1*	*CRA2*	*CRA3*	*CRA3*	*CRA4*	
Grading (mm)	*0–6*	*0–6*	*0–4*	*0–4*	*Average*	*4–20*	*4–20*	*4–10*	*10–20*	*4–20*	*Average*
K_2O (mg/kg)	596.8	149.9	232.4	227.6	301.7	249.5	140.9	164.5	151.0	56.6	152.5
Diss. rate (%)	7.65	1.43	4.94	2.30	4.1	2.66	1.61	2.95	2.67	0.51	2.1
Na_2O (mg/kg)	402.6	158.5	246.7	206.9	253.7	147.7	94.6	83.1	76.3	23.4	85.0
Diss. rate (%)	13.42	0.81	16.44	4.14	8.7	4.50	1.85	3.00	2.93	0.36	2.5
$Na_2O_{eq.}$ (mg/kg)	795.3	257.1	399.4	356.7	452.1	311.9	187.3	191.3	175.3	60.7	185.3
Diss. rate (%)	9.78	1.71	8.70	3.10	5.8	3.32	1.72	2.97	2.78	0.44	2.2

Table 12.6 Soluble silica (SiO_2/Na_2O) measured on the nine tested RA

	FRA1	*FRA2*	*FRA3*	*FRA4*		*CRA1*	*CRA2*	*CRA3*	*CRA3*	*CRA4*	
SiO_2/Na_2O	*0–6*	*0–6*	*0–4*	*0–4*	*Average*	*4–20*	*4–20*	*4–10*	*10–20*	*4–20*	*Average*
24h	0.01	0.02	0.02	0.01	0.02	0.16	0.02	0.13	0.20	0.02	0.11
48h	0.02	0.03	0.03	0.01	0.02	0.27	0.04	0.41	0.55	0.05	0.26
72h	0.03	0.05	0.05	0.02	0.04	0.55	0.06	0.81	0.94	0.07	0.49

Soluble alkalis have been measured in accordance to the French standard NF P18-544 2015 that consists to attack 500 g of materials during 7 h in boiling water saturated in lime. Results (Table 12.4) show that the fine aggregate release much more alkalis than the coarse aggregates. The smallest values obtained for the fine aggregates ($Na_2O_{eq.} \approx 250$–300 mg/kg) are close to the highest values measured on the coarse aggregates. Nevertheless, it represents a feeble dissolution rate of the total alkalis contained in each material (1.7%–9.8% for the FRA and 0.5%–3.3% for the CRA) as it is shown by their respective values.

RAs release very high contents in water-soluble alkalis compared to usual values measured on NA (Lavaud 2017). Water-soluble $Na_2O_{eq.}$ values are close to 20–40 mg/kg (fine) and 5–15 mg/kg (coarse) for alluvial aggregates from Paris Basin whereas they are close to 10–45 mg/kg (fine) and 2–15 mg/kg (coarse) for many limestones from the southern areas of France. RA values correspond much more (sometimes exceeding them) to the marine aggregates which can usually be close to 60–400 mg/kg (Table 12.5).

Soluble silica have been measured by following the protocol of the annex A included in the French standard NF P18–594 2015 that conduct to classify materials into three categories, NR, PR, and PR with pessimum effect (PRP). This test consists to place a small amount (25 g) of crushed sample in a molar soda solution (25 mL–1 M) during 24, 48, or 72 h. Titration of SiO_2 and Na_2O in the solution is done with an inductively coupled plasma mass spectrometry (ICP-MS). Results of soluble silica are explained through normalization to the soda, SiO_2/Na_2O (Table 12.6 and Figure 12.9).

The finest fractions of the RA release a very feeble part of silica (NR materials) on contrary to their coarse fractions that can give contrasted results. It is particularly marked for the three CRAs produced on platforms 1 and 3. The CRA corresponds to PR materials due to the presence of instable phases as indicated by the next tests. On contrary, FRA and CRA from platforms 2 and 4 present close values and are classified as NR.

In order to understand the respective origins of soluble alkalis and silica, those previous tests have been carried one more time (Tables 12.7 and 12.8) on the CRA artificially enriched (visual sorting) in one of their two constitutive fractions: ACP or NA. A sample only composed of broken tiles and bricks collected in CRA1 was also tested.

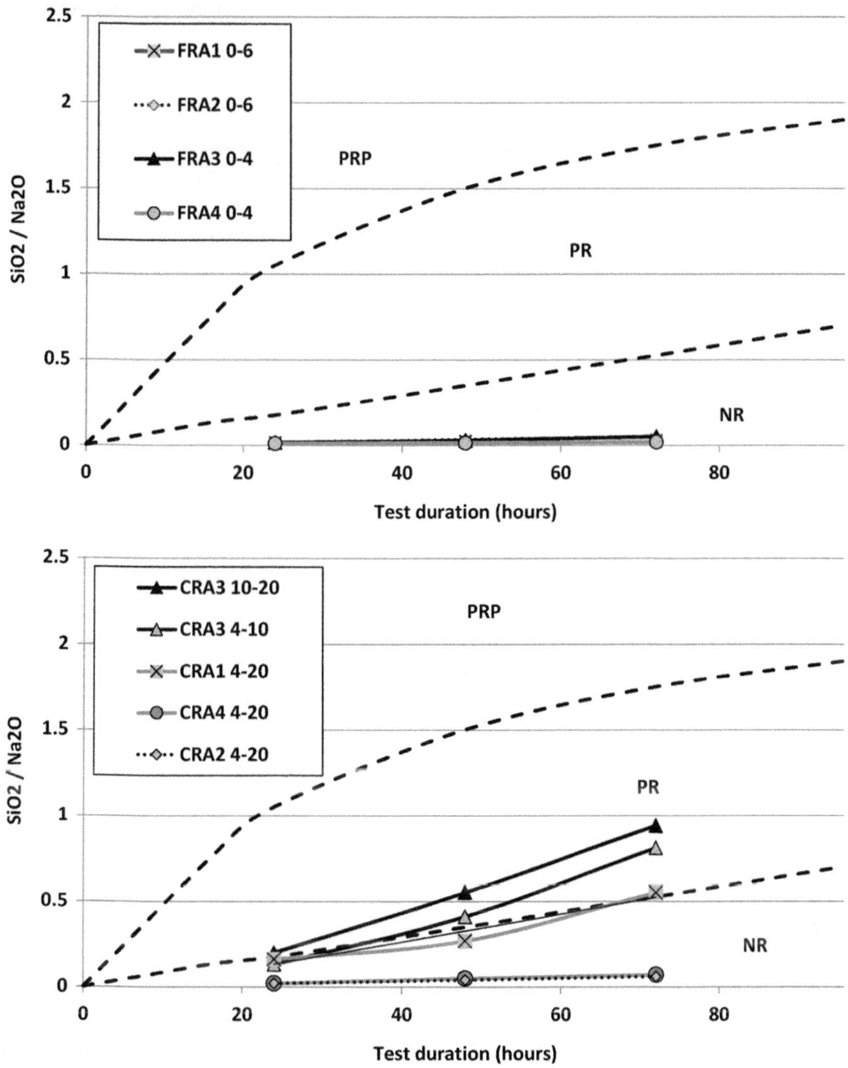

Figure 12.9 Soluble silica (SiO_2/Na_2O) measured after 24, 48, and 72 h.

Compared to the initial values (Tables 12.5 and 12.6), the contents in water-soluble alkalis strongly increase for the RA enriched in ACP in contrary to RA enriched in NA. The sample entirely composed of tiles and bricks releases a high content in water-soluble alkalis that partially correspond and explain the content of the original material CRA1.

The high content in soluble silica released by the CRA3 (10–20 grading) is essentially linked to its NA composition, rich in cherts. The test conducted on the same material enriched with ACP (excluding the presence of cherts) leads to feeble values that are close to the FRA. As for the soluble alkali tests, it appears for a second time that pollution in brick fragments could produce ASR due to its very high soluble silica content. This pollutant could be classified as PRP category.

These tests show the role of ACP in the release of soluble alkalis. Indeed, alkali compounds contained in ACP are less stable than alkalis contained into mineralogical structures (like feldspars) of the NA (except for altered minerals and structures). The concrete crushing

Table 12.7 Soluble alkalis (K_2O, Na_2O and $Na_2O_{eq.}$) and dissolution rates of the enriched RA

	CRA1	CRA2	CRA3		CRA4	
Grading (mm)	4–20	4–0	10–20		4–20	
Enrichment	Tiles/bricks	ACP	NA	ACP	NA	ACP
K_2O (mg/kg)	149.4	501.6	69.0	445.0	40.1	260.2
Na_2O (mg/kg)	64.5	346.7	24.2	110.8	19.6	69.9
$Na_2O_{eq.}$ (mg/kg)	162.8	676.7	69.6	403.6	45.9	241.1

Table 12.8 Soluble silica (SiO_2/Na_2O) measured on the enriched RA

	CRA1	CRA3	
SiO_2/Na_2O	4–20	10–20	
Enrichment	Bricks	NA	ACP
24h	1.41	0.49	0.03
48h	2.14	0.79	0.03
72h	2.03	1.05	0.04

process tends to concentrate ACP in the FRA (Hansen 1992), explaining their high contents in soluble alkalis. But pollutants like broken tiles and bricks can release a lot of soluble alkalis too. The soluble silica is essentially released by the unstable phases eventually contained in the NA fraction of some RA. The ACP releases a feeble amount of soluble silica and even nothing. Broken brick pollution is once again an important source of soluble silica.

12.3.3.3 Potential reactivity of recycled concrete aggregates

To complete the results of soluble silica test, (autoclave) mortar and (microbar) micromortar tests were carried out on the nine RAs selected for RECYBÉTON to determine potential reactivity of recycled concrete aggregates by using the French test NF P18-594 (2015).

First, (autoclave) mortar tests were carried out on 0.16–5 mm fraction obtained by crushing then sieving from CRA (and FRA if necessary). The mortar formulations had water/cement ratio equal to 0.5 and different cement/aggregate ratios equal to 0.5, 1.25, and 2.5. It should be noted that the standard protocol had to be adapted to consider the high water absorption of RA. A step of presaturation of sample was carried out following the protocol described by Delobel et al. (2016).

Each formulation was measured by the longitudinal expansion of three prisms (40 mm × 40 mm × 160 mm) after autoclave treatment for 5 h at 127°C. According to FD P18-542 (2015), giving the criteria for interpretation of the mortar test results, the prism should have less than 0.15% of expansion to be considered as acceptable regarding ASR.

Figure 12.10b shows that coarse aggregates of all RA were classified PR as almost all the expansion values were above the threshold. The results on CRA1 and both CRA3 were in accordance with those obtained by soluble silica test (Figure 12.9b) and characterization test. Indeed, CRA1 contained high part of brick fragments and both CRA3 had NA rich in cherts. On the other hand, the classification was different between the two tests for CRA2 and CRA4.

For all fine aggregates except FRA3, all the expansion values of mortar test (Figure 12.10a) were below the threshold giving NR classification like the result of soluble silica test (Figure 12.9). The possible presence of cherts of NA for FRA3 could explain this PR

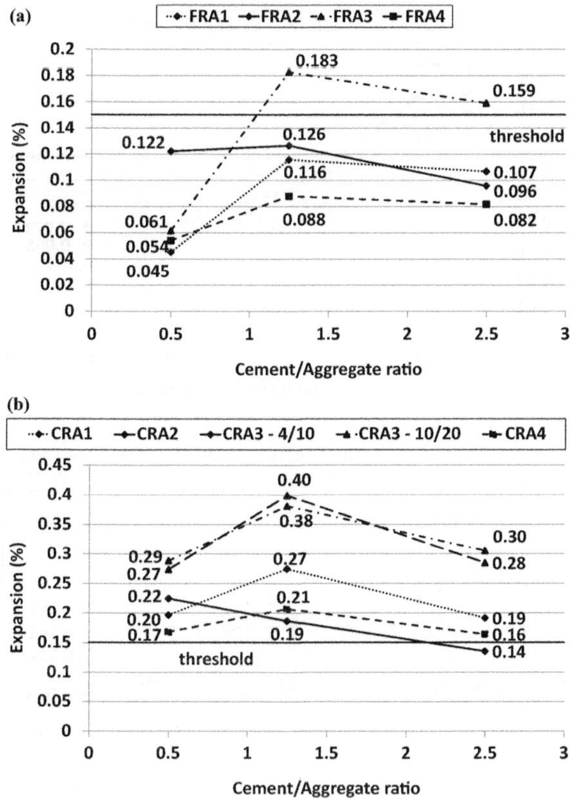

Figure 12.10 Expansion of mortar versus cement/aggregate ratio: (a) mortar from different FRA; (b) mortar from different CRA.

classification. The same conclusion was obtained with micromortar test (Figure 12.11a) and long-term concrete tests with A3 formulation in Figure 12.12.

Second, (microbar) micromortar tests were carried out on 0.16–0.63 mm fraction obtained by crushing then sieving from FRA and CRA. The micromortar formulations had water/cement ratio equal to 0.3 and different cement/aggregate ratios equal to 2, 5, and 10. It should be noted that the standard protocol had to be adapted to consider the high water absorption of RA. As for previous tests (autoclave), a step of presaturation of sample was carried out following the protocol described by Delobel et al. (2016).

Each formulation was measured by the longitudinal expansion of four prisms (10 mm × 10 mm × 40 mm) after 4 h of steam treatment then alkaline treatment for 6 h at 150°C. According to FD P18–542 (2015), giving the criteria for interpretation of the micromortar test results, the prism should have less than 0.11% of expansion to be considered as acceptable regarding ASR.

Figure 12.11 shows that fine and coarse aggregates of all RA were classified PR as at least one value was above the threshold. These results were contrary to the classification obtained by soluble silica test (Figure 12.9) for all FRA and CRA2 and CRA4. Only CRA1 and both CRA3 were in accordance with the results of soluble silica test. On the other hand, the PR classification of all CRA was the same between (autoclave) mortar and (microbar) micromortar tests (Figure 12.10b). Concerning FRA, only FRA3 obtained the same classification between both the tests (Figure 12.10a and Figure 12.11a).

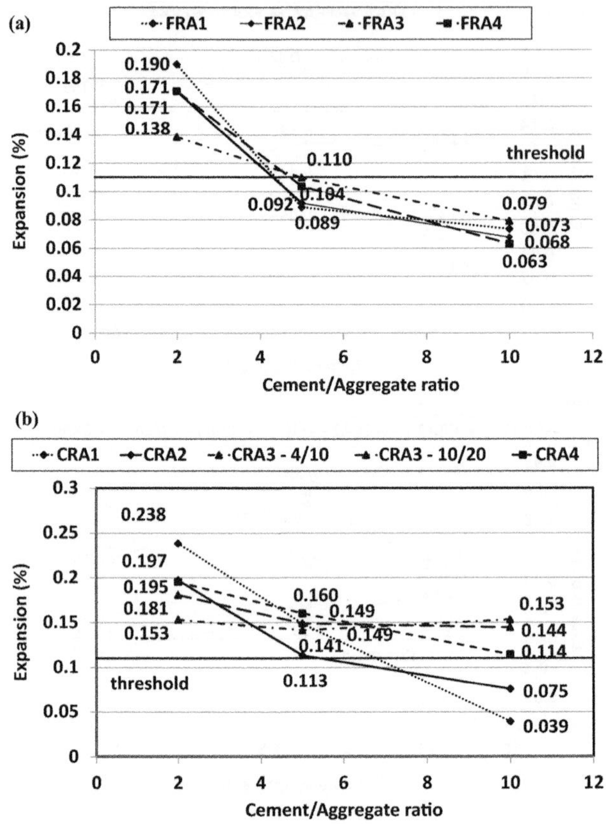

Figure 12.11 Expansion of micromortar versus cement/aggregate ratio: (a) micromortar from different FRA; (b) micromortar from different CRA.

The difference of case from FRA with micromortar test compared to the other tests (mortar test and soluble silica test) could be explained by the severe conditions of this test but also by the use of 0.16–0.63 mm fraction. Indeed, the crushing of 0.16–0.63 mm sample eliminated the finest fraction rich in cement past and thus concentrated the naturel aggregate part, which can contain reactive phases.

12.3.3.4 Concrete incorporating recycled concrete aggregates

In order to verify the main methods allowing the limitation of abnormal expansion due to ASR on concrete, the study focused on the recycled concrete aggregates and alkali contents and on the use of MK or GGBS as inhibitors of ASR.

Tests were carried out by using the French test NF P18-454 (2004), which studies the alkali reactivity of a given concrete formulation measured by the longitudinal expansion over time of three concrete prisms (70 mm × 70 mm × 282 mm) cured at 60 ± 2°C and 100% RH. According to FD P18-456 (2004), giving the criteria for interpretation of the performance test results, the concrete should have less than 0.02% of expansion at 5 months (20 weeks) to be considered as acceptable regarding ASR.

The concrete formulations are given in Table 12.9, with the information regarding the materials used in the mixtures. The RAs used for those concretes were FRA3 and CRA3 4–10,

Table 12.9 Formulations of concretes containing FNA and CRA (platform no. 3)

	Recycled aggregate content			*Alkali content*		*SCM effect*			
Concrete name	*A1*	*A2*	*A3*	*B1*	*B2*	*C1*	*C2*	*C3*	*C4*
Constituents (kg/m³)	30R-30R	0R-100R	100R-100R	0R-100R	0R-100R	0R-100R	0R-100R	0R-100R	0R-100R
Cement	321[a]	336[a]	381[a]	336[b]	336[b]	336[b]	280[b]	184[b]	336[a]
Limestone filler[c]	44	53	70	53	53	53	—	—	53
MK[d]	—	—	—	—	—	—	70	—	—
GGBS[e]	—	—	—	—	—	—	—	184	—
Natural fine aggregrate	482	698	—	698	698	782[f]	782[f]	782[f]	782[f]
Recycled fine aggregrate[g]	214	—	663	—	—	—	—	—	—
Natural coarse aggregrate	720	—	—	—	—	—	—	—	—
Recycled coarse aggregrate[h]	307	926	734	926	926	840	840	840	840
Total water	220	238	284	238	238	238	238	238	238
Superplasticizer[i]	1.64	2.18	2.78	2.18	2.18	2.18	2.62	2.18	2.18
Retarding agent	1.30	—	3.00	—	—	—	—	—	—
NaOH	0.6	0.6	0.7	0.2	4.5	4.5	5.0	5.9	0.6
Alkali content ($Na_2O_{eq.}$)	2.5	2.6	3.2	2.9	6.2	6.2	6.2	6.2	2.7
RA content (%/total aggregate)	30	57	100	57	57	57	57	57	57

a CEM II/A-L 42.5N (11% of limestone filler, 3,700 cm²/g, 0.59% $Na_2O_{eq.}$).
b CEM I 52.5R (4,500 cm²/g, 0.76% $Na_2O_{eq.}$).
c 4,600 cm²/g.
d Flash calcination, 16 m²/g, activity index (15%) at 28 days: 1.05.
e CaO/SiO_2: 1.14, 4,300 cm²/g, water-soluble $Na_2O_{eq.}$: 0.003%.
f For concrete of the series C, the natural sand was PR to alkalis, according to NF P18–594.
g 0/4 mm, absorption: 9.2%, water-soluble $Na_2O_{eq.}$: 0.04%.
h 4/20 mm, absorption: 6.2%, water-soluble $Na_2O_{eq.}$: 0.02%.
i Polycarbolylate ether.

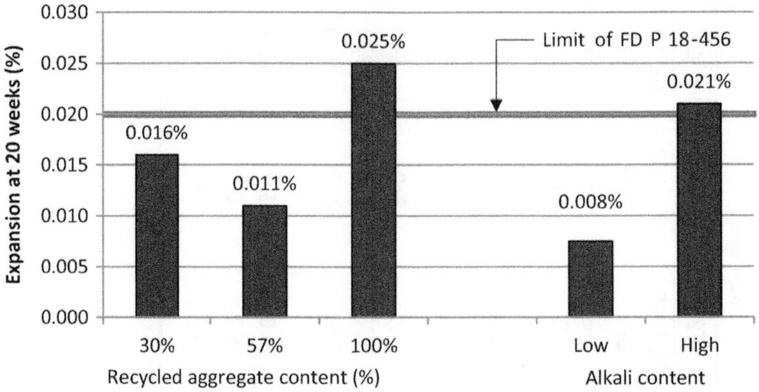

Concrete	A1	A2	A3	B1	B2
Fine recy.aggr.	30%	0%	100%	0%	0%
Coarse recy.aggr.	30%	100%	100%	100%	100%
Alk content	2.5 kg/m³	2.6 kg/m³	3.2 kg/m³	2.9 kg/m³	6.2 kg/m³

Figure 12.12 Expansion at 20 weeks of concrete cured at 60°C and containing RA, considering the effect of the content on recycled concrete aggregates and on alkali.

10–20 (platform no. 3 production) with the main properties already presented at the beginning of this section. It should be noted that the two coarser aggregates were classified as PR (Figure 12.9). NAs were all classified NR regarding ASR risks, except FNA used for concrete C1 which was PR. It should be noted that the strength class of the concretes was C35/45 (i.e., 35 MPa on cylinders or 45 MPa on cubes at 28 days), and that the hydric state of the FRAs and CRAs was always maintained at a value 1% higher than their absorption coefficients.

Figure 12.12 presents the expansion of concrete cured at 60°C after 20 weeks and shows the effects of RA content and of the alkali content in the mixtures. It can be seen that both the parameters are significant. On the one hand, increasing the RA content led to an increase in the expansion of the concretes. Up to 100% of coarse RAs could be used at alkali content below 3 kg/m³, but the use of 100% of RA (all the natural fine and coarse aggregate being replaced) should be avoided, even at alkali content around 3 kg/m³. It should be said anyway that this kind of concrete can hardly be cast, due to high water absorption of RAs. On the other hand, high alkali content involved expansion above the acceptable limit, as for NAs. Many authors propose a threshold effect of alkali, generally between 3 and 5 kg of alkali/m³ of concrete made of NAs (Oberholster 1983; Rogers and Hooton 1991; Hobbs 1993; Thomas et al. 1996; Shehata and Thomas 2000; Multon et al. 2008), below which ASR expansion is small. The French provisions for the prevention of ASRs recommended a maximum value of 3.5 kg/m³, when the binder used contains less than 60% of slag (FD P18-464 2014). It can be seen here that these values also stand for recycled concrete aggregates.

Figure 12.13 shows the effect of MK and GGBS on the reduction of expansion relative to concrete made of 100% cement (CEM I and CEM II, see Table 12.9), for high alkali content. It should be noted that the natural sand used in these formulations was reactive to alkalis (0.25% on mortars according to NF P18-594, superior to 0.15% to be considered as NR; the concrete C1 reached an expansion of 0.021%). These results confirm, on RAs concretes, that MK and GGBS, when used at sufficiently high levels, are able to significantly reduce the expansion of alkali-reactive formulations (expansion values given on the figure). In that case, the reductions of 52% and 43% for 20% of MK and 50% of GGBS, respectively, were enough to bring down the expansions at values under the limit fixed by the standard.

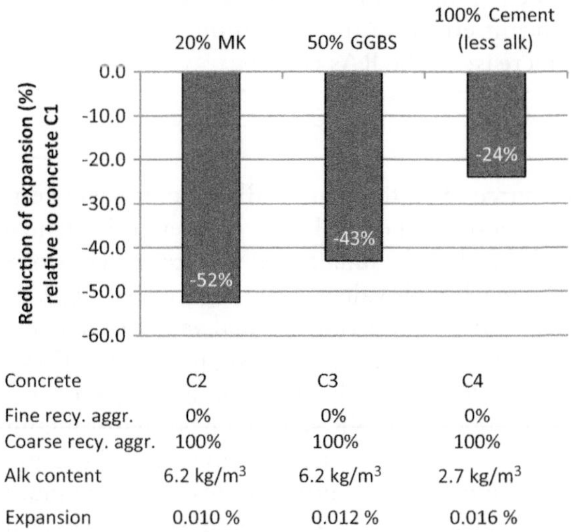

Figure 12.13 Reduction of expansion relative to concrete made of 100% cement (alkali content adjusted at 6.2 kg/m^3) of concretes containing 20% of MK and 50% of GGBS in replacement of the cement. Comparison with a concrete without MK or GGBS but containing less alkalis (2.7 kg/m^3 of alkalis). All the natural coarse aggregates were replaced by recycled coarse aggregates. The natural sand used in all formulations was reactive to alkalis.

The effect of MK and GGBS was even better here than the decrease of the alkali content in the formulation (2.7 kg/m^3 of alkalis), the reduction obtained reaching 24%.

Figure 12.14 summarizes all the expansion results in terms of RA content and alkali content. It is seen that a safe zone without supplementary cementitious materials (SCM) can be represented (although it should be defined more precisely), depending on both the parameters:

- More alkalis in the concrete will limit the replacement rate of NAs by RAs;
- More RAs in the concrete will imply a decrease in the alkali content to remain under the acceptable limit of expansion.

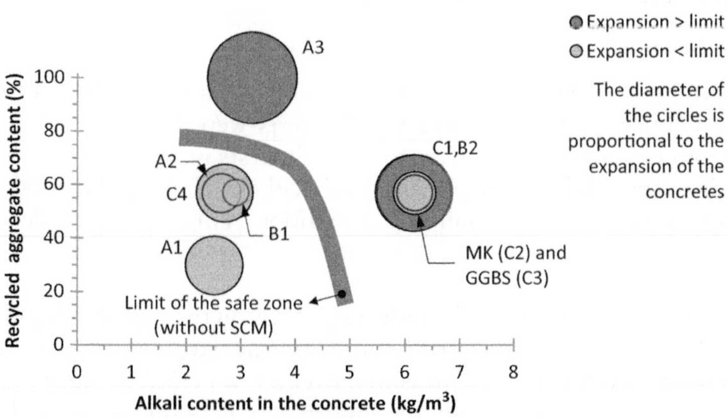

Figure 12.14 Expansion of the concretes (proportional to the diameter of the circles) containing different fractions of recycled concrete aggregates, at variable alkali contents. A1, A2, A3, B1, B2, C1, C2, C3, and C4 refer to the concrete names.

The use of SCM could be a good method to either accept an increase in the alkali content without having a significant effect on the expansion (Figure 12.14 see C–circles MK and GGBS) or to allow an increase in the RAs content.

12.3.3.5 Conclusion

RA can be a potential source of water-soluble alkali components especially the fine aggregates that contain an important part of ACP. The coarser grains tend to contain stable alkali components that are stabilized in the (unaltered) silicate minerals of the NA. Water-soluble alkali content of RA can reach high values >400 mg/kg ($Na_2O_{eq.}$). These values are higher than most of the NAs and close or superior to usual values of marine aggregates. The natural minerals contained in the RA can recover a significant fraction of unstable silica and/or structures that could develop ASR. This problem is more sensitive for the CRA because the fine fraction is often composed of NR silicate phase: natural (alluvial) quartz and ACP. Then potential reactivity of RA can be detected knowing those previous values or using specific test carried out on mortars, but autoclave test could be much more adapted to RA than microbar test. Water absorption of the RA must be necessarily considered in these types of tests.

The tests conducted on concrete (using RA classified as PR) show the possibility of using CRA up to 100%. A concrete mix containing substitution of NA with 30% fine and CRA also produce acceptable expansion (>0.02% after 5 months) regarding ASR. But the use of a mix that incorporate 100% FRA and 100% RA is not relevant. Experimental results permitted to define a safe zone corresponding to concrete mixes that provoke expansion values inferior to 0.02%. It is limited by a maximum RA content of 80% and a 3 kg of alkali/m^3. But beyond those limits, expansion can also be significantly reduced to acceptable values by introducing MK and GGBS at sufficient level. In any cases, these products can reduce expansion on concrete caused by ASR until 50% of the value.

All the results obtained in this study show that the current rules and recommendations to avoid ASR can be applied for RAC by considering few necessary evolutions on procedure and the variability of RA which can be more important than NA.

12.3.4 Effect of recycled concrete aggregate on the sulfate attacks in concrete

12.3.4.1 Introduction

This section presents the main results and conclusions obtained on RECYBÉTON French National Project concerning sulfate attacks. The tests were both carried out on concrete and mortars. Objectives of the study were to place the tests in unfavorable conditions, using RAC with high water-soluble sulfates with and without the experimental simulation of plaster pollution coming from the building demolition. This study included three majors' objectives:

- The first objective was to study the influence of RA on the risk of delayed ettringite formation (DEF). The tests were carried out on mortars steamed at 80°C. The evolutions of length of the specimens were measured over a test period of more than 450 days.
- The second objective was to study the influence of RAs on the risks of ISA related to ettringite formation. The tests were carried out on nonsteamed concrete cured in water at 20°C. The evolutions of length of concrete specimens were measured over a test period of more than 300 days.

- The final objective was to study the influence of RAs on the risks of ISA related to thaumasite formation. The tests were carried out on nonsteamed concrete cured in water at 4°C. The evolutions of length of concrete specimens were measured over a test period of more than 300 days.

12.3.4.2 Characterization of RAC and plasters' pollutions

12.3.4.2.1 Recycled aggregate concrete

Mineralogy and chemical composition of the different RAC have been presented in a previous section (see Chapter 3). It was shown that the RAC selected for RECYBÉTON mainly contains concrete fragments without visible plasters' pollution. Silicate and carbonate minerals are predominant both to XRD and microscopic identification.

In this study dedicated to sulfate attacks, the FRA from four platforms (no. 1, 2, 3 and 4) and the coarse RAC aggregates from platform no. 3 were tested. Their chemical composition showed the presence of small amounts of sulfate (SO_3) as well as alkaline compounds (K_2O, Na_2O). Let us recall the role of these alkaline compounds in the amplification of the ISA (Leklou 2008; Nguyen et al. 2013). The previous section dedicated to ASR also showed the great solubility of these alkaline elements, especially marked for the fine RAC from the platform nos. 2 and 3. The next table regroups all these different parameters (Table 12.10).

Water-soluble sulfate content present in the RAC was determined in accordance to the French standard NF EN 1744-1, §10. It consists of a blend 25 g of aggregate (crushed below 4 mm) and 1 L of water at 65°C (±5°C) under stirring. The next results give values of water-soluble sulfate for the RAC used. These tests (Table 12.10) reveal that RAC can release very variable proportion of sulfate (<0.08%–0.24%). The tests on RAC from the same platform (no. 3: C3 and F3) show that sand release more sulfate than coarse aggregate, probably because they contain more ACP which correspond to initial cement compounds.

In order to determine the origin of the sulfate, specific tests were carried out. It consists to enrich CRA (4/10 and 10/20, platform 1) with their own ACP and NA fractions. The respective enrichments were visually done considering the grains aspects. These tests show that CRA enriched in ACP releases at least twice more sulfates (Table 12.11) at 60°C (4/10) or 20°C (10/20) than RAC of same platform enriched in NA.

The sulfate dissolution was also studied, with variable temperature, type of stirring, water/RAC ratio, and so forth. For example, at 20°C (Table 12.12) without crushing, after 1 day, the sulfate dissolved represents nearly 20% of the soluble sulfate measured at 60°C

Table 12.10 Total contents (%) in Al_2O_3, SO_3, K_2O, and Na_2O for the FRA of the four platforms and for the CRA from platform no. 3

		FRA1	*FRA2*	*FRA3*	*FRA4*	*CRA3*	
Grading		*0–6*	*0–6*	*0–4*	*0–4*	*4–10*	*10–20*
Al_2O_3	Total (%)	4.54	5.58	2.47	7.76	2.47	2.39
SO_3	Total (%)	0.62	0.39	0.47	0.35	0.12	0.14
	Water-soluble (%)	0.20	<0.08	0.24	0.11	0.10	0.12
K_2O	Total (%)	0.78	1.05	0.47	0.99	0.56	0.56
	Water-soluble (mg/kg)	596.8	149.9	232.4	227.6	164.5	151.0
Na_2O	Total (%)	0.30	0.81	0.15	0.50	0.28	0.26
	Water-soluble (mg/kg)	402.6	158.5	246.7	206.9	83.1	76.3

Water-soluble alkaline components (mg/kg) are also expressed.

Table 12.11 Soluble sulfates in water, CRA3 (4/10) at 60°C or CRA3 (10/20) at 20°C manually enriched with ACP or NA

	60°C—CRA3 (4/10) Crushed 25 g (0–4)—water 1 L—time 15 min	*20°C—CRA3 2 kg (10/20) Water 4 L—time 7 days*
Enrichment in ACP	<0.08	0.036
Without sorting (reference)	0.10	0.065
Enrichment in NA	0.18	0.080

Table 12.12 Comparison of soluble sulfates in water at 60°C (NF EN 1744-1, §10.2) and at 20°C

		Soluble sulfate (%)			
CRA3	*Conditions*	*15 min*	*1 day*	*2 days*	*7 days*
4–10	20°C—CRA 2 kg (10–20)—water 4 L	—	0.025	0.038	0.055
	60°C—crushed 25 g (0–4)—water 1 L	0.10		—	
10–20	20°C—CRA 2 kg (10–20)—water 4 L	—	0.024	0.040	0.065
	60°C—crushed 25 g (0–4)—water 1 L	0.12		—	

(after crushing). Dissolution at 20°C in water is also a long process: between 1 day and 1 week, the quantity dissolved is multiplied by factor of 2–3.

The sulfate nature and localization have been determined by scanning electron microscopy (SEM) analysis. It confirms that sulfates are present in ACP and are combined with alumina and calcium. Sulfates in RAC are therefore associated to ettringite and/or monosulfoaluminate (or their carbonated product: limestone, gibbsite, and gypsum). For sample kept under water for 1 month, the surface exposed to water contains fewer sulfates than unexposed samples.

12.3.4.2.2 Plaster pollution

In the aim to study the possible sources of sulfates pollutants coming from the building's demolition, three usual plaster materials were characterized: one hydrated binder, usual plasterboard, and some bricks. Their respective mineralogy (XRD analyses) and chemical compositions (XRF analyses) are rather the same. They are composed of pure gypsum added with a feeble part of quartz (<1.5%). Impurities (Al_2O_3, MgO, and ZrO) represent less than 1%. The sulfate water solubility of these three products has also been measured, considering their high solubility. The values presented in Table 12.13 result from a modification of the standardized protocol used for RAC. Only 0.5 or 0.25 g of product has been introduced into 1 L of water. They show that the solubility increases inversely to the mass tested, due to the sulfate saturation that is easily obtained with these products. Plasterboard and bricks have similar values (40%–45%) slightly higher to the solubility of the binder (37%).

Considering those results, single representative plasters' pollution was prepared to study its effects on the mortars and concrete mixes. It is a homogeneous plaster powder obtained with 66.6% of plaster bricks and 33.3% of binder previously hydrated, finely crushed together to obtain the grading 0.1–2 mm. Water solubilities (with 0.5 and 0.25 g tested sample mass) were measured in the same conditions than the initial pollutants. The resulting values, 41.6% and 45.6%, are slightly higher than the brick ones (probably due to greater powder fineness), but they are in accordance with the values of its constituents (Table 12.13).

Table 12.13 Water solubility of plaster pollutants and final plaster pollution

	BA 13®		*Bricks*		*Binder*		*Plaster pollution = 2/3 bricks + 1/3 binder*	
Mass tested (g)	0.5	0.25	0.5	0.25	0.5	0.25	0.5	0.25
Water-soluble SO_4^{2-}	41.2	44.0	40.8	45.2	37.5	37.6	41.6	45.6
Average (%)	42.6		43.0		37.5		43.6	

Experimental pollution in plaster was then introduced in some mixes (in substitution of aggregates) in order to obtain 0.4%, 0.8%, or 1.2% of total water-soluble sulfate (including sulfate realized by RAC). Homogenization of the plaster pollution and aggregates of the mixes were obtained by a brief dry brewing in order to obtain a homogeneous product without modification of the constituent's grain sizes.

12.3.4.3 Overview of the main experimental aspects of the tests carried out during RECYBÉTON project

Three cements were selected to produce mortars and concretes. A first Portland cement CEM I produced with a clinker having a very high C_3A content, around 10% has been selected and designated "CEM I-10%". A blended cement compound with limestone filler (26%, without further addition), derived from the same clinker, was also selected and was designated CEM II. A second CEM I with clinker having a very low C_3A content (1%), as opposed to the first, has also been selected and was designated "CEM I-1%".

Two FRAs, FRA1 and FRA3 (platforms 1 and 3), were chosen to cast the mortars. Those materials respectively contain 0.2% and 0.24% in soluble sulfate as well as high contents in soluble alkali (Table 12.10). They were sieved to 2.5 mm to provide a size distribution close to the standardized sand (NF EN 196-1 2006), and their characteristics were then measured again (Table 12.14). The elimination of the coarser grains provokes an increase of the water absorption and water-soluble sulfates due to the higher relative percentage in ACP.

For the mortar, the water/cement ratio was 0.5 and the sand/cement ratio was 3. The mortar prisms (40 mm × 40 mm × 160 mm) were prepared according to the NF EN 196-1 standard. After casting, some mortar samples were treated with a heat treatment at 80°C for 72 h to evaluate the risk of DEF, (Taylor et al. 2001; Pavoine et al. 2012; Nguyen et al. 2013). The other specimens were cured directly in water at 20°C ± 1°C to follow the secondary ettringite formation.

12.3.4.4 Results related to the formation of ettringite

12.3.4.4.1 Analysis of mortar results incorporating FRA

The results obtained on mortar showed several important points. First, on the formation of secondary ettringite without heat treatment (NT) (Figure 12.15), we have shown that at more than 400 days of monitoring, unpolluted (CEMI-10%—FRA1—unpolluted [NT])

Table 12.14 Characteristics of the sand used in the mortars

	Density	*Ab*	*Water-soluble* SO_4^{2-}
FRA1 *sieved 0–2.5*	1.961	12.9	0.28
FRA3 *sieved 0–2.5*	2.070	9.9	0.27

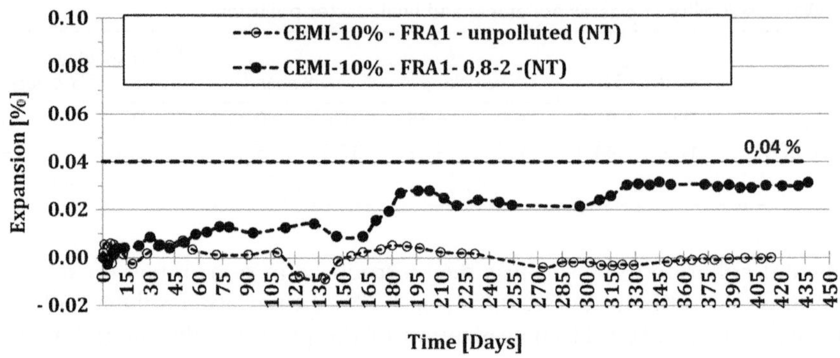

Figure 12.15 Expansions of mortars made with FRA1, CEM I-10% without heat treatment, with and without plasters' pollution (0.1/2). (Nonlinear scale).

expansions of mortar are still below the threshold of 0.04%. An expansion of 0.04% of the material is generally used for concrete as an arbitrary threshold for significant expansion. This same critical value is found in many studies (Pavoine and Divet 2007; Brunetaud 2005; Leklou 2008; Nguyen et al. 2013). For the performance test on concrete, developed in partnership between ATILH (*Association Technique de l'Industrie des Liants Hydrauliques*), IFSTTAR (*French Institute of Science and Technology for Transport, Spatial Planning, Development and Networks*) and CERIB (*Center for Studies and Research of the Concrete Industry*), this threshold of 0.04% was identified as a critical value below which concrete is considered suitable for use (Pavoine and Divet 2007). This threshold is also given in the French Publics Works Research Institute LCPC technical guide to prevent ISA (LCPC 2007). With plasters' pollution, the expansion is also below the threshold of 0.04% but it gradually increases [CEMI-10%—FRA1-0,8-2 (NT)].

The second important point is about the effect of a heat treatment. We have observed that mortars that have undergone heat treatment have a higher risk of DEF. The results on Figure 12.16 on mortars made with FRA3 and two CEM I cements (1% and 10% C_3A) showed that the C_3A content (1% or 10%) of the cement does not significantly affect the expansions regardless of the rate of plasters' pollution. A slight further expansion is observed with increasing C_3A content, but behaviors remain close.

Figure 12.17 presents the expansions of the mortars manufactured with FRA1, (CEM I-10%-FRA1). It is observed that the specimens of mortars made with FRA1 develop significant swellings. They stabilize on a plateau between 0.06% and 0.08% after 400 days of testing. The mortar specimens made with FRA3 (CEM I-10%) (Figure 12.16) showed much lower expansions and stabilized between 0.03% and 0.05% after 400 days. The content of soluble alkali can partly explain the difference in behavior between the mortars manufactured with FRA3 and those of FRA1. Their soluble sulfate levels are identical (Table 12.14), but FRA1 releases almost twice as much soluble alkaline (Table 12.10) 795.3 mg $Na_2O_{eq.}$/kg for 399.5 mg $Na_2O_{eq.}$/kg. Many authors have shown that the role of alkaline is very important because the more alkalis can be leached, the more the phenomena of DEF are amplified (Aubert et al 2009; Leklou 2008; Aubert et al. 2013). The soluble aluminate content of FRA may also explain this phenomenon as several authors have shown that the risk of DEF increases with the aluminate content (Odler et al. 1995; Hanehara et al. 2008a,b). The soluble aluminate content would be related to the number of expansive products formed, thereby influencing the degree of degradation. In absolute terms, FRA1 contains twice as much aluminum as that of FRA3 (Table 12.10), but it remains to be identified what quantities can

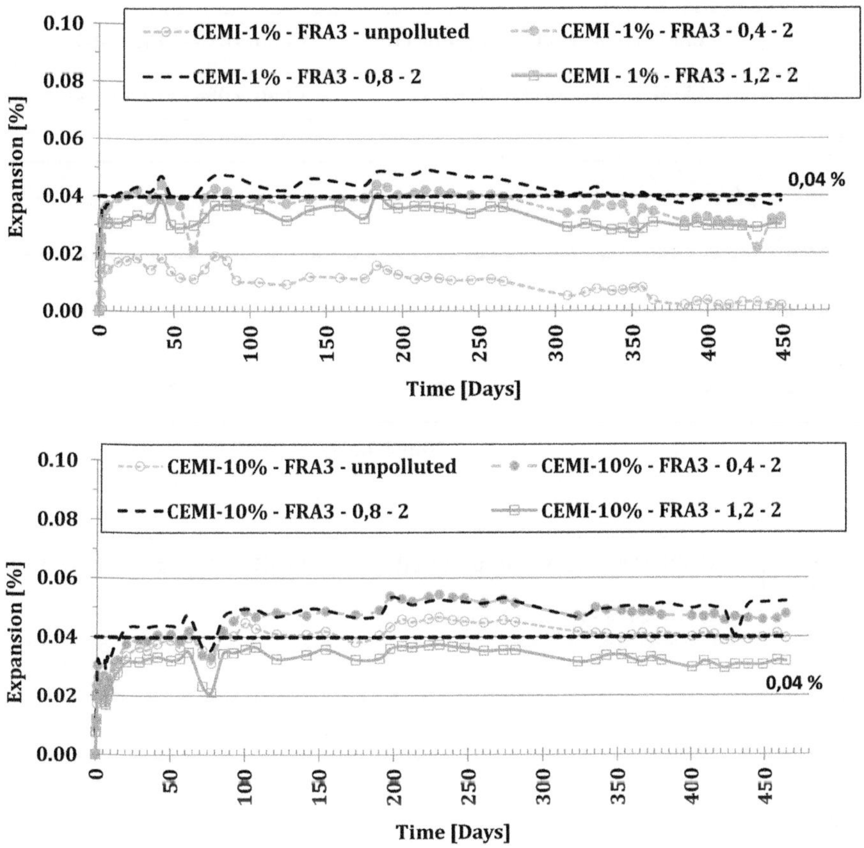

Figure 12.16 Expansions of the mortars (80°C during 72h for the hardening at early age) made with the FRA3, CEM I-1%, and CEM I-10% at variable plasters' pollution (0.1/2) contents. Hardening at early age: 80°C during 72h. Water storage at 20°C.

really be mobilized in these two sands. Figure 12.17 also shows the effect of the fineness (coarse and fine) of deconstruction plasters. It should be noted that the fineness (0/100 µm) of the deconstruction sulfates added in FRA3 does not seem to have a strong influence on the kinetics and the amplitude of the expansions, whatever the formulation. This phenomenon can be explained by the fact that the high sulfate fineness (0/100 µm) favors and accelerates the reaction between sulfates and aluminates in the first hours to form primary ettringite, thus limiting the amount of sulfates adsorbed by the C-S-H that will be reused later and after the thermal cure for the DEF (Yang et al. 1996; Divet et al. 1998; Taylor et al. 2001). However, swelling may occur over a longer period of time as shown in a study where expansions started after 5 years of conservation (Aubert et al. 2009).

12.3.4.5 Analysis of concrete results incorporating CRA

For the concrete, two mixes were used: 30% or 100% of CRA and 0% of FRA. Concrete compositions are detailed in Table 12.15.

Concrete samples' notation was defined as following: 0R-100R—CEM I-10%—P-0.27 (0R: content of FRA, 100R: content of CRA, CEM I-10%: type of cement and C_3A content, P: CRA origin, with P like Paris, 0.27: water-soluble sulfate content in CRA).

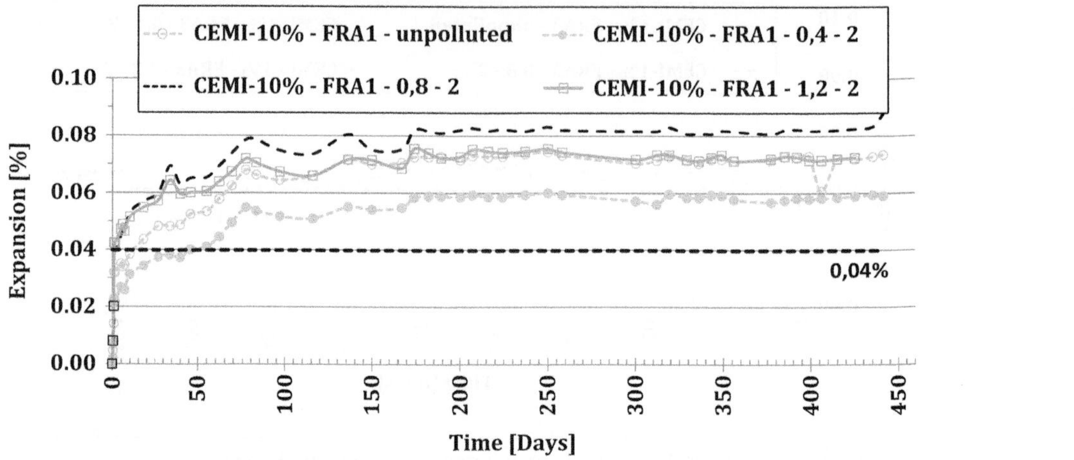

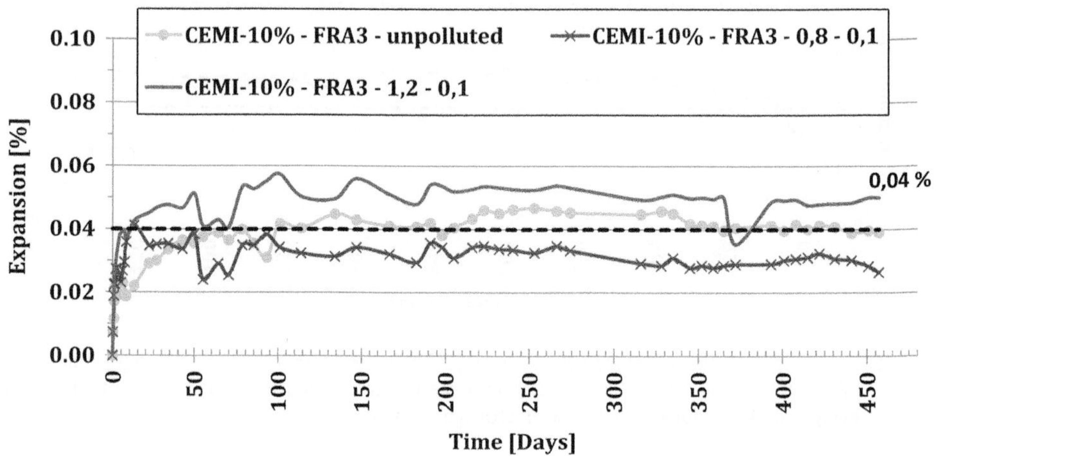

Figure 12.17 Expansions of the mortars made with the FRA1, (CEM I-10%-FRA1), at variable plasters' pollution (0.1/2) contents and expansions of the mortars made with the FRA3 at variable plasters' pollution contents and for plasters' fineness 0.1 mm (0/100 μm). Hardening at early age: 80°C during 72 h. Water storage at 20°C.

Table 12.15 Compositions of concretes

CRA content	*100%*	*30%*
Components	*Mass (kg/m³)*	
Cement	282	276
Limestone filler	31	31
Natural sand	806	813
Natural gravel 4/10	—	228
Natural gravel 6,3/20	—	462
CRA 4/10	163	—
CRA 10/20	701	296
Superplasticizer	1,167	1.14
Efficient water	164	161

Eight concrete mixes were studied for ettringite tests, with three sulfate contents in CRAs (0.27, 0.8, and 1.2), two cements (CEM I with 10% or 1% of C_3A) and two CRA contents (30% or 100%).

For ettringite tests, concretes were mixed and casted at 20°C. Due to their high water absorption, RACs were saturated during 24 h, with a mass of water corresponding to their water absorption +1%. After casting, samples were stored in mold for 24 h at 20°C, then in water until 28 days. After curing, samples were submitted to two cycles of 7 days of drying at 38°C ± 2°C and RH<30% and 7 days of immersion in water at 20°C ± 2°C. Expansion measurements were started after the last immersion period and carried out on prismatic samples (7 cm × 7 cm × 28 cm).

Three concrete mixes were studied for thaumasite tests, with three sulfate contents in CRAs (0.27%, 0.8%, and 1.2%), one cement (CEM II/B-LL), and one CRA content (100%).

Figure 12.18 shows expansion of concretes made with 30% or 100% of CRA Paris, and "CEM I-10%" or "CEM I-1%" cements. Concrete samples were stored in water for almost 1 year at 20°C. Expansion of concrete samples was very limited and was lower than the threshold of 0.04%. No significant differences were observed between these compositions, whatever the cement type and the CRA sulfate content.

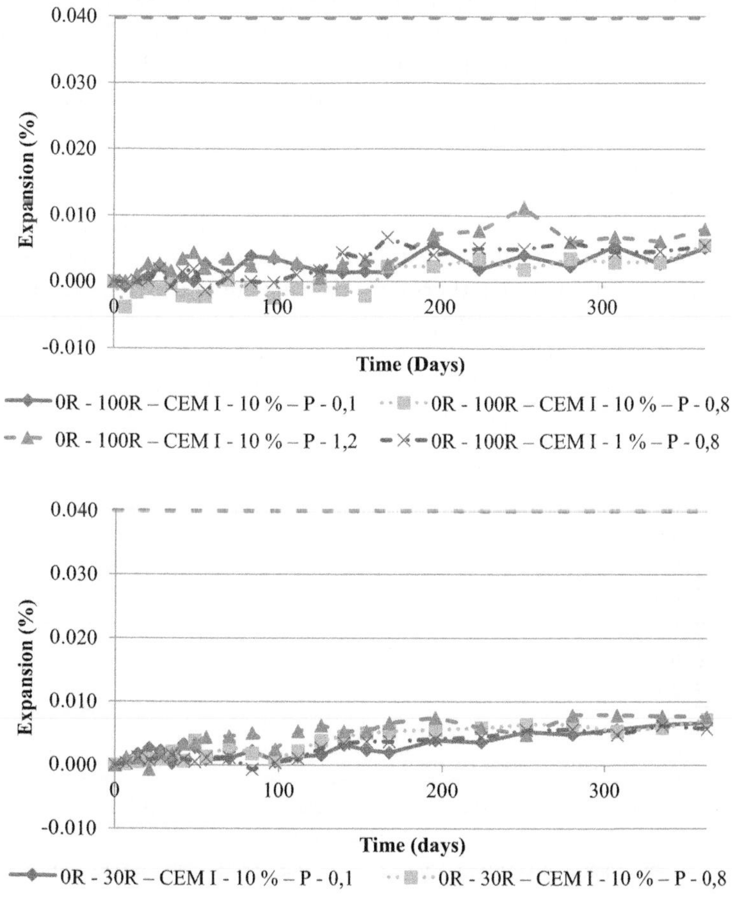

Figure 12.18 Expansions of concretes made with CRA of Paris, cements "CEM I-10%" and "CEM I-1%", and three sulfates contents. Hardening at a young age at 20°C followed by a water storage at 20°C.

SEM analysis was carried out on "0R-100R—CEM I-10%—P 1.2" sample at 1 year. This concrete composition contains the most important sulfate content from the CRA. Ettringite was observed in the microporosity of cement paste and less regularly, in air bubbles. The small quantity of ettringite observed during these examinations is in accordance with the very low swelling of concrete. The ettringite appears as a nonexpansive form.

Compressive strength has been determined on concrete cylinders (Ø11 cm, h22 cm) at 28 days and 1 year. Samples "0R-100R—CEM I-10%—P 0.27" present a compressive strength of 31.3 MPa at 28 days and 35.8 MPa at 1 year. These results agreed with expansion measurements and confirm the non-damage of concrete samples.

12.3.4.6 Results related to the formation of thaumasite

For thaumasite tests, concretes were mixed and casted at 20°C, with the same water saturation of CRA. After casting, samples were stored in mold during 24 h at 20°C, then in water at 4°C. Expansion measurements were started directly. Some reference samples were stored at 20°C in water.

Figure 12.19 shows the expansion of concrete samples stored in water at 4°C during 1 year. Concretes were made with 100% of CRA Paris, CEM II/B-LL cement, and three various sulfate contents. The more the sulfates content is, the more the samples are swelling. Indeed, after 1 year of immersion, expansion of concrete with 0.27% sulfates CRA is very low, and close to 0.01%, while swelling of concrete samples with 0.8% and 1.2% sulfates CRA are, respectively, close to 0.1% and 0.8%. Samples with 1.2% of sulfates were cracked at the end of the test, as illustrated in Figure 12.20.

SEM analysis was carried out on "0R-100R—CEM II/B-LL—P 1.2" sample at 28 days and 1 year. This concrete composition contains the most important sulfates content from the CRA. After 1 year of immersion, cement paste of this sample was cracked and was hand-friable. A lot of balls formed by crystals of thaumasite and ettringite were observed in air bubbles of the cement paste. Furthermore, thaumasite and ettringite crystals were also observed inside the cement paste microporosity (Figures 12.21 and 12.22).

Otherwise, compressive strength have been determined on concrete cylinders (Ø11 cm, h22 cm) stored 1 year at 4°C and 20°C. The 4°C samples of concrete "0R-100R—CEM II/B-LL—P 1.2" present a compressive strength of 8.3 MPa whereas the 20°C samples

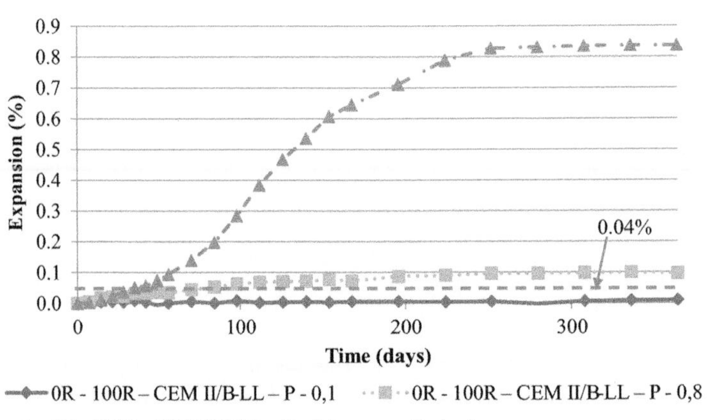

Figure 12.19 Expansion concretes made with CRA gravel of Paris, cement CEM II/B-LL, and three sulfates contents. Hardening at a young age: 20°C. Water storage at 4°C.

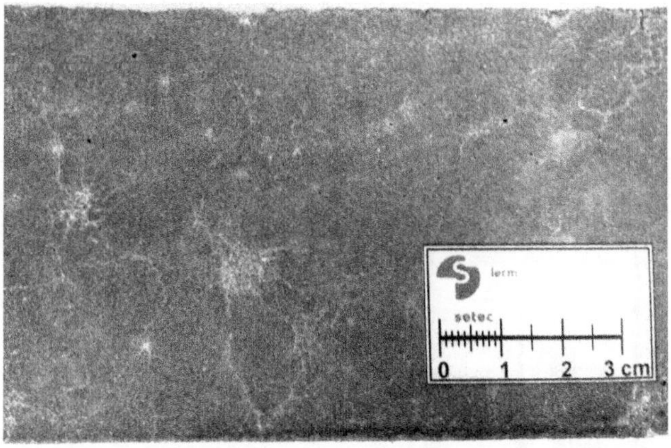

Figure 12.20 Photography of cracked surface of "0R-100R—CEM II/B-LL—P 1.2" sample at 1 year.

Figure 12.21 Scanning electron microscopy of "0R-100R—CEM II/B-LL—P 1.2" sample at 1 year. Detail of thaumasite and ettringite crystals in an air bubble of cement paste.

present a compressive strength of 21.8 MPa. This loss of strength is probably due to cracks affecting the 4°C samples (20°C samples were not cracked to the naked eye).

12.3.4.7 Conclusion

The use of RA with high water-soluble sulfates as replacement of NA in the production of concrete and mortar was analyzed. The following conclusions may be drawn from this investigation:

- Mortar tests showed that without thermal heating at early age, and even unfavorable factors (cement rich in C_3A, SO_4^{2}—total=0.8%), the mortars with FRA show a moderate expansion.
- When a temperature of 80°C during 72 h is applied for the hardening at an early age, the risk of swelling increases. Polluted mixtures of plasters (sulfate total between

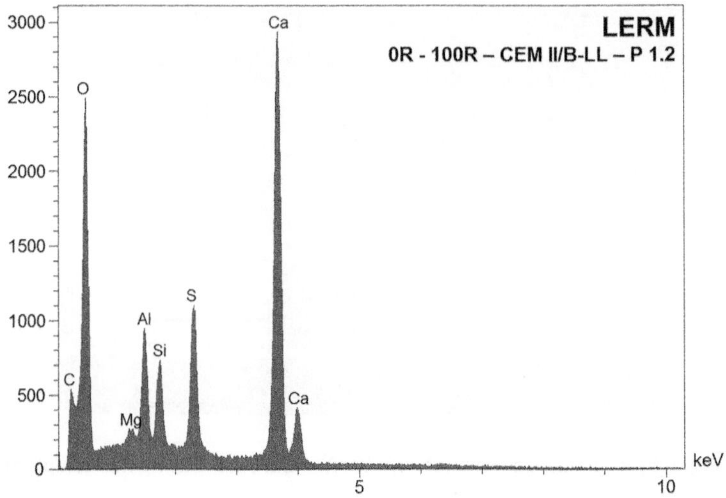

Figure 12.22 Elementary analysis by energy-dispersive X-ray spectroscopy of thaumasite and ettringite crystals.

0.4% and 1.2%) exhibit higher expansions that stabilize around 0.03%–0.05%. The presence of a high level of soluble alkali leads to values greater than 0.08%. Despite the levels of soluble sulfate close to 1%, DEF induced by RA is contained, the rate of dissolution of the sulfates (i.e., fineness) favoring the formation of nonexpansive ettringite at an early age. Finally, the effects of plasters' additions on the mechanical performance of mortars (Rc and Edyn) are very limited.

Ettringite and thaumasite formation is observed for concretes designed with 0.8% and 1.2% sulfates CRA (plaster pollution) stored in water at 4°C. These mineral crystallization cause an expansion of concrete, loss of strength, and cracks. Such results provide some answers to define RA sulfate content limit.

12.4 RESEARCH NEEDS

Future research needs relate to the following:

- The performance-based approach. The first question is to know if durability indicators, performance tests, and their threshold values established for NAC are relevant for RAC. For example, water porosity seems to be inadequate for RAC due to the higher porosity of RA. Another question is to determine the way to consider the specific variability of RA and its influence on durability properties of RAC.
- The durability modeling. Current models (carbonation and chloride propagation ones) are defined for NAC and are not calibrated for RAC. Some studies are needed to determine the eventual specific adaptation required for RAC modeling.

12.5 CONCLUSIONS

The PN RÉCYBETON and ANR ECOREB studies confirm the previous researches: durable concrete can be made using RA even at a high level of substitution. The physicochemical characteristics and the rate of substitution of RA are determinant factors for the durability properties of the new concrete.

There are many results in the literature, some of which may seemingly seem contradictory. This is explained by the diversity of RA used, the differences in methodologies used for keeping constant the composition parameters, and the diversity of the operating methods used.

The studies carried by PN RECYBÉTON on durability properties lead to the following conclusions:

- Because of the heterogeneity and variability of RA sources, more attention needs to be paid to the control of the regularity of RA characteristics. The "natural" variability of RA has to be considered when determining the measurement frequencies, as well as the threshold values.
- As for other high porous aggregates (lightweight aggregate for example), the porosity accessible to water appears to be a durability indicator not relevant in itself to sufficiently predict the risks of corrosion. It is more rigorous to consider at least the chloride diffusion coefficient and the resistance of concrete against carbonation penetration.
- Predictably the RAs decrease the performances of the concrete relating to the transfer properties with an intensity which depends on the intrinsic characteristics of RAs (porosity), the substitution rate, and the compactness of the cement matrix of the new concrete. By optimizing concrete mix (reduction of the *W/B* ratio in particular), it is easy to produce concretes that are as resistant as concretes constituted by NA.
- The frost resistance of RA depends on the characteristics of the original concrete. Concretes are resistant to freeze/thaw cycles with or without deicing salts provided to check the frost resistance of RA and to respect the rules of formulation (binder content, *W/B* ratio, entrained air content, and so forth).
- Concerning the risks of ASRs, RA can release significant amount of water-soluble alkalis, in particular, for the fine aggregates that contain an important fraction of ACP. RA can also contain unstable silica phases provided by some specific NA and possible pollutants (broken tiles). The studies also show that the main tests usually used for NA have to be adapted (great water-absorption of RA) and nevertheless some of them (microbar test) are not adapted to RA.
- For the risk of ettringite or thaumasite formation due to internal sulfate reactions, taking into account the results of the studies done and previous research [especially Orsetti (1997)] the proposed recommendations are the following:
  - Maximum water-soluble sulfate content for RA: 0.3%
  - Maximum water-soluble sulfate content for RA + NA: 0.2%.

Chapter 13

Behavior under fire

F. Robert
CERIB

A.L. Beaucour
L2MGC Université de Cergy Pontoise

H. Colina
ATILH

CONTENTS

Abstract

The use of recycled concrete aggregates in the casting of concrete elements is one of the ways to achieve more sustainable buildings. Nowadays, these new concretes are more and more characterized, allowing their incorporation in construction projects. However, the fire behavior remains a research area that has seen limited attention. This chapter aims to characterize the behavior of two specific concretes with 30% and 100% recycled coarse aggregate replacement rate (with 30% and 0% recycled sand replacement rate, respectively). It goes from the characterization of heated recycled concrete behavior up to the spalling occurrence assessment and fire resistance of large-scale elements. The spalling of concrete was evaluated in a comparative way on four slabs: 20 cm thick, 4.6 m long, and 1.5 m wide. In addition to the two formulas with recycled aggregate (RA) concretes (RAC), two other slabs with natural aggregates (NA, siliceous or limestone) were made. The thermal profiles within these slabs were also measured during the tests carried out under the conventional temperature time curve (commonly called ISO 834). The concrete mix with RAs has a higher moisture content than with NA (+ 0.5 to + 1.5%), which may explain some few localized spalling that are not detrimental to fire resistance. Furthermore, the thermal gradients inside the specimens show a lower thermal diffusivity for RAC, which has been confirmed by transient plane source (TPS) measurements through a lower thermal conductivity. Hot and residual thermomechanical characteristics were determined for the two concretes tested with RA on samples of diameter 10 cm and length 30 cm at 300°C and 600°C (according to the recommendations of RILEM). Lastly, the fire resistance of two beams with a total length of 4.2 m was also evaluated and analyzed in a comparative way with a design according to Eurocode 2-1-2 (the tests were carried out according to the reference NF EN 1365-3). The results show good agreement with the Eurocode 2-1-2 requirements.

13.1 INTRODUCTION

13.1.1 Literature review

Only a few authors have considered the performance of concretes including recycled aggregates (RAs) when subjected to high temperatures (Cree et al. 2013). When concrete is heated, several phenomena can occur, such as expansion of aggregates, shrinkage of cement paste, increase in vapor pressure, and cracking or spalling. In addition to potential differences in the mineralogy of aggregates, RAs contain mortar and thus have very different properties than natural aggregates (NAs): they are more porous, they have a higher water absorption coefficient, and they contain hydrates. Additionally, concretes with RA have in general two interfacial transition zones, because RAs have interfacial zones between the mortar and the original aggregate as well as between the RA and the new cement paste (Liu et al. 2011). These interfacial transition zones are known as being weak areas in terms of mechanical properties (Behera et al. 2014; Scrivener et al. 2004).

A few studies have been led on the subject of recycled concrete subjected to high temperatures, and from them, a few conclusions can be drawn (Zega and Di Maio 2006; Xiao et al. 2013b; Sarhat and Sherwood 2013; Vieira et al. 2011; Eguchi et al. 2007; Zega and Di Maio 2009; Xiao and Zhang 2007; Liu et al. 2016b). Just like natural concrete, recycled concrete exposed to high temperatures shows a decrease in its mechanical properties when compared with the properties before heating:

- the comparison between the evolution of residual mechanical properties of NA concrete (NAC) and that of RA concrete (RAC) leads to different conclusions depending on the studies;
- just as NA, RA from concrete made with calcareous aggregates shows better residual properties than RA obtained from concrete made with flint aggregates.

The thermal instabilities observed in these studies are limited to the appearance of cracks on cubical samples from 800°C (Xiao et al. 2013b) and from 600°C on prismatic samples. No explosive spalling has been witnessed in recycled concrete during ISO 834 heating (Xiao et al. 2013b; Sarhat and Sherwood 2013).

Among the various studies, there is a strong dispersion of results, in particular, for the evolution of the residual mechanical properties as a function of rate of substitution of aggregates. This dispersion is explained by the variety of heating rates and by the different mineralogical nature of investigated RA. Recent works compared the evolution of residual properties of a reference concrete and a concrete with 100% RA as coarse aggregates, the aggregates of the reference concrete being of the same mineralogical nature (silicocalcareous of Seine) as the original aggregates of RA (Laneyrie 2014). Results show that, in the absence of a contaminants (wood, asphalt, etc.) within the RA, the evolution of compressive strength with the temperature is the same for both types of concrete while the tensile strength of RA concretes decreases more rapidly with temperature. The higher number of interfaces in the recycled concretes enhanced the development of cracks and led to reduced tensile strength. Contaminants have a negative impact on the residual mechanical performances of RA concretes. The presence of noncementitious impurities in RA led to cracking, flaws, and porosity when they burned during heating (Laneyrie et al. 2016).

13.1.2 Aggregates, concrete mix design, and concrete mechanical properties

The NA and RA were provided by PN-RECYBÉTON (see Appendix). Coarse NAs are crushed calcareous aggregates from Givet. Two granular fractions are considered 4/10 and 6.3/20. Fine NA is crushed 0/4 silicocalcareous sand from Sandrancourt. RAs were produced in the platform of recycling from DLB Gonesse by crushing concrete waste of demolished buildings. Three granular fractions were used: 0/4, 4/10, and 10/20. Density and absorption coefficient of aggregates are given by Sedran (2017). A Portland cement (CEM II/A-L 42.5) and limestone filler with densities of 3.09 and 2.7 are used, respectively. To ensure high workability of all developed mixes, a superplasticizer was employed. The RAs are presaturated 1% above their absorption coefficient into barrels 48 h before mixing.

The assessment of spalling sensitivity of slabs, the fire resistance of beams and the hot and residual compressive strength of cylinders were investigated with 2 RA mixes coded C35/45 30S-30G and C35/45 0S-100G, respectively. In the nomenclature C35/45 xS – yG, "x" represents the replacement percentage by weight of natural sand by recycled sand and "y" represents the replacement percentage of natural gravel by recycled gravel. The mix design of investigated concretes was proposed by RECYBÉTON project (Sedran 2017). For the spalling study on slabs, two other reference concretes, C35/45 0S-0G (Givet) and C35/45 0S-0G (Seine), were manufactured, with Givet calcareous coarse aggregates and with Val de Seine siliceous coarse aggregates, respectively, according to the mix design given by Sedran (2017).

The study of thermophysical behavior was carried out on 110 mm × 220 mm and 150 mm × 300 mm concrete cylinders with three types of mixes, named C25/30 0S-0G (Givet), C25/30 30S-30G, and C25/30 0S-100G, respectively.

The mechanical properties of C35/45 and C25/30 concretes are given at 28 days in Table 13.1. Compressive strengths are measured on 110 mm × 220 mm cylinders.

Table 13.1 Mixes used in the different tests and their compressive strengths

	Compressive strength of concretes at 28 days (MPa)			
Type of tests	*Spalling*	*Fire resistance*	*Thermomechanical properties*	*Thermophysical properties*
C35/45 0S-0G Givet	44 ± 3			
C35/45 0S-0G Seine	36 ± 1			
C35/45 30S-30G	29[a] ± 1	32 ± 1	39 ± 2	
C35/45 0S-100G	40 ± 2	27 ± 4	34 ± 1	
C25/30 0S-0G Givet				29 ± 2
C25/30 30S-30G				28.5 ± 0.6
C25/30 0S-100G				25 ± 0.6

[a] 7 days

13.2 EXPERIMENTAL PROGRAM AND RESULTS

13.2.1 Thermophysical properties

13.2.1.1 Tests

The thermal properties (conductivity, specific heat) of concretes were measured with a Hot-Disk probe TPS1500. The system is based on the technology of the transient plane source (TPS). This method has been developed at Chalmers University of Technology by Gustavsson et al. (1994). A mica probe with radius 14.61 mm consisting of very fine nickel double spiral (thickness 10 µm) covered with two thin layers of electrically insulating materials is used. The maximum aggregate size (20 mm) remains smaller than the probe diameter. The probe is placed between two symmetrical 40-mm slices of concrete cylinders (110 mm × 220 mm). Each sample must have a flat surface to avoid contact defects with the sensor. Specimens were cured in plastic sealed bags with a wet towel for 90 days. Cylinder slices were predried at 80°C until a constant mass was obtained and then heated in a 5-L electric oven controlled by the Hot Disk software. The heating and cooling rates were 1°C/min. For each target temperature, measurements were taken under isothermal conditions at 30°C, 150°C, 200°C, 250°C, 300°C, 450°C, 500°C, 550°C, and 600°C during the heating phase and then during the cooling phase. Thermal conductivity and diffusivity are simultaneously obtained through a process iteration from one transient recording. The specific heat is deduced from diffusivity and conductivity. The results presented are the average values calculated from three tests.

The thermal response of concrete cylinders (150 mm × 300 mm) was also investigated during heating/cooling cycles at 0.5°C/min. Measurements of surface temperature, quarter-diameter, and middiameter temperature of the specimen has allowed determining the temperature difference (ΔT) between the middiameter (quarter-diameter) and the surface of the specimen as a function of surface temperature during heating–cooling cycles up to 600°C. The heating/cooling cycles were performed in a programmable electric furnace with size of 1,100 mm × 1,200 mm × 1,100 mm. The temperature rise is controlled by a regulator controller connected to a thermocouple in the oven. Thermocouples are used to monitor temperatures at the upper surface and inside the concrete specimen (this last thermocouple was embedded at the center of the sample).

13.2.1.2 Results

At room temperature, RA concretes display lower thermal conductivity values than reference concrete, 1.7 and 1.6 W/m/K for C25/30 30S-30G and C25/30 0S-100G instead of 2.1 W/m/K for C25/30 0S-0G. This would be explained by the higher porosity and paste volume fraction in RA concretes. Figure 13.1 shows that thermal conductivity decreased as temperature increased. In nonmetallic materials, the decrease of thermal conductivity with temperature is related to phonon scattering. This decrease was also partly linked to damage, since the residual values were lower than those observed for the nonheated samples. Some of the potential mechanisms of damage were as follows: (i) bound water departure since the thermal conductivity of water is higher (0.6 W/m/K) than that of air (0.02 W/m/K), (ii) the decrease of conductive bonds linked to decomposition of hydrates, and (iii) microcracking above 300°C. Conductivity of 0S-0G concretes decreases faster than RA concretes. Thus, the gap between the concretes reduces with temperature. It is indeed usual to observe that the most conductive concretes show a higher loss of conductivity with the rise in temperature (Xing et al. 2015; Niry 2015; Yermak 2015). The average loss of thermal conductivity for a temperature increase of 100°C is 0.17 W/m/K for C25/30 0S-0G and 0.13 and 0.12 W/m/K

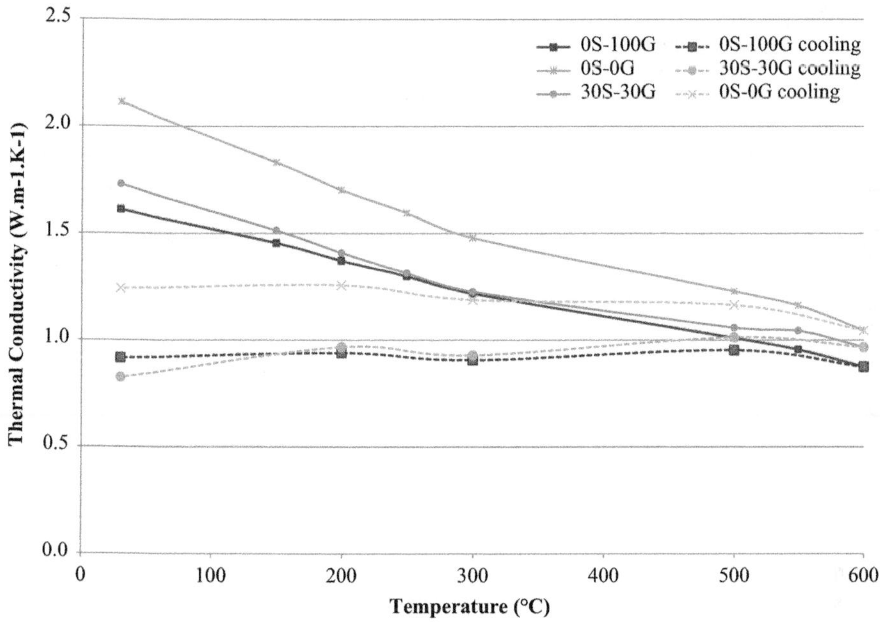

Figure 13.1 Thermal conductivity evolution of concretes with temperature, during heating and cooling.

for C25/30 0S-100G and C25/30 30S-30G, respectively. The conductivity at 600°C of the reference concrete represents 50% of its initial value and 55% for the two recycled concrete aggregates (RCA) concretes. These values are close and are within the range of values reported in the literature for most of the concretes. The measurements carried out during cooling show a hysteresis that highlights the irreversibility of the reactions leading to the damage of concrete (Yermak 2015; Yermak et al. 2017; Jansson 2004a,b).

Figure 13.2 shows the evolution of specific heat during heating up to 600°C and consequent cooling to ambient temperature. The thermal capacity values are measured in isothermal conditions, and specimens should achieve a hydrothermal equilibrium before measurement. The specific heat measured should be regarded as a fundamental specific heat, not considering the latent heat of different physicochemical transformations. RAs have no significant influence on the fundamental specific heat value of concretes. The measurement of thermal response of a concrete specimen will complete this analysis by evaluating the influence of RA on heat consumption related to physicochemical transformations. Results show that specific heat increases with temperature, and differences between the measured values of 0S-0G and 30S-30G are not significant. Between 500°C and 600°C, the specific heat of 0S-100G began to increase faster. This trend has also been observed on other types of concrete (Khaliq et al. 2011). The phenomenon of hysteresis is much less important than for conductivity. The increase of thermal capacity with temperature is essentially related to reversible phenomena. Specific heat is strongly dependent on atomic vibration, the main mode of absorption of thermal energy in solids. As temperature increases, the average energy of atomic vibration increases, thus leading to higher values of specific heat. Consequently, a residual measurement of heat capacity cannot be substituted for hot measurement.

The measurements, during the heating, of the surface (T_{surf}), quarter-diameter ($T_{1/4}$) and middiameter ($T_{1/2}$) temperatures of 150 mm × 300 mm concrete cylinders enable to compare the thermal response of three types of concretes, including the proportion of heat consumed

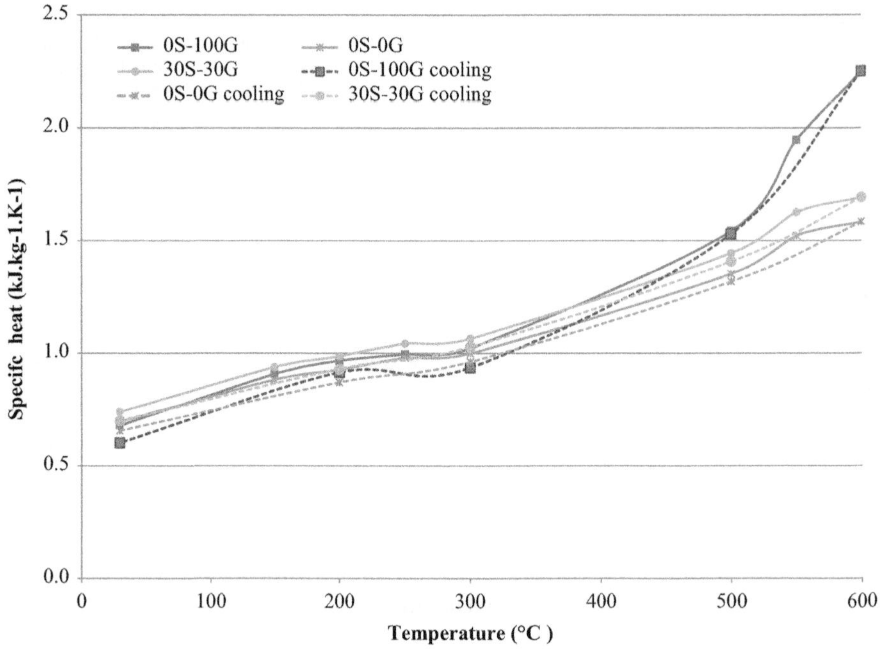

Figure 13.2 Specific heat evolution of concretes with temperature, during heating and cooling.

during the phase changes and chemical transformations. Figure 13.3 shows that the temperature differences ($T_{surf} - T_{1/4}$) and ($T_{surf} - T_{1/2}$) increased to a peak around 250°C and 270°C respectively, and then decreased slightly before a second increase around 600°C. The first peak is linked to evaporation of free water and decomposition of hydrates. A slightly higher maximum temperature difference for concretes containing RA, especially for the C25/30 30S-30G, is explained by the higher absorption coefficient of RAs (aggregates were presaturated before mixing) and the higher amount of cement paste. The two concretes 30S-30G and 0S-100G have almost the same volume of RA and cement paste. The highest peak observed for 30S-30G concrete may be related to the higher absorption coefficient of recycled sand compared with that of recycled gravel and consequently to the higher free water content. However, these differences of temperature remain rather low compared with the peaks of temperature recorded on higher-strength concretes containing a larger volume of paste.

13.2.2 Spalling tests and thermal profiles

The four slabs 4.6 m × 1.5 m × 0.2 m were subjected to the conventional temperature time curve for a period of 60 min (commonly called ISO 834). The test was carried out in accordance with the principles of NF EN 1363-1. The slabs have been stored during 90 days at 50% relative humidity (RH) and 23°C.

A manual measurement (with a ruler) of the spalling observed on the surface exposed to the fire of the slabs is carried out after the test (Figure 13.4). For each slab, the spalled surface and the maximum spalling depth of each zone are evaluated. The corresponding data are grouped in Table 13.2.

RAC showed some superficial and localized spalling (maximum depth of 3 cm over an area of about 10% of the total area). Their appearance may be related to a slightly higher

(a)

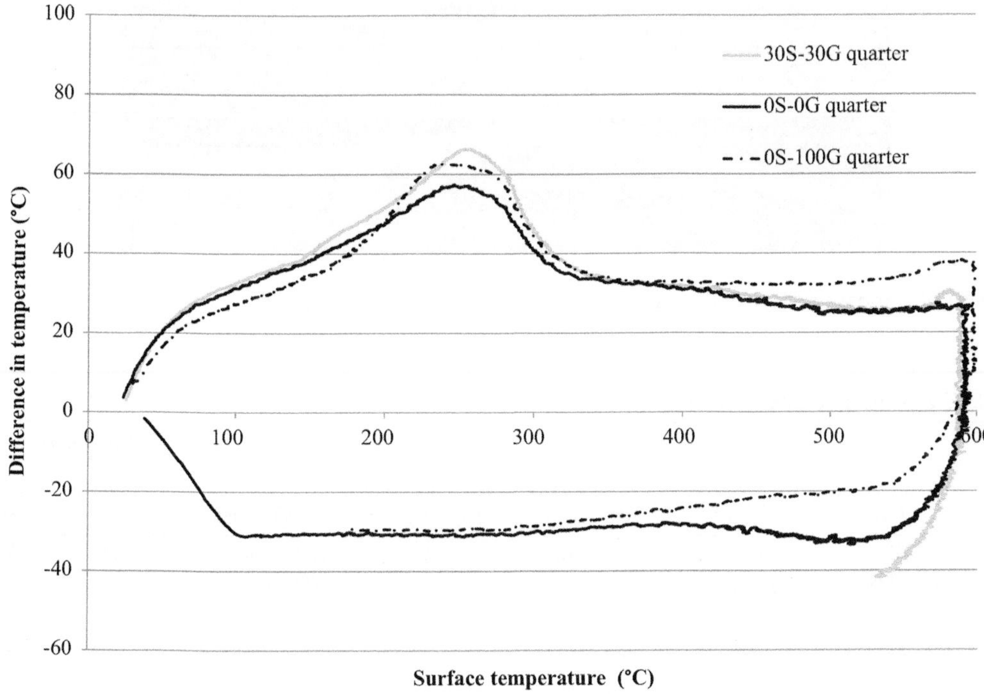

(b)

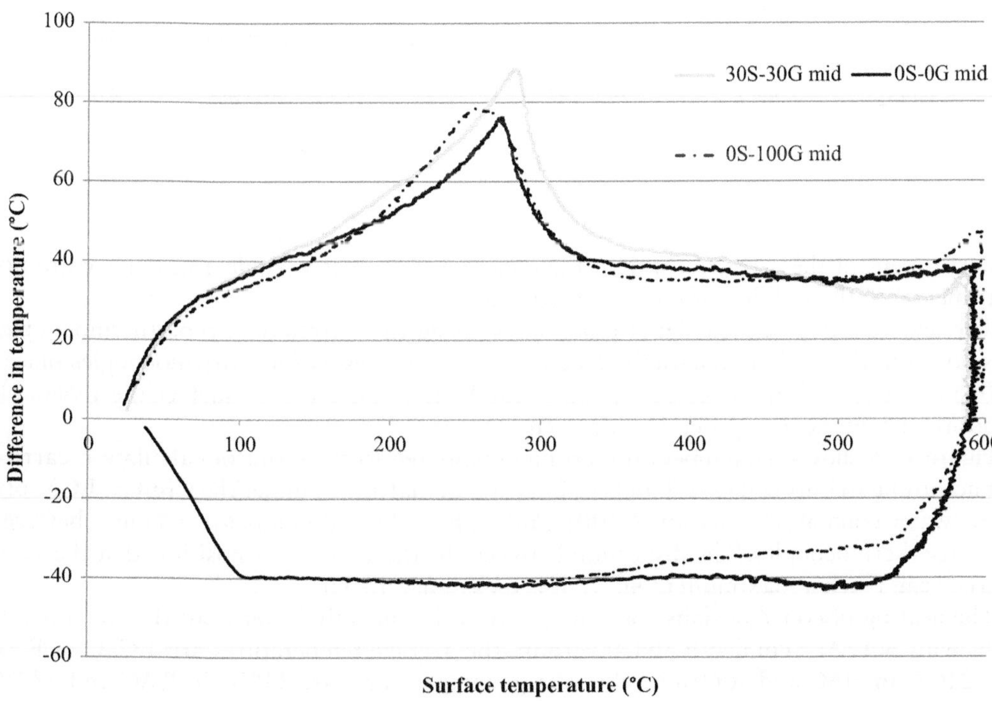

Figure 13.3 Difference of temperature ($T_{surf} - T_{1/4}$) (a) and ($T_{surf} - T_{1/2}$) (b) within the specimens during the heating—cooling cycle at 600°C (0.5°C/min).

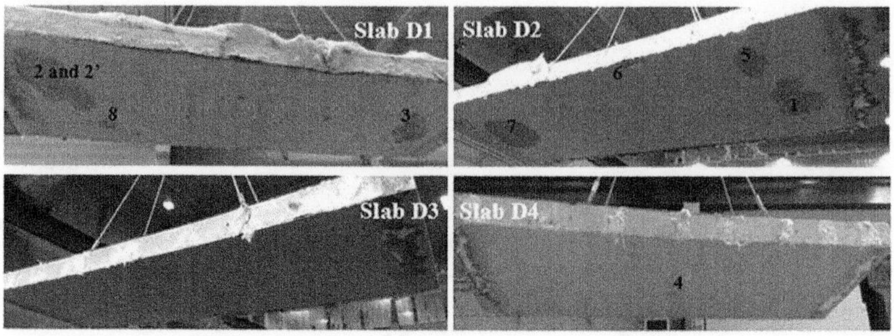

Figure 13.4 Exposed surface of the slabs after the test.

Table 13.2 Spalling results on slabs

						Moisture content	
Slab	*Mix*	*Zones*	*Spalled surface (cm × cm)*	*Maximal depth of spalling (cm)*	*Comment*	*Surface (%)*	*Heart (%)*
D1	C35/45 0S-100G	Zone 2 + 2'	40 × 78	3	Three main zones	5.5	5.8
		Zone 8	18 × 18	1			
		Zone 3	45 × 65	2.5			
D2	C35/45 30S-30G	Zone 7	80 × 65	3	Four localized zones	4.4	4.9
		Zone 6	25 × 10	2			
		Zone 1	35 × 20	3			
		Zone 5	55 × 36	2.5			
D3	C35/45 0S-0G Seine		—	—	Cracks at the surface	4.0	4.9
D4	C35/45 0S-0G Givet	Zone 4	25 × 18	2	One localized zone at the center of the slab	4.0	4.2

water content of these concretes than the concretes of NA (see Table 13.2). They are not detrimental to the stability of the tested element.

The temperatures of each slab are measured using three trees of K-type thermocouples positioned in the median longitudinal axis. Each tree consists of five thermocouples placed at 0 (on the exposed surface), 2, 3, 5, and 15 cm. In addition, a K-type surface thermocouple is positioned on the unexposed face on each slab.

Figure 13.5 shows the temperature profiles compared to the result of calculation carried out according to Eurocode 2-1-2 and the French national annex under the CimFeu EC2 software (water content 3%, density 2,300 kg/m^3). The indicated points are averages between the three thermocouples. The dispersion between the thermocouples positioned at the same altitude can reach a maximum of 40°C (it is on average 15°C).

The heating of concrete slabs containing RA is significantly lower than that of concrete slabs with NA. At 3 cm depth and at 30 min, the average temperatures are 182°C in RAC and 216°C in NAC and at 60 min, the average temperatures are 338°C in RAC and 370°C in NAC. This tends to indicate that the thermal conductivity of concretes with RA may be lower, and hence, the heat transfer calculation according to Eurocode 2-1-2 can then be applied.

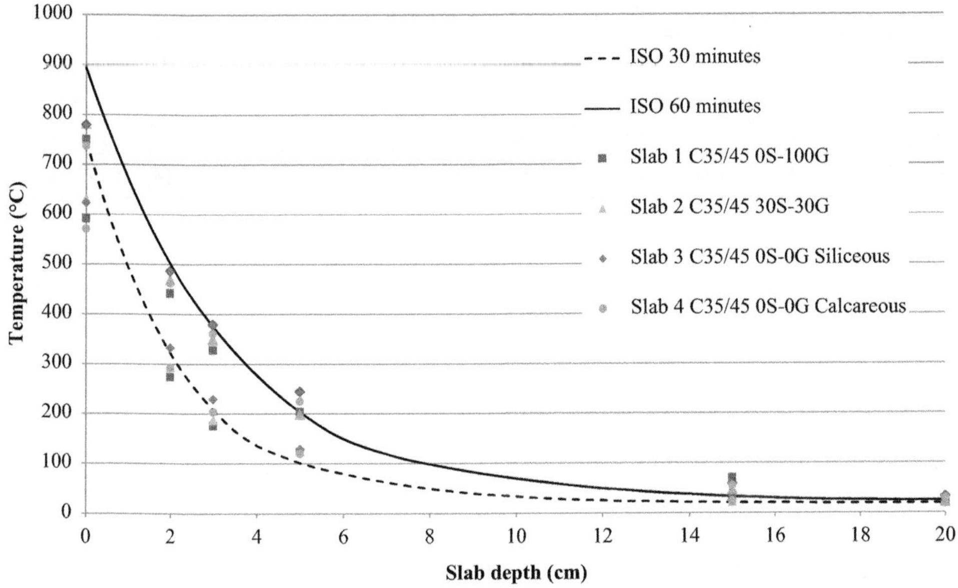

Figure 13.5 Temperature profiles in the slabs at 30 and 60 min.

13.2.3 Thermomechanical properties

The tests were carried out according to the RILEM TC 129-MHT recommendations (Rilem 1995) on cylindrical samples diameter 10 cm and height 30 cm. The samples have been stored 6 months at 50% RH and 23°C. A preload of 20% of the compressive strength measured at room temperature (20°C) is applied before the specimen is heated and is maintained throughout the test. The preloading rate is 0.5 MPa/s. The rate of temperature rise is 1°C/min. When the test temperature is reached, it is maintained during a stabilization step of

- 2 h for tests at 300°C;
- 1 h for the tests at 600°C.

When the test is carried out in a residual state, the test sample undergoes a cooling of 12 h after the stabilization stage is completed. When the stabilization stage is completed (or 12 h of cooling when the test is carried out in residual), the mechanical loading is increased at 0.5 MPa/s until the test sample fails.

The results at 20°C, 300°C, and 600°C and the corresponding ratios are presented in Table 13.3 and Figure 13.6. Four samples were tested at 20°C and two were tested for each point at high temperature (the scatter at high temperature is lower than at ambient temperature, and thus two samples are commonly tested at high temperature for the same point).

These results show that the coefficient $k_c(\theta)$ considering the decrease of characteristic strength f_{ck} of concrete with siliceous aggregates may be used for the design of structures using RCA. Furthermore, it appears that the 30S-30G performance is higher than the 0S-100G mix (about 10% on the relative values). These observations should be confirmed with other types of RA.

Table 13.3 Compressive strength (MPa) and ratios $[f_c(\theta)/f_c(20°C)]$

		A 300°C		*A 600°C*	
Mix	*A 20°C*	*Hot*	*Residual*	*Hot*	*Residual*
C35/45 0S-100G	34.4 ± 0.9 (1)	28.1 ± 0.3 (0.82)	29.4 ± 0.0 (0.85)	15.3 ± 1.0 (0.45)	12.5 ± 0.4 (0.36)
C35/45 30S-30G	39.3 ± 1.9 (1)	36.1 ± 0.3 (0.92)	30.8 ± 1.4 (0.78)	21.0 ± 0.3 (0.54)	14.8 ± 1.6 (0.38)

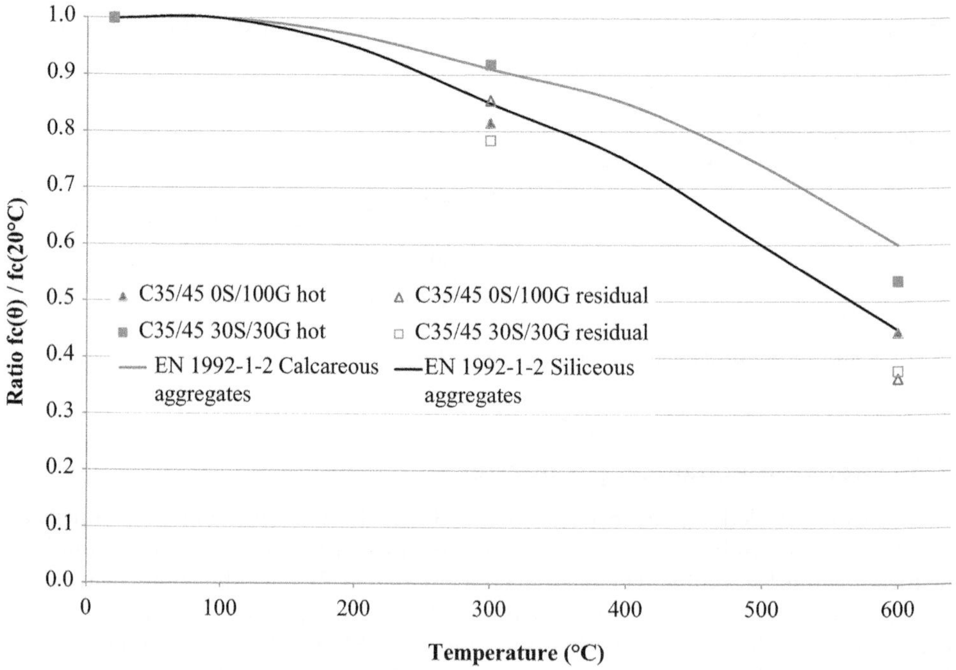

Figure 13.6 Ratio $f_c(\theta)/f_c$ (20°C) and comparison with EN 1992-1-2.

13.2.4 Fire resistance tests

13.2.4.1 Description of the tested beams

The beams (span 4.2 m; sections 30 cm × 40 cm) have been designed for an office building (G = self-weight + 1 kN/m²; Q = 1.5 kN/m²).

The cover of the transversal reinforcement is 10 mm under the beam; thus, the axis distance for longitudinal reinforcement is 26 mm.

Tabulated data of EN 1992-1-2 (Table 5.5) would lead to classify the beam R60.

During the fire test, the combination $G+0.5\ Q$ is applied (which does correspond to an applied isostatic moment 66.15 kN.m).

Two beams were tested at ambient conditions and two others were heated according to the conventional temperature/time curve for 60 min (NF EN 1363-1 and NF EN 1363-3) and then the load calculated according to the accidental combination (total load of 9.45 T with two symmetrical jacks positioned at 0.69 m from the center) was increased to failure (Figure 13.10).

The beams have been stored for 90 days at 50% RH and 23°C (Figure 13.7).

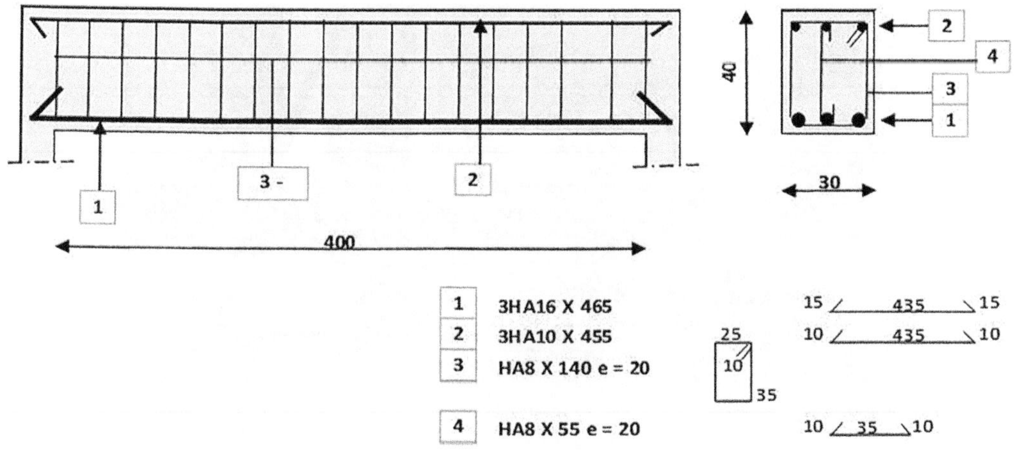

Figure 13.7 Beam reinforcement.

13.2.4.2 Test results of the beams tested at ambient conditions

The reinforced concrete beam C35/45 0S-100G and 30S-30G failed, respectively, at 18.1 T and 17.1 T (respective bending moments of 126.6 and 119.9 kN.m).

Analytically, by taking safety coefficients of the materials equal to 1, maximal bending moment was calculated at 108.3 kN.m.

13.2.4.3 Test results of the beams tested under fire

The reinforced concrete beam C35/45 0S-100G and 30S-30G failed, respectively, at 15.2 T and 16.6 T (respective bending moments of 106.4 and 116.2 kN.m).

The software CimFeu gives a resistant bending moment of 69.70 kN.m at 60 min.

The moments of resistance obtained experimentally (and the ratios $M_{R,fi}/M_R$) at the end of the tests are higher than those calculated according to Eurocode 2-1-2 (Figures 13.8–13.11).

Figure 13.8 Beam 0S/100G after the test.

Figure 13.9 Beam 30S/30G after the test.

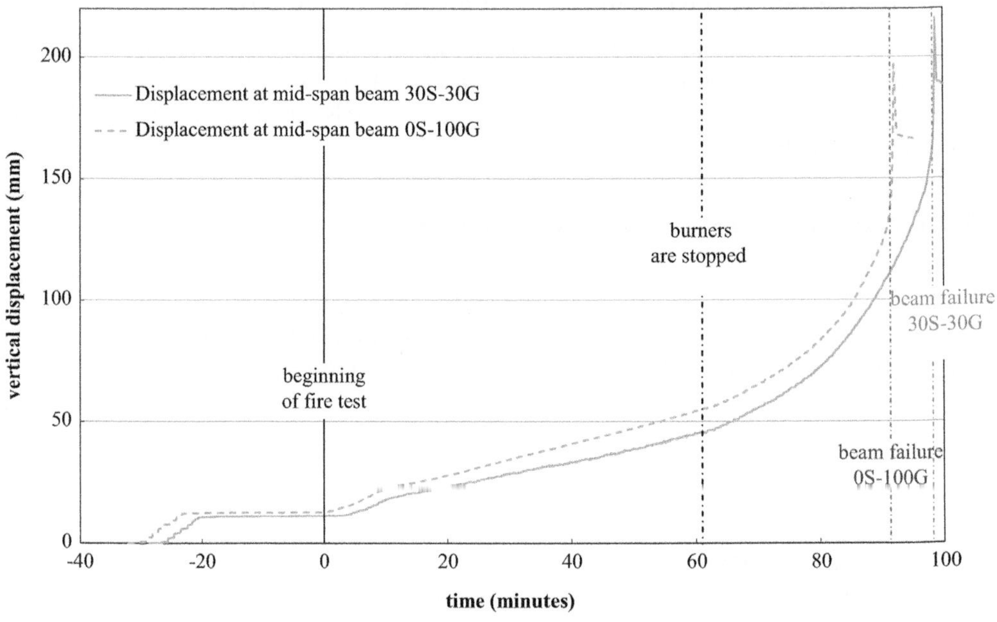

Figure 13.10 Vertical displacement at midspan of the beams during the fire test.

13.3 CONCLUSIONS

Two specific concrete mixes with 30% and 100% recycled coarse aggregate replacement rate (respectively, 30% and 0% recycled sand replacement rate) were made to study the behavior in the fire of concretes containing RA. C25/30 concretes were considered for testing thermophysical properties and C35/45 concretes for testing spalling, fire resistance, and thermomechanical properties. For spalling and thermophysical properties tests, reference concretes of the same resistance class were also manufactured. Four slabs were casted for

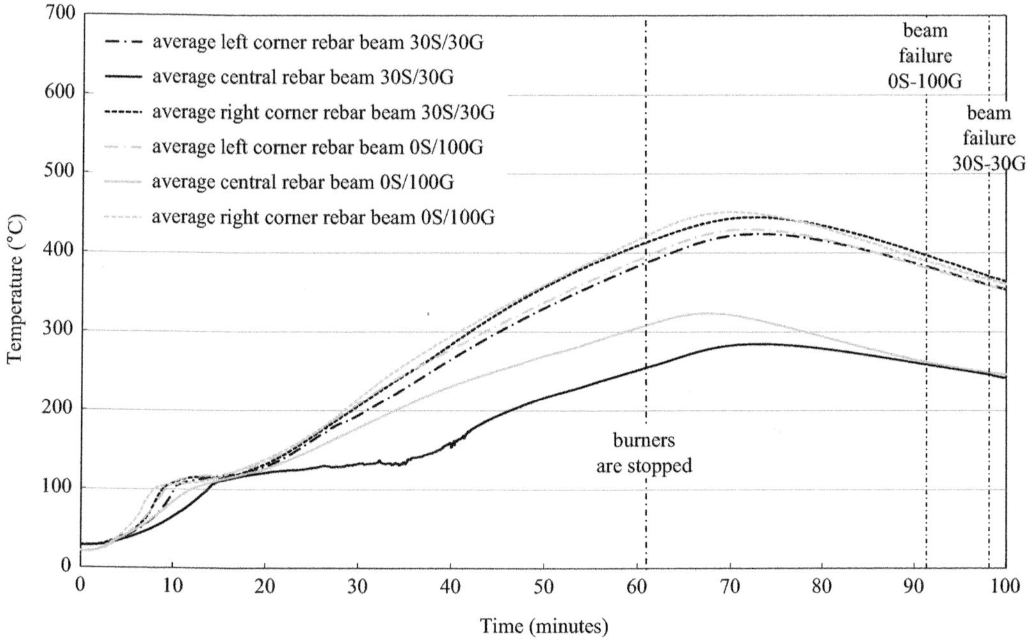

Figure 13.11 Temperature reinforcement in the beams.

spalling tests (where two reference slabs with different types of NA) and four beams for the fire resistance test (two for tests at ambient conditions and two for test at fire conditions).

The following conclusions can be drawn from this complete study:

- Concretes containing RA display lower thermal conductivity values than reference concrete, which is explained by the higher porosity and paste volume fraction in RCA concretes. Conductivity of 0S-0G concretes decreases faster with the increase of temperature and then the values of the tested concretes at 600°C are close. They are within the range of values reported in the literature for most concretes.
- Recycled concrete aggregates do not have an important influence on the specific heat value of concretes. The specific heat increases with the temperature, and differences between the measured values of 0S-0G and 30S-30G are not significant.
- Slabs containing RCA showed some superficial and localized spalling (maximum depth of 3 cm over an area of about 10% of the total area, not detrimental to the stability of the tested element), and maybe related to a slightly higher water content of these concretes than the concretes with only NA.
- The beams with concretes containing RCA, tested according to the conventional temperature/time curve for 60 min, present at the end of the tests experimental moments of resistance (and ratios $M_{R,fi}/M_R$) which are higher than those calculated according to Eurocode 2-1-2.

We can conclude that concretes containing RCA, compared with current concretes, have good fire behavior, in particular the 30S-30G concrete. These results are to be confirmed with other types of RA to introduce them in the specific standards (Eurocode 2-1-2).

Section V

Mix design of recycled concrete

T. Sedran
Ifsttar

This section starts with Chapter 14 that describes how the current standardization accounts for the recycled concrete aggregates (RAs) and their use in new concrete. The chapter focuses on the European context, and sometimes more specifically on the French one. It is observed that the standardization is often conservative and strongly limits the recycling rate. This can be explained by a lack of knowledge concerning the effect of RAs on concrete properties and particularly durability.

Without considering durability purposes at first, Chapter 15 is dedicated to mix design methods of concrete including RA. It first makes a summary of the particular properties of RA (water absorption, packing density, and mechanical strength) and the expected qualitative consequences on the concrete mix design. In the second part, it shows that for a simple set of requirements including the specification on compressive strength and slump, an empirical method of Dreux type, currently popular in France, can be adapted to roughly design concrete with recycled aggregrates, when the recycling rate is set a priori. Yet the main objective may not be a question of designing a concrete with a given recycling rate, but to optimize the recycling rate, so that the gain, either economic or environmental, is maximized. In that case, experimental methods are too cumbersome and imprecise. Thus, a more scientific approach is proposed to predict the optimal packing density and the mechanical properties for optimizing the RA content and selecting optimal recycling scenarios. These mix design processes must be completed by a prescriptive approach to ensure durability, considering the recycling rate, or better by a performance-based approach. Research summarized in Chapter 12 will help to identify criteria and thresholds adapted to the RAC durability in the future.

Chapter 16 is dedicated to a construction site realized in 2005 that appears as a pioneer in the French landscape, demonstrating the feasibility of a full and local recycling process, 7 years before RECYBÉTON started. It consists in the partial demolition of a concrete water treatment plant and its reconstruction with the RA produced on-site. The chapter describes the analysis of standards existing at the time, the characterization of aggregates, the detailed mix design process of a C25/30 concrete containing between 90% and 100% of RA (including fine fraction), and the results of quality control. At the moment of writing this book (2018), the structure behaves well.

Chapter 14

Specifications of concrete with recycled aggregates

W. Pillard
EGF.BTP

T. Sedran
IFSTTAR

P. Rougeau
CERIB

CONTENTS

Abstract

The specification of concrete, in terms of EN 206:2013, is defined as the final compilation of documented technical requirements given to the producer in terms of performance or composition, including the constituents. This standard considers the use of recycled concrete aggregates (RA) but in quite different ways in the various national provisions (see chapter 34). Nevertheless, there is a common way that is comprised of two aspects, the quality of the aggregates and the type of concrete. The present chapter details the standards currently in use in France dealing with the introduction of RA in concrete. A safe approach is adopted up to now. Due to the lack of knowledge concerning the durability of concrete including RA, recycling is limited for coarse aggregate, and in most cases prohibits the use of fine aggregate. The French national project

RECYBÉTON was thus organized to provide answers to a wide range of questions concerning RA and concrete including RA (RAC). Significant results were obtained to justify the evolution of the standards as the natural follow-up of the works.

14.1 INTRODUCTION

In the last decade, stringent environmental regulations and depletion of natural resources led to the need of using construction demolition waste (CDW) as aggregates for construction. The use of recycled concrete aggregate (RA) in new concrete is a way of maximizing the economic benefits of CDW. Nevertheless, it is well known that RA displays characteristics that are different from those of natural aggregates (NA). Consequently, the performances of concrete including RA concrete (RAC) are modified compared to normal concrete (see Sections I, III, and IV of this book). It is then necessary to account for the specificities of these RA in standards and utilization. For example, mechanical characteristics such as elastic modulus, shrinkage, and creep of RAC are modified, so that their use depends on the type of structures. Typically, in buildings, special design specifications allow to bypass the consideration of these characteristics. On the other hand, there are some structures (e.g., bridges) for which they are essential, and consequently, the designers need to know their evolution with the substitution rate of RA. More generally, in France, due to lack of knowledge that ensures the durability, the maintenance of safety, and in-use requirement of built elements, the use of RA is strongly limited by standards. So RA still remains employed in ordinary applications such as bases and subbases of road pavements, despite the potential shown by these aggregates in various experimental research projects performed worldwide and more permissive standards adopted in other countries.

In that context, one of the main goals of RECYBÉTON was to improve the knowledge of the properties and durability of RAC, to see how it is possible to change the French standards, ensure their use in structures, and give confidence to owners.

For a good understanding of the way RECYBÉTON was organized, it is necessary to have a global vision of the architecture of the existing French standards. This vision is also needed as a first step in RAC mix design process to establish a relevant set of requirements (see also Chapter 15). For these purpose, this chapter makes a survey of the French standards. It is completed by an important survey of European and international provisions, presented in Chapter 34.

14.2 STATE OF THE ART

For concrete structures, the global concept is the durability, e.g., the ability to answer, during the working service life, to the requirements for which they were designed. Figure 14.1 shows the interconnections between the European standards dealing with the design (Eurocodes), the concrete material (EN 206), the production of precast concrete products (EN 13369) and the execution of concrete structures (EN 13670). A suitable durability of the structure involves that all these texts need to be consistent with each other, so that specifications of concrete have to include all the relevant information coming from different standards.

In Europe, the general organization of standardization allows to have European standards accompanied by national provisions. For example, the EN 206 standard dealing with concrete is customized by NF EN 206/CN in France.

The following section details the different texts in their current version.

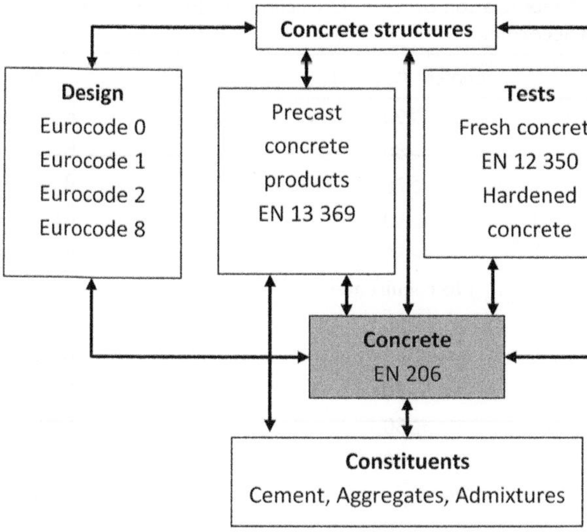

Figure 14.1 General organization of the European standards to ensure the durability of concrete structures.

14.2.1 Aggregates for concrete according to European standard EN 12620 and national provisions (NF P 18-545)

European standard specifying aggregates for concrete is EN 12620. It is a harmonized standard in which RAs are classified in terms of constituents (Rc, Ru, etc., see hereunder), and, like NA, of properties thresholds (according to test standards). Table 14.1 gives the normalized denomination of RA in which

- Rc stands for concrete, concrete from industrial products, mortars, and concrete masonry elements;
- Ru stands for unbounded aggregates, natural stones, and binder-treated aggregates;
- Rb stands for elements in fired clay such as bricks or tiles, elements in calcium silicate and foam concrete elements;
- Ra stands for bituminous materials;
- Rg stands for glass materials;
- FL stands for floating materials;
- X stands for the other constituents such as soils, ferrous or nonferrous metals, wood, plastics, and gypsum materials.

In France, the codification of RA is done according to the national version NF EN 12620 + A1 completed by the national provisions summarized in NF P 18-545. To be used in concrete, RA must at least comply with the specifications of these two standards, but additional constraints are added by NF EN 206/CN as will be seen further. The main test standards that present a specific interest regarding the classification of RA are as follows:

- EN 1097-6: Tests for mechanical and physical properties of aggregates—Part 6: Determination of particle density and water absorption;
- EN 1744-1: Tests for chemical properties of aggregates—Part 1: Chemical analysis (water-soluble sulfate);

Table 14.1 Normalized denomination of RA in EN 12620 on the basis of its composition according to NF EN 933-11

Constituents	*Content (% in mass)*	*Categories*
Rc	≥90	Rc_{90}
	≥80	Rc_{80}
	≥70	Rc_{70}
	≥50	Rc_{50}
	<50	$Rc_{declared}$
	No requirement	Rc_{NR}
Rc + Ru	≥95	Rcu_{95}
	≥90	Rcu_{90}
	≥70	Rcu_{70}
	≥50	Rcu_{50}
	<50	$Rcu_{declared}$
	No requirement	Rcu_{NR}
Rb	≤10	Rb_{10-}
	≤30	Rb_{30-}
	≤50	Rb_{50-}
	>50	$Rb_{declared}$
Ra	≤1	Ra_{1-}
	≤5	Ra_{5-}
	≤10	Ra_{10-}
X + Rg	≤0.5	$XRg_{0.5-}$
	≤1	XRg_{1-}
	≤2	XRg_{2-}
Constituents	*Content (cm³/kg)*	*Categories*
FL	≤0.2	$FL_{0.2-}$
	≤2	FL_{2-}
	≤5	FL_{5-}

- EN 1744-5: Tests for chemical properties of aggregates—Part 5: Determination of acid-soluble chloride salts;
- EN 1744-6: Tests for chemical properties of aggregates—Part 6: Determination of the influence of RA extract on the initial setting time of cement;
- EN 1097-1: Tests for mechanical and physical properties of aggregates—Part 1: Determination of the resistance to wear (micro-Deval);
- EN 1097-2: Tests for mechanical and physical properties of aggregates—Part 2: Methods for the determination of resistance to fragmentation.

14.2.2 Selection of concrete components and properties according to European standard EN 206:2013 and national provisions NF EN 206/CN: 2014

In EN 206 2013, the specifications are defined as the "final compilation of documented technical requirements given to the producer in terms of performance or composition". Typically, this includes any requirement for concrete properties needed for transportation after delivery, placing, compaction, curing, or further treatment. The specifications may include special requirements (e.g., to obtain an architectural finish).

Consequently, the specifier must ensure that all relevant requirements of concrete properties are included in the specification given to the producer. For this purpose, he has to consider the following:

- the environmental conditions to which the element is to be exposed (defined by exposure classes). The exposure classes give information to the producer in the way to have a mix design complying with requirements in terms of minimal cement content C, maximal free water-to-cement ratio (W/C), etc., to ensure the durability of concrete. In the French version of the standard with its national provisions (NF EN 206/CN), the tables of the national annex NA.F.1 (for ready-mixed concrete, site-mixed concrete, or industrial precast concrete products) and NAF.2 (for industrial precast concrete products only) give all the relevant information needed in terms of composition and properties (i.e., compressive strength class);
- the fresh behavior of concrete expressed in terms of workability classes. It has a direct relation with the execution of works according to NF EN 13670/CN and the choice of casting method (e.g., pumped concrete or self-compacting concrete). EN 206 2013 allows different ways to measure this property;
- the properties of hardened concrete mainly expressed in terms of classes of compressive strength for normal weight concrete. Compressive strength classes are defined in a large range, moving from C8/10 to C100/115, where the first number stands for the characteristic strength in MPa measured on Ø150 × 300 mm^3 cylinder and the second one for the characteristic strength in MPa measured on 150-mm^3;
- the method of placing and more especially the conditions of curing. Some concretes need a special attention during execution works because of their sensibility to the curing conditions, so the different curing classes defined in NF EN 13670/CN:2013 may have to be adapted;
- the dimensions of the element related to the heat development that could lead to internal delayed ettringite formation (DEF) if it is too high. The French experience dealing with DEF has led to national recommendations (Ifsttar 2017);
- the design life that is 50 years in EN 206 2013. The Fascicule 65 gives useful specifications for public works in France, for which this design working life is 100 years. Here, NA.F tables must be adapted or a performance-based approach must be used (see next section).
- any requirements related to the cover of reinforcement or minimum section width, e.g., maximum aggregate size in agreement with EN 1992-1-1 and EN 13670. These specifications depend on the design of the structure (according to Eurocodes) in which rules dealing with durability (prevention of steel reinforcement from corrosion) depend on the structural class of the structure and other dispositions such as quality of concrete and quality management of the project.
- any further restrictions on the use of constituent materials with established suitability (e.g., type of cement in special environments).

In EN 206 2013, concrete shall be specified either as

- designed concrete: concrete for which the required properties and additional characteristics are specified to the producer who has the responsibility to supply a concrete that satisfies all these properties;

- prescribed concrete: concrete for which the composition of the concrete and the constituents are specified to the producer who is responsible for providing a concrete complying with this specified composition.

In addition, both EN 206:2013 and NF EN 206/CN: 2014 give the specifications for suitability of constituents and especially for aggregates.

14.2.3 Use of RAs in concrete according to EN 206:2013 and national provisions

The EN 206:2013 allows the use of RAs, because it refers to suitability of European standard for aggregates for concrete (EN 12620) in which the classification of RAs is given (see Table 14.1); however, local suitability is to be established by reference to national provisions in accordance with local experience. Thus, in France, the NF EN 206/CN: 2014 standard imposes supplementary constraints as explained later.

First, three types of coarse RA are defined according to their codification by NF EN 12620 + A1 completed by the national provisions summarized in NF P 18-545 (see Tables 14.2 and 14.3) and the frequency of their control:

Table 14.2 Codification of coarse RA according to its composition

Code	*Main constituent (NF EN 12620 +A1)*	*Secondary constituents (NF EN 12620 +A1)*				*Type of test frequencies*	
						Temporal	*Quantitative*
CRB	Rcu95	Rb_{10-}	Ra_{1-}	$XRg_{0.5-}$	$FL_{0.2-}$	2/month	1/2,000 tons
CRC	Rcu90	Rb_{10-}	Ra_{1-}	XRg_{1-}	FL_{2-}		
CRD	Rcu70	Rb_{30-}	Ra_{10-}	XRg_{2-}	FL_{2-}		

Table 14.3 Codification of coarse RA according to their properties

Characteristic	*Test method*	*Code*	*Category*	*Type of test frequencies*	
				Temporal	*Quantitative*
Water-soluble sulfates	NF EN 1744-1 article 10.2	CRB, CRC CRD	$SS_{0.2}$ SSD $Vss_{0.7}$	1/week	1/1,000 tons
Determination of particle density	NF EN 1097-6	CRB, CRC CRD	≥ 2.0 t/m³ ≥ 1.7 t/m³	1/week	1/1,000 tons
Determination of the influence of RA extract on the initial setting time of cement	NF EN 1744-6	CRB, CRC CRD	A_{10} A_{40}	2/month	1/2,000 tons
Determination of particle shape—Flakiness index	NF EN 933-3	CRB, CRC, CRD	FI_{35}	1/month	1/4,000 tons
Methods for the determination of resistance to fragmentation–Los Angeles	NF EN 1097-2	CRB, CRC CRD	LA_{40} LA_{50}	1/2 months	1/8,000 tons
Determination of acid-soluble chloride salts	NF EN 1744-5	CRB, CRC, CRD	To be declared	2/month	1/2,000 tons
Water absorption measured at 24h (WA_{24})	NF EN 1097-6	CRB, CRC, CRD	To be declared	1/week	1/1,000 tons
Determination of active lime water-soluble alkalis	XP P 18-544	CRB, CRC, CRD	To be declared	2/month	1/2,000 tons

- Type 1: all the characteristics are CRB;
- Type 2: all the characteristics are CRB or CRC;
- Type 3: all the characteristics are CRB, CRC, or CRD.

The frequency of control can be either temporal or quantitative; however, in the first period of production (i.e., during at least the first 12 months with a minimum production of 10,000 tons), the producer must choose the frequency that gives the greater number of samples.

Then, RA can be used in concrete, provided the following items are respected:

- RA complies with NF EN 12620 + A1 and NF P 18-545;
- coarse RA is of type 1, 2, or 3;
- fine RA complies with Table 14.4 specifications;
- the concrete is not prestressed;
- the maximum recycling rates displayed in Table 14.5 are respected according to the exposure class.

Finally, it can be observed that the guidelines proposed by the NF EN 206/CN standard are restrictive and that recycling rates are quite limited (particularly for fine aggregate recycling and severe environments). This conservative approach is justified by the lack of knowledge about the durability of concretes including RA. However, it is interesting to note that EN 206:2013 allows the use of performance-based approach in addition to the prescriptive approach. The prescriptive approach expressed in tables NA.F gives yield values for the mix design (maximum water to cement ratio, minimum value of cement content C, etc.) and for the properties of concrete (compressive strength). The performance-based approach allows a

Table 14.4 Specifications for fine RA

			Type of test frequencies	
Characteristic	*Test method*	*Category*	*Temporal*	*Quantitative (t)*
Water-soluble sulfates	NF EN 1744-1	SSD Vss0.7	1/week	1/1,000
Determination of particle density	NF EN 1097-6	≥1.7 t/m^3	1/week	1/1,000
Determination of the influence of RA extract on the initial setting time of cement	NF EN 1744-6	A_{40}	2/month	1/2,000
Determination of acid-soluble chloride salts	NF EN 1744-5	To be declared	2/month	1/2,000

Table 14.5 Maximum substitution content of coarse and fine RA (in mass percentage of total coarse aggregate or fine aggregate, respectively)

	Exposure class according to NF EN 206/CN[a]			
Type of RA	*X0*	*XC1, XC2*	*XC3, XC4, XF1, XD1, XS1*	*Other exposure classes*
Coarse aggregate of type 1	60	30	20	0
Coarse aggregate of type 2	40	15	0	0
Coarse aggregate of type 3	30	5	0	0
Fine aggregate	30	0	0	0

[a] Definition of exposure classes: The higher the index, the most severe the environment. X0: No risk of corrosion or attack.

XC, corrosion induced by carbonation; XS, corrosion induced by chlorides from sea water; XD, corrosion induced by chlorides other than from sea water.

concrete recipe, once its durability is proven by tests or previous experience. This approach is a way to increase the acceptable recycling rate, but it is time and money consuming and can be justified only for large construction works. In France this concept is based on two approaches:

- the first one is called absolute approach and deals with the definition of durability indicators such as porosity, gas permeability, etc. (AFGC 2007). The limit values of these indicators depend on the service life and the environment of concrete. These rules could be applied to RAC, except the porosity requirement, which is likely to be too demanding when porous aggregates such as RA are used at a significant percentage;
- the second one is called equivalent approach and deals with the qualification of a concrete by comparison with a reference concrete already accepted by the NF EN 206/CN standard (FNTP 2009).

14.2.4 Alkali-reaction and freeze–thaw standards referenced in NF EN 206/CN: 2014

In addition of the general European standard EN 206:2013, French national provisions are referenced in NF EN 206/CN: 2014. Concerning the aggregates, it seems important to mention the technical texts dealing with the alkali-silica reaction and freeze–thaw resistance.

The document FD P 18-464 defines a set of national provisions and standards for aggregates and concrete to prevent alkali-silica reaction. For example:

- XP P 18-543: Aggregates—Petrographic investigation of aggregates relating to alkali-aggregate reactions;
- XP P 18-544: Aggregates—Determination of active lime water-soluble alkalis;
- FD P 18-541: Aggregates—Guide for the drafting of the quarry specification for the purposes of the prevention of deleterious effects of alkali-aggregate reactions;
- FD P 18-542: Aggregates—Criteria for qualification NAs for hydraulic concrete with respect to the alkali-reaction;
- NF P 18-594: Aggregates—Test methods on reactivity to alkalis;
- NF P 18-454: Concrete—Reactivity of a concrete formula with regard to the alkali-aggregate reaction—Performance test;
- FD P 18-456: Concrete—Reactivity of a concrete formula with respect to the alkali-aggregate reaction—Criteria for interpretation of the performance test results.

For freeze–thaw environment, national provisions are established based on a guide (LCPC 2003) and three test methods:

- XP P 18-420: Scaling test for hardened concrete surfaces exposed to frost in the presence of a salt solution;
- NF P 18-424: Freeze test on hardened concrete—Freeze in water—Thaw in water;
- NF P 18-425: Freeze test on hardened concrete—Freeze in air—Thaw in water.

Due to the heterogeneity of RA (in relation with their classification in EN 12620) and the possible presence of pollutants or other products than concrete, the question that may be raised is if the specifications on alkali-silica reaction given in FD P 18-464 are still relevant for RAC (see Chapter 12). In the same way, the high water absorption of RA could have an influence in the recommendations given in LCPC (2003) concerning freeze–thaw behavior.

14.2.5 Execution of concrete structures according to European standard NF EN 13670/CN and national provisions (DTU 21, Fascicule 65)

The European standard NF EN 13670/CN deals with the execution of concrete structures, including a section on "concrete operation", so that, according to Figure 14.1, concrete must comply with the NF EN 206/CN: 2014. Moreover, the user is the final specifier, in particular for the choice of the consistency classes in relation with the way the concrete will be poured. In France, the following national provisions complete the NF EN 13670/CN standard:

- Fascicule 65: Client General Specifications for public works—Execution of civil engineering concrete structures.
- NF DTU 21: Building works—Execution of concrete structures—Part 1–2: General criteria for the selection of materials.

There is no other limitation in NF DTU 21 than the ones given in NF EN 206/CN: 2014.

In Fascicule 65, some additional limitations are given:

- only RA of type 1 coming from the deconstruction of a civil engineering structure and for which the traceability is insured can be used;
- the compressive strength class of concrete is limited to C35/45;
- recycling is limited to exposure classes XC or XF1;
- the maximum rate of substitution (by mass) is 20%.

14.2.6 Reclaimed and RAs in structural precast concrete products according to European standard EN 13369 and national provisions

EN 13369 is the European standard that gives the common rules for structural precast concrete products. Structural precast products refer to this standard. For concrete composition and design, it refers to EN 206 and to EN 1992-1-1, respectively.

Reclaimed crushed aggregates and recycled coarse aggregates, mixed in concrete with other aggregates, are considered in EN 13369 since 2004. By definition:

- reclaimed crushed aggregates are obtained from precast concrete products of the same factory where the new concrete is produced. They belong to type 1 aggregate according to NF EN 12620 and NF P 18-545 standards
- recycled coarse aggregate assessed by the manufacturer come from an external source but are made of pure concrete debris.

The amount of reclaimed crushed aggregate can reach up to 10% in mass of the total content of aggregates in the concrete mix with no further testing on the hardened concrete other than the compressive concrete strength. However, when it is required and for specific applications, the amount of reclaimed crushed aggregates can be limited to 5% in weight.

This amount can be increased up to 20% if one of the following cases applies:

- the mechanical strength of the concrete product is determined by calculation, aided or not by full-scale testing, and all the hardened concrete properties relevant for calculation are determined by testing;
- the mechanical strength of the concrete product is determined by full-scale testing.

Amounts above 20% of reclaimed crushed aggregates can be used when all the hardened concrete properties relevant for calculation are determined by testing, and the mechanical strength of the precast concrete product determined by calculation is verified by initial full-scale testing. A maximum amount could also be determined by provisions valid in the place of use.

Recycled coarse aggregates from an external source made of pure concrete debris can be used under the same conditions as for reclaimed crushed aggregates, provided that the source and mix properties of the crushed concrete are known by the manufacturer.

Other recycled coarse aggregates, which do not correspond to pure concrete debris, should conform to the category RC_{90} according to EN 12620 to be used, and their amount into concrete mixes shall be limited to half of the percentages admitted for reclaimed crushed aggregates.

14.2.7 Concrete Eurocode: European standard EN 1992-1-1 and national provisions (NF EN 1992-1-1/NA)

Eurocode 2 (EN 1992) describes the principles and the requirements for safety and durability of concrete structures using the limit state concept. It only deals with requirements for mechanical strength and fire resistance. Properties of concrete are given in section 3 of Eurocode 2 part 1-1. This section includes compressive and tensile strength (f_{ck} and f_{ctm}), elastic modulus (E_{cm}), shrinkage (ε_{cs}), and creep (Φ (t,t_0)). It is well established that these three latter properties as well as f_{ctm} in a less extent may be modified, for the same compressive strength, when RA replaced NA in concrete (see Chapters 10 and 11). It should then be necessary to provide the Eurocode 2 with modified models to account for the influence of RA on these properties.

14.2.8 Survey of international standards and specifications

This section presents an overall review of how the production of concrete with RA is considered in normative standards of different countries. In de Brito and Sakai (2013), the authors provided a comprehensive compilation based on existing standards and specifications that allow the use of RA in the production of concrete. Their analysis covered a wide panel of countries (Brazil, Germany, Hong Kong, Japan, United Kingdom, the Netherlands, Portugal, Belgium, and Switzerland). During RECYBÉTON project, other countries were also under investigation like Austria, Italy, Norway, Sweden, China, USA, and Canada (Bodet et al. 2015).

The analysis of these standards shows that the use of RA generally obeys the following principles:

- it is a tradition in road applications for which the use of RAs is well controlled;
- there is a need to characterize their performances (level values and variability);
- there is a marked difference when considering recycled coarse and fine aggregates with, for the latter, limitations in their use in structural concrete for which risks can be higher.

It also shows that there are three ways to promote the use of RA:

- a general classification of aggregate (constituents, requirements, tests methods, and quality control);
- definition of possible substitution rates in concrete (depending on the types of concrete, exposure classes, and strength classes);
- adjustment of parameters used in design codes, considering the variations induced by RA (e.g., Young's modulus, shrinkage, and creep).

The survey underlines a marked disparity regarding the use of coarse RAs in structural concretes between different countries. Some of them, such as the Netherlands, Germany, Australia, United Kingdom, and Japan, are most advanced on the subject.

In the Netherlands, which is one of the most advanced countries in the use of RA in Europe, the rate of substitution is limited to 50% independently of the exposure classes. This rate could increase up to 100% if XS and XD classes are excluded. Strength classes of RAC can reach up to C35/45, and the use of fine RA is accepted.

In Germany, the rate of substitution can go up to 45% in volume for coarse aggregates and for exposure classes X0, XC, XF1, XF3, and XA1.

In United Kingdom, RAs are allowed until 20% with a maximum strength class C40/50 but limited to X0, XC, and XF1 environments.

In Australia, there is a global interest in recycling so that this country allows up to 100% for C25/30 concrete and up to 30% for C40/50 concretes, provided that the traceability is guaranteed and additional controls are carried out.

Some countries have introduced a safety coefficient in the calculation of concrete properties, such as elastic modulus, shrinkage, or creep coefficient, when using coarse RAs (e.g., the Netherlands, Switzerland, Spain).

Apart from Switzerland and Norway, countries prohibit the use of RA in prestressed concrete elements.

de Brito and Sakai (2013) identified a pattern for the different standards, considering the aggregates and their classification, the most relevant requirements, and the condition for using them in concrete. Most specifications classify RA in terms of composition.

Table 14.6 shows a comparison between three countries. To complete the classification, some supplementary requirements are specified focusing on minimum density, maximum water absorption, maximum chloride content, and maximum sulfate content (see Table 14.7). The use of RA is finally expressed in terms of maximum rate of NA replacement, exposure classes, and maximum strength classes (see Table 14.8).

14.2.9 Conclusion of the survey

The state-of-the-art review shows that RA and their use is beginning to be supported by documents and standards in many countries.

In France, the classification of RA seems rather complete even if the relevance of some characteristics still raise questions, as for the absorption or the content of sulfate soluble in water.

Table 14.6 Sample of an overview on RA composition (from de Brito and Sakai 2013), with RCA (recycled concrete aggregates) and MRA (mixed recycled concrete and masonry aggregates)

		Composition (maximum content %)					
Country	*Classification*	*Concrete*	*Masonry*	*Organic*	*Contaminants*	*Lightweight materials*	*Filler*
Brazil	RCA	>90	—	2	3	n.a	7
	MRA	<90	—	2	3	n.a	10
United Kingdom	RCA	>95	<5	1	0.5	n.a	
	MRA	n.a	1	1	n.a		
Portugal	RCA 1	>90	<10	0.2	1	n.a	
	RCA 2	>70	<30	0.5	1	n.a	
	MRA	>90	2	1	n.a		

n.a: not available.

Table 14.7 Sample of an overview on RA requirements

Country	*Classification*	*Minimum density (kg/m³)*	*Maximum water absorption (%)*	*Maximum chloride content (%)*	*Maximum sulfate content (%)*
Brazil	RCA	n.a	7	1	1
	MRA	n.a	12	1	1
United Kingdom	RCA	n.a	n.a	n.a	1
	MRA	n.a	n.a	n.a	n.a
Portugal	RCA 1	2,200	7		0.8
	RCA 2	2,200	7		0.8
	MRA	2,000	7		0.8

Source: From de Brito and Sakai (2013).

Table 14.8 Sample of an overview on RA field of application

Country	*Classification*	*Maximum replacement*		*Use conditions*	*Maximum strength class*
		Coarse (%)	*Fine (%)*		
Brazil	RCA	100	100	Nonstructural concrete	15 MPa
United Kingdom	RCA	20	0	X0, XC, XF1, DC-1	C40/50
Portugal	RCA 1	25	0	X0, XC, XS1, XA1	C40/50
	RCA 2	20	0		C35/45
	MRA	n.a	0	Nonstructural concrete	n.a

Source: From de Brito and Sakai (2013).

The recycling rates in concrete are also codified by the NF EN 206/CN: 2014 standard, but the values used are quite restrictive and conservative due to the lack of knowledge about the durability of RAC. However, some countries are more permissive, suggesting that these rates could be increased in future.

Finally, the influence of RA on the concrete properties of concrete at the hardened state, such as tensile splitting strength, elastic modulus, shrinkage, and creep, are not taken into account in Eurocode 2, whereas it is significant as soon as the replacement rate is higher than 10%–20%.

14.3 RECYBÉTON OUTPUTS

RECYBÉTON has been launched to promote the use of concrete RA in new concretes following the French experience but also the European and international ones. The different studies achieved during this project give enough information to consider that national or European standards could be improved in the future.

For example, it could be mentioned:

- Chapter 3 shows the applicability for RAs of the classical aggregate standards or the definition of thresholds (water absorption, water-soluble sulfates or acid-soluble chlorides, Los Angeles test);
- Chapter 12 shows the applicability of the French method to prevent alkali-silica reaction;
- the possibility to design RAC with replacement rates of NA by RA is higher than those proposed by the current French concrete standard NF EN 206/CN: 2014. Chapter 12

underlines the interest of the concept of performance-based evaluation allowed in the EN 206:2013 to reach this goal.

As this document is being written, it is interesting to mention another French National project that has been started, PerfDub, dealing with the development of performance-based approach, in which some studies on concrete with RA should be done. Experimental construction sites in Section VII also proved that recycling up to 100% is possible in some cases.

- Chapters 10 and 11 summarize interesting information concerning the influence of recycling rate on the evolution of hardened properties such as tensile splitting strength, shrinkage, elastic modulus, or creep behavior with the compressive strength of RAC. It could be integrated in the material models used in Eurocode 2 to better account for RAC in structural design.

14.4 CONCLUSION

The various European technical texts, such as Eurocode 2, EN 206, EN 13369, EN 13670, and EN 12620, are strongly interconnected and all of them contribute to define the conditions under which RA can be recycled into concrete. Thus, it is necessary that the specifications for a concrete used in a structure designed according to Eurocode 2, including aggregates complying with EN 12620 and the composition of which complies with EN 206 or with EN 13369 and executed according to EN 13670, are well considered in all standards.

In France, the classification of RA is quite complete through NF EN 12620 + A1 and NF P 18-545, and the NF EN 206/CN concrete standard allows the use of these RA. However, the constraints are strong in terms of RA quality, maximum recycling rate, and exposure class of concrete, which prevents a wide use of RA in concrete in practice. This conservative approach is explained by a lack of feedback on the use of RAC in structural applications, especially regarding their durability properties. Moreover, the influence of RA, for higher substitution rate, on the properties in the hardened state, such as tensile splitting strength, elastic modulus, shrinkage, and creep, is not taken into account in Eurocode 2, whereas it may be significant.

The survey realized within the RECYBÉTON project shows that several foreign countries are more permissive, suggesting that the recycling rates could be increased in France. In this context, RECYBÉTON has provided numerous studies dealing with RA characterization, fresh and hardened concrete properties, concrete durability, and experimental construction sites, which will be useful to give recommendations for further improvements of the overall chains of standards managing the use of RA in concrete.

Chapter 15

Adaptation of existing methods to incorporate recycled aggregates

T. Sedran
IFSTTAR

CONTENTS

Abstract

Many methods of concrete mix design, often empirical, exist worldwide. Because of their particular properties, presented in the other chapters of the present book, the introduction of recycled aggregates (RA) requires adaptations of these methods. This chapter makes first a summary of these particular properties of the RA and the expected qualitative consequences on the concrete mix design. In a second part, the chapter focuses on the adaptation of the Dreux-Gorisse method commonly used in France, to account roughly for the introduction of RA in concrete. Finally, the chapter shows the interest of a more scientific method of mix design to refine optimal recycling scenarios.

15.1 INTRODUCTION

Recycled aggregates (RAs) display a number of properties that differentiate them from natural aggregate (NA) and have consequences on the properties of concrete in which they are introduced. This has been described in previous chapters and is summarized in Section 15.2. To our knowledge, there is no recognized method to consider the specificities in the mix design of concrete. To be more precise, the question of the mix design of RA concrete (RAC) comes in two forms:

- optimization of concrete with an imposed recycling rate (usually by a standard or local guide);
- optimization of concrete whose recycling rate is sought to minimize local economic or environmental cost.

In the first case, empirical methods can be adapted, as proposed in Section 15.3, with the particular case of the French Dreux method. In the second case, it is necessary to use a more theoretical approach as described in Section 15.4.

15.2 FEATURES OF RAs AND GENERAL CONSEQUENCES ON MIX DESIGN

15.2.1 Water absorption

RA inherits higher water absorption from their source concrete (typically between 5% and 10%) when NA generally exhibits a lower one (1%–3%). This does not raise any particular problem in terms of mix design. It has however two main practical consequences.

The two main concrete properties, slump and compressive strength, are strongly controlled by free water content W_{eff}. This value corresponds to all the water available in the mixture (addition water, water supplied by the aggregates, and the admixtures) minus the water absorbed in the aggregates, which is assumed not to be available to lubricate the mixture nor to hydrate the cement. When replacing NA with RA, it must be considered that more water will be absorbed by the aggregates. It is therefore necessary to increase the total water when targeting a given W_{eff}. For example, in a mixture containing 1,400 kg/m^3 of RA with absorption of 7.5%, the absorbed water represents 105 L/m^3. So, for a target-free water content of 200 L/m^3, the total water must be 305 L/m^3. As for comparison, the water correction on a concrete with 1,700 kg/m^3 of NA with 1% water absorption would be only 17 L/m^3.

For concrete production, RAs are generally used in a state close to saturation. Indeed, when aggregates are dry, they do not absorb water instantaneously, which leads to an evolution of the free water in the first moments of concrete and thus to variations of workability (see Chapter 7). Moreover, when aggregate particles are not saturated, it is harder to measure precisely their real water content in the production plant. This is not acceptable from an industrial point of view.

15.2.2 Mechanical properties

NAs are essentially produced from rocks with more than 50 MPa in compressive strength, while RAs are generally crushed from concrete with a strength lower than this value. Moreover, crushing of RA may damage the remaining paste and lower its bonding to the original NA. As a consequence, for a given new paste quality (i.e., for the same W_{eff}/C ratio), the increase of recycling level generally leads to a decrease of compressive strength of new concrete (see Chapter 10). Because fine RA exhibits a greater remaining cement paste content, their negative effect is higher than with coarse RA. This can be compensated by a decrease of W_{eff}/C ratio, which means an increase of cement content and/or a decrease of free water, thanks to the use of a superplasticizer. It is worth noticing that replacing NA by

RA may sometimes increase the compressive strength for moderate strength concrete, when the NA is of poor quality (see Chapter 10), probably because RA exhibits good bonding with the new paste of RAC, due to the presence of remaining paste.

15.2.3 Packing density

Packing density of a granular material is defined as the ratio V_s/V_t, where V_s is the volume occupied by the skeleton after compaction while V_t is the volume of the container.

Concrete mix design methods around the world generally aim at increasing the packing density of the granular skeleton accounting for fine and coarse aggregates and sometimes binders and mineral additions (Caquot, Faury, Baron-Lesage, Dreux-Gorisse, ACI 211, etc.). Indeed, increasing the packing density decreases the volume of the voids between the skeleton. On the other hand, the free water added to concrete plays two roles: it first fills the voids between skeleton and then separates the grains from each other. Only this second part of water gives to the concrete its workability, while the first part is in some way useless. Hence, increasing the packing density of the skeleton allows decreasing the amount of "useless" water and consequently the free water, while keeping the same workability. Finally, because the compressive strength is a decreasing function of W_{eff}/C, a higher packing density leads to a lower cement content for a given target strength and workability and then a lower cost of the material.

RAs are crushed from concrete and made of a mix of NA and the remaining attached cement paste. Their surface is generally irregular and rough. Therefore, they generally exhibit a low packing density due to the friction between grains. For example, Dao (2012) and Dao et al. (2014) studied the packing density of RA, sourced from concretes with various compressive strength and made of two types of NAs: a siliceous semicrushed river aggregate (*S*) and a crushed limestone (*C*). The RAs were named second generation (G2) when the source concrete was made of 100% NA, and third generation (G3) when the source concrete included 100% of G2. From experimental values measured on fine aggregates on one side and coarse aggregates on the other, Dao deduced the residual virtual packing density for each size of grain β, with the Compressible Packing Model (CPM; de Larrard 1999a). The results summarized in Figure 15.1 show that globally the RAs have a packing density similar to that of a crushed limestone: lower than for more rounded aggregate and with values decreasing strongly when the size particle is decreasing. These results are confirmed with the RA obtained in RECYBÉTON (see Appendix). It can also be seen that virtual packing density is quite similar for coarse RA, while for fine RA, the range of value is greater.

The well-known consequences on mix design of concrete including RA are the following:

- Increasing the recycling rate generally increases the water demand of concrete (i.e., the free water needed to reach a target slump);
- Fine RAs have a stronger negative effect on water demand than coarse RA. Moreover, this effect is strongly dependent on the fine RA origin.

15.2.4 Summary on RA key properties

In conclusion, when replacing NA by RA, the morphology and roughness of RA make it necessary to increase the free water W_{eff} or to increase the plasticizer or superplasticizer content to maintain the slump value, while the low mechanical properties of RA generally lead to a decrease in W_{eff}/C ratio to maintain the compressive strength of concrete. Finally, the cement demand increases in a nonlinear way with the recycling rate, for a given slump

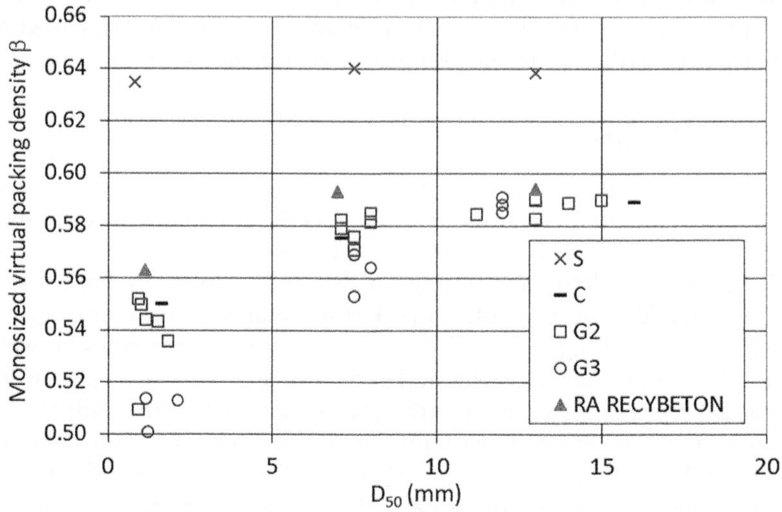

Figure 15.1 Monosized virtual packing density β. (Data from RECYBÉTON and from Dao 2012.) NA (S siliceous semicrushed river aggregate, C crushed limestone) and RA (G2 and G3) sourced from different concretes made of same NA.

and compressive strength. This effect is negligible for low recycling rates but important for higher ones. That is why, among other reasons, recycling rates are generally limited in codes (generally, no fine RA and a maximum amount of 30% of coarse RA).

Moreover, due to the high water absorption, it is mandatory (like for NA, but in a more crucial way) to correct the total water and thus the added water used to compensate the water absorbed by the RA to maintain the desired free water content.

15.3 ADAPTATION OF EXISTING MIX DESIGN METHODS

15.3.1 General classical approach

Concrete mix design methods aim to determine the composition having the lowest cost for a set of targeted properties. These methods follow more or less the same stages:

- Definition of a set of requirements;
- Determination of the W_{eff}/C ratio to reach the expected compressive strength and/or the required durability;
- Optimization of the granular skeleton compactness to minimize the water demand of the mixture;
- Adjustment of the paste and superplasticizer dosages to achieve the desired workability;
- Consideration of the possible use of a retarder to maintain workability.

In France, the Dreux-Gorisse method (Dreux and Festa 1995) is classically used to design concrete. This method is mainly empirical and based on reference optimum grading curves. These curves were obtained in the 1960s from an analysis of concrete mixes from the Paris region that performed well. Water content estimation and Y (ordinate of the breaking point on the optimum curve on Figure 15.2) are tabulated, and the values proposed in the original method must be adapted to local components.

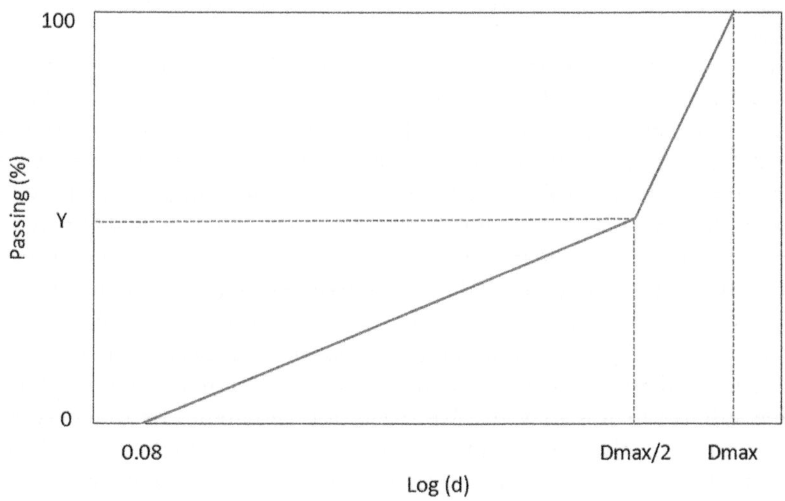

Figure 15.2 Shape of optimum aggregate grading curve in Dreux-Gorisse method.

In the following sections, we will detail each step of such a usual mix design method and explain the difficulties to adapt it to the use of RAs and show the need of a more rational mix design method.

15.3.2 Definition of a set of requirements

The set of requirements gives at least targets in terms of workability, design compressive strength at 28 days, and durability.

As noted, during the design of reference mixes for RECYBÉTON ((Sedran 2017) and Appendix) and in the specific study on fresh state behavior (see Chapter 7), the increase of fine RA content induces a significant loss of slump versus time, whereas it is moderate when coarse RAs are introduced. It is therefore important to address this concern during mix-design process.

To obtain values higher than the design compressive strength during the site job, a higher margin on mean compressive strength is aimed at during the mix design stage in laboratory. The difference is expected to cover the quality fluctuation of constituents and concrete composition variation (mainly water content). RAs have by nature less stable physicomechanical properties compared to NA. It may induce supplementary variations in RAC properties and should be accounted for when defining the mean compressive strength to target. As a first indication, a 50% increase of the security margin used for NA concrete (NAC) could be suggested when 100% of coarse RAC is used. But that still implies that the RA composition according to NF EN 933-1 is relatively stable, and W_{eff}/C is well controlled even if the water absorption of the RA may fluctuate) (see Chapter 29).

Durability of concrete can be addressed with a performance-based approach or a prescriptive approach (see Chapters 12 and 14). In Europe, the prescriptive approach is mainly dealt with, through EN 206 standard. It defines exposure classes and imposes prescriptions on concrete composition (typically on cement content and W_{eff}/C ratio) depending on the severity of the environment. These prescriptions were determined based on past experiences. The NF EN 206/CN, which is the French version of the code, accounts for the use of RA. But because we lack benefit of hindsight with RA, fine RAs are forbidden and replacement of coarse aggregate with RA is all the more limited as the environment is severe: 60% (in mass) for class X0, 30% for classes XC1 and XC2, 20% for classes XC3, XC4, XF1, XD1, and XS1, and 0% for the other classes (see Chapter 14). When going out the limits of NF EN 206/CN by increasing the rate of recycling, it is necessary to prove the durability of RAC, on a performance-based approach, as explained in AFGC (2007) or FNTP (2009) (see Chapter 14 for more details). Indeed, due to their specific nature and properties, the RA may have a strong impact on durability of RAC. RAs are more porous than NAs, and they predictably decrease the performance of concrete relating to transfer properties and may impact their freeze–thaw behavior. Due to their composite nature, RA can release significant amount of water-soluble alkalis, in particular for the fine aggregates that contain an important fraction of adherent residual cement paste. They can also contain unstable silica phases provided by some specific NA and possible pollutants. Consequently, special care is to be taken concerning the risk of alkali–silica reactions when using RA. Finally, the possible presence of water-soluble sulfates in RA may involve deleterious ettringite or thaumasite formation. Chapter 12 summarizes the main results obtained with RECYBÉTON and ECOREB projects concerning these issues and confirms that durable concrete can be made using RA even at high level of substitution provided to check the properties of RA and adapt the mix design (*W/B* ratio, entrained air content, etc.).

As shown in Chapter 10, for the same compressive strength, the splitting tensile strength of RAC may decrease with the RA replacement ratio. The decrease may typically reach around 13% to 17% for a 100% replacement ratio if the RAs are made of the same NA they replace. In other cases, the effect of RA on tensile strength may be neutral negative or even positive depending on the respective properties of RA and NA. In the same way, the reduction of elastic modulus may reach between 15% and 30%. This can be explained by the fact that RAs are softer than NA due to the paste remaining in RA and because RAC generally exhibits greater volume of paste than NAC for the same slump. The same causes explain that RAC may display shrinkage as great as twice that observed on NAC (Dao 2012) and a higher creep (see Chapter 11).

These properties may be important concerns: for example, splitting tensile strength for concrete road pavement or handling and storage phases for prefabricated pieces, elastic and delayed strain for prestressed concrete structures, shrinkage for cracking risk in restrained conditions (floors, etc.). In these cases, a particular focus is necessary if RA is used.

15.3.3 Definition of W_{eff}/C ratio

A general first step of mix design method is the determination of a target W_{eff}/C to ensure the compressive strength target.

Numerous models in literature express the compressive strength according to the following types of equation:

$$f_{cm} = k_g \cdot Rc_{28}^{ce} \cdot f\left(\frac{W_{eff}}{C}\right) \tag{15.1}$$

where

- k_g is a mechanical parameter accounting for the influence of aggregate, depending on its own strength and bonding quality with the cement paste. k_g also depends on the $f()$ function chosen;
- Rc_{28}^{ce} is the compressive strength class of cement at 28 days according to NF EN 197-1;
- $f()$ a decreasing function of W_{eff}/C ratio

For example, the Bolomey's equation used in the French Baron-Lesage's or Dreux' methods of mix design is given by equation 15.2, and Feret's one by equation 15.3.

$$f_{cm} = k_g \cdot Rc_{28}^{ce} \cdot \left(\frac{C}{W_{eff}} - 0.5 \right) \tag{15.2}$$

$$f_{cm} = k_g \cdot Rc_{28}^{ce} \cdot \left(\frac{1}{1+3.1\frac{W_{eff}}{C}} \right)^2 \tag{15.3}$$

A more sophisticated equation accounting for air content and aggregate skeleton packing density is proposed in Section 10.2.1 (see also Section 3.4).

The target W_{eff}/C ratio can thus be deduced from this kind of equation once the cement class and the mechanical performance of aggregate skeleton are known and the target compressive strength of concrete decided. The cement class is generally given by the producer, and the k_g term is fitted on experimental measurements of compressive strength on previous concrete mixes made of the same skeleton.

In Section 10.2.1, these equations proved to be still valid when RAs are introduced in concrete. For analysis purposes, it is assumed that the factor k_g of the overall aggregate skeleton is calculated according to equation 15.4, where $k_{g,j}$ is the mechanical parameter for the aggregate fraction j and VF_j its volume proportion in the skeleton.

$$k_g = \sum_j VF_j \, k_{g,j} \tag{15.4}$$

In conclusion for concrete including RA, once the $k_{g,j}$ terms for coarse RA and fine RA are determined, it is possible to choose the target W_{eff}/C ratio. These terms can be calibrated, like for NA, on previous mixes. This was done for the RECYBÉTON reference mixes (see Table 10.2 and (Sedran 2017)). They can also be estimated, thanks to Figure 10.2 adapted from Silva et al. (2014b), on the basis of oven-dried density and water absorption of RA or by the means of equation 15.7 (see also equation 3.11) based on the micro-deval of the RA.

In a second step, as explained in 15.3.1, the durability concerns impose a maximum acceptable W_{eff}/C, through the EN 206 standard for example or through performance-based tests, given the fact that durability is generally improved when W_{eff}/C decreases. For RAC, these limits are not clearly defined, but a certain amount of information can be found in Chapter 12.

The final target of W_{eff}/C is the minimum value between compressive strength and durability requirement.

15.3.4 Optimization of packing density

In the Dreux method, the Y value in Figure 15.2 is calculated according to equation 15.5, while $X = D_{max}/2$. D_{max} is the maximum size of aggregate skeleton (in mm), M_f is the fineness modulus of sand, and K a tabulated value depending on the nature of the aggregate (crushed or rolled, but without distinction between fine and coarse aggregate), the cement content, and the energy of vibration. Once the curve is thus defined, it is easy to choose the fine and coarse aggregate ratio to fit it.

$$Y = 50 - \sqrt{D_{max}} + K + (6M_f - 15) \quad (15.5)$$

The RECYBÉTON reference concrete mixes developed by Sedran (2017) and presented in the Appendix were developed with the CPM implemented in the BetonlabPro software (de Larrard 1999a; de Larrard and Sedran 2007). The interest of such approach will be discussed in 15.4 but is supposed to produce precise granular optimization, accounting for the real packing density and the grading curve of each aggregate.

Four particular optimum skeletons were calculated with this method (see Tables A.8 and A.9): one with fine and coarse NA (0S-0G), one with fine RA and coarse NA (100S-0G), one with fine NA and coarse RA (0S-100G), and one with fine and coarse RA (100S-100G). These optimal curves are presented in Figures 15.3 and 15.4, where they are fitted with bilinear curves like in Dreux method (breaking point at $d = 9$ mm).

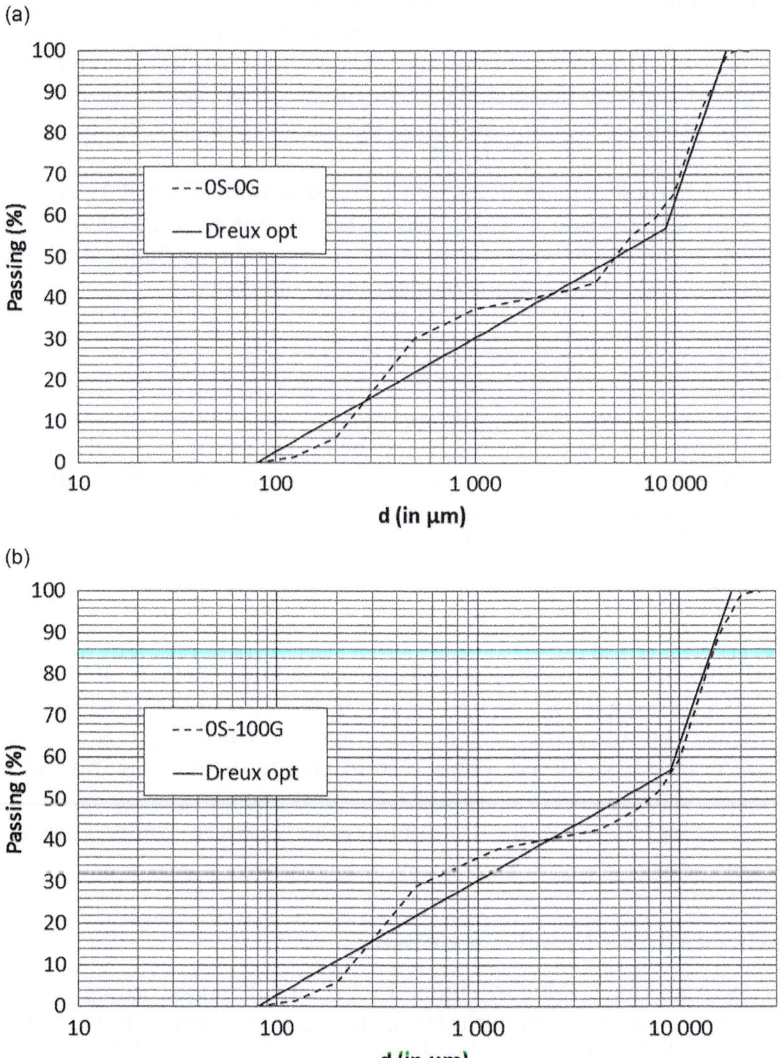

Figure 15.3 Optimum aggregate skeleton grading curves for concrete with fine NA and coarse NA or RA calculated with CPM model fitted by bilinear curves like in Dreux method.

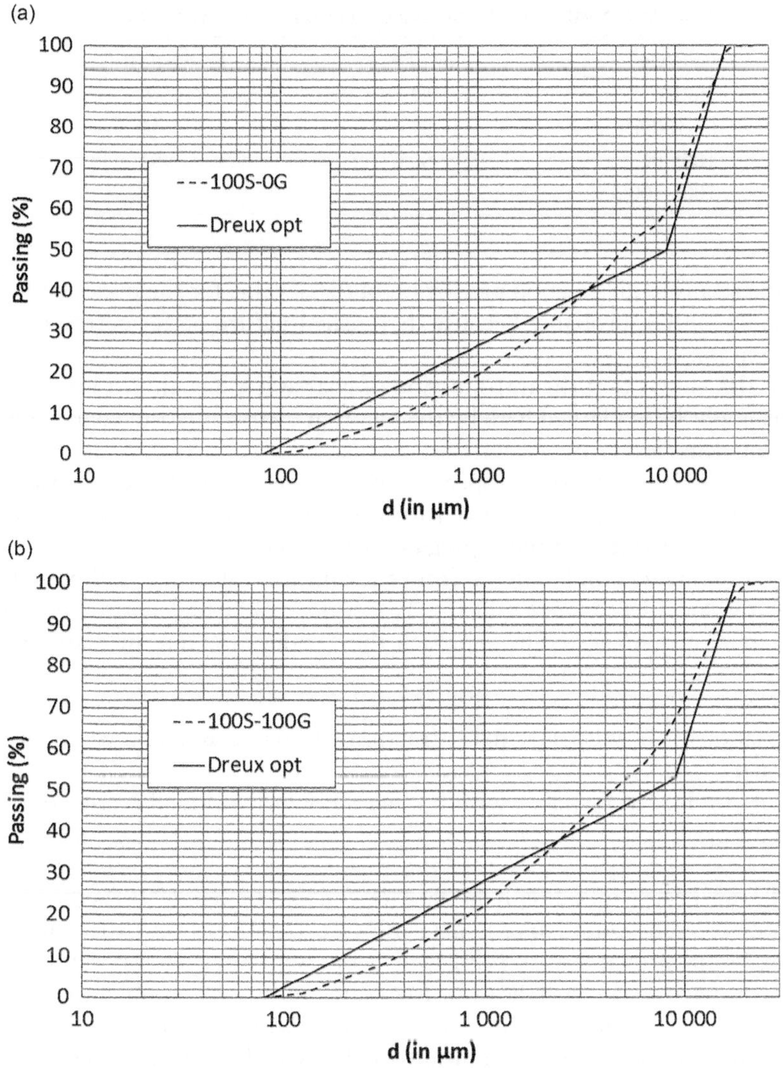

Figure 15.4 Optimum aggregate skeleton grading curves for concrete with fine RA and coarse NA or RA calculated with CPM model fitted by bilinear curves like in Dreux method.

On the other hand, the optimum curves according to Dreux method were calculated with equation 15.5 and the following constants: $D_{max}/2 = 9\,\text{mm}$, $K = 4$ (corresponding to concrete made of crushed aggregate, 350 kg/m^3 of binder and placed with low vibration), $M_f = 1, 9$ for fine NA and $M_f = 2, 9$ for fine RA.

The Y values of the breaking point obtained by two methods (fitting and calculation) are summarized in Table 15.1.

Table 15.1 first shows that even for the skeleton made with NA (0S-0G and 0S-100G), the Dreux method is not very precise and tends to underestimate the optimum content of natural sand. In that case, the high volume of fine NA might be explained by the good shape of the sand grains, which displays a particularly high virtual packing density (see Table A.7). Yet, the Dreux method does not account for the real packing properties of the aggregate (it is by the way a general criticism that can be made to the method). However, the

Table 15.1 Y values of the breaking point on Dreux-type curves

Concrete	*By CPM*	*By equation 15.5*
0S-0G	57	46
0S-100G	57	46
100S-0G	50	52
100S-100G	53	52

Dreux method seems to give a good prediction of the *Y* value for the mixes with the fine RA (100S-0G, 100S-100G), probably because its crushed nature better corresponds to Dreux method assumption.

Table 15.1 also shows that the introduction of coarse RA has almost no influence on the *Y* value. This might be explained here by the fact that coarse NA and coarse RA have quite similar packing densities.

In conclusion, the Dreux method appears to be sufficient to give a rough evaluation of the optimum curves for concrete containing 100% of RA, provided that the calculation of the *Y* value is done considering that the aggregates are crushed.

Based on these remarks and starting from the optimal value Y_{NAC} calibrated for a NAC according to the local experience and the Y_{RAC} value calculated with Dreux for a 100% recycled concrete, it is suggested to approach the *Y* values for intermediate recycling rates as following:

- No variation of *Y* with $\Gamma_{m,c}$ mass fraction of coarse RA related to overall coarse aggregate;
- $Y = Y_{NAC}\,(1 - \Gamma_{m,f}) + Y_{RAC}\,\Gamma_{m,f}$, with $\Gamma_{m,f}$ mass fraction of fine RA related to overall fine aggregate.

The French Baron-Lesage mix design method (Baron and Lesage 1976) is the same as Dreux one, except that the optimum curve is determined on the basis on LCL workability-meter measurements (same principle as in NF P18-452 but with a bigger apparatus). It consists in producing different concrete with the same free water and cement contents (chosen according to past experiences), but with varying coarse-to-fine aggregate ratio. According to granular concept, the ratio leading to the maximum workability is the optimum one (the workability could be measured by slump test, as well). The advantage of this method is that it allows a fine determination of the granular optimum. On the other hand, it requires numerous tests.

It can be applied as such to concretes containing RA, but only when the replacement rate of fine and coarse aggregates is preselected. Indeed, the method allows the optimization of only two components at once. Therefore, it is necessary to repeat the procedure when changing the recycling rates, which can be rapidly burdensome.

15.3.5 Water adjustments

The next step is to fix the amount of free water required to obtain the desired workability. The Dreux method provides a chart for approximating this quantity for a given slump and W_{eff}/C ratio.

This step is the weak part of the Dreux method because it does not explicitly consider the packing density of the optimized granular skeleton, but it takes little account of the presence of superplasticizer and accounts only for slump less than 12 cm. This is not satisfactory for the mix design of modern concretes, and concrete designers have generally adapted their own chart to their local NAs to calculate the necessary free water content. Obviously, the introduction of RA, which is known to lead to an increase in the water demand of the concrete, will not contribute to improving the accuracy of the method. At this stage, it therefore seems essential to go through slump tests in the laboratory.

So, to account for the RA effect, we tried here to estimate the magnitude of variations in free water demand for a constant slump with a constant superplasticizer dosage due to increasing recycling rates in fine or coarse aggregates. To do so, we compared theoretical recipes of concrete made with the components used by RECYBÉTON project (the cement was a CEM II/A). The mixes were calculated with BetonlabPro sofware (de Larrard 1999a; de Larrard and Sedran 2007). All the necessary data are summarized in Appendix. In the simulations, the cement, the limestone filler, and the superplasticizer were kept constant at 300 kg/m^3 and 50 kg/m^3 and 0.15% (% of dry extract relative to the cement weight), respectively. For each recycling rate, the aggregate skeleton was first optimized to reach the best packing density, and then the free water content was adjusted to reach a 15 cm slump. It was then possible to build the Figure 15.5 describing the theoretical increase of free water necessary when coarse RA and fine RA contents increase.

All these results can be summarized with the following equation:

$$\Delta W_{eff} = 10.2\,\Gamma_{m,c} + 14.6\,\Gamma_{m,f}^2 \qquad (15.6)$$

with $\Gamma_{m,f}$ mass fraction of fine RA related to overall sand, and $\Gamma_{m,c}$ mass fraction of coarse RA related to overall coarse aggregate.

This equation means that a supplementary 24.8 kg/m^3 of free water is needed when all the NAs are replaced by RA (at least in the case of RECEYBETON project). This trend is theoretical, as the concrete was designed with BetonlabPro. However, some concrete recipies were produced by RECYBÉTON in laboratory with the same components (see Tables A.8 and A.9 and Sedran 2017), which seems to show that the equation slightly overestimates the increase in water demand and that a maximum 18 kg/m^3 supplementary free water is needed

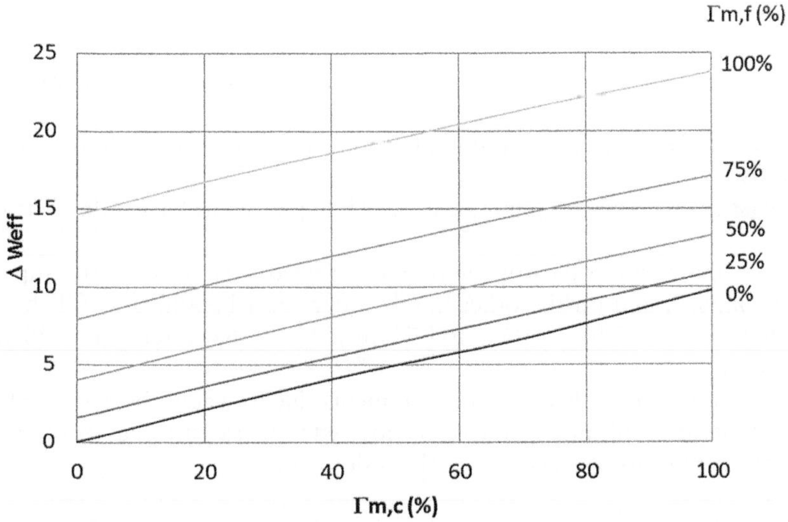

Figure 15.5 Increase of free water content demand for a 15 cm concrete compared to NAC when coarse RA replacement increases, for different fine RA replacement rates. $\Gamma_{m,f}$ mass fraction of fine RA related to overall sand, and $\Gamma_{m,c}$ mass fraction of coarse RA related to overall coarse aggregate.

with full recycling. So, the parameters of the equation were refitted on the experimental mixes while keeping its form.

Moreover, in this project, the NA displays virtual packing densities (see Table A.7) near to those measured on rounded aggregates, i.e., in the upper part of the range generally observed. So, the negative effect of recycling calculated in that case is probably a maximum value. Indeed, when the NAs are crushed aggregates, they may have virtual packing densities very similar to that of RA (Figure 15.1), and then recycling should have a very limited influence on water demand.

In conclusion, these observations can be summarized in the following set of equations illustrating the extreme case of NA nature. These can be a helpful guide to anticipate the increase of water demand with recycling rate when the water demand of the reference NAC is already known.

$$\begin{aligned}
&\Delta W_{eff} = 8\Gamma_{m,c} + 10\Gamma_{m,f}^2 \text{ when NA is rounded} \\
&\Delta W_{eff} = 10\Gamma_{m,f}^2 \text{ when fine NA is rounded and coarse NA is crushed} \\
&\Delta W_{eff} = 8\Gamma_{m,c} \text{ when fine NA is crushed and coarse NA is rounded}
\end{aligned} \quad (15.7)$$

with $\Gamma_{m,f}$ mass fraction of fine RA related to overall sand, and $\Gamma_{m,c}$ mass fraction of coarse RA related to overall coarse aggregate

For a given recycling rate and a given plasticizer content (selected according to past experience), a free water dosage is chosen. Since we know from the first stage the W_{eff}/C ratio, we deduce the cement mass. Knowing the densities of water and cement, we can calculate the volume of aggregates. From the previous step, the volume proportions of fine and coarse aggregates are known, and the masses of each granular fraction can therefore be deduced therefrom. It is recalled that it is necessary to take good account the density of the mixture NA+RA for each fraction of granular skeleton. The water to be added in the mix assuming the aggregates are dry is then calculated, accounting for the water absorption of aggregates. For example, let us consider that the previous calculations lead to the following masses:

- W_{eff} = 190 L/m³
- Fine RA = 670 kg/m³, in dry state. Fine RA is supposed to have a water absorption of 9%;
- Coarse RA = 740 kg/m³, in dry state. Coarse RA is supposed to have a water absorption of 6%;
- Superplasticizer = 4 kg/m³. Superplasticizer is supposed to have a dry extract of 30%

In that case, on one hand, a part of the free water needed is given by the superplasticizer (0.7 × 4 = 2.8 kg/m³), and on the other hand, water must be added to fill the porosity of aggregates (670 × 0.09 + 740 × 0.06 = 104.7 L/m³). Then, the water to be added in the dry mix is 291.9 L/m³.

Yet to avoid uncertainties in slump measurements due to water absorption by aggregates in the first instants after mixing, it is strongly advised to work with presaturated aggregates. In this case, the masses to be weighed are the following.

- Fine RA = 730.3 kg/m³, in saturated surface dry state, among which 60.3 kg comes from the absorbed water;
- Coarse RA = 784.4 kg/m³, in saturated-surface-dry state, among which 44.4 kg stands for the absorbed water;

- Superplasticizer = 4 kg;
- Water to add to the mix = 187.2/m^3

The free water is thus adjusted (and therefore the other constituents) on different mixes to achieve the desired slump. If the adjustment of water is not sufficient to achieve the required slump, the superplasticizer dosage can be increased.

A solution to compensate the increase of free water demand could be to keep the free water content constant (even if the added water increases to compensate the RA absorption) but to increase the superplasticizer content drastically. In that case, the plastic viscosity of the concrete may rise to unacceptable values, and it is worth checking if this solution is economic. However, the cement content may be kept constant for environmental reasons (see Chapters 30 and 33).

15.3.6 Other adjustments

As noted, during the design of reference mixes for RECYBÉTON (Sedran 2017 and Appendix) and in the specific study on fresh state behavior (see Chapter 7), the introduction of fine RA content induces, even at low replacement rate, a significant loss of slump versus time, whereas it is moderate when coarse RAs are introduced. The more the recycled fine aggregate, the greater is the slump loss. This effect is not due to the absorption of water, since the phenomenon occurs even when the aggregates are introduced into the mixer in a saturated state. It is a consequence of interactions with the superplasticizer, and the use of a retarder is therefore generally required as soon as RA fines are used. Retarder dosage can even reach the maximum value generally recommended by the supplier for the high rate of fine RA replacement.

To simplify the tests, the retarder dosage can be evaluated on mortar extracted from the studied concrete mix by retaining only the grains passing at 2 mm. The classical MBE (Schwartzentruber and Catherine 2000) or AFREM grout method (de Larrard et al. 1997) may be used, keeping in mind that it is advisable to use the fine RA in saturated condition, if any.

For high volume of fine RA, Chapter 7 shows that air content may increase up to 4% with some superplasticizer. This fact may lead to supplementary decrease of compressive strength, and if needed, a defoamer agent may be used.

Finally, if shrinkage is too high as expected with high amount of RA, the use of a shrinkage reducing agent may allow a 30% decrease of shrinkage (Dao 2012).

15.4 THE ISSUE OF RECYCLING RATE OPTIMIZATION

15.4.1 Contribution of a scientific mix design model

We have shown in the previous section that by adapting Dreux's empirical method, it was possible to design concrete with RA. The process is cumbersome because it requires many calibration tests on laboratory batches, but it can nevertheless prove sufficient to design a concrete, when the recycling rate is selected *a priori*.

As mentioned earlier, a scientific mix design concrete using NAs has been developed and implemented in a software (de Larrard and Sedran 2007). The method is based on a series of equations allowing to optimize the granular skeleton using the Compressible Packing Model (CPM) on one hand and to calculate a certain number of properties of use, such as slump, shear yield stress and viscosity, compressive strength, tensile strength, elastic modulus, on the other hand.

The CPM is applicable as it is for RA (Dao 2012; Dao et al. 2014; Pepe 2015; Sedran 2017). The frictional aspect of RA is then directly considered by means of the measurement of their packing density, which are an input data of the model. Similarly, the equations used to predict the slump or rheology of concrete can be used directly, and the negative impact of RA on these properties is considered though the packing density too. For the prediction of compressive strength, splitting tensile strength, or elastic modulus of concrete, models remain valid, provided that certain properties (k_g, k_t, E_g) of RAs are fitted (see Chapter 10). This set of equations is therefore adapted to design RAC. Even when it is still necessary to validate the mixes in the laboratory, it makes it possible to design concrete faster and more precisely than with the empirical methods, since it considers the real properties of the constituents (especially their packing densities). It also provides a more complete design since it calculates properties other than slump and compressive strength and is suitable for mineral additions (fly ash, silica fume, limestone filler, etc.).

This type of approach becomes essential when it is not a question of optimizing a concrete with a given recycling rate, but of optimizing the recycling rate so that the gain, whether economic or environmental, is maximized. This was the founding purpose of the RECYBÉTON project, and it is clear that this optimum does not necessarily correspond to the maximum recycling rate, since, for example, for high rates, the necessary increase in the cement content counteracts the saving on NAs. Indeed, to answer this question with the empirical Dreux method, it is necessary to multiply the batches to design concretes with variable recycling rates, which is prohibitive. The scientific method makes it very quickly with the help of a solver.

As an example, Figure 15.6 shows the cement consumption calculated with the scientific method for different recycling rates for a 10 cm slump and 35 MPa compressive strength concrete (Dao 2012). The calculations provide in the same way the quantities required of each constituent. It is then easy, by allocating financial or environmental costs to each constituent, to find the optimum dosage of RAs. Thus, if the NAs are locally rare or expensive,

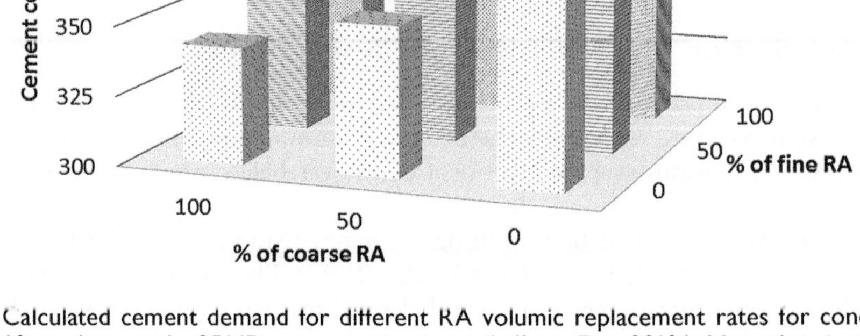

Figure 15.6 Calculated cement demand for different RA volumic replacement rates for concrete with a 10 cm slump and a 35 MPa compressive strength (from Dao 2012.). Note that, in this case, the replacement of the coarse fraction by RA decreases the cement demand.

there will be a tendency to put more RA. Similarly, the optimum RA content will be influenced by the nature of NAs they replace. If the NAs are locally of poor quality (low packing density and/or mechanical properties), the cement increase linked to the introduction of RA will be minimized and the optimum recycling rate will be greater.

In conclusion, the use of a scientific model such as that presented in this chapter makes it possible to explicitly take account of local conditions to optimize the recycling of RA in concrete.

15.4.2 Highlighting the interest of selective sorting

The mechanical properties of RA are improved and particularly the k_g term involved in the compressive strength calculation of RAC (see Chapter 10) when the compressive strength of the source concrete increases. Concrete from different sources are generally mixed on recycling platforms to produce an RA of "mean quality", but the question of the interest of selective sorting may be raised. The scientific mix design method can be used to quickly evaluate this scenario in terms of cost saving on RA production when using high-quality RA. Figure 15.7 summarizes the simulations made by Dao (2012) to produce a 10 cm slump and 35 MPa mean compressive strength concrete. The recycling rate was 100%, but partial replacement could also be calculated. The k_g terms or the RA were evaluated from their Micro-Deval (see equation 10.7). The figure shows significant cost differences, which may justify for important rehabilitation project, to make a selection of the concrete waste at the demolition stage.

Once again, the rational approach is then helpful to refine the recycling scenario.

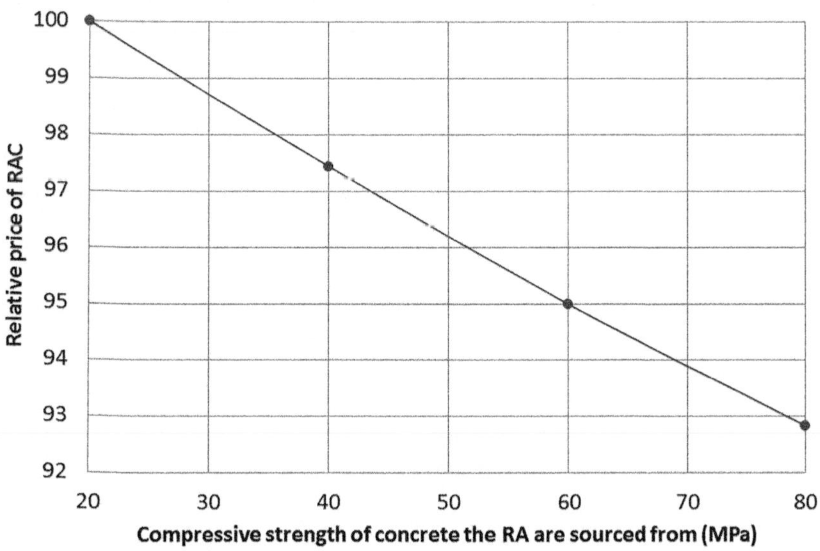

Figure 15.7 Theoretical relative cost for concretes made of 100% RA, with a 10 cm slump and a 35 MPa mean compressive strength, in function of the compressive strength of the concrete the RA are sourced from Dao (2012). The RAC made of RA crushed from 20 MPa concrete is taken as reference.

15.5 CONCLUSIONS

First, experience shows that it is possible to design concretes with high RA content with reasonable cement contents (see Chapter 20 and further).

The poor packing density of the RA, and in particular that of the fines linked to the crushing process inherent to their production and their frictional nature inherited from the residual cementitious paste, leads to increase the free water content of RAC. The total water content increases not only because of the increase of this free water but also largely to compensate the absorption of RA, which is generally much higher than that of NA (especially for fines). It is important to make a difference between these two effects and to clearly identify the free water content to control the mix design of concrete. This is true for NAC, but even more for RAC.

For a simple set of requirements including a specification on compressive strength and slump, we note that an empirical method of Dreux type conventionally used in France can be adapted to design roughly RAC, when recycling rate is set *a priori*. One of the precautions to be respected among others is to take a greater margin of safety on the average compressive strength to ensure the characteristic one, when RAs are introduced. This is related to the fact that RA is by nature more variable than NA for a given source.

Yet, the main objective may not be a question of designing a concrete with a given recycling rate but to optimize the recycling rate so that the gain, whether economic or environmental, is maximized. At least it was the founding purpose of the RECYBÉTON project, and it is clear that this optimum does not necessarily correspond to the maximum recycling rate since, for example, for high rates, the necessary increase in the cement content counteracts the saving on NAs in terms of sustainability.

In that case, experimental methods are too cumbersome and imprecise. More scientific models to predict the optimal packing density (de Larrard 1999a), and the mechanical properties (see Chapter 10), implanted in software like BetonlabPro (de Larrard and Sedran 2007) seem essential for optimizing the RA content and proposing optimal recycling scenarios.

In mix-design process, there is always a durability component. This is currently the main reason that is slowing down high rate recycling because of a lack of experience in the long term. We can use a prescriptive approach to ensure this durability, but considering the recycling rate, or better on a performance-based approach. Research summarized in Chapter 12 should in the future help to identify criteria and thresholds adapted to the RAC durability.

Chapter 16

Development, production, and control of a high replacement rate recycled concrete

S. Favre
Léon Grosse

G. Noworyta
CTG - Calcia

T. Sedran
IFSTTAR

CONTENTS

Abstract

In 2005, a concrete water treatment plant at Saint Cloud in the Paris region was partially demolished, and then rebuilt. At this occasion 6000 tons of 0/20 mm recycled concrete aggregates (RA) were produced on site. The public owner of the facility was very keen that all these materials be recycled on site to save natural resources, and more generally for the sake of eco-design. According to the contractor and the owner, the corpus of standards available at this time allowed the use at high rates of RA in concrete. This solution was selected for a complete recycling of the RA. This chapter describes the analysis of standards made at the time, the characterization of aggregates, the mix-design of a C25/30 concrete containing between 90 and 100% of RA, and the results of quality control. Seven years before RECYBÉTON started, this construction site appeared as a pioneer in the French landscape, demonstrating the feasibility of a full and local recycling process.

16.1 INTRODUCTION

In 2005, a concrete water treatment plant at St. Cloud in the Paris region was partially demolished and then rebuilt. On this occasion, the demolition of the roof of a water tank generated concrete wastes that were primary crushed on site in a 0/80 mm aggregate. A secondary crushing and sieving have finally resulted in about 6,000 tons of 0/20 mm recycled concrete aggregate (RA), stored in open air, on site (see Figure 16.1).

The public owner of the facility was very keen that all these materials be recycled on site to save natural resources and more generally for the sake of ecodesign. One solution to reach this objective was to use the RA massively into the new concrete when possible.

In this context, the contractor company (Leon Grosse) analyzed the standards available in 2005, concerning the use of RA in concrete (see Chapter 14), i.e.:

- NF EN 206-1 (2004)
- NF EN 12 620,
- XP P 18 545,
- Fascicule 65 A,
- Fascicule 74,
- BAEL 92 revised 99 (at this time, this document gave the rules to design reinforced-concrete constructions; since then it was replaced by Eurocode 2).

In 2005, in the French context, NF EN 206-1 stated that aggregates complying with XP P 18 545 and NF EN 12 620 requirements were suitable for use in concrete. Yet, these two latter standards covered RA in their scope. In fact, according to the NF EN 12 620, RA is an aggregate from the transformation of inorganic materials previously used in the construction and according to XP P 18 545, an aggregate was a natural, artificial, or recycled granular material. So, in 2005, it was concluded by the contractor and the public owner that RA could be authorized, even at a high rate of replacement, provided they complied with the following index classification according to XP P 18 545 (linked to NF EN 12 620).

According to XP P 18 545, the A-indexed aggregates were suitable for concrete with a compressive strength higher than 35 MPa dedicated to both infrastructures and buildings. Some characteristics could even be indexed B and at most two of them C or D, after studies and references. For the common concretes (item 10.7.3.), the C-indexed aggregates as well as those that had at most two D-indexed characteristics were suitable, relying studies and previous references.

According to Fascicule 65 A, the aggregates had to be B-indexed. For concrete with compressive strength lower than 35 MPa, some characteristics could even be C-indexed relying on local experience, studies, and previous experiences.

(a) (b)

Figure 16.1 Demolition of the water treatment plant and stockpile of 0/20 mm RA.

According to Fascicule 74 Section V 7, the concrete manufacturing complied with the Fascicule 65 A, and according to Section XIV "Tests and Evaluation", the requirements of Fascicule 65 A could be applied.

Finally, according to the BAEL 92, the concrete was composed of regular aggregates with a cement dosage at least equal to 300 kg/m^3.

In the absence of a normative prohibition concerning RA in concrete, the contractor then made a characterization of the RA stockpile. For this purpose only, it was necessary to sieve the 0/20 mm RA in two fractions: a fine 0/4 mm and a coarse 4/20 mm one. However, the raw RA was used in a single fraction to produce RA concrete (RAC) batches. All the results are summarized in the following tables. For further information, the water absorption of 0/20 mm was 8.9% and the specific gravity 2.29, while water absorption was 12% and 7.3% for 0/4 mm fraction and 4/20 mm fraction, respectively.

Table 16.1 shows that the fine to coarse aggregate ratio (CA/FA) of the raw RA is around 1.7, conveying that the fines particle content is relatively small. Moreover, it can be seen that the CA/FA displayed an important scatter. The consequences on concrete mix design of these observations will be discussed further on in the section (Tables 16.2 and 16.3).

Table 16.4 shows that the characteristics of RA are mainly A-indexed except for Los Angeles (B-indexed 4/20 mm fraction) and water absorption (D-indexed for fraction 0/4 mm). Then, both the contractor and the owner validated that the RA stockpile could be used to produce concrete. As a precaution, it was decided to limit the use of RA for C25/30

Table 16.1 Grading curve of several samples of the RA in the stockpile

Sieve (mm)\Sample #	*05 135 passing (%)*	*05 136 passing (%)*	*05-00462 passing (%)*
25		100	100
20	100	99	98
16	95	86	89
14	90.1	78.2	83
12.5	84.1	70.2	76
10	77.7	57.8	65
8	69.1	47.8	55
6.3	60.1	38.5	47
5.6	43.2	34.7	43
4	31.5	28.9	35
2	20.3	19.6	24
1	12.4	13.3	16
0.500	5.7	8	9
0.250	2.3	4.1	4
0.125	1.1	1.8	2
0.063	0.4	0.9	1.7

Table 16.2 Grading curve of 4/20 mm fraction of the RA

Sieve (mm)	4	5	6.3	8	10	12.5	16	20	25
Passing (%)	8.0	16.4	26.3	33.3	53.7	66.7	85.4	46.8	100

Table 16.3 Curve grading 0/4 mm fraction of the RA

Sieve (mm)	0.063	0.125	0.25	0.5	1	2	4	8
Passing (%)	6.0	9.4	17.6	33.3	52.7	75.5	99.6	100

Table 16.4 Summary of the physical and chemical characteristics of the different RA fractions

Test		*Value*	*Category according to NF EN 12620*	*Index according to XP P 18-545*
Fraction 4/20 mm				
Los Angeles (NF EN 1097-2)	LA	40	LA_{40}	B
Flakiness index (NF EN 933-3)	A	10	FL_{15} ($A \leq 15$)	A
Grading 4/20 mm (NF EN 933-1)	0.063 mm passing	1.4%	$f_{1,5}$ G_C 90/15	A
Fraction 0/4 mm				
Cleanliness (Equivalent sand) (NF EN 933-8)	SE	67	Declarative	A
Grading 0/4 mm (NF EN 933-1)	0.063 mm passing	6%	f_{10} G_f 85	A
Density and water absorption (NF EN 1097-6)	MVr Ab	2,03 t/m³ 11,9%	Declarative	— D
Chlorides soluble in water (NF EN 1744-1)	%	8.10^{-4}%	Declarative if the client asks	Declarative if >0.01
Sulfate dosage in water (NF EN 1744-1)	%	5.10^{-5}%	—	A
Organic pollutant affecting the cement setting (NF EN 1744-1)	—	No coloring	—	Results to declare
Prohibited impurities (NF EN 1744-1)	ImP	11.10^{-5}%	—	A

concrete dedicated to office building, thus excluding the concrete dedicated to the so-called "hydraulic parts", such as the storage and treatment tanks. The following section describes the mix-design process of the concrete including RA (RAC).

16.2 CONCRETE MIX DESIGN

16.2.1 Preliminary mixes

Around 6,000 tons of 0/20 mm were available, while the total volume of C25/30 yielded around 3,500 m³ in the project. So, a high recycling rate (90%–100%) was necessary to consume all RA stockpile within the construction.

The set of requirements for the concrete C25/30 was the following:

- Environment class: XF1 (according to EN 206—see Chapter 14);
- Slump: higher than 160 mm (S4) after 30 min;
- A compressive strength at early age (1 day) around 6–7 MPa to allow formworks removal after 16 h, for a daily rotation;
- A characteristic compressive strength at 28 days higher than 25 MPa.

The RA has a high water demand, so it was decided to prewet it at a water content above 12% to ensure that even the fine fraction 0/4 mm was saturated. Here, the aim was to avoid any stiffening of concrete just after mixing due to water absorption.

Three different cements were preselected and tested to get the best compromise between cost and compressive strength evolution (from 1 to 28 days): a CEM III 32.5, a slag-based CEM II 32.5 and CEM I 52.5, according to EN 197-1. Four polycarboxylate-type admixtures were also tested to get the best compatibility regarding the initial slump without segregation and slump retention during 30 min. Segregation was a main concern, considering the high value and scattering of CA/FA ratio. Moreover, the lack of fine particles in the RA leads to a low packing density of the granular skeleton and a high water demand for the concrete. Therefore, the introduction of natural 0/2 mm sand was tested at this stage to correct the RA grading.

Finally, a set of 11 batches were produced (among few adjustments batches) as presented in Table 16.5.

Noting that the slump of the mixture 1 is zero while that of the mixture 3 is 200 mm, we can deduce that the admixture 1 has a low compatibility with the cement CEM III. It was also observed on mixtures 10 and 11 that admixture 3 was as efficient as admixture 2 in terms of slump at 5′ but was slightly less effective in terms of slump retention at 30′. The same observation on slump retention could be made on admixture 4 when looking at mixtures 7 and 7b.

Concrete made of CEM III exhibited no compressive strength f_c at 1 day, while the one made of CEM II displayed a too low margin of security to allow formwork removal at 16 h, probably due to the presence of slag in both cements. Moreover, the increase of CEM II cement displayed a disappointing low increase of compressive strength between 7 and 28 days.

According to Fascicule 65A, it is necessary to obtain a mean value of f_c at 28 days that complies with the following requirements to ensure a concrete characteristic compressive strength of 25 MPa: $f_c > 25 + (R_c - R_{cmin})$ and $f_c > 1.1 \times 25$, with R_c the real strength of cement at 28 days (according to NF EN 196-1) and R_{cmin} the minimal guaranteed one. The difference between the mean and characteristic values is necessary to cover the unavoidable variations in concrete composition during the construction period. For the CEM I 52.5, $R_c = 64.4$ MPa according to Table 16.6 and $R_{cmin} = 52.5$ MPa. This leads to an f_c goal value of 36.9 MPa. So, it can be seen that both requirements in terms of f_c at 28 days and 1 day were satisfied with CEM I.

In conclusion, admixture 2, CEM I 52.5, and concrete No. 19 were selected for further investigation at this preliminary stage. Nevertheless, it is noteworthy that the introduction

Table 16.5 Summary of the preliminary mixes (in kg/m³)

Type of cement	*CEM III 32.5*				*CEM II 32.5*			*CEM I 52.5*			
No. of mix	01	02	03	04	09	10	11	07	07b	08	19
Cement (kg)	295	331	371	334	361	343	337	345	334	362	335
RA 0/20 mm (kg)	1,631	1,553	1,571	1,544	1,530	1,542	1,558	1,586	1,536	1,534	1,550
Natural 0/2 mm (kg)									85		
Admixture type	No. 1	No. 2	No. 2	No. 2	No. 2	No. 2	No. 3	No. 4	No. 2	No. 2	No. 2
Admixture content (%)	2	1	1.5	1	0.9	1.2	1.0	2	1	1	1
Free water (kg)	177	199	178	202	198	198	194	182	176	199	201
Slump at 5′ (mm)	0	220	200	220	200	210	210	210	220	220	220
Slump at 30′ (mm)	0	190	80	160	140	120	80	130	160	170	160
f_c 1 day (MPa)				0	7	7		23	25	21	24
f_c 2 days (MPa)				6,5	18	19		32	33	28	
f_c 7 days (MPa)				22	23	23		42	42	38	40
f_c 28 days (MPa)								48	47		45

Table 16.6 Properties of CEM I cement

Specific gravity	3.15
Blaine specific surface (cm^2/g)	4,100
R_c at 1 day (MPa)	23
R_c at 2 days (MPa)	36
R_c at 7 days (MPa)	51.5
R_c at 28 days (MPa)	64.4

Table 16.7 Properties of mix No. 19 (see Table 16.5): nominal and derived

Mix No.	*19 (Nominal)*	*20 (No. 19 + 10 L)*
Slump at 5′ (mm)	220	240
Slump at 30′ (mm)	160	210
R_c 1 j	24	19.5
R_c 7 j	40	36.5
R_c 28 j	45	42

of 0/2 mm proved to be efficient in reducing the water demand for a given slump, as demonstrated by mixtures 7b and 19. This point was exploited for final concreting on site.

16.2.2 Derived mixes

To ensure the robustness of a concrete recipe during the construction period, Fascicule 65A suggests to verify it respects the requirements even with ±10 L/m³ variations of free water. This was measured on mixture 19 selected on previous step, but only with 10 L more water (by lack of time). Table 16.7 shows that, even for the derived mix, the concrete is satisfactory and is still a C25 (or even a C30).

16.2.3 Supplementary evaluations

Figure 16.2 confirms that, as expected (see Chapter 11), the introduction of a high rate of RA leads to high values of total shrinkage, thus increasing restrained shrinkage cracking risk (although it was shown on a posterior construction site that such a high shrinkage can be found without special cracking—see Chapter 22). Then, a particular attention must be paid to concrete curing when possible (on horizontal parts, for example).

The temperature elevation of concrete No. 19 was evaluated in a quasi-adiabatic box supposed to be representative of the elements that were cast, with a maximum thickness limited to 20–60 cm (see Figure 16.3). The figure shows that even if a CEM I 52.5 is used the temperature elevation is expected to be moderate in the structure. There is thus no risk of internal delayed ettringite formation as the maximum concrete temperature remains below 60°C (Ifsttar 2017) with a limited risk of cracking due to thermal gradient and shrinkage.

Finally, the risk of alkali-silica reaction of concrete mix No. 19 was evaluated, following the procedure proposed in FD P 18-464 (2014) standard. According to this procedure, the prevention level B was selected for the present project, matching the category II of risk of disorder (unacceptable) and the environmental class 2 (with hygrometry above 80% or in contact with water).

In a first step, RA 0/20 mm was detected to be potentially reactive, on a microbar test carried out according to NF P18-594 (2015) standard. Then, the Na_2O_{eq} alkaline balance

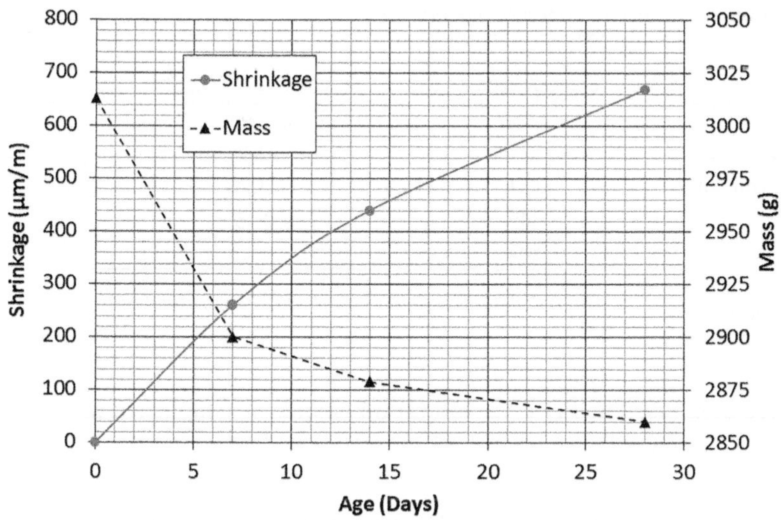

Figure 16.2 Mean total free shrinkage and mass loss measured on three 7 cm × 7 cm × 28 cm samples of mix No. 20 (see Table 16.7, this derived mix was selected as it was the "worse" case expected).

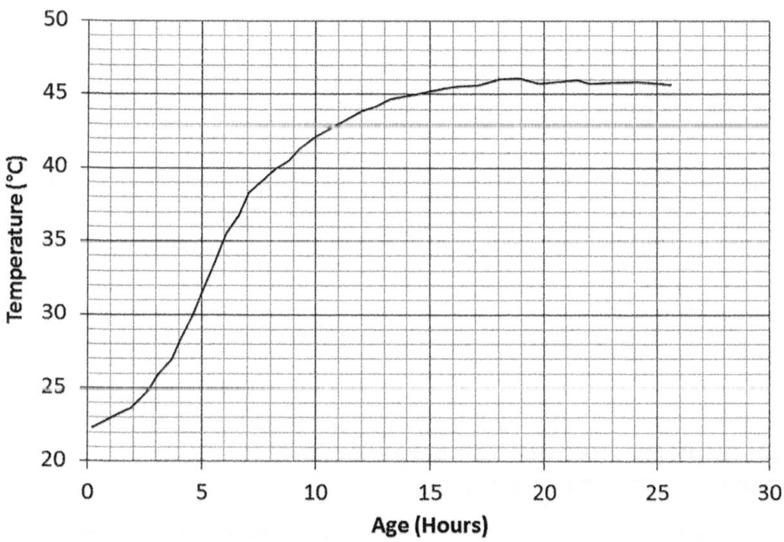

Figure 16.3 Temperature elevation of concrete mix No. 19 (see Table 16.5) in a quasi-adiabatic box. Test stopped when the temperature reached its maximum value.

of the concrete was calculated to be 2.58 kg/m^3 (see Table 16.8). According to FD P 18-464 (2014) standard, it is a sufficient condition to exclude the risk of alkali reaction that this value is lower than $T_m = 3.5/(1 + 2V_c)$, where V_c is the coefficient of variation of active Na_2O_{eq} in cement. This was the case, with a T_m value of 3.08 kg/m^3 ($V_c = 6.8\%$ in that case). Nevertheless, to ensure these results and considering the lack of experience with RA, it was decided to carry out a supplementary verification through a swelling test on the concrete mix, according to NF P18 454 (2004) standard. The mean average swelling measured on three samples was 0.01%, that is lower than 0.02% after 2 weeks, and no individual

Table 16.8 Alkaline balance of concrete mix No. 19

	Dosage (kg/m³)	*Active alkali Na_2O_{eq} (%)*	*Active alkali Na_2O_{eq} (kg/m³)*
RA 0/20 mm	1,550	0.0099[a]	0.15
CEM I 52.5	335	0,54[b]	1.81
Admixture content	3.51	<3.2[b]	<0.11
Total water	339	<0.15[c]	<0.51
Free water	201		

[a] measured.
[b] from datasheet.
[c] from EN 1008 standard.

value exceeded 0.025%. The FD P 18-456 (2004) standard established that in that case the concrete can be considered as nonreactive.

16.3 MANUFACTURING AND TESTING OF RAC ON JOBSITE

16.3.1 Control of RA water content on site

As explained earlier, RAs have a high water demand, so it was decided to prewet it at a water content around 13% to ensure that even the fine fraction 0/4 mm was saturated. The aim was to avoid any stiffening of concrete just after mixing due to water absorption, thus limiting the variability of workability on site.

To do so, a garden-like watering system was installed near the stock piles at the concrete plant, which operated a few minutes per hour from the middle of the day until the evening, to maintain the humidity almost constant, even in the first 10–15 cm on the surface that was subject of evaporation. This simple device proved to be efficient to fix this issue. To ensure that the water content remained in the range of 13% ± 0.5%, a daily control was carried out directly on the stockpile with a microwave sensor, weekly calibrated through a drying test on a RA sample. Moreover, the workability was systematically controlled with the mixer wattmeter (a value of 30 was aimed at) as a supplementary insurance that the free water was kept constant.

16.3.2 Final adjustments of the recipe during concrete implementation

Once the mix design procedure presented earlier was completed, a site suitability test was made in real conditions with mix No. 19, at the beginning of concrete works. This test shows that the mix was not satisfactory (segregation and bad aspect at the fresh state). It was then decided to test the mix A (see Table 16.9) derived from mix No. 19 by increasing the paste volume (with almost the same free water to cement ratio) and decreasing the admixture content. This recipe was used for the implementation of the first concrete walls (around 600 m³), but it was not satisfactory to have such a high cement content and it still displayed a slight proneness to segregation leading to random emergence of honeycombs.

As analyzed during the preliminary mix design, this was probably due to the lack of fine particles in the RA. It was then decided to correct the bottom part of the grading curve with the natural 0/2 mm sand used in Mix 7b (see Table 16.5). The concrete B was then selected including 10% of natural 0/2 mm to continue the works. This mixture gave satisfactory results in terms of workability (see Figure 16.4) and surface quality (see Figure 16.5) as well.

Table 16.9 Final recipes and results of suitability tests

	Mix A	*Mix B*
Date of trial	20 April 2005	6 May 2005
CEM I 52.5 N (kg)	389	320
RA 0/20 mm (kg)	1,456	1,480
Natural sand 0/2 mm (kg)	—	150
Admixture 2 (kg)	2.5	3.5
Free water (kg)	228	185
Free water/cement	0.59	0.58
Total water (kg)	358	324
Concrete temp./ext. temp.	24°C/17°C	25°C/19°C
Water content of RA (%)	13%	13%
Wattmeter value	30	30
Air (%)	1.1%	1.4%
Slump at 5′	200	205
f_c at 1 day	—	—
f_c at 2 days	26	24
f_c at 7 days	32	32.1
f_c at 28 days	39	40.4
f_c at 90 days	—	48.2

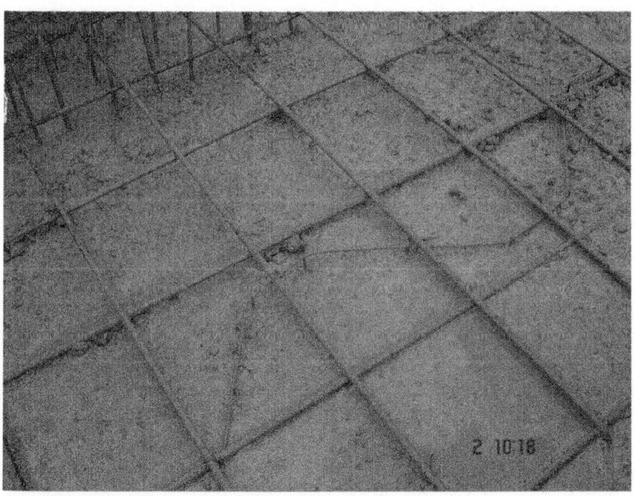

Figure 16.4 Aspect of the fresh concrete.

16.3.3 Concrete strength follow-up

During the works, samplings were carried out every 250 m^3 to follow up the compressive strength at 7, 28, and 90 days. Results are summarized in Table 16.10. At 28 days, the compressive strength was in the range of (30.1, 40.4 MPa) with a mean value of 34 MPa and a standard deviation of 3 MPa at 28 days. Then, the C25/30 requirement was matched although the variability was somewhat high.

(a) (b)

Figure 16.5 Aspect of harden concrete surfaces.

Table 16.10 Follow-up of compressive strength during concreting works

		Mean f_c (MPa)		
Dates	*Test Ref.*	*7 days*	*28 days*	*90 days*
29 Apr	1,309	32.4	36.8	39.8
25 May	1,399	29.0	33.7	38.4
2 June	1,423	—	34.5	—
2 June	1,422/1,423	30.6	34.3	—
16 June	1,468	27.9	32.7	37.6
4 July	1,531	32.1	40.4	41.7
19 July	1,555	31.1	36.6	41.8
29 Aug	1,645	25.9	31.7	32.9
19 Sep	1,709	24.6	31.0	34.0
26 Sep	1,745	27.4	31.1	34.0
21 Oct	1,860	32.4	35.4	39.8
10 Nov	1,980	25.9	30.1	32.8
13 Nov	2,023	27.6	34.0	48.2

16.4 CONCLUSIONS

In 2005, after the demolition of a water treatment plant, 6,000 tons of 0–20 mm RA were produced on site. The public owner of the facility was keen that all these materials be recycled on site to save natural resources, and more generally, for the sake of ecodesign. The solution adopted by the contractor was to recycle the RA at high rate in 3,500 m^3 of C25/30 MPa concrete dedicated to office buildings part of the construction. This project leads to the following remarks:

- in 2005, the corpus of standards available at this time allowed this achievement. Nowadays, the standards and particularly the NF EN 206/CN are more restrictive (see Chapter 14);

- RA displayed a high water absorption that made it necessary to prewet the stockpiles using a watering system, to ensure that the aggregates were saturated. Then, post-batching water absorption by the RA and corresponding workability changes could be avoided. A 13% water content was aimed at to ensure a full saturation of all aggregate particles;
- the RA was produced in a single fraction and important variations in coarse-to-fine particles ratio were observed. Moreover, the grading curve lacked fine particles (CA/FA around 1.7). It appeared necessary to correct the grading curve with at least 10% of natural 0/2 mm sand, to avoid segregation and improve the fresh behavior of concrete;
- even if only C25/30 concrete was aimed at, a CEM I 52.5 cement had to be used as preliminary trials with a CEM III 32.5 or a CEM II 32.5 with slag and gave a too low compressive strength at 1 day, incompatible with formwork removal at 16 h;
- it would have been possible to produce a C30 or a C35 concrete with a better control of RA grading (in two fractions for example) and the water content as well;
- up to now, the concrete behaves well after more than 12 years.

This construction site is the first documented one, where a (almost) 100% recycled concrete was used in France. Moreover, it appears as a pioneer case of application of current concepts of "short circuit" and "rebuilding the city on the city" (see Chapter 33).

Section VI

Reinforced recycled concrete

B. Fouré

Consultant, Bougival

This section deals with the behavior of reinforced Recycled Aggregates Concrete in structural members and compares it with reinforced Natural Aggregates Concrete behavior for the same concrete compressive strength.

Chapter 17 considers the topic of bond between RAC and reinforcement bars. The test results show that there is no significant difference in the ultimate bond strength between RAC and NAC. However, the ratio between bond and tensile strength may be higher for RAC than for NAC, due to the smaller tensile strength of RAC.

Dealing with members in compression, Chapter 18 is related to short reinforced columns. Their ultimate limit state (ULS) of strength is not affected by the use of RAC instead of NAC, despite the differences in modulus (smaller for RAC) and creep factor (higher for RAC). It is different for slender reinforced columns because, as shown in Chapter 19, these parameters can affect significantly the ULS buckling load and the conventional limits of slenderness below which it is allowed to neglect the second-order moment in the calculation.

For members in bending, covered in Chapter 20, the replacement of NAC by RAC does not affect the ULS of strength. However, at the serviceability limit state (SLS), smaller modulus and higher creep factor of RAC lead to higher deflections. The crack spacing and crack widths of RAC beams are only marginally different from those of NAC. For structural members where the deflection is a sensitive parameter (for instance, at SLS or ULS of buckling), it is advisable to measure the modulus and the creep factor.

When the failure occurs in shear, Chapter 21 shows that the possible difference in ultimate strength of beams without transverse reinforcement results primarily from the difference in tensile strength. The ratio between the shear strengths of RAC and NAC may be assumed proportional to the ratio of their tensile strength in the case of a brittle failure. For a non-brittle failure of beams with high flexural reinforcement, the ultimate mechanism implies not only the tensile strength but also the aggregate interlock in the crack interfaces, which is smaller for RAC; the ratio between the shear strengths would be on the order of magnitude of the square ratio of the tensile strengths. In beams with stirrups, the truss ultimate mechanism can form as well in RAC as NAC. The ultimate strength is not affected by the change from NAC to RAC, perhaps except the limit compressive strength of the concrete struts.

Chapter 17

Bond of reinforcement in recycled aggregate concrete

E. Ghorbel and G. Wardeh
University of Cergy-Pontoise

B. Fouré
Bougival

CONTENTS

Abstract

The aim of the present work is to investigate the bond strength and cracking properties of recycled aggregate concretes (RAC) with steel reinforcement. For this purpose, six recycled concrete aggregate mixtures and two reference conventional concrete mixtures (NAC) with C25/30 and C35/45 strength classes and S4 class of workability were chosen. Specimens were fabricated with different incorporation ratios of fine and coarse recycled aggregates or only coarse recycled aggregates. A total of 96 pull-out specimens were prepared using 10mm and 12mm diameter deformed steel bars concentrically embedded in concrete cylinders with two embedment lengths of five and ten times the rebar diameter. To study the cracking behavior, four 200x300x3000mm reinforced concrete beams were prepared with two ribbed bars of 12mm in diameter. The mechanical properties of the studied concretes were characterized in terms of compressive strength, splitting tensile strength, and Young's modulus. Concerning reinforced elements, two types of tests were performed: direct tension for pull-out tests and 4-point bending for beams. Test results for pull-out specimens showed that for the same strength class the bond strength and related failure mechanisms remain the same and the obtained values are, at least, five times higher than the predicted values by Eurocode2 when safety factors are considered. For flexural members, RAC showed more cracks (in number) and smaller spacing than NAC. Flexural tests showed, however, that crack openings remain in the same order of magnitude whatever the class of compressive strength (C25/30 or C35/40). These results show only slight differences between RAC and NAC which may be due to the difference in tensile strength and allow concluding that the Eurocode2 (EC2) predictions for the control of cracking remain valid for the studied RAC as well as the bond characteristics, without questioning significantly the security level that it provides.

17.1 INTRODUCTION

The bond between concrete and reinforcing bars is one of the main features determining the performance of reinforced concrete members. The earliest research on the effect of recycled aggregates (RAs) on the bond characteristics was published by Xiao and Falkner (2007). Using pull-out tests, they investigated three different recycled coarse aggregate replacement ratios being 0%, 50%, and 100% (i.e., effective replacement ratios (see definition in Section 17.2.2) of 0%, 34%, and 70%) with two types of steel rebar (plain and deformed). They found that, under the equivalent mix proportion (constant cement and sand contents and water/cement ratio), the bond strength between the RA concrete (RAC) and plain rebar decreases with an increase of replacement percentage, whereas for the deformed rebar, it has no obvious relation with the replacement ratio despite the reduction in compressive strength.

Butler et al. (2011) studied the influence of replacing natural coarse aggregate with recycled concrete aggregate on concrete bond strength with reinforcing steel. They also concluded that there is no significant difference between the bond strength of RAC and the conventional one.

Prince and Singh (2013, 2014) evaluated the bond strength from 90 pull-out tests carried out on 8, 10, 12, 16, 20, and 25 mm diameter deformed steel bars concentrically embedded in RAC. The studied mixtures were designed using equivalent mix proportions with coarse recycled concrete aggregate replacement levels of 0%, 25%, 50%, 75%, and 100% (equivalent replacement ratios of 0%, 13%, 26%, 39%, and 52% according to the definition in Section 17.2.2) by keeping cement and natural sand contents as well as water-to-cement ratio. This mix proportioning method also led to a decrease in the compressive strength as a function of substitution ratio.

The bonded length was five times the rebar diameter and was so selected to avoid, yielding of the steel bar under pull-out load. The authors observed a slight increase in the bond strength proportional to the replacement ratio and explained this fact by the effect of the internal curing action of the recycled particles. It has been conservatively suggested that anchorage lengths of deformed steel bars embedded in RAC may be taken to be the same as that in natural aggregate (NA) concrete (NAC).

Seara-Paz et al. (2013) studied the bond behavior of RAC by replacing different percentages of natural coarse aggregate with recycled coarse aggregate (20%, 50%, and 100%). Pull-out tests were carried out on 10-mm diameter deformed steel bars and an embedment length of five times the rebar diameter. Based on the obtained results, it was found that compressive strength and bond stress decrease with the increase of the percentage of recycled coarse aggregate used. The decrease in bond strength was about 13% for 100% replacement ratio.

The results obtained by Breccolotti and Materazzi (2013), Kim and Yun (2013, 2014), and Guerra et al. (2014) are consistent with the earlier results where no significant effect of RAs on the bond characteristics of concrete is found. It is worthily noticed that all these studies were conducted by replacing one type of aggregate that is mainly the coarse aggregates.

The WRAP's tests (WRAP 2007) involve a class C35 concrete with either NAs or 60% of recycled coarse aggregates (i.e., an effective ratio of about 30%); the compressive strength f_{cm} was practically the same for RAC and NAC, about 42 MPa. Splitting strength $f_{ctm,sp}$ of RAC was about 6% smaller than that of NAC, for the same f_{cm}. Deformed bars Ø16 mm with a small cover 1.6 Ø and a large embedment length 14 Ø were used. It is probable that the ultimate bond f_{bm} is governed by the small cover and that the bar does not yield in spite of the rather long embedment. The mean f_{bm} is 5% higher for RAC than for NAC. Assuming the compressive and splitting strength to be the same in the pull-out specimens as in the specific studies for f_c and $f_{ctm,sp}$, the ratio f_{bm}/f_{ctm} would be about 10% higher for RAC than for NAC.

This study aims to characterize the bond behavior of six RACs corresponding to two classes of compressive strength: C25/30 and C35/45 (Sedran 2017). These mixtures, used in all studies conducted by the national project RECYBÉTON, were designed from two reference mixtures by substituting partially or completely NAs with materials from the recycling of concrete while keeping the same class of compressive strength and S4 class of workability (see Appendix). Hence, the mixture proportions change within each series unlike studies available in the literature that keep the binder volume constant (Gomez-Soberon 2002a; Evangelista and de Brito 2007a).

17.2 EXPERIMENTAL PROGRAM

17.2.1 Raw materials

Raw materials used in this study were selected and provided by the technical support of the national project RECYÉTON to all associated partners (Sedran 2017). The materials were the following (see Appendix for more details):

- A Portland cement type CEM II/A-L 42.5 with a specific density of 3.09. The compressive strength at 28 days, measured according to the standard NF EN 196-1 (2006) is higher than 53 MPa;
- Limestone fillers, with a bulk density, measured according to standard NF EN 1097-7, are about 2.7 and the Blaine surface is equal to 500 m²/kg;
- a superplasticizer;
- Three NAs and three RAs (one sand and two coarse aggregates).

For pull-out specimens, deformed steel bars with $\Phi = 10$ mm and $\Phi = 12$ mm diameter were used. The two types of rebars are named, respectively, HA10 and HA12 and were made from 500 MPa nominal yield strength steel.

17.2.2 Concrete mixtures

The two reference concrete mixes C25/30 and C35/45 developed for the National Project RECYBÉTON were chosen (Sedran 2017) (see Appendix). To compare the results of this study with those of the literature in a uniform manner, we propose to introduce the equivalent replacement ratio, named Γ, which is the ratio between the weight of recycled aggregates and the total weight of aggregates in the mixture. It can be given by

$$\Gamma = \frac{\sum M_{RA}}{\sum M_{(NA+RA)}} \tag{17.1}$$

where M_{RA} and $M_{(NA+RA)}$ are the weight of RAs and the total weight of NA and RAs, respectively. Hence, $\Gamma = 1$ when both recycled sand and RAs are used. The equivalent replacement ratio for all mixtures is given in Table 17.1.

Table 17.1 Mix proportions of concrete mixtures

Constituents (kg/m³)	C25/30-0S0G	C25/30-30S30G	C25/30-0S100G	C25/30-100S100G	C35/45-0S0G	C35/45-30S30G	C35/45-0S100G	C35/45-100S100G
Γ (%)	0	30	52	100	0	30	50	100

17.2.3 Material test methods

Concrete cylinders with 220 mm height and 110 mm diameter were produced and cured in water at room temperature for 28 days for each of the mix designs, to verify the mechanical properties of studied concretes. Uniaxial compression and tensile splitting tests were performed using a servohydraulic machine with a capacity of 3,500 kN by imposing stress increment rate of 0.5 MPa/s for compression test and 0.05 MPa/s for splitting tests. Moreover, the dynamics modulus of elasticity was determined using E-Meter MK II device. Each test was repeated three times, and results shown in Section 17.3 are the averages of obtained values.

With the aim to characterize the mechanical behavior of the steel, tensile tests were performed on steel bars using a 250 kN testing machine. The results show that the yield strength, f_y, is about 550 MPa, where the ultimate strain is approximately 10% and the ultimate tensile strength exceeds 650 MPa.

17.2.4 Bond test specimens

Pull-out specimens were cast in 110 mm × 150 mm cylindrical molds with reinforcing steel positioned at the center. To ensure that the rebar was in a straight vertical position, a specific molding table was designed.

To characterize the influence of the embedded length on the failure mode, the test specimens were prepared by varying the following parameters:

- Two ribbed bar diameters, $\Phi = 10$ mm and $\Phi = 12$ mm. The two types of rebars are named respectively as HA10 and HA12.
- Two embedded lengths ($Lb = 5\Phi$ and $Lb = 10\Phi$).

For each configuration, three specimens have been manufactured. The steel bars were first cut to a length of 42 cm and were partially disconnected from the concrete using polyvinyl chloride (PVC) tubes injected with silicon so that the concrete/steel embedded length was to conform to the requested one. The bars were then introduced into the mold so that the effective anchorage length is positioned at midheight of concrete cylinders (Figure 17.1a).

Two 200 mm × 300 mm × 3,000 mm beams of the series C25/30 and two of the series C35/45 were also prepared. The flexural reinforcement was two ribbed bars with diameter 12 mm, and the shear reinforcement was made with deformed bars of 6 mm diameter while the secondary reinforcement was ensured with bars of 8 mm diameter (Figure 17.1b).

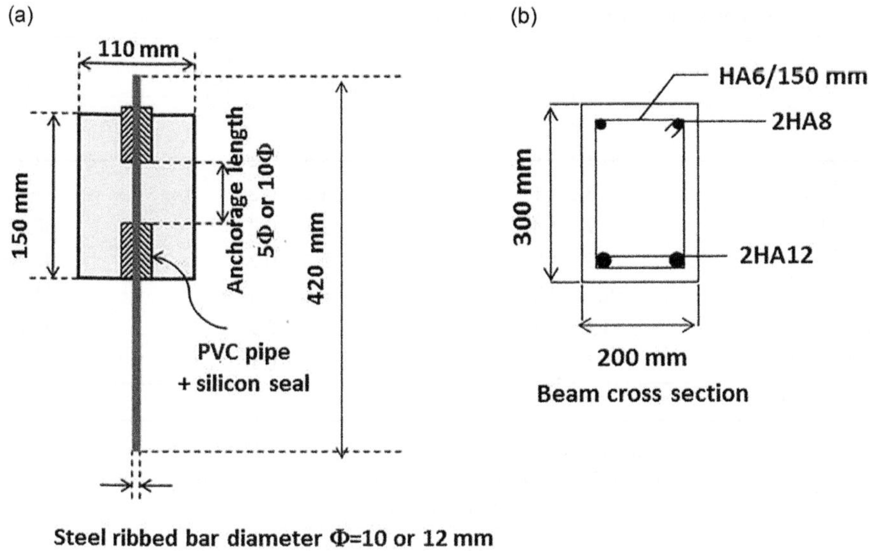

Figure 17.1 (a) Details of pull-out specimens. (b) beams cross section.

17.2.5 Tests setup

The pull-out tests were performed using a 250 kN Perrier testing machine and a dedicated sample supporting frame rigidly connected to the machine. The test was performed by pulling the embedded rebar downward out of the specimen according to the following protocol:

- For the first part of the test (preloading), the press was controlled with a constant;
- 0.11 kN/s load rate;
- After the preloading phase, the press was controlled with a constant 0.05 mm/s displacement rate.

The loaded-end slip was measured with the help of a linear variable differential transducer (LVDT) while the slip of the free end was measured using a camera coupled to an image analysis tool software DEFTAC.

Beams were loaded up to failure using a 350 kN capacity hydraulic 3R machine. The four-point bending tests were carried out at a constant displacement rate of 1 mm/min. Two concentrated loads were applied at 250 mm from midspan at the top of the beams. The deflection was measured at midspan using a displacement transducer (LVDT) placed on the bottom of beam specimens.

17.3 MECHANICAL PROPERTIES OF CONCRETES

Mean values and standard deviations obtained from three specimens tested at 28 days for the eight different concretes are summarized in Table 17.2. The results show that all concretes reach the target strength classes C25/30 and C35/45 as defined by the National Project RECYBÉTON (PN RECYBÉTON 2011). It can be observed that the splitting tensile strength remains almost constant for the same class of compressive strength with a slight decrease when the granular skeleton is completely recycled (C25/30-100S-100G and

Table 17.2 Mechanical properties of concretes at 28 days

Mix	*Compressive strength f_{cm} (MPa)*	*Splitting tensile strength f_{ctm} (MPa)*	*Dynamic elastic modulus E_d (GPa)*
C25/30-0S0G	30.8 ± 1.3	3.1[a] ± 0.9	40.8 ± 0.6
C25/30-0S100G	30.3 ± 0.4	3.4 ± 0.2	32.9 ± 0.3
C25/30-30S30G	30.8 ± 1.1	3.3 ± 0.2	34.4 ± 0.4
C25/30-100S100G	29.9 ± 1.4	2.8 ± 0.3	30.1 ± 0.2
C35/45-0S0G	41.5 ± 1.2	3.6 ± 0.2	42.1 ± 0.7
C35/45-0S100G	38.1 ± 2.2	3.4 ± 0.4	31.8 ± 0.4
C35/45-30S30G	40.3 ± 1.4	3.7 ± 0.2	35.7 ± 0.9
C35/45-100S100G	37.3 ± 1.3	3.4 ± 0.1	30.7 ± 0.5

[a] Probably underestimated.

Table 17.3 Relative variation of the splitting strength with replacement ratio

RA ratio	*30%*	*52%*	*100%*
C25/30	+7%	+11%	–8%
C35/45	+4%	0%	+1%

C35/45-100S-100G). Regarding the elastic modulus, it can be observed that it decreases when the replacement ratio increases for both series C25/30 and C35/45.

For the same compressive strength f_{cm}, the relative variation of the splitting strength is given in Table 17.3.

The mean correlation between the tensile strength $f_{ctm} = 0.9\ f_{ctm,sp}$ and the compressive one is $f_{ctm} \approx k f_{cm}^{2/3}$, with $k = 0.29$ for C25 and $k = 0.27$ for C35, which is slightly sur-safe as compared to EC2 formula $(0.275)\ f_{cm}^{2/3}$. It may be remarked that the other studies performed within the frame of the national project RECYBÉTON generally show a decreased $f_{ctm,sp}$, for the same f_{cm}, when the ratio of RA increases.

17.4 RESULTS OF PULL-OUT TESTS

17.4.1 Load versus slip curves

Force-loaded end displacement curves, obtained experimentally for both series C25/30 and C35/45, are shown in Figures 17.2 and 17.3, where each curve is the average of three pull-out tests. The analysis of these curves shows that the bond-slip behavior of RAC is similar to the behavior of reference concrete. At the beginning of loading, the force–displacement relationship is linear until a critical force (about 60% of the maximum force). During this phase, microslip occurs between the reinforcement and the concrete, and then microcracks appear at the rebar/concrete interface; as the load increases and reaches the maximum force, the slip increases, and the behavior becomes nonlinear. After the maximum load is reached, we can distinguish two types of postpeak behavior depending on the rebar diameter and the embedded length:

- A softening behavior with a residual bond capacity of about 40% of the maximum bond stress was observed during the test, for the smaller bar diameter and embedment length;
- Rebar yielding for the longer embedment length.

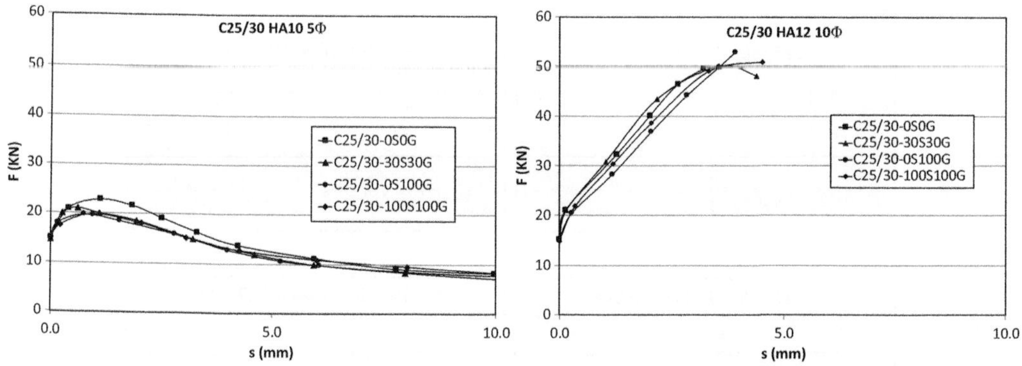

Figure 17.2 Force-loaded end displacement of the C25/30 series for the different bar diameters and embedded lengths.

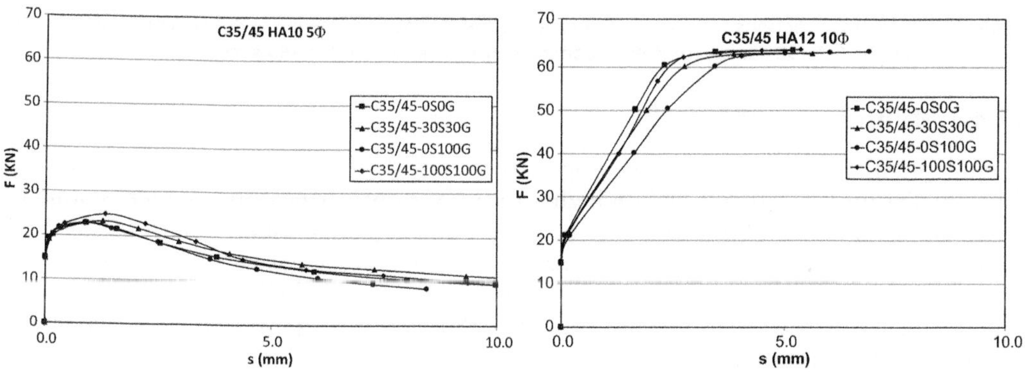

Figure 17.3 Force-loaded end displacement of the C35/45 series for the different bar diameters and embedded lengths.

The results also show that

- Two C25/30-30S-30G HA12 10Φ specimens have shown a softening behavior.
- For C35/45 specimens, a softening behavior is observed for anchorage length equal to 5Φ whatever the bar diameter while a rebar yielding occurs when the embedded length is equal to 10Φ.
- The ultimate bond strength seems not to vary significantly with the ratio of RA.

17.4.2 Macroscopic observations

At the end of each pull-out test, the specimen was broken into two parts to observe the interface between the rebar and the surrounding concrete. Figure 17.4 displays pictures of the interface for each configuration. It shows that when the anchorage length is 5Φ, the dominant failure mode is the complete shearing of concrete at the top surface of the rib whatever the type of concrete and the strength class. These results are supported by the softening behavior of force–displacement curves. For the configuration HA12 10Φ, the prints of the rebar ribs are clearly visible on the concrete. However, for the configuration HA10 10Φ, the situation is intermediate between the two previous ones.

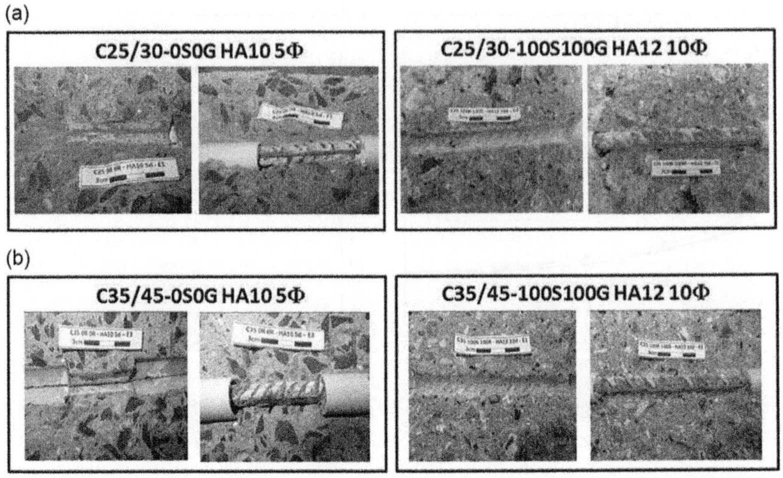

Figure 17.4 Split cross sections for both C25/30 and C35/45 series specimens. (a) Rebar–concrete interface for the C25/30 series. (b) Rebar–concrete interface for the C35/45 series.

17.4.3 Bond strength

Considering the variation of f_{cm}, the bond strength of RAC compared with NAC is about 10% smaller for C25/30 and slightly higher for C35. Considering only the highest amount of replacement, namely100%, the ratio between bond f_{bm} and tensile strength f_{ctm} for RAC compared with NAC is rather constant for C25/30, slightly increasing for C35/45. In addition, for a given bar diameter, the ultimate bond decreases if the embedded length increases because the ultimate force is anchored on a length less than l_b; so, the stress f_{bm} calculated on the overall length is underestimated.

17.4.4 Comparison with EC2 prevision

Eurocode EN 1992-1-1 (2004) gives a good model for bond, which includes all the parameters that are important to consider. For the comparison to tests results, it is necessary to write it in a convenient manner. EC2 considers a conventional value of the ultimate bond strength f_{bd}, including only a limited number of parameters (EC2 equation (17.2)).

$$f_{bd} = \eta_0 \eta_1 \eta_2 f_{ctd} \tag{17.2}$$

For deformed bars, $\eta_0 = 2.25$; for good bond condition $\eta_1 = 1$, other conditions $\eta_1 = 0.7$. Depending on the diameter Ø, size effect: $\eta_2 = 1.32 - \frac{\varphi}{100} \leq 1.0$

The EC2 design calculation is done with $\gamma_C = 1.5$ and α_{ct}, a coefficient taking account of long-term effects on the tensile strength (recommended value $\alpha_{ct} = 1$), with $f_{ctk,0.05} = 0.7 f_{ctm}$ and $f_{ctm} = 0.3(f_{ck})^{2/3}$ for $f_{ck} \leq$ C50/60 and $f_{ck} = f_{cm} - 8\,\text{MPa}$.

Thus, a conventional value, called the required anchorage length, is defined for a steel stress σ_{sd} equation (17.3):

$$l_{bd,rqd} = \frac{\varphi \sigma_{sd}}{4 f_{bd}} \tag{17.3}$$

Then, the design anchorage length is defined (EC2 equation (17.4)):

$$l_{bd} = \alpha_1\alpha_2\alpha_3\alpha_4\alpha_5 l_{bd,rqd} \tag{17.4}$$

For the straight bars, $\alpha_1 = 1$; effect of the cover c, $\alpha_2 = 1 - 0.15\left(\frac{c}{\varphi} - 1\right)$ between 0, 7, and 1; without transverse reinforcement $\alpha_3 = 1$; nonwelded mesh of bars, $\alpha_4 = 1$; without the confining effect of a transverse compression, $\alpha_5 = 1$. So

$$l_{bd} = \left[\frac{\alpha_1\alpha_2\alpha_3\alpha_4\alpha_5}{\eta_0\eta_1\eta_2}\right]\left(\frac{\varphi\sigma_{sd}}{4f_{ctd}}\right) = \frac{1}{\Psi_k}\left(\frac{\varphi\sigma_{sd}}{4f_{ctd}}\right) \tag{17.5}$$

For the comparison to test, the effective bond strength f^*_{bd} corresponding to l^*_{bd} is defined:

$$l^*_{bd} = \frac{\varphi\sigma_{sd}}{4f^*_{bd}} \tag{17.6}$$

that is

$$f^*_{bd} = \left[\frac{\eta_0\eta_1\eta_2}{\alpha_1\alpha_2\alpha_3\alpha_4\alpha_5}\right] f_{ctd} = \Psi_k f_{ctd} \tag{17.7}$$

It is necessary to suppress the safety factor γ_c and to replace characteristic value by mean value: $f_{ctd} = f_{ctk}/\gamma_c$; f_{ctk} replaced by f_{ctm} and $\gamma_c = 1$. Furthermore, it is necessary to suppress a hidden factor γ_R that takes into account the scatter of experimental results used to establish the formula, passing from a mean value f^*_{bm} to the characteristic one f^*_{bk}. Using the global term Ψ_k from equation (17.7), its mean value is $\Psi_m = \gamma_R \cdot \Psi_k$. After the test results used in the reference (Eligehausen and Balazs 1993), a value $\gamma_R \approx 1.4$ may be used. Finally,

$$f^*_{bm} = \Psi_m f_{ctm} = \left[\gamma_R\left(\frac{\eta_0\eta_1\eta_2}{\alpha_1\alpha_2\alpha_3\alpha_4\alpha_5}\right)\right] f_{ctm} \tag{17.8}$$

The comparison to the present tests is done using the measured splitting strength $f_{ctm,sp} = \frac{f_{ctm}}{0.9}$ instead of the predicted one, which is derived from the measured compressive strength f_{cm} through the relationship $f_{ctm} = 0.3 f_{cm}^{2/3}$. The comparison is performed with $\eta_2 = 1$, $\alpha_2 = 0.7$. Thus,

$$f^*_{bm} = 4.5 f_{ctm} = 4.0 f_{ctm,sp} \tag{17.9}$$

or

$$f^*_{bm} = 1.35 f_{cm}^{2/3} \tag{17.10}$$

Neglecting the marginal underestimation of f_{bm} due to the yielding of some bars, the overall mean experimental results f_{bm} and their comparison to the equations (17.9) and (17.10) are displayed on Table 17.4.

Table 17.4 Comparison between experimental results and EC2 prevision without safety factors

Class	*Mix*	*Bond strength* f_{bm} *(MPa)*	*Mean test/ equation (17.9)*	*Relative variation*	*Mean test/ equation (17.10)*	*Relative variation*
C25/30	0S0G	13.7	1.10	—	1.03	—
	30S30G	12.2	0.92	−16%	0.92	−11%
	0S100G	12.6	0.93	−15%	0.96	−7%
	100S100G	12.3	1.10	0%	0.94	−9%
	Mean (deviation)		1.01 (11.7%)		0.96 (8.7%)	
C35/45	0S0G	15.1	1.05	—	0.93	—
	30S30G	14.5	0.98	−7%	0.91	−2%
	0S100G	14.8	1.09	+4%	0.97	+4%
	100S100G	15.0	1.10	+5%	0.99	+6%
	Mean (deviation)		1.06 (9.6%)		0.95 (8.9%)	

The prediction by the formula of EC2 without safety factors is on the whole satisfactory, irrespective of RAC or NAC, which is due to the small variation of f_{ctm} with the type of concrete. The scatter is smaller for equation (17.10) based on f_{cm} than for equation (17.9) based on f_{ctm}, which is probably due to a greater uncertainty on f_{ctm}. The difference in mean value for equations (17.9) and (17.10) is of course due to the discrepancy between the numerical factor in formula giving f_{ctm} versus $f_{cm}^{2/3}$ in EC2 and its experimental values.

The design values for all mixtures are also reported in Figure 17.5, where it can be observed that all experimental bond stress values are close to the value calculated using equation (17.10).

17.5 RESULTS OF FLEXURAL MEMBERS

Load–midspan deflection relationships obtained from the tests are illustrated in Figure 17.6, where each curve consists in three lines with different slopes. From the tests, it is shown that the bearing capacity is identical for both natural and recycled concrete beams. However, the stiffness of NA beams is slightly higher. The loss of stiffness can be explained by the decrease of elastic modulus and the more pronounced cracked state of beams made with RAC.

The observed failure mode of beams is related to the steel yielding. Concrete crushing occurred after the yielding of the steel under load points. By the end of the test, inclined cracks, due to the shear force, appeared near supports. All crack patterns appeared at lower loads for RAs beams compared with NA ones. Moreover, it can be noticed that the number of cracks increases when recycled gravels are used. This cracked state is accompanied by a decrease in the spacing between cracks and an increase in the width of the cracked zone and finally an increase in the height of the cracks (Table 17.5).

During the test, an acquisition system (camera + software) allowed to follow the opening of the first vertical flexural crack of C25/30 beams (see Figure 17.7). From this figure, it can be observed that, for each beam, the crack opening is almost identical during loading. Strain gauges placed at the extreme fiber in compression with the midspan of beams showed that, for a given load, the strain of RAC is slightly higher than the strain of the NA one. Finally, it may be concluded that EC2 provisions for cracking analysis remain valid for concrete incorporating RAs.

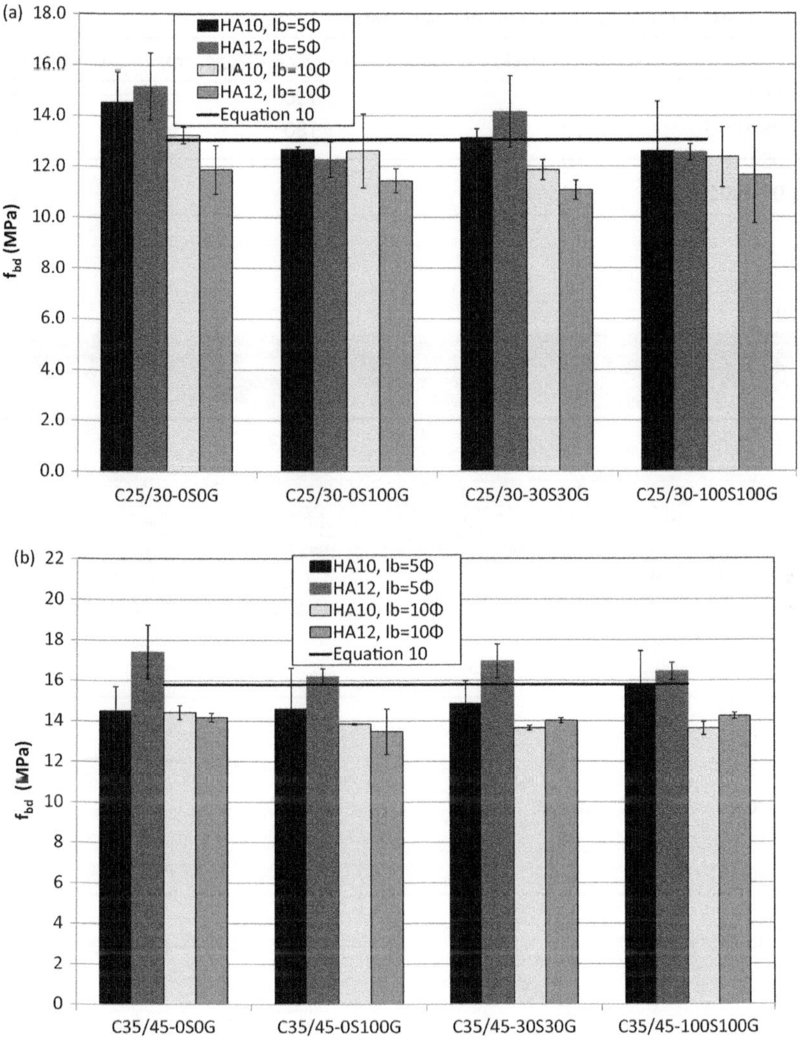

Figure 17.5 Average bond strength for studied concretes. (a) Bond stress for the series C25/30. (b) Bond stress for the series C35/45.

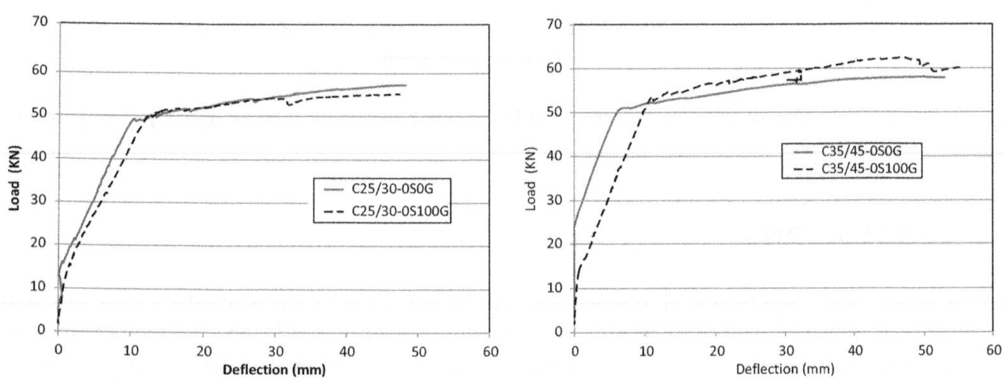

Figure 17.6 Load–deflection curves.

Table 17.5 State of cracking at the end of beam tests

Beam	*Number of cracks*	*Width of the cracked zone (cm)*	*Crack spacing (cm)*
C25/30-0S-0G	13	150	11.5
C25/30-0S-100G	18	180	10.0
C35/45-0S-0G	10	100	10.0
C35/45-0S-100G	14	125	9.0

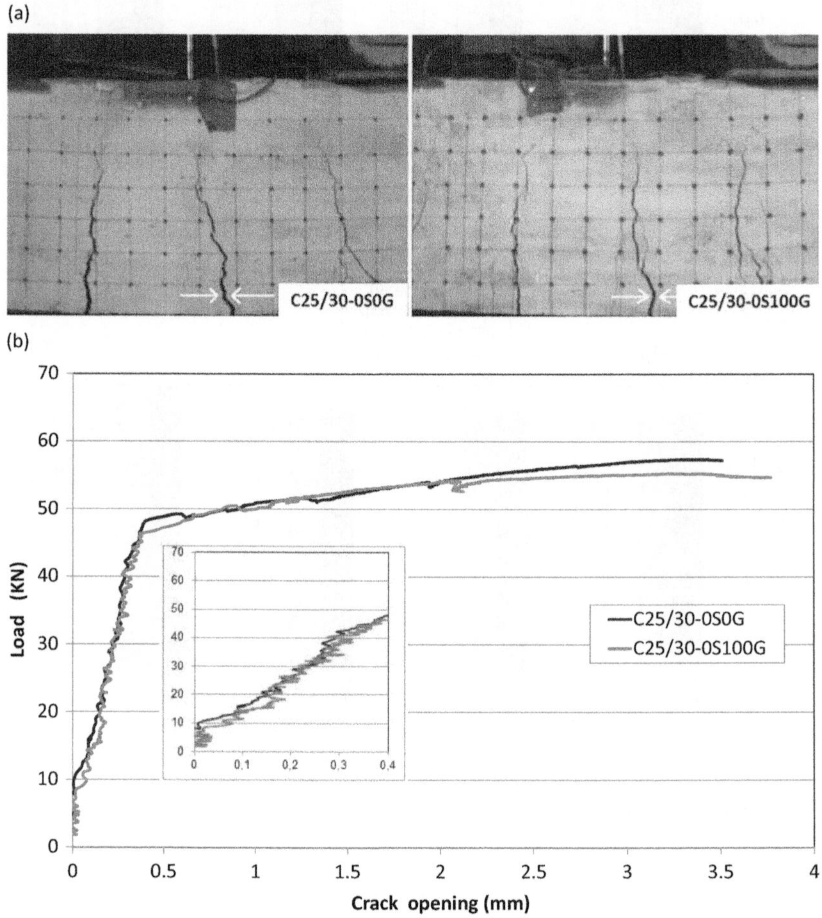

Figure 17.7 Flexural behavior (a) Crack patterns at failure load, (b) Flexural crack opening during loading.

17.6 CONCLUSIONS

In this work, the experimental results related to the bond characteristics and cracking behavior of RACs are presented. Eight mixes with C25/30 and C35/45 strength classes and S4 class of workability were studied. Based on experimental results, the main conclusions are as follows:

- The bond strength of RAC remains of the same magnitude order as that measured for the reference NAC, as far as the tensile strength remains with a less than 10% reduction from the reference. The bond-slip behavior and the associated failure mechanisms remain the same for RAC and NAC;
- In all cases, the measured values are equal to the values calculated by the formula of EC2 without safety factors and by multiplying the characteristic value by 1.4 to obtain the mean value. They are therefore at least five times higher than the design values;
- RAC members showed more cracks than conventional concrete members, thus a smaller spacing;
- Crack openings remain of the same order of magnitude as in the case of normal concrete.

This work was restricted to bars of relatively small diameters (10 and 12 mm), and it would be important to extend this study to other reinforcement sizes. The obtained results constitute a significant database on the bond behavior of RAC, which is exploitable for modeling work.

Chapter 18

Members in compression—Short columns

R. Boissière and F. Al-Mahmoud
Université de Lorraine

A. Hamaidia
Université de Jijel

B. Fouré
Bougival

CONTENTS

Abstract

The present study deals with the mechanical behavior of reinforced concrete columns made from Recycled Coarse and Fine Aggregates replacing the natural ones. These columns are loaded in eccentric static compression with different replacement ratio in their mix, with the same compression strength class. An experimental program followed by an analytical analysis is developed. Four columns are tested: one with natural aggregate (NA) as a reference, and three with recycled aggregates (RA) and different replacement ratios (gravel and sand) as defined by the French National Project RECYBÉTON. The columns are instrumented with LVDT and strain gauges to measure concrete and steel strains. Results indicate that a low RA replacement ratio leads to negligible change in the ultimate strength of RA columns compared to the NA one. On the other hand, a ratio of 100% induces a slightly different mechanical behavior. Finally, the analytical study confirms that EUROCODE 2 standard allows designing short columns in a safe way, irrespective of the use of recycled or natural aggregates.

18.1 INTRODUCTION AND BIBLIOGRAPHY

The study of the properties of recycled aggregate (RA) and the basic properties of RA concrete (RAC) has been going on for the past few decades (Yagishita et al. 1994; Ajdukiewicz and Kliszczewicz 2002; Otsuki et al. 2003; Domingo-Cabo et al. 2009; Fathifazl et al. 2011b; Kou et al. 2012; Kou and Poon 2012; Kikucki et al. 1988), leading a number of countries to establish standards or recommendations supporting their use. From these studies, the full and partial replacement of natural aggregate (NA) with RA has been proven to be a feasible option. However, few investigations have been carried out in the field of structural behavior (behavior under flexure conditions, shear, compression, torsion, etc.). The earliest research on structural performance of RAC was published in Japan (Yagishita et al. 1994). This study discusses the shear performance of reinforced concrete beams with RA. Zhuang (2007) performed an analysis of the influence of varying RA contents on the complete stress–strain curve, peak stress, peak strain, elastic modulus, and compressive strength. It was concluded that the stress–strain curves of RAC exhibits the same tendencies than those of the conventional concrete. Although the ratios of the prism compressive strength to the cube compressive strength and the peak strain for RAC are higher, the elastic modulus is lower than those of the NA concrete (NAC) of same strength. The compressive behavior of reinforced concrete columns that contain RA was investigated by Choi and Yun (2012). The quality and quantity of RAs were studied. They found that, as the percentage of replacement of RA increases, the maximum axial load capacity decreases by approximately 6%–8% compared to columns with NA with the same mix proportion, due to a decrease of f_c. The authors concluded that, in general, the axial behavior exhibited in RAC columns is comparable to the one found in conventional reinforced concrete columns that are made with NA. A study was conducted by Liu et al. (2010) on the behavior and mechanical performance of RAC with very short concrete columns (geometrical slenderness 4) from the same compressive strength class (45/50 MPa). They conducted tests on six NAC and RAC, where the replacement ratios were 0% and 100%, with eccentricity values of 60 and 150 mm. The results obtained showed that the behavior and failure mechanism of RAC columns are similar to those of NAC columns. The bearing capacity of NAC columns was higher than the RAC columns, probably because the compressive strength of concrete is higher. The authors mention the variability of the ultimate load due to the uncertainties on the compressive strength and eccentricity. Zhou et al. (2010) studied the behavior of RAC columns rather short (slenderness 9) under large eccentric compressive loads with different replacement ratios of RA. They found that the smaller the load that is applied to a recycled concrete column, the more RA can be used with no changes in mechanical behavior up to 80% RA. However, the ductility of RAC columns is shown to be slightly higher than that of NAC columns. They suggest that concrete structures that contain 50% RA could have practical applications in engineering. Based on a review of the existing literature, the use of RACs is very recent and few researchers have studied the subject. These studies are not sufficient for making clear the behavior of structural members, leading to establishing the design method. Indeed, very few results are available in literature regarding using recycled fine aggregates, and very few tests have been conducted under eccentric compression loading for concrete members that contain RA, and always conducted on short columns. The following study was undertaken to fill this gap and deals with testing of the mechanical behavior of reinforced concrete short columns made from recycled coarse and fine aggregates. Also, the coincidence of experimental results with code predictions is studied to check the applicability of current design provisions for recycled concrete beams.

18.2 EXPERIMENTAL PROGRAM

An experimental program followed by an analytical analysis was developed. The columns were loaded in eccentric static compression with different coarse and fine aggregate replacement ratios in their mix proportions with the same compressive strength class. Four columns were tested: a control one with NA as a reference and three with RA composed of different replacement ratios (gravel and sand) as defined by the French National Project RECYBÉTON (see Appendix).

18.2.1 Geometry of the columns

The experimental program consisted of four square columns tested under eccentric loads. All columns have cross-sectional dimensions of 150 mm × 150 mm and height of 1,200 mm. Longitudinal steel reinforcements were 4 Ø10 mm deformed bars with yield stress of 560 MPa. Stirrups were 6 mm in diameter with a constant spacing of 150 mm. To avoid premature failure, column ends were confined internally by reducing spacing of ties from 150 to 50 mm. Concrete external cover was 15 mm. All tested columns were loaded as pin-ended columns with a load eccentricity of 15 mm. The columns were designed to study the effect of different replacement ratios of RA. The dimensions and reinforcement details are shown in Figure 18.1.

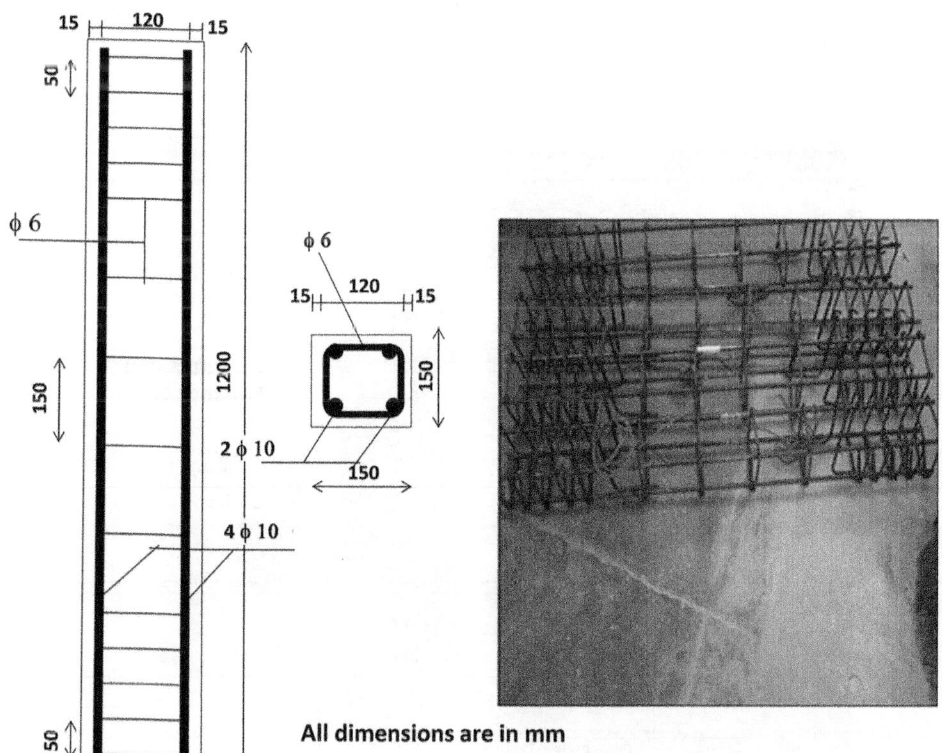

Figure 18.1 Reinforcement arrangement along the column height.

18.2.2 Eccentric loading device

Eccentric compression tests were carried out using a hydraulic testing machine with a capacity of 2,000 kN. To avoid premature failure and ensure a good distribution of the load inside the columns, both column ends were confined by external steel caps height 150 mm cross section 200 mm × 200 mm (Figure 18.2). Those caps were constructed using steel plates of 10 mm thickness. A square steel plate was fixed in the upper part of the testing machine to apply the eccentric load by steel cylinders placed between this plate and the cap to give target eccentricity in both upper and lower ends of tested columns. Another circular steel plate was attached to the testing machine base in contact with the metallic cap of the lower part; the system can be considered simply supported (hinged columns). The eccentricity was chosen equal to one-tenth of the side of the concrete specimen (15 mm). The purpose is to minimize the relative effect of uncertainty of the eccentricity by imposing a controlled shift that allows studying the effect of bending moment, contrary to the pure axial compression with zero eccentricity. Figure 18.2 shows the developed eccentric loading mechanism for the study. The distance between the hinges is l_0 = 1,280 mm, so the geometrical slenderness is equal to 8.5.

18.2.3 Casting and curing

All the columns were cast in the laboratory and were compacted using an electrical internal vibrator. After casting, the molds were covered with wet burlap and plastic sheets to ensure concrete curing and to control the moisture. The specimens were demolded after 24 h, but their moist curing continued till 7 days. After that, the moist curing was discontinued and the concrete specimens were stored in laboratory conditions.

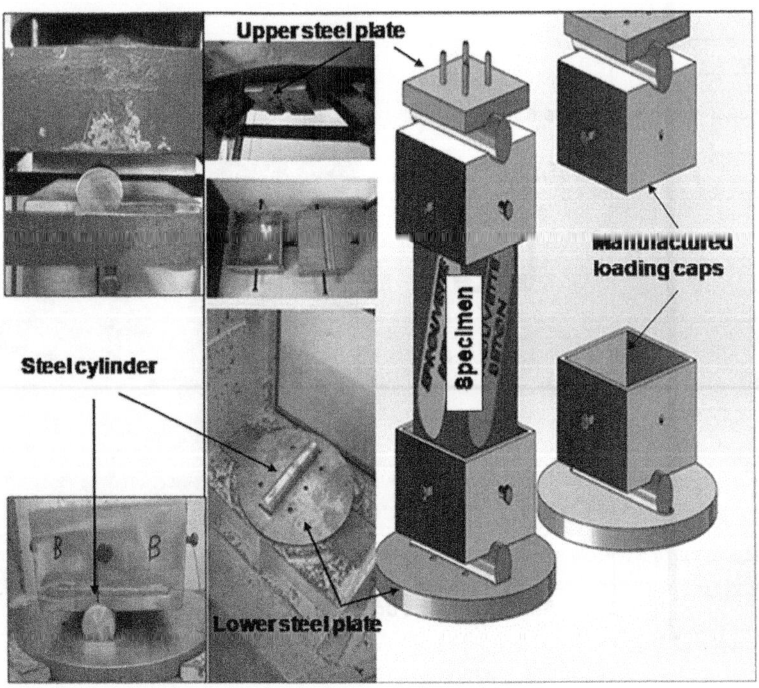

Figure 18.2 Developed eccentric loading device.

18.2.4 Instrumentation

To control strains and displacements of the column, strain gauges and linear voltage differential transducers (LVDTs) were used. Three strain gauges were glued on concrete surface at both tension and compression sides and in the middle of the lateral faces. Steel bar strains were measured by means of two strain gauges glued at mid-height. Two vertical LVDTs were installed on both tension and compression sides at mid-height (Figure 18.3). A horizontal LVDT was placed perpendicularly to the column, to measure the transverse displacement in the middle of the column. The horizontal displacement at the ends was neglected. The columns were subjected to a quasi-static compressive loading. The total loading time was between 8 and 11 min and can then be considered as quasi-instantaneous. Figure 18.3 shows details and setup for tested columns.

18.3 MATERIAL PROPERTIES

18.3.1 Concrete mixes

Four C25/30 concretes were made. The mixes include a reference concrete with C25/30 NAC and three RACs. The concrete mixes must be of class S4 (concrete slump between 160 and 210 mm).The designation of each mix is the following, i.e., Column xS−yG:

- x is the replacement ratio of recycled fine aggregate, S;
- y is the replacement ratio of recycled coarse aggregate, G.

RAs were used with a moisture content corresponding to absorption plus 1% (in absolute value). The water content of 9.9% for the recycled fine aggregate 0/4, 6.6% for the recycled coarse aggregate 4/10, and 6.8% for the recycled coarse aggregate 10/20. The RAs were stored in barrels for at least 2 h, after homogenization by rolling. Mixes studied in this part are given in the appendix. It is noted that the proportions of the constituents vary from one mix to another. In addition, the cement content increased as the substitution ratio increased to achieve the target compression strength at 28 days.

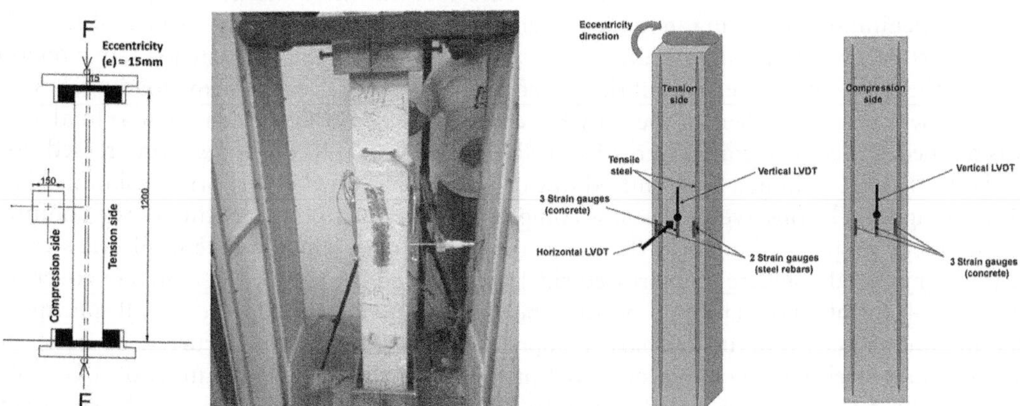

Figure 18.3 Experimental setup, LVDT, and gauge installation.

Table 18.1 Mechanical properties of the concrete for all columns

Concrete column	Compressive strength f_{cm} (MPa)	Splitting tensile strength f_{ctm} (MPa)	ψ	Elastic modulus E_{cm} (GPa)	k
0S-0G	28.4	2.9	0.276	24.3	8,000
30S-30G	27.9	2.9	0.274	22	7,200
0S-100G	28.7	2.7	0.264	21.5	7,100
100S-100G	29.0	2.5	0.238	17.2	5,600

18.3.2 Concrete strengths

Standard cylinders (diameter = 160 mm, height = 320 mm) were cast from the same concrete batch as the columns. The properties of the hardened concrete (compressive strength, tensile strength, and instantaneous elastic modulus) were measured at 28 days on the concrete cylinders. The specimens were removed from their molds 24 h after casting and stored for 28 days in laboratory conditions ($T = 20°C$, relative humidity = 60%). The tensile strength was obtained using splitting test. The concrete strains were measured using an extensometer with three LVDTs located on three generators at 120°. This instrument was used to measure the static secant modulus of elasticity for each tested sample. Table 18.1 shows the mechanical properties of the concrete mixes measured at 28 days for all specimens. In this table, k and ψ correspond to the coefficients for the predicted tensile strength ($f_{ctm} = \psi \cdot f_{cm}^{2/3}$) and elastic modulus ($E_{cm} = k \cdot f_{cm}^{1/3}$). Thence, it is clear that for the same strength f_{cm} the tensile strength f_{ctm} and the modulus E_{cm} decrease when the ratio of RA increases, as was found by other partners in the RECYBÉTON project. Both splitting tensile strength and Young's modulus appear to be significantly lower compared to what was expected from the compressive strength.

18.4 TEST RESULTS

18.4.1 Cracking and failure modes

Figure 18.4 shows the tested elements after failure. Since the columns were loaded in an eccentric compression way, they remained mainly in a compressive state with a dissymmetry between the left and right sides as shown on Figure 18.4 (the most and less compressed sides).

At the beginning of loading, no cracks were observed on the external surface of the columns. Vertical cracks appeared in all columns at about 65% of the ultimate load, in a region close to the most compressed side of the upper part of the specimens. During loading, lateral displacement became higher for the fully RA column (100S-100G). When the maximal load was reached, transversal cracks were observed in the central zone of the less compressed side of the 100S-100G and in a zone shifted towards the upper third of the other columns. The plastic hinge can be observed in the less compressed side due to the buckling of the column. These cracks propagated towards the most compressed side. The increase in plastic strains of the compressed concrete (compressed side) led to a sudden loss of the concrete cover. In Figure 18.4, the observed failure is beyond the ultimate state and occurred for all specimens due to concrete crushing in the most compressed side. The failure occurred by concrete crushing and steel bars buckling, occurred in the upper third of the columns 0S-0G, 30S-30G, and 0S-100G. For the column 100S-100G, buckling occurred almost at the specimen center. As can be seen in Figure 18.4, for all cases the buckled steel bars are curved toward the external side of the column.

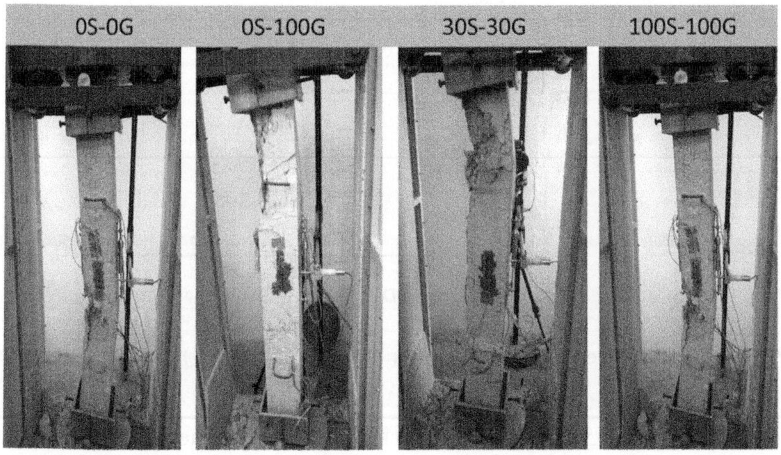

Figure 18.4 Columns after failure.

18.4.2 Ultimate state

The ultimate loads and corresponding deflections for all columns are presented in Table 18.2.

Table 18.2 shows that the measured bearing capacities ranged from 514 kN (30S-30G) to 444 kN (100S-100G). The maximal difference was 15% depending on the replacement ratio. It seems that a low replacement ratio of NA by RA (<50%) induces limited changes in bearing capacity (less than 10%). Regarding the fully RAC column (100S-100G), a significant decrease is observed but remains in the scope of what is expected. These differences between the columns in terms of ultimate load do not allow drawing conclusions, as only one column of each concrete type has been tested. There is a little decrease in load capacity for RAC columns that probably does not affect expected requirements and the provisions. This conclusion will be discussed in Section 18.5. Lateral horizontal displacements of the columns were measured during the test. Figure 18.5 shows the horizontal displacement in the middle of the column versus applied load. Two phases can be distinguished: first, the displacement is almost linear, and then it increases rapidly due to plastic strain in the compressed concrete and compressed bars yielding for 30S-30G and 100S-100G columns (see Section 18.4.3). Figure 18.5 shows that, globally, columns made from RAs 30S-30G, 0S-100G, and 100S-100G exhibit more lateral displacement than the control column (0S-0G). Displacements at the load of 400 kN were 1.1, 2.25, 1.2, and 3.3 mm for the columns 0S-0G, 30S-30G, 0S-100G, and 100S-100G, respectively. This result is in agreement with the conclusion by Zhou et al. (2010). Indeed, the results of their tests showed a higher transversal displacement for RAC column (including sand recycling) than a natural one.

Ultimate deflections shown in Table 18.2 are 2, 4.48, 2.96, and 4.85 mm for 0S-0G, 30S-30G, 0S-100G, and 100S-100G columns, respectively. These values correspond to an increase of +124%, +48%, and +143% for RAC columns compared with NAC columns. These deflections lead to additional bending moments due to second-order phenomena.

Table 18.2 Ultimate loads for all columns

Column	*0S-0G*	*0S-100G*	*30S-30G*	*100S-100G*
Ultimate load n_u (kN)	502	475	514	444
Ultimate deflection (mm)	2.0	3.0	4.5	4.8

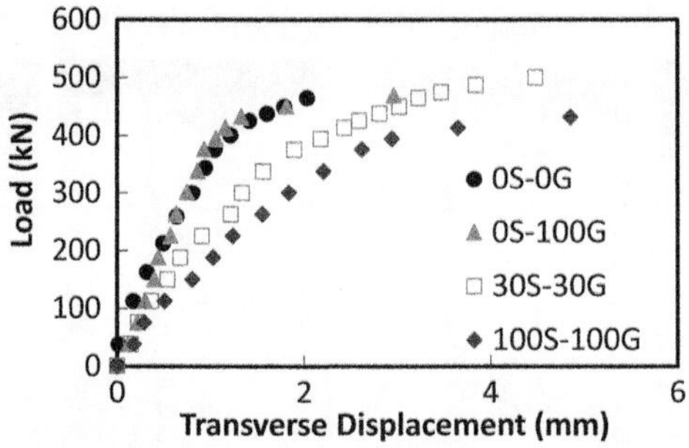

Figure 18.5 Middle height transverse displacement for columns 0S-0G, 0S-100G, 30S-30G, and 100S-100G.

The initial eccentricity is 15 mm and the horizontal displacement is 4.85 mm at most, the bending moment is hence expected to be raised up to 30% during the test due to second-order eccentricity. However, its effect on the ultimate load is much less than 30%.

18.4.3 Concrete strains

Strains in concrete were assessed, and Figure 18.6 presents the concrete strains measured on the most compressed side versus applied load (for 50% of ultimate load, 400 kN, and ultimate load) for 0S-0G, 30S-30G, 0S-100G, and 100S-100G columns. It can be noticed that the strain measurements exhibit the same trend as the one observed for transverse displacement. Strains measured from 0S-0G and 0S-100G columns are close to each other. The 100S-100G column showed an important difference compared to the other columns for the same load. Strains measured at 400 kN are 1,150, 1,500, 1,250, and 2,150 microstrains

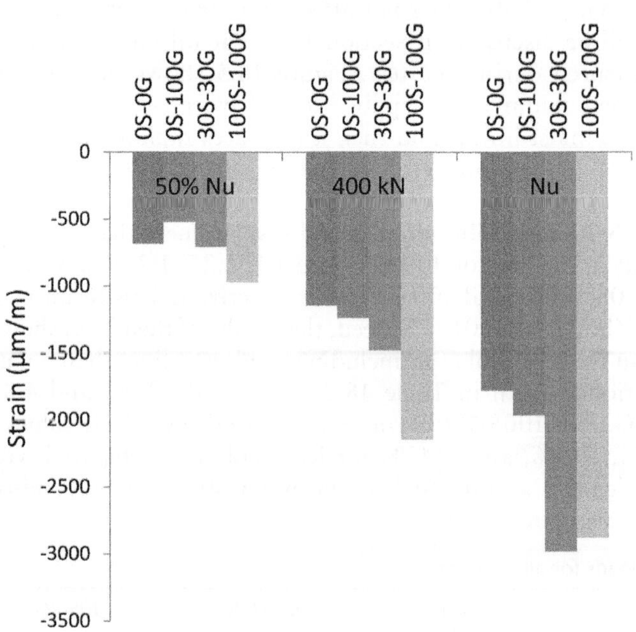

Figure 18.6 Concrete strains for the most compressed side for 0S-0G, 0S-100G, 30S-30G, and 100S-100G columns at 50% of ultimate load (n_u), 400 kN, and ultimate load (n_u).

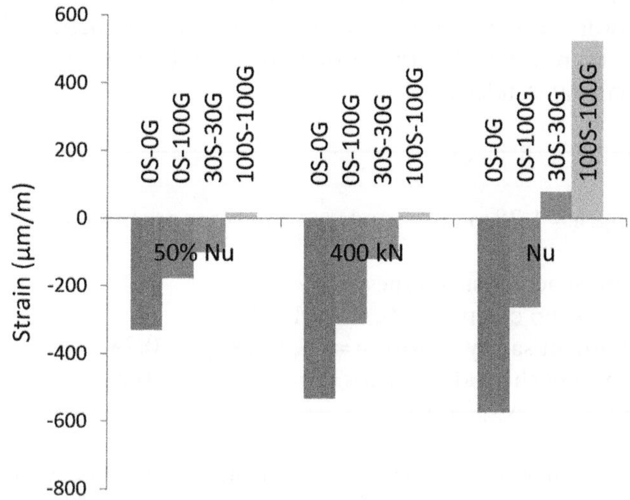

Figure 18.7 Concrete strains in the less compressed zone for 0S-0G, 0S-100G, 30S-30G, and 100S-100G columns at 50% of ultimate load (50% n_u), 400 kN, and ultimate load (n_u).

for columns 0S-0G, 30S-30G, 0S-100G, and 100S-100G, respectively. These values correspond to a difference of +30, +7, and +87% compared to 0S-0G column. Figure 18.7 presents the concrete strains measured on the less compressed side for tested columns versus the applied load (for 50% of ultimate load, 400 kN, and ultimate load). This figure shows that for 0S-0G, 30S-30G, and 0S-100G columns, this side is in compression during the test, whereas 100S-100G column is already in low tension at 50% of the ultimate load (200 kN), what may be due to an unintended larger eccentricity that was probably induced by the settings of the column inside the end caps. This larger eccentricity was also checked through numerical simulations. It can also be noted that 30S-30G and 100S-100G columns exhibit a quick increase in their "tension" strain at the end of the test, between 400 kN and failure. Observations seem to show that the displacements and strains measured are different, depending on the substitution rate. The mechanical behavior of the columns with at least 30% of recycled sand exhibits higher compliance than the natural one (0S-0G column) due to lower modulus and higher peak strain of their concrete. According to the strain–load curves, the ultimate displacement increased following this order 0S-0G, 0S-100G, 30S-30G, and 100S-100G.

18.5 STRUCTURAL ANALYSIS

In this section, comparison is done with the design calculations according to the Eurocode 1992-1-1(EC2), to check whether the use of recycled concrete still fits with the requirements. Even if the second-order effect is small, the calculation is done by the so-called Faessel method (Robinson and Modjabi 1968) or "model column method", which is roughly equivalent to the general method of EC2 (§ 5.8.6).

The design eccentricity must be determined. It contains three terms: the eccentricity due to external loads, the geometrical imperfection, and the second-order eccentricity equal to the deflection at mid-height. The first one is known following the set up: $e_0 = 15$ mm. The second one e_i is equal to the maximum between 20 mm (EC2, National Annex) and $L_0/400$

(EC2, § 5.2.7 a). Here: $e_i = 20$ mm, thence, the first-order eccentricity is 35 mm. The third one is the second-order eccentricity that is active for slender structures. The EC2 formula (§ 5.8.3.1 (1)) used to know whether the second order can be neglected is the following condition on the mechanical slenderness:

$$\lambda = l_0/i < \lambda_{lim} = 20 \cdot A \cdot B \cdot C/\sqrt{n}$$

with l_0 buckling length = 1.28 m, as boundary conditions are hinges

$i = \sqrt{(I/S)} = 0.15$ m → actual slenderness $\lambda = 29.6$
$A = 1/(1 + 0.2\ \varphi_{ef})$ → no creep $\varphi_{ef} = 0$, then $A = 1$
$B = \sqrt{(1 + 2\ \omega)}$; without safety factors $\omega = (A_s\ f_y)/(A_c\ f_c) \approx 0.28 \rightarrow B = 1.25$
$C = 1.7 - r_m$ → ratio of the end moments $r_m = 1$, then $C = 0.7$
Relative axial force: $n = N_{Ed}/(A_c\ f_c)$

In this study, the experimental ultimate load n_u is about 500 kN, with $f_c \approx 28$ MPa, i.e., $n_u \approx 0.79$. The corresponding slenderness limit can be calculated as $\lambda_{lim} = 20\ A\ B\ C/\sqrt{n_u} \approx 24$. The actual slenderness is slightly higher. Thus, the second-order effect included in n_u is not negligible, even if it is rather small.

Then, the eccentricity taken into account for the design EC2 calculation is equal to $e = e_0 + e_i + e_2$, with e_2 depending on loading. As a consequence, the use of a calculation software is compulsory to solve this request. For the studied case, the one developed by Thonier has been used. It gives a critical design load $N_{Rd} = 235$ kN, including safety factors.

Figure 18.8 shows the different ultimate loads measured in experimental tests compared with the design according to EC2. The margin between N_{Rd} and n_u lies between 1.9 and 2.2. It is worth to notice that the decrease of the ultimate load n_u due to second-order moment is far less than the increase of the moment itself. The material safety factors $\gamma_c = 1.5$, $\gamma_s = 1.15$ and the geometrical imperfection e_i being introduced into the calculation of N_{Rd}, the margin of safety between N_{Rd} and n_u (assumed to be a mean value) should include the load factor $\gamma_F = 1.5$ and a factor γ_R that divides the mean value n_u of the ultimate load to obtain the characteristic one. From the smallest ratio $n_u/N_{Rd} = 1.9$, it follows that γ_R must be larger than about 1.3, which is probably realistic. These results confirm the ones from Zhou et al. (2010), which underline the fact that the mechanical behavior is not influenced by the addition of RA up to 50%.

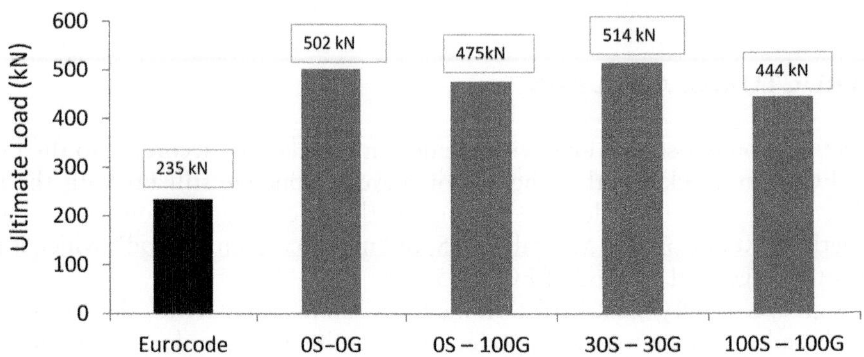

Figure 18.8 Ultimate load for each column tested in comparison with design EC2 prediction.

18.6 CONCLUSION

This study deals with the mechanical behavior of RAC with four mixes having the same mechanical compressive strength class fc28 (but there are slight differences in fc28 between the mixes). First observations tend to show that a low replacement ratio (30S-30G) leads to little change in the ultimate load compared to a control column without RA. It is also noticed that a high replacement ratio of coarse aggregates accompanied with natural sand (0S-100G) does not fundamentally modify the ultimate load n_u of the column (decrease of 5% compared to the control column). In both previous cases, observations suggest that measured strains are different and vary depending on the replacement ratio. Anyway, it would be difficult to draw an obvious conclusion for the 0S-100G and 30S-30G even if the strains measured for the two specimens are larger than the ones for the 0S-0G column. On the other hand, a total replacement of the natural aggregates by recycled ones, particularly the use of recycled sand significantly increases the structural compliance. A fall of the elastic modulus and higher strains and displacements is noted; thence, higher second-order moment for RAC columns may lead to a slight decrease of the ultimate load. These conclusions should be confirmed as only one specimen was tested for each formulation (uncertainty about a characteristic value of n_u). Accordingly, there are still uncertainties about the eccentricity that could lead to larger changes. Calculation of the ultimate design load N_{Ed} according to EN 1992-1-1 shows that the uncertainties on the materials, eccentricity, and characteristic value of n_u seem to be adequately covered by the safety factors γ_c, γ_s and the geometrical imperfection e_i, irrespective of the nature of the aggregates, either natural (NA) or recycled (RA). Obviously, this conclusion was expected for the ultimate limit state (ULS) of short columns. Furthermore, their serviceability limit state is not questioned by the use of RA, even if their modulus and creep factor are higher than that of NAC. On the contrary, the behavior of slender columns deserves special consideration, at least for their ULS, which is the subject of Chapter 19.

Chapter 19

Members in compression—Slender columns

B. Fouré
Consultant

CONTENTS

Abstract

This chapter points out that the buckling behaviour of slender columns may differ significantly when RAC is used in place of NAC, contrary to short columns (chapter 18). The main parameters responsible for the difference are discussed: smaller elastic modulus for RAC; consequently, modified stress-strain law in compression (a model valid for RAC as well as NAC is proposed, which needs only to know the strength and the modulus); creep higher for RAC. Considering the lack of tests of slender reinforced RAC columns, an analogy is done with LWC (Light Weight Concrete) and some comparative tests on slender reinforced LWC and NAC columns are presented, only under short term loading. Furthermore, some parametric calculations by means of the so-called model column method are done. They show that the decrease in buckling load of RAC slender columns with respect to NAC will remain moderate for short term loading (effect of modulus), but may be important for long term loading (cumulative effects of modulus and creep). Finally, the modifications of the EC2 calculation rules at the ULS that would be necessary for RAC are analyzed, such as the limit of slenderness below which the second order effect can be neglected.

19.1 INTRODUCTION

As has been stated in Chapter 18, the behavior of short reinforced columns made of recycled aggregate (RA) concrete (RAC) exhibits only minor differences with that of natural aggregate (NA) concrete (NAC); thence, the ultimate limit state (ULS) calculation rules need not to be modified. On the contrary, slender reinforced columns exhibit a different behavior when RAC is used and may need adaptation of the calculation rules valid for NAC. This difference is mainly due to the smaller Young's modulus of RAC (and thence the modified stress–strain law in compression) and to the higher creep, which result in larger deflection and second-order moment and reduction of the ultimate buckling load. The more or less conventional limit of slenderness between short and slender columns should be modified accordingly.

From the author's knowledge and after a state-of-the-art review, no tests of reinforced RAC slender columns have been performed. But an analogy can be made, to a certain extent, with reinforced lightweight concrete (LWC) columns, for which some buckling tests are reported. The following analysis should thus be checked, so that distinctive provisions between NAC and RAC are appropriately validated.

19.2 MAIN PARAMETERS TO BE CONSIDERED

19.2.1 Elastic modulus

As showed in Chapter 10, the elastic modulus E_c of RAC will be always lower than that of NAC, the larger decrease being about 30% when the ratio of RA is 100% for RAC tested in the French project RECYBÉTON, as well as in some foreign studies.

19.2.2 Stress–strain law in compression

It is needed specially to calculate the ultimate buckling load by the general method (§ 5.8.6 of (EN 1992-1-1 2004), referred to as EC2 subsequently).

Generally speaking, this law (Figure 19.1) is defined by four parameters at least: strength f_c, tangent modulus E_{c0}, peak strain ε_{c1}, and a point on the falling branch (for example, the strain ε_{cu} corresponding to the stress 0.6 or 0.7 f_c). For a given compressive strength f_c, it is physically obvious that the smaller the modulus E_{c0}, the larger the peak strain ε_{c1}. So, the value of ε_{c1} cannot be a function solely of f_c (as in EC2 and other codes and models), but must also depend on the modulus. Such a model was proposed by Fouré (1996, 2005) and Fouré et al. (1996) and is valid for NAC, RAC, LWC, and high strength concrete and ultrahigh performance concrete. In this last situation, the model properly describes the tendency towards linear behavior when the strength increases to very large values; for example, the more performant reactive powder concretes (Dugat et al. 1996). Furthermore, it is generally agreed that the falling branch recorded in an axial compressive test does not

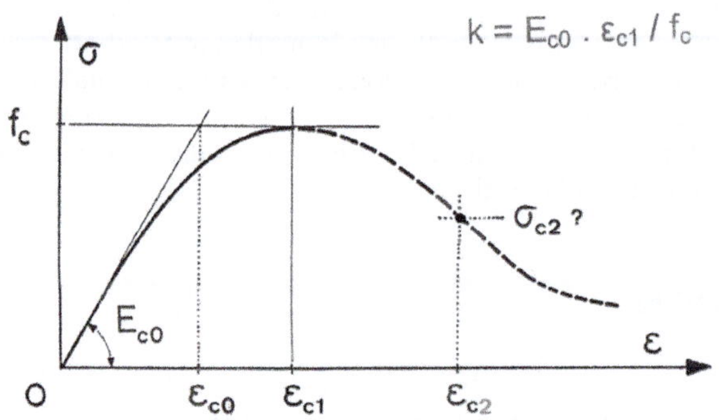

Figure 19.1 Schematic compressive stress–strain curve and main parameters.

have a well-established physical meaning and depends greatly on the test machine and its servocontrol device. So, the value of ε_{c2} for $\sigma_{c2} = 0.7\, f_c$ was derived from the ultimate state of strains in tests of unreinforced columns under eccentric compression. For practical purpose, the model avoids to need calibration based on recording the entire stress–strain curves by means of difficult and expensive tests.

To comply with the EC2 relation between the modulus E_{cm} and the strength f_{cm}, the model is slightly modified for the peak strain:

$$\varepsilon_{c1} = \left[1+0.15k_0/\left(f_c^{\,2}+800\right)\right]\left(f_c^{\,0.7}/k_0\right)\left(f_c \text{ in MPa}\right)$$
$$k_0 = 1.05\,E_c/f_c^{\,0.3} \text{or } k_0 = E_{c0}/f_c^{\,0.3} \tag{19.1}$$

$$\varepsilon_{c2} = \left(1+20/f_c\right)\varepsilon_{c1} \tag{19.2}$$

For RAC as other types of concrete, the relative amount of peak strain increase is smaller than the modulus relative decrease. For example, equations 19.1 and 19.2 are applied to the test results of Cergy's University (Chapter 10) in Table 19.1, where the measured values of f_{cm}, E_{cm}, ε_{c1}, and ε_{c2} are given as well as the ratio test/calculation for the two latter.

The model gives reasonably satisfactory results for ε_{c1}. As may be anticipated, the scatter is greater for ε_{c2}, due to the somewhat unreliable nature of the values measured in axial compression and to the rather oversimple character of equation 19.2. However, it seems that there is a tendency for RAC to give comparatively higher values than NAC, i.e., the postfailure ductility being a little higher.

19.2.3 Creep

As shown in Chapter 11, the creep of RAC is always higher than that of NAC. But the amount of increase varies from one study to another, and is not necessarily proportional to the replacement ratio of RA. Especially, the effect of fine RA may be different from that of coarse RA. The few tests performed for the project RECYBÉTON and the review of other published results show that the long-term creep strain can be increased from 50% to far over 100%, for a ratio of RA 100%. But for a given increase of creep strain ε_{cc}, the increase of the creep factor $\varphi = \varepsilon_{cc}/\varepsilon_i$ will be less, because the instantaneous strain ε_i at loading of RAC is larger than that of NAC, due to the smaller modulus.

Table 19.1 Cergy's compressive tests

Class/Mix	f_{cm} *(MPa)*	E_{cm} *(GPa)*	k_0	ε_{c1} *(10^{-3})*	*Test/calc. equation 19.1*	ε_{c2} *(10^{-3})*	*Test/calc. equation 19.2*
C25/0S-0G	33.8	36.9	13,470	1.67	0.94	2.41	0.85
C25/30S-30G	33.5	29.5	10,800	1.90	0.95	3.04	0.95
C25/0S-100G	32.4	28.1	10,390	1.94	0.96	3.56	1.09
C25/100S-100G	28.6	25.3	9,710	1.95	0.95	3.80	1.09
C35/0S-0G	38.6	42.3	14,840	1.73	1.01	2.43	0.93
C35/30S-30G	40.2	34.9	12,100	1.97	1.03	3.41	1.19
C35/0S-100G	39.1	34.0	11,880	2.11	1.09	3.90	1.33
C35/100S-100G	35.7	30.7	11,030	2.23	1.12	4.24	1.37
Mean/Std. deviation					1.01/7%		1.10/17%

Comparison of the measured values of εc_1 and εc_2 to the calculated ones.

19.3 ANALOGY WITH LWC COLUMNS

19.3.1 Analysis of test results

To obviate the lack of tests for slender RAC columns, it is possible to take advantage of some LWC column tests, as far as only short-term loading is considered. Indeed, compared to NAC, LWC exhibits a smaller modulus and the corresponding modification of the stress–strain law, as do RAC. The buckling behavior is modified accordingly.

Some tests were performed by Fouré (1986) for which more details are given by Kavyrchine et al. (1976). The columns section is 0.29 m wide and 0.145 m high; its mechanical ratio of reinforcement is about 0.2. The main results for the material are given in Table 19.2 for all the concrete batches.

For a same strength f_{cm}, the relative decrease of the tensile strength from NAC to LWC is about 10%. It corresponds rather well to the class 1,8 of EC2, Section 11, which gives a reduction factor $\eta_1 = 0.4 + 0.6\ (\rho/2{,}200) \approx 0.89$ (for the limit value of the class $\rho = 1{,}800\,\mathrm{kg/m^3}$). The decrease of modulus E_{cm} is about 30%. This is the same order of magnitude than the largest decrease observed for RAC in the studies of the project RECYBÉTON (for a ratio of RA 100%), and also the same as the value predicted by EC2, Section 11, which gives a reduction factor (for the limit value $\rho = 1{,}800$) $\eta_E = (\rho/2{,}200)^2 \approx 0.67$. The measured peak strain ε_{c1} is well predicted by the model referred to in Section 19.2.2. The increase of ε_{c1} for LWC over NAC is about 25% for the same f_{cm}.

Only two columns with the highest geometrical slenderness 25 and relative eccentricity 0.15 ($e_0 = 22\,\mathrm{mm}$) are considered here, one with NAC (IU 1, concrete mass $\rho \approx 2{,}300\,\mathrm{kg/m^3}$) and the other with LWC (IU 3, $\rho \approx 1{,}700\,\mathrm{kg/m^3}$). The main results of the buckling tests, ultimate load N_u, and deflection d_u are given in Table 19.3.

For a same strength f_{cm}, the decrease of the ultimate load for LWC with respect to NAC would be only about 14%, whereas the second-order moment would be almost 30% higher

Table 19.2 Material tests results for NAC and LWC

Column	*Strength* f_{cm} *(MPa)*	*Flexure* $f_{ctm,fl}$ *(MPa)*	*Modulus* $E_{cm,0}$ *(GPa)*	k_0	*Peak strain* ε_{c1} (10^{-3})	*Calculated* ε_{c1} *equation 19.1*	*Test/ calculation*
IU 1 (NAC)	40.5	2.7	31.3	10,310	2.15	2.11	1.02
IU 5 "	38.3	2.6	35.0	11,720	2.05	1.94	1.05
IU 2 (LWC)	37.2	2.2	21.4	7,230	2.55	2.62	0.97
IU 3 "	36.4	2.3	21.6	7,350	2.45	2.56	0.96
IU 4 "	37.3	2.1	21.7	7,330	2.55	2.58	0.99
IU 6 "	36.6	2.3	20.7	7,030	2.65	2.64	1.00
IU 7 "	39.5	2.3	21.6	7,170	2.70	2.66	1.02
IU 8 "	41.3	2.0	21.4	7,010	2.60	2.74	0.95
Mean/Std. dev.							0.99/3.4%

Table 19.3 Main results of the columns tests—calculated values of N_u

Column	*Geometric slenderness*	*Load* N_u *(kN)*	*Deflection* d_u *(mm)*	*Peak strain for calculation (affinity* δ*)*	*Calculated load (kN)*	*Test/ calculation*
IU 1 (NAC)	25	753	29	0.0020 (1.00)	729	1.03
IU 3 (LWC)	25	629	35	0.0023 (1.15)	641	0.98
				0.0026 (1.30)	629	1.00
				0.0029 (1.45)	620	1.01

than that of NAC. Of course, this is valid only for these particular values of slenderness and eccentricity, but indicates that the increase of second-order effect for RAC compared to NAC may not be neglected. This conclusion is drawn for short-term loading only. However, the EC2 does not modify the requirements for buckling when applied to LWC. This is probably because the effect of the decrease of modulus is balanced by that of the creep factor that is decreased in the same proportion (EC2: same factor as η_E).

In the present case, according to EC2 formula (5.13)—see Chapter 18 and Section 19.5 hereafter—but without the safety factors, the limit of mechanical slenderness λ under which the second-order effect can be neglected is about 30, i.e., geometrical about 8.5. The tested columns are far above this limit for NAC. A question is raised: is it necessary to modify this limit for RAC, and how to do it?

19.3.2 Calculation with an affinity on strain ε_c

The ultimate load N_u of the NAC column can be predicted by the so-called Faessel method (Faessel et al. 1971) or EC model-column method (see also Chapter 18). It uses the stress–strain law of Sargin, with $\varepsilon_{c1} = 0.002$ and $E_{c0} = 1{,}000\ f_c$ for short-term loading, which is convenient for the NAC column IU 1. For the RAC column IU 3, a factor of affinity $\delta > 1$ is applied on the strain: $\varepsilon_{c1} = 0.002\ \delta$, thence on the entire curve (E_{c0} divided by δ). Three values of δ are considered: 1.15 (ratio of the peak strain RAC/NAC), 1.45 (ratio of the modulus NAC/RAC), and 1.30 (intermediate value). The results for N_u in Table 19.3 show that, in this particular case, there is no important difference between the three calculations and that the affinity governed by the modulus is the better. The calculation of N_u using the factor of affinity δ = NAC modulus/RAC modulus will be probably always safe.

19.4 PARAMETRIC STUDY OF THE DECREASE OF BUCKLING LOAD FROM NAC TO RAC

There are mainly two features that lead to an increase of strains, and hence of the deflection and the second-order moment.

- The decrease of the modulus is characterized by the ratio $\alpha = E_c(\text{RAC})/E_c(\text{NAC}) < 1$. Correlatively, the peak strain increases with a relative amount, a little less than $1/\alpha$.
- The increase of the creep factor φ is characterized by the ratio $\beta = \varphi(\text{RAC})/\varphi(\text{NAC}) > 1$. For buckling calculations, φ is the long-term value for an age of loading t_0: $\varphi(\infty; t_0)$; generally t_0 = 28 days. The EC2 allows to modify the stress–strain law by an affinity factor:

$$1+\varphi_{\text{ef}} = 1+\varphi(\infty;t_0)\,M_{0\text{Eqp}}/M_{0\text{Ed}}$$

$M_{0\text{Eqp}}/M_{0\text{Ed}}$ being the ratio of the moments due to quasi-permanent loads (qp) and design loads (d). For the following calculations, a value of 0.7 was assumed.

A limited number of calculations were done to compare the buckling load N_u of either RAC or NAC rectangular columns, by means of the method quoted earlier (Faessel et al. 1971), with different values of the geometrical slenderness l_0/h, the relative eccentricity e_0/h, the mechanical ratio of reinforcement ω, and assuming that the peak strain is multiplied by $1/\alpha$ for RAC (assumption on the safe side).

For short-term loading, the results show that the decrease of buckling load due to the smaller modulus of RAC will be important only for high slenderness, small eccentricity, and small reinforcement. For example, the decrease is about 25% for $\alpha = 0.7$, $l_0/h = 40$, $e_0/h = 0.1$, $\omega = 0.1$; with $\omega = 0.6$, the decrease is only about 10%. Possibly, it could be often neglected.

For long-term loading, the results of the calculations are not always reliable because of some hazardous extrapolations. However, it can be ascertained that, due to the combined effects of modulus and creep, the decrease could be in certain cases very important and rarely neglected, if simultaneously α is small and β high. For example, with the same values of the parameters as mentioned earlier and a large value for the increase of the creep factor, $\beta = 2.5$ (which will be probably on the safe side), the decrease of the buckling load is only about 25% for $\omega = 0.6$, but much greater for $\omega = 0.1$.

19.5 POSSIBLE MODIFICATIONS IN EC2

19.5.1 Slenderness criterion for isolated members (EC2, §5.8.3.1)

The slenderness limit under which the second-order effect can be neglected is defined by the formula (EC2, 5.13):

$$\lambda_{lim} = 20 \cdot A \cdot B \cdot C/\sqrt{n}$$
$$A = 1/(1 + 0.2\,\varphi_{ef})$$
$$B = \sqrt{(1 + 2\omega)} \text{ Mechanical ratio of reinforcement } \omega = A_s{\cdot}f_{yd}/A_c{\cdot}f_{cd}$$
$$C = 1.7 - r_m,\ r_m = M_{01}/M_{02} \text{ ratio of the end moments}$$
$$n = N_{Ed}/A_c{\cdot}f_{cd}$$

The factor A depends on creep; the factors B and C do not depend on the type of concrete. So, for the same creep factor φ_{ef} but different modulus, RAC and NAC columns should have the same slenderness limit, which is probably not acceptable. A safe solution may be to multiply A by α (see Section 19.4).

19.5.2 Creep (EC2, §5.8.4 (4))

Among the three conditions to satisfy for neglecting the creep effect ($\varphi_{ef} = 0$), the second on the slenderness might be modified by reducing its value in the same ratio as the slenderness limits of EC2, §5.8.3.1:

$$\lambda \le 75\alpha\left[1+0.2\varphi_{ef}(\text{NAC})\right]/\left[1+0.2\varphi_{ef}(\text{RAC})\right]$$

19.5.3 Nominal stiffness (EC2, §5.8.7.2) and nominal curvature (EC2, §5.8.8.3)

Modifications might be necessary, but rather difficult to define without doing some parametric studies.

19.5.4 Lateral instability of slender beams (EC2, §5.9)

The slenderness limit of formulas (EC2, 5.40 a and b) might be multiplied by the ratio of the slenderness limits, as in Section 19.5.2:

$$l_{0t}/b \le \left[(50\,\text{or}\,70)/(h/b)^{1/3}\right]\alpha\left[1+0{,}2\varphi_{ef}(\text{NAC})\right]/\left[1+0{,}2\varphi_{ef}(\text{RAC})\right]$$

It will be necessary to verify that such a modification is not oversafe.

19.6 CONCLUSION

There are clear trends for RAC to have a smaller modulus and a larger creep factor than NAC. The consequence is a decrease of the ultimate buckling load of reinforced RAC slender columns with respect to NAC. For short-term loading, the decrease will be moderate, which is confirmed by some test results on reinforced LWC columns that have an analogy with RAC (such tests are lacking for RAC). For long-term loading, the decrease may be important, but its amount is difficult to quantify, due to the lack of tests and the insufficient reliability of some of the simplified calculations that have been done.

The calculation of the buckling load by the so-called general method for NAC in EC2 does not cause any problem for RAC as far as its compressive stress–strain law and its creep factor are known. An empirical but very general model is proposed for the compressive law, which needs to know only the strength and the modulus, as affected by the RA ratio.

Other parts of the EC2 design rules for buckling (such as the limit of slenderness below which the second-order calculation can be omitted) need probably to be modified for their application to RAC, in cases where significant decrease of the modulus and increase of the creep factor with respect to NAC will be confirmed.

Chapter 20

Beams under bending

H. Mercado-Mendoza, K. Apedo and P. Wolff
Université de Strasbourg/INSA

B. Fouré
Bougival

CONTENTS

Abstract

This chapter reports a study carried out in order to: *(i)* evaluate the impact of the utilization of recycled concrete aggregates on the flexural behavior of reinforced concrete beams and *(ii)* assess the applicability of the Eurocode 2 to estimate the parameters characterizing the mentioned behavior. Four-point bending tests were performed on beams manufactured with four different concrete mixes. Three of them contained recycled aggregates while the fourth one was a reference containing only natural aggregates. The different mechanical properties of concrete and steel used to manufacture the beams were measured through standard tests. Deflection, crack spacing and crack width along the beams as well as loading force and displacement were monitored during the bending tests. The recorded parameters were compared to the values determined using the Eurocode 2 prescriptions. An impact of the use of recycled aggregates was detected on all the serviceability-limit-state parameters that were investigated. The Eurocode 2 specifications gave quite accurate estimations of the experimental deflections, independently of the recycled aggregate content, provided that the actual mechanical properties of the materials are used as inputs. As for the cracking parameters, the Eurocode 2 expressions seem to overestimate the actual measured values on the beams containing recycled aggregates. A theoretical analysis of the Eurocode 2 cracking model is made so as to explain this discrepancy. Finally, no noticeable effect of the incorporation of recycled aggregates on the flexural strength and the failure mode of the beams was demonstrated. Accordingly, the Eurocode 2 approach applies well in order to estimate the failure parameters of recycled aggregate concrete beams tested in this work.

20.1 INTRODUCTION

Several studies have been carried out dealing with fresh and hardened recycled aggregate concrete (RAC) properties (Xiao et al. 2006b; Otsuki et al. 2003; Domingo-Cabo et al. 2009; Guerra et al. 2014; Wardeh et al. 2015a), see also Chapter 10. However, research on RAC structural members appears to be rather scarce. In particular, as far as reinforced concrete beams are concerned, along with the pioneering works of Mukai and Kikuchi (1988) and Yagishita et al. (1994), a limited number of studies are found. These studies (Arezoumandi et al. 2015; Ajdukiewicz and Kliszczewicz 2007; Bai and Sun 2010; Knaack and Kurama 2015; Wardeh and Ghorbel 2015c) state in general that the ultimate strength of flexural members is practically not affected by the incorporation of recycled concrete aggregates (RA). Also, the experimental results presented in those works—for different reinforcement ratios of the members—seem to evidence that the inclusion of RA has an impact on the deformation parameters of the members considered. Thus, an increase of deflection during bending, somehow proportional to the RA content, is ascertained. Notwithstanding, controversy raises in relation to these deflection-related observations (Fathifazl et al. 2009; Ignjatovic et al. 2013).

Concerning cracking of RAC beams, even fewer and contradictory information has been found, especially from a quantitative point of view. Arezoumandi et al. (2015), Kang et al. (2014), and Wardeh and Ghorbel (2015c) state that the crack spacing diminishes with the RA content, whereas Bai and Sun (2010) conclude that RA inclusion does not significantly modify the aforesaid spacing. Sato et al. (2007) and Zhang and Zhao (2016) are among the rare groups to propose accurate measurements of crack widths on RAC members submitted to bending. Nevertheless, the former present quite scattered results, which moreover shows no clear trend about the influence of RA utilization. The latter express their results in terms of the sum of the widths of several cracks altogether, while no information about the measured maximum and average crack widths is given.

Furthermore, to envisage the feasibility of a normalized design of RAC structural members, a comparison between the experimental data and the predictions derived from standard code prescriptions is requested. Again, to our knowledge, a lack of attempt in this sense prevails. The comparison made by Sato et al. (2007) between their experimental measurements of cracking parameters and the estimated values following the Japanese specifications (JSCE 2002) showed a considerable discrepancy. Arezoumandi et al. (2015) compared their experimental deflection and ultimate strength data with the estimations of both the Eurocode 2 (NF EN 1992-1-1 2005) and the American (ACI 318R-11 2011) standards. Again, poor congruity was found thereby.

Accordingly, this section reports a study carried out within the project RECYBÉTON on (i) the impact of RA utilization on the behavior of structural members in bending and (ii) the applicability of the Eurocode 2 specifications to estimate the different parameters related to RAC beams under bending. With this aim, full-scale flexure tests were performed on reinforced beams. Those were manufactured using three different concrete mixes, containing different amounts of RA, together with a fourth formulation reference containing only natural aggregates. Several parameters were monitored during the tests and were compared, with the corresponding values estimated following the Eurocode 2 stipulations.

20.2 EXPERIMENTAL STUDY

20.2.1 Definition of concrete types

Four types of concrete were used for this study. The main purposes leading to their choice were as follows: to have a reference mix containing only natural aggregates, to dissociate

Table 20.1 Concrete types

Mix designation (class C25/30)	Mass substitution (%) Recycled sand	Recycled coarse aggregates	Overall mass substitution (%)
0S-0G	0	0	0
30S-30G	30	30	30
0S-100G	0	100	52
100S-100G	100	100	100

the effects of sand and fine gravel, and to test the concrete with only recycled aggregates. Mixes tested previously within a complementary study of the RECYBÉTON project (cf. Chapter 17) were also considered. The four concrete mixes, in terms of the content of recycled aggregates, are presented in Table 20.1. These concretes belong to the Eurocode 2 C25/30 strength class (commonly used in the civil engineering domain).

20.2.2 Bending test characteristics

The Eurocode 2 prescriptions on flexural members at serviceability limit states (SLSs) are mainly developed on the basis of reinforced concrete section under pure bending. Thus, a four-point bending test was carried out in this work on four reinforced concrete beams made of the aforementioned concrete mixes.

To keep a representative scale of civil engineering constructions, 5-m long beams were tested (cf. Figure 20.1). Accordingly, the beams had a cross section of 40 cm × 15 cm so as to have appropriate slenderness (with regard to expected cracking and deflection parameters) and shape (to avoid any risk of lateral buckling).

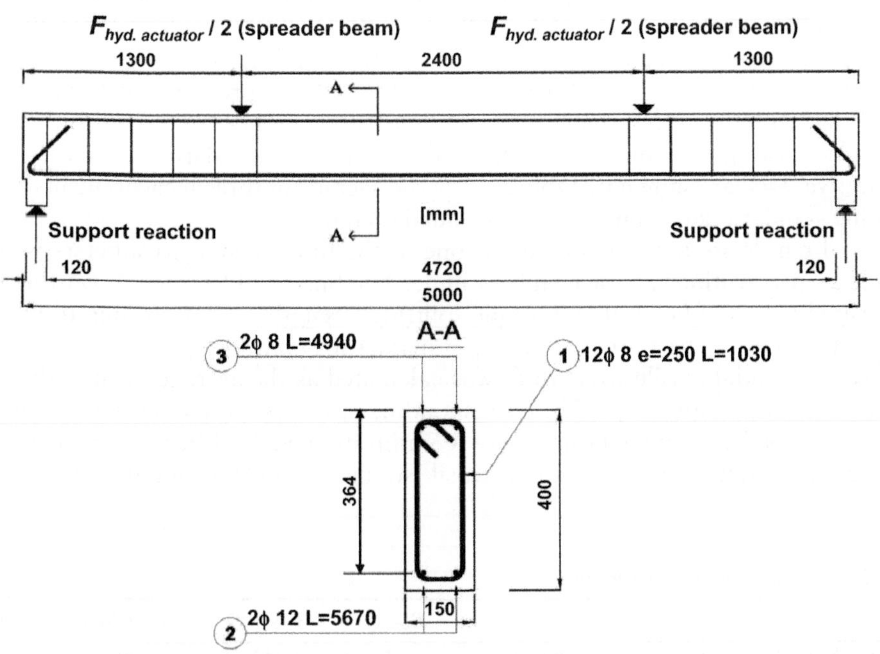

Figure 20.1 Four-point bending test and reinforcement configuration.

In practice, the use of over reinforced beams (brittle failure due to concrete compression) is quite uncommon because, in this type of members, steel reinforcement is not used in an optimal way (cf. for instance, Granju (2012) and Ricotier (2012)). Furthermore, the influence of recycled aggregates on the behavior of concrete under compression is discussed elsewhere in the RECYBÉTON project (Chapter 18). Hence, for the work described here, underreinforced beams were used. Moreover, this type of members (which fail after rebar yielding) allowed to point out parameters related to the SLSs, since a moderate reinforcement ratio was used.

Throughout the definition of reinforcement characteristics, two essential criteria were followed, namely (i) the ratio between the bending moment supposed to generate the first crack in concrete and the bending moment which should produce the lowest reference width of a crack according to the Eurocode 2 and (ii) the ratio between the bending moment that should cause the collapse of the member and the bending moment which should produce the highest reference crack width prescribed by the Eurocode 2.

Thus, the longitudinal reinforcement of the samples consists of two ribbed 12-mm diameter rebars (cf. Figure 20.1), which account both for the nonbrittleness condition and the minimal reinforcement with respect to the crack control.

As for the shear reinforcement, to avoid any correlation between crack spacing and the position of the stirrups, the latter was essentially placed outside the zone between the loading points of Figure 20.1 (the so-called pure bending zone). This shear reinforcement was defined according to the Eurocode 2 specifications. Nevertheless, the strength of the unreinforced central zone was verified so as to withstand the shear force generated by the self-weight of the beam.

20.2.3 Materials

The four types of concrete that were used in this study were designed, mixed, and cast following the requirements of the RECYBÉTON project, in which reference mix proportions and a preparation procedure were defined (Sedran 2017) (see Appendix). Table 20.2 shows the characteristics measured on fresh concrete and the standards that were followed.

For each concrete mix, at least three specimens for each test presented in Table 20.3, were cast (together with the corresponding beam) to measure a number of mechanical properties 28 days after casting. The measured values, along with the standard followed to perform the respective tests, are shown in Table 20.3 as the average of three measurements. The corresponding standard deviations are also included therein.

The axial tensile strength of concrete is one of the fundamental parameters within the Eurocode 2 specifications about members under bending at SLSs. Hence, on the basis of the measured values of flexural and tensile splitting strengths and according to the related relationships given in the Eurocode 2: $f_{ct,fl} = \max\{(1.6 - h/1{,}000)f_{ct}; f_{ct}\}$ and $f_{ct} = 0.9f_{ct,sp}$, respectively, the axial tensile strength f_{ct} was calculated as the average of the values resulting from these two relationships. The corresponding f_{ct} values are also shown in Table 20.3.

With the aim of determining the impact of RA utilization on the different concrete mechanical properties, the ratios between both the tensile strength f_{ct} and the modulus of elasticity E_{cm}

Table 20.2 Fresh concrete characteristics

Mix		*0S-0G*	*30S-30G*	*0S-100G*	*100S-100G*
Slump (NF EN 12350-2 2012)	*h* (mm)	67	113	65	168
Air content (NF EN 12350-7 2012)	*A* (%)	1.4	1.7	2.5	3.1

Table 20.3 Concrete mechanical properties (in parentheses: standard deviation)

Mix		0S-0G	30S-30G	0S-100G	100S-100G
Bulk density	ρ_a (kg m^{-3})	2,346 (20)	2,251 (8)	2,188 (9)	2,041 (19)
Compressive strength (NF EN 12390-3 2012)	f_{cm} (MPa)	30.1 (2.6)	27.6 (0.4)	26.4 (0.2)	23.7 (2.1)
Tensile splitting strength (NF EN 12390-6 2012)	$f_{ctm,sp}$ (MPa)	2.8 (0.3)	2.5 (0.2)	2.2 (0.1)	2.0 (0.1)
Flexural strength (NF EN 12390-5 2012)	$f_{ctm,fl}$ (MPa)	4.6 (0.2)	3.9 (0.2)	4.1 (0.1)	3.2 (0.2)
Calculated tensile strength	f_{ctm} (MPa)	2.8	2.4	2.3	1.9
Secant modulus of elasticity (prEN 12390-13 2013)	E_{cm} (GPa)	31.8 (1.1)	23.0 (1.3)	24.2 (1.1)	15.3 (1.1)
	$(f_{ctm}/f_{cm}^{2/3})/10^{-1}$	2.9	2.6	2.6	2.3
	$(E_{cm}/f_{cm}^{0.3})/10^4$	1.1	0.9	0.9	0.6

and the compressive strength f_{cm} (affected by the exponents from Eurocode 2, Table 3.1) were calculated for different tested formulations (cf. Table 20.3). When comparing these ratios, one notices that the impact of RA seems even more pronounced on f_{ctm} and E_{cm} than on the compressive strength (see Chapter 10). In fact, in our case, both ratios significantly decrease as a function of the overall recycled aggregate content. It is worth mentioning that both the tensile strength and the modulus of elasticity are important factors when it comes to the assessment of serviceability parameters such as cracking and deformation (see Section 20.3).

Tensile strength tests were performed on samples of the ribbed steel rebars following the ISO 6892-1 (2009) standard. The results are summarized in Table 20.4 as the mean values of three measurements together with the corresponding standard deviations.

20.2.4 Bending test protocol

Load was applied to the tested beams using a 640 kN hydraulic actuator that was fastened to an autostable pendular gantry (cf. Figure 20.2). In addition, a spreader beam was allowed to exert the load along two lines of contact, according to the configuration of the four-point bending test (cf. Figure 20.1). The supports of the beam under test as well as its load-contact mechanisms were hinged-type ones.

During the execution of the bending test, in addition to the force developed by the hydraulic actuator and the correspondent displacement, other parameters characterizing the specimen response in bending were monitored by means of electrical resistive and inductive sensors. On the one hand, the deflection of the beam, the strain at its top concrete fiber, and the strain of its reinforcement bars were measured at midspan. On the other hand, the widths of the different cracks along the beam were also recorded by means of inductive clip-on sensors.

With the aim of validating the measurements of deflection and crack widths mentioned earlier, digital image correlation technique was used. In fact, through computer processing

Table 20.4 Reinforcement steel mechanical properties (in parentheses: standard deviation)

Modulus of elasticity	E_s (GPa)	201 (5)
Yield stress	$f_{0.2k}$ (MPa)	555 (9)
Tensile strength	f_t (MPa)	624 (12)
$k = f_t/f_{0.2k}$	—	1.12 (0.01)
Elongation at maximum force	ε_{uk} (%)	5.44 (0.35)

Figure 20.2 Bending test setup.

of images taken during the bending tests, this technique has permitted to determine the fields of strain and displacements along the specimens.

The experimental procedure followed to perform the bending tests comprised several phases, as described hereafter.

- *Precracking:* first, an initial low-level load was applied to the specimen, within the precracking stage, to produce barely visible cracks along the pure bending zone. Consequently, several inductive sensors, used to measure the width of the sharpest cracks, were positioned.
- *Initial cracking:* second, two loading steps, similar to those shown in Figure 20.3 for the final cracking phase described later, were applied throughout an initial cracking phase. With reference to the widest crack along the beam, the load level of each plateau generated a 0.2- and a 0.3-mm wide crack, respectively.
- *Permanent loading:* after the end of the initial cracking phase, a permanent loading phase took place. This phase was intended to assess, to a certain extent, the impact of sustained loads on the response of members in flexure, especially regarding the SLS parameters. Thus, the hydraulic actuator was immobilized (imposed displacement) for approximately 50 h at the position reached at the end of the aforementioned second plateau, to reproduce a permanent load during this time. Then, the load was completely removed.
- *Final cracking:* next, the two loading steps mentioned earlier (initial cracking phase) were applied again, followed by a third one corresponding to a 0.4-mm wide crack (cf. Figure 20.3). Indeed, within the Eurocode 2 prescriptions related to the SLS, three main reference crack widths ($w_{k,ref}$) are dealt with (NF EN 1992-1-1 2005, §7.3.1 and §7.3.3), namely 0.2, 0.3, and 0.4 mm.
- *Failure:* following the three final loading steps, the last phase of the bending test, characterized by the failure of the beam, was launched. In this phase, the load was gradually increased until it visibly began to reach a final plateau, where the measured force markedly attained an upper limit, even though the actuator continued to move on before the test ended (cf. Figure 20.3).

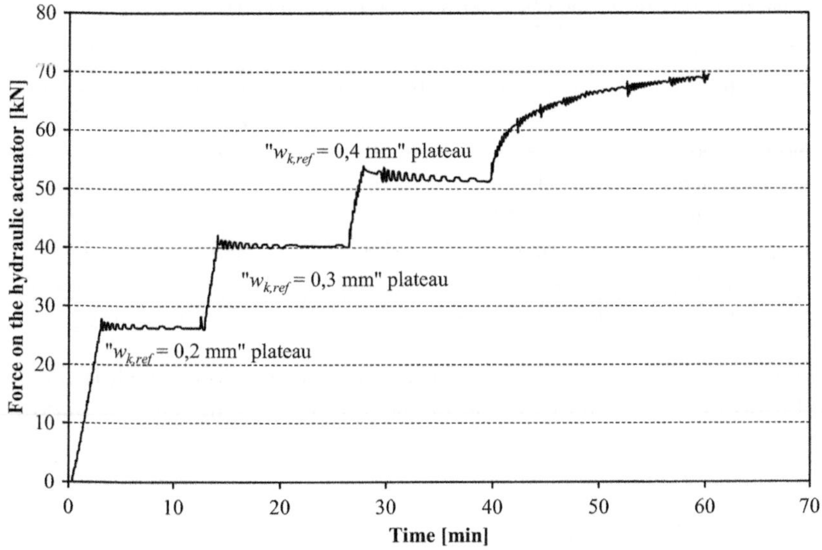

Figure 20.3 Bending test procedure—typical final cracking and failure phase.

20.3 RESULTS AND DISCUSSION

One should note that all the remarks stated in this document are subjected to precaution, because only one specimen was tested for each concrete mix.

20.3.1 Spacing between the cracks

A typical lateral view of the test specimens (pure bending zone) at the end of the bending test is shown in Figure 20.4. It allows to observe the characteristic cracking configuration due to bending. Accordingly, the mean spacing ($s_{r,mean,exp}$) between the different vertical cracks, within the pure bending zone, was determined by taking into account the main cracks (the sharpest ones on Figure 20.4) that were only present throughout the whole initial and final cracking phases (cf. Section 20.2.4), i.e., for service loading levels. Also, the maximal spacing between two contiguous cracks ($s_{r,max,exp}$) and the number of cracks inside the aforesaid zone were recorded. These parameters are summarized in Table 20.5 (in parentheses: standard deviation) for the different tested beams. The general trend of these results seems to correspond with those reported in previous studies (see references in Section 20.1) i.e., the spacing between the cracks diminish with the overall recycled aggregate content. As seen hereafter, this is plausibly due to the increase of the ratio of bond over tensile strength with the overall recycled aggregate content.

The results displayed in Table 20.5 suggest that the experimentally measured values of the spacing between the cracks, and in particular, the mean spacing $s_{r,mean,exp}$, seem to decrease with the overall RA content.

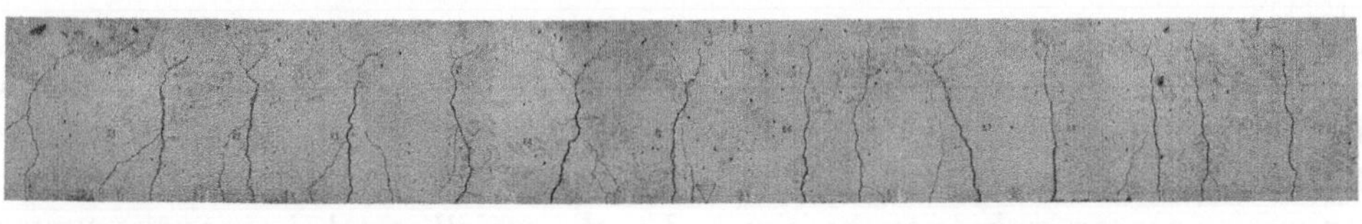

Figure 20.4 Typical cracking configuration.

Table 20.5 Registered and calculated spacings between cracks

Beam concrete mix	*0S-0G*	*30S-30G*	*0S-100G*	*100S-100G*
$s_{r,mean,exp}$ (cm)	20.2 (5.8)	16.6 (4.4)	15.6 (5.1)	14.5 (2.8)
$s_{r,max,exp}$ (cm)	27.0	26.0	26.4	20.0
No. of cracks	12	14	15	17
$s_{r,max,EC2,bend}$ (cm)	21.2			
$s_{r,max,EC2,bend}$[a] (cm)	9.0 + 12.2			
$s_{r,max,EC2,k2}$ (cm)	30.0	29.9	29.9	29.7
$s_{r,max,EC2,k2}$[a] (cm)	9.0 + 21.0	9.0 + 20.9	9.0 + 20.9	9.0 + 20.7

[a] The values corresponding to the two terms of equation (20.1) are shown separately.

When cracking along a beam is governed by rebars in tension (which is the case in this work), the Eurocode 2 prescribes the following equation to calculate the maximum crack spacing (NF EN 1992-1-1 2005, equation (7.11)):

$$s_{r,max} = k_3 \cdot c + k_1 \cdot k_2 \cdot k_4 \cdot \frac{\phi}{\rho_{p,eff}}, \tag{20.1}$$

where c is the cover to the longitudinal reinforcement, ϕ is the nominal diameter of the longitudinal reinforcement bars, $\rho_{p,eff}$ is the reinforcement ratio of the effective area of concrete in tension (cf. NF EN 1992-1-1 2005, equation (7.11)).

Within the expression given by equation (20.1), the first term allows for the ineffective anchorage zone of the rebar near the cracks (where there is a lack of support for the concrete struts that allow the anchorage), with k_3 specified as a constant. The second term accounts for the interaction of the rebar and its surrounding concrete stresses, from one crack to the contiguous one, where the tensile strength is attained. Thus, the product $k_1 \cdot k_4$—whose influence on the calculated spacing is detailed hereafter—concerns the relationship between the tensile strength of concrete and the rebar-concrete bond stress.

With regard to k_2, it is a coefficient that accounts for the shape of the diagram of tensile stress in concrete between the cracks. The Eurocode 2 prescriptions seem to be rather ambiguous as for the value that should be taken for this coefficient. On the one hand, "for bending" (which is actually the case in this study), a value of 0.5 is specified. On the other hand, "for local areas" the prescribed value of k_2 is to be calculated from the largest and the smallest stresses in the effective area of concrete in tension. Hence, as shown in Table 20.5 for each specimen tested in this work, two different values of maximum crack spacing were calculated according to the two mentioned approaches, i.e. $s_{r,max,EC2,bend}$ ($k_2 = 0.5$) and $s_{r,max,EC2,k2}$ ($0.5 < k_2 < 1$), respectively. Note that these calculations were performed using the experimentally determined characteristics of the materials (cf. Section 20.2.3).

The 0S-0G beam (which contains only natural aggregates) is taken as a reference, and one can notice from Table 20.5 that the calculated value $s_{r,max,EC2,bend}$ (stipulated by the Eurocode 2 for members in bending) is very close to the experimental mean spacing between cracks $s_{r,mean,exp}$. Furthermore, the maximum spacing between two contiguous cracks $s_{r,max,exp}$ seems to be well predicted by the value determined rather on the basis of the tensile stress diagram $s_{r,max,EC2,k2}$.

Regarding the specimens containing recycled aggregates, it can be seen from Table 20.5 that the Eurocode 2 relationships seem to overestimate the values of spacings between cracks measured during the bending tests, and, in particular, the mean spacing between cracks $s_{r,mean,exp}$ (as evidenced by the number of cracks recorded in each case). Moreover, this overestimation appears to be more pronounced as the overall recycled aggregate content is higher. Thus, at first sight, the cases of 30S-30G and 0S-100G seem to be similar (cf. Tables 20.1 and 20.5). However, one should note that, by reason of the slightly lower value of the mean spacing between the cracks of 0S-100G beam with reference to that of 30S-30G beam, the former has one crack more than the latter.

In an attempt to explain the trends exhibited in Table 20.5, one should recall the aforestated $k_1 \cdot k_4$ product of equation (20.1). As a matter of fact, one can demonstrate that the spacing between the cracks along a reinforced concrete member in bending is (i) directly proportional to the tensile strength of concrete and (ii) inversely proportional to the bond stress that the rebar/concrete interface is able to generate during flexure. Actually, the $k_1 \cdot k_4$ product stands for the ratio between the aforementioned tensile strength and bond stress. Now, by comparing Tables 20.3 and 20.5, one can remark that the variation of the tensile strength of the different concretes follows an identical trend (diminishes with an increase of the overall recycled aggregate content), from a qualitative point of view, as that of the spacing between their cracks. Above all, with regard to another study carried out within the RECYBÉTON project (cf. Chapter 17), one could infer that the bond stress does not vary following the aforestated trend. The results presented therein seem to suggest that the variation of the bond stress appears to be practically negligible with regard to the tensile strength presented here. As a consequence, it seems conceivable to ascribe the decrease of spacings between the cracks along the beams (as a function of their overall recycled aggregate content) highlighted in Table 20.5, to the diminution of the tensile strength of the corresponding concretes.

The approach stipulated in the Eurocode 2 (NF EN 1992-1-1 2005, §7.3.4) consists in assigning a constant value to the $k_1 \cdot k_4$ product of equation (20.1) (and so implicitly to the aforesaid ratio between the tensile strength and the bond stress), irrespective of both the concrete and the concrete/steel interface mechanical properties. In that sense, this approach likely induces that the diminution of tensile strength of concrete with its recycled aggregate content is probably neglected. As a plausible result, the spacing between cracks evaluated through the equation (20.1) overestimate the actual measured values, as stated earlier. Nevertheless, it is worth noticing that statistical studies dealing with crack spacing and widths of natural-aggregate concretes (cf. for instance, Robinson and Morisset (1969); their conclusion can be extended in a large field of concrete strength) showed little influence of the concrete properties on those parameters.

The observed tendency of the crack spacing to decrease when the ratio of RA increases was also reported by several other authors (cf. Section 20.1).

20.3.2 Crack widths

The evolution of the crack widths along the 0S-0G beam, measured using the inductive sensors mentioned in Section 20.2.4, as a function of the load applied on the specimen is shown in Figure 20.5. Therein, the continuous-line curves represent the behavior of the widest crack along the beam during the initial cracking phase ($w_{max,exp,INI}$) and the final cracking one ($w_{max,exp,FIN}$). Also, the dashed-line curves $w_{mean,exp,INI}$ and $w_{mean,exp,FIN}$ show the mean value of the crack widths measured, respectively, during the two mentioned phases within the pure bending zone. These measurements—as well as those corresponding to the deflection of the specimen considered—were compared to those obtained, thanks to the Digital Image Correlation technique, and the congruency was quite satisfactory (the discrepancies are within the range of accuracy of the experiments).

Following the Eurocode 2 prescriptions, the crack width (w_k) is to be calculated using NF EN 1992-1-1 (2005, equation (7.8)):

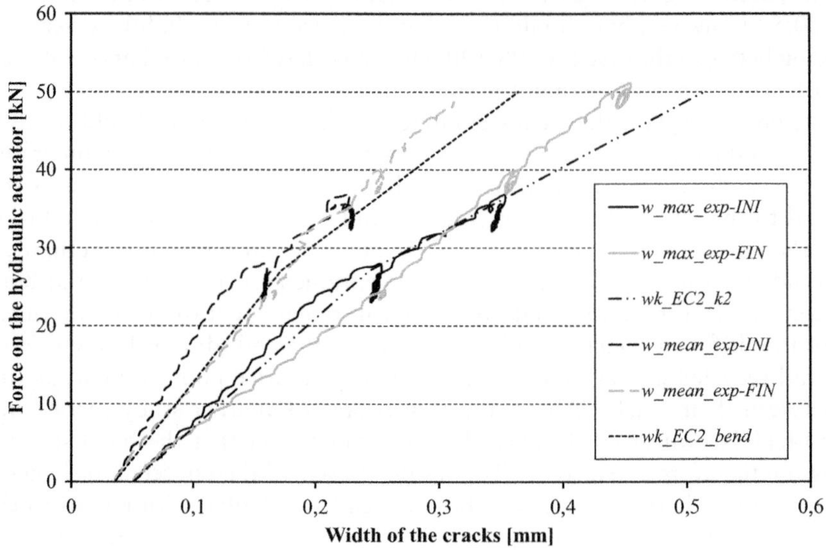

Figure 20.5 Crack widths—0S-0G beam.

$$w_k = s_{r,max}\left(\varepsilon_{sm} - \varepsilon_{cm}\right), \tag{20.2}$$

where $s_{r,max}$ is the maximal crack spacing treated in the previous section and the second factor stands for the difference of elongation between the rebar and its surrounding concrete from one crack to the contiguous one. The Eurocode 2 prescribes to obtain this strain difference as follows (NF EN 1992-1-1 2005, equation (7.9)):

$$\varepsilon_{sm} - \varepsilon_{cm} = \frac{\sigma_s - k_t \dfrac{f_{ct,eff}}{\rho_{p,eff}}\left(1 + \alpha_e \cdot \rho_{p,eff}\right)}{E_s} \geq 0.6\frac{\sigma_s}{E_s}, \tag{20.3}$$

where σ_s is the stress in the rebar at a given load level, k_t is an adjustment coefficient coming from the modeling of the behavior of the partially cracked bent member, $f_{ct,eff}$ is the effective tensile strength of concrete at the moment of the crack width calculation (equal to f_{ctm} if concrete is at least 28 days old), and α_e is the ratio between the steel and concrete moduli of elasticity E_s/E_{cm}.

As stated in equation (20.2), the theoretical crack width is a function of the maximal crack spacing $s_{r,max}$ [cf. equation (20.1)]. So, as proceeded in the previous section, two different curves, $w_{k,EC2,bend}$ and $w_{k,EC2,k2}$, have been determined using, respectively, $s_{r,max,EC2,bend}$ and $s_{r,max,EC2,k2}$ for the calculation of the crack width at each level of loading held by 0S-0G beam. These two graphs are also included in Figure 20.5. The actual measured properties of the materials, presented in Tables 20.3 and 20.4, were used to carry out the corresponding calculations.

Note that the theoretical $w_{k,EC2,bend}$ and $w_{k,EC2,k2}$ curves were determined taking account of the test specimen self-weight (whose impact is nonnegligible), whereas, for unavoidable technical reasons, the measurements giving rise to the experimental curves do not include the self-weight effect. In fact, the cracks along the beam were visible only after a certain precracking load was applied, which allowed to place the corresponding sensors. Therefore, once the precracking load was completely removed, the self-weight of the beam gave ineluctably a zero level for the sensors. Hence, for the sake of comparison, the experimental curves shown in Figure 20.5, as well as those concerning crack widths (Figures 20.6–20.8)

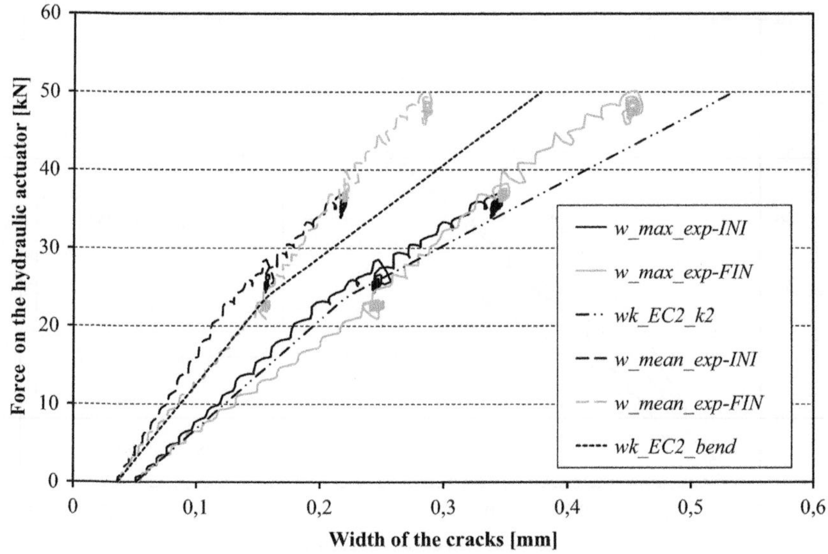

Figure 20.6 Crack widths—30S-30G beam.

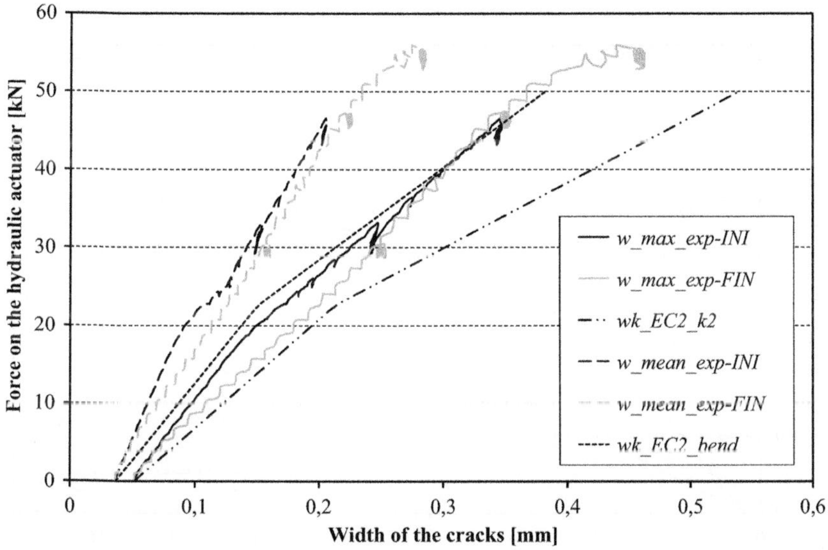

Figure 20.7 Crack widths—0S-100G beam.

and beam deflections (Figures 20.9–20.12) hereafter, were shifted with and offset on the abscissa so as to correspond with the theoretical ones with a similar shape. This supposes to admit linearity for low levels of loading.

As noticed in the case of crack spacings, as far as the reference beam is concerned, Figure 20.5 exhibits very good agreement between the estimations of the Eurocode 2 and the experimental data, especially during the initial cracking phase. In a logical way, as $s_{\mathrm{r,max,EC2,k2}}$ matches the measured maximal crack spacing, $w_{\mathrm{k,EC2,k2}}$ retraces the widest crack measurements very well. In the same way, the mean value of the measured crack widths is estimated by $w_{\mathrm{k,EC2,bend}}$ in a satisfactory way. In spite of a change in the shape of the curves

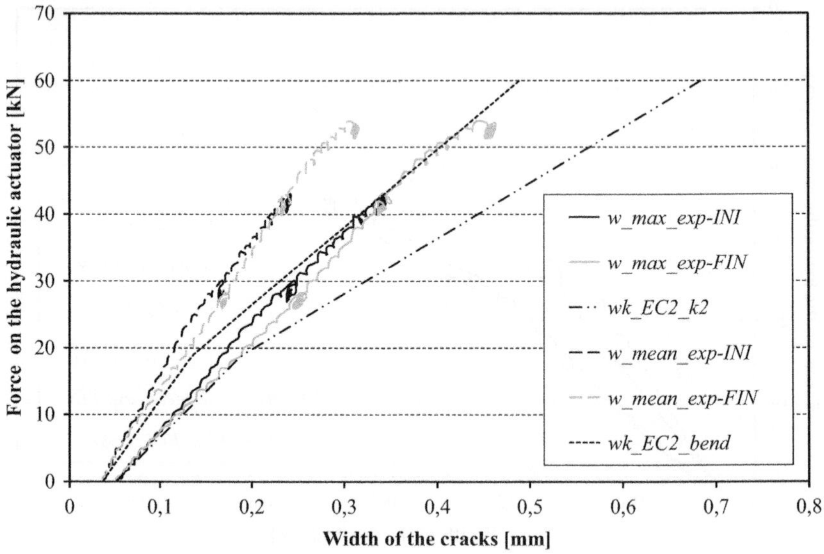

Figure 20.8 Crack widths—100S-100G beam.

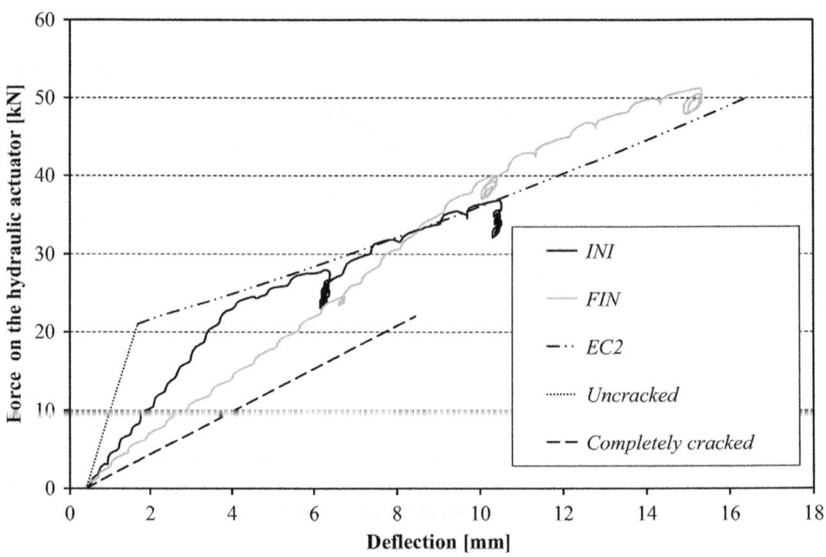

Figure 20.9 Deflection—0S-0G beam.

related to the final cracking phase, the estimations remain adequate, in particular, in terms of the orders of magnitude.

Analogously to Figure 20.5, the response of the specimens containing recycled aggregates are shown in Figures 20.6–20.8 for 30S-30G, 0S-100G, and 100S-100G, respectively. For the sake of comparison between the different mixes, the measured values shown in Figures 20.5–20.8 for a given load level of 40 kN (high enough to emphasize the distinctive responses during the final cracking phase) are shown in Table 20.6. As a matter of fact, the 30S-30G beam (RA total content of 30%) has a similar experimental behavior with

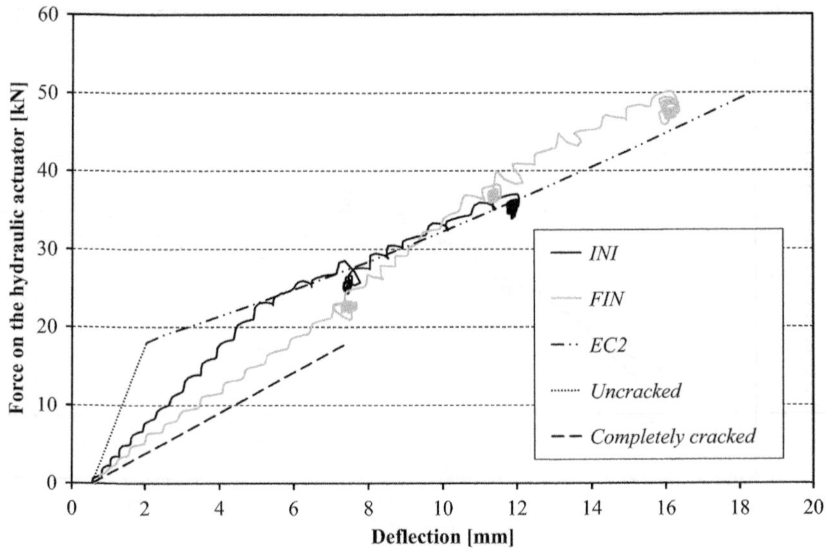

Figure 20.10 Deflection—30S-30G beam.

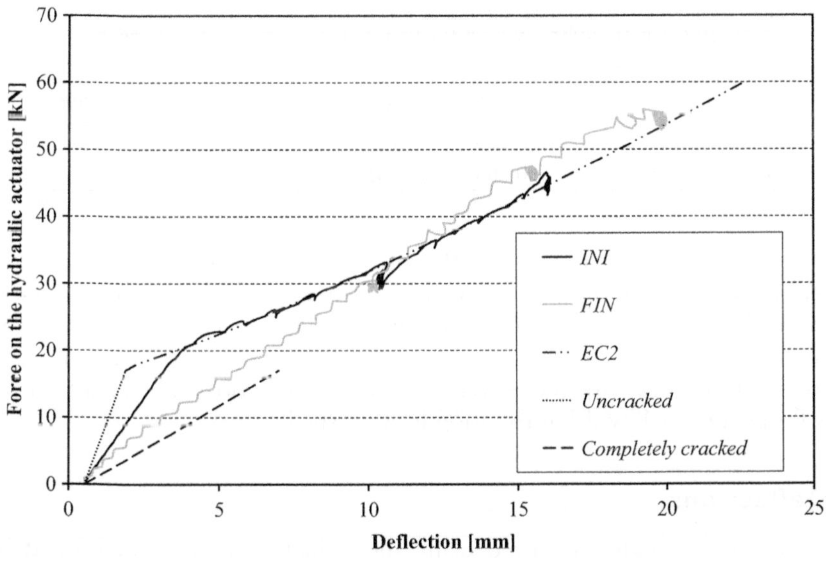

Figure 20.11 Deflection—0S-100G beam.

respect to that of the natural concrete beam 0S-0G. The 0S-100G beam (RA total content of 52%) shows smaller crack widths (maximal and mean values) than those of the other three beams. One could possibly ascribe this observation to the fact that, as mentioned earlier, the 0S-100G beam possesses one more crack than the 30S-30G beam, whereas their concrete mechanical properties and spacings between cracks are quite similar (cf. Tables 20.3 and 20.5, respectively). Finally, the 100S-100G beam shows an intermediate response.

In a logical way, the crack width being directly proportional to the crack spacing (equation (20.2)), similar commentaries to those stated for the latter also apply with regard to

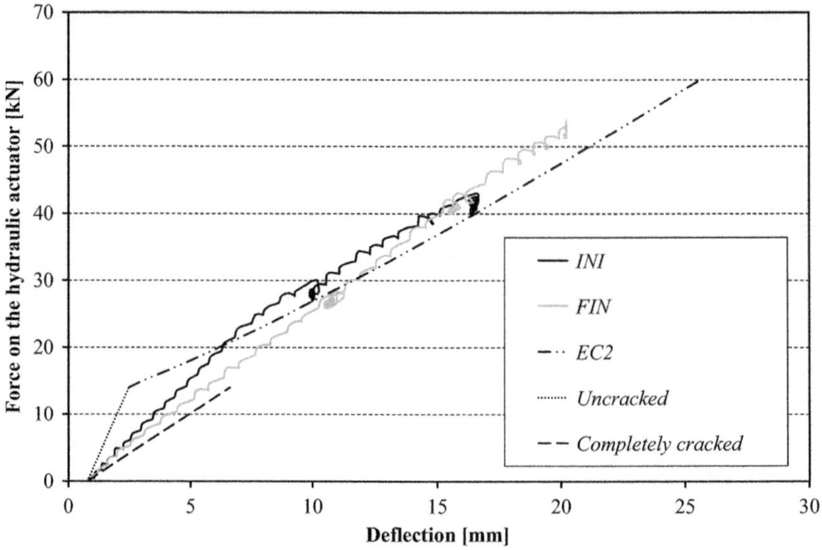

Figure 20.12 Deflection—100S-100G beam.

Table 20.6 Crack widths (within the final cracking phase) at an applied load of 40 kN

Beam mix designation	*0S-0G*	*30S-30G*	*0S-100G*	*100S-100G*
$w_{mean,exp\text{-}FIN}$ (mm)	0.256	0.228	0.193	0.228
$w_{max,exp\text{-}FIN}$ (mm)	0.363	0.361	0.302	0.332

Figures 20.6–20.8. Indeed, on the basis of the results obtained for the reference beam 0S-0G, it seems that the calculation of the crack widths, by means of the Eurocode 2 prescriptions, tends to overestimate the actual experimental values. This overestimation appears to be all the more significant that the total recycled aggregate content grows. Meanwhile, the substitution of the natural sand with the recycled one up to 30% does not seem to have a major impact, as far as the crack width estimation is concerned.

20.3.3 Deflections

The deflections at midspan, measured using the inductive sensors mentioned in Section 20.2.4, as a function of the load applied on 0S-0G, 30S-30G, 0S-100G, and 100S-100G beams are shown in Figures 20.9–20.12, respectively. Therein, the continuous-line black curve stands for the evolution of the deflection during the initial cracking phase of the bending test, whereas the continuous-line gray graph corresponds to the final cracking phase. As made for the crack widths in the previous section (Table 20.6), the deflections measured at a value of 40 kN during the final cracking phase have been displayed in Table 20.7. As reported in previous works (see references in Section 20.1), the deflections showed in Figures 20.9–20.12 and in Table 20.7 increase with the augmentation of the overall recycled aggregate content up to about 50% more for 100S-100G. Indeed, this observation appears to be in accordance with the mechanical properties (moduli of elasticity) of the different concretes tested in this study (cf. Table 20.3).

Table 20.7 Deflections (within the final cracking phase) at an applied load of 40 kN

Beam mix designation	*0S-0G*	*30S-30G*	*0S-100G*	*100S-100G*
Measured deflection (mm)	10.4	11.9	12.9	15.2
Calculated deflection (mm)	11.9	13.8	13.9	16.5

The Eurocode 2 stipulates that, "in most cases", "it will be acceptable" to use the following expression to estimate the deflection (f) of a member subjected to a bending moment higher than the one which, in theory, generates the first crack in concrete along the beam (NF EN 1992-1-1 2005, §7.4.3):

$$f = (1-\zeta) f_{\mathrm{I}} + \zeta \cdot f_{\mathrm{II}}, \tag{20.4}$$

where f_{I} and f_{II} are the deflections calculated, admitting either that the totality of the member remains uncracked (state I) or that the concrete in tension along the member is entirely (at each section along the beam) cracked (state II), respectively.

As for the distribution coefficient ζ, it is given by (NF EN 1992-1-1 2005, equation (7.19)):

$$\zeta = 1 - \beta \left(\frac{\sigma_{\mathrm{sr}}}{\sigma_{\mathrm{s}}} \right)^2, \tag{20.5}$$

where β is a coefficient that allows for the impact of either sustained or cyclic loads on the deformation parameters of the beam, σ_{sr} is the stress in the rebars at the moment when the loading applied on the beam produces the first crack along it and σ_{s} is the rebar stress at the load level under consideration.

Accordingly, the graphs representing the deflection at midspan were determined by means of the earlier-stated expressions, with the experimentally measured properties of the materials (Tables 20.3 and 20.4) as inputs, and through the classical formulae of the Strength of Materials. The corresponding curves for each specimen tested in this work are included in Figures 20.9–20.12 (EC2 label), and the calculated values for a given load level (40 kN) during the final cracking phase are presented in Table 20.7. Thus, the curves corresponding to both ideal deflections were also displayed on Figures 20.9–20.12 (*Uncracked* and *Completely cracked* labels) so as to visualize the experimental behavior of the beams tested in this work, in relation to these theoretical states.

As shown in Figures 20.9–20.12 for the initial cracking phase, the Eurocode 2 specifications appear to be suitable to assess the deflections occurring during the bending tests of the different specimens scoped out in this study. This observation seems to remain true independently of the recycled aggregate content in reinforced concrete, as far as the reinforcement and geometry features detailed earlier (cf. Section 20.2.2) are concerned. Likewise, even though the final cracking phase curves are somewhat deviated, the estimations keep satisfactory, especially regarding the orders of magnitude, provided that the actual mechanical properties of the materials measured experimentally are used. According to the discussion presented here, Silva et al. (2016) proposed an approach to consider the impact of recycled aggregates incorporation in concrete on the calculation of deflection of reinforced concrete beams. In essence, this approach consists of calculating a reduced RAC modulus of elasticity from the measured compressive strength and to perform the deflection calculation by means of the Eurocode 2 expressions using this reduced modulus of elasticity.

Notwithstanding, it is worth to note that the Eurocode 2 approach used here tends to underestimate the measured deflections within both initial and final cracking phases for the low-level loading range (up to about 20–25 kN, depending on the concrete type). Yet, as this underestimation applies to the natural-aggregate beam (Figure 20.9), it should not raise questions about the applicability of the Eurocode 2 prescriptions herein to members containing recycled aggregates. Furthermore, because deflections measured during the initial cracking phase are tangibly underestimated, explanation should be searched elsewhere. Indeed, one could think that because the initial cracking phase was performed after a precracking one (cf. Section 20.2.4), i.e., that the specimens giving rise to the initial cracking phase curves were actually previously cracked, these cracked specimens could undergo larger deflections than those that originally uncracked beams would do. Nevertheless, the fact that the underestimation in question does not show when it comes to the assessment of the crack widths (cf. Figure 20.5) could hint at another hypothesis, namely that the model used to calculate the deflections might need an enhancement or a complement to estimate them throughout the whole range of loading levels, including those in the low-level loading range stated earlier.

To sum up, the underestimations of the measured deflections, within the low-level loading range part of the graphs in Figures 20.9–20.12, could likely be misleading since the tested elements were not in an uncracked state (precracked beams, cf. Section 20.2.4). On the other hand, the calculated deflections seem to allow appropriate estimations of the measured values, regardless of the recycled aggregate content, as far as serviceability levels of short-term loads are concerned (cf. Section 20.3.5 for sustained loads). In a logical way, this accuracy appears to diminish gradually with the transition towards the failure of the tested beams (indeed, the Eurocode 2 model for deflection is not supposed to deal with the ultimate limit states (ULSs) behavior). Finally, since an important difference prevails between the measured deflections of the different tested members according to their recycled aggregate content, for elements whose deflections need to be accurately controlled, it would be advisable to measure the modulus of elasticity and the creep factor.

20.3.4 Failure phase

Considering the reinforcement characteristics of the elements tested in this study, that is underreinforced beams (cf. Section 20.2.2), their expected mode of failure was a ductile one. Namely, this progressive failure was confirmed during the bending tests for all the samples, as presented in Figure 20.13. Therein, the force/downward displacement of the hydraulic actuator is plotted. Indeed, the displacement values being too large to be measured by the deflection inductive sensors, these were withdrawn from the testing zone at the end of the last plateau of the final cracking phase.

Figure 20.13 allows to see that the four tested beams behave in a similar way, with reference to their collapse phase (ductility plateau at around 70 kN), independently of their recycled aggregate content. This agrees with the results obtained by other authors. One can single out a noticeable change of slope of the graphs at a measured force of about 60 kN. This likely corresponds to the yielding of the steel reinforcement. Indeed, on the basis of the cracked cross-section characteristics and the measured mechanical properties of the materials, the average steel strain for the tested beams at 60 kN gives 2,686 μm/m; whereas the experimentally determined steel yield strain (Table 20.4) is 2,757 μm/m.

As a matter of fact, since the beams tested were underreinforced, their failure was supposed to take place due to the yielding of the steel rebars. A verification of the Eurocode 2 specifications related to the ULS of members subjected to flexure (NF EN 1992-1-1 2005, §6) was carried out, as follows (cf. Table 1.8). For each specimen, the stress ($f_{s,calc}$) and the strain ($\varepsilon_{s,calc}$) in rebars due to the maximal measured load (F_{exp}) at the end of the flexure tests (Figure 20.13) were calculated (considering the compressive strength of concrete measured experimentally).

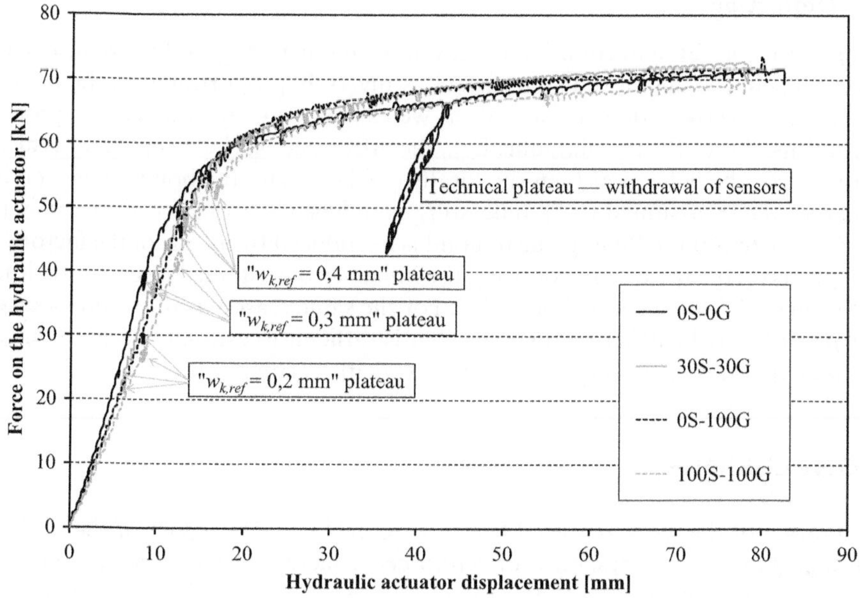

Figure 20.13 Measurements on the hydraulic actuator—final cracking and failure phases.

Then, on the basis of the experimental curves resulting from the tensile strength tests performed on steel rebars, the stress $f_{s,exp}$, corresponding to $\varepsilon_{s,calc}$, was determined. Finally, the experimental ($f_{s,exp}$) and calculated ($f_{s,calc}$) stresses were compared, showing a very good agree ment, as shown in Table 20.8. This allowed to validate the applicability of the Eurocode 2 specifications to the estimation of the strength of the beams containing recycled aggregates.

20.3.5 Specific influence of the permanent loads at SLS

20.3.5.1 Crack width

According to Eurocode 2, equation (7.8), the second term in the mean strain $\varepsilon_{sm} - \varepsilon_{cm}$, which considers the tension stiffening, is proportional to the factor k_t equal to 0.6 for short-term loading or 0.4 for sustained or cyclic loading. Even if there is no significant difference on bond between RAC and natural aggregate concrete (NAC) for instantaneous loading, the value $k_t = 0.4$ may be different for RAC due to the difference in the creep factor φ. But the value of k_t for RAC is difficult to estimate, because there are no experimental results of comparative tests between RAC and NAC under sustained loading and, furthermore, because for NAC, the values of β are not explicitly related to the creep factor φ.

Table 20.8 Stress in steel at the ULS

Beam mix designation	*0S-0G*	*30S-30G*	*0S-100G*	*100S-100G*
F_{exp} (kN)	72.3	72.8	71.8	70.0
$f_{s,calc}$ (MPa)	605	610	605	595
$\varepsilon_{s,calc}$ (%)	3.01	2.71	2.60	2.34
$f_{s,exp}$ (MPa)	616	613	612	609
$\lvert f_{s,calc} - f_{s,exp}\rvert / f_{s,exp}$ (%)	1.8	0.5	1.2	2.3

20.3.5.2 *Deflection*

According to the format of Section 20.3.3, given the moment M_{Ed} and the reinforcement A_s, the deflection f_I in the uncracked state, which is inversely proportional to the modulus E_c, will be always correctly evaluated for RAC as well as for NAC by means of the proper value of E_c. By means of the effective modulus $E_{c,eff}$ the difference on the creep factor is also taken into account. For the deflection f_{II} in the totally cracked state, the bond between reinforcement and concrete is assumed to be null, so f_{II} is independent of E_c and inversely proportional to E_s. The tension stiffening due to bond is introduced by means of the factor ζ, which in turn depends on the factor β introducing the influence of sustained or cyclic loading ($\beta = 1$ for short term or 0.5 for sustained or cyclic). For the same reason as in the previous section, the value $\beta = 0.5$ may be different for RAC, but is difficult to estimate because, once more, there are no experimental results and β is not explicitly related to φ.

20.4 CONCLUSIONS

The main target of the work reported here was to evaluate the impact of the utilization of recycled aggregates on the response of reinforced concrete elements subjected to flexure and to assess the applicability of the Eurocode 2 prescriptions to their design. Therefore, a campaign of bending tests up to failure was carried out on full-scale underreinforced beams.

The spacings between cracks seem to decrease with the total recycled aggregate content. This may be a consequence of a ratio of bond over tensile strength a little higher for RAC than NAC, probably due to a similar bond but a little lower tensile strength for RAC. Also, the crack widths along the specimens appear somewhat to diminish with the recycled aggregate content. The measured and the calculated spacings were compared. They showed a very good correspondence as far as the natural-aggregate reference beam was concerned. Nevertheless, an overestimation of the experimental crack spacings was ascertained in the case of the elements containing recycled aggregates. A conceivable source of explanation lies in the fact that some intrinsic material parameters, e.g., the tensile strength of concrete, are neglected in the corresponding Eurocode 2 expressions. The overestimation also concerns the assessment of the crack widths along the beams that were tested. In a logical way, this seems to be the consequence of the aforementioned crack spacing misestimation.

The deflections at midspan increase with the content of the recycled aggregate, hence following an inverse trend with reference to that of the modulus of elasticity of concrete. The calculated deflection values by means of the Eurocode 2 formulae are in very good agreement with those arising from the bending tests, regardless of the recycled aggregate content, as far as serviceability levels of short-term loads are concerned. Notwithstanding, an underestimation of the measured deflections was noticed within the low-level loading domain of the bending tests. It appears to be plausible to impute this underestimation, which could likely be misleading, to the flexural test protocol, which included a precracking phase. In any case, the applicability of the Eurocode 2 rules does not seem to be compromised, since the underestimation concerns the natural-aggregate beam as well. The agreement between the calculated and the experimental deflections appears to decrease, in all cases, with the transition to the failure phase of the tested members. In fact, the deflection model of the Eurocode 2 is not intended to account for the ULS behavior of flexural members. It is also worth mentioning that, since a clear increase of the deflection with the recycled aggregate content was ascertained, the Eurocode 2 specifications related to the omission of the deflection calculation should be checked out.

Moreover, it was shown that the Eurocode 2 expressions allow a satisfactory prediction of the bearing capacity (ULS) of the elements tested in this work, independently of their recycled aggregate content.

It is worth mentioning that all the earlier-stated calculations were performed using the actual experimentally measured mechanical parameters of the materials. However, only short-term loading was involved in these tests. It is important to consider that the creep of RAC is higher than that of NAC. For the calculations according to Eurocode 2, the coefficients accounting for permanent loading k_t (for crack width) and β (for deflection) must be modified. The Eurocode 2 conditions to omit these calculations should also be revised. In the case of structural members for which the deflection is a sensitive parameter, it would be advisable to measure the modulus of elasticity and the creep factor.

Finally, it shall be noted that the aforementioned conclusions pertain to the experimental results ensued from only one specimen tested for each type of concrete and to the single configuration of the steel reinforcement placed therein. Consequently, precaution should be taken when extrapolating these conclusions.

Chapter 21

Members in shear

G. Wardeh and E. Ghorbel
University of Cergy-Pontoise

B. Fouré
Bougival

CONTENTS

Abstract

The results from an experimental work on the shear behavior of concrete beams with natural aggregates (NA) or 100% recycled aggregates (RA) are presented. Full scale 200x250x1900 mm reinforced concrete beams without stirrups were manufactured from two mixtures with C35/45 target class of compressive strength and S4 class of workability. The beams were tested under 4 points bending for a shear span-to-depth ratio (a/d) equal to 1.5 or 3.0. The mechanical properties of the two mixtures were characterized in terms of compressive strength, splitting tensile strength and modulus of elasticity. The experimental results show that, for the same class of compressive strength, the shear failure mechanisms in recycled aggregates concretes (RAC) are the same compared to the natural aggregate concretes (NAC) while the shear strength is lower. The decrease in the shear strength is directly related to the decrease in the splitting tensile strength of the RAC compared to the NAC, when the failure is brittle. When it is not brittle, this decrease is significantly higher than that of the tensile strength. In terms of load–deflection response, RAC beams exhibited higher defection compared to the NAC beams. Fracture surfaces of beams with a/d = 3 were observed by using an optical microscope for both NAC and RAC. It was found that the failure surface of NAC beams occurs at the interface between the aggregates and the mortar, whereas it occurs by the rupture of recycled aggregates with a denser cracking network in the case of RAC beams. The experimental shear strengths were compared with the shear provisions of EN 1992-1-1which may need adaptation to be applied to RAC. The final conclusions are supported by the results of other published tests results.

21.1 INTRODUCTION

This study aims to experimentally characterize the effect of recycled aggregates (RAs) on the shear capacity of reinforced members without stirrups and to study the applicability of Eurocode 2 (EC2) (EN 1992-1-1 2004) for the prediction of the bearing capacity of members made with RA concrete (RAC) with and without transversal reinforcement.

Shear failure of reinforced concrete beams is a complex phenomenon due to the effect of too many parameters. Main factors influencing the shear capacity of beams are as follows: the shear span-to-depth ratio (*a*/*d*), the longitudinal reinforcement ratio (ρ_l), the compressive strength of concrete (f_{cm}), the aggregate size, and the transverse reinforcement ratio (ρ_t). Depending mainly on the shear span and transverse reinforcement ratios, various failure modes exist, such as diagonal failure, flexural failure, and deep beam failure. Research programs have been reported in the literature on the shear capacity and the shear failure mechanisms of beams made with concrete, incorporating mainly recycled coarse aggregates. Only some of them allowed to draw reliable conclusions, principally, on the concrete part of the shear strength V_{Rc} (Arezoumandi et al. 2014; Etxeberria et al. 2007; Fathifazl et al. 2011; Gonzalez-Fonteboa et al. 2007; Schubert et al. 2012; Yun et al. 2011). Some of these conclusions are contradictory, but generally V_{Rc} is found to be equal or lower for RAC compared with natural aggregate (NA) concrete (NAC), due to possible differences in tensile strength and roughness of the surface of cracks (aggregates interlock). The failure can be brittle or not, depending on the shear span-to-depth ratio *a*/*d* and the amount of flexural reinforcement. The tests on beams with stirrups are scarcer, but the stirrups term of strength V_{Rs} seems to be independent of RAC. These results scattered and sometimes at first sight is contradictory were considered insufficient to conclude in the frame of this National Project. Having in mind that the resisting truss mechanism associated to the transverse reinforcement cannot be significantly modified by the use of RA, it was decided to study only the concrete contribution to shear strength in beams without stirrups. By considering the variability of the strength, it was decided to test three identical beams for each concrete mix.

The conclusions issued from these and other published test results are presented and complemented by considerations on the strength due to the transverse reinforcement and on the limit strength of the concrete struts in the truss mechanism. Furthermore, an analogy is done with lightweight aggregate concrete (LWC).

21.2 EXPERIMENTAL STUDY

21.2.1 Materials and methods

21.2.1.1 Materials

The mixtures C35/45 (f_{ck} = 35 MPa) 0R-0R with only NAs and C35/45 100R-100R with total replacement by RAs developed for the National Project RECYBÉTON were chosen for this study. The mix proportions are recalled in the appendix.

21.2.1.2 Test specimens

Concrete cylinder specimens 110 mm × 220 mm were manufactured to determine the compressive strength, splitting tensile strength, and dynamic modulus of elasticity.

Six 200 mm × 250 mm × 1,900 mm beams without stirrups were constructed for each concrete type. For each series, three beams were tested with a shear span-to-depth ratio $a/d = 1.5$ and another three with $a/d = 3.0$. For all tested beams, the longitudinal reinforcement was four 16-mm diameter deformed bars. The secondary reinforcement was two 8-mm diameter longitudinal bars on the compressed side and 6-mm diameter stirrups, four in the central part of the beam and one at each end anchorage of the flexural bars (Figure 21.1). There are no stirrups in the shear spans. For all tested beams, the longitudinal reinforcement ratio is $\rho = A/bd = 1.78\%$.

All beams and specimens were made, cured, and tested in the structural laboratory of the University of Cergy-Pontoise. The beams and specimens were kept for 1 week in the frameworks and were humidified for 1 week after demolding. Afterwards, they were preserved at room temperature with a relative humidity of about 50% in the laboratory until the test at 28th day.

21.2.1.3 Testing procedures

Uniaxial compressive and splitting tensile tests were performed using a servohydraulic INSTRON equipment with a capacity of 3,500 kN. Experiments were performed at a stress rate of 0.5 MPa/s compression and 0.05 MPa/s traction. Beams were loaded up to failure using a 350 kN capacity hydraulic equipment. The four-point bending tests were carried out under displacement control at a constant rate of 1 mm/min. The position of the load was varied to obtain an effective shear span-to-depth ratio of 1.5 or 3.

The deflection was measured at midspan using a displacement transducer (LVDT) placed at the bottom of beam specimens. The area next to the supports was filmed to locate the beginning as well as the evolution of shear cracking.

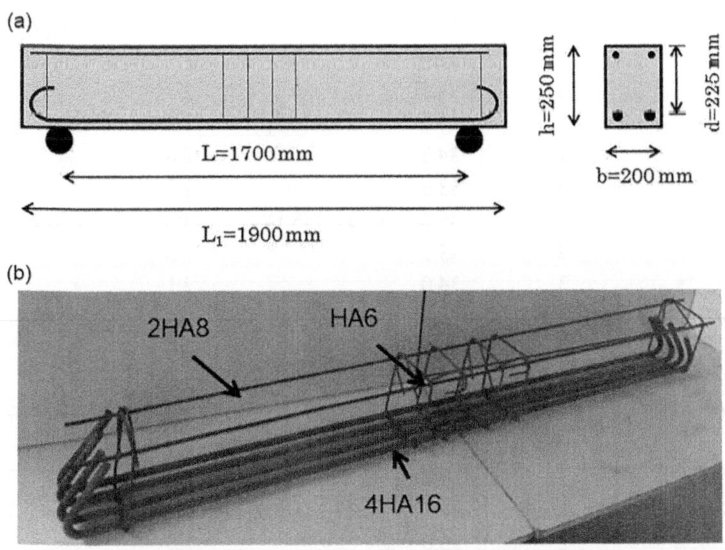

Figure 21.1 (a) Dimensions and (b) reinforcement layout of the tested beams.

21.2.2 Experimental results

21.2.2.1 Mechanical properties of concrete mixes

Mean values and standard deviations obtained for both concrete mixes are summarized in Table 21.1. It can be observed that all the properties (compressive strength, splitting tensile strength, and elastic modulus) decrease when the granular skeleton is fully recycled. The decrease is about 8%, 27%, and 37%, respectively. The variation in f_{cm} between RAC and NAC may be corrected assuming correlations in $f_{cm}^{2/3}$ for $f_{ctm,sp}$ or $f_{cm}^{2/3}$ for $E_{cm,dyn}$. For the same compressive strength, the decrease is about 18% for the tensile strength and 22% for the modulus.

21.2.2.2 Response of the tested beams

For all tested beams, the first cracks, called flexural cracks, appear in the zone between the two loading points, and then the failure occurs at the formation of a diagonal crack near one of the supports. Two types of responses are distinguished as a function of the ratio *a*/*d*:

- When *a*/*d* = 3.0, the failure is brittle and occurs abruptly with a large opening of the diagonal crack;
- When *a*/*d* = 1.5, the element continues to resist after the formation of the diagonal crack up to failure.

Cracking and shear failure are shown in Figure 21.2. In terms of crack formation and propagation, the behavior of NAC and RAC beams is practically identical. The mode of failure is by shear compression when *a*/*d* = 1.5, whereas it is by diagonal tension for *a*/*d* = 3.0.

Table 21.1 Mechanical properties of concrete mixes at 28 days (between brackets: standard deviation)

	Mix (C35/45)	*Specimen*	f_c *(MPa)*	f_{cm} *(MPa)*	$f_{ct,sp}$ *(MPa)*	$f_{ctm,sp}$ *(MPa)*	E_{dyn} *(GPa)*
Series 1	0S-0G	1	35.3	37.3 (4.1)	3.2	3.2 (0.09)	42.1 (0.7)
		2	42.0		3.1		
		3	34.5		3.3		
	100S-100G	1	33.4	33.6 (0.5)	2.5	2.6 (0.09)	30.7 (0.5)
		2	34.3		2.6		
		3	33.6		2.5		
Series 2	0S-0G	1	35.0	37.4 (5.1)	3.5	3.5 (0.4)	—
		2	43.3		3.8		
		3	34.0		3.1		
	100S-100G	1	33.4	35.6 (1.9)	2.4	2.7 (0.4)	—
		2	36.5		2.5		
		3	37.0		3.1		
Series 3	0S-0G	1	34.0	36.9 (4.4)	3.6	3.5 (0.07)	—
		2	42.0		3.4		
		3	34.8		3.5		
	100S-100G	1	31.5	33.3 (1.8)	2.6	2.6 (0.1)	—
		2	35.0		2.5		
		3	33.3		2.7		

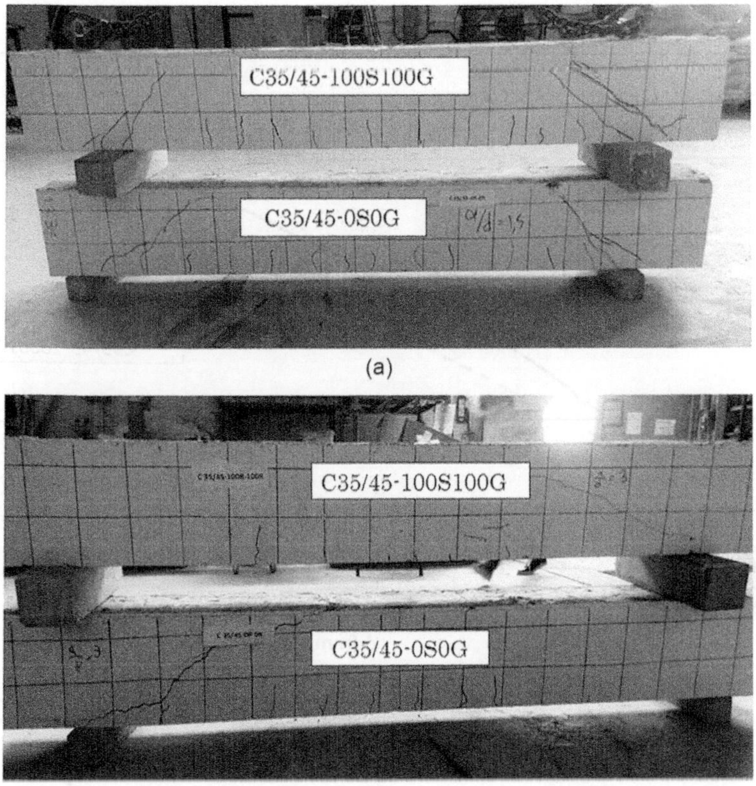

(a)

(b)

Figure 21.2 Crack patterns of the tested beams.

21.2.2.3 Load deflection curves

Load deflection curves for all tested beams are shown in Figure 21.3. The behavior is linear before the onset of the first flexural crack, obviously with a smaller slope for the RAC beams due to the smaller modulus. After cracking, the slope of the curve changes whatever the mixture, NAC or RAC. For $a/d = 3$, the cracked flexural stiffness, which is mainly determined by the reinforcement, is predominant, and this slope has close values for the two types of RAC and NAC. This difference in slope is larger for $a/d = 1.5$, because the shear term of deflection has a larger relative amount and relies mainly on concrete compressive strains that are larger for the RAC beams.

21.2.2.4 Analysis of the fracture surface

The surfaces of shear fracture for C35/45-0R-0R and C35/45-100R-100R beams tested with $a/d = 3.0$ have been observed macroscopically using an optical stereoscopic microscope with 20 times magnification. It was found that the failure surface of NAC beams occurs at the interface between the aggregates and the mortar, whereas it occurs by the rupture of RAs for RAC. The analysis also reveals that the fracture surface of RAC is smoother than the fracture surface of NAC, which means that the contribution of bridging and branching phenomenon is reduced during the stress transfer between the two lips of the crack (Figure 21.4).

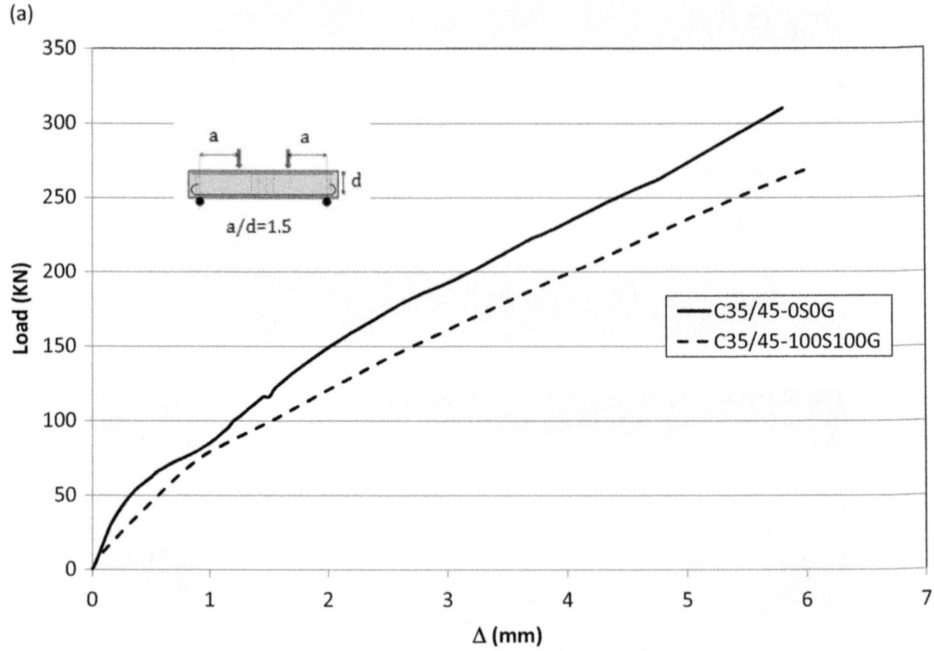

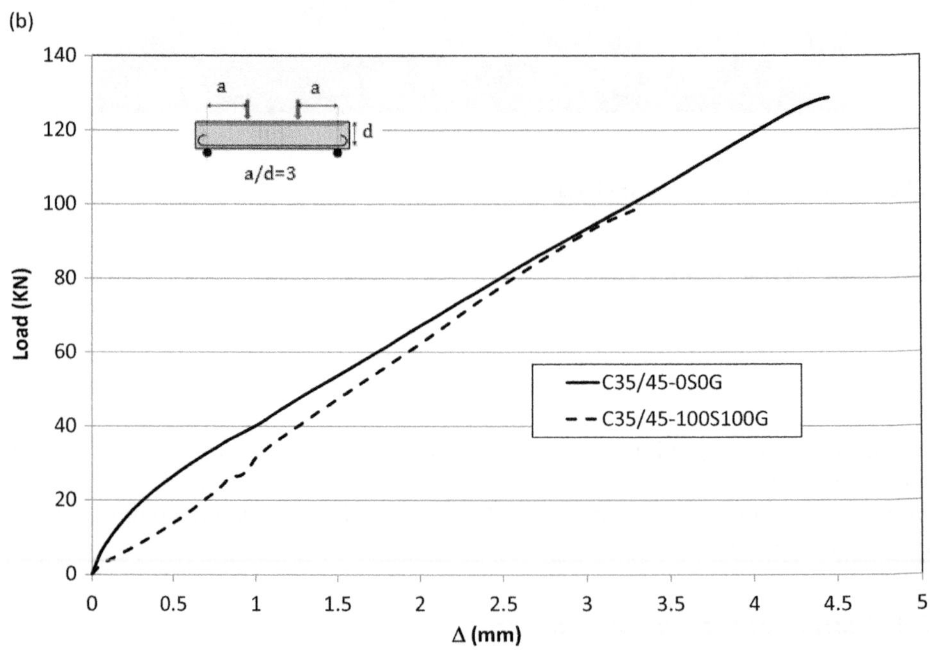

Figure 21.3 Force–deflection curves. (a) a/d=1.5, (b) a/d=3.0.

The microscopic observations showed that the crack network is denser for RAC than NAC beams. The cracks pass through the RAs while they remain localized in the paste around NAs. Finally, the crack width is larger for RAC than for NAC. For both materials, no interfacial debonding was observed between new mortar and RAs.

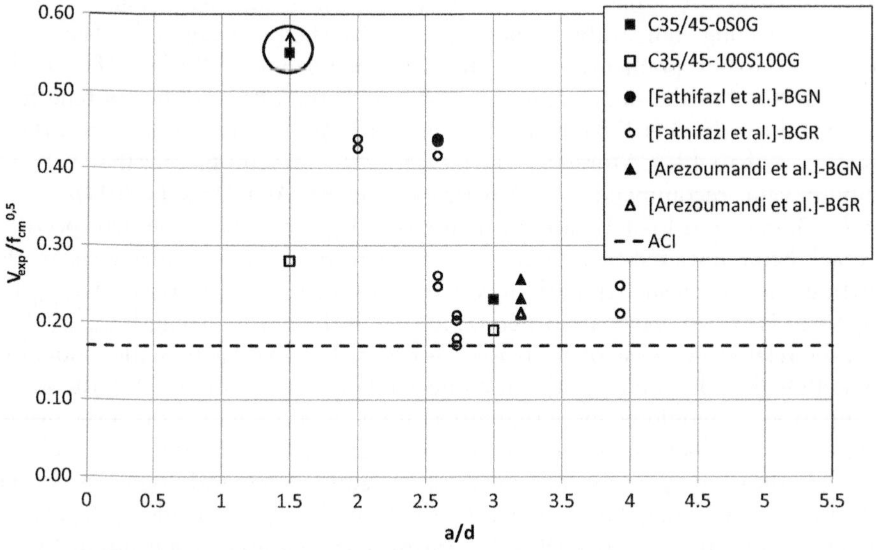

Figure 21.4 Comparison between the results of this study and other authors (Arezoumandi et al. 2014; Fathifazl et al. 2011) for NAC and RAC.

21.3 ANALYSIS OF THE ULTIMATE STRENGTH

21.3.1 Present study

The ultimate force F_{exp} and shear strength V_{exp} of all tests are summarized in Table 21.2. For the NAC beams with $a/d = 1.5$, the maximum jack force 300 kN did not allow to attain the shear failure; thus, the ultimate shear force is larger than 150 kN.

The ultimate experimental shear stress was calculated using the following equation:

$$v_{exp} = V_{u,exp}/b_w \cdot d \tag{21.1}$$

Table 21.2 Experimental results from four-point bending tests (shear failure)

Mix (C35) Relative shear span	*Beam n°*	*Failure force F_{exp} (kN)*	*Mean F_{moy} (standard deviation)*	*$V_{u,exp} = F_{exp}/2$*	*$v_{exp} = V_{exp}/b_w\, d$ (MPa)*	*$\frac{v_{exp}}{\sqrt{f_{cm}}}$*	*Mean*
0S-0G *a/d* = 1.5	1	>300		>150	>3.33	0.55	>0.55
	2						
	3						
0S-0G *a/d* = 3.0	1	127	123 (3)	63.6	1.41	0.24	0.23
	2	121		60.6	1.35	0.23	
	3	120		60.0	1.33	0.23	
100S-100G *a/d* = 1.5	1	150	153 (5)	75.0	1.67	0.27	0.28
	2	160		80.0	1.78	0.29	
	3	150		75.0	1.67	0.27	
100S-100G *a/d* = 3.0	1	100	99 (—)	50.2	1.12	0.19	0.19
	2	98		49.0	1.09	0.19	
	3	—		—	—	—	

Moreover, this stress, v_{exp}, has been divided by the root of the mean compressive strength f_{cm}, as $v_{exp}/\sqrt{f_{cm}}$, having comparable values corrected for the variation of f_{cm}. The proportionality to $\sqrt{f_{cm}}$ was used to predict the strength V_{Rc} by the formula of Rafla (1971) and Kordina and Blume (1985). For the same compressive strength, the ultimate shear strength $V_{u,exp}$ of RAC beams with $a/d = 3$ is 17% lower than that of NAC beams; for $a/d = 1.5$, the relative decrease is more than 49%. Moreover, the ratio $v_{exp}/\sqrt{f_{cm}}$ remains greater than 0.17, which is the minimum value recommended by American standard (ACI 318R-11 2011).

For $a/d = 3$, cracking load V_{cr} and ultimate load $V_{u,exp}$ are the same. Irrespective of the critical crack being initiated by a flexural crack or formed directly during shear, this load is proportional to the tensile strength of concrete. So the relative decrease in $V_{u,exp}$ for RAC with respect to NAC (–17%) is almost equal to that of the tensile strength $f_{ctm,sp}$ (–18%). For $a/d = 1.5$, the relative decrease of shear force for RAC has two parts, with a global value of more than 49%. For the same f_{cm}, the first one is related to V_{cr} (about17%); the second one, from V_{cr} up to $V_{u,exp}$, would be more than 40%. It involves the activation of the mechanisms of (1) aggregates interlock along the critical crack, (2) dowel effect of the flexural reinforcement, and (3) shear-compression resistance of the zone in flexural compression. The ultimate effect of interlock may be estimated by the formulas for shear along the construction joints in § 6.2.5 of Eurocode (EN 1992-1-1 2004). The cohesion term is proportional to the tensile strength f_{ct} (so, a decrease of about 15% from NAC to RAC). The friction term is proportional to the friction coefficient (design value μ with a safety factor; tan φ without); it varies according to the roughness of the surface. It may be assumed that the crack surface is rough for NAC ($\mu = 0.7$; $\tan \varphi \approx 1$) and smooth ($\mu = 0.6$; $\tan \varphi \approx 0.8$; decrease of the actual tan φ of 20%) or very smooth for RAC ($\mu = 0.5$; $\tan \varphi \approx 0.6$; decrease 40%); this last assumption seems to be a better one to explain the decrease of $V_{u,exp} - V_{cr}$.

21.3.2 Other test results

All the experimental works used rectangular beams. Gonzalez-Fonteboa et al. (2007) carried out tests on beams with or without stirrups, a relative shear span $a/d = 3.3$, with a constant longitudinal reinforcement. Concrete of mean strength $f_{cm} = 40$ MPa was used, NAC or RAC, with an RA ratio of about 25% (coarse only). The compressive strength f_{cm} is slightly variable; the tensile strength f_{ctm} seems not to vary with the RA content for the same f_{cm}. For the beams without stirrups, the shear strength V_{Rc} seems not to vary due to RA, with the failure being a brittle one. Etxeberria et al. (2007) performed tests on beams with or without stirrups, a relative shear span a/d a little less than 3.3, with a constant longitudinal reinforcement. Concrete of mean strength $f_{cm} = 41$ MPa was used, NAC or RAC with a ratio of RA from 14% to 62% (coarse only). The compressive strength f_{cm} is slightly variable; for the same f_{cm}, the tensile strength f_{ctm} increases with the RA content, up to 22%. Nevertheless, for the beams without stirrups, the shear strength V_{Rc} decreases down to 14%. Failure is rather brittle. Schubert et al. (2012) performed tests on beams without stirrups, a relative shear span of $a/d = 3.5$, with a constant longitudinal reinforcement. Concrete of mean strength $f_{cm} = 36$ MPa was used, NAC or different RAC mixes with 100% of RA (coarse + sand). The strength f_{cm} is slightly variable; for the same f_{cm}, the tensile strength f_{ctm} increases with RA while shear V_{Rc} would be slightly decreasing, with a brittle failure. Fathifazl et al. (2011) carried out tests on beams without stirrups, a relative shear span a/d variable between 1.5 and 4, with varying longitudinal reinforcement and a varying height of the section to study the size effect. Concrete used was NAC or RAC with a ratio of RA of 41% or 47% (coarse only). Two types of aggregates were used (crushed limestone or river gravel), from the same source for NA and RA. The compressive strength f_{cm} is significantly different for RAC (mean 45 MPa) than NAC (mean 36 MPa); for the same f_{cm}, the tensile strength f_{ctm} of RAC would be about 10% lower than NAC. The shear strength V_{Rc} is the same for RAC as NAC with crushed limestone and is significantly lower (about 30%) with river gravel. Probably, two types of failure, brittle or not, are involved. Yun et al. (2011) performed tests on beams without stirrups, a relative shear span $a/d = 5.1$, with a constant longitudinal reinforcement. Concrete of mean strength $f_{cm} = 33$ MPa was used, NAC or RAC with a ratio RA from 15% to 52% (coarse only) was used. The compressive strength f_{cm} is variable; for the same f_{cm}, and the tensile strength f_{ctm} increases with the RA content. The failure is brittle. Arezoumandi et al. (2014) performed tests on beams without stirrups, a relative shear span a/d about 3, with a variable longitudinal reinforcement. Concrete of mean strength $f_{cm} = 34$ MPa was used, NAC or RAC with a ratio of RA 52% (coarse only) was used. The compressive strength f_{cm} is slightly variable; for the same f_{cm}, the tensile strength f_{ctm} decreases for RAC. Depending on the amount of flexural reinforcement (ratio ρ_l), the failure is brittle, with a small decrease of V_{Rc} for RAC compared with NAC, or nonbrittle with a much larger decrease of V_{Rc}.

21.3.3 Synthesis

As far as V_{Rc} is concerned, these results confirm that it is necessary to distinguish between the brittle or nonbrittle failure, as is clearly shown by the test results given in Section 21.3.1. To quantify the possible difference in V_{Rc} between RAC and NAC, it is difficult to do a synthesis of all the results, for several reasons:

- The great scatter of the results themselves (on f_{cm}, f_{ctm}, V_{Rc});
- The uncertainty on the type of failure, either brittle or not;
- In certain series, it is impossible to compare RAC and NAC directly, because several parameters vary simultaneously. In this case, the comparison is done by the intermediate of the calculated values of V_{Rc} using the formula of Rafla (1971) and Kordina and Blume (1985), which includes all the parameters of influence:

$$V_{Rc} = 0.54 f_{cm}^{1/2} \rho_l^{1/3} d^{-1/4} k_a b_w d \tag{21.2}$$

where b_w and d are cross-section dimensions in mm, f_{cm} is the mean compressive strength expressed in MPa, ρ_l is the longitudinal reinforcement ratio in %. d is expressed in cm and the factor $d^{-1/4}$ describes the size effect. k_a is a factor without unit and can be expressed as follows:

$$k_a = 6 - 2.2\frac{a}{d} \text{ for } \frac{a}{d} < 2$$

$$k_a = 0.795 + 0.293\left(3.5 - \frac{a}{d}\right)^{2.5} \text{ for } 2 < \frac{a}{d} < 3.5 \text{ and}$$

$$k_a = 0.9 - 0.03\frac{a}{d} \text{ for } \frac{a}{d} > 3.5$$

The respective ratios V_{Rc}(test)/V_{Rc} (calculated) for RAC and NAC are then compared.

A correlation is searched between the ratio of shear strength $S = V_{Rc}(\text{RAC})/V_{Rc}(\text{NAC})$ and the ratio of tensile strength $T = f_{ctm}(\text{RAC})/f_{ctm}(\text{NAC})$ for the same f_{cm}. T implicitly includes the effect of the ratio Γ of RA, assuming that a correlation between f_{ctm} and Γ is given elsewhere. The individual results or the mean result of several identical tests are balanced as follows: weight 1, direct comparison between RAC and NAC for the same value of the parameters; weight 3/4, reliable indirect comparison; weight 1/2, less reliable indirect comparison. Even with this treatment, the scatter remains high, a part of which is due to the uncertainty of the classification into brittle or nonbrittle failure. Furthermore, the field of variation of the experimental values of S and T are small. So, it is impossible to succeed in finding even the simplest linear regression between S and T. So the conclusion is based only on global mean values:

- Brittle failure (diagonal tension), mean $T \approx 1$, mean $S \approx 0.95$. A slight decrease of V_{Rc} for RAC even if the tensile strength is unchanged seems unlikely. The highly reliable results from Cergy tests (three identical tests for the same parameters), supported by some other published results, allow to conclude the proportionality between S and T.
- Nonbrittle failure (shear compression), mean $T \approx 0.9$, mean $S \approx 0.8$, i.e., $S \approx T2$.

21.3.4 Analogy with LWC

The analysis of shear behavior of RAC compared with NAC may take advantage of an analogy of RAC with LWC, which also exhibits a decrease of tensile strength with respect to NAC for the same f_{cm}, due to cracks passing through the aggregates. As a function of the density ρ, the EC2 Section 11 gives the factor of reduction $\eta_1 = 0.4 + 0.6\,\dfrac{\rho}{2{,}200}$ for tensile strength (formula 11.1), assuming a density $\rho = 2{,}200\,\text{kg/m}^3$ for NAC. In formula (11.6.2), it gives two reduction factors for shear strength of LWC beams without stirrups and no axial force compared to NAC. One (η_V) is applied to V_{Rc}, the other ($\eta_{v,min}$) to $V_{Rc,min}$ (see also § 4) with $\eta_v = \left(\dfrac{0.15}{0.18}\right)\eta_1 = 0.833\,\eta_1$ and $\eta_{v\,min} = \left(\dfrac{0.03}{0.05}\right) \approx 0.86$

It is worth to recall that v_{min} is likely related to the critical shear crack formation (i.e., V_{cr}), while V_{Rc} corresponds to the complete mechanism with aggregates interlock and so on. Taking for example, the light weight concretes LWC with $\rho = 1{,}800\,\text{kg/m}^3$ (EC2 for a density class of 1.8) suggests $\eta_1 = 0.89$, $\eta_{V,min} = 0.86$, i.e., slightly less than η_1; $\eta_V = 0.74$, i.e., slightly less than η_1^2.

21.4 COMPARISON WITH EC2 PROVISION FOR V_{RC}

The theoretical shear force using EC2 (EN 1992-1-1 2004) is calculated for all beams using the following equation:

$$V_{Rd,c} = \left[C_{Rd,c}k\left(100\,\rho_1 f_{ck}\right)^{1/3}\right]\cdot b_w d \tag{21.3}$$

with a minimum of

$$V_{Rd,c} = \left[0.035\,k^{3/2} f_{ck}^{1/2}\right]\cdot b_w d \tag{21.4}$$

where $C_{Rd,c}$ = $0.18/\gamma c$ (with $\gamma_C = 1.5$ in most cases); $k = 1 + \sqrt{(200/d)} \leq 2$, with d in mm; $\rho_l = A_{sl}/b_w \cdot d$ (A_{sl} area of the tensile reinforcement, b_w and d, respectively, width and effective height of the cross section); f_{ck} characteristic compressive cylinder strength of concrete at 28 days in MPa (for the tests: $f_{ck} = f_{cm}$ measured—8 MPa; only the mean values for either NAC or RAC beams are considered here).

Table 21.3 EC2-predicted ultimate shear force according to equation (21.3), compared to test values (kN)

Concrete mix (C35)	f_{cm} *(MPa)*	f_{ck} *(MPa)*	*a/d*	$V_{u,exp}$	*(A)* $V_{Rd,c}$	*(B)* $V_{Rd,c}$
0S-0G	37.3	29.3	1.5	>150	39.2	63.7
			3.0	61.4		
100S-100G	34.5	26.5	1.5	76.7	37.9	64.1
			3.0	49.6		

The results are given in Table 21.3 (column A); of course, the safety margin with respect to the experimental values is smaller for RAC than for NAC regardless of shear span-to depth ratio. The equation (21.3) is also applied without safety $f_{ck} = f_{cm}$ and $\gamma_c = 1$ (column (B)). In this latter case, $V_{Rd,c}$ is lower than $V_{u,exp}$ for $a/d = 1.5$ and higher for $a/d = 3.0$. But in the case $a/d = 1.5$, EC2 considers a reduced value of the acting force, equivalent to an enhanced value of resistance.

21.5 OTHER CONSIDERATIONS FOR THE CALCULATIONS OF SHEAR STRENGTH

21.5.1 Reinforcement term V_{Rs}

As was likely to be anticipated, the published test results do not show that the truss model may not be applicable. So the calculation of the shear strength due to the transverse reinforcement V_{Rs} remains unchanged. The only question raised is that of the minimum value of the angle θ of the compressive strut. There are too few test results to allow this point to be analyzed.

21.5.2 Limit strength of the concrete struts

The reduced concrete compressive strength $\nu 1 \cdot f_{cd}$ introduced in the limit shear force $V_{Rd,max}$, according to EC2, results from the behavior of the concrete struts that are subjected to a longitudinal compression and a transversal tensile force. If the tensile strength f_{ct} of RAC is likely to be less than NAC, it seems logical to suppose that the reduction factor $\nu 1$ will be lower for RAC than NAC. In the absence of test results, the analogy with LWC may be taken as background; in EC2 Section 11, $\nu 1$ is multiplied by $\eta 1$ (i.e., proportionally to f_{ct}).

21.6 CONCLUSIONS

Full-scale beams constructed without stirrups of NA concrete and 100% recycled coarse and fine aggregates concrete have been tested for failure in shear. Based on the results of this study and on other published results, the following conclusions are drawn:

- The behavior of the RAC and NAC beams is identical in terms of cracking morphology and crack propagation.
- RAs may decrease the shear strength of beams compared to beams constructed with NAs, which is consistent with a possible reduction of the tensile strength. The amount of reduction varies with the character of the failure (brittle or not) depending on the

shear span-to-depth ratio and other parameters. For brittle failure, the decrease is proportional to that of the tensile strength; for nonbrittle failure, its order of magnitude would be the power 2 of the previous factor.

- EC2 conservatively predicts the shear strength of the beams tested here, with a lower safety margin for RAC than for NAC.
- Shear fracture is due to the detachment of NA for NAC beams, while it occurs by the rupture of RA for RAC beams. The consequences of this difference on tensile and shear strengths for RAC may be considered similar to that of LWC compared with NAC.

Section VII

Experimental construction sites

P. Dantec
Consulting Engineer

The organization of experimental constructions in various areas aimed to prove the operational feasibility of the incorporation of recycled aggregates in concrete mixes, to build a parking lot (Chapter 22), parts of a bicycle path bridge (Chapter 23), an archive's building (Chapter 24), parts of an office building (Chapter 26), industrial facilities in a ready-mixed concrete plant (Chapter 25), and precast concrete products (Chapter 27).

The effect of partial or total substitution of natural aggregates with recycled ones (coarse and fine aggregates) on the properties of manufactured concretes is explored. The effects on the production methods of the recycled aggregates, those of concrete mixes, and their implementation are also described at each stage. Quality control of the recycled aggregates related to the quality of the components, grading curves and physical properties, and mechanical and chemical characteristics are carried out by following the standards tests and acceptable thresholds of specifications. The impact of the introduction of recycled aggregates on the mechanical properties of recycled concrete aggregates was assessed investigated through the following properties: shrinkage, strength, and elastic modulus, which were taken into account when this was relevant in structural design. In parallel, this impact was also assessed through the degradation level of transfer properties by measuring durability-related properties. Visual inspections were carried out at the end of the construction stage, and when this was possible, at later age, they were used to evaluate the possible effect of recycled aggregates on the quality of the execution and in particular on the concrete facings of formed and unformed surfaces.

Then provisions applicable for worldwide and particularly in France about both material quality control and the execution of buildings and civil engineering structures are presented (Chapter 28). The primary standard regulations are provided, along with a number of examples drawing upon the outcomes of several experimental worksites conducted within the framework of the RECYBÉTON national project. Finally, it is worth recalling that 7 years before the start of RECYBÉTON, a construction was carried out in Paris where several thousands of cubic meters of RAC were cast at a high replacement rate. The material development aspects are related to those in Chapter 16.

Chapter 22

Slab-on-grade

Chaponost

F. de Larrard, E. Garcia, and T. Dao
LafargeHolcim R&D

C. De Sa and F. Benboudjema
LMT

D. Rogat
Sigma Béton

CONTENTS

Abstract

This section presents the first experimental site carried out in France on recycling concrete into concrete, within the RECYBÉTON National Project. A 2,000 m^2 parking lot was built, divided into six sectors made up with various concrete mixes. All mixtures matched the same set of specifications, except the replacement rate, which ranged from 0% to 100% of recycled aggregates (fine and coarse fractions). A comprehensive characterization campaign was carried out on the six concrete mixes, mainly devoted to fresh concrete properties, strength, E-modulus, and shrinkage. The usual trends—increase of cement content for high replacement rate, decrease of E-modulus, and increase of shrinkage—were noted. A comprehensive finite-element modeling of the slab-on-grade was performed, based on measured concrete parameters complemented by a number of data taken from the literature and from experimental results of laboratories participating with the national project. The objective was to anticipate cracks originating in a combination of thermal and hygral effects. The simulations only predicted the appearance of cracks in the joints of the slabs, and in a laboratory sample of fully recycled concrete used to perform a ring cracking test. The construction site went smoothly with no difference between the mixes, as reported by the practitioners. No crack appeared out of the joints sawn every 5 m according to usual procedures, as predicted by the simulations. From this experimental site, it was concluded that the production and casting of recycled concrete did not require any significant change in the current practices. As for the risk of shrinkage-induced cracking, it only grows in case of very high replacement rates. Even in such a case (slab-on-grade cast in winter, with moderate thickness), it was possible to place a fully recycled concrete, with a higher cement dosage but without crack. To better assess the risk in other cases, more experimental and numerical creep data have to be generated. Three years after this project completion, some supplementary slabs were cast with concrete incorporating recycled cement (i.e., a cement where 15% of the raw materials were substituted by ground recycled concrete). In terms of fresh and hardened mechanical properties, as well as behavior during and after the casting phase, no difference was noted by comparison with a control concrete batched with a regular Portland cement.

22.1 INTRODUCTION

This construction site, located in Chaponost (Rhône, France, in the southern Lyon's suburb), was the first one organized within the RECYBÉTON national project. Here, the goal was to produce recycled concrete in an industrial environment (a ready-mix concrete plant near Lyon) and to carry out the usual procedures in terms of batching, transporting, and casting. A 2,000-m^2 area was dedicated to build a parking lot. The questions addressed by this experiment were the following:

- For such a construction, is it necessary to change the usual practice, from the structural design to the finishing procedures, when a part of the aggregate is replaced by recycled concrete aggregates in the slab concrete?
- If the answer is yes, where is the border between moderately recycled concrete (where "business-as-usual" habits can be kept) and highly recycled concrete, which should be considered as a particular material?
- Knowing that recycled concrete generally displays a higher drying shrinkage, when compared with a natural aggregate concrete of the same grade, would this peculiarity entail a higher cracking risk?

To address these queries, a comprehensive team was formed within RECYBÉTON, encompassing the ELTS company (Entreprise Lyonnaise de Travaux Souterrains, acting as the owner of the parking lot and the contractor), Filliot Eurovia (delivering recycled aggregate), the Lafarge group (now LafargeHolcim, having developed the mixtures and delivered the concrete mixes for the site), the CEREMA (Centre d'études et d'expertise sur les risques, l'environnement, la mobilité et l'aménagement, in charge of quality control for the owner), Sigma Bétons (for the mix characterization campaign), and LMT ENS Cachan (Laboratoire de Mécanique et Technologie, Ecole Normale Supérieure de Cachan, for the modeling and simulation of cracking risks).

22.2 THE CHAPONOST PROJECT

22.2.1 Genesis

ELTS is a contractor specialized in underground civil engineering works. For hosting a part of its trucks and machines float, ELTS decided to build a parking around a new building. Acting both as a (private) owner and a contractor, ELTS could carry out a project where current standard limitations could be overcome without any risk of litigations. As ELTS was convinced of the interest of developing the recycled concrete technology in terms of economy and environment, the company accepted to take the risk of building with up to 100% of recycled aggregates in the mix. Filliot, a well-established recycler in the Lyon's area, was keen to demonstrate the possibility of using its product not only in conventional markets as road subbases but also in higher added value applications as structural ready-mix concrete. The LafargeHolcim group had long claimed its wish to take part in this trend; therefore, the Chaponost site provided a unique opportunity to demonstrate its ability to produce recycled concrete at a much higher level when compared with what was allowed by current standards. ELTS ordered a design from its local consultant. With a 25-MPa design strength, the structural calculations led to a slab-on-grade with an 18 cm thickness, divided in 5 m × 5 m pads identified by saw cuts on one-third of their full depth.

22.2.2 Mixture design and properties

The recycled aggregate originated from a single lot prepared by Filliot for the Chaponost project. The full characterization was performed according to the aggregate tests required by NF EN 206/CN (de Larrard et al. 2014). The fine aggregate displayed water absorption of 9.4% (according to NF EN 1097-6) and matched the requirements of the concrete standard. As for the coarse aggregate, it was classified as Rcu_{95} (with a total amount of recycled concrete and virgin aggregate of 98%), but due to its content of 1.3% of asphalt leading to a Ra_{10-} label, the final category was a type 3 according to NF EN 2016-1.

All concretes had to match the same set of specifications detailed in Table 22.1. This set refers to EN 2016-1. It contains a provision of minimum entrained air. As for the replacement rate, although the standard does not allow any recycled material of this type of environment (XF2), here the amounts of fine and coarse recycled aggregates were chosen between the three following values: 0%, 30%, and 100%. The recipes were adjusted to provide comparable slump values and compressive strength at 28 days. Owing to the lower packing ability of recycled aggregates (as compared to natural ones), the water amount increased with the recycling rate. To maintain the water/binder ratio under the maximum specified level, the cement amount was augmented (only in mixes containing 100% of coarse recycled aggregate). Regarding slump retention, it was necessary to add a retarder in the recycled mixtures, even when the aggregates were previously saturated. As a matter of fact, it turns out that recycled fines can bring a perturbation in the cement/superplasticizer system (see Chapter 7). The concrete recipes and properties appear in Table 22.2, as produced on site.

Table 22.1 Concrete specifications

Concrete type (according to NF EN 206-1)	*Slump between T_0 and T_0 + 1 h 30 (mm)*	*Minimum compressive strength (MPa)*	*Minimum dosage in equivalent binder (kg/m³)*	*Maximum water/equivalent binder ratio*	*Entrained air volume (%)*	*Recycled aggregate rate (%)*
C25/30 XF2 D20 S4	$150 \leq S \leq 230$	25	300	0.55	≥4	0–100

Table 22.2 Concrete mix proportions and properties. XS-YG: X% of recycled fine aggregate, Y% of recycled coarse aggregate

Constituents (kg/m³)/recipes	*REF*	*30S-0G*	*0S-30G*	*30S-30G*	*0S-100G*	*100S-100G*
11/22 natural aggregates	720	752	509	506	0	0
4/11 natural aggregates	286	174	139	139	0	0
0/4 natural aggregates	790	541	789	506	772	0
4/20 recycled aggregates	0	0	282	280	778	725
0/4 recycled aggregates	0	235	0	234	0	654
Cement CEM II (Portland limestone cement)	302	306	305	308	346	390
Plasticizer	2.57	3.65	2.6	2.62	2.94	3.32
Retarder	0	1.54	0.88	1.54	1.04	1.95
Air-entraining agent	0.91	0.46	0.76	0.61	0.35	0.4
Water	173	188	178	192	205	260
w/c (free water/cement ratio)	0.52	0.54	0.53	0.54	0.54	0.53
Slump (mm) (NF EN 12350-2)	200	200	200	200	190	190
Air (%, at the concrete plant) (NF EN 12350-7)	7.2	8.5	7.6	7.6	5.5	9
Air (%, in the lab) (NF EN 12350-7)	5.8	7.3	6	7	3.5	6
Krai test (ASTM C 1579, %)	n.d.	238	75	181	200	274
Shrinkage from 0 to 24h (10^{-6})	1,580	2,420	2,130	2,280	3,200	3,090
Compressive strength at 1 day (MPa) (NF EN 12390-3)	7.2	7.8	6.7	6.7	10.8	13.6
Compressive strength at 7 days (MPa) (NF EN 12390-3)	22.7	25	25.6	22.2	33.1	25.6
Compressive strength at 28 days (MPa) (NF EN 12390-3)	31.1	31.3	32.1	29.1	40.1	33.3
Splitting tensile strength at 28 days (MPa) (NF EN 12390-6)	2.8	3.2	2.9	2.8	3.3	3.2
Shrinkage from 1 to 90 days (10^{-6}) P18 457	397	430	469	492	545	838
E-modulus at 28 days (GPa) ISO 6783	30	28	28	23	25	21
Ring cracking test (local procedure)	No	No	No	No	No	Yes

In spite of a lower dosage of air-entraining agent, the air volume was increased by the presence of recycled sand, leading to somewhat excessive values, but this phenomenon did not markedly affect the compressive strength, which was well above the 25-MPa specification. The 0S-100G mix exhibited a higher strength, because of its lower-than-expected air content.

All hardening or hardened concrete measurements were performed on samples taken at the concrete plant and brought to the Sigma Beton's laboratory. The sensitivity to plastic shrinkage was assessed through Krai tests (ASTM C 1579). It appears that recycled mixtures are more prone to this problem. However, given the season, plastic shrinkage did not affect the parking lot (see Section 22.2.3). Drying shrinkage, as measured on 7 cm × 7 cm × 28 cm specimens, was strongly affected by the replacement rate, as does the E-modulus but in an opposite trend. As compared to the reference mix, the fully recycled concrete had a double shrinkage (see shrinkage data at 90 days for REF and 100S-100G), and about one-third less in terms of E-modulus (see E-modulus data at 90 days for REF and 100S-100G). A restrained ring cracking test was carried out on all mixtures (see Figure 22.1) following a local procedure. Only the fully recycled mix showed a crack, having occurred at an age of 4 days.

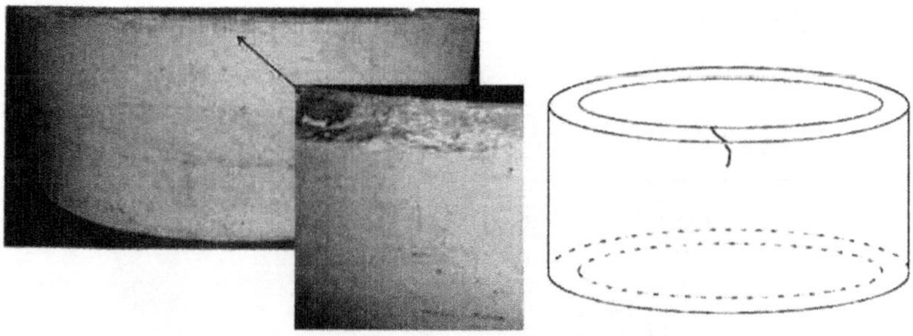

Figure 22.1 Cracking of the 100S-100G ring specimen.

22.2.3 Parking layout and construction process

Figure 22.2 shows the general layout of the parking lot. A small zone on the right-hand side was cast in advance with the 100S-100G mix, to check its suitability using a conventional casting process. Then, the six zones were laid down within a week in December 2013. The weather was cold and misty. During construction of the D6 zone, a light rain fell on the site.

No special remark came from the production of concrete, which was performed as usual whatever the recycling rate. Although difficulties in handling recycled aggregates (especially sand) were previously reported, no difficulty arose regarding the aggregate flowing process in hoppers. The minimum mixing time of 55 s was observed according to the current standard requirement for concrete-containing admixtures.

Figures 22.3–22.5 show the various stages of construction. The slabs were manually finished and then troweled with a "helicopter"-type machine. No curing of any type was carried out. On site, the crew did not report any difference between the mixes, which were all fluid and easy to trowel. The only difference consisted in the aspect of fully recycled concrete surface. Here, some wood chips were visible, unlike the other zones where no alien particles were visible.

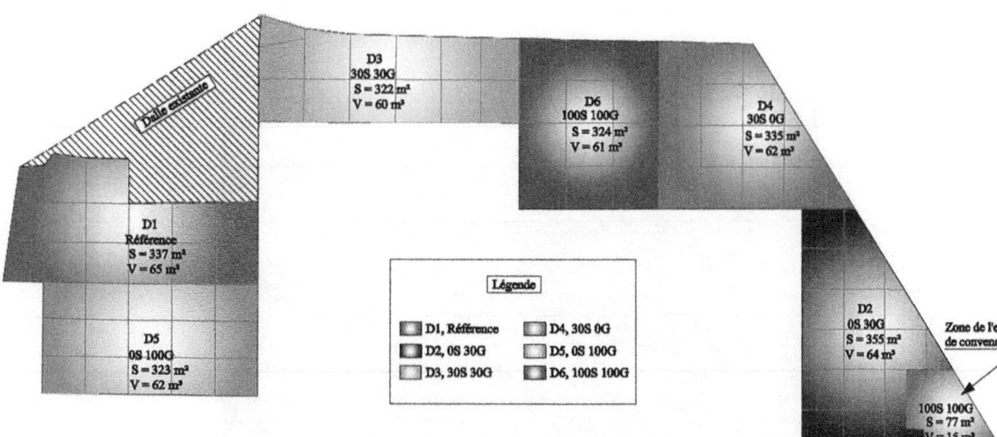

Figure 22.2 The Chaponost parking lot (de Larrard et al. 2014).

Figure 22.3 View of the site before concrete casting (de Larrard et al. 2014).

Figure 22.4 Casting of fresh concrete (de Larrard et al. 2014).

Figure 22.5 View of the site after concrete casting (de Larrard et al. 2014).

Figure 22.6 In-place cracking of a joint observed after 4 months [Picture Lafarge LCR].

22.2.4 Observations

The site was subject to various visits after construction. The main aim was to detect a possible cracking of the concrete slabs, generated by the restrained shrinkage of concrete. As a matter of fact, no crack was observed between the sawn joints. Examination of some joint sections at the edge of the pads allowed identifying few little cracks (see Figure 22.6).

22.3 MODELING OF CRACKING RISK

Cracking risk can be due to different degradation mechanisms associated to different scales of observation:

- at a macroscopic scale, hydric and thermal gradients between the surface and the core of the slab will induce strain gradients and so stress gradients (tension at the surface can lead to cracking when this tension overcomes the tension strength). Moreover, hydric restriction by the ground could prevent slab deformation and lead to through crack and debonding between slab and ground.
- at a mesoscopic scale, cement paste deformation is restrained by the aggregates (more rigid), leading to debonding at the cement paste/aggregate interfaces (circumferential cracks) and to the growth of intergranular cracks (radial cracks).

To consider these different mechanisms and to predict and compare the risk of cracking for the different recycling mixes, thermohydromechanical simulations were carried out at LMT by C. De Sa and F. Benboudjema using the finite element code CAST3M developed by the French Atomic Energy Commission (Cast3M).

22.3.1 Formulation and identification of models for the various transfers and mechanical properties

The hydration process is first described by the concept of chemical affinity, based on the Arrhenius' law. To describe the temperature field, the heat equation is solved, with a source term originating in cement hydration. Boundary conditions are of the convective type; the heat flow is assumed to be proportional to the difference between ambient and surface temperature.

For the humidity transfers, only the liquid flow is considered, with a diffusion law accounting for a desorption isotherm. From the humidity field, the free drying shrinkage deformation is deduced through an empirical model, assuming a linear dependency between humidity and shrinkage. Autogenous shrinkage, although being minor in this range of water-binder ratio, is accounted for, assuming a linear dependency with the cement degree of hydration. The thermal strain is deduced from the local temperature by using a dilation coefficient.

The compressive strength and E-modulus are controlled by the degree of hydration. Basic creep modeled by a classical chain of Kelvin-Voigt models is taken, in series with a dashpot. The stiffness of the springs, as the viscosity of the dashpots, is assumed to depend on hydration degree, according to a model proposed by de Schutter (1999) and Stefan et al. (2010). The same model is used for tensile and compressive creep deformations, although this assumption is contrary to most experiments. Then comes the drying creep, which is supposed to be proportional to drying shrinkage, according to Bazant and Chern (1985). The total deformation is eventually the sum of the elastic and thermal strains, plus the autogenous/drying shrinkage and the basic/drying creep strains. When the calculated tensile stresses equal the strength, cracking is described with a modified Mazars damage model (Mazars 1984).

The identification of the models was carried out based on the following data:

- original strength data could allow the calibration of the degree of hydration law;
- the shrinkage tests, which included weight loss measurements, were used to calibrate the drying shrinkage law. At this stage, the shrinkage prismatic samples were simulated through a finite-element model. Comparison between simulations and measurements appear in Figure 22.7;
- for the basic creep model, original experimental data were lacking, except two mixes that were tested at Ecole Centrale de Nantes (see Chapter 11). The model was calibrated with these experimental data for these two mixes and for the other ones, basic creep was assumed to be inversely proportional to the Young modulus.
- typical data taken from the literature were used for the rest of the numerous adjustable parameters. For example, to capture the effect of recycling rates on drying creep, total creep experiments (Gomez-Soberon 2002b) were simulated to make the calibration of different parameters. The Chaponost formulations were then simulated and compared to Gomes-Soberon's ones (see Figure 22.8);

The whole modeling process is described in detail in the original RECYBÉTON report (de Sa and Benboudjema 2016).

22.3.2 Simulation of the ring cracking test

Application of the earlier models to the ring cracking test described in Figure 22.1 led to the damage field displayed in Figure 22.9. For all mixes except the fully recycled one, a limited damage appeared corresponding to a thin skin cracking. However, for the 100S-100G mix, the damage was much more developed, corresponding to crack developing in one-third to

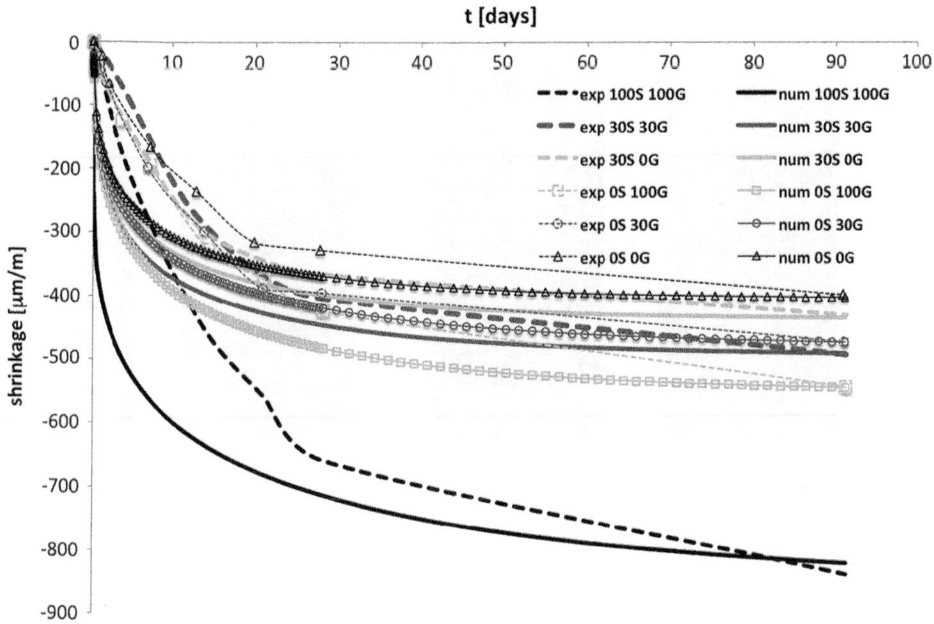

Figure 22.7 Calibration of drying shrinkage.

one-half of the ring thickness. This prediction is consistent with the observation reported in Figure 22.1, where a crack started from the top exterior ridge of the ring, crossing the whole thickness but stopping before reaching the bottom of the specimen.

22.3.3 Simulation of the slabs

Three-dimensional simulations were run on models forming one-fourth of six 5 m × 5 m slabs cast with the various mixes, supported on a 1-m thick soil layer (see Figure 22.10).

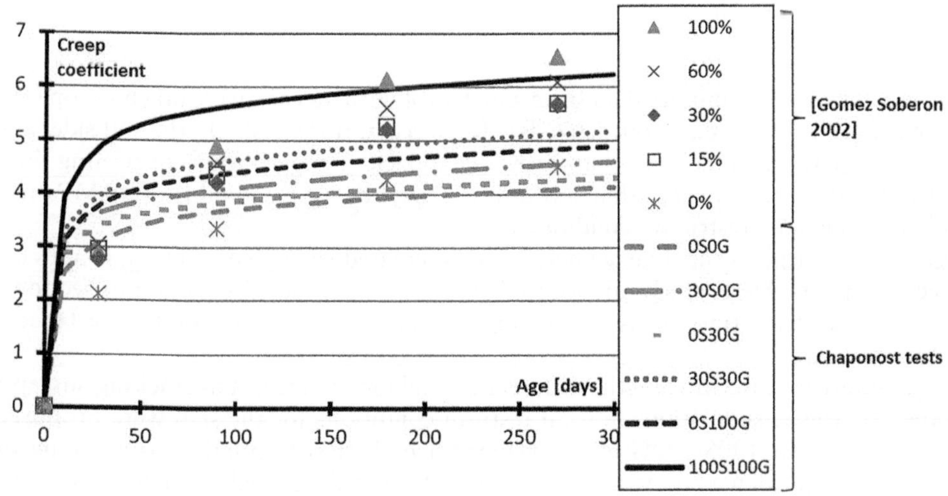

Figure 22.8 Creep coefficients (experimental and modeling).

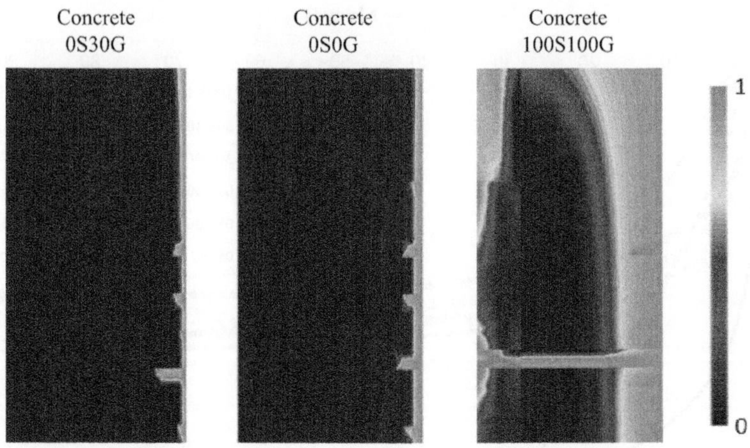

Figure 22.9 Simulated damage of rings.

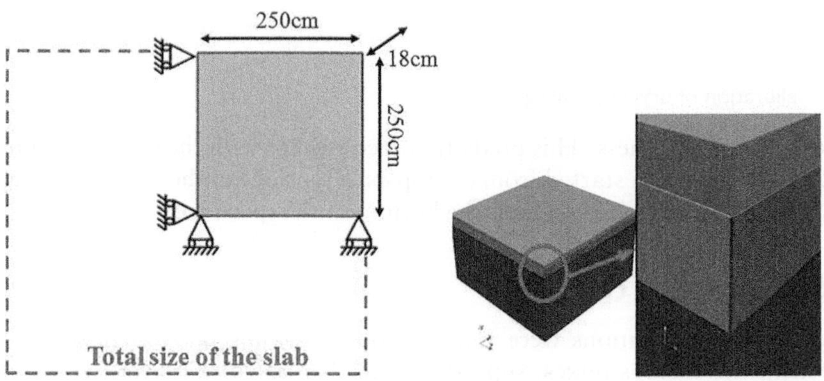

Figure 22.10 3D slab-on-grade mesh () with detail on joints. The top and the bottom layer stand for the concrete slab and the soil, respectively.

Real temperature and humidity data from a meteorological agency were taken as top boundary conditions (considering a convective law for thermal transfers). On the soil side, a 98% humidity value was assumed. Mechanically, due to lack of experiments concerning the soil, a perfect bond was assumed with a soil characterized by an E-modulus of 40 MPa (corresponding to the more restrictive conditions).

Figure 22.11 displays the damage field in six simulated slabs. No cracking is anticipated between the joints. Here, cracks are located, owing to the sudden reduction of section that induces a stress concentration. As for the ring test, simulations are in good accordance with the observations.

To quantitatively discriminate the different formulations in terms of cracking susceptibility, other 1D elastic simulations in total restrain conditions for the slab were proposed by LMT, showing that 100S-100G mix developed a highest susceptibility to cracking (indicated by an index calculated by dividing tension stress σ_t developed by material strength f_t, see Figure 22.12). The other compositions presented similar values of cracking index and not higher than the reference one.

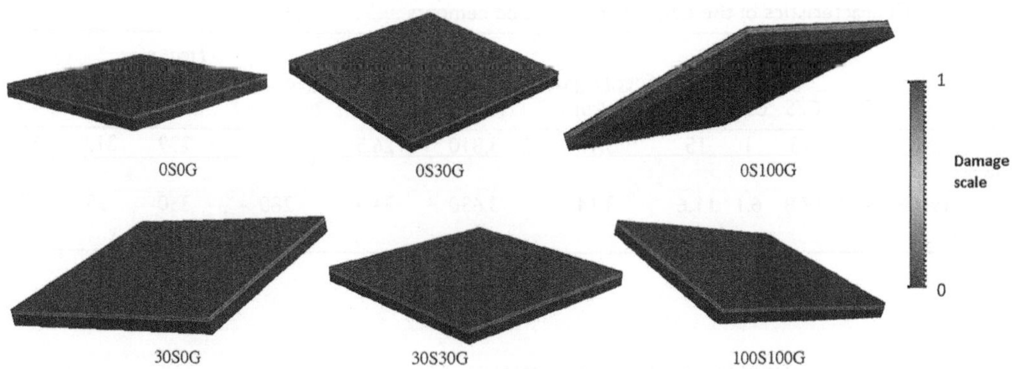

Figure 22.11 Simulated damage of slabs.

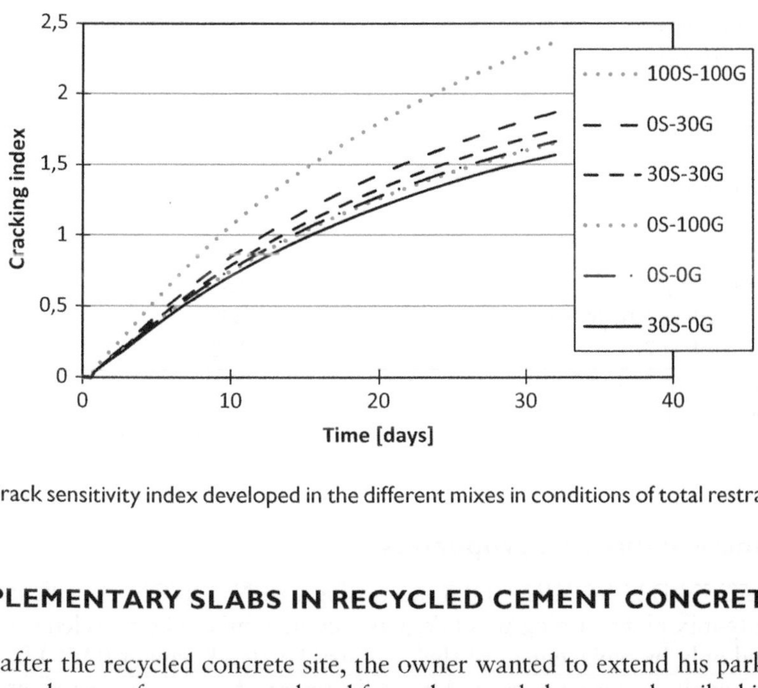

Figure 22.12 Crack sensitivity index developed in the different mixes in conditions of total restrain for the slab.

22.4 SUPPLEMENTARY SLABS IN RECYCLED CEMENT CONCRETE

Three years after the recycled concrete site, the owner wanted to extend his parking lot, and accepted to test the use of concrete produced from the recycled cement described in Chapter 4. Let's recall that this cement's clinker incorporated 15% of ground recycled concrete in its raw materials. Approximately, 5.8 tons of this cement, produced by the Vicat company, was delivered to the LafargeHolcim's Herriot ready-mix concrete plant (located in Lyon).

22.4.1 Cement incorporating ground recycled concrete as an alternative raw material

The development phase of the recycled cement is detailed in Chapter 4. Its mains physical, chemical, and mechanical properties are reported in Table 22.3. No significant difference can be detected between the two cements, apart from the water demand, which is higher for the recycled cement (although the Blaine specific surface is lower).

Table 22.3 Characteristics of the control and recycled cements

	C3S	C2S	C3A	C4AF	*Specific gravity (g/cm³)*	*Blaine SS (cm²/g)*	*Water demand (%)*	*Setting time (mn)*	*Heat of hydration at 41 h*	Rc_2 *(MPa)*	Rc_{28} *(MPa)*
OPC CEMI 52.5	66	13	1	15	3.18	3,810	26.5		299	31.8	61
CR1 recycled cement	61.8	17.8	6.1	11.6	3.14	3,630	34.9	280	350	35	58

Table 22.4 Characteristics of the produced concretes

Constituents (kg/m³)/recipes	*Control concrete*	*Recycled cement concrete*
11/22 virgin aggregates	720	720
4/11 virgin aggregates	290	290
0/4 virgin aggregates	790	790
CEM I 52.5 N CP2 cement	300	—
CEM I 52.5 N recycled cement	—	300
Plasticizer (%)	2.55	2.55
Air-entraining agent (%)	1.05	1.05
Water	178	187
w/c (free water/cement ratio)	0.59	0.62
Slump (mm)	220	195
Air (%, in the lab)	3.5	2.8
Air (%, at the concrete plant)	5.0	6.0
Compressive strength at 1 day (MPa)	6.9	6.8
Compressive strength at 7 days (MPa)	23.2	25.3
Compressive strength at 28 days (MPa)	29.4	29.2
Splitting tensile strength at 28 days (MPa)	2.67	2.80

22.4.2 Concrete mix and properties

A C25/30 I 52.5N D22 S4 XF2 CL0.4 (according to EN 206/CN classification) concrete from the ready-mix plant catalog was taken as a control mix. The recycled cement mixture was generated only by replacement of the ordinary Portland cement (OPC) by the recycled cement, keeping all dosages constant. After adjustment of water content to provide a comparable consistency, the recycled cement concrete appeared to have a slightly higher water/cement ratio. However, fresh concrete as well as hardened concrete properties were very similar. The workers did not detect any difference during the casting process, while the technician in charge of sampling the mixes found a slightly stickier consistency. Mix designs and fresh and hardened state properties are presented in Table 22.4.

22.4.3 Laying of supplementary slabs at Chaponost

Slabs-on-grade were cast with the same thickness and same technology as those built 3 years before. To date, neither degradations nor differences were reported, which could be attributed to the different cements used.

22.5 RESEARCH NEEDS

Regarding the use of recycled concrete, the Chaponost construction site did not raise any significant roadblock overcoming the practical implementation of concrete recycling into practice. The mix-design process was quite straightforward, except that the admixture management raised two difficulties. The slump retention was partially puzzled by the introduction of recycled (fine) aggregate. Research is needed in the field of organomineral chemistry of the cement–superplasticizer interactions, to understand the role of recycled aggregate. Likewise, the negative effect of recycled sand on air content requires further study. By the way, this effect was noticed on site, but not during the previous laboratory trials, in which a different delivery of recycled aggregates was used. This means that the perturbing role of recycled aggregate on fresh concrete rheology could depend on slight variations of secondary parameter.

In terms of material characterization, the modeling process suffered from a lack of creep data. A large investigation on recycled concrete creep is necessary, tackling the effects of several parameters as the replacement rate, the age of concrete at loading, and the effect of humidity. From such a complementary characterization, reliable simulations could be performed aiming at cracking risk assessment in specific projects.

Regarding concrete incorporating recycled cement, it should be important to check that there is no more effect of incorporation of recycled concrete into the clinker on durability-related properties than in fresh and mechanical hardened concrete properties.

22.6 CONCLUSIONS

The Chaponost parking lot was the first experimental construction site carried out within RECYBÉTON. From a practical viewpoint, the aim was to check that recycled concrete could be produced and cast without changing the usual procedures. On the design and mechanical behavior side, the challenge was to see to what extent a high-rate replacement of virgin aggregates by recycled concrete aggregates could be carried out without creating shrinkage-induced cracking in large slab-on-grade constructions. Four years after construction, the following conclusions are drawn:

1. A type 3 (according to EN 206-1 standard) recycled aggregate was used and did not lead to any identified difficulty, in spite of the recommendation of current standards to prefer type 1 or 2 recycled aggregate in recycled concrete;
2. Six recipes were developed, with replacement rate ranging from 0% to 100%, including the fine fraction. They were subject to a large characterization test series (except the creep behavior that could not be studied). Usual trends were found, with a decrease of E-modulus and an increase of drying shrinkage at various stages. To maintain compressive strength, the cement dosage was increased only for the mixes containing more than 30% of recycled aggregates;
3. Ring cracking tests were carried out in the laboratory. The 100% recycled concrete was the only one to display an increase in cracking tendency, as compared to the control mix;
4. The construction of the parking lot, with six zones corresponding to each concrete recipe, could develop smoothly. No difference between mixtures was noted by the work crew during casting and finishing. Some wood chips appearing at the surface of the fully recycled concrete were the only obvious difference between the zones;
5. A comprehensive finite-element modeling was carried out for the cracking process of the in-place slabs, accounting for the various mix properties. The shrinkage and ring

lab tests were simulated for calibration or prediction purpose. Creep results from the literature were used to calibrate the creep model, while site-related data were sufficient to calibrate the other models;

6. As expected, it was found that, for an 18-cm thick slab, drying shrinkage is the main motor for cracking, thermal effects playing a minor role. The simulations could reproduce the cracking of the fully recycled concrete ring, unlike the other samples that were supposed to only display fine skin cracking;
7. Both experience and simulation did not show any crack on site, except in the joints sawn every 5 m in the slabs. The cracking tendency only grows for replacement rates higher than 30%. Event at 100%, in spite of the high cement paste amount and for the environment conditions of the site, cracking is not an issue with recycled concrete. This is to the authors' knowledge the first study demonstrating this important fact;
8. Although some creep tests were performed after this site within RECYBÉTON (see Chapter 11), a comprehensive creep study of recycled concrete would be desirable in the future, where parameters such as design strength, replacement rate, and age at loading would be systematically studied. Based on such a campaign, the cracking risk could be more deeply assessed in less favorable cases (thicker concrete pieces, dryer environment, etc.).
9. The practical use of cement made from partially recycled clinker neither displayed any difference in terms of concrete properties nor in terms of constructability, as compared to ordinary Portland cement. Keeping in mind that this recycled cement is in fact already allowed by EN 197 standard, this is an encouraging result. Once the durability brought by this cement has been checked, there will be no barrier other than economical or environmental (see Chapter 33) for a dissemination of this new practice.

ACKNOWLEDGEMENTS

The authors thank the ELTS company (Chaponost, Rhône) for having provided the opportunity to organize this experimental construction site, Vicat, for the production of recycled cement and Lafarge France for the production of recycled concrete.

Chapter 23

Civil engineering structure

Nîmes Montpellier rail bypass site

I. Moulin and E. Perin
Lerm Setec

O. Servan
Setec TPI

M. Verbauwhede
Bouygues TP

CONTENTS

Abstract

This section presents the construction of a conventional truss bridge of the Nîmes Montpellier rail bypass project using recycled coarse aggregates (RA). This full-scale experiment is combined with laboratory study. For this structure, a C35/45 concrete was required to comply with an environmental class XC4 XF1 defined in NF EN206/CN. Two recycled concrete mixes were designed: the first one was limited to 20% RA according to NF EN206/CN for this kind of environmental class, and the second one explored the impact of a higher content of 40% RA replacing a totally natural 6/14 coarse aggregate. Chemical, mineralogical, and physical properties of the selected RAs were determined. Concretes were studied in laboratory as usual for this kind of structure; in addition, further characterizations were performed (water porosity, oxygen permeability, carbonation depth, and shrinkage assessments) to evaluate the impact of RA ratio on concrete durability. Laboratory studies reveal that fresh concrete and hardened concrete properties of both 40% and 20% RA mixes were in agreement with requirement specification. The bridge was built with the 20% RA mix and standard concrete mix. No particular difficulties were observed neither in concrete production nor in its placement. One sacrificial wall element to allow core sampling was cast at the same time as the structure to follow concrete properties over time. More than 2 years after construction, no particular defaults were observed, and the measurements of durability parameters on core samples were as expected. This experience shows that RAs could be generalized even in engineered structures without major difficulties, probably, even at higher than 20% content as proved by the positive results obtained for the 40% mix.

23.1 INTRODUCTION

Recycled aggregate (RA) concrete is allowed in France within some limits defined by NF EN 206/CN (2014). However, such recycled concrete is barely used all the more in engineering structures, and feedback is limited. Setec group is in charge of the construction management of Nîmes Montpellier rail bypass project. Thanks to its innovation approach (Seteclab initiative) RECYBÉTON national project partners, Oc'Via Company and Bouygues Company, were put through. This collaboration has led to the implementation of a RA concrete in one of the bridges of the project. The goal is to demonstrate the suitability of RA concrete and its implementation for an engineering structure. Another outcome of this project will be to better identify hindrances to the development of recycled concrete in such context.

This full-scale experiment is combined with laboratory study to evaluate the impact of RA proportion on the concrete properties. Coarse aggregates were substituted by 20% type 1 of RA, which is the maximum allowed proportion by the NF EN206/CN standard and also by 40% in a prospective way. Special emphasis is also placed on concrete durability, an important aspect for the possible use of RAs in engineering structures.

23.2 GENESIS OF THE PROJECT AND SITE DESCRIPTION

The Nîmes Montpellier rail bypass project is a new high-speed railway line in the south of France, which requires the construction of 188 civil engineering structures. Its delivery is scheduled for 2017.

The project is financially and contractually structured as a public–private partnership between RFF (Réseau Ferré de France: French network rail company) and Oc'Via. Working with the engineering companies SETEC and SYSTRA, Oc'Via is in charge of the design and construction of the railway line and must ensure its operation and maintenance.

The joint SETEC/Oc'Via decision to build a structure using RAs was put into effect in February 2014. The selected structure is a conventional rigid-framed bridge (about 12 m long and 7.5 m wide) located in Nîmes in the south east of France. This bridge allows a bicycle path to cross over Valdebane creek. The bicycle path largely follows the new railway line and is an integral part of the Nîmes Montpellier bypass project. Nevertheless, the structure is outside the railway zone. Concrete must be compliant with NF EN 206/CN (2014) standard.

The selected structure was scheduled for completion in June 2014; therefore, project members have had a tight delay to design and approve the concrete mixture in respect of French regulation (Fascicule 65 2014).

The whole crossmember and one return wall (called M1D in top view, Figure 23.1) were casted with RA concrete, giving 42 m^3 for a total of 100 m^3.

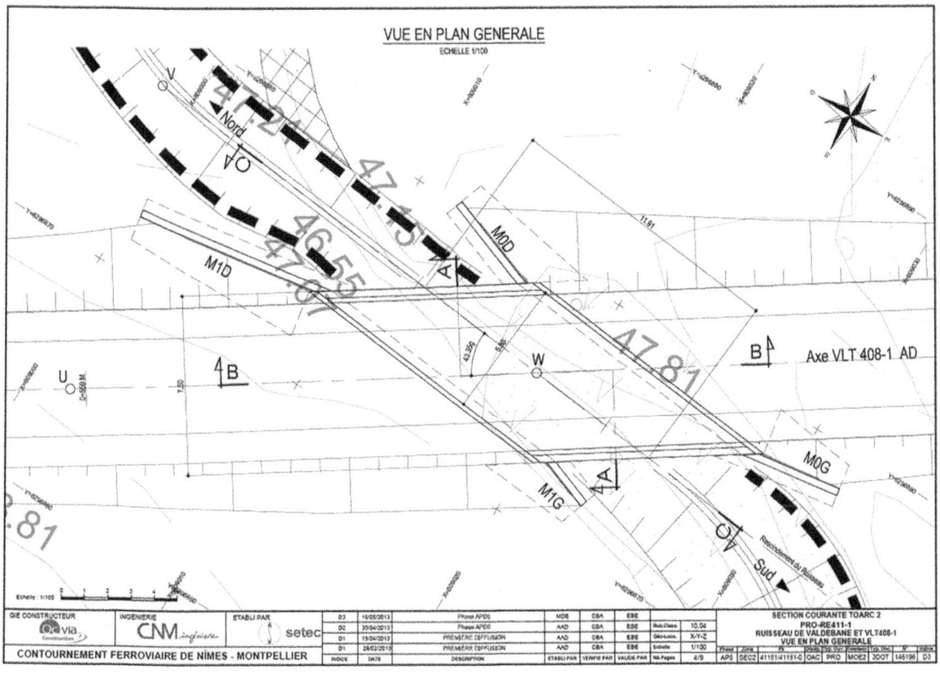

Figure 23.1 Top view of selected structure.

23.3 MATERIALS AND MIXTURE DESIGN

23.3.1 Recycled aggregate

23.3.1.1 Supply

The RA was from a single lot (50 tons) prepared by LRM Company and dedicated to the project. The platform was located 20 km of Unibeton mixing plant, which was in charge of concrete production. Recycled 6/20 mm aggregate was produced by crushing concrete block to reduce their size and allow steel elimination. The RA (Figure 23.2) was classified type 1 according to NF EN 206/CN (2014) standard (all parameters were CR_B).

Figure 23.2 Stock of RA prepared by LRM. (Photo credit: Lerm.)

23.3.1.2 Aggregate characteristics

Specific gravity, absorption coefficient, and water-soluble content were determined on ten samples. Other characterizations required by NF EN 12620 standard were carried out on an average representative sample. Attention has been paid to potential alkali aggregates reaction problems. To this end, petrographic analysis, active alkalis content, and dimensional stability tests under alkaline conditions according to XP P 18-594 standard were performed. RA characteristics are summarized in Table 23.1. All properties matched the requirement of concrete standard, and note that the RA displayed water absorption of 6.1%–7.3% with an average of 6.6%. Petrographic analysis showed different nature of rocks, which indicate that RA was from various old concrete. It can be estimated that RA contains near 4% of reactive silica, a range that could lead to potential alkali aggregate reaction. However, dimensional stability tests under alkaline conditions allow classifying the RA as nonreactive.

The RA is classified as category D due to its absorption coefficient and its sulfate content, but all other parameters belong to category A (according to NF EN 12620).

23.3.2 Concrete mix-design

At the design stage, C35/45 concrete was needed to comply with environmental class XC4 XF1 defined in standard (NF EN 206/CN 2014). Concrete mixes selected for the overall project were more efficient and suited for more aggressive environment, in particular, to saltwater environment. Two recycled concrete mixes were designed from one of those mixes (called standard mix) and studied in laboratory:

- recycled concrete mix noted 40% RA, which allows replacing an all-size fraction of coarse natural aggregate,
- recycled concrete mix noted 20% RA, for which RAs was limited to 20% according to NF EN 206/CN (2014).

Due to potential alkali aggregates reaction, alkali supply was controlled using CEM I 52.5 and ground granulated slag mix.

Table 23.1 RAs properties

Physical characteristics		
Specific gravity (kg/m^3)	NF EN 1097-6	2,350–2,420
Absorption coefficient (%)	NF EN 1097-6	6.6
Micro Deval (%)	NF EN 1097-1	29
Los Angeles (%)	NF EN 1097-2	29
Chemical characteristics		
Chloride content (%)	NF EN 1744-1 article 7	<0.01
Sulfate content (%)	NF EN 1744-1 article 12	0.34
Total sulfur content (%)	NF EN 1744-1 article 11	0.14
Organic matter	NF EN 1744-1 article 15-1	Negative
Water-soluble sulfate content (%)	XP P18-581	0.11–0.15
Alkali aggregate reaction tests		
Active alkalis (%)	LPC no 37 test (LCPC 1993)	0.0111
Petrology	NF EN 932-3	Limestone
Dimensional stability test under alkaline conditions	XP P 18-594 article 5.1	Nonreactive

Table 23.2 Concrete compositions

Concrete materials	*Std.*	*20% RA*	*40% RA*
CEM I 52.5N SR3 CE PM (kg/m^3)	320	320	320
Ground granulated slag (kg/m^3)	60	60	60
0/4 natural sand (kg/m^3)	860	850	840
6.3/16 natural coarse aggregate (kg/m^3)	340	270	190
11/22 natural coarse aggregate (kg/m^3)	595	450	330
6.3/20 RAs (kg/m^3)	—	180	350
Water-reducing plasticizer 1 (kg/m^3)	2.66	2.66	2.66
Water-reducing plasticizer 2 (kg/m^3)	0.76	0.76	0.76
Effective water/equivalent binder	0.44	0.44	0.44

Standard concrete mix and recycled concrete mixes are detailed in Table 23.2. It may be noted that no water was added and additive ratio was not modified.

23.4 LABORATORY TESTS

23.4.1 Concrete mixes properties

Hardening concrete and mechanical properties measurements were performed by Unibéton and managed by Bouygues TP. Laboratory tests were in accordance with the requirements of French legislation and specific requirements of French network rail company. Results are shown in Table 23.3. It may be noted that aggregates were not previously water saturated.

S4 slump-class (160–220 mm) was maintained for 2 h as required without changing recipes. In spite of RA introduction, the compressive strength at 28 days was not significantly modified.

Durability tests were performed by Lerm laboratory on samples taken at the concrete plant, except for shrinkage evaluation measured on $7 \times 7 \times 28\,cm^3$ specimens made at Lerm and stored at 20°C and 50% HR. Mass and length were measured at different due dates up to 1 year. To evaluate, at least partially, the importance of autogenous shrinkage (on hardened concrete), the dimensional stability of a second set of specimens embedded in an aluminum adhesive film was followed (first measurement being made at 24 h). Figure 23.3

Table 23.3 Concrete properties

Concrete mixes	*Std.*	*20% RA*	*40% RA*
Fresh concrete			
Slump T0 (mm)	230	235	235
Slump T60 (mm)	225	220	235
Slump T120 (mm)	190	215	220
Specific density (kg/m^3)	2,356	2,340	2,323
Entrained air (%)	2.0	2.9	2.1
Hardened concrete			
Compressive strength at 7 days (MPa)	44	38	41
Compressive strength at 14 days (MPa)	—	47	47
Compressive strength at 28 days (MPa)	55	57	55
Tensile strength at 28 days (MPa)	4.0	3.8	3.6

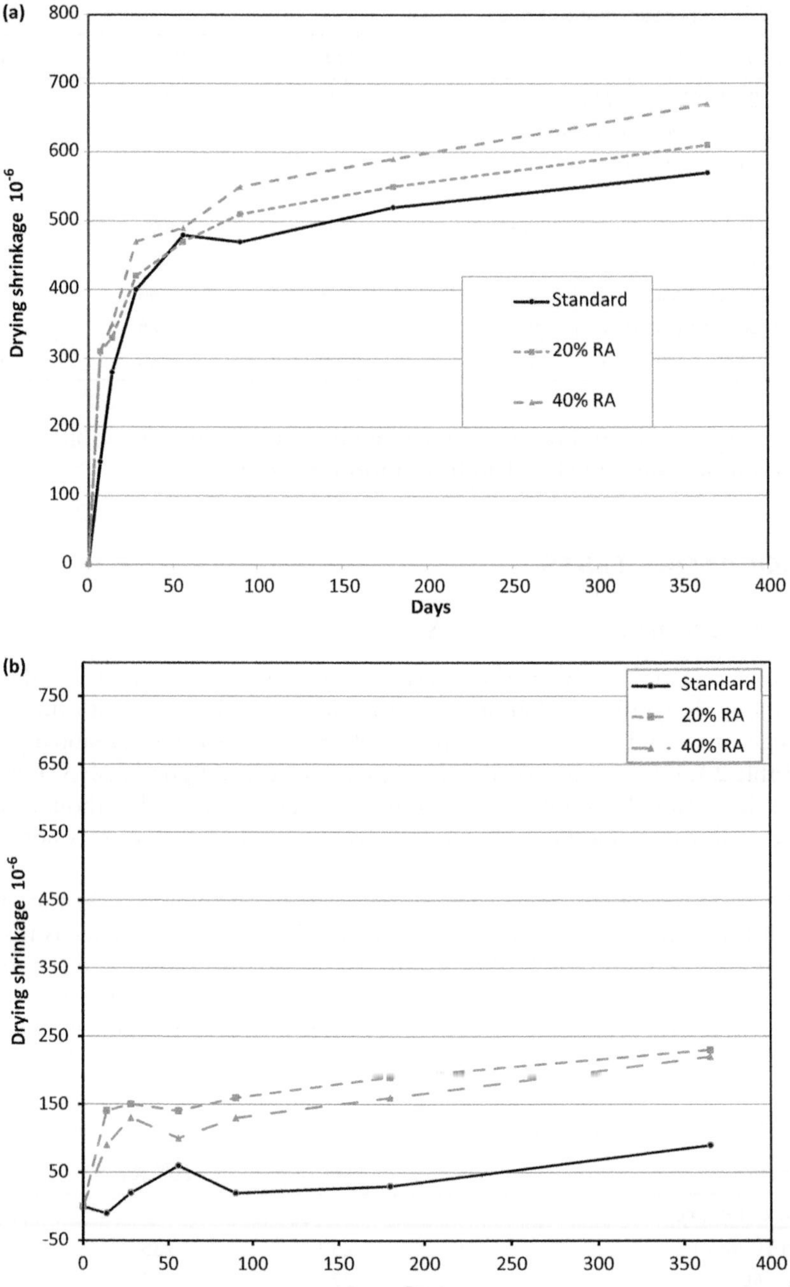

Figure 23.3 Drying shrinkage: (a) specimens stored in 20°C and 50% HR, (b) specimens stored in 20°C and 50% HR with a protective waterproof.

illustrates the total drying and autogenous drying shrinkage over time of 20% RA, 40% RA, and standard concrete.

Durability indicators (water porosity, oxygen permeability, apparent chloride diffusion coefficient, and carbonation depth) were performed on concrete specimens cured in water at 20°C during 90 days.

Table 23.4 Durability properties of concrete (averages over three measurements)

Mix	*Std.*	*20% RA*	*40% RA*
Water porosity at 90 days (%) NF P 18-459	13.1	14.6	14.4
Oxygen permeability at 90 days ($10^{-18}m^2$) XP P18-463	49	213	149
Apparent chloride diffusion coefficient at 90 days ($10^{-12}m^2/s$) XP P18-462	8.6	9.6	8.3
Carbonation depth at 90 days (accelerated test) (mm) XP P 18-458	7.8	7.1	—
Alkali aggregate reaction test (week 52) NF P18 454 (%)	—	0.016	0.019
Drying shrinkage from 1 to 90 days (10^{-6})	470	510	550
Autogenous shrinkage from 1 to 90 days (10^{-6})	20	160	130

Water porosity was measured by imbibition under vacuum and hydrostatic weighing according to French standard NF P 18-459.

Oxygen permeability measurement was performed on previously dried concrete disc (diameter of 95 mm and thickness of 50 mm) according to the French standard XP P18-463.

Apparent chloride diffusion coefficient was performed according to the French standard XP P18-462: concrete discs (90 days moist cured) are placed between two compartments, the first one containing a 0.1 M NaOH+0.5 M NaCl solution and the second one a 0.1 M NaOH solution. Chloride migration was accelerated in an electric field, and the apparent diffusion coefficient was calculated based on the penetration chloride front.

Carbonation depth was measured after an accelerated reaction through an exposure of a gas mixture containing 50% of CO_2 according (90 days) to XP P18-458 standard.

All results are detailed in Table 23.4.

As expected, porosity and permeability values are higher for RAs concretes but differences are not so large, and RA concretes are still classified as moderately durable concrete. It could be noted that permeability measurements for RA concrete are widely distributed; therefore, the difference between 20% and 40% RA is not significant, considering the standard deviation for these parameters.

Use of RAs in concrete led to a marked increase of autogenous shrinkage; however, less impact on drying shrinkage is noted. This effect may be due to the delayed water absorption of RAs.

23.5 EXECUTION OF THE WORK

23.5.1 Production of concrete

The concrete containing 20% RAs (20% RA) has been chosen for the construction of the bridge. Concrete was produced by the Unibéton mixing plant and was transported for 15 km by a truck mixer. No particular preparations were performed on RA before their use (like prewetting). No difficulties have been reported at the production stage. The concrete was casted on 27 June 2014 for return wall and 17 July 2014 for the crossmember. The weather was sunny and hot.

23.5.2 Delivering and placing concrete

The implementation of the RA concrete occurred exactly as standard concrete. Figures 23.4 and 23.5 show different stages of the casting (construction). The concrete was cured by water sprinkling and a protective covering soaked with water to avoid premature dehydration. No particular difficulties appeared during delivering and placing concrete, and RA concrete behaved exactly as standard concrete.

Figure 23.4 Casting of 20% RA concrete. (Photo credit: Lerm.)

Figure 23.5 Leveling of 20% RA concrete. (Photo credit: Lerm.)

23.6 OBSERVATIONS

As all structures of the Nîmes-Montpellier bypass project, the bridge was visited and inspected after casting. No particular defaults were observed. In the framework of PN RECYBÉTON, a specific follow-up action was adopted in addition to classical visits. Throughout construction, one sacrificial concrete wall element was cast at the same time as the actual structure and under the same circumstances (formula, personnel, and conditions). This wall allows core sampling and to measure the concrete characteristics at different ages after the construction period in addition to visual observations. Two campaigns were planned: the first after 1 year and the second after 3 years. Concrete samples allow to measure water porosity, oxygen permeability, and carbonation depth. Figures 23.6–23.8 show the 1-year campaign

Figure 23.6 Core sampling in the sacrificial wall at 1 year. (Photo credit: Bouygues.)

Figure 23.7 View of the sacrificial concrete wall after core sampling. (Photo credit: Bouygues.)

Figure 23.8 Core sample and carbonation depth test. (Photo credit: Lerm.)

Table 23.5 Concrete properties

Mix 20% RA	*Design stage*	*1 year*
Water porosity at 90 days (%) NF P 18-459	14.6	14.4
Oxygen permeability at 90 days (10^{-18}m^2) XP P18-463	213	141
Carbonation depth at 90 days (accelerated test) (mm) XP P 18-458	7.1	—
Carbonation depth (mm) NF EN 14630 (6 cores)	—	4 (2.5–6)

and core sampling. Table 23.5 provides results of the 1-year campaign in comparison with properties measured at 90 days during laboratory study. All parameters are within expected value regarding previous laboratory tests. One-year permeability value is smaller than previous laboratory results, which confirm that heterogeneity of RAs concrete makes it difficult to measure oxygen permeability with a good reproducibility.

23.7 CONCLUSIONS

A conventional rigid-framed bridge of the Nîmes-Montpellier bypass project was partly constructed with RA concrete. The aim was to demonstrate that RA concrete could be used without changing mix design and implementation processes while maintaining the durability requirements specific for engineering structures. Laboratory studies revealed that the use of 20% or 40% of RA in concrete allowed maintaining the durability indicators in the expected range. Although drying shrinkage is favored by RAs, it stays within acceptable values. No particular preparations of the RAs was needed, and its implementation was straightforward. More than 2 years after construction (Figure 23.9), no particular defaults were observed, and measurements of durability parameters on core samples (water porosity, oxygen permeability, carbonation depth, etc.) were as expected. This experience shows that recycled aggregates could be generalized even in engineering structures without major difficulties, probably even at higher than 20% content as proved by positive results obtained for the 40% mix with type 1 RA. In this case, a particular attention must be paid to any potential degradation of creep modulus.

Figure 23.9 View of the bridge after 2.5 years. (Photo credit: Lerm.)

ACKNOWLEDGEMENTS

The authors thank Oc'Via for having provided this opportunity, Setec group for its financial support, LRM for providing RAs and Unibeton for the production of concrete.

Chapter 24

Administrative archives building

CD77

A. Ben Fraj
Cerema

S. Decreuse
CEMEX

CONTENTS

Abstract

This section presents the first experimental building, based on recycled aggregate concrete (RAC), carried out in France, within the RECYBÉTON national project. This building, with 40 m^3 of concrete, is particularly loaded (750 kg/m^2) and is intended for the archive of documents. Two concrete mixtures were manufactured: The first one with natural aggregate concrete (NAC) and the second (RAC) containing 30% of recycled fine aggregates and 50% of recycled coarse aggregates. A first laboratory characterization campaign was carried out by CEMEX company to measure the slump, the shrinkage, the compressive strength, and the Young's modulus of concrete. Then, tests of suitability were conducted by the Center for studies and expertise on Risks, Environment, Mobility, and Urban and Country Planning (Cerema), mainly devoted to fresh concrete properties (slump) and its compressive and tensile strengths. Results show the necessity of increasing cement content (+5%) and using superplasticizer, in RAC mixture, to reach the same specifications of NAC. An increase of shrinkage (+40%) and a decrease of elastic modulus (−17%) were measured when recycled aggregates were used. For building, the concrete was implemented five times. For every concrete delivery, an experimental characterization was done, through slump, compressive, and tensile tests. Drying shrinkage and durability properties (water porosity, gas permeability, and carbonation depth) were measured once, on both RAC and NAC specimens. On construction site, measured slump, compressive, and tensile strengths were the same as for RAC and NAC. Shrinkage, water porosity, and gas permeability increased, respectively, by 10%, 20%, and 80%, when recycled aggregate was incorporated. The carbonation depth, after 1 month, is 4 mm higher in RAC. The detailed inspection of the building carried out after construction and after 3 months do not reveal any defaults or micro-cracks that could be caused by recycled aggregates. The observed defaults (flatness, apparent steel, segregation) were due to the used framework and the lack of vibration.

24.1 INTRODUCTION

Today, sustainable development and ecology related to construction materials are strategic issues. Recycling concrete into concrete provides a promising alternative to depleting natural resources. In this context, Seine-et-Marne department (CD77) wants to be a territory leader in ecoconstruction and ecorenovation and federates various local actors within ecoconstruction projects. CD77 building, located in Mitry-Mory (Seine-et-Marne, France, in the northeast Paris' suburb), is the third experimental site built within RECYBÉTON national project. It also seems to be the first building in France based on recycled aggregates. The building has an area of 40 m² and is intended to store archives, which is the critical point in this project. Indeed, the applied load is of 750 kg/m². This peculiarity induced some particular dispositions in slab sizing, considering:

- The allowable deformation: the load should not damage movable elements for storing archive;
- Creep: the use of recycled aggregate induces a decrease of long-term modulus of recycled aggregate concrete (RAC).

For a better knowledge of recycled aggregate effect, durability tests and detailed inspection were carried on RAC specimens and CD77 building, respectively.

To achieve our goal, a comprehensive team was formed within RECYBÉTON, gathering the CD77 (acting as the owner), design and control offices, a construction company, CEMEX (for recycled aggregates and concrete production and delivery) and the Cerema (in charge of design and quality control, durability tests, and inspection).

24.2 CD77 BUILDING PROJECT

24.2.1 Presentation

The building was to be built in Mitry-Mory (Seine-et-Marne, in Île-de-France region) with an area of 40 m² (Figure 24.1). It was not supposed to be accessible to people, and it was annexed to an existing building ("Solidarity" house). The construction, intended to store archives, has shallow foundations, reinforced concrete slab and panels (with three windows) and a connection with existent building (canopy). A reinforced concrete ramp would be built to evacuate disabled persons.

From a technical viewpoint, a small quantity of concrete (40 m³) is needed, which justifies the use of a single RAC mixture, where 30% of fine aggregates and 50% of coarse aggregates are substituted by recycled ones: 13, 7, 13, and 7 m³ are the concrete volumes for foundations, slab, panels, and floor, respectively. The required characteristic compressive strength at 28 days is 25 MPa.

24.2.2 Materials and mixture design

The used cement is CEM II/A-L 42.5 R CP2, where calcareous 0/4 sand and 4/20 aggregates are used as natural materials. To improve the rheology of NAC, fly ash and plasticizer are

(a)

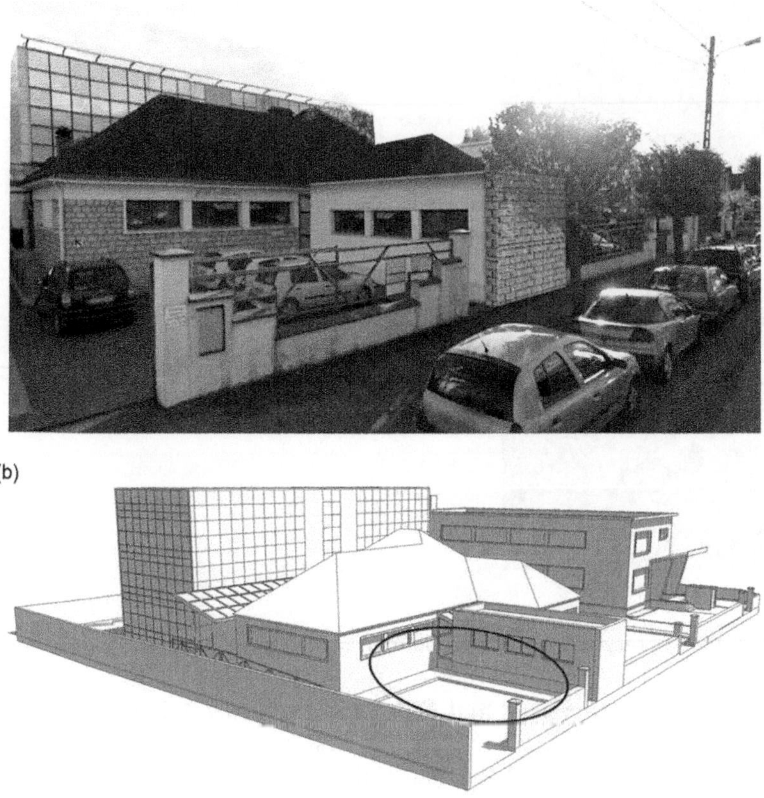

(b)

Figure 24.1 Building's location. (Source: CD77.)

used. The latter is replaced by a superplasticizer for RAC. Recycled aggregates, used as partial substitution of natural ones, are produced by CEMEX and described later.

24.2.2.1 Recycled aggregates

After demolishing, concrete blocks are transported to recycling plant or transformed in mobile platform (Figure 24.2). They first undergo crushing, reducing their size and allowing the elimination of steel reinforcement. Aggregates then undergo a secondary crushing before sorting. Pollutants, such as wood and plastic, are removed.

In our case, demolition concrete has been collected on Paris' harbors by CEMEX from various demolition sites. The demolition concrete has been transferred to Bouafles' quarry (Normandie) by barge (an empty barge that had brought natural aggregates to Paris has been used. So, no extra energy was needed to transport the material to be recycled). Recycled aggregates have been produced in Bouafles' quarry, using

- Hydraulic shears to remove the steel frames from the demolition concrete
- Jaw crusher to produce 0/100 mm recycled concrete material that is separated through a screening device in 0/40 mm and 40/100 mm.
- The 40/100 mm size fraction is crushed in a gyratory crusher and passed on a screen to produce 0/4 mm and 4/20 mm.

Figure 24.2 Production of recycled aggregates. (Source: CEMEX.)

Fifty tons of recycled 0/4 mm sand and 150 tons of recycled 4/20 mm aggregates are produced. According to NF EN 206/CN (2014), these materials are type 1, and their category is CR_B. The physical properties of natural and recycled materials are summarized in Table 24.1. As shown, old mortar decreased the density of recycled fine and coarse aggregate by 16%. The presence of this mortar and its high porosity are at the origin of high water absorption for recycled materials: 8 times and 15 times higher than that of natural coarse and fine aggregates, respectively.

24.2.2.2 Concrete mixture

As the building is located in Île-de-France region, a moderate freezing (XF1) and high risk of corrosion by carbonation (XC4) are considered, according to NF EN 206/CN (2014). The minimum characteristic compressive strength (MPa) and maximum water-to-equivalent binder ratio are, respectively, 25 and 0.6. For easy implementation, S4 slump-class is required, which corresponds to a slump of 160–220 mm. It should be noticed that it

Table 24.1 Physical properties of natural (NA) and recycled (RA) aggregates

Materials	*Density (-)* *NF EN 1097-6*	*Water absorption (%)* *NF EN 1097-6*
0/4 mm natural sand	2.63	0.52
0/4 mm recycled sand	2.22	8.3/7.3[a]
4/20 mm natural aggregates	2.67	0.66
4/20 mm recycled aggregates	2.28	5.5

[a] First value is measured based on the 0/4 mm size fraction, and the second value is measured based on the 0.063/4 size fraction.

Table 24.2 NAC and RAC mixtures

Constituents	*NAC (reference)*	*RAC*
Aggregates		
0/4 mm natural sand (kg)	826	543
0/4 mm recycled sand (kg)	—	233
4/20 natural coarse aggregates (kg)	1,041	489
4/20 mm recycled coarse aggregates (kg)		489
Binders		
CEM II/A-L 42.5 R CP2 (kg)	243	255
Fly ash (kg)	61	61
Admixtures		
Air entraining agent (%)	0.15	
Plasticizer (%)	0.60	
Viscosity agent		0.33
Superplasticizer (%)		0.80
Effective water (L)	166	155
Equivalent binder (kg)	280	292
W/equivalent binder	0.59	0.53
Aggregate-to-sand ratio (A/S)	1.26	1.26

was necessary to replace the plasticizer by a superplasticizer and a viscosity agent in RAC mixture to maintain the required rheology (slump). Both recycled and natural aggregate concrete (NAC) mixtures are presented in Table 24.2.

24.2.3 Preliminary study and suitability tests

A preliminary study was carried out on NAC and RAC, by CEMEX laboratory. This study concluded the necessity to increase by 10% the cement content (Table 24.2) of RAC mixture to achieve the same compressive strength as the reference formula. The use of a superplasticizer in RAC mixture was necessary to improve its rheology.

Figure 24.3 presents the slump's evolution versus time (min) for the studied concretes. Except the period between T0 and T30, for which the slump of RAC decreases rapidly, the slump variation is quite the same for both concretes. Indeed, recycled aggregates are not saturated, and their high porosity results in high water absorption after the first contact of different materials in mixer. This phenomenon decreases the concrete slump rapidly. At T90, the measured slump is the same for both concretes, which shows the beneficial effect of using superplasticizer in RAC mixture. The suitability tests, carried by Cerema, showed the same trends. The measured slump was 185, 185 and 120 mm, at T30, T60, and T90, respectively.

To measure the drying shrinkage, 7 cm × 7 cm × 28 cm specimens were manufactured. Every shrinkage value is the average of three measurements. In Figure 24.4, the drying shrinkages of RAC and NAC at 2, 7, 14, and 28 days are drawn. As can be shown, the shrinkage at 28 days of RAC is 40% higher than that of NAC. The old mortar in recycled aggregates and the cement content of RAC explain this difference.

Table 24.3 summarizes the mechanical properties of studied concretes at 28 days. The obtained results confirm the expected trends. Due to their low elasticity, recycled aggregate decreases the elastic modulus of concrete by 17%. The characteristic compressive strength of RAC is 4 MPa lower than that of NAC. However, both formulated concretes have the required characteristic compressive strength for their use in building construction (>25 MPa).

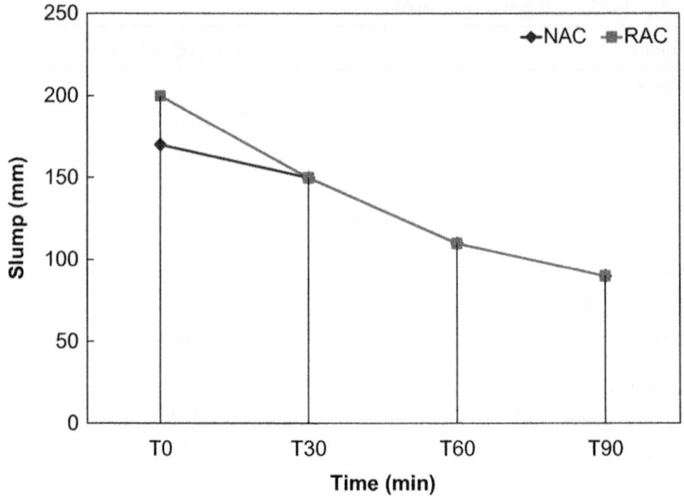

Figure 24.3 Rheology of NAC and RAC.

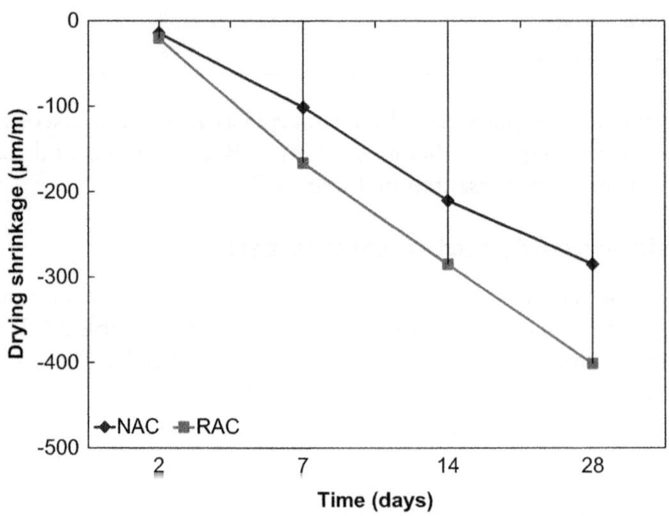

Figure 24.4 Drying shrinkage of NAC and RAC.

Table 24.3 Mechanical properties of NAC and RAC

Properties	*NAC*	*RAC*
Rc_28 days (MPa)	38.9	34.7
E_28 days (GPa)	32.5	27.1

24.2.4 Construction process

As discussed earlier, the critical point of this project is the service load of 750 kg/m^2. In fact, specific dispositions have to be taken according to EN 1992-2-1 standard. The allowed deflection was calculated as L/500 (L/250 in other cases), to avoid damaging the movable elements for archive storage. Otherwise, by considering the decrease of creep modulus when recycled aggregates are used, an equivalent coefficient of 27 (15 for NAC) was considered in calculation of steel reinforcement section (de Larrard et al. 2014). Therefore, the calculated steel section should be doubled (2 × ST25), as shown in Figures 24.5–24.9.

24.2.5 Experimental site concrete properties

For every casting, an experimental campaign is carried out, by Cerema Île-de-France, on manufactured and delivered concrete. First, the slump is measured to check the required rheology; its slump class. Then, cylindrical samples were prepared for mechanical (compressive and tensile) tests. For drying shrinkage and durability (water porosity, gas permeability, and carbonation depth) tests, only slab concrete (RAC) and that for ramp (NAC) were characterized. Eco-Materials Laboratory (LEM) of Cerema Île-de-France was in charge of preparing, conditioning, and measuring these properties.

Figure 24.5 Steel reinforcement of slab. (Source: Cerema.)

Figure 24.6 Pouring of ready mixed concrete. (Source: Cerema.)

Figure 24.7 Leveling of fresh concrete. (Source: Cerema.)

Figure 24.8 Vibration of fresh concrete. (Source: Cerema.)

In Table 24.4 all measured properties for different delivered concretes are presented. In the last column, the properties of NAC for ramp casting are listed.

Except for casting concrete of 22 October 2015, the measured slump is quite the same (200 mm → S4). Indeed, at this date, it was cold (13°C), wet (88%), and raining. The required characteristic compressive strength (25 MPa) was achieved for all mixtures, and tensile strength was the same for both RAC and NAC.

As expected, the drying shrinkage of RAC is more important than that of NAC. This is due to the presence of old mortar in recycled aggregates and the additional cement in RAC mixture to reach the required compressive strength. However, the shrinkage increment (+12%), when recycled aggregates are incorporated in site concrete, remains slight in comparison with that was measured in the preliminary study (+40%).

Even if cement was added to the RAC mixture, which decreased its water-to-cement ratio and then the porosity of new cement paste, the total porosity of RAC is 20% higher than

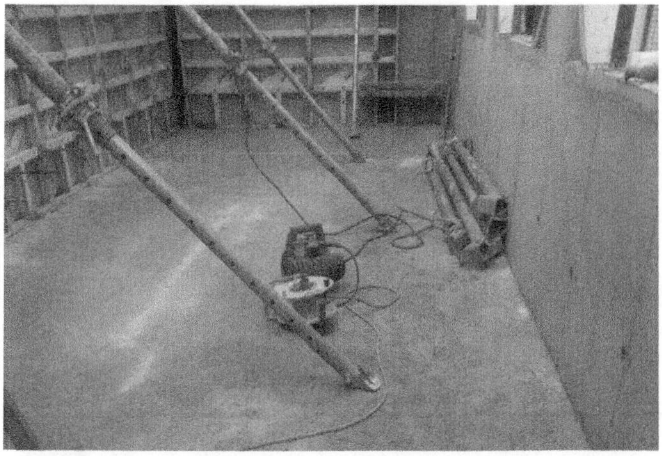

Figure 24.9 Framework of steel reinforced concrete panels. (Source: Cerema.)

Table 24.4 Construction concrete properties

	RAC				*NAC*
	Casting date				
Properties	*6 October*	*9 October*	*15 October*	*22 October*	*30 October*
A (mm)	205	200	200	240	205
Rc_{28} (MPa)	38.4	35.0	36.2	34.5	33.3
Rt_{28} (MPa)	3.4	3.4	3.3	3.2	3.3
$\varepsilon_{28\,days}$ (µm/m)	418	n.d.	n.d.	n.d.	373
P_{water_90days} (%)	18	n.d.	n.d.	n.d.	15
K_{gas_90days} ($\times 10^{-18}$ m^2)	180	n.d.	n.d.	n.d.	100
X_d (mm)	6	n.d.	n.d.	n.d.	2

that of NAC. Recycled aggregates are porous, which increases the porosity of RAC greatly. This increase affects gas permeability and carbonation depth. Indeed, recycled aggregate could be seen as composite material composed of a conventional aggregate, an old mortar, and Interfacial Transition Zones (ITZ) between these constituents and between each of them and the new surrounding cement paste matrix. These ITZs, particularly old ones, are considered as preferred path for aggressive agents. The obtained results agreed with previous researches, where carbonation depth increased with increasing recycled aggregate content (Silva et al. 2015).

24.2.6 Observations

Concerning the implementation of concrete, workers confirmed that the recycled concrete behaves exactly as the conventional one.

On 3 December 2015, a first inspection of building (Figure 24.10) was done after removing the framework. This inspection was carried out according to the technical instruction of Transport Ministry for Bridges' supervision and maintenance, delivered in 19 October 1979 and revised in 2010. Facings were also examined, according to the French standard (FD P 18-503 1989) criteria.

Figure 24.10 Building after framework removing. (Source: Cerema.)

After the initial inspection, the following observations are noticed:

- Two vertical cracks were observed: the first one has 0.4 mm width, 20 cm length, and is located in the northern wall. The second one has 0.1 mm width and is located in the eastern foundation. Both cracks could be the consequence of restrained shrinkage and casting default;
- Rust trace on floor intrados and western wall (Figure 24.11): the use of unclean framework seems to be the origin for this casting default;
- Apparent steel reinforcement: the concrete cover for reinforcement was not respected, which explains this observation;
- Segregation (Figure 24.12): the nonsealing of framework and the lack of vibration were the explanations for this phenomenon;

Figure 24.11 Rust and bubbling. (Source: Cerema.)

Figure 24.12 Segregation phenomenon. (Source: Cerema.)

- Flatness defaults (Figure 24.13): located on northern and eastern walls, these defaults are the consequence of framework mounting default;
- There are no undesirable materials (wood, plastic, etc.) on concrete surface.

Mentioned observations show clearly that inspected defaults are attributed to the construction and equipment quality. Hence, the use of recycled aggregates is not at the origin of these defaults.

Figure 24.13 Flatness default. (Source: Cerema.)

The second part of inspection, carried out by Cerema Île-de-France, deals with the observations of facings. Three parameters were considered: flatness, bubbling, and color. It was shown

- P3-flatness: the global flatness is 2–5 mm and the local one is 1–2 mm.
- E(3-3-3)-bubbling: the first index indicates that the maximal bubble surface is $0.3\,cm^2$, with a maximal depth of 2 mm and maximal bubbling surface of 2%, and the second corresponds to concentrated bubbling zones lower than 5% while the latter characterizes localized defaults.
- T3-color: the difference of color (gray) between adjacent zones and extreme ones is, respectively, 1 and 2.

These observations correspond to acceptance criteria for thin facings, according to article "fine concrete facing" of Fascicule 65 (2014).

24.3 RESEARCH NEEDS

The construction of the archive building did not raise any significant difficulty. The use of recycled aggregates did not change the practical implementation of concrete. Except the used admixture (superplasticizer), the mix design process was quite the same for RAC and NAC.

The applied service load was the critical point of this project, and considering the decrease of the long-term modulus of concrete, when recycled aggregates were added, was necessary. For best estimation of this parameter, further research on concrete creep is necessary. The effect of several parameters on creep, like *W/C*, the replacement rate of natural aggregates by recycled ones, and the age at loading application should be studied.

24.4 OUTLOOK FOR NEW PROJECTS

As the construction of the archive building did not raise any significant difficulty, CD77 has decided to start another bigger project. A new college, based in Montevrain (Seine-et-Marne department), was built in 2018. CD77 requires the use of recycled aggregates in concrete. This building will soon be opened to the general public; therefore, the maximum content of recycled aggregates in the concrete will be limited to 20% (according to the current standards).

Even if RECYBÉTON will be ended when the worksite will start, the project team has been asked to help in writing the specifications for the public tender project.

24.5 CONCLUSIONS

For this project, the challenge was to build a construction with a high amount of recycled aggregates, overcoming the current standard limitations, without changing mix design and implementation processes. One year after its construction, the following conclusions about CD77 building can be drawn:

- The produced recycled aggregates are of good quality. They pertain to the type 1, according to NF EN 206/CN (2014) standard;

- Superplasticizer was used in recycled aggregates mixture to improve its rheological behavior. At T90, the same slump was measured for both RAC and NAC.
- Usual expected trends were found (decrease of elastic modulus, increase of shrinkage) when recycled aggregates were incorporated. For the same required compressive strength, a slight increase of cement content, by 5%, was necessary for the RAC mixture.
- To take into account the high service load, the allowable deflection was considered as L/500. This particular disposition should protect the movable elements, for archive storing, from damaging.
- The decrease of long-term modulus, when recycled aggregates are used, induced a doubling of steel reinforcement section in the slab;
- From a practical viewpoint, the implementation of RAC and NAC was the same, and no difficulty was encountered on this project site;
- Detailed inspection of the building showed some casting and labor defaults. No default was attributed to the use of recycled aggregates.

ACKNOWLEDGEMENTS

The authors thank CD77 for this opportunity to build a construction with high level of recycled materials, CEMEX for the production of recycled aggregates, and RAC and Cerema for design and quality control.

Chapter 25

Industrial applications

Walls and sidewalks

P. Vuillemin
EQIOM

A. Cudeville
CLAMENS

CONTENTS

Abstract

This section presents one of the first experimental construction trial site that was carried out in 2014/2015 within the RECYBÉTON National Project/PN RECYBÉTON.

The objectives were as follows:

- To produce recycled aggregates (RA) having sufficient characteristics to be used in structural concrete, such as recycled gravels (RG) and recycled sands (RS).
- To study various concrete mix designs with a replacement ratio of 0% up to 100% of the "traditional" aggregates, or natural aggregates (NA), by recycled aggregate (RAs).
- To select a mix design, enabling to produce a C25/30 concrete for structural use.
- To produce the concrete in a ready mix plant, using only the existing equipment.
- To place the concrete on a construction site in reinforced concrete elements, such as walls, and architectonic sidewalks.

The relevant conclusions are that if it can be recommended to use recycled concrete coarse aggregates with quite a high replacement ratio, the use of recycled concrete fine aggregates/sands is more delicate.

The use of superplasticizer allows to adjust the workability and the rheology of the recycled aggregate concrete (RAC).

In our study, the RAs are high-quality ones, but a larger use of recycled sand would lead to poor durability characteristics: shrinkage, water porosity, and gas permeability, increase quickly with the percentage of sand substitution.

However, these results have to be considered with care because we were aiming aimed to produce C25/30 concrete for rather small constructions. High-durability objectives do not concern our study and mean values for potential durability evaluation are promising.

We draw the attention of future users of RAC to anticipate on the number of storage silos that the RMX (ready-mix) plant can handle, in order to use 4 to 6 different sizes of aggregates.

25.1 PRESENTATION OF THE PROJECT

Running parallel with laboratory studies, the national project RECYBÉTON asked for volunteers to carry out field tests using recycled aggregate concrete (RAC).

25.1.1 Clamens

Clamens is a leading company in recycling demolition waste, producing yearly 600,000 tons of recycled materials (Figure 25.1).

Three plants are used:

- A crushing and screening device that breaks the concrete blocks with a percussion crusher was used to generate crushed all in materials 0/31.5.
- In the recycling installation of concrete muds, the materials are washed under water, sieved, and cycloned to separate the aggregates contained in the slurry.
- The third plant allows to treat the recycled all in 0/31.5 aggregate by adding a few percentage of hydraulic binder to improve their mechanical properties. This material is devoted to pavement applications.

However, there is a large gap between the requirements of earthworks/roadworks, and concrete for buildings, in terms of physical, mechanical, and chemical characteristics. Therefore,

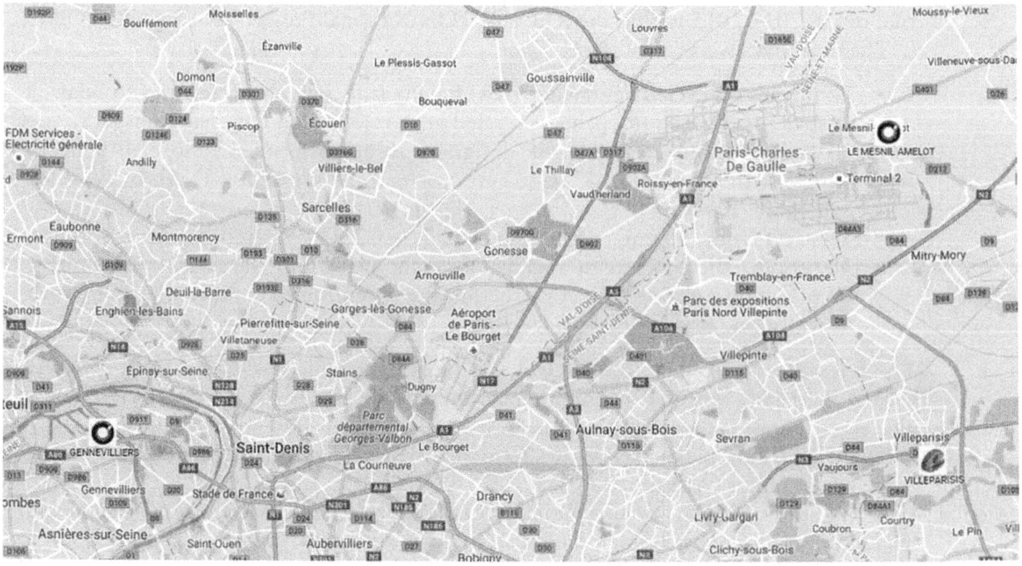

Figure 25.1 Locations of Clamens and Holcim/Eqiom plants. (Source: Clamens.)

Figure 25.2 Clamens platform at Villeparisis (77)—Photo: Clamens.

Clamens had to start producing gravels and sands coming from selected demolition waste and properly treated in terms of cleanliness and sulfate content (Figure 25.2).

25.1.2 Holcim Bétons

Holcim Bétons (renamed Eqiom Bétons at end of 2015) is a major actor of ready-mix (RMX) in France, producing 3 million m^3 of concrete per year and is very active in the implementation of solutions linked with sustainable development.

Their study of RAC consisted mainly in an extensive campaign of laboratory tests, with eight mix designs of C25/30 concrete, ref. F0–F7.

F0 is the reference concrete, daily used on sites and containing only natural aggregate (NA).

F1–F7 mixes integrate various quantities of recycled aggregate (RA), from 0% to 100%.

For all mixes, the compressive and tensile strengths, the 90-minutes rheology, the shrinkage, the gas permeability, porosity, and apparent chloride diffusion coefficient, were measured.

In addition, the elastic modulus of concrete was measured for the mixes that seemed the most suitable for use in reinforced concrete (F2/F3/F4/F6).

25.2 MATERIALS, MIXTURE DESIGNS, AND PROPERTIES

The RMX plant of Roissy was chosen because of its close location with Clamens plant in the North of Paris.

That plant is currently delivering concrete for large civil engineering projects of Aéroports De Paris/ADP, and the current NAs are of high standard quality, namely crushed limestone sand and gravels from Carrières du Boulonnais/CB (Figure 25.3).

25.2.1 Fabrication of RAC at Clamens plant

Clamens started to store pieces of concrete demolition waste, such as blocks of concrete coming from concrete pile *cut-off*. Close attention was paid to ensure that the concrete wastes contain as little plaster as possible, as measured by the sulfate content.

The pictures shown later illustrate the various stages of RA production (Figures 25.4–25.7).

Figure 25.3 Roissy RMX plant—Photo: Holcim.

Figure 25.4 Construction waste is stored on the platform—Photo: Holcim.

Figure 25.5 For RECYBÉTON Project, blocks of concrete pile cut-off were selected to produce RAs—Photo: Holcim.

Figure 25.6 Left: recycled sand to be used in earth works. Right: recycled washed sand to be used in structural concrete—Photo: Holcim.

Figure 25.7 Blocks of concrete-containing plaster are not considered for use in RAC.

For commercial reasons, Clamens produced one fraction of 0/4 mm sand and two fractions of 4/10 mm and 10/20 mm gravels (Table 25.1).

Due to the precautions taken by Clamens to produce "good aggregates", the analysis shows that both the sand and gravels can be classified as the highest class CRB, type 1, of RAs, as per the French standard (NF EN 206/CN 2014). The water absorption coefficients range from 5.8 to 6.6 and are low values for RACs that offer frequently values of 7–9.

The crushed materials were washed before the final screening, leading to clean gravels and sand. For the sand, both the methylene blue test (0.4), the fines content (0.7%), and the fineness modulus (3.2) are good indicators of high-quality recycled sand. However, the French standard restricts the use of RAs at 30% in XC1 exposure and class C25/30 concrete, which explains that the selected mix was F3 S30 G30 with 30% of replacement of NA by RA.

Table 25.1 Laboratory analysis and classification of Clamens recycled aggregates according to NF EN 206/CN (2014) standard

Test	*Standards*	*Recycled sand 0/4*		*Recycled coarse aggregates 4/10*		*Recycled coarse aggregates 10/20*	
Los Angeles	NF EN 1097-2 (%)					37	LA40
Micro-Deval	NF EN 1097-1 (%)					29	—
Fines content	NF EN 933-1 (%)	0.7	—	0.1	—	0.3	—
Flakiness index	NF EN 933-3 (%)					1.8	FI40
Fineness modulus	NF EN 12620 (-)	3.2	—				
Organic pollutants	NF EN 1744-1 (-)	Negative	—				
Water absorption	NF EN 1097-6 art. 8 and 9 (%)	5.8	—	6.4	—	6.6	—
Specific density	NF EN 1097-6 art. 8 and 9 (Mg/m^3)	2.33	≥2	2.25	≥2	2.18	≥2
Methylene blue test	NF EN 933-9 (g)	0.4	—				
Classification of constituents	NF EN 933-11 (%)			100	Rcu_{95}	100	Rcu_{95}
Total sulfur as S	NF EN 1744-1 art 11 (%)	0.8	—	0.14	—	0.13	—
Active alcalines Na_2O eq	LPC 37 (LCPC 1993) (%)	0.0196	—	0.0247	—	0.0369	—
Soluble sulfates in water as SO_4	NF EN 1744-1 art. 10.2 (%)	0.02	SS0.2	0.04	SS0.2		
Soluble chlorides in acid	NF EN 1744-5 (%)	0.03	—	0.014	—	0.017	—
Reduction of setting time (constituents)	NF EN 1744-6 (min)	0	A10	10	A10	10	A10
			CR_B		CR_B		CR_B

Nevertheless, we could have increased the percentage of recycled coarse aggregates to 50% or 60% with satisfactory results (see laboratory tests) (Figure 25.8).

25.2.2 Laboratory study of the various concrete mix designs

For availability reasons, the laboratory studies were performed at the laboratory of Amiens/Camon in Hauts-de-France. Eight mix designs were studied and tested as follows.

The reference traditional mix F0, which is daily used at Roissy RMX plant, offers a very high strength for reasonable cement content, thanks to its high-quality aggregates.

The seven mixes of RAC are named by the respective content of RAs.

For example: F3 S30 G30 means that 30% of recycled sand (RS) was replacing 30% of VS and that 30% of RA was replacing 30% of NA.

All mixes were kept with the same cement dosage of 280 kg/m³ and the same water content of 168 liters, which corresponds to a water/cement ratio of 0.6.

Figure 25.8 Big bags of RAs 10/20, 4/10, and 0/4—Photos: Holcim.

The consistency and rheology were adjusted to obtain a slump test value of S4 during 6090 minutes, by modifying the dosage and type of admixtures. The water reducer plasticizer was added at a percentage of 0.35 of the cement weight and the superplasticizer was added with a percentage varying from 0.5% to 1.05% depending on the difficulty to keep a sufficient working consistency.

It appears clearly that the recycled concrete sand has a great impact on the properties of concrete, and all mixes containing more than 30% of recycled sand are adversely affected (Table 25.2):

- decrease of the compressive strength, tensile strength, and elastic modulus;
- increase of the shrinkage with the percentage of substitution of sand (see Figure 25.9);
- increase of the porosity and gas permeability and therefore lower long-term durability.
- the porosity is found to be highly dependent on the presence of RA, with a value of 15.7 compared to 13.8 for NA in mix F3.

25.3 CONSTRUCTION OF WALLS AND SIDEWALKS AT GENNEVILLIERS

25.3.1 Production of concrete

The RMX plant had first to prepare a mix of 33% of recycled gravel (RG) 4/10 mm and 67% of RG 10/20 mm, because only one storage bin was available for the use of RGs.

Table 25.2 Results of laboratory tests on eight mix designs

Mix	*F0 standard*	*F1 S30G0*	*F2 S0G30*	*F3 S30G30*	*F4 S0G100*	*F5 S100G100*	*F6 S10G10*	*F7 S100G0*
No sample	150,639	150,632	150,633	**150,634**	150,635	150,636	150,637	150,638
NA-FA-0/4 crushed limestone Ferques (kg)	924	647	924	**647**	924		832	
NA-CA-4/20 crushed limestone Ferques (kg)	999	999	699	**999**			899	999
RA-FA-0/4 recycled Clamens (kg)		277		**277**		924	92	924
RA-CA-4/10 recycled Clamens (kg)			100	**100**	333	333	33	
RA-CA-10/20 recycled Clamens (kg)			200	**200**	666	666	67	
Cement CEMII/B-S 42.5 N CP1 NF Héming (kg)	280	280	280	**280**	280	280	280	280
Water (efficient) (L)	168	168	168	**168**	168	168	168	168
ChrysoDelta CER %	0.35	0.35	0.35	**0.35**	0.35	0.35		0.35
Chryso Optima 175%	0	1.05	0.5	**0.8**	0.5	0.8	0.5	0.9
Air temperature T_0 (°C)	16	13.8	15.4	**15.3**	15.2	14.1	15.7	11.4
Concrete temperature T_0 (°C)	17.1	15.5	17	**17.1**	17.7	15	16.7	12.4
Slump T0 (mm)	165	190	190	**180**	200	200	160	180
Slump T30 (mm)	135	180	180	**180**	200	200	130	180
Slump T60 (mm)	100	165	180	**160**	185	200	115	170
Slump T90 (mm)	80	160	175	**120**	185	180	90	170
Density (kg/m³)	2,409	2,376	2,362	**2,340**	2,272	2,112	2,361	2,278
Air content (%)	1.5	2.3	1.8	**2.3**	1.7	6	2.3	2.5
Rc 1 day (MPa)	4.05	5.8						
Rc 7 days (MPa)	33.2	34.3	32.6	**32.9**	25.8	17.3	28.1	23.1
Rc 28 days (MPa)	49.3	47.9	43.8	**43.8**	33.5	24	40.5	33.2
Flexural strength (MPa)	3.56	3.27	3.47	**3.29**	2.8	2.1	2.9	
Shrinkage 28 days (μm/m)	271	377	236	**346**	407	663	361	
Elastic modulus 28 days (MPa)			39,722	**38,047**	32,482	39,200		
Porosity (%)	13.8	14.1	15	**15.7**	18.6	21.5	14.9	

Altogether, Clamens supplied the plant with 100 tons of RS 0/4 mm and 200 tons of RG 4/10 and 10/20 mm.

We decided to carry out the construction trials using a concrete containing 30% of RS and 30% of RG, with mix design ref. F3. The reason is that RG has a "lower" impact on the properties of concrete than RS, which has a large impact in terms of rheology, strength, over cost, and ultimately durability. We could have afforded to use RG up to 50% of replacement. According to the French standard NF EN 206/CN (2014), the F3 mix is labeled as follows: "BPS NF EN 206/CN I IB Voile C25/30 CEM II/B-S 42,5 N CE CP1 NF D_{max} = 20 mm recyclé S3 affaissement 100 à 150 mm XF1-XC3-XC4-XD1 CL0,40". The concrete was mixed during 58 seconds to reach a flat wattmeter curve and to comply with the French standard (55 seconds minimum mixing time).

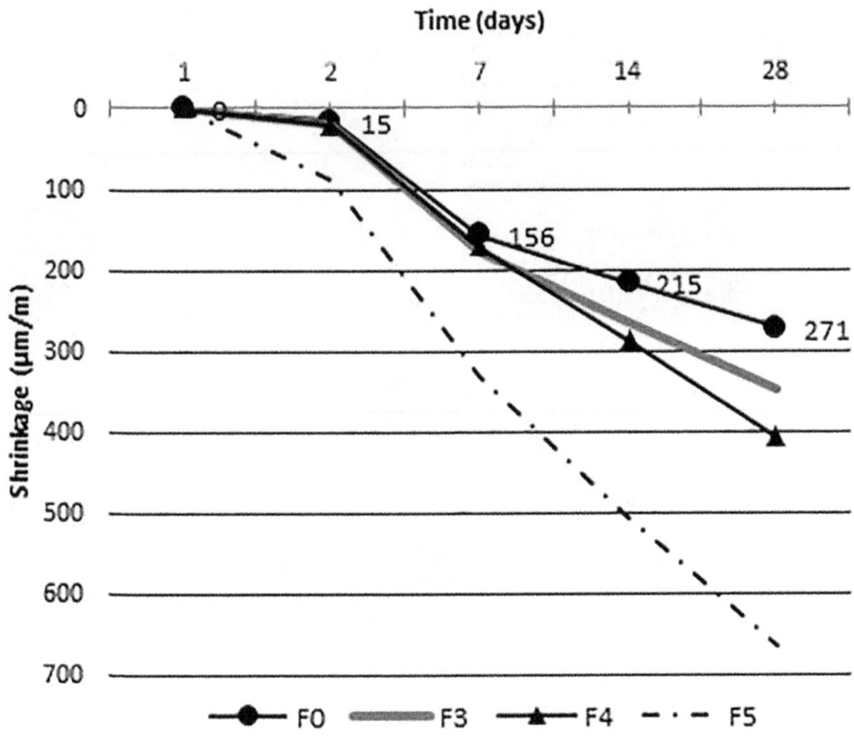

Figure 25.9 Shrinkage vs. time.

25.3.2 Delivery and placement of concrete

The concrete was then transported by truck mixers of individual capacity of 7.5 m^3, to the construction site of Gennevilliers, which is located 30 km far from Roissy plant. The transportation time using congested A1 highway and north ring of Paris ranges from 1 to 2 hours; therefore, the rheology of the concrete was considered to ensure a delivered S3 concrete by using a mix of plasticizer and a superplasticizer.

It took 1 hour to place the concrete on site by casting reinforced concrete walls and sidewalks. For the sidewalks, yellow pigments were added to the concrete to obtain the "Sahara Earth" yellow color.

The concrete was easy to place. The air temperature was quite fresh (4°C–8°C) in early February. The contractor did not report any problems linked with RAC during the casting stages.

A site inspection in October 2016 shows that the early 2015 cast reinforced concrete structures do not reveal any problems. The facings of the walls are satisfactory, with very little air bubbles. On the sidewalks, there is only a transverse crack located, where a joint was forgotten to be sawn (Figures 25.10 and 25.11).

25.4 CONCLUSIONS

Recycled concrete aggregates produced from all but the poorest quality original concrete can be expected to pass the same tests required of conventional aggregates.

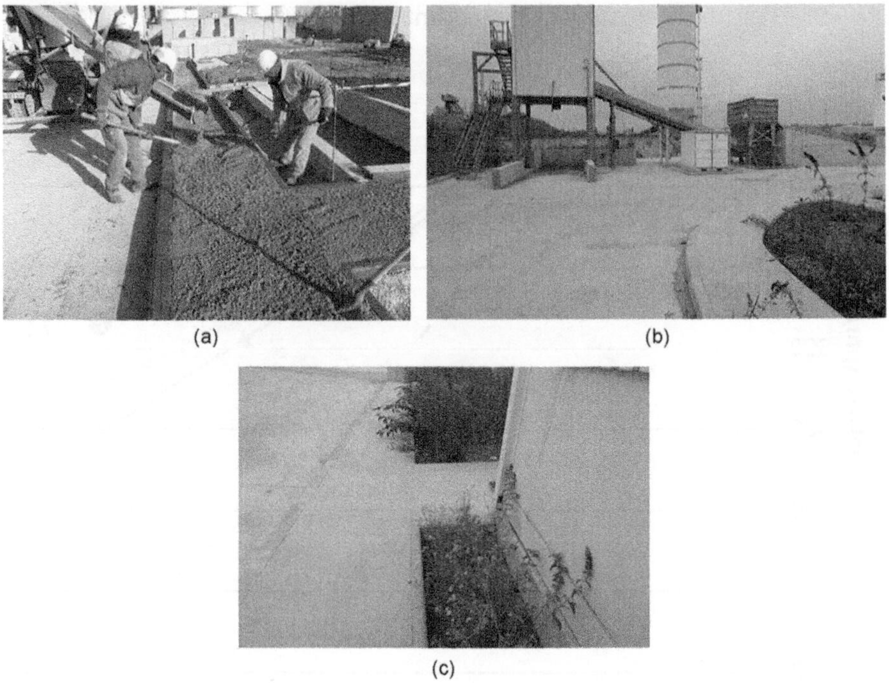

Figure 25.10 Construction of sidewalks with colored RAC—"ocre savane" pigments—Photos: Holcim.

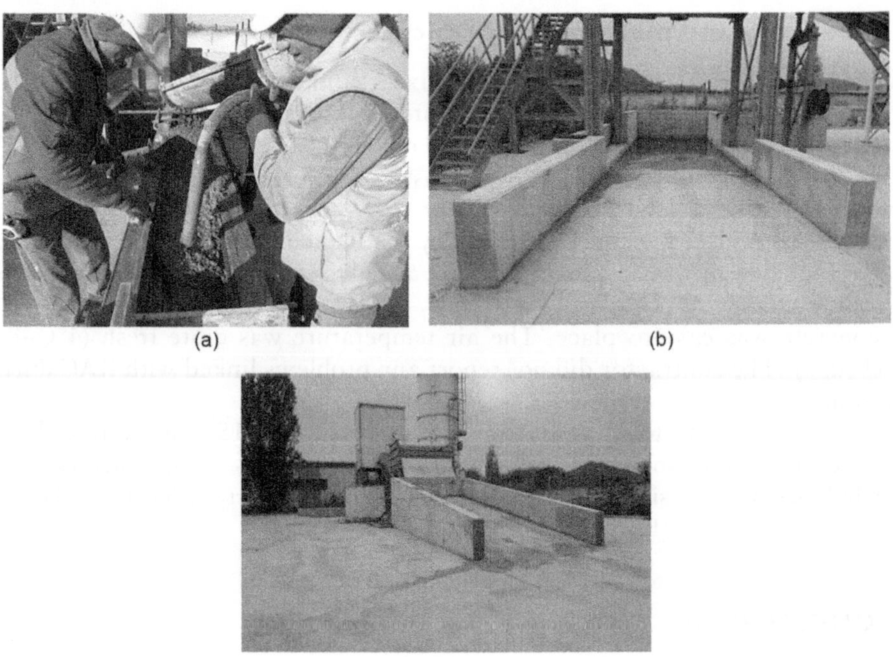

Figure 25.11 Construction of various walls using RAC—Photos: Holcim.

Recycled concrete aggregates contain not only original aggregates but also hydrated cement paste. This paste reduces the specific gravity and increases the porosity and therefore the absorption of RAC.

It is generally accepted that, when natural sand is used, up to 30% of natural crushed coarse aggregate can be replaced with coarse RA, without significantly affecting the mechanical properties of the concrete. As replacement amounts increase, drying shrinkage and creep will increase and tensile strength and modulus of elasticity will decrease. However, compressive strength is not significantly affected.

If the strength and workability are affected, it is quite easy to increase the paste/cement content and the dosage of plasticizer and superplasticizer.

In our case, the RAs come from a well-organized selective demolition (blocks of concrete pile cut-off), which leads to high-quality sand and gravel. This is why we used 30% of recycled sand in our mix. More generally, we will have aggregates of less quality, and the tendency will be to sacrifice the sand that will create an imbalance to the recycling company.

An increased use of RAC is more a strategy and a wish. We could remember famous words like "Just do it" or "Yes we can" to illustrate that the technical outlets are solvable.

Renovation works represent a large part of the public and housing works and huge quantities of old concrete will be available.

ACKNOWLEDGEMENTS

The authors thank the laboratory engineers and technicians from the Materials Applications Center and the Holcim Quality Department, the batcher of Roissy RMX plant, and the contractor Gameiro at the construction site.

Chapter 26

Office building

Compression slab

R. Deborre

Nacarat (groupe Rabot Dutilleul)

E. Garcia

Lafarge Bétons France (groupe LafargeHolcim)

CONTENTS

Abstract

This chapter presents the joint points of view of a property developer, Nacarat, and an industrial concrete manufacturer, the Lafarge Bétons company, on the experimental use of recycled aggregates when building an office block in the Paris region. The first successful experiment was carried out in the North of France in 2015 involving the re-use of crystallized blast furnace slag coming from the industrial sector. In 2016, the Rabot Dutilleul group wanted to extend this experimentation in the Paris region, this time with aggregates from the building and public works sector, i.e., made from demolition concrete. The selected project was a standard office building under construction. This choice was not made with a view to changing the design but to integrate the experiment into a "standard" building site. The contractors were quite willing to cooperate, provided the poured concrete fully complied with all current standards and that the previously negotiated cost price would not be modified. The choice made soon after the discussions was to carry out experimental testing on a compression slab. The supplier of ready-mix concrete (RMC) already had all the materials required for carrying out the experiment and for testing the recycled aggregate concrete. This amounted to a volume of 80 cubic metres formula incorporating recycled aggregates compliant with the current concrete standard).

The RMC Lafarge Bétons concrete plant possessed recycled aggregate type 1 AGGNEO BPE 6/20, compliant with current aggregate and concrete standards from the Lafarge Granulats recycling facility in the Paris suburbs, i.e., in Gennevilliers (92230).

26.1 PRESENTATION OF THE D1 BUILDING PROJECT, Parc de l'étoile

The construction work was undertaken in May 2016 on the initiative of Nacarat, the property development subsidiary of Rabot Dutilleul, in Villeneuve-la-Garenne in the Paris region. In cooperation with the French National R&D Project RECYBÉTON, Nacarat undertook to use recycled aggregates (RA) in an office building project under construction, within the limits authorized by current rules and regulations.

26.1.1 The main features of the project

The main features of this project are listed as follows:

- Address: Parc de l'étoile, bâtiment D1, 54 avenue du Maréchal LECLERC, 92390 Villeneuve-la-Garenne
- Type of building: ground floor+two levels over a semiburied basement.
- Use of the premises and delivery date: handover scheduled for January 2017 to Hauts-de-Seine department, the owner and occupier.
- Description of the premises: 1,825 m^2, RT2012 certified, 69 offices, three meeting rooms, two restrooms, one PMI unit (maternal and child welfare services) on the ground floor
- Owner and developer: Nacarat
- Design Architect: Preconcept Architectes
- Main works contractors: Artis Construction; Sofialex
- Concrete supplier: Lafarge Bétons France, a member of the LafargeHolcim group (Figures 26.1 and 26.2).

Figure 26.1 The "Parc de l'étoile" building D1 completed construction. (Photo by Nacarat.)

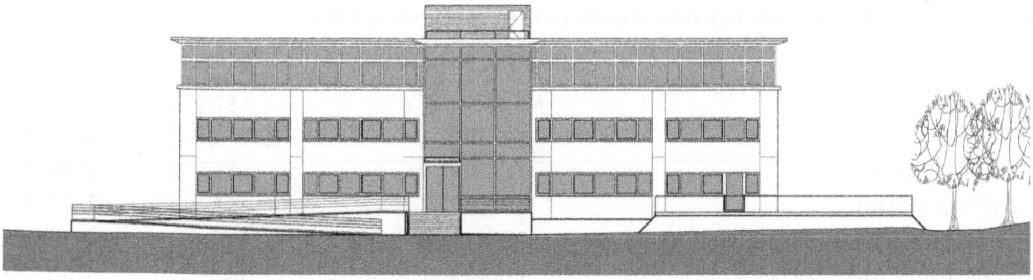

Figure 26.2 East side elevation of the "Parc de l'étoile" building D1. (Drawing by Preconcept Architectes.)

26.1.2 Selecting the work area

The advantage of selecting a building currently under construction was the time saved for the experimentation. Indeed, the RA concrete (RAC) was poured in May 2016, and the decision to do so had been taken in March 2016.

The main disadvantage was the limited perimeter in which it was possible to study concrete slabs made from recycled materials: construction was already in progress when the decision was taken, with RECYBÉTON, to study the subject of RAC: thus, the design had not accounted for this; the foundations had already been poured; the site works were progressing at the same time as the formulation tests were carried out, etc.

Lastly, the current policy of the Rabot Dutilleul Group is to comply fully with current standards and to remain in the frame of the contracts already signed with all the site parties. No significant extra cost was noted, which would otherwise have made the experimentation unfeasible in terms of the economic balance of the project. Moreover, jeopardizing the satisfactory completion of the building was out of the question given that the client (Hauts-de-Seine department), though highly interested in the circular economy, was not asking for any experimentation on this issue in this building.

Having noted these constraints, Nacarat soon agreed with the building contractor and the main works Sofialex company that the concrete supplier, Lafarge Bétons, would be able to supply the required concrete incorporating RA.

26.2 MATERIALS AND MIXTURE DESIGN

In this program, 80 m^3 of concrete, where 20% of its coarse aggregates mass was RA, were poured into the compression slab on the second level above the ground floor (terrace). RA was got from deconstruction sites located in the Paris region. The Lafarge Bétons Paris Region concrete batching plant, the project's concrete supplier, was located in Nanterre near Paris.

26.2.1 Recycled aggregates

The RMC plant had a supply of AGGNEO BPE 6/20 No. 1 type RA, which complied with the (NF EN 12620+A1 2008; NF P18-545 2011) standards and was got from the Lafarge Granulats recycling plant in Gennevilliers (Tables 26.1 and 26.2).

The AGGNEO RA used in this project was made from deconstruction concrete material from work sites in the Paris region located within a radius of a few kilometers of the Gennevilliers plant. Its production requires the controlled sorting of waste, in line with a

Table 26.1 6/20 RA characteristics with classification of constituents CRB

Test	*Standard or method*	*Results*	*Dates*
Water absorption	NF EN 1097-6	3.57%	06/08/2015
Specific density	EN 1097-6	2.32 mg/m^3	06/08/2015
Los Angeles (10/14)	EN 1097-2	32	21/07/2015
Micro Deval (10/14)	EN 1097-1	22	21/07/2015
Reduction of setting time (constituents)	EN 1744-6	6 min	21/07/2015
Recycled agg.—floating particles	EN 933-11	$FL_{0.2}$	21/07/2015
Recycled agg.—bituminous materials	EN 933-11	Ra_1	21/07/2015
Recycled agg.—masonry elements	EN 933-11	Rb_0	21/07/2015
Recycled agg.—concrete and mortar	EN 933-11	Rc_{97}	21/07/2015
Recycled agg. RA—glass	EN 933-11	Rg_0	21/07/2015
Recycled agg.—unbound aggregate	EN 933-11	Ru_2	21/07/2015
Recycled agg.—others	EN 933-11	X_0	21/07/2015
Recycled agg.—glass and others	NF EN 12620	$XRg_{0.5}$	21/07/2015
Chlorides	EN 1744-1. art 7.8 or 9	None	21/07/2015
Soluble chlorides in acid	NF EN 1744-5	0.01%	21/07/2015
Total sulfur as S	EN 1744-1 art 11	0.12%	21/07/2015
Soluble sulfates in water as SO_4	EN 1744-1 art 10.2	SSB or SSC	21/07/2015
Alkali-réaction	P 18-542	PR	21/07/2015
Active alcalines (Na_2O Eq)	LPC 37 (LCPC 1993)	0.0104%	21/07/2015

Source: Lafarge Granulats France.

Table 26.2 6/20 RA granular distribution

Granular class 6.3/20					*NF P 18455 article 10 (obtained results 21/07/2015)*							
	d/2	*d*			*D/1.4*		*D*		*1.4D*	*2D*		
	3.15	6.3	10	12.5	15	16	20	22.4	28	40	f	FI
VSS	5	15			70		98					20
VSI		0			40		83		98	100		
moy X_f	2	3	16	36	51	65	88	96	100	100	0.6	10

Source: Lafarge Granulats France.

precise set of product specifications. Using such aggregates helps to reduce the consumption of natural resources while multiplying opportunities to reuse the waste generated by deconstruction concrete for constructing buildings inside urban areas.

26.2.2 Concrete mixture

Formulating Aggneo RA is a process well known to Lafarge Bétons France, which produces several hundred thousand cubic meters of such concrete annually, thanks to its network of 250 concrete batching plants.

As Lafarge Holcim say: "Yesterday, we indeed supplied SOFIALEX with AGGNEO BPS C30 CEMI+Cv D20 S3 XC4 XF1 concrete comprising 20% of RA. Twenty percent is the maximum ratio authorized by the (NF EN 206/CN 2014) standard for XF1 exposure with a RA of type 1 (CR_B). The client had asked for concrete with small dimensions (D_{max} 14 mm). He knew the RA had a diameter of D_{max} 20, but the 20% ratio was not a problem for him" (Table 26.3).

Table 26.3 Mix design of BPS C25/30 CEMI+CV D22 S3 XFI recycled concrete

Characteristics	*Technical data*
Exposure class	XF1
Compressive strength 28 days on cylinders	25 MPa
Dosage of equivalent binder	≥273 kg/m³ for D_{max} = 22.4 mm
Ratio of W_{eff}/equivalent binder ratio	≤0.6
Consistency	S3
Type of sand	Siliceous-calcareous, partly crushed 0/4 alluvial, corrected with an alluvial correcting sand 0/1
Type of aggregate	Siliceous-calcareous 8/22 alluvial partly crushed
	6/20 recycled type 1 (20% of coarse aggregate content)
Type of binder	CEM I 52.5N OPC + fly ash (activity index 0.6)
Type of admixture	Water-reducing plasticizer
Origin of water	Decanted water, EN 1008 compliant

Source: Lafarge Bétons France.

26.3 EXECUTION OF THE WORKS AND FEEDBACK

26.3.1 Execution of the work

The terrace slab was poured in May 2016 using a pump and workers to adjust its thickness (Figures 26.3 and 26.4).

26.3.2 Feedback from the various participants

26.3.2.1 Sofialex company

"The pouring and execution did not show any significant difference compared with conventional concrete. However, the quality of the finish was less good than with conventional concrete (plastic shrinkage)".

Figure 26.3 Pouring the slab with a pump. (Photo by NACARAT.)

Figure 26.4 Execution of the slab and adjusting its thickness. (Photo by NACARAT.)

26.3.2.2 *Nacarat: the owner developer*

"We were worried beforehand, because the client had not been warned of this experimentation, even though it complied with the standards. Then we felt relieved or even frustrated because this concrete had a very conventional aspect and there was no "visible" difference due to the experimentation".

26.3.2.3 *Lafarge Bétons: the supplier of concrete*

Lafarge Bétons France was pleased to be a partner, at the request of Nacarat, in a project that promoted the circular economy in a local context. As part of this project, Lafarge Bétons France, with RECYBÉTON and other partners, poured concrete whose RA content reached a ratio of up to 100% in another RECYBÉTON Experimental Construction Site (Chapter 22).

26.4 RESEARCH NEEDS

It seems that no extra research is required to deploy the use of RA for using concrete in buildings that meet the existing standards and seem compatible with the economic realities of work sites.

26.5 OUTLOOK FOR THE NEW PROJECT

The circular economy is a major issue in the construction of new buildings. Concrete may still have a "significant" role to play, with other low-CO_2 construction materials, provided we maximize the use of reused or recycled materials.

This experiment suggests that the technical challenge is certainly not as great as generally feared (for cultural reasons?) regarding recycling across the entire real estate value chain. Perhaps, it is much simpler than expected? This seems to be the case.

Therefore, according to us, it seems most appropriate, particularly in the Paris region, to ask our contractors, subcontractors, and industrial partners to supply concrete containing maximum amount of recycled materials without having any impact on the cost base.

Finally, should a final customer ask us to carry out an experiment that goes beyond existing standards, we would be tempted to undertake this venture in a fully transparent way with our key partners involved in our ecoconstruction system.

26.6 CONCLUSIONS

The aim of this project was to experiment the use of RA in a concrete slab used in a real estate development in the Paris region at the request of the owner. The limits set were not to go beyond existing technical standards and to leave the budget unchanged compared with a conventional project.

The concrete with RA was supplied by Lafarge Bétons France with the following technical features: BPS C30 CEMI+Cv D20 S3 XC4 XF1 AGGNEO with 20% of coarse RA (maximum ratio authorized by the (NF EN 206/CN 2014) standard for XF1 exposure with a type 1 RA). The origin of the RA supply did not pose any problems to the contractors.

On the other hand, the initial reluctance of the project operators was quite significant. We believe this was due to cultural obstacles resulting from the preconceived notion that the quality of recycled materials is inferior. Even when the concrete used meets the current standards, there remains some fear that "recycled material" will be visible and not appealing. In this case, once the "recycled" concrete had been poured, the same operators felt reassured technically and commercially.

ACKNOWLEDGEMENTS

The authors wish to thank the RECYBÉTON National Project, which suggested this experiment. We also wish to thank our operational colleagues and the contractors involved, who helped us complete the project successfully and put their trust in us.

Chapter 27

Recycling in the precast concrete industry

P. Francisco and P. Rougeau
CERIB

CONTENTS

Abstract

Some precast concrete plants manufacture products with recycled aggregates (RA) for many years. Part of them are gained by washing fresh concrete, but traditionally, RAs are got from concrete products after being tested under the factory production control or because a nonconformity is identified. This represents some percent of the whole mass produced and concerns both small unreinforced products, such as blocks, flags, kerbs, and large reinforced products such as pipes, beams, and building elements. After being crushed, sorted, and tested, these RAs are used to produce new precast concrete products by replacing a part of original natural aggregates. This crushed RA concrete are from internal source. To increase the ratio of RA, other sources can also be used: concrete products or concrete coming from construction and demolition waste. These crushed recycled concrete aggregates are from external source. Such sources may present variable characteristics such as maximum aggregate size, water absorption, or fine content that have to be monitored to respect the constant performance required on final concrete products. Requirements on crushed RA concrete and possibilities of use are described in the European standard entitled "Common rules for precast concrete products" and adjusted by provisions valid in the place of use. Practical examples from some French companies are presented.

27.1 INTRODUCTION

Some precast concrete plants manufacture products with recycled aggregate (RA) for many years. This applies both for use in nonreinforced products such as blocks, flags, kerbs, and for use in reinforced products such as building elements.

The concrete products industry is fully committed to addressing the challenges of the circular economy, including the sustainable use of resources. A wide variety of applications exist concerning the use of secondary raw materials in concrete (Petitpain et al. 2017). Their use is almost always technically possible and more and more often economically viable. Secondary raw materials contribute to a rational use of natural resources with respect to product performances. Several precast concrete plants manufacture products with RAs for many years.

RAs essentially consist of two categories:

- Concretes from the industrial processes of precast concrete plants;
- Concretes resulting from the demolition of structures or construction sites.

27.2 RECYCLING CONCRETE FROM THE INDUSTRIAL PROCESSES OF PRECAST CONCRETE PLANTS

Industrial sectors generate materials that can become secondary raw materials. In the precast industry, they are essentially inert materials such as concrete and aggregates and some others like most of plants such as paper, cardboard, plastics, wooden pallets, scrap, etc. Concrete and aggregates represent significant volumes for the entire precast industry. They are close to 500,000 tons per year and can represent for a large plant up to 2,000 tons per year (Bresson 2003). However, due to the multiplicity of products and components manufactured by the factories, as well as the size of the production volumes produced, the quantities produced per plant vary between 200 and 2,000 tons per year. This represents on average about 1.7% of the mass of concrete produced, considering both uses in nonreinforced products and reinforced products.

The origin and shape of these materials vary widely, as they are generated during each stage of production or storage. These include

- Residues from the cleaning of mixers, transport buckets, machinery hoppers, molds, machines, and workshops;
- Rejected or unused mixes;
- Defective fresh products removed from the machine and broken products during handling, storage, or loading;
- Products destroyed during the control tests;
- Residues of unusable aggregates;
- Powdered residues resulting from surface treatments such as sandblasting;
- Solid residues from the clearing of settling ponds.

At present, there are three ways of treating these materials with a view to their conversion into recyclable aggregates in concretes or roadworks:

- Internal treatment at the plant with fixed or semipermanent installations allowing the production of RA in the manufacture of concrete products;
- Temporary treatment by crushing-screening campaigns of a large stock of material stored at the plant site;
- Transport to fixed installations specialized in the treatment of demolition materials (public works or recycling companies, quarries).

Internal treatment is obviously the best way of protecting the environment, because it eliminates many transports. It can also be the most economical if the processing costs (depreciation of the installation, treatment management, operation, and maintenance) remain below the costs of the aggregates saved, avoided transport, and evacuation costs. To do this, it is necessary to set up an organization and material means adapted to the types and volumes produced, as well as to the production conditions and the configuration of the plant.

For example, a precast concrete company carries out concrete slabs (garden slabs, slabs for sealing terrace roofs, etc.). Manufacturing rejects are stored safe from weather (Figure 27.1).

Figure 27.1 Manufacturing rejects to be crushed and recycled. (Photo by MARLUX France.)

Once or twice a year, when the amount is enough (about 1,000 tons), a crushing campaign is conducted. The same kind of crusher used for roads purpose is routed over the production site. The crushing operation lasts for 3 days. Aggregate sizes are adjustable from 0 to 6 mm to coarser fractions. Those new aggregates are used to perform mass concrete products for paving purpose.

27.3 RECYCLING CONCRETE FROM CONSTRUCTION AND DEMOLITION WASTE

Close to most large cities, there are now platforms specialized in the treatment of demolition or deconstruction materials. These sites receive all types of inert materials (reinforced concrete, asphalt mixes, gravel) directly from construction sites or platforms for grouping and sorting construction materials, as well as concretes that are not recycled in precast concrete plants.

These sites are managed by specialized companies (public works, recycling or quarrying companies), using large equipment on large platforms (on old quarries or brownfields) to allow separate storage according to the different inert materials. These facilities are capable of handling large volumes (200,000–300,000 tons per year), including fixed or semimobile installations with powerful impact crushers (200–400 kW), vibrating screens, sorting stations (for the removal of wood, plastics, glass), magnetic separators (for the removal of steel) and sometimes sand washing equipment (or air separators).

For example, precast concrete company use RA concrete (RAC) from electrical poles. According to progressive burying electric power airlines, the French electrical network company (ERDF: Electricité Réseau Distribution France) launched several campaigns of concrete electrical poles' recycling, with the collaboration of French counties' departmental councils and companies or recycling platforms.

The departmental council of Sarthe and ERDF led a common approach of recycling concrete poles. Each year in Sarthe, 3,500 poles are transformed into rubble for future use in construction sites. The recycling site of concrete pole for electrical line stands is able to separate steel and concrete from the pole to reuse completely those concrete electrical poles.

This recycled material is then used in backfilling sliced, layout industrial platforms, and sublayers of roads. Finally, all components that made the "waste" were converted into raw materials for new types of applications.

Figure 27.2 Precast concrete company (located in Normandy) uses aggregates from recycling of concrete electrical poles to make wall elements for façades. (Photo by CMEG.)

In the Tarn-et-Garonne, a recycling company set up in Maine-et-Loire handles collection and transformation steps of concrete poles. So, after each stand drop off operation, either after burying the network or as part of stands replacement, useless poles are carried toward three storage areas provided for this aim in the county.

Thus, depending on the location of ongoing work in the county, poles need to be recycled are carried toward storage areas provided for this purpose. In 2014, across the county, more than 100 poles to a total volume of more than 100 tons were recycled by the precast concrete company (Figure 27.2). First, poles are treated to separate steel and concrete, and then steel is recycled in foundry for reuse while concrete is reduced to rubble by a finishing treatment.

Materials are turned into raw materials for new applications, especially in public works (backfilling, road sublayers, etc.).

Industrial aims are as follows:

- Reduce the environmental footprint of concrete products;
- Anticipate future regulations;
- Be proactive with customers.

The particle size of RAs collected is 4–14 mm (sand being used for other applications). They are used by 20% of total aggregate amount (Figures 27.3 and 27.4). Each year, this precast concrete company recycles almost 400 tons of RAs.

27.4 RECYCLING OF CONCRETE INTO LARGE SIZE CONCRETE BLOCKS

Since 1999, a French company provides support, treatment, and marketing of materials and inert minerals. This company has five production sites located in four departments in Brittany. Each year, it values more than 20,000 tons of materials, including concrete from the demolition of buildings and infrastructure (Figure 27.5).

Figure 27.3 Precast product with RAs. (Photo by CMEG.)

Figure 27.4 Building using precast concrete products with RAs. (Photo by CMEG.)

Figure 27.5 Inert wastes from buildings demolition and ready-mixed concrete's drum returns. (Photo by INERTA.)

Figure 27.6 Aggregate after crushing operation. (Photo by INERTA.)

Platforms of this company are equipped with crushers, sifters, shovels, and loaders. They produce aggregates for road sublayer or new concrete uses (Figure 27.6). Ferrous materials are valued in foundries.

This company produces especially large size concrete blocks (1.6 m × 0.80 m × 0.80 m) with an approximate weight of 2.4 tons each (Figure 27.7). These plain concrete blocks are equipped by a lifting device. They fit one into another to make a wall. They are used by this company's customers to mark out storage areas or to build shoring walls.

These blocks are made of traditional constituents (cement, aggregates) that are incorporated in varying amounts depending on the application of RAs from demolition or even Ready-Mixed Concrete's drum returns. Blocks may also incorporate inert industrial coproducts coming from foundry for example.

They can be produced either in Saint Avé's factory or directly on site using a movable facility. The stationary "Blend" concrete-mixing plant allows making concrete blocks where industrial coproducts or recycled concrete deposit are located. Transportation is thus minimized.

Figure 27.7 Concrete blocks made from RAs. (Photo by INERTA.)

27.5 CONCLUSION

Both nonreinforced products such as blocks, flags, kerbs, and reinforced products such as building elements can be produced using RAs.

The practical examples presented here show a large range of origins compatible with the requirements on precast concrete products defined in current European or French standards.

This knowledge in combination with results from National Project for Research and Development RECYBÉTON will increase the fraction of RAs in concrete by extending the field of the traditional practice.

ACKNOWLEDGEMENTS

The authors would like to thank MARLUX France, CMEG, and INERTA for information and pictures delivered.

Chapter 28

Quality control in the French context

A. Ben Fraj
Cerema

P. Dantec

CONTENTS

Abstract

This chapter presents the provisions applicable in the world and particularly in France as regards both material controls and the execution of buildings and civil engineering structures. The primary standard regulations are provided, along with a number of examples drawing upon the outcomes of several experimental worksites conducted within the framework of the RECYBÉTON national project (NP). Special controls dedicated to the incorporation of recycled aggregates are exposed at the various construction stages: design, execution, and acceptance of works.

28.1 QUALITY REQUIREMENTS BASELINE

28.1.1 Introduction

The national complement to the European standard addressing the execution of concrete structures, NF EN 13670/CN (2013), encompasses the fields covered by the two documents applicable to concrete buildings and civil engineering facilities, namely: DTU (Unified Technical Document) No. 21 (NF DTU 21 2017), "Execution of concrete structures"; and≈the new Fascicule 65 (Fascicule 65 2017) titled "Execution of reinforced and prestressed concrete civil engineering structures" released in 2014. These documents have been updated and are now being published to acknowledge the impact of NF EN 13 670/CN (2013), which imposes writing execution specification in compliance with the targeted quality level, depending on the requirements of the particular execution class selected for the structure. The nature of controls performed varies with respect to the selected execution class, which in turn depends on the scope of the structure or portion thereof, as well as on the "criticality" of its execution in terms of anticipated functions.

In Execution Class 3, beyond self-inspection and in-house controls performed by the builder itself, a more extensive control may be mandated by national regulations or project specifications. Such an extensive probe may also be conducted by another firm, known as an independent external control body.

The quality control is a key parameter in concrete structures, particularly when recycled aggregates are used, and in the absence of traceability. Thereby, the use of recycled aggregates leads to more vigilance and more tests. In this context, the first part of this chapter deals with the quality requirement and control for buildings and civil engineering structures. It focuses on French experience and gives an overview of standardization at an international scale. In the second part, a description of special controls and observations dedicated to the incorporation of recycled aggregates in concrete is presented. This description is carried out at various construction stages for several experimental worksites conducted within the framework of the "RECYBÉTON" national project.

28.1.2 Building: NF DTU 21, technical specifications, project owner, and manager

The NF DTU 21 (NF DTU 21 2017) specifies increasing levels of control depending on both the magnitude of the works and the potential for special structural components. Three categories of worksites are distinguished herein:

- Category A: Small-scale construction project comprises at most two stories above the ground floor plus a basement level; this category refers in particular to detached single-family dwellings or attached townhomes built in small numbers.
- Category B: Medium-sized project comprises structural elements that are of widespread use and loaded under normal conditions. This category would specifically apply to buildings limited to 16 stories, a major single-family subdivision, or a basic industrial complex. The quantity of concrete poured remains below 5,000 m^3.
- Category C: Large-scale building project contains commonly sized structural elements with standardized loadings. This category targets buildings taller than 16 stories, industrial or commercial warehouses subjected to heavy loads or intense traffic, as well as expansive sports facilities.

Particular structural components PA, PB, and PC used in projects of categories A, B, and C would include

- large overhangs,
- load transfer floors or components designed to withstand heavy loads,
- slender columns,
- long-span floors,
- structures needing the use of sophisticated execution techniques,
- structures whose required 28-day characteristic design strength for concrete cast in situ is at least 35 MPa.

These elements must be highlighted on the structural drawings.

The execution controls adopted for special structural elements PB and PC cannot be weaker than those ascribed to their corresponding structural category, i.e., B or C.

Control levels are calibrated based on the requirements of the given execution classes of Standard NF EN 13670/CN (2013). More specifically, unless otherwise indicated in the contractual documents, projects belonging to category A, B, or C correspond, respectively, to execution class 1, 2, or 3. For the PA structural category, the execution control will be no lower than level 2.

The imposed concrete controls are based from both the material producer's controls and controls organized by sampling or batches assigned in the dedicated contractual documents. They are conducted by the contractor and interpreted in accordance with the provisions stipulated in Appendix A (standards) of the NF DTU 21 Codes of Practice.

As regards structural facing, the NF DTU 21 refers to the requirements stated (FD P 18-503 1989). Such requirements concern flatness and texture (bubbling and localized defects).

The technical controller is assigned to prevent the various uncertainties potentially encountered when building structures. This controller is commissioned by the project owner, which provides him with an opinion regarding problems within the technical realm. This opinion is related to problems relative to both structural robustness and personal safety; it is mandatory for all facilities open to the public as well as tall buildings and buildings containing underground structural elements or foundations located in seismically active zones. Even where it is merely optional, this control is often requested by the project owner as a prerequisite for construction insurance coverage or to verify the durability of the given project.

The technical controller is mainly commissioned to verify compliance with the requirements of standards at all stages of the construction process.

Relations between the public project owner and private project management consultant are framed in the so-called "MOP" Law (Law on public project contracting) (MOP law no. 85-704 of 12 July 1985), which serves to define the respective primary missions.

28.1.3 Civil engineering structure: Fascicule 65, use of an external control body

Fascicule 65 (Fascicule 65 2017), titled "Execution of concrete civil engineering structures" is applicable to both concrete structures and the temporary structures needed to complete their execution successfully. This Fascicule establishes the general set of technical specifications and lays out the general requirements applicable when building structures for a 100-year facility use period. Each contract concluded is complemented by a set of the special technical specifications (or Cahier des Clauses Techniques Particulières (CCTP) in French).

Currently, the new version of Fascicule 65 (Fascicule 65 2017) must be made contractually binding and included in the project contract appendices being furnished. Execution oversight is addressed in Chapter 4, which mandates the set of documentation to be submitted by the contract recipient at the various stages of the construction process: preparation, execution, and completion. This documentation package is composed of a set of execution specifications (calendar, design notes, drawings, sketches, etc.) and a set of high-quality drawings and environmental protection plans. The various subparagraphs of the Fascicule list the execution and environmental protection procedures that the contractor must provide.

Quality management is based on the provisions stipulated in execution class 3 of Standard NF EN 13670/CN (2013), whose requirements are twofold: (i) for construction products, compliance with standards, and the European Conformity ("CE" marking protocol) and (ii) for execution, both an in-house control organized by the contractor awarded the contract and an external control assigned to the project manager. Tables compatible with those included in Standard NF EN 13670/CN (2013) summarize the material and product controls (design and information testing) as well as the execution controls (trial and control testing). Custom controls can also be added to the technical clauses specific to the project contract, depending on the techniques slated to be implemented. A dedicated control plan establishes, for the given building operation, the breakdown between the various internal and external control steps being scheduled.

Moreover, quality management relies on a list of breakpoints stipulated in the Fascicule and then complemented and adapted to the construction project specifics, as set forth in the technical clauses (CCTP). The breakpoints relative to concretes consider the following: concrete specifications, design testing, validation of trial testing, ultimate acceptance of the specimen element, acceptance of the mixing plants, authorization to pour a part of the structure, and the acceptance of the building facings.

Removal of these breakpoints by the project manager helps justify compliance of control results with respect to contractual requirements, based on the actions completed by the contractor for in-house controls and by the project manager for external controls.

The various articles of Fascicule 65 (Fascicule 65 2017) state the applicable execution specifications and provisions outlined in the areas of design, trial, and control, with emphasis on concretes and facings (see Chapter 8). The key contents of Chapter 8 are listed as follows:

- Compliance of concretes with Standard NF EN 206/CN (2014) is imposed.
- The concretes used in each part of the structure correspond to a group of exposure classes and, as such, must satisfy all the requirements inherent to each class.
- The concrete may be prescribed either by mixture composition specifications through applicable threshold values, as determined by the exposure class or by the performance-based specifications tied to the exposure classes. In this case, partial waivers are issued for mixture composition requirements; moreover, depending on the maximum measured durability indicator values, it becomes possible to reduce the cover to reinforcement.
- All concrete components must be compliant with current standards, in particular with (NF EN 12620 + A1 2008) and (NF P18-545 2011) as regards aggregates. It is authorized to use type 1 recycled coarse aggregate, according to standard NF EN 206/CN (2014), originating from the demolition of civil engineering structures, whose traceability is guaranteed for concretes of strength classes weaker than C35/45 (i.e., in class XC1, XC2, XC3, XC4, or XF1), with a 20% maximum substitution rate.
- The concrete is subjected to a design test based on a use reference or mixtures produced in the laboratory, thus making it possible to verify the performance and robustness of the nominal mix design compared with both market specifications and the anticipated execution method.
- The contractor furnishes a concreting schedule listing the provisions adopted to ensure good execution quality and durability, in addition to the conditions relative to the execution of the facings.
- The concrete undergoes a trial test, conducted in a mixing plant under the same conditions as the target project, more specifically with the same components, production

equipment, transportation, and casting conditions. The results obtained during this trial test, including if applicable acceptance of the specimen element (facing compliance), constitute a breakpoint.

- Production takes place at a ready-mix concrete plant that has been granted the right to affix the corresponding French Standardization on Ready Mixed Concrete, known in France as NF-BPE mark (structural concrete certification) or its equivalent with additional requirements, as indicated in the contractually binding Appendix B of the Fascicule. In the case of production in situ or at a precast concrete plant, the mixing plant must satisfy the specifications set forth in Appendix B.
- The concretes undergo compliance inspection at the time of casting by the contractor, on the basis of a division contractually defined for each structural part. Any given batch is considered to be compliant if the results obtained satisfy both the required consistency values and characteristic strength values, according to a statistical interpretation based on the number of samples per batch. As for durability-related values, compliance criteria are specified, as are the values for heat-treated concretes.
- Each sampling entails execution of a consistency measurement and the preparation of three specimens for the purpose of determining the 28-day compressive strength and an air content measurement for concretes mixed with an air-entraining agent.
- Execution specifications are defined for casting steps involving: vibration, construction joint treatment, execution of nonformed surfaces, formwork removal and cradling removal, pouring special concrete, curing, and protection.
- As regards the facings, sidewalls, and nonformed surfaces, these requirements are based on existing regulations and, especially for facings, Fascicules (FD P 18-503 1989) and FD CEN/TR 15739 (2010). The facing types are classified as single, thin, or elaborate.

28.1.4 Recycled concrete: aggregates and concrete standards

28.1.4.1 At the international scale

The use of recycled aggregates in roads and concrete become an international preoccupation, as shown in standards and methodology guides collected in RECYBÉTON. Forty documents from 20 countries of different continents, excepting Africa, were analyzed. The most experimented countries in recycling are Netherlands, Germany, Austria, United Kingdom, Japan, United States, and Canada (Quebec province). They are concerned with half of the analyzed documents. The report “Synthesis of international documents on recycled aggregates use in concrete” produced in RECYBETON details all international practices. The following conclusions can be drawn:

- In Germany, the best recycled aggregates types are used, and the substitution rates varied between 25% and 45% for exposure classes X0, XC, XF1, XF3, or XA1.
- In Belgium, only aggregates with best quality are incorporated with a rate of 20% in C16/20 concrete for exposure classes X0 and XC1.
- In Denmark, only exposure classes X0 and XC1 are concerned. The incorporation rate is 30% for fine aggregates and 100% for medium and coarse aggregates. Their use is limited for C30/37 concrete, and there are supplementary requirements in calculation codes (only 10% of fine aggregates and 20% of coarse aggregates, to control shrinkage, creep, and elastic modulus impact).
- In Austria, the use of recycled aggregates containing bricks in concrete is authorized.
- In Spain, a draft regulation aims at limiting the rate of recycled aggregates to 20%, considering a correction of shrinkage, creep, and elastic modulus coefficients.

- In Italy, the incorporation rate varies between 5% and 100%, for concretes C8/10–C45/55. Additionally, some precisions about their origin are required.
- In Norway, the recycled aggregates should come from controlled constructions before demolition. For required water resistance, freezing–thawing, and chloride resistances, the recycled aggregate properties must be specially controlled.
- In Netherlands, a great panel of possibilities exists; the substitution rate of 50% is used for all exposure classes, and it can reach 100% if the classes XD and XS are excluded. The resistance class is limited to C35/45. For more than 50% of recycled aggregates, the design rules have to be checked. In Netherlands, recycled fine aggregate is also authorized.
- In United Kingdom, two kinds of recycled aggregates are distinguished, depending on their origin: masonry recycled aggregates (RA) and recycled concrete aggregates (RCA). RCA is authorized up to 20% for maximum resistance class of C40/50 in X0, XC, and XF1 environments. The use of RA should be approved by a certification organization. There is no restriction if the resistance class does not exceed C16/20. It should be noticed that the Waste & Resources Action Programme (WRAP quality protocol) is followed to leave the waste status.
- In Russia, the authorized substitution rate is 100% for C16/20 concrete. This rate decreases to 10% for higher resistance class, and a durability test is required in freeze–thaw environment.
- In Sweden, the practice is similar to that of Norway
- In Switzerland, the use of recycled aggregates containing at least 25% in mass Rc constituents is authorized for environments X0 and XC. In this case, the designation of "natural aggregates" is authorized, even in the presence of recycled aggregates. The environments are limited to X0 and XC1 for recycled aggregates containing at least 5% mass of recycled bricks. Their use in another exposure classes requires preliminary studies.
- In Japan, the classification of recycled aggregates depends on their physicochemical properties (absorption and impurity), and their use is the function of the concrete type in the structure.
- In China, there are three classes of aggregates (three coarse and three fine aggregates). Similarly, in Japan, the use of recycled aggregates depends on concrete resistance class, and a maximum rate of 50 is applied if no technical data is available.
- In Australia, two classes of aggregates are considered, respectively, for concrete and road use. The maximum authorized rate is 100% and 30%, respectively, for C25/30 and C40/50 concretes.
- In New Zealand, the recycled aggregate classification is similar to that in United Kingdom.
- In Brazil, the substitution rate is limited to 20% for maximum resistance class C40/50.
- In United States and Québec, the use of recycled aggregates concerns particularly road applications.
- Except in Switzerland and Norway, the use of recycled aggregates is not authorized for prestressed concrete.

28.1.4.2 At the national scale

The two reference standards for aggregates are NF EN 12620 + A1 (NF EN 12620 + A1 2008) and NF P18-545 (NF P18-545 2011).

Standard NF EN 12620 + A1 (2008) comprises the European specifications, along with the lists of characteristics cataloged under the CE marking. Standard NF P18-545 (2011) is voluntarily applied and indicates European specifications for various uses. Moreover, this standard establishes the link between the NF EN Product Standards and the specifications targeting aggregates.

These two standards enable classifying recycled aggregates according to the various cri teria outlined as follows:

- Classification of components according to Standard NF EN 933-11 (2009) on the basis of manually sorted test specimen results and dividing them into a list of components;
- Soluble sulfate rate measurements in water, according to Standard NF EN 1744-1+A1 (2014);
- Modification of the concrete setting time, according to Standard NF EN 1744-6 (2007);
- Measurement of soluble chlorides in acid, according to Standard NF EN 1744-5 (2007).

The national complement for concrete standard (NF EN 206/CN 2014) stipulates the provisions applicable for the use of recycled aggregates in France. Three types of recycled gravel have been defined based on their classifications and associated testing frequencies:

- Type 1: all characteristics defined herewith are CR_B;
- Type 2: all characteristics defined herewith are CR_B or CR_C;
- Type 3: all characteristics defined herewith are CR_B or CR_C or CR_D.

During the initial period, which corresponds to production extending over a time span of at least 12 months and a quantity of at least 10,000 tons, the minimum frequency for sampling and testing recycled aggregates must match the contents of Tables NA.2–NA.4 of Standard (NF EN 206/CN 2014), by selecting the frequency yielding the greatest number of samples. Subsequently, during the continuous production regime, producers respect at least one of the testing frequencies (temporal or quantitative) listed in Tables NA.2–NA.4 of Standard (NF EN 206/CN 2014) (see Tables 28.1–28.3). This step complies with the provisions set forth in the producer's Aggregate Production Control Manual upon completion of the initial period.

The technical product data sheets drawn up by the aggregate producer serve to verify the compliance of recycled aggregates with respect to the standardized specification requirements.

Table 28.1 Table NA.2 of Standard (NF EN 206/CN 2014)—constituents of recycled aggregates

Code	Main constituents category, NF EN 12620 + A1	Secondary constituents: Categories NF EN 12620 + A1				Types of test frequency[a]: Temporal	Types of test frequency[a]: Quantitative (in tons)
CR_B	Rcu_{95}	Rb_{10-}	Ra_{1-}	$XRg_{0.5-}$	$FL_{0.2-}$	2/month	1/2,000
CR_C	Rcu_{90}	Rb_{10-}	Ra_{1-}	XRg_{1-}	FL_{2-}		
CR_D	Rcu_{70}	Rb_{30-}	Ra_{10-}	XRg_{2-}	FL_{2-}		

Note: Glossary dedicated to recycled aggregates: Rc: concrete, concrete products, mortar, concrete masonry elements contained in a recycled aggregate; Ru: unbound aggregates, natural stone, aggregates treated with hydraulic binders contained in a recycled aggregate; Rcu corresponds to Rc + Ru; Rg: glass contained within a recycled aggregate; Rcug corresponds to Rc + Ru + Rg; Ra: bituminous material contained in a recycled aggregate; Rb: elements containing burned clay (bricks and tiles), elements made of calcium silicate, non-floating cellular concrete contained in a recycled aggregate (Rb_1 means Rb less or equal to 1% of the total aggregate mass); X: clays, soils, ferrous and non-ferrous metals, wood, plastic, non-floating rubber, plaster contained in a recycled aggregate; XRg corresponds to X + Rg; FL: floating material (expressed in volume terms) contained in a recycled aggregate.

[a] Any lot of recycled aggregates whose production is below the test frequencies of the Table NA.2 of Standard (NF EN 206/CN 2014) must be subjected to a minimum test of classification of constituents. The Standard NF P 18-545 defines the notion of lot.

Table 28.2 Table NA.3 of Standard (NF EN 206/CN 2014)—conventional standardized characteristics of recycled gravels

				Types of test frequency[a]	
Property	*Testing method*	*Code*	*Category*	*Temporal*	*Quantitative (in tons)*
Sulfate soluble in water	NF EN 1744-1, article 10.2	CR_B, CR_C CR_D	$V_{SS0.2}$ NF P 18 545, code SS_D $V_{SS0.7}$	1/week	1/1,000
Specific gravity	NF EN 1097-6	CR_B, CR_C CR_D	≥2.0 t/m³ ≥1.7 t/m³	1/week	1/1,000
Influence on initial setting	NF EN 1744-6	CR_B CR_C, CR_D	A_{10} A_{40}	2/month	1/2,000
Flakiness coefficient	NF EN 933-3	CR_B, CR_C, CR_D	FI_{35}	1/month	1/4,000
Los Angeles	NF EN 1097-2	CR_B, CR_C CR_D	LA_{40} LA_{50}	1/2 month	1/,8000
Chlorides soluble in acid	NF EN 17₄4-5	CR_B, CR_C CR_D	To be declared	2/month	1/2,000
Water absorption at 24h (WA_{24})	NF EN 1097-6	CR_B, CR_C, CR_D	To be declared	1/week	1/1,000
Releasable alkalis according to LPC no 37 (LCPC 1993)	pr XP P 18-544	CR_B, CR_C, CR_D	To be declared	2/month	1/2,000

[a] Any lot of recycled aggregates whose production is below the test frequencies of the Table NA.3 of Standard (NF EN 206/CN 2014) must be controlled for every characteristic. The notion of lot is defined according to NF P 18-545.

Table 28.3 Table NA.4 of Standard (NF EN 206/CN 2014)-characteristics of recycled sands

			Types of test frequency[a]	
Property	*Testing method*	*Category*	*Temporal*	*Quantitative (in tons)*
Sulfate soluble in water	NF EN 1744-1, article 10.2	NF P 18-545, code SSD $V_{ss0.7}$	1/week	1/1,000
Specific gravity	NF EN 1097-6	≥1.7 t/m³	1/week	1/1,000
Influence on initial setting	NF EN 1744-6	A_{40}	2/month	1/2,000
Chlorides soluble in acid	NF EN 1744-5	To be declared	2/month	1/2,000

[a] Any lot of recycled aggregates whose production is below the test frequencies of the Table NA.4 of Standard (NF EN 206/CN 2014) must be controlled for every characteristic. The notion of lot is defined by the Standard NF P 18-545.

28.2 QUALITY CONTROL ON EXPERIMENTAL CONSTRUCTION SITES

28.2.1 Concrete and materials testing

28.2.1.1 Recycled coarse aggregates

In the various experimental projects, recycled aggregates undergo testing in accordance with the concrete and aggregate standards. The required specifications were verified, and the obtained results served to classify coarse aggregate into type 1, 2, or 3 (Tables 28.4 and 28.5).

Table 28.4 Example of experimental construction site Chaponost—properties of recycled aggregates (Lafarge laboratory, Vitry-sur-Seine)

Property	*Associated standards*	*Results*	*Associated code*	*Associated category*
Sulfate soluble in water	NF EN 1744-1	0.06	CR_B, CR_C	$SS_{0.2}$
Specific density	NF EN 1097-6	2.61	CR_B, CR_C	≥2.0 t/m^3
Absorption with fines	NF EN 1097-6	9.4	CR_B, CR_C, CR_D	Declarative
Influence on initial setting	NF EN 1744-6	15	CR_C, CR_D	A_{40}
LA (10/14)	NF EN 1097-2	29	CR_B, CR_C	LA_{40}
Alkalis	LPC no 37 (LCPC 1993)	0.022	CR_B, CR_C, CR_D	Declarative

Table 28.5 Example of experimental construction site Chaponost—properties of recycled aggregates according to NF EN 933-11 (NF EN 933-11 2009)

	Rcu	*Rc*	*Ru*	*Rb*	*Ra*	*Rg*	*X*	*FL*
4/20 recycled	98%	87%	11%	1.50%	1.30%	0%	0.20%	0.49 cm^3/kg
Classification		Rcu_{95}		Rb_{10-}	Ra_{10-}		$XRg_{0.5}$	FL_{2-}
		CR_B		CR_B	CR_D		CR_B	CR_C

Given the classification outcome of Chaponost project gravel, a type 3 could be determined. It is noticeable that, if the aim is to stick to current standards, these aggregates would not have been accepted. However, no problem was encountered at the construction site and after (Chapter 22). Perhaps, the importance of concrete deformability increases with the replacement rate, higher than reported in the state of the art (Chapter 10) is due to the nature of this specific RA.

As for the fine aggregates, the practical testing campaign findings are compliant with the specifications found in Table NA.4 of Standard (NF EN 206/CN 2014) (Table 28.6).

28.2.1.2 Concrete tests

Concrete controls are carried out by the concrete producer at frequencies stipulated by both the Standard (NF EN 206/CN 2014) and the user (contractor), while following the frequencies listed in the NF DTU 21 Codes of Practice and Fascicule 65, respectively, for buildings and civil engineering structures. The contractual documentation or technical clauses provide the necessary complements for these controls. Regarding engineering structures, further controls are conducted by the external body, notably during critical construction steps to remove any breakpoints.

Table 28.6 Example of experimental construction site Chaponost—properties of recycled sand (Lafarge laboratory, Vitry-sur-Seine)

Property	*Associated standards*	*Results*	*Associated category*
Sulfate soluble in water	NF EN 1744-1	0.15	$SS_{0.7}$
Specific density	NF EN 1097-6	2.55	≥1.7 t/m^3
Absorption with fines	NF EN 1097-6	9.8	—
Influence on initial setting	NF EN 1744-6	20	A_{40}

In the case where durability indicator thresholds are specified in the contract, the controls pertain to the compliance verification not only of all results obtained at the various development stages but also at the production of the concrete mixture (mixture development, on-site trials, control during construction).

These controls target the characterization of both fresh and hardened concrete properties, in addition to the determination of durability properties as needed.

28.2.2 Quality control of concrete mixtures

Compliance with production requirements is verified by checking that the ready-mix concrete plant has been granted the right to use the NF BPE mark (in accordance with the NF033 (NF 033 revision 24 2016) structural concretes certification regulation) or an equivalent. In other cases, a description of the production control system, as intended in Standard (NF EN 206/CN 2014), must be made available and accompanied by a statement of the controls performed, thus making it possible to ensure concrete compliance as per this same standard.

For concretes dedicated to civil engineering structures, the mixing plant must verify the additional specifications listed in Appendix B of Fascicule 65. In the case of casting in situ or in a precast plant, the mixing station must satisfy the specifications of the regulation in Appendix B of Fascicule 65. Verifying that this requirement has been met typically entails a quality system audit, which may then be followed up by conducting tests over the course of the actual production run. During such an audit, it may prove worthwhile to verify the recycled aggregate storage conditions.

At the time of delivery, comparison with the delivery note enables ensuring that the initial order has been accurately filled. For civil engineering projects covered under Fascicule 65, the delivery slip is complemented by a weight receipt, which allows controlling for compliance with production requirements, notably in terms of component concentrations and the E_{eff}/L_{eq} ratio (water binder ratio). This provision is also imposed on category C building projects that extend to special elements PA and PB (see Section 28.1.2).

28.2.3 Quality control before casting: site trials

Before casting the structure, a trial test is organized to verify that design characteristics are satisfied for the concrete produced, under the conditions and using the equipment specified for the given project (Table 28.7).

- Slump (mm) (NF EN 12350-2 2012)
- Rc: compressive strength at 1, 7, and 28 days (MPa) (NF EN 12390-3 2012)
- Rt: splitting tensile strength at 1, 7, and 28 days (MPa) (NF EN 12390-6 2012).

Table 28.7 Example of tests' program for trial tests in the case of archives' building—Mitry-Mory

Tests	*CEMEX concrete and laboratory*	*Cerema (construction site)*
Temperature of concrete	1 measure t_0	1 measure t_0
Consistency	1 test $t_0 + t_{30}$	1 test $t_0 + t_{30}$
Rc_{2d}	3 specimens t_{30}	—
Rc_{7d}	3 specimens t_{30}	—
Rc_{28d}	3 specimens t_{30}	3 specimens t_{30}
Rt_{28d} (splitting)	3 specimens t_{30}	3 specimens t_{30}

Note: t_0 and t_{30} mean when concrete was available, and 30 min after, respectively (see Chapter 24).

28.2.4 Control of concrete mixtures

These controls concern the verification of fresh and hardened concrete properties, in addition to the measurement of durability indicators and mechanical characteristics, if applicable. The list of tests commonly practiced is indicated in the following.

It should be pointed out that the quality of results obtained from these tests must be backed by certification or accreditation of the participating laboratories. For example, the COFRAC (French Accreditation Committee) guarantees compliance with requirements imposed by Standard NF EN ISO/CEI 17025 (2005) and ensures the reliability of test results according to this standard.

28.2.4.1 Compressive strength, slump, and air content

- Slump (mm) (NF EN 12350-2 2012)
- Air (%) (NF EN 12350-7 2012)
- Rc: compressive strength at 1, 7, and 28 days (MPa) (NF EN 12390-3 2012)
- Rt: splitting tensile strength at 1, 7, and 28 days (MPa) (NF EN 12390-6 2012).

28.2.4.2 Durability

- Water porosity at 90 days (%) (NF P18-459 2010)
- Gas permeability at 90 days ($\times 10^{-18} m^2$) (XP P18-463 2011)
- Apparent chloride diffusion at 90 days ($\times 10^{-12} m^2/s$) (XP P18-462 2012)
- Carbonation depth at 90 days (accelerated test) (mm) (XP P18-458 2008)

28.2.4.3 Mechanical characteristics: modulus and shrinkage

- 28-Day Young's modulus (E) (GPa) (ISO 1920-10 2010)
- Shrinkage from 1 to 28 days (%) (NF P18-427 1996)

28.2.4.4 Control program

A control program must be established both to declare compliance of the control batches on the basis of contractual specifications and verify the concrete properties at the time of casting (Table 28.8).

Table 28.8 Example of control program in the case of archives' building—Mitry-Mory (see Chapter 24)

Tests	*Concrete producer control (laboratory and site)*	*External control (site)*
Temperature of concrete	1 measure on concrete plant	1 measure
Consistency	1 test/batch	1 test/batch
Rc_{2d}	3 specimens on site	
Rc_{7d}	3 specimens on concrete plant + 3 specimens on site	
Rc_{28d}	3 specimens on concrete plant + 3 specimens on site	3 specimens on site
Rt_{28d} (splitting)	3 specimens on site per batch	3 specimens on site
Total shrinkage (NF P18-427)	1 test	1 test
Elastic modulus (28 days)	3 specimens	3 specimens
Water porosity (90 days)		2 specimens
Gas permeability (90 days)		2 specimens
Accelerated carbonation (90 days)		2 specimens

28.2.5 Control of facings and initial inspection

After formwork removal, the facings must be examined so as to verify their compliance with the acceptance criteria, as per the regulation applicable to the type of structure built and to the contractual specifications, which must also indicate the batches subjected to verification. For buildings, this regulation is the NF DTU21 Codes of Practice (2017). For civil engineering structures, this regulation is Fascicule 65 (Fascicule 65 2017).

Depending on the type of structure, it may prove necessary to perform a detailed initial inspection for recording the flaws and imperfections observed at the end of execution. Such a detailed inspection is mandatory for engineering structures and serves to compile a state of reference essential for monitoring the structure over time.

28.3 GUIDANCE FOR RECYCLING COARSE AGGREGATE

28.3.1 Rate of incorporation in accordance with the standardized values

The maximum substitution rates are provided in Standard (NF EN 206/CN 2014) (see Table 28.9). These maximum rates are limited by the provisions contained in Fascicule 65 (Fascicule 65 2017): use of type 1 recycled coarse aggregate, according to Standard (NF EN 206/CN 2014), derived from the demolition of engineering structures, whose traceability is covered for concretes weaker than the C35/45 strength class (i.e., XC1, XC2, XC3, XC4, or XF1), with a maximum substitution rate of 20%.

For projects satisfying the maximum substitution rate authorized by current regulations, it is necessary to verify that the control-related provisions have been implemented by the concrete supplier, the contractor, and the project owner and manager (see Table 28.10).

28.3.2 Rate of incorporation above standardized values

Feedback from experimental projects suggests the possibility of increasing the substitution rate without being subjected to any noteworthy impacts on the concrete mix design or on the casting and final quality of structures. In practice, this rate may range from 30% to 40% for the most common exposure classes and for type 1 recycled aggregates.

Should a substitution rate exceeding the values imposed by Concrete standard (NF EN 206/CN 2014) be employed, complemented by the prescriptions listed in Fascicule 65 (Fascicule2014) for civil engineering structures, it would become necessary to carry out the following additional controls, beyond those outlined in Section 28.3.1 (Table 28.10):

Table 28.9 Maximum substitution rate (in %) depending on recycled aggregate type (NF EN 206/CN 2014)

	Exposure class			
Type of recycled aggregate	X_0	XC_1, XC_2	$XC_3, XC_4, XF_1, XD_1, XS_1$	*Other exposure classes*
Recycled coarse aggregate type 1	60	30	20	0[a]
Recycled coarse aggregate type 2	40	15[a]	0[a]	0[a]
Recycled coarse aggregate type 3	30	5	0	0
Recycled fine aggregate	30	0	0	0

[a] For asphalt concretes, the use of 20% of reclaimed asphalt aggregates coming from base or surface layers or airfields pavements is allowed when their traceability is known.

Table 28.10 Control of constituents and concrete

	Recycled aggregates	*Concrete supplier*	*Contractor*	*Project owner*	*Project manager*
Building DTU21	Product data sheet of aggregates with characteristics according to NF EN 12620 +A1 NF 18545 NF EN 206/CN	Respect of controls on constituents and concrete according to NF EN 206/CN	Control, depending on construction and particular structure category Respect of sampling frequency and the contract's specifications	According to contract's specifications (MOP law)	Verification of repository conformity at the phase of control, conception and realization and site control Reception control
Bridges Fascicule 65	Ditto + NF Mark (structural concretes certification)/ its equivalent or a control procedure	Ditto+verification of requirements' respect of Appendix B of Fascicule 65	External control according to Tables 4.A and 4.B of Fascicule 65 Decomposition in structure parts according to contractual specifications	Ditto + external control according to Tables 4.A and 4.B of Fascicule 65 Lifting of breakpoints	Initial and final detailed inspection of execution

- Characterization of mechanical properties: modulus, shrinkage;
- Verification of incorporation of mechanical characteristics in the design calculation notes for the portion of the structure under consideration, along with any ultimate impacts on the rate of reinforcements;
- Concrete control plan adapted to the particular structure and calibrated to meet the requirements of execution class 3 of standard NF EN 13670/CN (2013);
- Characterization of the relevant durability indicators regarding both structural exposure and risks of concrete deterioration.

28.4 CONCLUSION

The controls with the requirements included in standards or regulations, associated with the production of recycled aggregates and the execution of structures, can effectively guarantee quality and durability, provided that the requirements have been appropriately implemented. These regulations specify controls for the operations of recycled aggregate production, concrete production, and concrete casting. Such controls enable verifying compliance with the defined properties at each stage of the process. Moreover, the setup constitutes a bona fide "toolbox", whose effective use depends on the unique features and criticality of the portion of the structure to be built. This organization of structural controls must be adapted to each operation and targeted by a control program in place before commencing the works, in associating all the actors in the construction process. The traceability of controls and respect of test procedures also guarantee the reliability of results and the compliance of structures relative to the contractual specifications.

It is of great interest to precise that the use of recycled aggregates leads to more vigilance and thus more control tests, particularly in their production stage, to guarantee a good concrete quality and durability. Nevertheless, the recycled aggregates concrete is subjected to the same control tests than natural aggregates one.

Section VIII

Sustainability of concrete recycling

A. Feraille
Ecole des Ponts ParisTech

In a more constrained context in terms of sustainable development and circular economy, it is important to evaluate from an environmental point of view the concrete recycling process. This is the scope of Chapters 29–33.

First of all, nothing is possible if the resource of recycled aggregate (RA) is not well identified. Thus, Chapter 29 is dedicated to qualitative and quantitative characterization of the resources on materials able to be recycled in concrete in France. Scenarios are drafted to provide the life cycle analysis (LCA) that is presented in Chapter 30. This study aims to compare the environmental impacts of recycled aggregate concrete to normal aggregate concrete. A multicriteria analysis of the factors impacting the LCA of such concretes (cement content and cement type, aggregate substitution rate, transport distances of aggregates and concrete) is proposed in a case study.

The release of substances from construction materials into the soil and groundwater is a growing concern which has been addressed in the framework of the European regulation on construction materials. Chapter 31 answers the following two questions: How different could be the leaching characteristics of recycled concrete compared to concrete prepared with only natural aggregates? and Do recycled concrete comply with existing threshold limit values for the environmental assessment of construction materials? Chapter 32 describes how the multi-recycling may change the concrete properties after more than one recycling cycle. The results show a deterioration of concrete with the increase of recycling cycles and confirm the interest of limiting a recycling rate to control the performance of concrete through the multi-recycling. It also suggests that in the long run, a separation of the original natural aggregate from the cement paste is desirable to limit the amount of cement paste in multi-recycled concrete.

The construction industry organization is complex, including many actors, for the success of environmental, economic, and societal preservation, and a coordination between actors is necessary. Thus, Chapter 33 considers how the development of the recycling of concrete demolition waste to produce new concrete could be a major component of sustainable construction.

Chapter 29

Resource and variability of recycled aggregate in France

L. Mongeard
ENS Lyon

D. Collonge
LafargeHolcim France

F. Jezequel
SIGMA BETON

CONTENTS

Abstract

This chapter is dedicated to qualitative and quantitative characterization of the resource of reusable materials in concrete in France. It originates from two sources:

- a statistical study based on the exploitation of data issued from the reports written by a French regional organization acting in the frame of the departmental construction and public works waste management plan. This study allows to approach a national consistent evaluation by waste types clarifying several evolution tracks related observed practices;

- a recycling platforms supervision – 13 representative sites of national diversity – showing medium course of recycled aggregates over several months with a focus on two sites during 24 months.

The first study integrates the conclusion of the platforms monitoring in order to approach the ways of actions to increase recycling in concrete considering the current practice and its knowledge. Furthermore, a geographical characterization of twelve of the main French urban areas is made by the inventory of recycling sites and their spatial distribution in relation to concrete production units (ready-mix and precast concrete factories). Thus, scenarios are prepared to provide the life cycle analysis that has been carried out on four towns (Lyon, Lille, Strasbourg, Bordeaux, see chapter 30). Finally, recycling target published in the 2015 Energy Transition Law has been added with a specific development linked to the increased risk of competition in the recycling between public works (road subbases and embankments) and concrete utilizations.

29.1 INTRODUCTION

The RECYBÉTON project aims not only at increasing the knowledge of the recycling of concrete but also at better understanding the recycling context in France of inert waste, in terms of flow and quality by nature of materials. Recent figures on construction and demolition waste (CDW) show that its recovery is very advanced; however, misconceptions still circulate. It is therefore necessary to specify the national resource available to produce recycled aggregates (RA) for concrete, and to answer the following question: can recycling practices established since a long time and required for road subbases or embankments are transposable to the development of concrete in terms of quality? The location, nature, quantity, and quality of these materials sources are keys to determine the extent to which sustainable recycling in concrete could make sense. These issues are addressed in the current chapter.

29.2 RESOURCES OF INERT MATERIALS THAT ARE RECYCLED IN CONCRETE IN FRANCE (VOLUMES AND LOCATIONS)

This evaluation is based on waste studies made by a French regional organization (Cellules Economiques Régionales de la Construction) (CERC 2018) in the frame of diagnostics for departmental—or regional—waste management plans in public works and construction fields. These local studies, mainly at departmental scale, have been built following the same rigorous methodology. They give a detailed account of wastes provided by the 32 French departments, which have already published their studies at the end of March 2015; a national assessment is also extrapolated from it to estimate the potential concrete recyclable resources (Figure 29.1).

29.2.1 Volumes

General figures relating to the waste production from public works and construction activities are published at the national scale (ADEME—Agence de l'environnement et de la maîtrise de l'énergie, Environment Ministry (ADEME 2016a)), but the scheduled updates are not frequent. The key figures of 2016 about waste edited by ADEME remains on mineral waste study of 2012, wrote in accordance with rubble and spoil investigation made in 2008 for public works and construction field (CGDD-SOeS 2015). The last survey, managed on 2014 data by the Ministry, which had given rise to a synthetic publication by Service

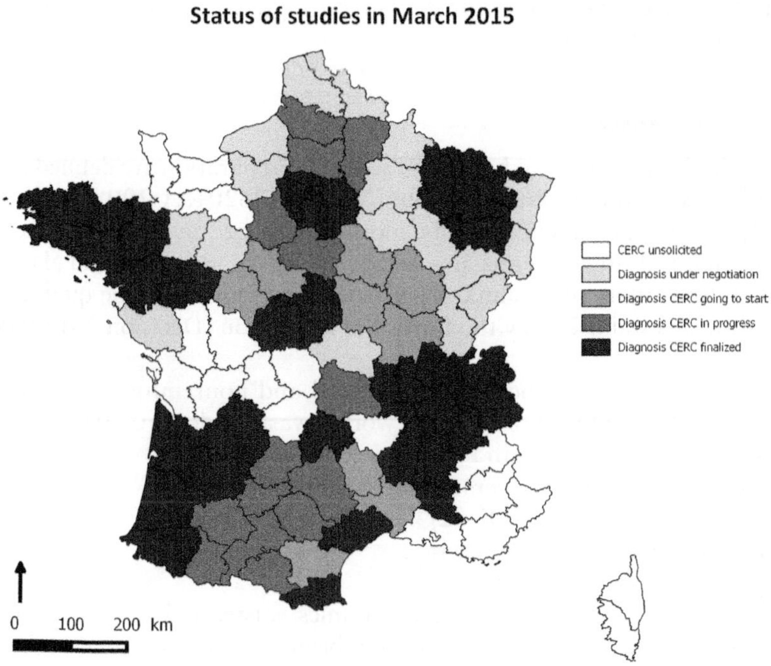

Figure 29.1 Study sources. Black represents the departments whose published reports provide the evaluation.

de l'Observation et des Statistiques (SOeS) in March 2017 (CGDD-SOeS 2017) has just been published in October 2018.

As part of RECYBÉTON, it seems interesting to consider an inventory of recyclable materials in concrete from detailed and recent sources, cross-checked and validated by local actors in public works and recycling. Although the coverage of available studies does not seem complete across France, the robustness of the CERC diagnosis seemed fundamental for us to identify the best trends for this inventory in France. In fact, the CERCs carry out surveys matching material flows with local practices from sites receiving waste (revenues and volumes), public works, demolition, construction, and local authorities. Understandably, the balance between the local trend resulting from the analysis of the CERC questionnaires and that of the national reports was checked as and when the publications of the Ministry of the Environment were published. This study first relies on the situation of 2008 with an initial estimate of more than 237 million tons of inert waste from decommissioning works in France (CGDD-SOeS 2010), which was used as a basis from 2010. An update was published in 2015 reducing this estimate to 231 million tons (CGDD-SOeS 2015). In 2017, on the data of the new national survey based on the year 2014, a drop to 215 million tons of inert was communicated with 61% recovery (reused, recycled, or used as backfill in the quarries) (CGDD-SOeS 2017). The figures and the study charts have been updated. The scope and considerations of the 2014 study are unchanged (in this document, the estimates for 2008 and 2014 are presented in the form of a range). More recently, in January 2018, the CERC GIE (Groupement d'Intérêt Economique) published major trends in waste surveys in 71 French departments (see CERC website [CERC]. The observations made in 2014 in 32 departments are confirmed with the same proportions in the typology of deconstruction waste. Very recently, the Ministry's detailed report published in October 2018 cross-checks this information and gives an estimate of the recovery rate at 69%, close to the target set by the European Directive.set by the European Directive [(CGDD-Service de la Donnée et des Etudes Stattistiques SDES 2018].

29.2.2 Waste typology

The accepted CDW components to be used in concrete preparation are defined and described in standards for recycled concrete aggregate [NF EN 12620+A1 2008] and concrete [NF EN 206 2014]: concrete, concrete elements, mortar, parts of concrete work (Rc), aggregates without binder, natural stones, aggregates with hydraulic binder (Ru), and glass (Rg). These component families—more or less mixed with other elements—can be qualified in the different types of waste established by the European regulation [Decision 2014/955/EU]:

- concrete waste: generally including reinforced and nonreinforced concrete, concrete products, mortar, and part of concrete works (waste code 17 01 01),
- bricks (code 17 01 02), tiles, and ceramics (code 17 01 03),
- mixtures of concrete, bricks, ceramics, etc. (code 17 01 07),
- nonreyclable glasses (code 17 02 02)
- and soils and stones (code 17 05 04)

In the CERC surveys, the bricks, tiles, and ceramics waste, and the nonrecyclable glasses (which could cause problems in the chemical equilibrium of concrete) are marginal. The other wastes—bituminous aggregates, soils—are unfit for concrete. Within soils and stones, which represent the majority of the construction, demolition, and public works waste (two thirds of inert waste), it has been possible to discriminate the part of recycled all-in material and the part of natural stones that make a secondary resource for recyclable aggregates in concrete.

The assessment of the potential resource led towards a global environmental logic to identify all the available materials eligible for building needs; then, ideally, the local constraints of place, time, and waste quality have to allow an optimization according to building sites and recycling rules.

Figure 29.2 synthesizes the inert waste proportion for the investigated departments: it is an average picture based on 2014 data, updated with last figures (CGDD-SOeS 2017) and recalculated for the streams plotted in the CERC reports. Indeed, the 2014 study was exclusively based on rates of waste in transit through recycling platforms. The recent communication from the CERC GIE confirms these proportions (see CERC website [CERC]).

In the context introduced by the law on the energy transition for green growth [Law No. 2015-992], it is important to know all the waste in their respective proportions (knowing that the differences between the initial extrapolations of the 2014 report and the latest trends are insignificant).

29.2.3 Streams assessment method

Different approaches have been carried out to schematize the statistic behavior of the French departments and propose a relevant methodology for extrapolating the local data to the whole metropolitan territory. This reasoning led in terms of trend. The national resource is thus estimated to produce 200 million tons (Mt) of inert waste maximum, which is, certainly, lower than Ministry of Environment data (215–231 Mt), but it can be possible to explain: the national survey identifies nonauthorized industries that are not visible in the CERC surveys; a consolidation of the local approaches inputs a statistic reducing bias. It is necessary to be cautious in the potential secondary resources' identification in an environment where national volume estimations are regularly challenged.

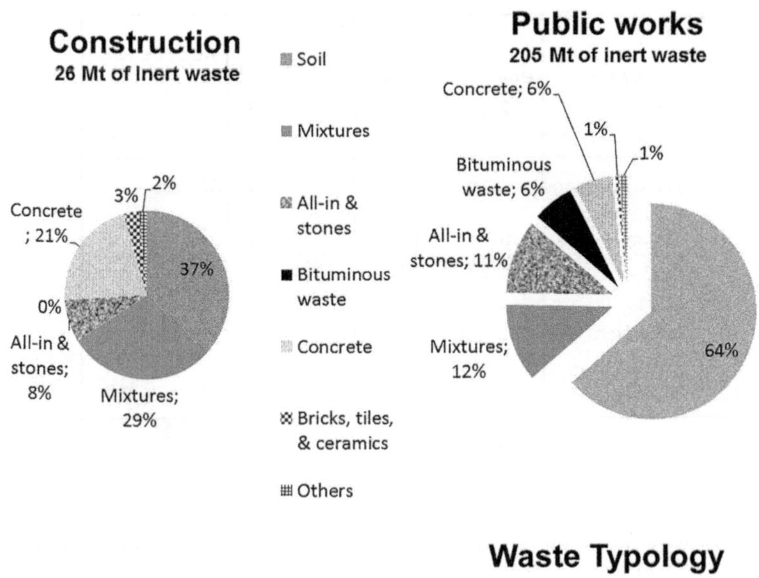

Figure 29.2 Inert waste proportion by source—public works and construction. (Results based on 32 French Departments, around 84 million tons of inert).

Recycling effective hypothesis about each three selected categories (concrete, mixed concrete waste, all-in/stones) has been considered due to the fact that 100% of material category is never recyclable. These hypotheses are based on a follow-up realized on recycling platforms (Jezequel 2013) (see chapter 29.3.1, p. 476 and following pages) and according to waste experts. The following values have been taken: 60% for concrete, 30% for mixed inert waste, and 75% for all-in and stones.

The CERC surveys on recycling platforms have inventoried all waste per type without distinction of source. A part of concrete waste does not come from deconstruction but from scraps of concrete units (mobile or fixed) and prefabricated plants that are not checked by the CERC methodology. Also, based on 2014 data (to be consistent with previous balance sheets), 20 million tons of precast concrete and 37 million m^3 of ready-mix concrete generated (with an average rate of 3% waste) a mass of concrete waste estimated at 3.2 million tons ((RMC volume × volumetric mass of concrete + 20 Mt) × 3% of waste). Of these 3.2 million tons, 60% is hardened concrete waste, about 2 million tons, with the rest being noninert sludge that is difficult to recycle.

29.2.4 Assessment of the global potential resource

Considering these remarks, the state of the surveys and the data checking in the 215–231 million tons of inert wastes generated from demolition, public works, and concrete industries (ready-mixed and precast), it is observed that (see Figure 29.3)

- the total of concrete waste, of mixture with concrete, all-in material, and stones are about 71 million tons;
- about 38 million tons of these waste constitute the secondary resource potentially recyclable in concrete, whose 12–21 million tons are concrete wastes and mixtures of concrete-based materials.

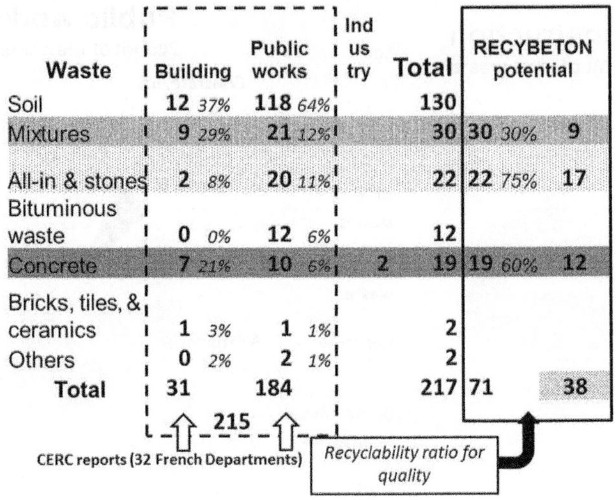

Figure 29.3 Potential resource of inert wastes for recycling aggregates in concrete. (Data SOeS 2014/2017—UNICEM 2014).

29.2.5 Industry analysis and evolution margins in construction

Figure 29.4 summarizes material flows in construction by source and family of use (from the 2014 national data of the Ministry (CGDD-SOeS 2017) and [Union Nationale des Industries de Carrières Et de Matériaux (UNICEM)], where the gray arrows represent recycling:

- first step: quarries extraction
- second step: aggregate production from primary resource quarries completed with secondary resources (industry and construction and demolition (C&D) - waste)

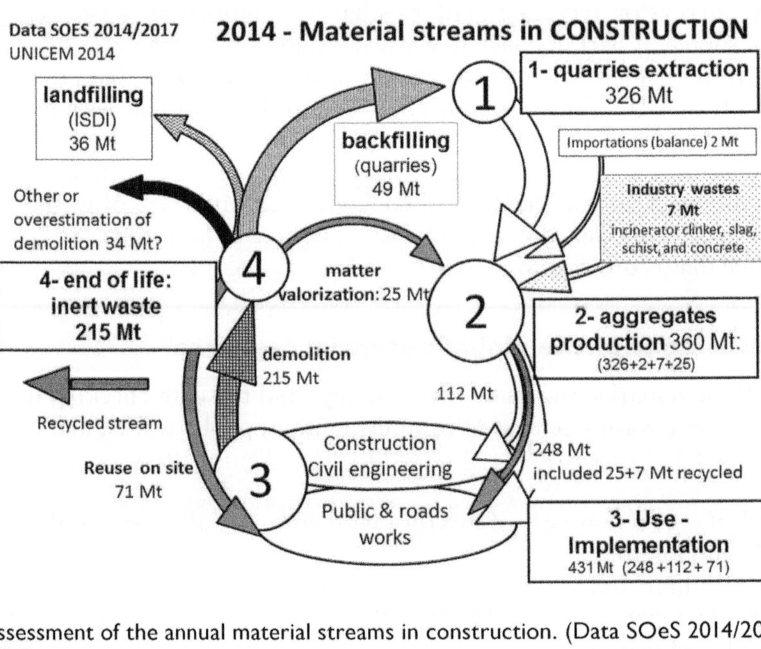

Figure 29.4 Assessment of the annual material streams in construction. (Data SOeS 2014/2017—UNICEM 2014).

- third step: implementation in civil engineering structures and construction, where it is possible to distinguish the part only built from concrete.
- fourth step: end of life (demolition) with inert wastes directly reused on works (short cycle and less transport), recycled on platforms or in quarries to become secondary aggregates, recovered in quarries in backfilling, or stored in landfilling (ISDI = inert waste landfill).

This figure brings out that recycling practices work, but the concrete structures do not benefit from them.

The demolition and industry wastes to recycling steps (see Figure 29.5) show the progress margins, knowing that, still today, the "concrete" sector only has few materials (less than a million tons in 2014):

29.2.6 Consequences of the energy transition for green growth law: local practices at the scale of the construction sites

The use of CDW on construction works provides the first stream of recycling operations: it is globally estimated at 71–81 Mt of inert wastes includes not only soils (45–50 Mt) but also concrete, mixtures of concrete-based materials as well as all-in material and stones (22–25 Mt in total). The nearness makes these streams preferential with respect to public works activities. In addition, the goals of the Energy Transition for Green Growth Law of 17 August 2015 [Law n° 2015-992] will strengthen the need for maintaining all these wastes as far as possible available for these works.

Among the measures described by this law in the Environmental Code and public market rights, it is necessary to consider matter and waste resulting of construction, demolition, and maintenance of road works, where the state and communities are the contracting authorities:

- 70% of the waste (generated from road works) has to be recycled or recovered by 2020 (in terms of the [Directive 2008/98/CE]).
- 60% of implemented material in road works arise come from recycling and reuse till 2020, with 20% minimum for surfaces layers and 30% for roadbeds.

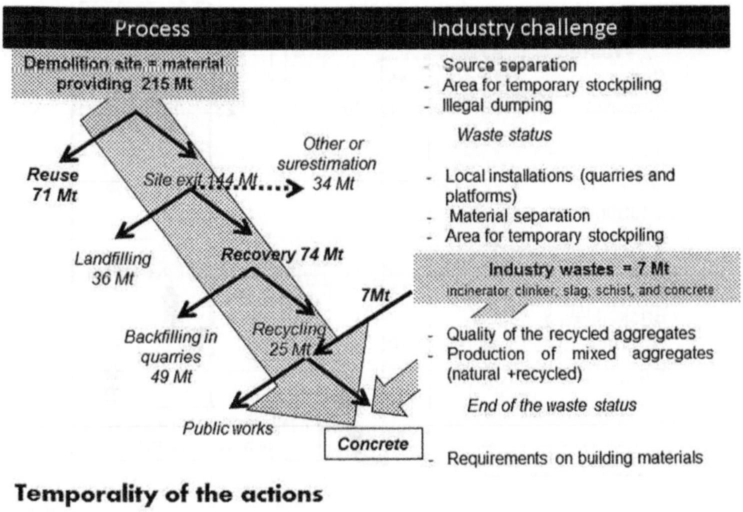

Figure 29.5 Stages towards recycling of inert wastes in concrete. (Data SOeS 2014/2017—UNICEM 2014).

29.2.7 Evaluation of the total effective resource

Therefore, potential resource figures for the concrete industry (Figure 29.3) need to be reevaluated without the portion that will be used for public works for the actual resource values available for recycling into concrete, as shown in Figure 29.6.

Reconciliation of these actually available quantities (25 Mt) with those currently used in structures and construction (112 Mt—Figure 29.4) results in an overall theoretical substitution potential of 22%. This approach is very much theoretical, because it would mean that materials and waste are distributed homogeneously throughout the territory: we will see in the next chapter that this is far from being the case. In addition, two parameters must always be considered: the schedule of operations and the configuration of the works. The organization of the works neither allows the use of inert wastes generated nor the access to adequate spaces if the materials are adapted and available.

Moreover, civil structures work related to road works—currently out of scope of laws requiring a recycling rate—such as crash barriers, bridges, tunnels, foundations, and works intended to water collection are mainly made with concrete, represent 39 Mt in 2014. Globally, these structures are a potential market of RA concrete, representing additional benefits, thanks to the time and transport savings if recycling operations and concrete production can be organized near the works. The will to promote recycling in the concrete industry and proximity logic could then converge on requirements leading the incorporation of recycled materials from the worksites themselves. Nevertheless, that assumes evolutions of methods in technical management of works as well as adjustments in the production of concrete.

To complete this prospective analysis, we should project this image of current flows in the long term and consider the changes in our lifestyles and our habitats imposed by energy issues and a frugal management of raw materials and space. For example, densification of the habitat may lead to a reconstruction of the city itself, which should lead to an increase in the deconstruction wastes from the cities—knowing that we have seen that these are the most mixed sources. Likewise, changes in modes of travel will induce changes in road networks by remobilizing the materials that constitute them. Therefore, we perceive that the evolutions are not so simple. On the other hand, it is necessary to insist on the point that

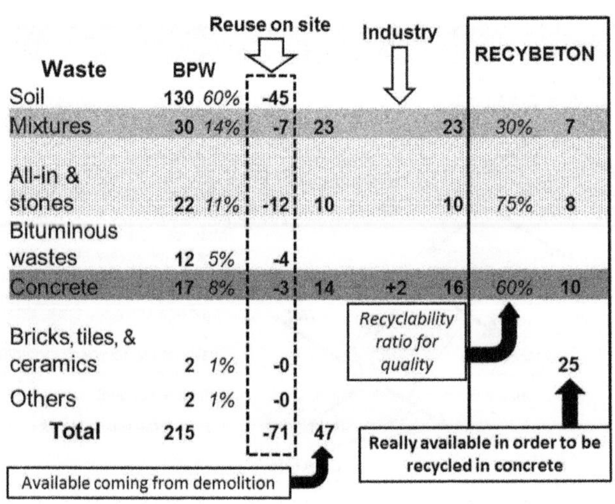

Figure 29.6 Available effective resource for RAs within concrete after works reusing. (Data SOeS 2014/2017—UNICEM 2014).

appears clearly in the study (Figure 29.6): the first way of progress for RA in the concrete is in the increase of rates of recyclability of the different families of waste.

29.2.8 National territory: need to optimize the recycling sites distribution

29.2.8.1 Department analysis

Looking forward to a future website [Materrio] that could include all the information on recycling in France (one of the points of the commitment for green growth signed in 2016 between the Ministries and UNICEM [Engagements pour la croissance verte]), the study is based on the CDW platforms of the website of the French Building Federation. The number of recycling and storage platforms located less than 30 km from urban areas was identified. A great heterogeneity between the departments appears. Some departments do not have a recycling platform for inert waste within 30 km of their main agglomeration. Not surprisingly, the most equipped departments with recycling sites show a higher recovery rate, with the waste resource exceeding the local economic threshold compared to the market price of natural aggregates (NA; Figure 29.7).

Recycling is based on the number of platforms adapted to the volumes of waste in a given area. A Geographic Information System (GIS) tool approach was used as part of a study (thesis currently being published) to provide a territorial analysis on optimizing the location of recycling sites. This type of study is to try to anticipate the flow of materials according to the perspectives of works while integrating the sources of natural materials in a complementary vision of the materials as well as the settlements of production of concrete (mixing unit ready to employment, prefabricated unit). Unfortunately, the GIS tool could only be tested in two French regions (Northern France and Greater Lyon) and was not easy to develop mainly because of data collection difficulties (some related to confidentiality or strategy).

29.2.8.2 Analysis regarding the 12 main cities in France

Analysis has also been carried out on 12 main French cities, focusing on four of them to contribute to a life cycle analysis study carried out by Strasbourg University (Idir et al. 2015) by providing local and relevant data (see chapter 30.3). For each urban area, data regarding local production (quarries, recycling platforms) and local concrete plants (precast and ready-mix) have been collected.

Data studies allow observing the perimeter where the identified and listed sites are distributed in each considered urban area, depending on waste inert volumes—real or estimated—for each department (Figure 29.8). For instance, this analysis shows that Strasbourg city is poorly equipped in terms of proximity platforms compared to waste production, probably due to the proximity of NA quarries.

29.2.9 Conclusion on resources of inert materials

- Waste streams are mainly under external influence: demolition works, road and building construction. A stricter control of practices in waste management would send the streams to the authorized sites (backfilling, landfilling, sorting, and recycling): currently, 10% of whole inert waste is dropped in illegal landfills.
- Then, the study has emphasized that these waste streams are unequally distributed on territories: a dozen of urban areas focuses the main demolition resources. They constitute the favorable space to recycling and, more specifically, they can kick-start the supply

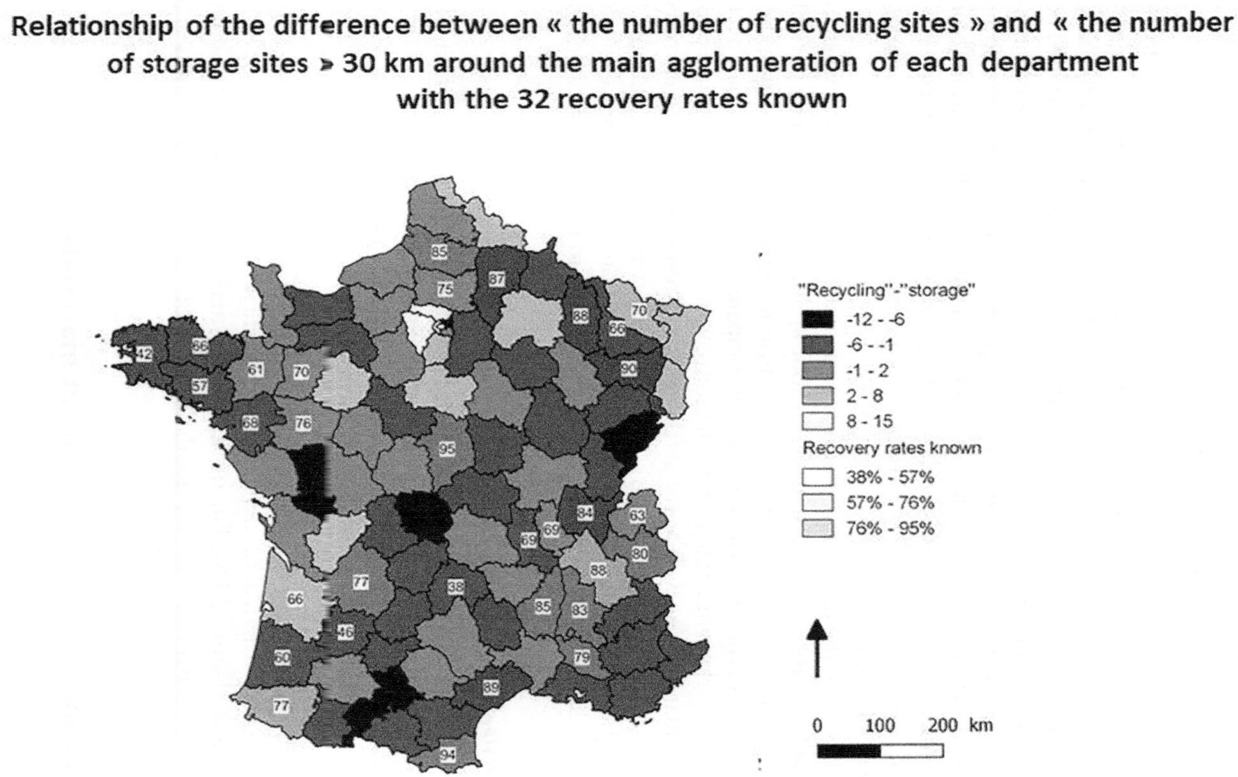

Figure 29.7 Department approach by the number of sites (recycling–landfilling) and the recovery rate.

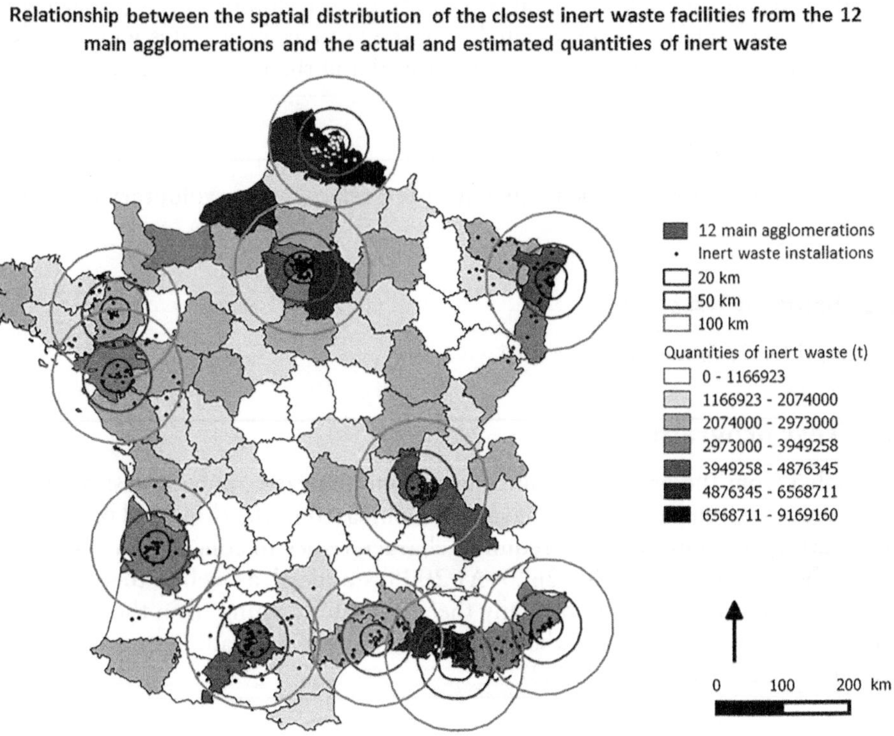

Figure 29.8 Spatial distribution of the closed inert facilities from the 12 main urban areas, represented by black (or white) dots.

of concrete sector. The spatial simulations show that the present network—quarries and recycling platforms—is working and will be better with the expected normative advances, allowing integration of RA more easily into industrial processing. In contrast to road sector, concrete industry structure is based mainly on industrial units (precast and ready-mix plants). This is why technical evolutions need time for their implementation.

- As for NAs, the recycling market remains and will remain local, depending on the players established according to the resources, their location, the volumes involved, business strategies, equipment, and technical capabilities. This makes sense because it is heavy materials whose environmental footprints are dominated in transport. The result of all these aspects is that the price is one of the keys for diversification toward the concrete industry.

29.3 RECYCLING PLATFORMS SURVEY IN FRANCE—GEOGRAPHICAL AND TEMPORAL QUALITY STUDY

The study presented here deals with the variability of the different characteristics of RA available in the French market and its suitability to be incorporated in concrete production. The first part of the study concerns the geographical variability of RA. In a second part, the variability over time of the RA characteristics is considered.

Thirteen recycling platforms spread throughout France have been studied for 16 different productions, which have led to the study of 45 size fractions of aggregates (from fine to coarse aggregates (CAs)). The goal is to draw the variability of RA production to have parameters allowing to estimate the quality of the produced concretes with recycled aggregates (RAC) in terms of durability and mechanical and rheological characteristics.

29.3.1 RA platforms

The 13 recycling platform participants have been selected on a voluntary basis (see Figure 29.9).

29.3.2 Studied RAs

The material list is shown in Table 29.1.

29.3.3 Test program

RAs sampled from the production sites have been analyzed applying test methods related to their use in concrete. The results have been compared to the threshold of the French [NF P 18-545 2011] standard, and the required characteristics in concrete have been measured following the European [NF EN 12620+A1 2008] standard, as well as the specific RA characteristics. It allows to compare not only the aggregates coming from the same or different platforms between them but also those of the different sizes grading or arising from different treatments (Tables 29.2 and 29.3).

29.3.4 Geographical diversity results

According to NF P18-545 (art.10), 15 are classified in f_A code with a fines content ≤10 (11 for materials with $D > 4$ mm), and one recycled sand is classified in f_B code with a fines content between 10 and 16. The figures given later give an image of the main results, with the

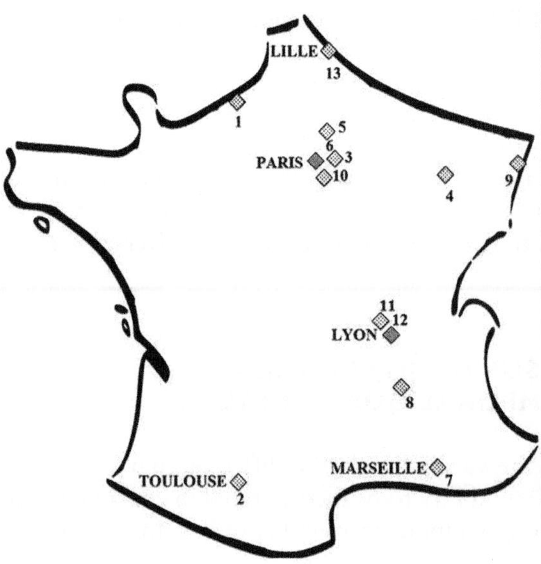

	Location
1	TOURVILLE (76)
2	ROQUES SUR GARONNE (31)
3	VALPARISIS (77)
4	NEUVES MAISONS (54)
5	LOUVRES (95)
6	GONESSE (95)
7	GARDANNE (13)
8	L'ARMAILLER ISDI (26)
9	NIERDERHAUSBERGEN (67)
10	SYNEOS NEUILLY (93)
11	GENAY (69)
12	VILLEFRANCHE SUR SAONE (69)
13	FRETIN (59)

Figure 29.9 Mapping of the 13 recycling platforms taking part on studies; on a gray background, the two platforms were surveyed for 2 years.

Table 29.1 List of the tested RAs according to the different platforms

Platform	*Batch number (–13_1–16)*	*Received aggregate*	*Tested aggregate*	*Comment*
1	1_1	FA 0/4 (input 0/63)	0/4	Input plan aggregate: 0/63 mm
		CA 4/12,5 (input 0/63)	4/12,5	
	1_2	FA 0/4 (input 4/63)	0/4	Input plan aggregate: 4/63 mm
		CA 4/12,5 (input 4/63)	4/12,5	
	1_3	FA 0/4 (input 20/63)	0/4	Input plan aggregate: 20/63 mm
		CA 4/12,5 (input 20/63)	4/12,5	
2	2_4	All-in 0/31,5 CNL	0/4	Made in laboratory
			4/10	
			10/20	
3	3_5	Washed FA 0/4	0/4	Washed aggregates
		Washed CA 4/10	4/10	
		Washed CA 10/20	10/20	
	3_6	Raw all-in 0/31,5	0/4	Sieved in laboratory
			4/10	
			10/20	
4	4_7	Recycle FA 0/4	0/4	—
		Recycle CA 4/10	4/10	
		Recycle CA 10/20	10/20	
5	5_8	Crushing FA 0/6,3	0/6,3	—
		Crushing CA 6,3/12,5	6,3/12,5	
		Crushing CA 6,3/20	6,3/20	
6	6_9	Recycle FA 0/4	0/4	—
		Recycle CA 4/10	4/10	
		Recycle CA 10/20	10/20	
7	7_10	Concrete FA 0/4	0/4	—
		Concrete CA 4/10	4/10	
		Concrete CA 10/20	10/20	
8	8_11	Recycle all-in 0/20 (R21020)	0/4	Sieved in laboratory
			4/10	
			10/20	
9	9_12	Recycle all-in 0/22,4	0/4	Sieved in laboratory
			4/10	
			10/20	
10	10_13	Recycle FA 0/5	0/5	—
		Recycle CA 6,3/10	6,3/10	
		Recycle CA 12,5/31,5	12,5/31,5	
11	11_14	All-in 0/20 GRM	0/4	Sieved in laboratory
			4/10	
			10/20	
12	12_15	Recycle all-in 0/22,4	0/4	Sieved in laboratory
			4/10	
			10/22,4	
13	13_16	Recycle FA 0/6,3	0/6,3	—
14		Recycle CA 6,3/14	6,3/14	
15		Recycle CA 6,3/20	6,3/20	

Table 29.2 List of recycled FA tests and testing standards

Test	*Norms*	*Tested fraction*
Size grading by sieving	NF EN 933-1	0/*D* supplied or made in laboratory
Methylene blue test	NF EN 933-9-Août 1999	0/2 mm
True density and water absorption	NF EN 1097-6 §9	0.063/4 mm
0/*D* water absorption	NF EN 1097-6 §9 complétée par la note (1) du tableau 54 de la NF P18-545	0/4 mm
Soluble sulfates in RAs water	NF EN 1744-1 §10.2	0/4 mm
Soluble sulfates in acid	NF EN 1744-1 §12	0/4 mm
Soluble chlorides in acid	NF EN 1744-5	0/4 mm
Effect of RAs extract on initial setting cement	NF EN 1744-6	0/4 mm

Table 29.3 List of recycled CA tests and testing standards

Test	*Norms*	*Tested fraction*
Size grading by sieving	NF EN 933-1	*d*/*D* supplied or made in laboratory
Flattening	NF EN 933-3	4/*D*
Classify of recycled constituting gravel essay	NF EN 933-11	8/*D* of the coarsest gravel available of each production
Los Angeles essay	NF EN 1097-2 §5	6,3/10 mm and 10/14 mm
True density and water absorption	NF EN 1097-6 §8	6,3/10 mm and 10/20 mm
Soluble sulfates in RA water	NF EN 1744-1 §10.2	*d*/*D* of the coarsest gravel available of each production
Soluble sulfates in acid	NF EN 1744-1 §12	*d*/*D* of the coarsest gravel available of each production
Soluble chlorides in acid	NF EN 1744-5	*d*/*D* of the coarsest gravel available of each production
Effect of RAs extract on initial setting cement	NF EN 1744-6	*d*/*D* of the coarsest gravel available of each production

different thresholds in reference of standards for recycled products (product standards and testing standards) (Figure 29.10).

For CAs, 23 of the 29 samples have a fines content less than or equal to the threshold allowed for use in concrete of 1.5 (GrA code of NF P18-545), five could be used (<4), and only one exceeds the value of four and needs a methylene blue test on the 0/125 µm fraction (Appendix A of standard NF EN 933-6, as proposed in reference (1) of table 45 of NF P18-545) (Figure 29.11).

Twelve of the 16 recycled fines aggregates (FA) are classified in P_A code (lower than or equal to the only threshold admitted for use in concrete of 1.5% according to article 10 of NF P18-545). They can therefore be used in concrete. The nonclassifiable recycled FAs (fines content between 4.4% and 10.9%) are not washed. So we can imagine the setting of a specific treatment (washing plant, upfiller selector) that could reduce the fines content and increase the clearness characteristics of the RA in order to use them in concrete.

The measured flatness coefficients point out that the whole tested recycled CA has a suitable shape to be used in concrete. In addition, Los Angeles coefficients give a good strength for the tested CA.

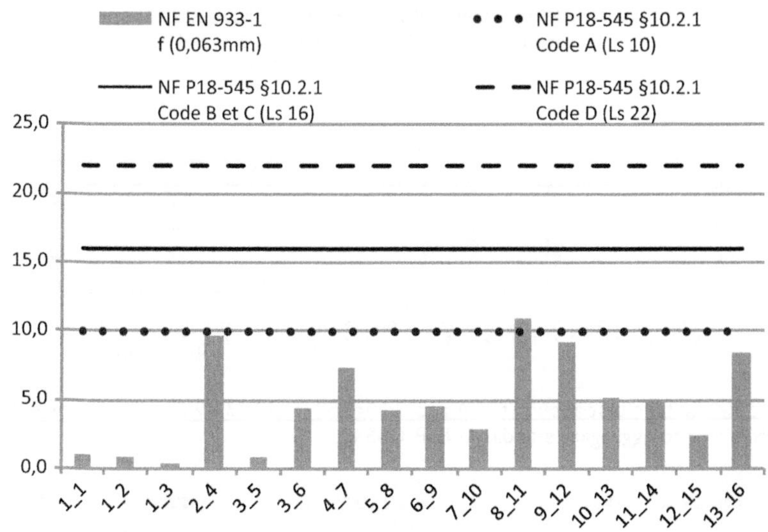

Figure 29.10 Content of particles passing the 63-µm sieve (%) in FA according to NF EN 933-1 and the corresponding threshold according to aggregate standard NF P 18-545.

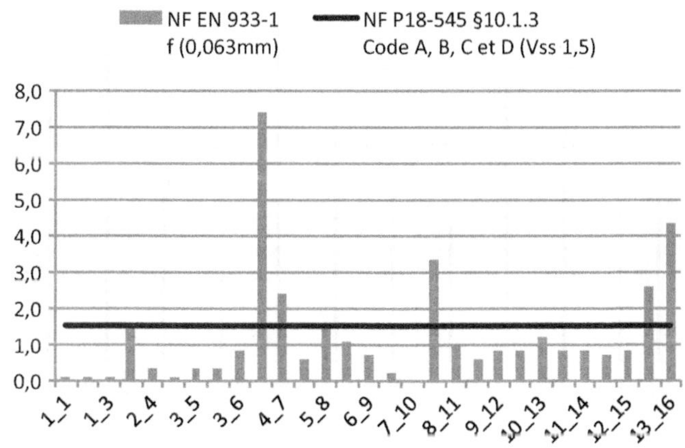

Figure 29.11 Fines content (%) in CA according to NF EN 933-1 and the corresponding threshold according to aggregate standard NF P 18-545.

The water absorption coefficient in RA is a sensitive issue, particularly for FA, at the sight of the high values that are measured during the survey (Figure 29.12). The use in concrete of RA will demand to be careful with this characteristic because of his importance in the physics of concrete (mainly rheology and resistance) and therefore prevents disorders in order to ensure durability in concretes. The comparison between the values of water absorption coefficient obtained on the fractions 6.3/10 mm and 10/14 mm does not show any significant difference. On the other hand, the FA values are higher than those measured on the recycled CA, whether or not the sand fines are considered (Figure 29.13). The presence of a higher proportion of cement paste in FA than in CA probably explains this result.

The results of the components classification tests of the recycled CA and those of the content of soluble sulfates in RA water (Figure 29.14) are good in most of the cases. The less satisfactory values remind the importance of source sorting in recycling treatments to reduce the adverse components in concrete aggregates.

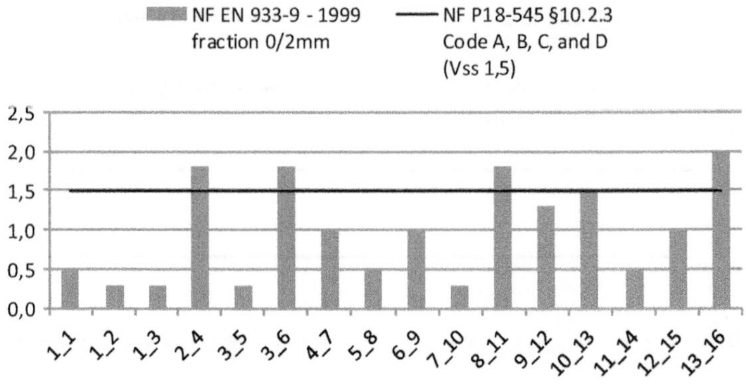

Figure 29.12 Methylene blue test applied to FA according to NF EN 933-9 and the corresponding threshold according to aggregate standard NF P 18-545.

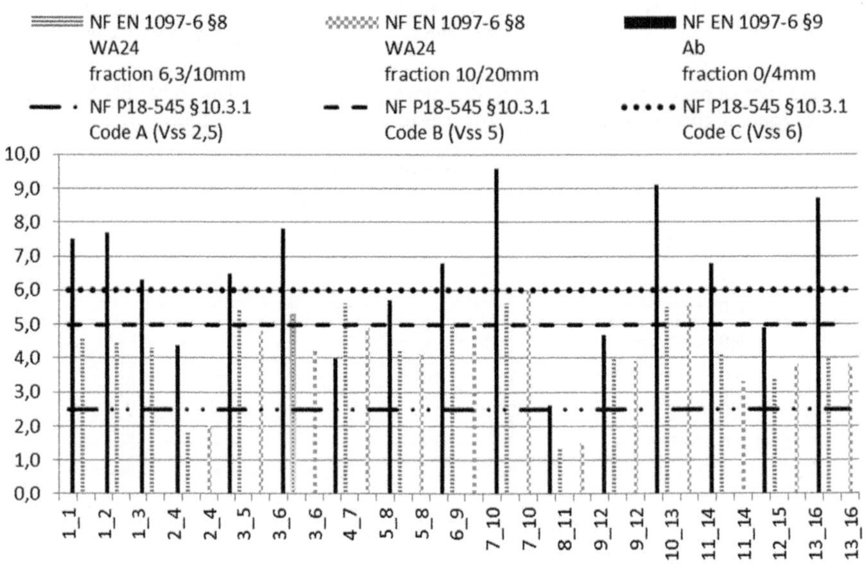

Figure 29.13 Water absorption coefficient in FA (black) and in 6.3/10 mm and 10/20 mm (dotted and hatched lines) according to NF P18-545 and testing standard NF EN 1097-6.

All the results of tests for determining the influence of RA extract on the initial setting time of the cement is less than or equal to the most restrictive threshold of 10 min. The RAs are all classifiable, according to article 10 of NF P18-545, in code A_B. The test results show a tendency towards a setting retardation effect in view of the 22 negative values obtained against nine positive values and one null value. The comparison between the values obtained on the recycled sands and those obtained on the recycled gravel does not make it possible to highlight logic for this characteristic (Figure 29.15). All these results do not show disadvantages for concrete use.

29.3.5 Survey of the variability of two platforms for 2 years

Following the previous protocol (see Section 29.3.3), the variability of RAs during a significant period focused in the survey of two of the tested platforms: one situated in Gonesse (95)

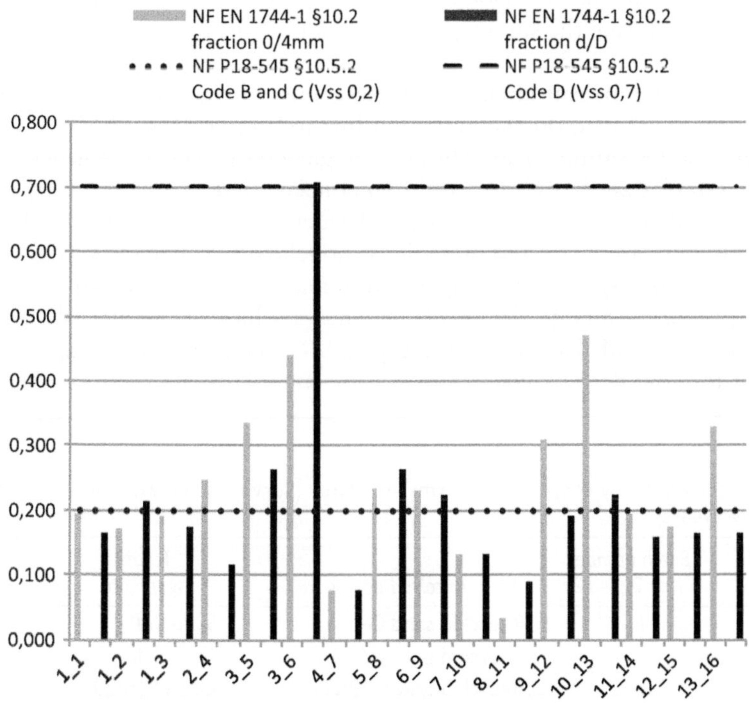

Figure 29.14 Content of soluble sulfates (SS) in RA water (SS in %) according to NF EN 1744-1 and thresholds of the standard NF P18-545

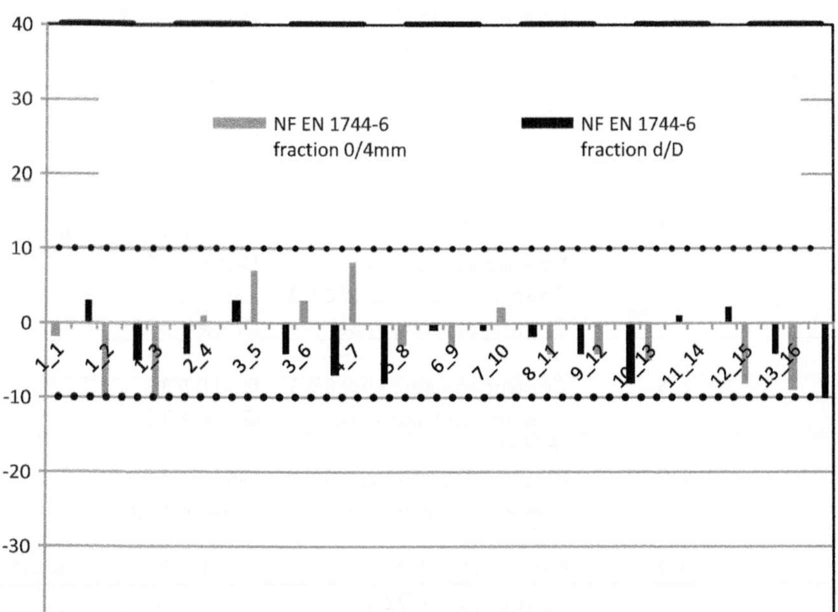

Figure 29.15 Effect of RAs extract on initial setting cement (time in minutes) measured according to the testing standard NF EN 1744-6 and corresponding thresholds in NF P18-545.

and the other in Villefranche-sur-Saône (69), which have been relocated during the first half of 2014 in Anse (69) during the survey (same company, same inert wastes sources, and same staff).

On the whole of the measured characteristics, most of the recycled FA generated from Gonesse platform are graded on D code, according to Section 10 of the NF P18-545 (2011) standard. Only one FA sample is graded as B. Regarding recycled CA generated from this platform, four samples are B, 6C, 1D, and two nonclassified.

In the case of Villefranche-sur-Saône/Anse platform, most of the recycled FA are ranked D, one sample B, one sample C, and only one as nonclassified. The recycled CA of this platform are B (three samples), C (four samples), and nonclassified for six samples (Table 29.4).

During the platforms survey, the grain size analyses have sometimes showed high fines content as well as off-graded screening (Figures 29.16 and 29.17). The shape of the recycled CA, following the low flatness coefficient, seems suitable to be used in concrete because

Table 29.4 List of tested RA by recycling platforms and dates (Survey of the variability of two platforms for two years)

Platform number (new reference)	*Batch number (1 or 2_1–12)*	*Received aggregate*	*Tested aggregate*	*Comment*
6(1)	1_1	Recycled sand 0/4	Sand 0/4	—
		Recycled gravel 4/10	Gravel 4/10	
		Recycled gravel 10/20	Gravel 10/20	
	1_2	Crushing concrete sand 0/6,3	Sand 0/6,3	—
		Crushing concrete gravel 6,3/20	Gravel 6,3/20	
12(2)	2_1	Recycle all-in 0/22,4	Sable 0/4	Made in laboratory
		Recycle all-in 0/22,4	Gravel 4/10	
		Recycle all-in 0/22,4	Gravel 10/22,4	
6(1)	1_3	Crushing concrete sand 0/6,3	Sand 0/6,3	—
		Crushing concrete gravel 6,3/20	Gravel 6,3/20	
	1_4	Crushing concrete sand 0/6,3	Sand 0/6,3	—
		Crushing concrete gravel 6,3/20	Gravel 6,3/20	
12(2)	2_2	Recycled all-in 0/22,4	Sand 0/6,3	Made in laboratory
		Recycled all-in 0/22,4	Gravel 6,3/20	
6(1)	1_5	Crushing concrete sand 0/6,3	Sand 0/6,3	—
		Crushing concrete gravel 6,3/20	Gravel 6,3/20	
	1_6	Crushing concrete sand 0/6,3	Sand 0/6,3	—
		Crushing concrete gravel 6,3/20	Gravel 6,3/20	
	1_7	Crushing concrete sand 0/6,3	Sand 0/6,3	—
		Crushing concrete gravel 6,3/20	Gravel 6,3/20	
12(2)	2_3	Recycled all-in 0/22,4	Sand 0/6,3	Made in laboratory
		Recycled all-in 0/22,4	Gravel 6,3/20	
	2_4	Recycled all-in 0/22,4	Sand 0/6,3	Made in laboratory
		Recycled all-in 0/22,4	Gravel 6,3/20	

(Continued)

Table 29.4 (Continued) List of tested RA by recycling platforms and dates (Survey of the variability of two platforms for two years)

Platform number (new reference)	*Batch number (1 or 2_1–12)*	*Received aggregate*	*Tested aggregate*	*Comment*
6(1)	1_8	Crushing concrete sand 0/6,3	Sand 0/6,3	—
		Crushing concrete gravel 6,3/20	Gravel 6,3/20	
12(2)	2_5	Recycled all-in 0/22,4	Sand 0/6,3	Made in laboratory
		Recycled all-in 0/22,4	Gravel 6,3/20	
6(1)	1_9	Crushing concrete sand 0/6,3	Sand 0/6,3	—
		Crushing concrete gravel 6,3/20	Gravel 6,3/20	
12(2)	2_6	Recycled all-in 0/22,4	Sand 0/6,3	Made in laboratory
		Recycled all-in 0/22,4	Gravel 6,3/20	
6(1)	1_10	Crushing concrete sand 0/6,3	Sand 0/6,3	—
		Crushing concrete gravel 6,3/20	Gravel 6,3/20	
12(2)	2_7	Recycled all-in 0/22,4	Sand 0/6,3	Made in laboratory
		Recycled all-in 0/22,4	Gravel 6,3/20	
6(1)	1_11	Crushing concrete sand 0/6,3	Sand 0/6,3	—
		Crushing concrete gravel 6,3/20	Gravel 6,3/20	
12(2)	2_8	Recycled all-in 0/22,4	Sand 0/6,3	Made in laboratory
		Recycled all-in 0/22,4	Gravel 6,3/20	
6(1)	1_12	Crushing concrete sand 0/6,3	Sand 0/6,3	—
		Crushing concrete gravel 6,3/20	Gravel 6,3/20	
12(2)	2_9	Recycled all-in 0/22,4	Sand 0/6,3	Made in laboratory
		Recycled all-in 0/22,4	Gravel 6,3/20	
12(2)	2_10	Recycled all-in 0/22,4	Sand 0/6,3	Made in laboratory
		Recycled all-in 0/22,4	Gravel 6,3/20	
12(2)	2_11	Recycled all-in 0/22,4	Sand 0/6,3	Made in laboratory
		Recycled all-in 0/22,4	Gravel 6,3/20	
12(2)	2_12	Recycled all-in 0/22,4	Sand 0/6,3	Made in laboratory
		Recycled all-in 0/22,4	Gravel 6,3/20	

of their relatively cubic form. Only one result of the methylene blue test is nonstandard. It concerns a recycled FA sample prepared in a laboratory.

The recycled CA consists mainly of (95%) elements from concrete and stones (Rc + Ru). They contain very few elements of fired clay (Rb) and thus largely comply with the most restrictive threshold of 10%. Glass and other elements (X + Rg) are present in proportions lower than the most restrictive threshold of 0.5% for 23 samples and only one between 0.5% and 1%. None of the 24 samples contain floating materials (FL) meeting the most restrictive threshold of 0.2 cm^3/kg. Half of the production of recycled gravel has bitumen content of less than 1%, and the other half remains below the second threshold of 10%.

In conclusion, the tested recycled CAs are clean without floating components and very few residual materials: 11 of the 24 samples are classified in CR_B code, according to article 10 of NF P18-545 (2011) standard; 13 others are classified in CR_C code, and only one is out of code. The aggregate ranking depends mainly on its bituminous content. It is time to wonder

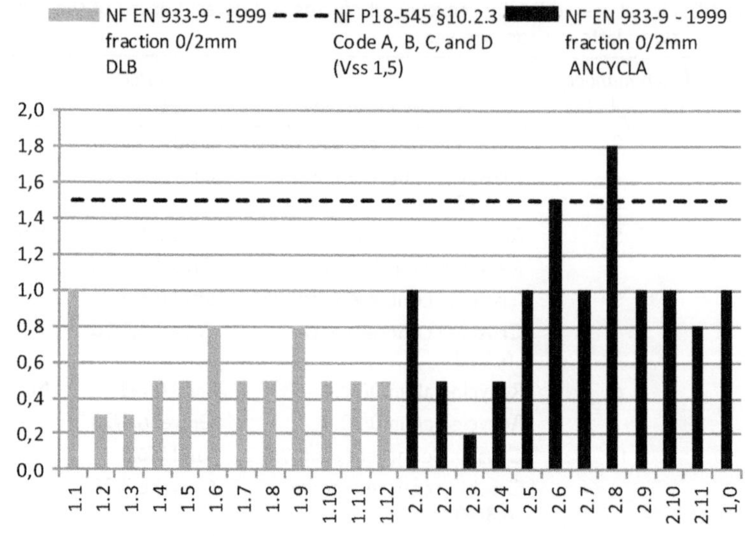

Figure 29.16 Methylene blue test applied to FA coming from Gonnesse and Villefranche-sur-Saône/Anse platforms according to the testing standard NF EN 933-9, and threshold of NF P18-545.

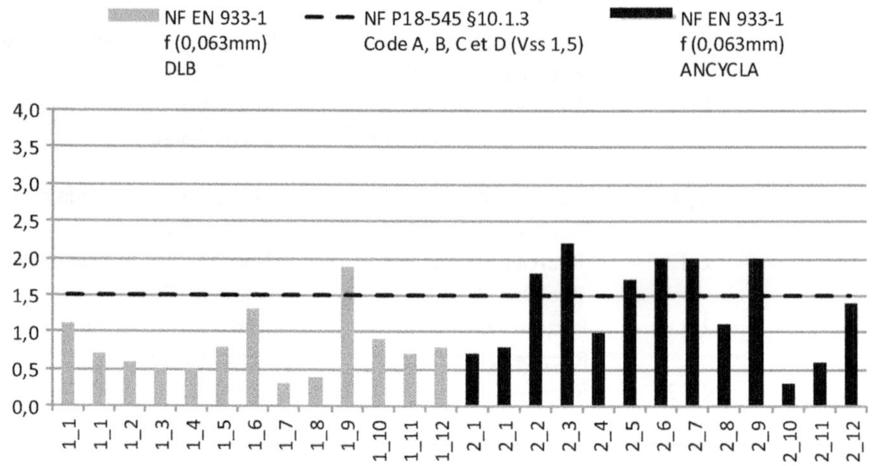

Figure 29.17 Fines content (%) in CA coming from Gonnesse and Villefranche-sur-Saône/Anse platforms according to the testing standard NF EN 933-1 and threshold of NF P18-545.

whether the thresholds are not too restrictive and whether it is necessary to revise the influence of bituminous rate in RA on concrete manufacturing.

The recycled CA has a good strength suitable for use in concrete according to NF P18-545 (2011) standard: the Los Angeles coefficients are near the threshold given by article 10 of this standard separating the codes A and B. Better results were obtained with 6.3/10 mm than for 10/14 mm CA: is variability arising from the strength of the origin material or the cement past ratio in the two tested fractions?

It remains to be seen whether the variability of these results is to be attributed to the variability of the raw material or to the account of the test method that poses difficulties in producing in the presence of fines, as is the case for the natural sands that are filled. In any case, it will be necessary to remain attentive to this characteristic that risks to limit or even prohibit the use of RAs for the realization of certain concretes.

The water absorption index of RA is one of the major parameters influencing its ranking. The obtained values—particularly with recycled FA—are higher than for NA (Figure 29.18). It is a general rule that water absorption index of recycled FA is higher than that of recycled CA, and also their specific density is lower than that of recycled CA. Probably, this could come from a higher amount of cement past in recycled FA than in recycled CA. It can be noticed that the presence of very fine particles (<63 µm) increase not only the water absorption index of recycled FA but also their statistical spread. It remains that the influence of the nature of the fines in the variability of these results, and therefore, the variability of the raw material or that related to the test method are understood. In fact, in the presence of fines, as is already noted in the case of added filler FA (natural), there are difficulties in carrying out the tests. In any case, it is necessary to be attentive to this characteristic, which may limit the use of RA in concrete manufacturing.

Another key factor is the content of water-soluble sulfates in RA, often close to the most restrictive threshold of the 10th article of the NF P18-545 (2011) standard, even sometime exceeding for recycled FA (Figure 29.19).

The NF P18-545 (2011) standard does not propose a threshold for the content of soluble chlorides in acid for aggregates. In this study, two results differ widely from the others

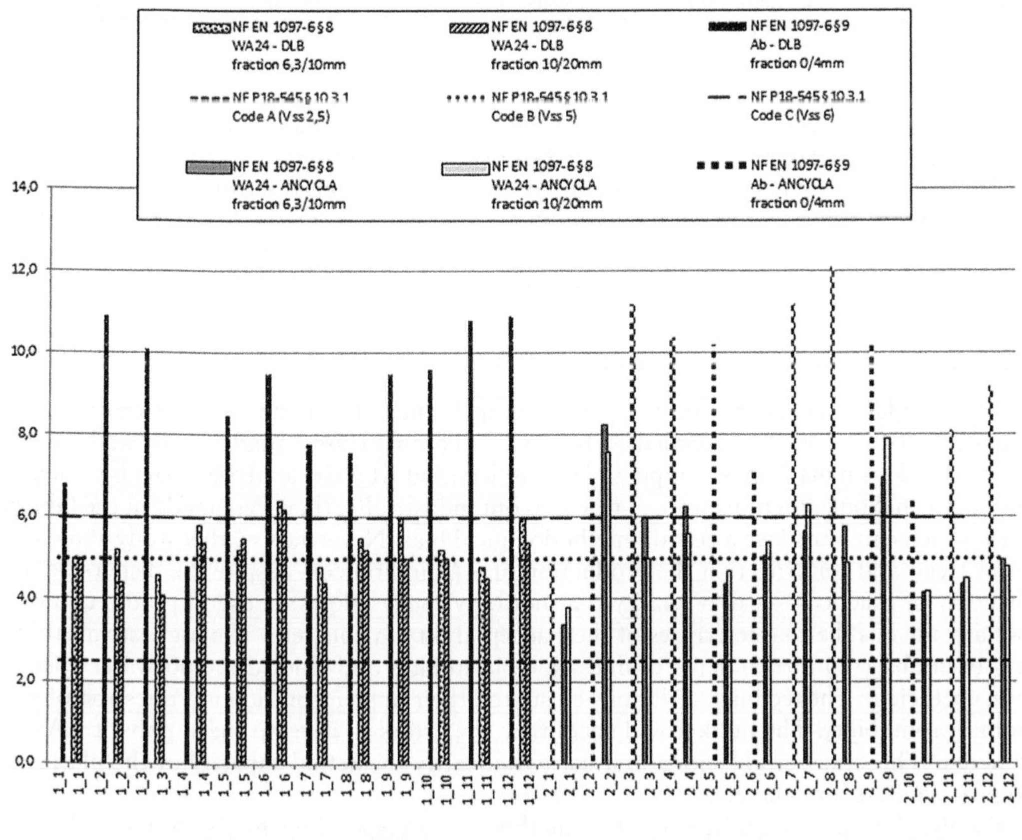

Figure 29.18 Water absorption coefficient of RAs—sand (dark), 6.3/10 mm and 10/20 mm (gray) (%) coming from platforms. According to the testing standard NF EN 1097-6 and thresholds of NF P18-545.

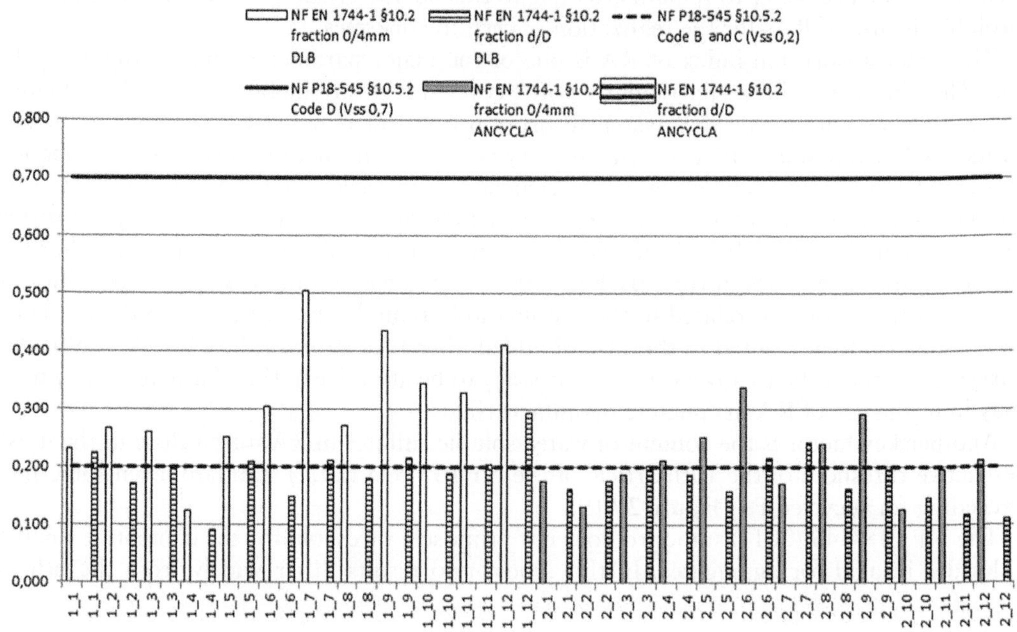

Figure 29.19 Content of soluble sulfates in RA water (SS in %)—FA coming from the two surveyed platforms, according to the testing standard NF EN 1744-1, and thresholds of NF P18-545.

(Figure 29.20), probably because of either pollution of the RA or a problem during handling. The question of the direct applicability of this test to RA is raised.

The determination tests of the effect of RA extract on the initial setting of cement (Figure 29.21) show a trend to delay the hardening of concrete. The methodology used for these tests is based on a water extract arising from the RA rinse and not on the RA itself; perhaps, it would be better to test the initial setting of cement directly with them.

29.3.6 Conclusion on variability of RAs

As a reminder, this geographic (13 platforms in France) and temporal (2 years) monitoring study focuses on RA intended to be used in common uses: roads, beds, and various networks. This remark makes it possible to understand why CA analyses show bitumen on half of them. Some fractions were not carried out industrially; they were made in the laboratory, which can introduce a certain methodological bias. Nevertheless, this study shows the same trend and confirms that the production of RAs in France—despite the dispersion and diversity of practices—is of a quality that not only today allows its use in road techniques but already has the characteristics of their acceptability in concrete. The highest variability has been observed for: contents of fines, soluble sulfates and chlorides, and water absorption coefficient. The technical solutions exist, and their implementation is a question of economic profitability. The market will encourage the actors to develop the supply of RAs for concrete. This study establishes a first basis to identify the actual capacity of the platforms to produce RAs for concrete.

Finally, during this study, some test methods—corresponding to the test standards—showed that they would require clarification or adjustments (fines in recycled FAs, chlorides, bitumen content, etc.).

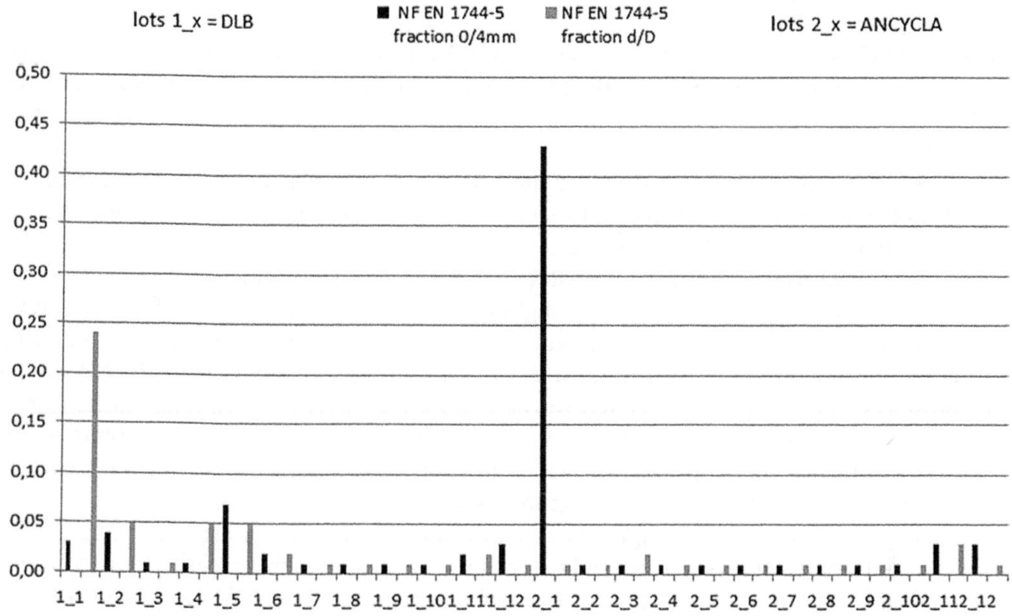

Figure 29.20 Content of soluble chlorides in acid coming from the two surveyed platforms, according to the testing standard NF EN 1744-5.

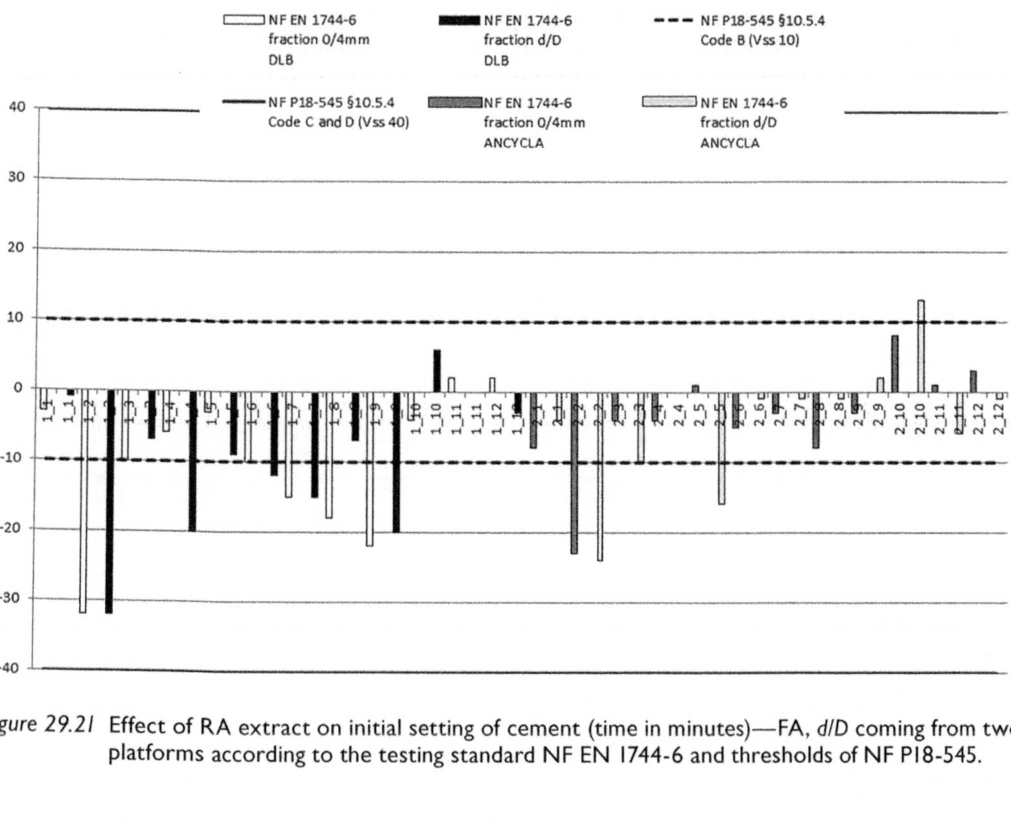

Figure 29.21 Effect of RA extract on initial setting of cement (time in minutes)—FA, *d/D* coming from two platforms according to the testing standard NF EN 1744-6 and thresholds of NF P18-545.

29.4 CONCLUSION

The study of the potential resource of recyclable materials in concrete in France showed that it is more limited than one might think at first sight. The mechanisms that will promote this kind of materials follow multiple factors:

- The monitoring carried out across the 13 French platforms shows that, where the demolition activity is sufficient for recycling to work, quality is already present to meet the new market for RA for concrete.
- An optimization of the quality of RA first requires a continuous and rigorous organization (sorting, storage, etc.) throughout the entire chain of works from the diagnosis stage before demolition, naturally by the worksites, up the transfer to recycling platforms for screening, crushing, selective separation, and stocking of the RA finally produced. These steps are related to the volumes involved: appropriate equipment to treat the type and quantity of waste in the right place at the right time could be the best summary. This optimization will contribute to the improvement of the recycling rate mentioned at the beginning (Figure 29.6).
- As in the manufacture of aggregates from natural deposits, waste recycling (CDW and industry waste) is a group whose progress benefits each specialty. For example, new soil treatment techniques or all-in material may allow RA substitutions today produced from concrete waste, whose quality could be better valued in RAC. And it is understood that it is the progress of all the recycling players, whatever their fields, to participate for the improvement of potential sources of RA for concrete.

Beyond these technical factors, others relate to the governance of the resources themselves, such as regional career plans, regional waste plans, etc. These tools are being put in place in a new dimension of sustainable development desired by the legislator. Now, the stakeholders—communities, companies, representatives of states, and citizens—will be able to develop local action plans and appropriate strategies through these territorial management tools in terms of recycling, to set objectives adapted to local situations according to the resources, the works, and constructions.

Chapter 30

Life cycle analysis of recycled concrete

S. Braymand
Université de Strasbourg

A. Feraille
Ecole des Ponts ParisTech

N. Serres
INSA Strasbourg

R. Idir
Cerema

CONTENTS

Abstract

The aim of this study is to confirm the eco-respectability of recycled aggregates concrete (RAC). Comparisons are performed according to the normative criteria of Life Cycle Assessment (LCA) methodology applied to construction materials (EN 15804 standard). Thus, this study aims to compare the environmental impacts of RAC (Recycled Aggregate Concrete) as compared to NAC (Natural Aggregate Concrete). A multicriteria analysis of the factors impacting the LCA of such concretes is proposed in a case study (cement content and cement type, aggregate substitution rate, transport distances of aggregates and concrete). The first focus is the study of the influence of the concrete composition (quantity of RA, the cement dosage, and the water content). Then, optimized formulations of concrete with close mechanical strengths are performed in order to obtain the same cement content and the same efficient water content. Four substitution rates are tested: 0S0G; 10S10G; 30S30G; 100S100G. In order to study the sensitivity to transport, several transport distances of NA and RA are proposed. It allows studying the integrated management of demolition waste streams in several towns. Finally, LCA of experimental construction sites are performed.

The study of the concrete composition influence indicates that the use of recycled aggregates in concrete formulations leads to an increase of environmental impacts at different levels, when the standardized mechanical strength required is reached by the cement content adjustment. Moreover, with cement content similar for the RAC and the NAC (optimized formulations), when the situation of the area leads to the proximity of the concrete production site with the production sites of NA or RA, the use of RA cannot be justified by a sufficient decrease in transport in view of its resulting impacts. Concerning LCA of concrete for experimental sites, the origins of aggregates (RA or NA) play a relatively low role, due to the reduced part in the slab studied of RAC with high RA content.

Finally, the environmental benefit of the use of RA is ensured by calculating the avoided burdens in addition to an LCA, which quantifies the damage of the manufacturing of such concrete (the amount of saved NA or the consumption of the area of landfill avoided).

30.1 INTRODUCTION

Environmental consideration at the level of products, services, and companies becomes a factor of added value creation. In this context, the Life Cycle Assessment (LCA) is used more and more for strategic management and innovation. LCA is a standardized method for quantifying the potential environmental impacts of goods, processes, and services throughout their life cycle. LCA allows global and multicriteria evaluation (Feraille-Fresnet 2016).

As it is known, construction industry has significant impacts on the economy, environment, and society (Geng et al. 2017; Zuo et al. 2012). Due to their number, built concrete structures take a significant share of environmental responsibility. Thus, construction and demolition wastes constitute one of the largest waste streams in the developed countries (Coronado et al. 2011; Vandecasteele et al. 2013; Torgal 2013). Construction industry is also a high consumer of resources needed for construction materials (Limbachiya 2004; Oikonomou 2005). In a global circular economy context, resources economy and wastes management incite to the recovery of construction and demolition wastes as recycled aggregates (RAs), to be reintroduced into new concrete.

The ecofriendly evidence of RA concrete (RAC) is often touched on without being systematically demonstrated. Indeed, recycling demolished concrete is not an "environmentally friendly" option by default. It must be verified because recycled concrete aggregate (RCA) production and formulation of concrete with such an aggregate can cause damage to the LCA of the finished product (Torgal 2013).

The objective of this section is to study whether RAC is more ecofriendly than natural aggregate (NA) concrete (NAC) according to the normative criteria of LCA applied to construction and building materials. The studied RAC and NAC are only devoted to structural use. An analysis of transportation sensitivity completes the comparison between RAC and NAC LCA.

A global analysis of the benefit due to the use of RAC could be proposed by taking into consideration the avoided impacts, provided that LCA indicators stay low. In this regard, other environmental criteria or indicators are also proposed in this work to include for RAC the evaluation of the aggregate resource preservation (environmental and societal criteria), the cost of materials transportation (by calculation of transported quantity in t.km, economic criterion), and the storage site avoidance contributing to a reduction of area consumption (environmental and societal criteria).

30.2 LCA OF RAC: STATE OF THE ART

30.2.1 LCA: an environmental assessment method

LCA is a method that permits environmental balance calculation. This method has been created for industrial products (for example, the first study was carried on a conditioning of Coca Cola® in 1969) and based on standard ISO 14040 (Gomes et al. 2013). A wide number of LCA studies of sustainable building (traditional or not) is provided in literature and articles on building. LCA has gained a rapid growth over the past 15 years (Geng et al. 2017; Vitale et al. 2017; Anand and Amor 2017; Kofoworola and Gheewala 2008; Ji et al. 2016; Asif et al. 2007). Among these studies, some of them focus on specific component/construction materials (Ortiz et al. 2009; Huntzinger and Eatmon 2009; Josa et al. 2004), such as concrete (Park et al. 2012; Wu et al. 2014; Van den Heede and De Belie 2012; Purnell and Black 2012; Marinkovic 2013; Hájek et al. 2011; Kawai 2011). Only the LCA of RAC is discussed here.

LCA has two basic principles:

- It considers all the steps of a product life cycle (from cradle to grave)
- It conducts a multicriteria analysis (various impact indicators are calculated, and these indicators can be imposed by standards or can vary following the scope of the study).

This method is divided into four parts [NF EN ISO 14040 2006], as illustrated in Figure 30.1:

- Definition of the goal and scope.
- Life cycle inventory: consisting in the data collection.
- Indicators calculation using different methods (Jolliet et al. 2010).
- Interpretation.

The goal and scope find out first about the definition of the performance characteristics of the product described by a measurable and clearly defined notion called

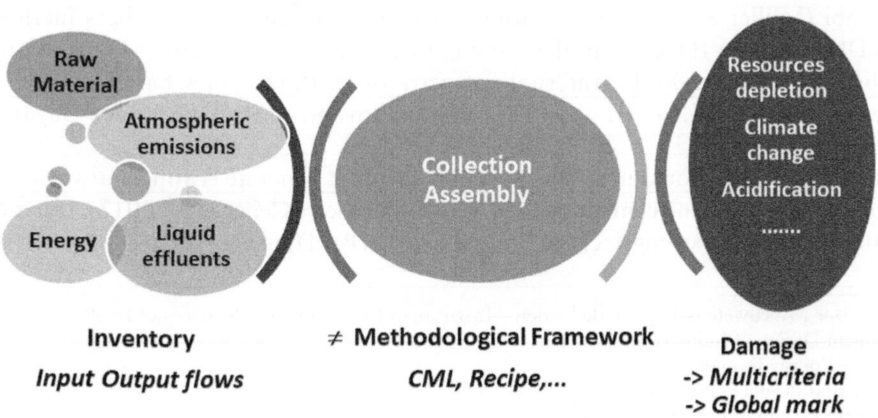

Figure 30.1 LCA approach principle.

the Functional Unit, and second, the reference flow that refers to the quantity of the product and consumables used by this product, necessary to ensure the needs of the functional unit.

LCA method is at present adapted to the domain of construction and is in the case of our study based on the standard NF EN 15804 (2012). For this case study CML,[1] EDIP,[2] and CED[3] are used as impact calculation methods. How these methods are adapted to the case studies will be explained in Section 30.3.

30.2.2 LCA of RAC: state of the art

Several studies in the literature have as scope the environmental assessment of RAC, but not all of them apply it following the well-recognized and standardized LCA methodology according to NF EN ISO 14040 (2006). For instance, some researchers focus on a specific methodology for quantifying the embodied energy or the gas emission (Wijayasundara et al. 2017; Teh et al. 2017). Moreover, even when the standardized LCA method is applied, LCA can be applied with many options (Marinković et al. 2017). These various options lead to differences in many aspects and make the results hardly comparable. Among these parameters, which can vary in the application of the LCA methodology from one to another, the following points are found: system boundaries: life cycle perimeter ("cradle to gate" or "cradle to grave") and allocation issues for recycling; functional unit choice (FU); LCI modeling (Life Cycle Inventory : normalization and aggregation for indicators calculation); database collected, etc. Differences with the same application of ISO 14040 standard following recommendations can also be observed, as described in the European standard (NF EN 15804 2012) because the databases or system boundaries are not the same (Braga et al. 2017; Serres et al. 2016). Van Den and De Belie (2012) consider three main points that influence LCA: the definition of the FU connected to concrete composition, the quality of the data collected, and the impact assessment method.

Most authors consider a cradle to gate system for the boundaries of the system (Marinković et al. 2017; Marinkovic and Radonjanin 2013; Serres et al. 2016; Braga et al. 2017; Kleijer et al. 2017; Cuenca-Moyano et al. 2017). Cradle to grave or cradle to cradle systems are sometimes proposed (Ding et al. 2016; López Gayarre et al. 2016).

Various life cycle impact assessments methods are employed, where the mainly used method is CML (see Figure 30.1) (Evangelista and Brito 2007b; Marinković et al. 2017; Serres et al. 2016; Braga et al. 2017), which can be applied with SimaPro (Serres et al. 2016), Gabi (Müller et al. 2015), or other software programs. Some others methods like EDP,[4] EDIP,[5] BEEP,[6] IPCC[7] can also be employed (Serres et al. 2016; Ding et al. 2016; Evangelista and Brito 2007b). Internal software or methodologies based on addition of LCI are also applied, (Kleijer et al. 2017; Cuenca-Moyano et al. 2017; López Gayarre et al. 2016).

Concerning the functional unit, the one cubic meter of concrete is almost always the reference. Sometimes, the same cement content is the reference (Kleijer et al. 2017), other times it is the same mechanical strength class, limited to 40 MPa (Ding et al. 2016; Marinković et al.

[1] Centrum voor Milieuwetenschappen de Leiden—Institute of Environmental Sciences of Leiden.
[2] Environment Design of Industrial Product.
[3] Cumulative Energy Demand.
[4] Environmental Product Declaration.
[5] Environment Design of Industrial Product.
[6] Bati Environnement Espace Pro.
[7] Intergovernmental Panel on Climate Change.

2017; Braga et al. 2017), and rarely the both cement content and mechanical strength are used as reference (Serres et al. 2016). In this last case, the cement content and the mechanical strength were fixed references; however, additional admixture and different aggregate content were used.

The most known discrepancy in the application of LCA concerns the databases collected. Even if Ecoinvent is a well-recognized database in Europe, it is always used complementary of local national databases (Portuguese, Serbian, French, etc.) obtained from suppliers or universities for LCI of some materials (cement, aggregate, admixture). Sometimes, only an internal database is used (Kleijer et al. 2017; López Gayarre et al. 2016; Braga et al. 2017). The update of the database is also important. Especially concerning the transport, the EURO 3 process (harmful process) can be used (Müller et al. 2015).

One of the questions taken up by authors is how to allocate the environmental impact of recycling and, as a consequence, the LCA determination of RA (Marinkovic and Radonjanin 2013; Müller et al. 2015; Vrijders and Wastiels 2017). In this case study, this question won't be addressed because the LCA is performed using the local databases of the French suppliers, as explained further.

There is also another source of discrepancy in results issued from research. The concrete compositions induce a high difference in the conclusions of the authors concerning the benefit of the use of RA. More precisely, the amount of cement influences the LCA results directly. Some authors refer that NAC and RAC present similar environmental impacts as long as the cement content variation is low, whereas with higher cement content in RAC, its impact is greater than that of NAC (Braunschweig et al. 2011; Müller et al. 2015). To obtain the mechanical strengths of RAC close to NAC ones using the same cement amount, the water-to-cement ratio should be decreased, and a superplasticizer should be used to solve workability problems. In that case, environmental performances won't be deteriorated for RAC (Marinković et al. 2017). Indeed, one of the main contributors for global warning and energy consumption in concrete is cement proportions (Ding et al. 2016).

In the same way, there is a limited number of studies with a high substitution rate of NA by RA. Most of the time, rates are below 50% (Knoeri et al. 2013; Evangelista and Brito 2007b). Sometimes, a 100% substitution rate of coarse NA by coarse RA has been studied (Marinković et al. 2017; Vrijders and Wastiels 2017). Most rarely, a total substitution of NA by RA (CA and FA) is proposed for LCA evaluation (Serres et al. 2016).

In the end, the third source of discrepancy is the limit distance for transporting RA or NA. So, the real environmental impact of NAC or RAC is different in all regions (Kleijer et al. 2017). The transport distances of aggregates considered in literature studies can reach 100 km. A reduced distance (Ding et al. 2016) or even zero distance (Evangelista and Brito 2007b; Vrijders and Wastiels 2017) is often considered for RAs, which leads to a reduction in impacts. If long delivery distances for NA exist, RAC is an opportunity for lowering the environmental impacts of concrete on condition that recycling plants are located near concrete plants (Ding et al. 2016; Marinkovic 2013). With close transportation distances and equal cement content, a decrease of impacts can then be observed due to the modification of other parameters of formulation (water/cement ratio, aggregate content, etc.) (Serres et al. 2016; Müller et al. 2015).

Considering the state of the art, it can be concluded that it is impossible to draw general conclusions on comparison of the ecoefficiency of RAC and NAC (Vieira et al. 2016; Marinković et al. 2017). To complete LCAs, some authors proposed to consider avoided burdens. With these additional indicators, improvements in environmental behavior can be observed with impacts reduced to 70% and 80% (Knoeri et al. 2013; Ioannidou et al. 2015; Ding et al. 2016; Habert et al. 2010).

30.3 RECYBÉTON'S OUTPUT

30.3.1 A case study: influence of RAC constituents and transportation on LCA

LCA of RAC in comparison with NAC was realized in several contexts to identify the influence of RAC constituents and RAC transportation on these LCA.

A first focus consisted in identifying the influence of the composition parameters of the concrete, in particular, the RA contents, the cement dosage, and the water content. In this part, CEM II cement was used.

Then, to have the same cement content for close mechanical strengths, concrete compositions with constant formulation in volume (NA or RA content, W/C, cement dosage) and with a minimum strength required were studied (Deodonne 2015). In this part, CEM I cement was used to obtain the same mechanical strengths with a constant cement content.

To study the integrated management of demolition waste streams in different areas, several transport distances of NAs and RAs were proposed for diverse cities.

At last, some experimental construction sites described in Chapters 20 and 22 were assessed by the same methods, and the role of RA use was analyzed. In that case, real transportation distances were taken in account for LCA calculation. In this part, CEM II cement was used.

30.3.1.1 LCA study framework

30.3.1.1.1 Homogenization of LCA data and methods used

The same data and calculation method between the laboratories involved in this study were used. According to standard (NF EN 15804 2012), the environmental analysis proposed here is done on the step of production, implying steps which must be considered (cradle to gate). The other steps (construction on site, use phase, end of life) would be considered while doing an LCA on the entire perimeter but are excluded in this study.

30.3.1.1.1.1 DATABASE AND SOURCES

Professional data were used if they are available; if not, the Ecoinvent database (Ecoinvent 2011) was used. The data used will be detailed in the following.

30.3.1.1.1.1.1 MANUFACTURE OF COMPONENTS

- Cement: data from ATILH[8]—Technical Association of Hydraulic Binders (June 2011).
- Aggregates: data from UNPG[9]—National Union of Aggregate Producers (May 2011).
- Plasticizer, retarder: data from SYNAD[10]—National Association of Concrete and Mortars Admixtures (March 2006).

The production of RA considers the processing of materials from demolition, shaping, or earthworks for the calculation of flows for RA. Thus, any material from one of these processes is considered in this study. Upstream transport of deconstruction materials to the recycling site is not considered in this study.

[8] http://www.infociments.fr/ciments-chaux-hydrauliques/caracteristiques-applications/les-ciments/declarations-environnementales-produits-ciments-courants-francais.

[9] http://www.unpg.fr/accueil/dossiers/environnement/analyse-de-cycle-de-vie-des-granulats/.

[10] http://www.synad.fr/.

All the activities of the production site are considered for NA: clearing, discovery, and exploitation of the production site, processing and marketing of aggregates, as well as redevelopment of the site after the extraction of rocks.

30.3.1.1.1.1.2 Transports and Electricity

Ecoinvent processes were used for transport and electricity, such as "electricity, medium voltage, FR (France)", and the type of transport used is described in Table 30.1.

30.3.1.1.1.1.3 Mixing Process of Concrete

A mixing process was created using the following data (Table 30.2). It was assumed that the mixing conditions are the same for the studies based on the compositions of concrete and for one of the experimental site study.

30.3.1.1.1.2 CALCULATION METHODS

This work is based on standards (NF EN 15804 2012) and (XP P01/064/CN 2014). Then, three calculation methods, such as CML, EDIP, and CED, were used, and they were adapted to the French standardization (NF EN 15804 2012) with the following impact indicators (Table 30.3). Two software were used: OpenLCA for the study on experimental sites and SimaPro for the studies on transportation and composition influences.

30.3.1.1.2 Compositions and properties of RAC: LCA goals and scopes

The same RA and NA were used for the variable cement content concrete and the constant cement content concrete. Fine NA is produced at Sandrancourt (78),[11] and coarse NA is

Table 30.1 Type of transports used

Components	*Type*	*Processes*[a]
Aggregates	>32t	Transport, freight, lorry >32t
Cement	>32t	Transport, freight, lorry >32t
Admixture	<16t	Transport, freight, lorry 7.5–16t
Concrete	16–32t	Transport, freight, lorry 16–32t
Experimental sites	Road	Transport, freight, lorry 16–32t
Experimental sites	Train	Transport, freight train, Europe without Switzerland

[a] Ecoinvent processes used are EURO 4 (corresponding to emissions standards) and RER (corresponding to European region).

Table 30.2 Data used to create a mixing process

Mixer (m^3)	2
Mixing time (s)	55
Real capacity (m^3/h)	55
Power (kW)	220
Energy consumed (MJ/m^3)	14.4

[11] Yvelines French Department.

Table 30.3 Impact indicators taken into account

Environmental impact indicator	*Unit*	*Method*
Consumption of energy resources • Renewable energy • Nonrenewable energy	MJ	Cumulative energy demand (CED)
Depletion of abiotic resources	kg Sb eq	Impact-oriented characterization (CML 2001)
Hazardous waste	kg	Environmental design of industrial products (EDIP)
Nonhazardous waste	kg	Environmental design of industrial products (EDIP)
Radioactive waste	kg	Environmental design of industrial products (EDIP)
Climate change	kg CO_2 eq	Impact-oriented characterization (CML 2001)
Acidification potential	kg SO_2 eq	Impact-oriented characterization (CML 2001)
Stratospheric ozone depletion	kg CFC eq	Impact-oriented characterization (CML 2001)
Photochemical oxidation	kg C_2H_4 eq	Impact-oriented characterization (CML 2001)
Eutrophication	kg PO_4^{3-} eq	Impact-oriented characterization (CML 2001)

produced at Givet (08).[12] RA is produced at Gonesse (95).[13] For the following studies, real transport distances were not taken in account, but more realistic distances were used.

The codes of the concrete formulations (xS–yG) are based on RA contents (sand and gravel). Cement, water, and admixture contents depend on the step of the study as described later.

30.3.1.1.2.1 RAC AND NAC WITH VARIABLE CEMENT CONTENT

In this first part, the study focuses on the influence of the composition parameters of the concrete, in particular, the content of RA or NA, the cement dosage, and the water content. For this concrete, the objective was to ensure the same compressive strengths for RAC and NAC (C25/30 for concrete presented here). Compositions of concrete are defined in Table 30.4. The declared unit for this analysis was "production of 1 cubic meter of concrete on a ready-mix concrete plant".

30.3.1.1.2.2 RAC AND NAC WITH CONSTANT CEMENT CONTENT

Optimized formulations of concrete using superplasticizers were performed to propose the same cement content for mechanical strengths obtained, which were included between 29 and 33 MPa (Braymand et al. 2017b). This aim requires, using at a laboratory scale, a CEM I cement. Four substitution rates were tested: 0%, 10%, 30%, and 100% (see Table 30.5). For this study, the substitution rates were calculated in volume. Substitutions included sand and gravel (fine and coarse aggregates). The declared unit for this analysis was "production of 1 cubic meter of concrete on a ready-mix concrete plant" and a step "transportation on construction site" is added in the LCA calculation.

30.3.1.1.2.3 EXPERIMENTAL CONSTRUCTION SITES

Two experimental sites, detailed in Sections 20 and 22 (a site at Chaponost, near Lyon, in the French department 69 and another in Seine et Marne, the French department 77), were

[12] Ardennes French Department.

[13] Val-d'Oise French Department.

Table 30.4 Compositions of RAC and NAC with variable cement content

	Constituent (kg/m³)	*C25/30-0S-0G*	*C25/30-0S-30G*	*C25/30-0S-100G*	*C25/30-30S-0G*	*C25/30-30S-30G*	*C25/30-100S-100G*
Formulation expressed in dry mass for aggregates	Added water	190	210	244	213	228	303
	Cement CEM II	270	276	282	276	277	326
	Efficient water/cement	0.6	0.7	0.7	0.7	0.6	0.6
	Calcareous filler	45	31	31	31	31	50
	Natural sand	780	813	806	549	500	—
	Recycled sand 0/4	—	—	—	235	218	673
	Natural gravel 4/10	267	228	—	190	171	—
	Recycled gravel 4/10	—	—	163	—	145	304
	Natural gravel 6,3/20	820	462	—	829	552	—
	Recycled gravel 10/20	—	296	701	—	167	442
	Superplasticizer	1.31	1.51	1.40	1.16	1.08	1.18
	Setting time retarder	0	0	—	1.1	1.1	2.6
	Efficient water (kg/m³)	180	185	189	185	185	199

Table 30.5 Compositions and properties of RAC and NAC with constant cement content

Component (kg/m³)	*0S/0G*[a]	*10S/10G*[a]	*30S/30G*[a]	*100S/100G*[a]
Efficient water	169	169	169	169
Efficient water + absorption water	182	192	213	284
Cement CEM I	260	260	260	260
NA	1,906	1,715	1,334	—
RCA	—	153	458	1,527
Superplasticizer	1.92	1.95	2.08	2.34
1 day Comp. strength (MPa)	9.1	11.2	13.5	6.8
28 days Comp. strength (MPa)	32.0 ± 1.1	34.9 ± 2.0	33.10 ± 1.15	29.0 ± 1.0

[a] Substitution rate in volume.

modeled. Only the results obtained on Chaponost site are presented here. The LCA results were similar for the two sites.

Chaponost experimental site consists of six slabs plus one zone for the suitability tests (with the same composition as D6 slab), as shown on Site ground plant in Chapter 22 (de Larrard et al. 2014). The FU was fabrication of the slabs. The various compositions of concrete and the volumes of each slab are detailed in Table 30.6.

30.3.1.1.3 Transportation circuits

30.3.1.1.3.1 RAC WITH VARIABLE CEMENT CONTENT

For this part, short circuits for the transport of aggregates were taken as working hypothesis, as shown in Table 30.7.

Table 30.6 Compositions and volumes of each concrete slab casts on Chaponost site

	D1	*D4*	*D2*	*D3*	*D5*	*D6 + conv.*
Components (kg/m³)	*REF*	*30S–0G*	*0S–30G*	*30S–30G*	*0S–100G*	*100S–100G*
NA	1,796	1,467	1,437	1,151	772	—
RA	—	235	282	514	778	1,379
Cement	302	306	305	308	346	390
Plasticizer	2.57	3.65	2.60	2.62	2.94	3.32
Retarder	—	1.54	0.88	1.54	1.04	1.95
Water	173	188	178	205	205	260
Surface (m²)	337	355	322	335	323	324
Volume (m³)	65	64	60	62	62	62

Table 30.7 Total quantity of transportation (t.km) of each composition

Formulation	*Transportation quantity (t.km)*
C25/30-0S-0G	19.8
C25/30-0S-100G	20.7
C25/30-0S-30G	18.7
C25/30-100S-100G	23.8
C25/30-30S-0G	20.1
C25/30-30S-30G	21.7

30.3.1.1.3.2 RAC WITH CONSTANT CEMENT CONTENT

An average (fixed) value of 50 km was used for the "cement transport" distance, for each considered city. Concerning the admixture (superplasticizer), a distance of 165 km was chosen (distance to go to the RMC (ready mix concrete) plant). These values were defined according to expert opinion. For aggregates and concrete, several values of transport distances were proposed and defined as follows. They were limited to a maximum value of 35 km from the aggregate site production and the ready-mix concrete plant.

These distances are calculated for several cities (Strasbourg, Lyon, Lille, and Bordeaux) and for several circuits. The definition of circuits and one of these circuits are illustrated in Figure 30.2.

These circuits illustrate current practices (i.e., production of each aggregate on its own site before transportation to the RMC plant: circuit 1) and also the possibility to premix NA and RA on NA production sites. In this last case, the RAs can be produced on the recycling platform site (circuit 2) or on the NA production site (circuit 3). In this way, these combinations of circuit, cities, and substitution rates lead to a parameter sensitivity analysis. For this sensitivity study, the distances to be considered were calculated from the mean of the three nearest distances of each stage of the circuit (with a maximum of 35 km).

30.3.1.1.3.3 EXPERIMENTAL CONSTRUCTION SITES

The concrete plant is located in Sérézin, the cement plant in Val d'Azergues, RA comes from Pierre Benite and NA from Petite Craz de St. Laurent de Mure and St Bonnet. All these cities are located in the French department 69. The transportation distances are detailed in Table 30.8. The transportation is carried out by lorry:

To calculate impacts due to transportation, the transported mass is required, as estimated in Table 30.9 using theoretical density.

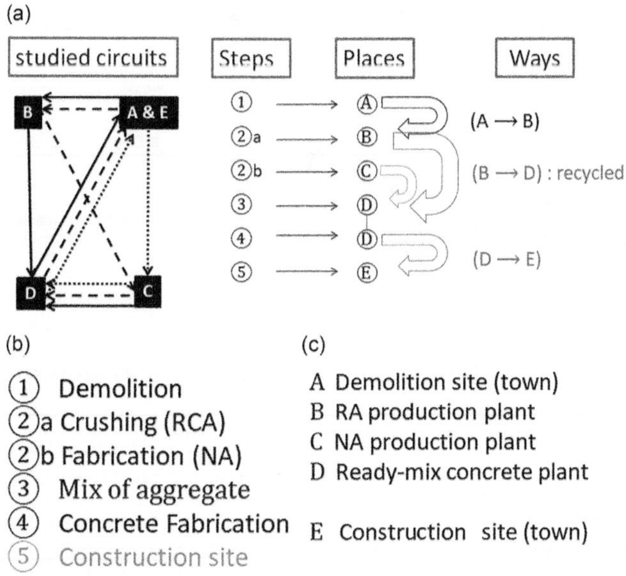

Figure 30.2 Transport circuits framework—an example of circuit.

Table 30.8 Transportation distances of each component

Val d'Azergues → Sérézin/cement → RMC plant	55 km
Pierre Benite → Sérézin/RA → RMC plant	15 km
St Bonnet → Sérézin/NA → RMC plant	20 km
Petite Craz → Sérézin/NA → RMC plant	20 km
Sérézin → Chaponost/RMC plant → construction site	22 km

Table 30.9 Estimated transported masses

	Theoretical density (kg/m³)	*Mass transported (kg)*
REF	2,274	137,930
0S30G	2,206	131,200
30S30G	2,183	121,740
30S0G	2,202	125,798
0S100G	2,105	123,752
100S100G	2,035	113,887

30.3.1.2 LCA of RAC with variable cement content

Figure 30.3 shows that the use of RAs in concrete formulations increases environmental impacts at different levels. This result is due to the increase of the cement content in the samples tested on this first part, because a standardized mechanical strength was required. As an example, the part of each process is illustrated in Figure 30.4. Thus, it is essential to remember that this penalizing effect linked to the increase in the rate of substitution cannot be generalized to all RAC. The observed effects and their analysis are valid only for the compositions studied.

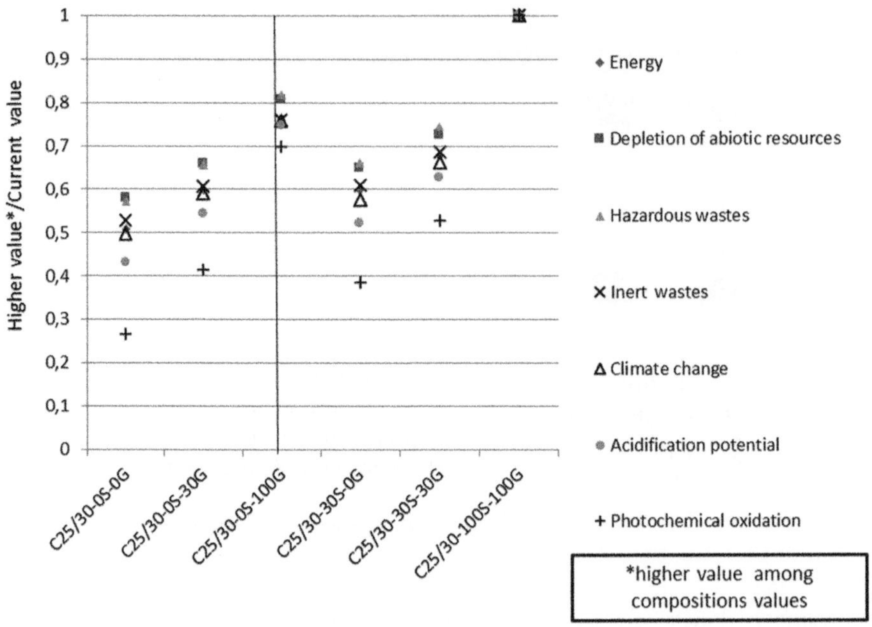

Figure 30.3 LCA of variable cement content RAC.

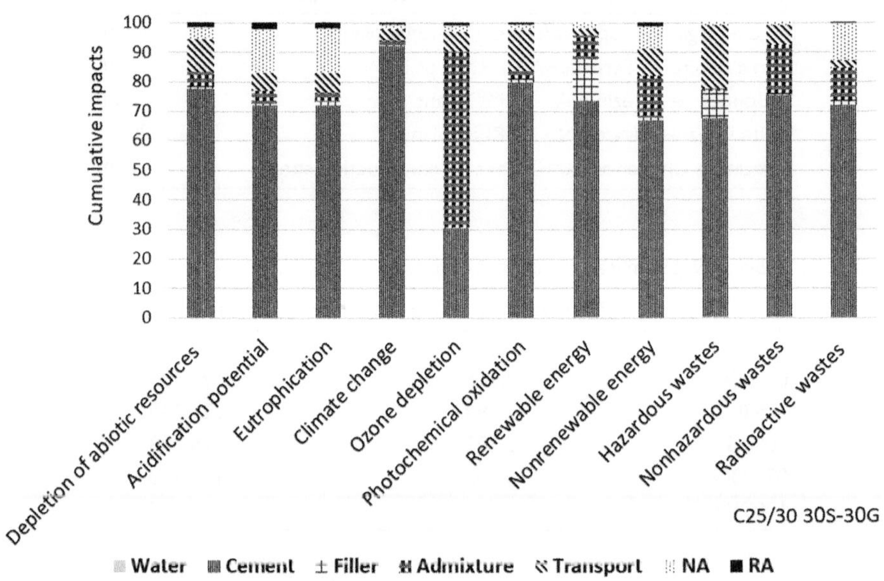

Figure 30.4 LCA—part of each process—case of 30S/30G.

30.3.1.3 LCA of RAC with constant cement content

Before quantifying the influences of substitution rate in the function of transport distances, it was necessary to calculate the LCA of referent concrete established without transportation of aggregates (0S0G to 100S100G, circuit 0) or transportation of concrete (Table 30.10). No significant differences are observed for RAC and NAC, as shown in Figure 30.5.

Table 30.10 LCA results for referent RAC with constant cement content

		Circuit 0—without aggregate or concrete transport			
Impact indicator	*Unit*	*0S/0G*	*10S/10G*	*30S/30G*	*100S/100G*
Depletion of abiotic resources	kg Sb eq	5.70×10^{-1}	5.70×10^{-1}	5.71×10^{-1}	5.72×10^{-1}
Acidification potential	kg SO_2 eq	6.89×10^{-1}	6.85×10^{-1}	6.79×10^{-1}	6.56×10^{-1}
Eutrophication	kg PO_4^{3-} eq	9.60×10^{-2}	9.52×10^{-2}	9.35×10^{-2}	8.77×10^{-2}
Climate change	kg CO_2 eq	2.39×10^{2}	2.39×10^{2}	2.39×10^{2}	2.39×10^{2}
Stratospheric ozone depletion	kg CFC-11 eq	1.39×10^{-5}	1.40×10^{-5}	1.41×10^{-5}	1.43×10^{-5}
Photochemical oxidation	kg C_2H_4 eq	5.56×10^{-2}	5.56×10^{-2}	5.56×10^{-2}	5.55×10^{-2}
Renewable energy	MJ	2.80×10^{1}	2.78×10^{1}	2.77×10^{1}	2.69×10^{1}
Nonrenewable energy	MJ	2.26×10^{3}	2.25×10^{3}	2.24×10^{3}	2.20×10^{3}
Dangerous waste	kg	6.43×10^{-2}	6.39×10^{-2}	6.31×10^{-2}	6.02×10^{-2}
Nondangerous waste	kg	4.65	4.67	4.78	4.95
Inert waste	kg	4.80×10^{-1}	4.78×10^{-1}	4.75×10^{-1}	4.62×10^{-1}
Radioactive waste	kg	7.40×10^{-3}	7.33×10^{-3}	7.20×10^{-3}	6.70×10^{-3}

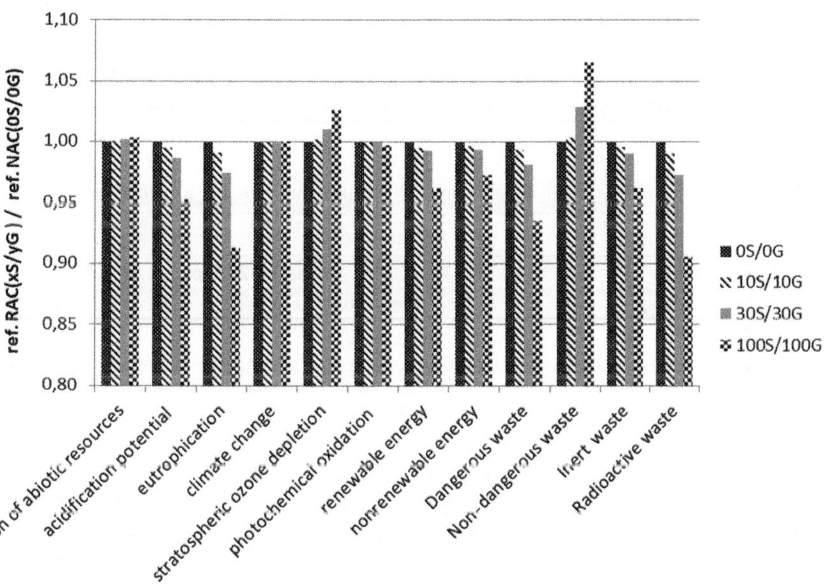

Figure 30.5 Influence of substitution rate on LCA.

When the composition is formulated at a constant volume dosage, the use of RAs does not improve or deteriorate the impact indicators. It confirms the dilution effect due to the cement share, which reduces the differences observed between the aggregates of different origin, all the more so because the cement used in this study is a CEM I cement.

Furthermore, LCA of the aggregates do not show significant variations according to their origins.[14] Only a few indicators (atmospheric acidification, eutrophication, photochemical ozone) indicate that the use of massive rocks aggregates is significantly detrimental. For all the others, value indicators of RCA are slightly higher than the ones of NA. As the concrete

[14] http://www.unpg.fr/accueil/dossiers/environnement/analyse-de-cycle-de-vie-des-granulats/.

was formulated with equivalent aggregates volume, the mass of aggregates used in RAC is lower than the mass used in NAC (see Table 30.5). This induces a compensation of slightly higher values cited earlier.

30.3.1.4 Sensitivity to transport of LCA

The sensitivity analyses of transport to LCA is divided into three steps:

- Influence of the substitution rate of RA for the most disadvantageous combination town/circuit (whole combinations of circuits for each town and for each composition).
- Influence of the circuit choice for a substitution rate of 30% and disparities between the advantageous and disadvantageous circuit. A 30% substitution rate was chosen because it corresponds to the upper rate allowed in the standards nowadays.
- Disparity of the substitution rate influence for a selected circuit between the four towns.

  The highest values are observed with the town of Lille, which presents the highest disparity between circuits (see Table 30.11).

The results of benchmarking of material composition influences and transportation influences on LCA of RAC and NAC are presented in Figure 30.6 for the negative combination presented in Table 30.11. Transport contribution is evaluated by comparison of LCA of transport versus LCA of total process. Total process incudes LCA calculated previously (LCA

Table 30.11 Combination inducing the highest and the lowest t.km transported—Lille

RAC/NAC	*Highest t.km*	*Lowest t.km*
0S/0G	Circuit 1, 2, 3 (equal, no RA)	
10S/10G	Circuit 3	Circuit 1
30S/30G	Circuit 2	Circuit 1
100S/100G	Circuit 3	Circuit 1

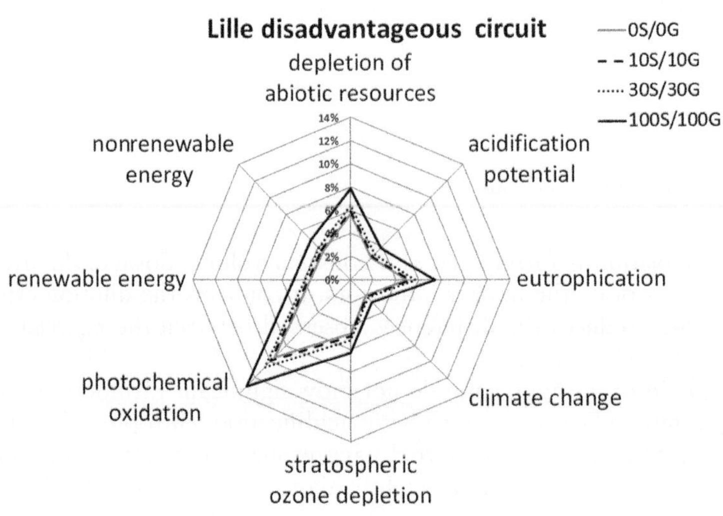

Figure 30.6 Transport contribution vs. total LCA.

of RAC/NAC with constant cement content) added with LCA of aggregate transportations (NA+RA) and with concrete transportation from the ready-mix concrete plant to the town.

At first, it has to be noticed that transport contribution on LCA is low, except for some waste indicators shown in Table 30.12. Second, even in the case of these "high" waste indicators, using RCAs in substitution of NAs does not change the transport contribution on LCA significantly (maximum increase of 10% between 0S0G NAC and 100S100G RAC). It is important to note that the results are dependent on the used database. As explained previously, Ecoinvent is used while data from the construction sector are not available. Figure 30.7 shows the way Ecoinvent LCA process linked to the transport is constructed.

The depreciation of infrastructure, fuel production, and emissions related to the transport phase itself is included in this module, and it is known that, especially, depreciation of infrastructure can induce important dangerous or inert waste indicators for LCA of transport (Figure 30.6). But this "high value of the part of transport for these two indicators has to be correlated to the absolute value in kg": there is a 10–100 factor between dangerous or inert waste value and nondangerous waste value. It puts into perspective the previous analyses.

In the second step, to complete the previous analysis, the part of the transport for the different circuits is presented in Figures 30.8 and 30.9 in the case of Lille and Lyon for the 30S/30G RAC. It indicates that, for this substitution rate, the choice of the circuits does not influence significantly the part of transport, even in the case of Lille, which presents the highest disparity between the circuits. In the case of Lyon, the choice of circuit has no influence.

In the same way, Table 30.13 indicates differences of transport LCA values between the highest and the lowest affecting circuit for each substitution rate and LCA values of transport for the highest affecting circuit. Part of the LCA of cement (at the used content) is also presented and compared to transport LCA values of the highest affecting circuit and composition (100S/100G—circuit 3). These results compared with those of Table 30.10 indicate that the influence of circuit choice stays reduced compared to the influence of cement on LCA indicators. Figure 30.10 illustrates the part of each process for a negative combination circuit and town and confirms the results of Table 30.13. The part of cement reaches to 60%–80% for main indicators.

As presented in Table 30.5, cement used for this part of study was CEM I, and this choice, required to reach performances, accentuates the part of cement.

Table 30.12 Transport contribution versus total LCA—wastes indicators—Lille

Part of transport (%)	*Dangerous waste*	*Nondangerous waste*	*Inert waste*	*Radioactive waste*
0S/0G	48	14	53	0.8
10S/10G	49	14	54	0.9
30S/30G	52	15	57	1
100S/100G	58	18	63	1.3

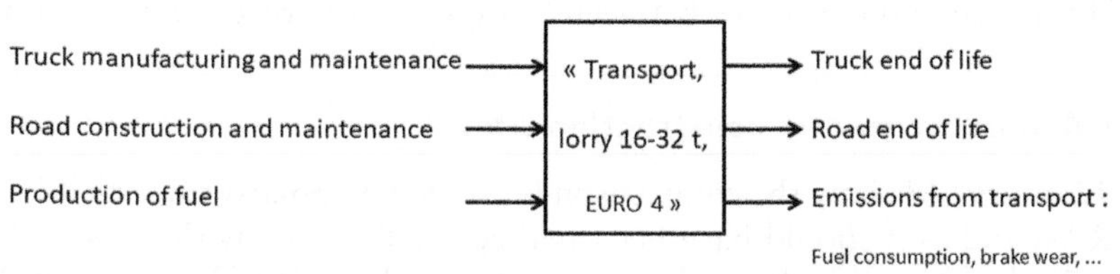

Figure 30.7 Description of Ecoinvent module "Transport, lorry 16–32t, EURO 4".

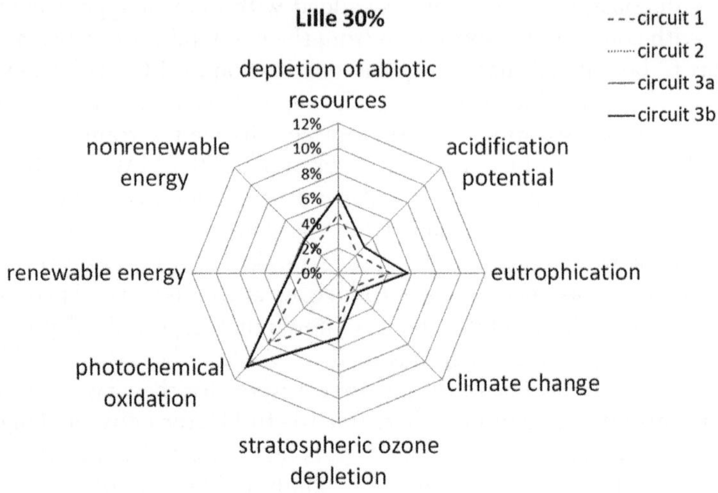

Figure 30.8 Transport contribution vs. total LCA—Lille—different circuits.

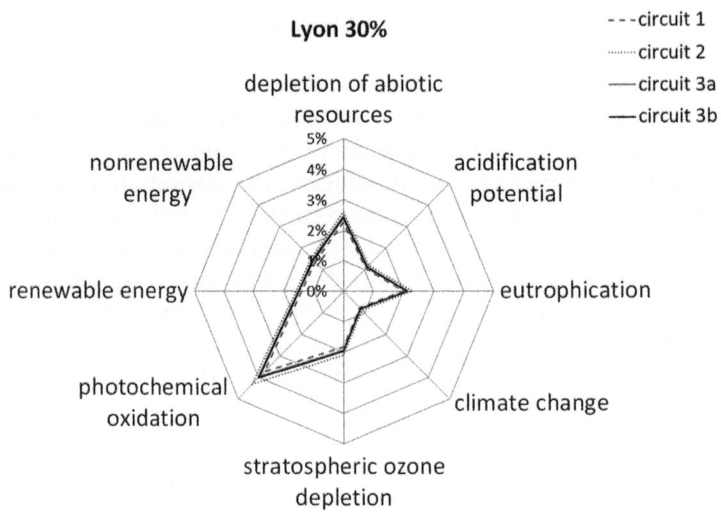

Figure 30.9 Transport contribution vs. total LCA—Lyon—different circuits.

Finally, the third step, to compare the influence of the situation territory, presents the part of transport for the first circuit, the four towns and the four compositions (see Figure 30.11). Only four indicators are represented here to facilitate the comprehension. It can be observed that the indicator evolution as a function of substitution rate is not the same depending on the town. This is due to the distance between NA quarry and ready-mix concrete plant.

30.3.1.5 LCA of experimental construction sites

Figures 30.12 and 30.13 show the comparison between Chaponost sites, as it has been realized (with RAs) and as it should have been realized (REF) only with NAs and the part of each process in the two cases. It can be observed that these differences are really low. For atmospheric acidification, climate change, and eutrophication, cement takes the majority of

Table 30.13 Comparison of highest and lowest circuits and ratio of transport versus cement

Lille town			*Highest scenario for transport–Lowest scenario for transport*				
Indicators	*Unit*	*260 kg of cement CEM I*	*10S/10G*	*30S/30G*	*100S/100G*	*Highest[a] LCA transport*	*Ratio (%): Highest[a] LCA transport/mass of cement*
Depl. abiotic resource	kg Sb eq	4.88×10^{-2}	3.67×10^{-3}	1.09×10^{-2}	3.42×10^{-2}	5.03×10^{-1}	9.70
Acidification potential	kg SO_2 eq	2.62×10^{-2}	1.96×10^{-3}	5.84×10^{-3}	1.83×10^{-2}	5.89×10^{-1}	4.44
Eutrophication	kg PO_4^{3-} eq	7.04×10^{-3}	5.29×10^{-4}	1.58×10^{-3}	4.94×10^{-3}	7.57×10^{-2}	9.30
Climate change	kg CO_2 eq	6.62	4.94×10^{-1}	1.47	4.61	2.29×10^{2}	2.89
Strat. ozone depletion	kg CFC-11 eq	9.59×10^{-7}	7.21×10^{-8}	2.15×10^{-7}	6.73×10^{-7}	1.0810^{-5}	8.88
Photochem. oxidation	kg C_2H_4 eq	8.31×10^{-3}	6.36×10^{-4}	1.89×10^{-3}	5.94×10^{-3}	5.08×10^{-2}	16.36
Renewable energy	MJ	1.42	1.06×10^{-1}	3.15×10^{-1}	9.88×10^{-1}	1.69×10^{1}	8.39
Nonrenewable energy	MJ	1.13×10^{2}	8.38×10^{1}	2.52×10^{1}	7.89×10^{1}	1.83×10^{3}	6.16
Dangerous wastes	kg	8.46×10^{-2}	6.22×10^{-2}	1.93×10^{-2}	6.07×10^{-2}	3.45×10^{-2}	245.13
Nondangerous wastes	kg	1.07	7.82×10^{-1}	2.48×10^{-1}	7.78×10^{-1}	1.60	66.95
Inert wastes	kg	7.76×10^{-1}	5.64×10^{-1}	1.81×10^{-1}	5.68×10^{-1}	2.06×10^{-1}	376.70
Radioactive wastes	kg	8.47×10^{-5}	6.33×10^{-5}	1.87×10^{-5}	5.87×10^{-5}	5.48×10^{-3}	1.55

[a] The highest LCA transport is obtained for the 100S/100G RAC and the circuit 3.

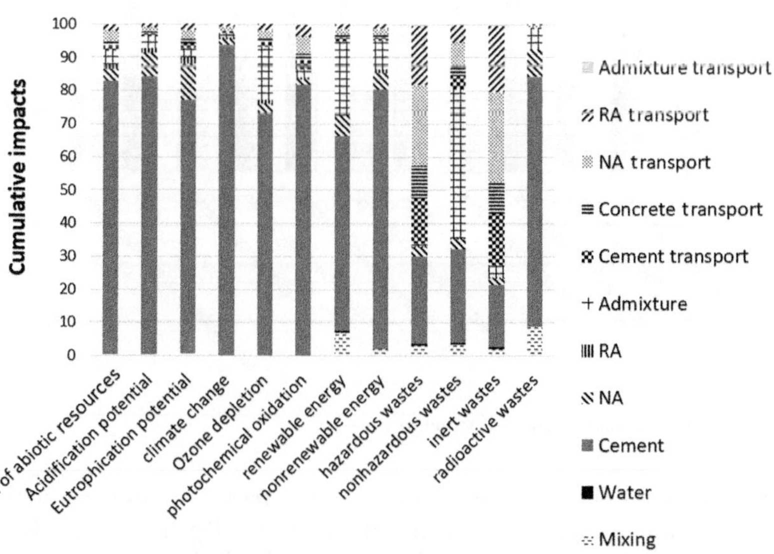

Figure 30.10 Part of each process in the case of the circuit 2—Lille—30% RA.

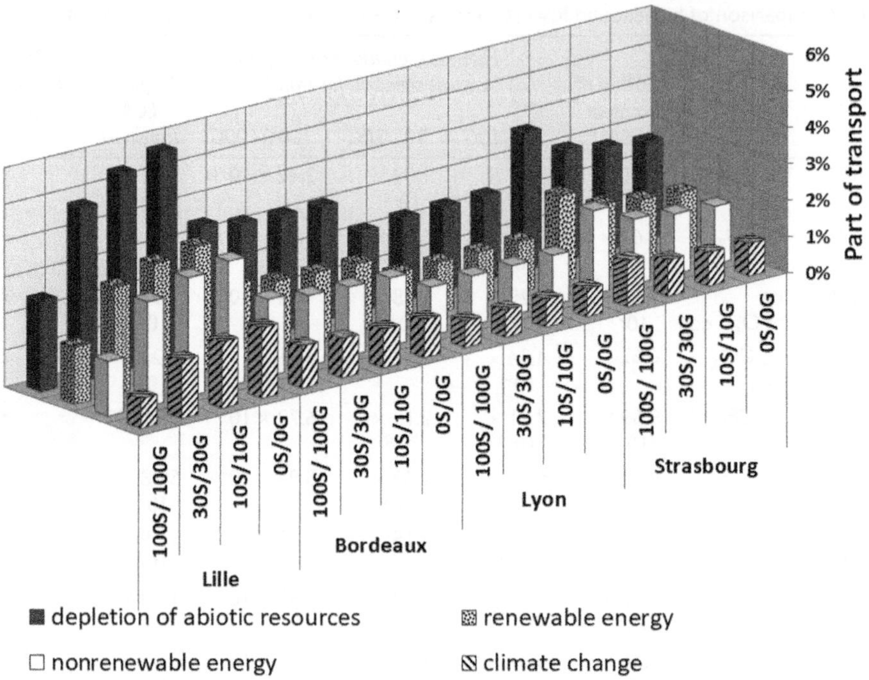

Figure 30.11 Part of transport for town and substitution rate—circuit 1.

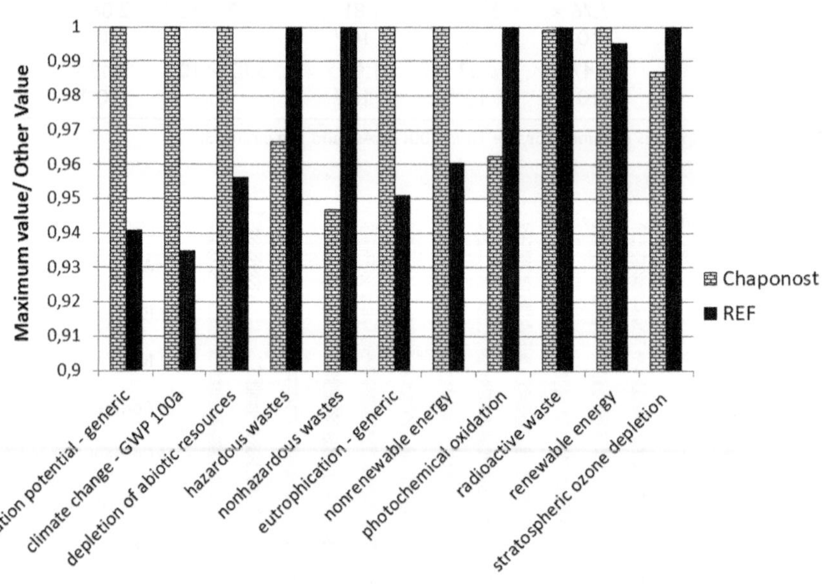

Figure 30.12 Environmental comparison—Chaponost site.

the impact. Aggregates play a relatively low role. Electricity represents a majority of radioactive waste, renewable energy, and stratospheric ozone depletion; it is also an important part of hazardous waste and nonrenewable energy. This is due to a specific compilation of the electricity process; the value obtained with open LCA is the total value of electricity

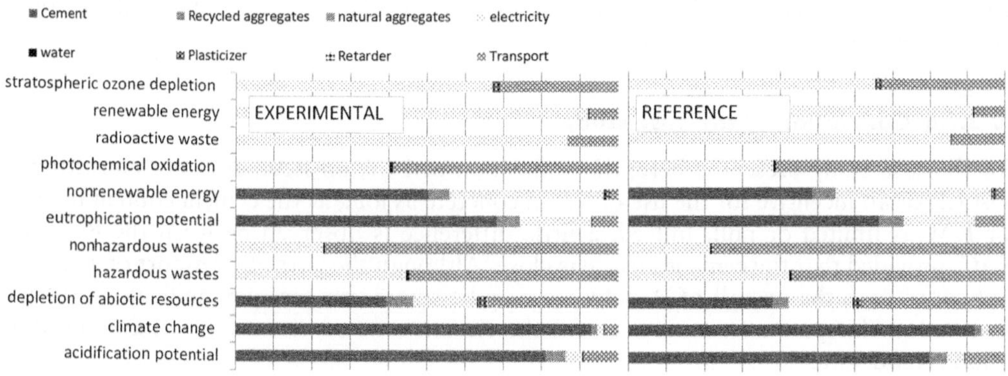

Figure 30.13 Part of each process in the case of the work site made with a reference concrete (without RAs) and in the case of the work site made with RAs.

consumed by all processes. The calculation is different with SimaPro in the previous part of the case study. SimaPro allocates the electricity on each separate process (aggregate, cement, and mixing). Finally, transport represents the majority of impacts for hazardous and nonhazardous waste and photochemical oxidation. As explained in the previous Section 30.3.1.4, considering the depreciation of infrastructures as it is done in Ecoinvent database induces important waste indicators. If it was not the case, transport would not probably represent the majority of impacts for hazardous and nonhazardous wastes. The trends for the two figures are quite similar. This is due to the fact that aggregates, whether NA or RA, have a small contribution to different environmental impact indicators and to the fact that transport distances are similar. Even if the cement content is not the same according to the substitution rates, it differs significantly for the slab D5 and especially for D6 (see Table 30.6 and Figure 30.14). D6 represents only 16.5% of the volume built.

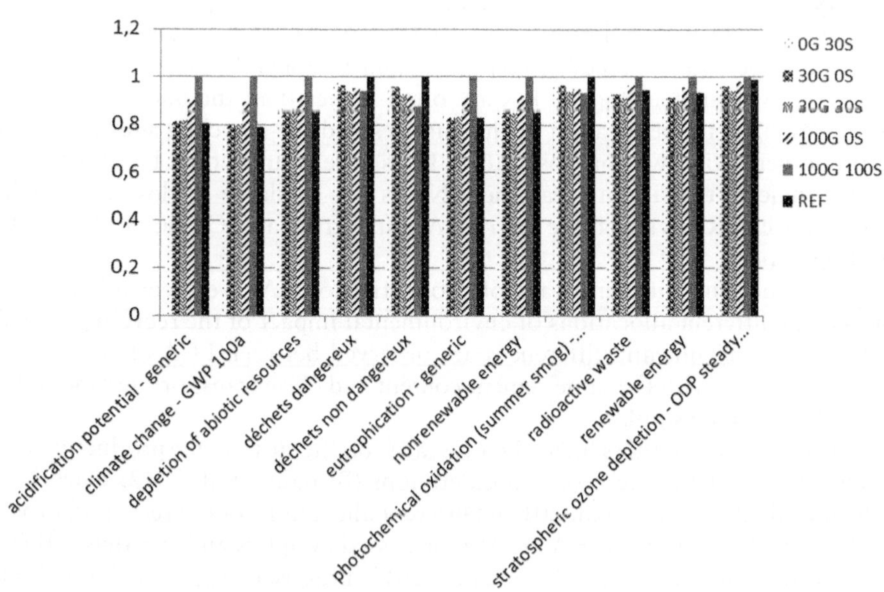

Figure 30.14 Influence of slab composition on LCA.

The effect of cement content increase is thus less sensitive than the effect observed previously (see Section 30.3.1.2).

30.3.1.6 Conclusion of the case study

For the case study presented here; the realization of LCA on RAC does not constitute an obstacle or an incentive for the use of RA in concrete as long as the cement content increase for RAC is limited or null. Indeed, a little difference is observed between the RAC and NAC, provided that the cement dosage is close. Although the part of transport of aggregate and concrete in the overall LCA outcome may be significant for some indicators, the gap observed in this part of transport versus the circuits studied or in terms of compositions remains moderate.

These conclusions in term of no benefits observed linked with the use of RA in concrete can be different from those obtained in some other researches (Knoeri et al. 2013; Vrijders and Wastiels 2017; Serres et al. 2016). In their studies, these authors conclude a beneficial use of RA, but these improvements are mainly due either to avoided impacts (Knoeri et al. 2013), nor to favorable allocation of the environmental impact of recycling on aggregate (Vrijders and Wastiels 2017), or to compositions with the same cement content but not the same quantity of aggregates (Serres et al. 2016).

Conversely, the conclusions of many authors are similar to those of this case study concerning the influence of cement content. If the cement content in RAC is larger than in NAC, the LCA demonstrates that environmental indicators are greater in the case of RAC due to the high contribution of cement to all impact categories (Braunschweig et al. 2011; Weil et al. 2006; Marinkovic and Radonjanin 2013; Müller et al. 2015). The choice of concrete formulation in the case of "RAC with the same cement content" presented in this study is also proposed in another research (Marinković et al. 2017).

Concerning the influence of transport, to enhance the interest of RAC, authors often consider a long delivery distance for NA contrary to RA (Marinkovic and Radonjanin 2013; Vrijders and Wastiels 2017; Ding et al. 2016; Evangelista and Brito 2007b). With this hypothesis, LCA results indicate that many environmental impacts (energy use, global warning, eutrophication, acidification, and photochemical oxidant) depend on the transport condition (type and distances) (Marinkovic and Radonjanin 2013). In the case study presented here, the distances for the RA are often higher than the ones for NA, and this situation corresponds to the location of the quarries and the recycling sites for the studied regions. If another region was proposed (e.g., Paris), the results could be different. Müller et al. (2015) indicate that in an area where NA is not available at close range while the transport distance of RA is short, the use of RA could, from an LCA point of view, become an interesting option.

In this case study, no different ways of production of RAs were considered, and as a consequence, no different allocations of environmental impact of the recycling on RA were proposed. Thus, no significant differences are observed between LCA of RA and LCA of NA. It is obvious that, with the same cement content and close transport distances, LCAs of RAC and NAC are quite similar.

On this purpose, some authors focus their research on the sustainable production and recycling of aggregate, including territory considerations (Blengini et al. 2012; Faleschini et al. 2016; Jullien et al. 2012; Silva et al. 2017). Different allocations of the recycling process and of the part to be attributed to the RA are also discussed (Vrijders and Wastiels 2017).

As other authors indicate (Marinković et al. 2017; Braunschweig et al. 2011; Weil et al. 2006; Vieira et al. 2016; Van den Heede and De Belie 2012), the conclusions of this case study are valid only under the allocations, assumptions, databases, and compositions proposed here. It would be dangerous to draw general and universal conclusions given that LCA results depend on many factors detailed in the state of art.

30.3.2 Supplementary impact indicators (avoided burdens and economic indicators)

30.3.2.1 LCA's limits in this case study

As shown previously (Section 30.3.1), the LCA method does not show significant differences between RA and NA when using the indicators' impact recommended in the standard (NF EN 15804 2012). Particularly, the indicator of resource depletion is not representative of the territorial context. It is a global indicator; however, the resource in NAs is not the same all around the territory. According to CML method, depletion of aggregate should be evaluated by silicon inventory, well this resource is considered with a very little depletion. As a result, no significant difference is observed between the composition for this inventory (and for calcium also). Moreover, the land consumption due to waste storage is not evaluated in NF EN 15804 (2012) standard.

30.3.2.2 Environmental and societal indicators

To confirm that RAC is ecofriendly, other comparisons should be performed. For example, criteria "preservation of NA" and "waste storage avoided" could promote the use of RA in concrete. For instance, if an impact indicator different from the normative ones such as the quantity of NAs that is avoided to be extracted from the environment is considered, it can be observed that the Chaponost site has induced 260 t of aggregates that were not taken from the ground. Even if this indicator is not standardized in the sense of EN 15804, it represents a reality of construction works.

A similar indicator is used in the ecocomparator SEVE[15], made for road professionals. Some other indicators are proposed in literature as the CMR indicator (kg) (consumption of primary mineral resources) (Ding et al. 2016) or new indicators for resources accessibility assessment. In this last case, the indicator considers the impact of other parameters, social and anthropogenic factors, like the proximity to an urban area (Ioannidou et al. 2015; Habert et al. 2010).

A second criterion "waste storage avoided" was estimated with a supposed RA apparent density equal to 1,200 kg/m^3. For instance, in the case of Chaponost site, 260 m^3 of wastes were not stored. This criterion is also to get closer to the decrease of uncontrolled landfills. It should be noted that, to respect the LCA standard for considering this criterion, the building or road deconstruction step should be included in the processes (functional unit). Similar criteria are proposed in literature (Knoeri et al. 2013; Mousavi et al. 2017).

30.3.2.3 Assessment of t.km in aggregate transportation: an economic indicator dependent on territoriality

A different analysis independent from the used LCA method could also be proposed. In this way, the following analysis was made for RAC with constant cement content transported following the circuits described Figure 30.2. As shown in Table 30.14, a comparative analysis of circuits in t.km (see Section 30.1) indicates, for the majority of the cities studied, that the circuit which consists in separately preparing the RA and the NA on their respective

[15] http://www.usirf.com/les-actions-de-la-profession/developpement-durable/eco-comparateurseve/.

Table 30.14 t.km calculation of each circuit for the four towns studied

t.km	*Strasbourg*	*Lyon*	*Lille*	*Bordeaux*
Disadvantageous circuits: maximum values for each town				
0S0G		*Circuits 1–2–3*		
	16.033	13.925	40.574	19.107
10S10G	*Circuit 3*	*Circuit 2*	*Circuit 3*	*Circuit 3*
	17.707	14.733	42.737	21.333
30S30G	*Circuit 2*	*Circuit 2*	*Circuit 2*	*Circuit 2*
	20.582	16.345	46.969	23.026
100S100G	*Circuit 1*	*Circuit 3*	*Circuit 3*	*Circuit 3*
	25.108	17.620	59.137	26.982
Advantageous circuits: minimum values for each town				
0S0G		*Circuits 1–2–3*		
	16.033	13.925	40.574	19.107
10S10G	*Circuit 3*	*Circuit 1*	*Circuit 1*	*Circuit 1*
	16.577	13.867	38.047	19.989
30S30G	*Circuit 3*	*Circuit 1*	*Circuit 1*	*Circuit 1*
	17.665	13.754	33.007	22.049
100S100G	*Circuit 3*	*Circuit 1*	*Circuit 1*	*Circuit 1*
	21.477	13.357	15.362	40.574

platform and then transporting them to the site of the RMC plant is the least penalizing (circuit 1). The choice of preparing a premix on the NA production plant leads to an increase in the t.km value. Depending on the towns, the most disadvantageous solution would be to crush the RA on the site of the recycling platform before being transported to the site of the NA (circuit 2) or that of crushing the RA on the site of NA production (circuit 3). It has to be noticed that the conclusions of this analysis are similar to the ones obtained in Section 30.3.1.4.

The specific case of Paris was not studied here owing to fluvial transportation of aggregates that are often used in Paris town, where the circuit modeling would be different.

Some towns present special configurations. For instance, the circuit 1 for Strasbourg is not the least penalizing, whereas it is the one that considers fewer steps. Lille has a different territorial distribution that leads to high gaps of t.km values between the different circuits with a clear preference for the circuit 1. For this town, very few NA sites are located in the near periphery downtown, with a high RA ratio to reduce the impact of transport. Thus, the increase of RA content penalizes whatever the circuits except for Lille (and Lyon to a lesser extent) in the case of the least penalizing circuit. This is due to the remoteness of the NA sites. Except for the city of Lille for which circuit 1 is preferential, the gaps of t.km values between circuits do not exceed 15% for premixtures.

$$\text{t.km} = M_{\text{RA}} \times D_{\text{RA}} \times M_{\text{NA}} \times D_{\text{NA}} \times M_{\text{C}} \times D_{\text{C}} \tag{30.1}$$

with M_{RA} = RA mass; D_{RA} = RA distance transport;
M_{NA} = NA mass; D_{NA} = NA distance transport;
M_{C} = concrete aggregates mass; D_{C} = concrete distance transport

A universal conclusion cannot be announced whatever the town studied. A territorial study is essential to determine the advantageous/penalizing circuit at the rate of substitution envisaged. This "economic" criteria could be included in an LCCA (Life cycle cost assessment), but the approach to compare environmental and economic impact is not much proposed by authors (Braga et al. 2017; Geng et al. 2017).

30.4 RESEARCH NEEDS

To complete the case study proposed earlier, it could be interesting to study the possibility of a new circuit including premixing of NA and RA on the recycling platform site. A different distribution of RA sites could also be considered, and an extension of the sensitivity to transport study to particular configurations leading to an increase in the transport distances would make it possible to determine whether one circuit becomes critical compared to another. In the study presented here, configurations with a lack of NAs and different transportation ways have not been studied, and it might be interesting to do so. In that case, a model based on equation 30.1 usable for one circuit (premixing on recycling platform) could be proposed for a defined concrete composition; the limit value of substitution rate could be determined beforehand. In the same vein, a comparison with the territorial situations of other large European cities would widen the scope of this study.

A short circuit corresponding to a demolition/recycling/reuse on the same site should also be considered.

RECYBÉTON project proposes to use cement incorporating recycled concrete fine fraction (see Chapters 4 and 5). LCA analysis of such concrete should be proposed in the future.

30.5 CONCLUSIONS

The results of LCA carried out in agreement with the standard (NF EN 15804 2012) do not demonstrate the ecorespectability of the RAC proposed in this study, whatever the rate of substitution used. This is due to several reasons. First, by applying the recommendations of this standard, RA and NA do not show any significant differences in their LCA. Then, some of the compositions are formulated at variable cement content, because it is usual to formulate concrete with cement content increasing with the rate of substitution of NA by RA, to guarantee a required mechanical strength. This results in an increase of impact indicators, in particular because of clinker content. Finally, even if the cement content is the same for the RAC and NAC, when the situation of the region area leads to the proximity of the city (concrete production) and of the production sites of NA and RA, the use of RA cannot be justified by a decrease in transport and its resulting impacts. Nevertheless, if the concrete compositions are formulated at a constant cement dosage, it has been shown that the optimum substitution rate of the RA and the favorable transport circuit can be estimated by an analysis according to the t.km criterion. The results of the indicators according to LCA will then be proportional to the values of t.km.

It should be noted that the ecorespectability of RAC is not questioned in this study. The environmental benefit of the use of RA is ensured by calculating avoided burdens (the preserved amount of NA or the avoided consumption of ground by waste materials landfilling). The ecorespectability of a concrete should be analyzed using a global approach that considers the LCA indicators to determine the damage and the calculation of supplementary impacts. In addition, if the analysis of the territory indicates that the NA production sites are further away (significantly), the delivery of the RA to the concrete production sites will be beneficial for the LCA.

Finally, the conclusions presented here cannot be generalized to other studies in which concrete compositions, transport distances, and, in particular, the LCA method would be different.

Chapter 31

Leaching of recycled aggregates and concrete

E. Vernus and L. Gonzalez
PROVADEMSE

D. Blanc
INSA de Lyon

R. Bodet
UNPG

J.-M. Potier
SNBPE

CONTENTS

Abstract

The release of substances from construction materials into the soil and groundwater is a growing concern that has been addressed in the framework of the European Regulation on construction materials. This issue is particularly obvious when waste is used as a component of construction materials. The environmental assessment of dangerous substances emissions from construction products has been discussed by the CEN (European Committee of Normalization) normative ad hoc technical committee TC351, establishing harmonized leaching procedures for testing construction materials.

The main issues addressed within the RECYBÉTON project on this topic are "How different could be the leaching characteristics of concrete containing recycled aggregates

(RAC) compared to concrete prepared with only natural aggregates (NAC)?" and "Do RAC comply with existing threshold values for the environmental assessment of construction materials?"

During the last two decades, several scientists, particularly within ISCOWA organization (International Society for the Environmental and Technical Implications of Construction with Alternative Materials), have studied the environmental assessment of construction materials containing waste in order to assess the environmental acceptability of using alternative materials in construction. Recycled aggregates (RA) are, with municipal waste incineration bottom ash and coal fly ash, one of the three main waste concerned by these scientific studies. In the framework of the RECYBÉTON project, the main leaching characteristics of RA and RAC described in the literature have been drawn up. Leaching tests have been conducted on concrete prepared according to the formulations of the RECYBÉTON project (see Appendix). The main goal was to compare the leaching characteristics of RAC with those of NAC.

France has not yet established any testing procedure or threshold values to assess the environmental acceptability of construction materials containing waste. Hence, the leaching characteristics of the RECYBÉTON materials have been determined by means of the Netherlands and German procedures for construction materials. The compliance with the local environmental acceptability conditions defined in these countries was verified.

31.1 INTRODUCTION

The use of waste materials and industrial by-products in the construction industry has been studied and developed since 40 years. Some of these works have led to great technical and commercial success, such as the use of coal fly ash, blast furnace slag, or silica fume, in the cement and concrete industry with a positive impact in concrete formulations. The technical properties and the economic opportunity have been the main topics of interest for the development of waste recycling in construction materials for many years. The environmental and health aspects of waste utilization in construction emerged progressively in some countries and, particularly, in Europe with the Construction Product Directive replaced by the Construction Product Regulation (CEE 2011).

Some of the construction and demolition waste (CDW) have interesting technical properties that make them useful as secondary raw materials. However, they could also contain some potentially harmful constituents, and their safety in construction materials must be assessed.

Existing methods of determining leaching characteristics of construction materials incorporating CDW are described, and the main results of scientific works on leaching behavior of recycled aggregate (RA) and recycled aggregates concrete (RAC) are presented. The experiments conducted within the French project RECYBÉTON (recycler le béton dans le béton) to assess the environmental acceptability of the use of RA instead of natural aggregate (NA) in concrete materials are then descripted and the conclusions drawn.

31.2 STATE OF THE ART

Despite the fact that the release of dangerous substances from construction materials into the soil, groundwater, marine waters and surface waters has been specified since 1989 in the third basic requirement of the Construction Product Regulation and that some methods of determination of emission and release have been harmonized by the European normalization committee, only few European member states have established national legislation on the environmental acceptability of using CDW in construction.

These countries are Netherland, Belgium, Denmark, Finland, Sweden, Norway, Austria, and Germany. CDWs are the most commonly secondary raw materials used in earth work (such as road embankment). France has published national guidelines for the use of alternative materials in road construction (CEREMA 2015), and nowadays is preparing a new one for their use in construction (excepted road construction).

Several scientific studies about the release of constituents from construction materials containing waste are being conducted in view to define the basis of national regulation. Most of these papers are published by north European researchers.

These studies aim to better understand the effect of incorporating RAs in concrete, especially on the increase of the amount of metal trace elements (Cr, Pb, As, Ba, Cu, Mo, Sb, Ni, Zn) and anions (sulfates, chlorides) and consequently their potential release.

To characterize the behavior of these elements, leaching tests have been chosen, taking into account specific parameters such as the influence of pH, the liquid-to-solid ratio, equilibrium concentrations, contact time, etc.

To analyze the results of these tests, the authors often used geochemical modeling tools. This type of modeling allows predicting the reactions occurring in the leaching of elements that depend on their speciation in solution and solid phase.

31.2.1 Review of laboratory testing literature

The main studies on concrete containing recycled or NA conducted at the laboratory scale assess the influence of pH on leaching behavior and the middle/long-term diffusion-controlled leaching.

The tests performed (Chen et al. 2013; Galvin et al. 2012; Galvin et al. 2013; Engelsen et al. 2010; Engelsen et al. 2009) focused on the dependence towards the pH on the leaching behavior of elements contained in RAC and NA concrete (NAC).

Engelsen et al. (2009, 2010) showed that the amount of metal trace elements leaching was low when the pH value is high in either tested material because of the interactions with hydrate constituent. For each oxyanion-forming element, the lowest solubility was found at pH 4–6 and the highest solubility at pH 8–11. At pH higher than 12, these oxyanions could be integrated not only in the structure of the hydrated phases but also in other phases such as ettringite, hydrocalumite, and alumina, ferricoxide monosulfate (AFM), limiting the leaching process (Chrysochoou and Dermatas 2007). For each cation-forming elements, the leaching was the lowest at pH 7–10 and it increased at pH below 7.

The leaching behavior of elements contained in concrete has been modeled by Engelsen et al. (2010) in order to identify the retention mechanisms. The results showed that there was no difference between the leaching behavior of NAC and RAC.

(Galvin et al. 2014), used the Dutch Diffusion test [NEN 7375 2004] to measure the elements released by diffusion. This is the main mechanism of release showed in monolithic materials.

Galvin et al showed that the behavior of metal cations and oxyanions is independent of the type of concrete or the percentage of replacement of NAs. The main release mechanisms identified were diffusion for oxyanions, with the exception of Cr and Mo for which the preponderant mechanism is depletion as for Zn and Cu (most mobile species). The diffusion curves are similar regardless of the type of concrete and the replacement percentage. The only differences observed in release were directly related to changes in the formulation pH values.

31.2.2 Literature review of pilot testing

Very few studies upon pilot testing of RAC used in construction work have been reported in the literature. The main published works on pilot-scale experiments with RA are those of Chen et al. (2012) Engelsen et al. (2012), and Mulugeta et al. (2011). They studied the use of these aggregates in base course or subbase course in road construction. This type of use does not comply with the scope of RECYBÉTON, so the results of these tests are presented only for information.

The RAs were used as granular unbound materials, thus the pH value of the water in contact with this material was different than the one of water in contact with RAC in the cement-bound material.

Chen et al. (2012) placed three lysimeter cells under the base course of a road section to study the percolating solution and the leaching behavior. The three cells had different proportions of RA from 0% and 50% to 100%. After 7 months of monitoring the exposed cells, the pH varied between the cells from 6.5 to 8.4. These differences were explained by the residual alkali present in RA, which are partially neutralized by the depleting effect of infiltrating water and the weathering effect of dissolved CO_2 in the percolated solution. The authors estimated that there were no significant differences in the pollutants release between the three pilot devices.

The leaching behavior of the oxyanions, such as, Cr, and Se, was highlighted because they tend to have a higher solubility on the front of carbonation as was already observed by Mulugeta et al. (2011).

Engelsen et al. (2012) observed the behavior of major and trace elements of RA used in an asphalt-covered road subbase and in an uncovered road subbase. The results were compared with a reference-covered road subbase made with NA.

The initial interstitial water of the aggregates was characterized by pH 13 in the three experiments because of the alkaline nature of the road base layer. The pH value tended to decrease more rapidly in the uncovered road than in the covered one. During the first 100 days of monitoring, the leaching of Cr, Cu, and V was significantly higher in the constructions made with RAs than those made with NAs. After that time, concentrations of every elements decreased below the Norwegian acceptance criteria.

The leaching behavior of the elements contained in the aggregates complied with the observed behavior as a function of the pH value in batch experiments.

Finally, a seasonal effect has been observed probably due to the influence of deicing salt agents increasing the leaching of Cr and Mo.

31.2.3 Main results of the review

From the laboratory to pilot-scale experiments recently published on the leaching behavior of trace elements contained in RA, it is clear that pH is the predominant impacting parameter. Incorporation of RA in concrete itself does not induce an increased emission of pollutants in water in contact with the materials.

However, the influence of pH on the solubility of elements forming cations such as Cu, Ni, Pb, and Zn is different from that of the pH on the solubility of elements forming oxyanions such as, Cr, Mo, Sb, Se, and V.

These studies concluded that the contact of the materials with water is more impacting on the release of elements than the amount of RA contained in the concrete.

The environmental acceptability of using RAC in the different tested scenarios has been assessed on the basis of different criteria for each author from landfill criteria to drinking water quality criteria and specific national threshold values. These assessments tend to conclude that the level of water contamination with trace elements contained in the RAs is very low.

31.3 RECYBÉTON'S OUTPUTS

31.3.1 Compliance testing of concrete with various amounts of RA

The first experiments conducted under the RECYBÉTON project on the leaching characteristics of RAC consisted of comparing the release of elements from concrete incorporating various amounts of RA according to the European standardized protocol CEN EN 12457-2 (CEN 2002). This leaching test is a one-stage batch test at a liquid-to-solid ratio of 10 L/kg dedicated to materials with particle size below 4 mm. The fragmented material is brought into contact with water on a rotating shaker during 24 h at room temperature (20°C ± 5°C)

These results can be used to assess the environmental acceptability of RAC reuse in road construction at its end of life according to the French guidelines "Acceptability of alternative materials in road construction—environmental assessment" (SETRA 2011) and its specific declination on materials from building and public works (CEREMA 2015).

31.3.1.1 Influence of the amount of RAs

The three different tested formulations were made of

- A reference sample containing only NA (C25/30 0S 0R)
- A test sample containing 30% of recycled coarse aggregates (CA) and no recycled fine aggregates (FA) (C25/30 0S 30R)
- A test sample containing 100% of recycled CA and 100% of recycled FA (C25/30 100S 100R)

  The results of the leaching test according to the standard EN 12457-2 are presented in Figure 31.1. The result shows that the incorporation of RA does not influence the pH value of the solution in contact with the fractionated material.

The release of trace elements and anions in the test solution presented in Figures 31.2 and 31.3 is expressed as a soluble mass fraction of the tested material. Very few differences are observed between the leaching behavior of different formulations. Antimony, arsenic, cadmium, mercury, nickel, and selenium have not been detected in any eluate. The graphs

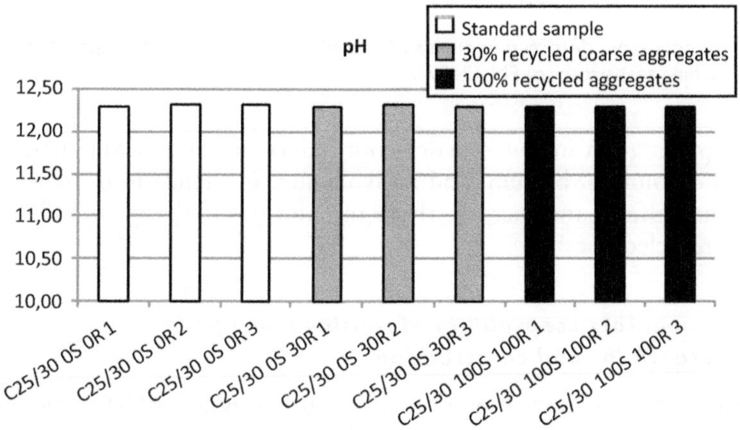

Figure 31.1 pH values of eluate of tested concrete samples prepared in accordance with the leaching test protocol EN 12457-2.

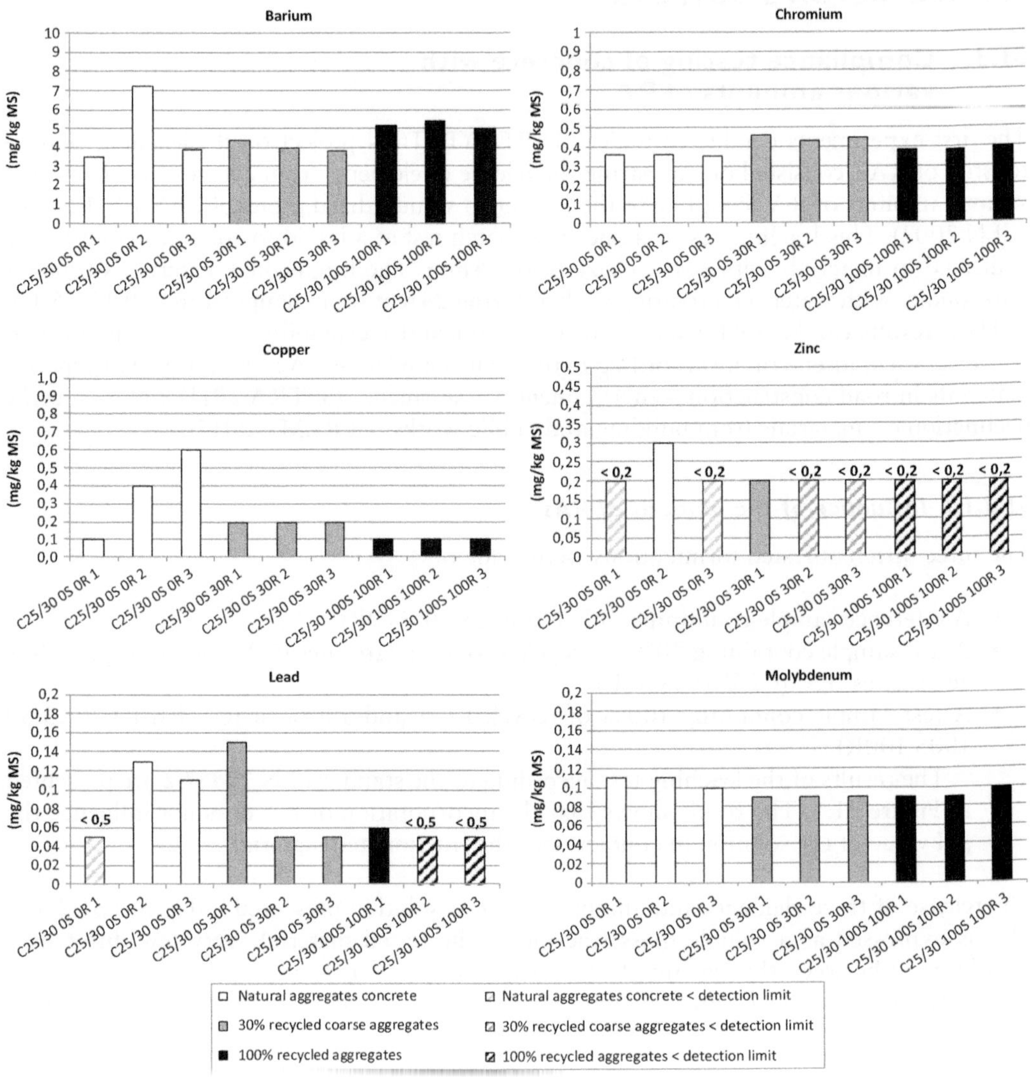

Figure 31.2 Mass of pollutants detected in eluate of tested concrete samples prepared in accordance with the leaching test protocol EN 12457-2.

indicate that supplying RA in concrete does not induce any significant change in the leaching behavior of chromium, barium, and molybdenum but tends to decrease the release of copper, lead, and sulfates and increase those of chlorides (only when 100% of aggregates are replaced of recycled ones).

31.3.1.2 Environmental acceptability of reusing end-of-life concrete RA in road construction

Despite the main targeted application of concrete RA in RECYBÉTON project is not road construction, the leaching characteristics of these materials have been compared to the threshold values defined in France to assess the environmental ability of alternative materials to be used as road construction materials (SETRA 2011; CEREMA 2015). It could be

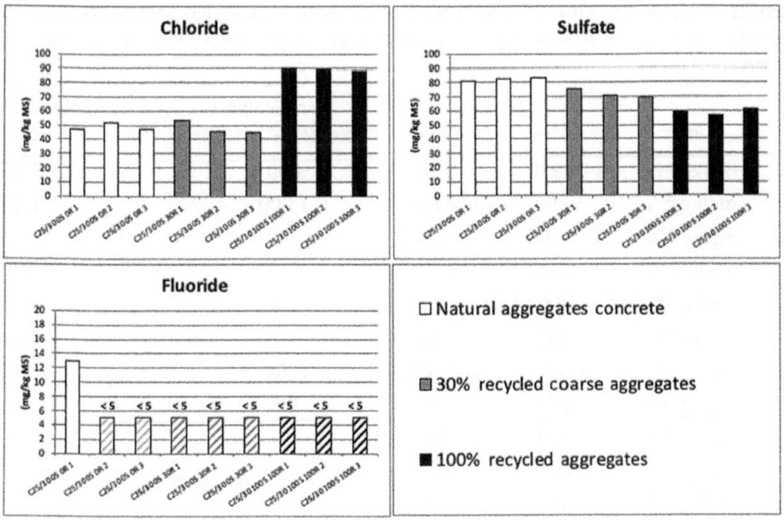

Figure 31.3 Mass of pollutants detected in eluate of tested concrete samples prepared in accordance with the leaching test protocol EN 12457-2.

considered as an interesting way to assess the reuse or recycling of these materials at the end of their life.

Table 31.1 shows the leached amount of each material compared to the French threshold values for using demolition waste in road surface course (as defined by the Guide CEREMA, (CEREMA 2015)) and to the waste landfill compliance criteria (as defined by the European Commission and transposed in France by the 12/12/2014 Decree) for inert waste.

These results show that all the tested formulations comply with the threshold values defined in the specific guide for the use of demolition materials in road surfaces (CEREMA 2015). All formulations except one (one of the three reference formulations) comply with the limit values for disposal of these materials in an Inert Waste Storage Facility. Only chromium has a leachable fraction for all formulations that is relatively close to the compliance criteria for Inert Waste Storage Facility.

Consequently, these concrete materials, at the end of their lifetime (as construction materials), could be reused in road construction regardless of their RA incorporation rate, under the conditions established on RECYBÉTON program.

31.3.2 Dynamic leaching characteristics of concrete with various amounts of RA

The last experiments carried out in the RECYBÉTON project on leaching characteristics of RAC were focused on the release dynamics of elements from concrete incorporating various amounts of RA. These experiments were conducted according to the European standardized protocol CEN TS 16637-2 (CEN 2014). This leaching test is a seven-step batch test under a specific liquid renewal program (from 6 hours to 20 days) at a liquid-to-surface ratio of 80 L/m^2 dedicated to monolithic products. The test portion is brought into contact with water without shaking at room temperature (19°C–25°C).

The leaching behavior of the concrete samples has been compared with the compliance criteria defined in the Netherlands and Germany to assess the environmental acceptability of construction materials. These criteria refer to leaching test protocols specific for each country.

Table 31.1 Results of elements released quantities from concrete materials tested compared with the French limit values for using demolition waste as materials for road surface course and inert waste landfill

Parameters	*Units*	*Min*	*Max*	*Inert landfill acceptance criteria*[a]	*CEREMA application guide using type 3 concrete category*[b]
Chloride	mg/kg MS	29.4	90.4	800	1,000
Sulfate	mg/kg MS	<50	83.4	1,000	1,300
Fluoride	mg/kg MS	<5	**13**	10	13
Antimony	mg/kg MS	<0.02	<0.02	0.06	0.08
Arsenic	mg/kg MS	<0.05	<0.05	0.5	0.6
Barium	mg/kg MS	2.72	7.27	20	25
Cadmium	mg/kg MS	<0.005	<0.005	0.04	0.05
Chromium	mg/kg MS	0.31	0.46	0.5	0.6
Copper	mg/kg MS	0.01	0.6	2	3
Molybdenum	mg/kg MS	0.09	0.11	0.5	0.6
Nickel	mg/kg MS	<0.05	<0.05	0.4	0.5
Lead	mg/kg MS	<0.05	0.15	0.5	0.6
Selenium	mg/kg MS	<0.1	<0.1	0.1	0.1
Zinc	mg/kg MS	<0.2	0.3	4	5
Mercury	mg/kg MS	<0.002	<0.002	0.01	0.01

[a] Limits to be complied with for elimination in landfill for inert waste.
[b] Limits to be complied with for using materials from demolition waste as road surface course.

31.3.2.1 Influence of the amount of RAs on leaching dynamics

The leaching dynamics of RAC containing 30% of recycled CA and 100% of recycled FA and CA determined by using the European test protocol CEN TS 16637-2 (CEN 2014) were observed and compared with a reference NAC.

The results of these tests are shown in Figure 31.4. The pH did not significantly change with time, and there were no visible differences between the reference and tested materials.

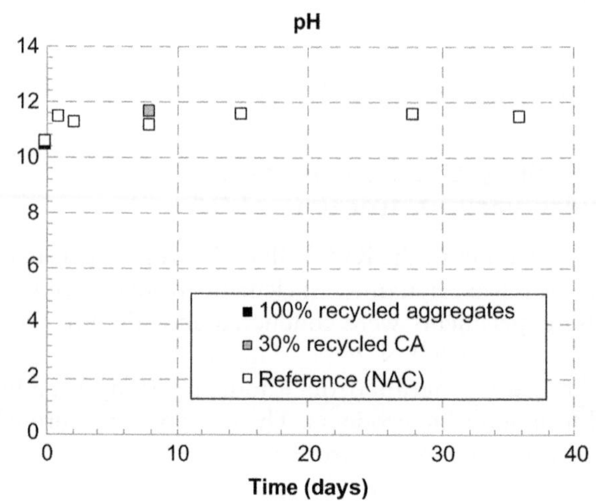

Figure 31.4 pH values of eluates collected during dynamic leaching test CEN TS 16637-2.

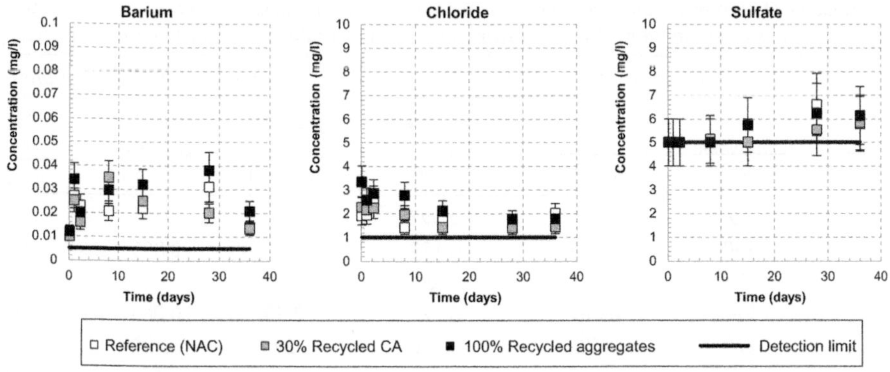

Figure 31.5 Concentration values of eluates collected during dynamic leaching test CEN TS 16637-2.

As shown in Figure 31.5, only barium, chlorides, and sulfates were measured in the eluates. The concentrations detected were not significantly different between the studied materials.

The observed release of pollutants from RAC was lower than those observed by other authors (Galvin et al. 2014), who found leaching differences due to dissimilarities of pH between the tested formulations.

31.3.2.2 Environmental acceptability of RAs toward existing thresholds values of construction materials

The testing protocol applied by the Dutch Soil Decree for molded building materials is based on the diffusion test referred as NEN 7375 (2004), which consists of a leaching test for monolithic sample.

Particular conditions applied according to the NEN 7375 (2004) protocol are characterized by a "Volume-to-Surface" ratio of approximately 100 L/m^2 and a test duration of 64 days.

The German testing protocol applied under the Deutsches institute für Bautechnik (DiBt) for concrete and concrete constituents is based on the long-term static test draft standard of the German Committee for Reinforced Concrete (DAfStb), which consists of a leaching test for monolithic sample.

Particular conditions applied according the DAfStb Draft Standard protocol are characterized by a "Volume-to-Surface" ratio of 80 L/m^2 and a test duration of 56 days.

The results of application of these tests on reference concrete and recycled concrete containing 100% of recycled FA and CA are shown in Table 31.2. They are expressed as cumulative mass values of elements extracted per unit area of exposed material (mg/m^2).

Results indicate that reference and recycled concrete comply with Dutch and German limit values determining the environmental acceptability of construction materials.

The experiments carried according the Dutch Diffusion test (NEN 7375 2004) reveal that sulfates release from 100% recycled concrete was two times higher than that from the reference concrete. Contrarily, the selenium and bromide release from 100% recycled concrete was much lower than that from the standard reference.

31.4 RESEARCH NEEDS

Due to the application of standard protocols to assess the RAC leaching behavior, the impacts of some real-life exposition parameters, such as weathering (exposure to atmospheric carbon

Table 31.2 Results of elements released quantities from concrete materials tested compared to the Dutch and German limit values of environmental acceptability of construction materials

	Dutch soil decree (NEN 7375)			*German DiBt (DAfStb draft protocol)*		
Element	*Limit value (mg/m²)*	*Reference concrete (mg/m²)*	*100% recycled concrete (mg/m²)*	*Limit value (mg/m²)*	*Reference concrete (mg/m²)*	*100% recycled concrete (mg/m²)*
Antimony	8.7	0.081	0.097	—	—	—
Arsenic	260	0.115	0.139	5	0.095	0.082
Barium	1,500	7.30	3.80	—	—	—
Cadmium	3.80	0.081	0.071	2.40	0.063	0.054
Chromium	120	4.75	4.46	24	3.126	2.720
Hexavalent chromium	—	—	—	4	2.130	2.271
Cobalt	60	0.145	0.071	24	0.090	0.131
Copper	98	2.40	2.23	24	1.301	1.101
Tin	50	0.403	0.353	—	—	—
Mercury	1.40	0.083	0.072	—	—	—
Molybdenum	144	0.924	0.774	—	—	—
Nickel	81	0.806	0.706	24	0.625	0.544
Lead	400	0.781	0.949	12	1.380	1.473
Selenium	4.80	2.19	0.177	—	—	—
Vanadium	320	0.852	1.634	—	—	—
Zinc	800	2.02	1.766	150	1.563	1.360
Bromide	670	202	177	—	—	—
Fluoride	2,500	202	177	—	—	—
Chloride	110,000	1,025	1,111	—	—	—
Sulfate	165,000	2,557	5,002	—	—	—

dioxide), have not been observed in these studies. Several pilot-scale experiments have shown that weathering could induce important pH variations over the surface of concrete materials and hence result in an increase of pollutant leaching potential.

This subject should be studied in further work that may be initiated in the continuation of the RECYBÉTON project, considering the carbonated level of RAs.

31.5 CONCLUSIONS

The scientific papers review of the RAC leaching behavior, and the results of the experiments conducted within the RECYBÉTON project show that the contribution of RA in concrete formulation does not result in increased release of pollutants as far as the pH conditions are maintained.

Waiting for a French environmental assessment procedure for the incorporation of CDW into building materials, the observation of the cumulative release of pollutants under the Dutch and German assessment of environmental acceptability of construction materials show that, even if RAC is made of 100% of recycled FA and recycled CA, the material complies with the threshold values and could be used.

ACKNOWLEDGEMENTS

These experiments were conducted, thanks to RECYBÉTON financial and scientific support, Thierry Sedran and his team from IFSTTAR, who prepared the different samples of test materials. This section was written based on the work conducted by Lorena Gonzalez from PROVADEMSE and thanks to the reviewing done by Denise Blanc from the DEEP laboratory of INSA Lyon and Adelaïde Feraille from the NAVIER laboratory of ENPC.

Chapter 32

Towards a multirecycling process

T. Sedran

Ifsttar

D. T. Dao

LCR (LafargeHolcim Research Center)

A. Feraille

Ecole des Ponts ParisTech

CONTENTS

Abstract

This chapter describes, through a theoretical approach, how the multi-recycling may change the concrete properties after more than one recycling cycle. By using different empirical models linking the concrete mix proportions to some of their properties, the partial recycling and the number of recycling cycles effects on different properties are considered. While the results show a fast deterioration of concrete after several cycles with full recycling rate, they confirmed the interest of limiting the recycling rate to preserve the performance of concrete through several recycling phases.

32.1 INTRODUCTION

In the previous sections, the properties of recycled concrete aggregates (RA) were properly studied (see Section I) as well as their influence on concrete during a first recycling cycle (see Sections III–V). These sections confirm that recycling RA into concrete is feasible with precautions while satisfying the performance criteria (both mechanical and durability ones) and the economic issue as well.

However, within a long-term vision of recycling, one should now wonder if multirecycling is feasible, and if so, what is the best scenario in terms of recycling rate. To answer these questions, Dao studied the evolution of the properties of RA and RA concrete (RAC) through two recycling cycles (Dao 2012; Dao et al. 2014). A first series of concrete, named B1, with various compressive strength, was made of different natural aggregate (NA) called G1 (rounded siliceous, crushed calcareous, and a mix of the two previous types). The RA obtained by crushing B1 concrete was named G2, and a second generation of concrete B2 was made with 100% of G2. A new generation of RA named G3 was crushed from B2, and a third generation of concrete B3 was made with 100% of G3. From this huge campaign, Dao proposed different empirical models linking the properties of RA to the concrete they are sourced from as well as models to predict the properties of RAC from the properties of RA. All these models can be found in Dao (2012) and are partially presented in Chapters 10 and 15. By sequencing these models one after the other, it is theoretically possible to predict the properties of RAC and RA during multiple cycles of recycling.

The present chapter summarizes these models and shows their application through four recycling cycles, starting with the NA selected in RECYBÉTON project (see Appendix). Influence of partial multirecycling and target compressive strength is also presented.

32.2 MULTI-RA PROPERTIES PREDICTION

The following paragraphs describe how to predict the relevant properties of RA (G_n) from those of the concrete (B_{n-1}) it is crushed from, which are used to predict the properties of the concrete of the next generation (B_n) (Dao 2012).

32.2.1 Residual parking density

The residual packing density β (virtual packing density of a monosized class of grain) of each constituent is needed to optimize the granular skeleton of a concrete and calculate its slump, thanks to the Compressible Packing Model (CPM, see (de Larrard 1999) and Chapter 15). It is governed by the shape and surface texture of aggregate particles and is deduced from the actual packing density measured on each constituent (coarse and fine aggregates, binders, etc.).

From his experiments, Dao shows that the β values of RA depends mainly on the mean size of the constituent. There is no effect of NA they are made of, of the compressive strength of the concrete they are crushed from, and of the number of recycling cycles (see Figure 15.1). This is probably due to the presence of remaining paste on the RA. For the selected crushing and sieving procedures, Dao obtained the following equation:

$$\beta = 0.525 + 0.055 \log(d_{50}) \tag{32.1}$$

where d_{50} is the sieve diameter having 50% of passing, in mm.

32.2.2 Aggregate abrasion coefficient (micro deval) and $k_{g,g}$ parameter

The $k_{g,g}$ parameter of CA (coarse aggregate) describes its influence on the compressive strength of a concrete made of it. The same parameter $k_{g,s}$ is determined for FA (fine aggregate) (see Chapter 10). Dao shows that, for RA, $k_{g,s}$ can be approximated by 4.2 whatever the number of recycling cycles or the concrete the RA is crushed from. He also determined (see equation 10.7, Chapter 10) that $k_{g,g}$ for medium and coarse RA can be calculated from its resistance to abrasion (Micro-Deval test (MDE), according to NF EN 1097-1 (2011)). The MDE itself depends on the compressive strength of the previous generation concrete B_{n-1} from which it is crushed and the abrasion resistance of the NA G1 used in the first cycle, according to the following equation:

$$\mathrm{MDE}_{G_n} = 23.69 - 0.21 fc^{c}_{B_{n-1}} + 0.58\ \mathrm{MDE}_{G_1} \tag{32.2}$$

where

$fc^{c}_{B_{n-1}}$ is the compressive strength of the previous generation concrete B_{n-1} at crushing time, in MPa;

MDE_{G_1} and MDE_{G_n} are the MDE value of first-generation (NA) and nth generation RA, respectively.

32.2.3 Density

The different fractions (fine, medium, and coarse) of the multi-RA G_{n+1} are composed of the different fractions of RA G_n and of the new paste (including cement and eventually mineral additions) P_n used to produce the concrete B_n, as described in Figure 32.1.

Hence, the density of the different classes of multi-RA G_{n+1} may be described by the following equation:

$$\rho_{G_{n+1},i} = \left(1 - P_{n,G_{n+1},i}\right)\rho_{G_n} + P_{n,G_{n+1},i}\ \rho_{P_n} \tag{32.3}$$

where

$\rho_{G_{n+1},i}$ is the density of the i-fraction of RA G_{n+1}, in kg/m³;

$P_{n,G_{n+1},i}$ is the volume proportion of the new paste P_n contained in the i-fraction of RA G_{n+1}, in percent.

From his experimentation, Dao concluded that, after crushing and sieving, the RA G_{n+1} contains less new paste P_n than its parental concrete B_n, probably due to loss of paste as dust

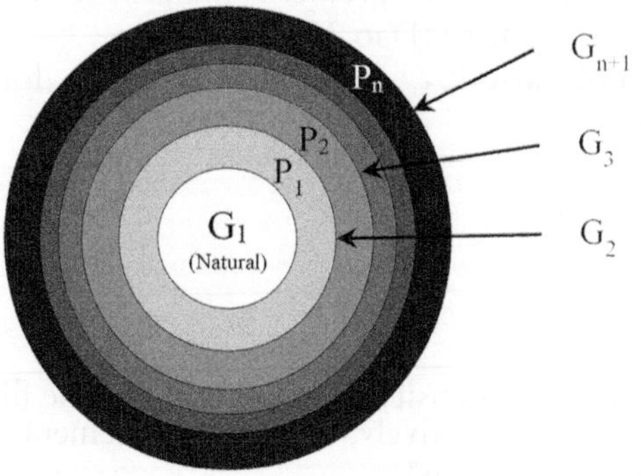

Figure 32.1 Composition of RA.

Table 32.1 Differences in terms of new paste content in RA versus its original concrete $\left(P_{n,G_{n+1},i} - P_{n,B_n}\right)$

i-fraction	G_2 vs. B_1	G_{n+1} vs. B_n
Coarse aggregate (11/22 mm)	−8%	−10%
Medium aggregate (6/11 mm)	−10%	−13%
Fine aggregate (0/6 mm)	−2%	−11%

in the process. The absolute average value (in percent) of decrease in paste content is given in Table 32.1 for each fraction,
where

ρ_{G_n} is the mean density of the aggregate skeleton in concrete B_n. It is supposed that the proportions of the different fractions of G_n in B_n remain constant when producing G_{n+1};

ρ_{P_n} is the mean density of the new paste P_n concrete B_n.

32.2.4 Water absorption

The water absorption (WA, in mass percent) is related to the porosity accessible to water p (in volume percent) according to the following equation, where ρ_D and ρ_w are the respective densities of the aggregate and water, in kg/m^3:

$$\mathrm{WA} = p\frac{\rho_w}{\rho_D} \tag{32.4}$$

By applying the same arguments as for the density, the porosity accessible to water of multi-RA can be determined from the following equation:

$$p_{G_{n+1},i} = \left(1 - P_{n,G_{n+1},i}\right) p_{G_n} + P_{n,G_{n+1}}\, p_{P_n} \tag{32.5}$$

where $p_{G_{n+1},i}$ is the porosity of the i-fraction of RA G_{n+1} in percent and p_{G_n}, p_{P_n} are, respectively, the porosity accessible to water of RA G_n, and of paste P_n, both contained in the concrete B_n, in percent.

32.3 MULTIRECYCLED CONCRETE PROPERTIES PREDICTION

The following sections describe how to predict some properties of concrete B_{n+1} from the properties of RA G_n they are made of (Dao 2012).

The volume proportion of paste P_{n,B_n} in concrete B_n made, for example, of cement and limestone filler is given by

$$P_{n,B_n} = \left(\frac{M_{c,B_n}}{\rho_c} + \frac{M_{lf,B_n}}{\rho_{lf}} + \frac{M_{weff,B_n}}{\rho_{cw}} + V_{a,B_n}\right) \tag{32.6}$$

where

ρ_c, ρ_{lf}, ρ_w are, respectively, the density of cement, limestone filler, and water, in kg/m^3;

M_{c,B_n}, M_{lf,B_n}, M_{weff,B_n} are, respectively, the mass of cement, limestone filler, and free water in concrete B_n, in kg/m^3;

V_{a,B_n} is the volume of air in concrete B_n, in m^3/m^3.

32.3.1 Density

The density of concrete B_n is described by the following equations:

$$\rho_{B_n} = \left(1 - P_{n,B_n}\right)\rho_{G_n} + P_{n,B_n}\,\rho_{P_n} \tag{32.7}$$

where

ρ_{B_n} is the density of concrete B_n, in kg/m^3;
P_{n,B_n} is the volume proportion of the new paste P_n, as defined in equation 1.6;
ρ_{G_n} is the mean density of the skeleton made of the different classes of RA G_n in the concrete B_n, in kg/m^3;
ρ_{P_n} is the density of the new paste P_n, contained in the concrete B_n, in kg/m^3. This value is calculated according to equations 32.8 and 32.9, where h_c is the cement final degree of hydration computed by the Waller's empirical model cited (de Larrard 1999).

$$\rho_{P_n} = \frac{M_{\mathrm{lf},B_n} + M_{c,n}\left(1+0.23h_c\right)}{P_{n,B_n}} \tag{32.8}$$

$$h_c = 1 - \exp\left(-3.38\frac{M_{c,B_n}}{M_{W_{\mathrm{eff}},B_n}}\right) \tag{32.9}$$

32.3.2 Porosity accessible to water

The porosity accessible to water of concrete B_n can be calculated according to the following equations:

$$p_{B_n} = \left(1 - P_{n,B_n}\right)p_{G_n} + P_{n,B_n}\,p_{P_n} \tag{32.10}$$

where

p_{G_n} is the mean porosity of the skeleton made of the different classes of RA G_n in the concrete B_n, in percent;
p_{P_n} is the porosity accessible of the new paste P_n contained in the concrete B_n, in percent. This value can be calculated with the following equation:

$$p_{P_n} = 0.94\frac{\left(V_{a,B_n} + 100\frac{M_{W_{\mathrm{eff}},B_n}}{\rho_W} - \left(\frac{0.21}{\rho_w} - \frac{0.086}{\rho_c}\right)h_c\,M_{c,B_n}\right)}{P_{n,B_n}} \tag{32.11}$$

32.3.3 Overall paste content

The total paste volume of concrete B_n is composed of the different layers of paste accumulated on aggregate G_n through the different recycling cycles and the new paste directly added in B_n. This value is needed to calculate shrinkage of concrete (see later).

$$V_{p,B_n} = 1{,}000\left(P_{n,B_n} + \sum_{j=1}^{n-1}\left(\prod_{i=1}^{j}\left(1 - P_{n+1-i,\,B_{n+1-i}}\right)\right)P_{n-j,G_{n-j+1}}\right) \tag{32.12}$$

where

V_{p,B_n} is the total volume of paste concrete B_n;

P_{i,B_i} is the volume proportion of new paste P_i in concrete B_i;

$P_{i,G_{i+1}}$ is the mean volume proportion in G_{i+1} of new paste P_i included in B_i from which G_{i+1} are recycled.

32.3.4 Elastic modulus

A model to predict the elastic modulus of concrete from one of the aggregate skeleton and the one of the paste is cited in de Larrard (1999) and presented in Chapter 10 (see Section 10.2.3). Dao shows that this model could be applied considering that

$$E_{G_{n+1}} = 0.65E_{B_n} + 0.35E_{G_n} \tag{32.13}$$

where

E_{B_n} is the elastic modulus of concrete B_n, in GPa;

E_{G_n} and $E_{G_{n+1}}$ are the elastic modulus of the aggregate G_n and G_{n+1}, in GPa, respectively.

32.3.5 Drying shrinkage

The paste volume of RA increases with the number of recycling cycles, leading to soften the RA and increase the total shrinkage. Thus, Dao proposed and validated the following simplified model to predict the drying shrinkage of the B_n concrete at 290 days, where the main parameter is the total paste volume in the concrete (including paste around the RA):

$$\varepsilon_{B_n}^{tot} = -0.0026\left(V_{p,B_n}\right)^2 + 3.83V_{p,B_n} - 613.61 \tag{32.14}$$

where

$\varepsilon_{B_n}^{tot}$ is the total shrinkage of concrete B_n, in µm/m, measured according to NF P 18-427 (1996) on $7\times7\times28\,cm^3$ samples;

V_{p,B_n} is the total volume of paste in a cubic meter of concrete B_n, calculated according to equation 32.12, in liters.

32.3.6 Gas permeability

Dao measured the gas permeability of different recycled concrete with CEMBUREAU apparatus and proposed the following equation:

$$k_{int,B_n} = 4\times10^{-21}\left(V_{k,B_n}\right)^2 - 3\times10^{-18}\,V_{k,B_n} + 6\times10^{-16} \tag{32.15}$$

where

k_{int,B_n} is the apparent gas permeability of concrete, measured on Ø11 cm cylinders, in m^2.

V_{k,B_n} is the total volume of clinker paste in a cubic meter of concrete B_n, in liters. This volume can be calculated in a similar way as the total paste with equations 32.6 and 32.12, but neglecting the limestone filler if any added, and accounting only for the clinker part of the cement if the cement is a composite one (CEM II/A-L…).

32.3.7 Carbonation

Finally, Dao shows that carbonation depth of a multirecycled concrete B_n mainly depends on the mass of new clinker added to concrete. The RA G_n appears to have no direct influence on the carbonation depth. This was explained by the fact that the RAs are probably already carbonated when introduced to the concrete. Nevertheless, RA has an indirect effect, as the cement content must generally be increased to counteract the lower mechanical properties and maintain the compressive strength of concrete. Dao finally fits his experiments with the following equation:

$$e_{B_n} = -0.0657 M_{k,B_n} + 39.144 \tag{32.16}$$

where

e_{B_n} is the carbonation depth in concrete B_n, measured on Ø11×7cm³ samples after 28 day of exposure, in mm;

M_{k,B_n}, the mass of the new pure clinker contained in a cubic meter of B_n concrete, in kg/m³.

32.4 EXAMPLE OF APPLICATION OF THE MODELS TO FOUR RECYCLING CYCLES

The previous sets of equations were used together with the CPM (de Larrard 1999) implemented in BetonlabPro software ((de Larrard and Sedran 2007), see also Chapter 15) to predict the evolution of concrete properties through four recycling cycles. The properties of RA were deduced from the concrete of previous generation with equations given in Section 32.2. Then, the compressive strength of the next-generation concrete was calculated according to the model presented in Chapter 10 (see equation 10.4), the slump with the BetonlabPro software, and finally the other properties with equations of Section 32.3 (see also Chapter 15). These calculations were iterated for the different cycles.

32.4.1 Set of requirements

Two families of concrete (C25 and C55) were simulated, whose requirements are summarized in Table 32.2. Moreover, for each family, two recycling rates were tested: 0S-30G (no natural FA replaced by recycled FA and 30% in mass of natural CA replaced by recycled CA) and 100S-100G (total replacement). The aggregate skeleton was optimized for each combination of RA and NA. For the four families thus obtained, the RA was recycled from a concrete of the same family, starting for NA concrete (NAC).

32.4.2 Materials

The G1 comprises three fractions of NAs selected in RECYBÉTON project (see Appendix). The theoretical recycled G_n was assumed to be separated in three fractions and have the

Table 32.2 Characteristics of C25/30 and C55/67

Concrete family	*C25*	*C55*
Design compressive strength at 28 days (MPa)	25	55
Mean compressive strength at 28 days (MPa)	30	65
Slump (cm)	20	20
Exposure class (NF EN206/CN, see Chapter 14)[a]	XC1	XF1

[a] According to NF EN206/CN, exposure class XC1 imposes that $C_{eq} \geq 260$ kg/m³ and $W_{eff}/C_{eq} \leq 0.65$, with C_{eq} the equivalent binder content and W_{eff} the free water content. XF1 exposure class XC1 imposes that $C_{eq} \geq 280$ kg/m³ and $W_{eff}/C_{eq} \leq 0.60$.

same grading curves as the RA of the RECYBÉTON project (see Appendix). The other properties were calculated from the equations in Section 32.2.

The same CEM II/A LL 42.5 cement as in RECYBÉTON project was selected for the C25 concrete family (see Appendix). For the C55 family, a theoretical cement was created with all the same properties but with a compressive strength of 64 MPa and a saturation amount of superplasticizer of 0.4% (in dry extract).

The same superplasticizer as in RECYBÉTON project was used (see Appendix) for all mixes with a fixed dosage of 0.15% in C25 and 0.32% in C55.

The same limestone filler as in RECYBÉTON project was finally selected (see Appendix) for all mixes with a ratio LF/(LF+C) lower than 0.15 in C25 and 0.20 in C55.

32.4.3 Simulation results

The following figures summarize the results obtained with the simulations. Cycle 1 stands for the production of first-generation concrete made of NA. Recycling starts with cycle 2.

Figure 32.2 shows that the cement demand increases mainly in the first cycle of recycling but then remains stable for subsequent cycles. This is mainly related to the fact that the

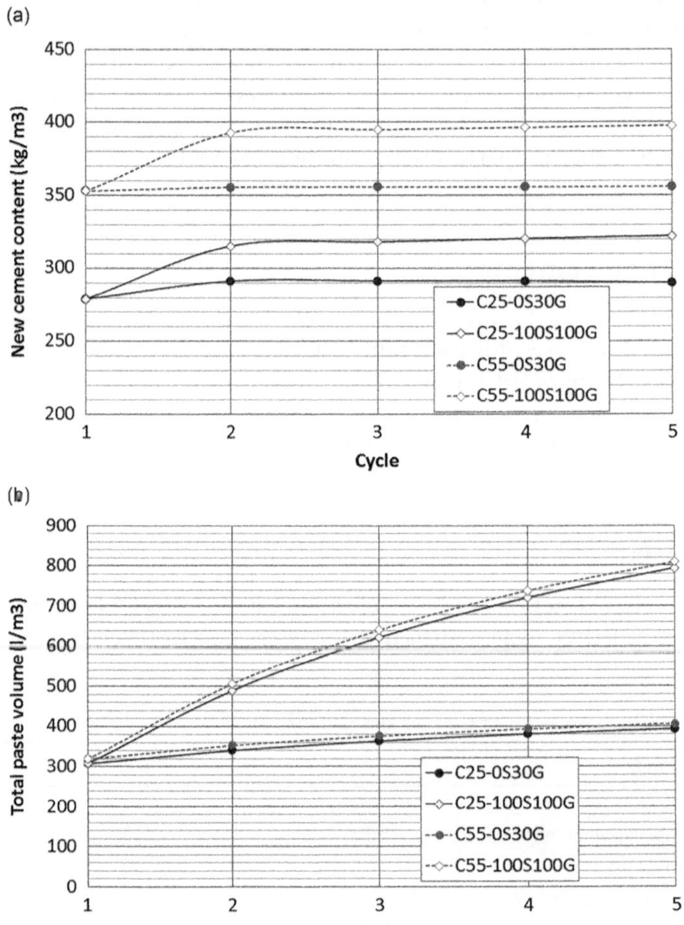

Figure 32.2 New cement content and overall paste volume of multirecycled concretes.

term k_g depends only on the strength of the original NA for a given family, as the compressive strength of concrete is constant (see Section 32.2.2). There is therefore no overconsumption linked to multirecycling. On the other hand, the total volume of paste (residual paste on RA and new paste) increases during cycles, in particular, for 100% recycling. In the latter case, the volume becomes excessive and greatly worsens the increase in porosity, gas permeability, and shrinkage and the decrease of elastic modulus with the cycles (Figures 32.3–32.5). On the contrary, multirecycling does not cause any problem with respect to carbonation since there is an even decrease of carbonation depth at 28 days during cycles.

All these trends are similar whatever the aimed level of compressive strength is.

32.4.4 Environment issues

The multirecycling process was evaluated from an environmental point of view using life cycle assessment method according to NF EN 15804 (2012), as described in Chapter 30. The concrete mixes defined previously were considered; as well the following mean

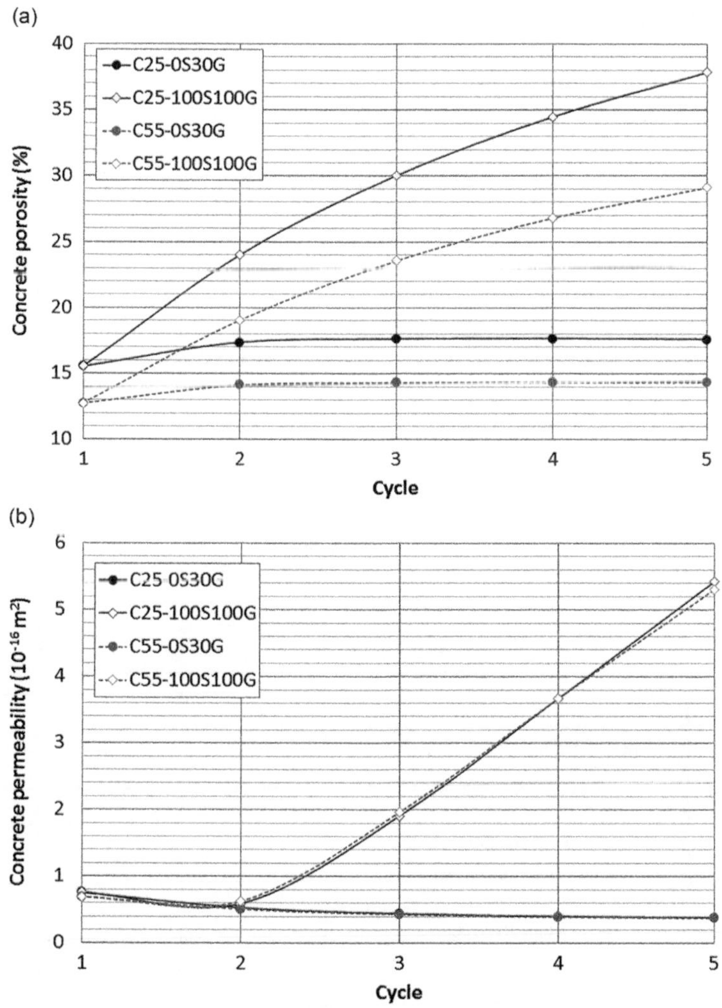

Figure 32.3 Water porosity and gas permeability of multirecycled concretes.

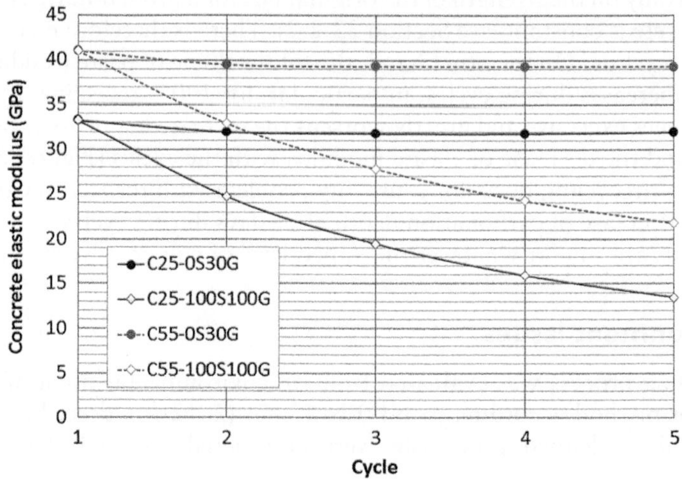

Figure 32.4 Elastic modulus of multirecycled concretes.

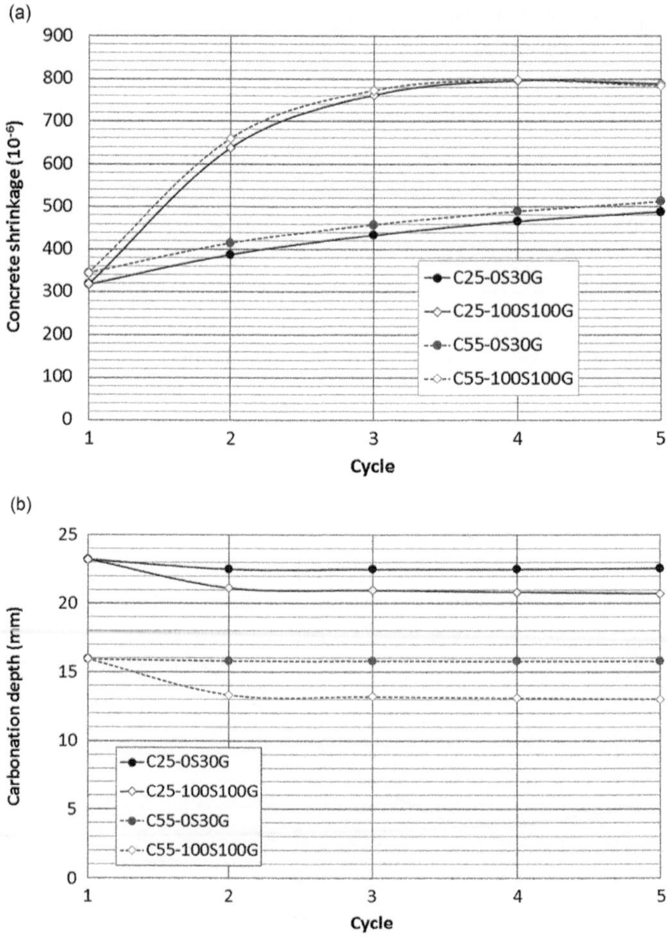

Figure 32.5 Shrinkage and carbonation depth of multirecycled concretes.

transportation distances, recently adopted by the French union of ready-mix concrete in the software Betie (SNBPE 2011): 124 km by road, 18 km by train, and 26 km by river for the cement; 22 km by road and 23 km by river for the NAs.

Figure 32.6 shows the results obtained for the concrete C25-100S-100G recycled four times, considering the same transportation distances for RA and NA. It can be seen that differences are not meaningful and that for just about all impact indicators, except three of them, multirecycling induces a decrease. Focusing on climate change, we can note that this impact indicator increases mainly with the first recycling cycle (cycle 2). This is explained by the increase of cement content that has an important impact on this indicator. For the following cycles, the cement content remains nearly constant, and thus is the climate change indicator. For nonrenewable energy indicator, the trend is the same because it also depends strongly on the cement content (cement induces important nonrenewable energy). Regarding depletion of abiotic resource, it increases between the normal concrete (cycle 1) and the first recycling (cycle 2) due to the fact that, as mentioned in Chapter 30, this impact indicator is more important (about a factor of three) for 1 ton of RA than for 1 t of NA. This impact indicator decreases for the following recycling cycle because the use of aggregates diminishes.

It can be assumed that the use of RA directly on site could be generalized in the future. Figure 32.7 summarizes the results of the calculations on the same concrete C25-100S-100G, with no transportation affected to RA. The figure shows that, in that case, the first recycling cycle (cycle 2) generally leads to a strong decrease of impact indicators. For example, wastes and photochemical oxidation indicators display a decrease as high as 40%. During the following cycles, the indicators almost remain constant. Climate change and nonrenewable energy indicators are the exceptions, which remain almost constant for the cycles. Thus, being able to recycle on site and avoid transportation is a track to dig. The current situation is probably between the two cases studied here and is shown in Figures 32.6 and 32.7: RAs are used while their transportation distance is less than that of NAs. The same trends were observed on C25-0S30G, C55-0S30G, and C55-100S100G concrete.

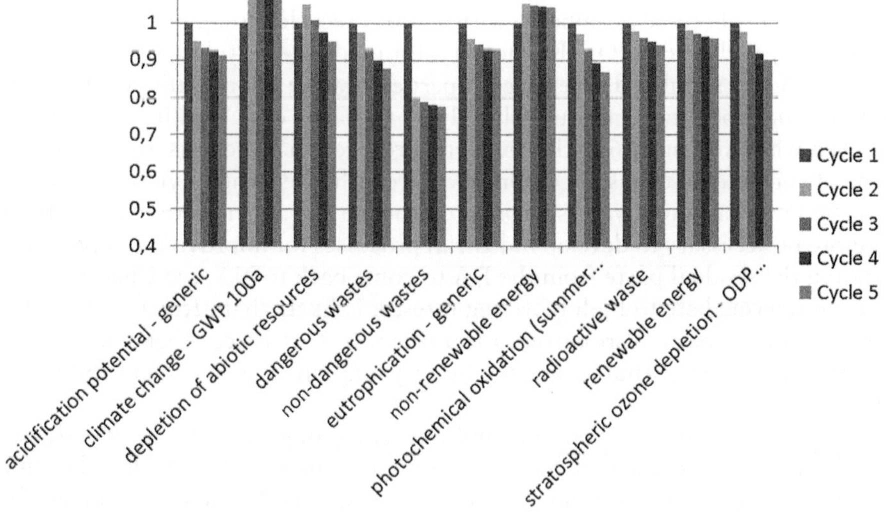

Figure 32.6 Environmental impact indicators of concrete C25-100S-100G, assuming the same transportation distance for NA and RA.

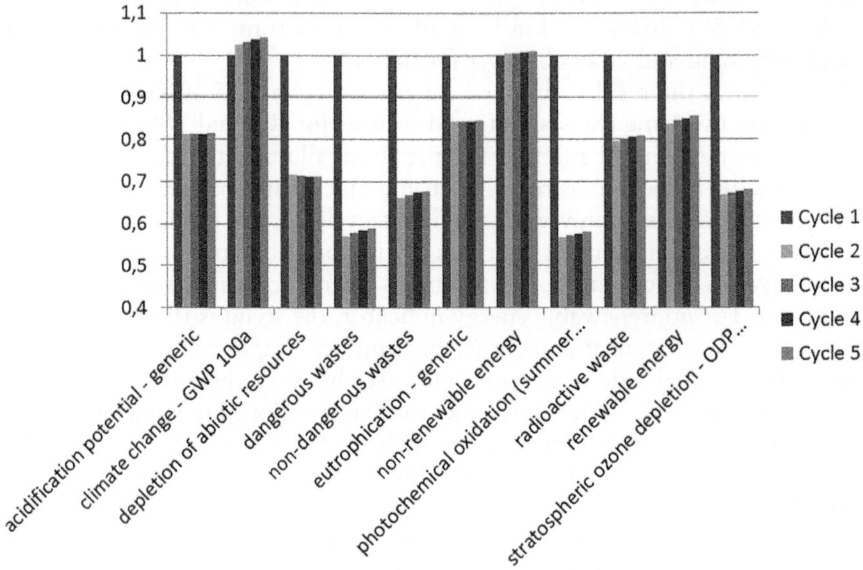

Figure 32.7 Environmental impact indicators of concrete C25-100S-100G, considering that RAs are used on site.

32.5 CONCLUSIONS

The multirecycling approach presented here is theoretical and based on semiempirical models validated on two cycles of recycling. It makes it possible to study different multirecycling scenarios. Four recycling cycles were simulated here.

The results show that cement demand increases mainly during the first cycle (cycle 2), but then remains constant during the following cycles. However, the increase of the total paste volume (paste remaining around RA and new paste) with the number of cycles leads to an increase of the porosity, gas permeability, and shrinkage and a decrease of elastic modulus. These issues quickly become prohibitive with a 100% recycling rate, because of the degradation of durability and rigidity. On the contrary, for moderate recycling rates (a 30% replacement rate of CA was simulated here), the properties relating to durability change a little and the resistance to carbonation increases slightly. In the latter case, multirecycling seems quite feasible, even when aiming at 65 MPa concrete, provided that the RAs are crushed from the same type of concrete (in terms of compressive strength). Yet, with a view to achieve a truly circular economy, that is to say, by stopping to produce NA and adopting 100% recycling, it will therefore be necessary to develop industrial processes for improving RA properties (e.g., by separating the residual paste from the RA to come back to NA, see Chapter 2).

In cases of intermediate to high recycling rates, it is likely that after a few cycles (two or three) the volume of paste also becomes too large. In that case, other recycling channels could be more appropriate than concrete like recycling fine RA in cement kilns, as studied in Section II.

On an environmental purpose, the multirecycling improves or, at worst, keeps the impact indicators almost constant compared with a single recycling cycle. The simulations also confirm the interest of recycling aggregates on site to eliminate the impact linked to transportation.

Chapter 33

Discussion

Conditions for a sustainable process

F. Buyle-Bodin
University of Lille

S. Decreuse
CEMEX

A. Feraille
Ecole des Ponts ParisTech

F. de Larrard
LCR, LafargeHolcim

X. Chomiki
Conseil départemental de Seine & Marne

CONTENTS

Abstract

Conditions for a sustainable process of recycling old concrete to produce new concrete are nearly similar with that claiming for other industrial products. This chapter focuses on specificities of concrete. The first one is the fact that this material is the most widely used in the world. However, the construction industry organization is complex, including many actors. For the success of environmental, economic and societal preservation, a coordination between actors is necessary.

The introduction aims to place the process of recycling of concrete in the general question, how to rebuilt the city on the city.

In the state of the art, the recycling stakeholders' ecosystem is firstly analyzed, then the decision making about recycling is detailed.

A recall is proposed, presenting the consequences of the use of RA on different characteristics of concrete. Then the preliminary questions are presented: for what type of construction, for what usage, with what specified strength regarding to site conditions, with what level of acceptable risks.

The question of the availability of the resource of RA is discussed in terms of distance to concrete production site, type of products, and quality of resource. Then the legal and political environment is developed: regulations, standard and incentives, in France, Europe and in the world. The links with the global question of the Circular Economy and the short loop are analyzed.

The last paragraph aims to discuss the decision-making frame. What are some questions to address? What are the technical, environmental and economic conditions?

33.1 INTRODUCTION

The recycled concrete aggregates (RA) produced by the demolition of old buildings seem to be an option to construct new buildings and to preserve the environment by the reuse of inert materials, minimizing the use of natural materials and the disposal in landfills. Numerous countries are multiplying studies concerning this subject and evolving their regulations consequently. These days, it seems fundamental to go further in the environmental protection by minimizing the impact of road transport through the implementation, on the same urban site, of the following process: in a first phase the demolition work, then the rubble processing, followed by the storage of RA, and finally the building's reconstruction.

In fact, the increase of the recovery rates of RAs seems globally meeting the current environmental constraints. However, the transportation of construction and demolition waste (CDW) from the site to the recycling plant and their potential return are still problems, which are not correctly considered in environmental assessments. The research of the shortest loop is being developed in France.

It is necessary to understand the limitations to this development. If the aim of the short-loop financial viability is obvious, the complexity of the organization of building projects in urban refurbishment is the first lock to remove. The various roles of the multiple operators of the project do not comply with the global coherence provided by short-loop circuits.

At present, the evolutions in terms of availability of natural resources and sites of landfills are favorable to recycling if transportation distances are acceptable economically and environmentally.

However, the disposal of demolition products is diffuse, unlike the quarries of natural aggregates (NAs). The flow depends on the importance and on the calendar of demolition works, and on the nature of demolition sites.

33.2 STATE OF THE ART

33.2.1 The recycling stakeholders' ecosystem

The first stakeholder is the contracting authority (or the owner) of the building operation located at the old construction place. For him, the demolition is at present a loss of time and money. He can entrust this phase to a specialized contractor (named demolition contractor) who describes the deconstruction works, selects specialized companies, and takes on the demolition and the distribution of the demolition products. The main stake is to clean the land and then allow a fast reconstruction.

The demolition company takes the property of the demolition waste, and simultaneously, the responsibility of the waste management. He must respect various regulations aiming to respect the health and safety of the workers and the preservation of environment.

The management of inert mineral waste passes by different ways:

- Reuse on site of some rough waste as backfill after simple grading;
- Transport of waste to recycling plants and then transferring its environmental responsibility at good financial conditions;
- Transport of waste to landfill at bad financial conditions.

Nevertheless, the financial conditions depend first on the quality of the waste after preliminary sorting (for the second way, allowing the acceptance of the waste by the recycling plant and then its price, for the third way, inducing the class of the landfill and then the price). Second, the distances between the three sites (demolition site, recycling plant, and landfill plant) must be considered. In conclusion, the role of the demolition companies is crucial. That explains why the localization of the demolition sites has been considered in Life Cycle Assessment (LCA). In addition, the quality of the preliminary on-site sorting is of great importance.

The third stakeholder is the recycling company. Its part is to receive CDW, accept them after control (then becoming the owner of the waste), and convert them into construction materials (end-of-waste status). These recycled materials are sold to construction companies or to concrete producer (ready-mix concrete or precast products). However, a deficient sorting can induce a poor valorization of the outgoing products. The main applications are of low or medium value, as in embankment and road works, whereas valorizations are possible towards nobler materials like concrete. Technologies of sorting on site are under development but remain expensive in the medium term without warranty of economic equilibrium. This equilibrium can be improved by reducing the distances between the sites (recycling plant, road and/or building construction site, and concrete batching plant). Moreover, the preservation of environment is also improved (see Chapter 30).

The last stakeholder is the owner of the new construction in which the recycled materials will be used. He selects construction companies, which take on the construction of the new buildings. Moreover, the choice of the construction materials is a matter for these companies. The main stake is to build at the best price respecting technical prescriptions.

To close the loop and develop the circular economy, it would be desirable that the same entity manages both demolition works and construction works. Even if he delegates some tasks to contractors and subcontractors, he might have a complete sight of the entire loop and could optimize the global cost simultaneously to the environmental impacts. For example, he could accept a higher price for demolition if it is accompanied with a better sorting. Then, he could set the use of RA for concrete works.

In other cases, the construction company contractor may propose a variant of concrete including RA.

33.2.2 Decision making about recycling

The decision loop includes the stakeholders previously presented. The contracting authority is the main decision maker. He must be convinced of the environmental stakes of the construction sector, particularly the shortage of nonrenewable resources and the increase of construction waste flows. Different tools and guides, produced by collectivities and social and environmental associations, can help make the decision.

The demolition contractor has to explain to the contracting authority the need of the quality of the demolition work and of the sorting of demolition waste. He considers the territorial conditions of the industrial organization (demolition site, recycling plant, and depot site). Specialized subcontractors can help him. His action should induce a better quality of the loop until RA production. At this part of the loop, professional guides give help as the one to be produced by RECYBÉTON.

The construction contractor has to explain to the contracting authority the need of the use of RA concrete (RAC) to build new constructions. With this agreement, he can prescribe various RAC for different parts of the construction. Of course, he also considers the territorial conditions of the industrial organization (recycling plant, NA quarries, and concrete batching plant). Specialized subcontractors and produces of construction materials can help him. It is certain at this part of the loop that professional guides are welcome (Figure 33.1).

33.3 RECALL

In this section, the main features of RAC, as compared to a conventional one, are recalled, as a summary of the matters developed in Sections III–VI. Here it is assumed that the recycling process is performed through incorporation of crushed concrete aggregates into new concrete. The case of recycled cement is not yet so developed to be considered here. Moreover, no specific changes in concrete properties are anticipated from the use of recycled cement, as shown in Section II.

Therefore, in RAC, the first difference is the fact of using a supplementary constituent, which is one or several fractions of RA, used in combination with NA in most cases. This new material is not traceable in most cases, and requires a suitable quality control plan to ensure that the content in pollutants remains below standard thresholds. Being significantly more porous than NA, RA needs a proper moistening system to avoid an important suction of water just after concrete batching. The development of a suitable RAC mix design requires some trial batches before starting an industrial production, and cannot be carried out by a simple replacement of a part of the NA in an existing formulation, even if the changes on concrete properties are minor as long as the replacement level is low (say less than 20% of coarse aggregate (CA) fraction).

The obtained RAC, for the same level of initial workability and compressive strength at 28 days, differs of conventional concrete in a number of extents. With increasing replacement levels, starting from low percentage of coarse aggregate to high values and partial or total replacement of the fine aggregate (FA) fraction:

- The cement demand is equal or higher. In case of a marked increase, the CO_2 balance of the solution can be affected, but potential savings from lower transport distance should also be accounted for;

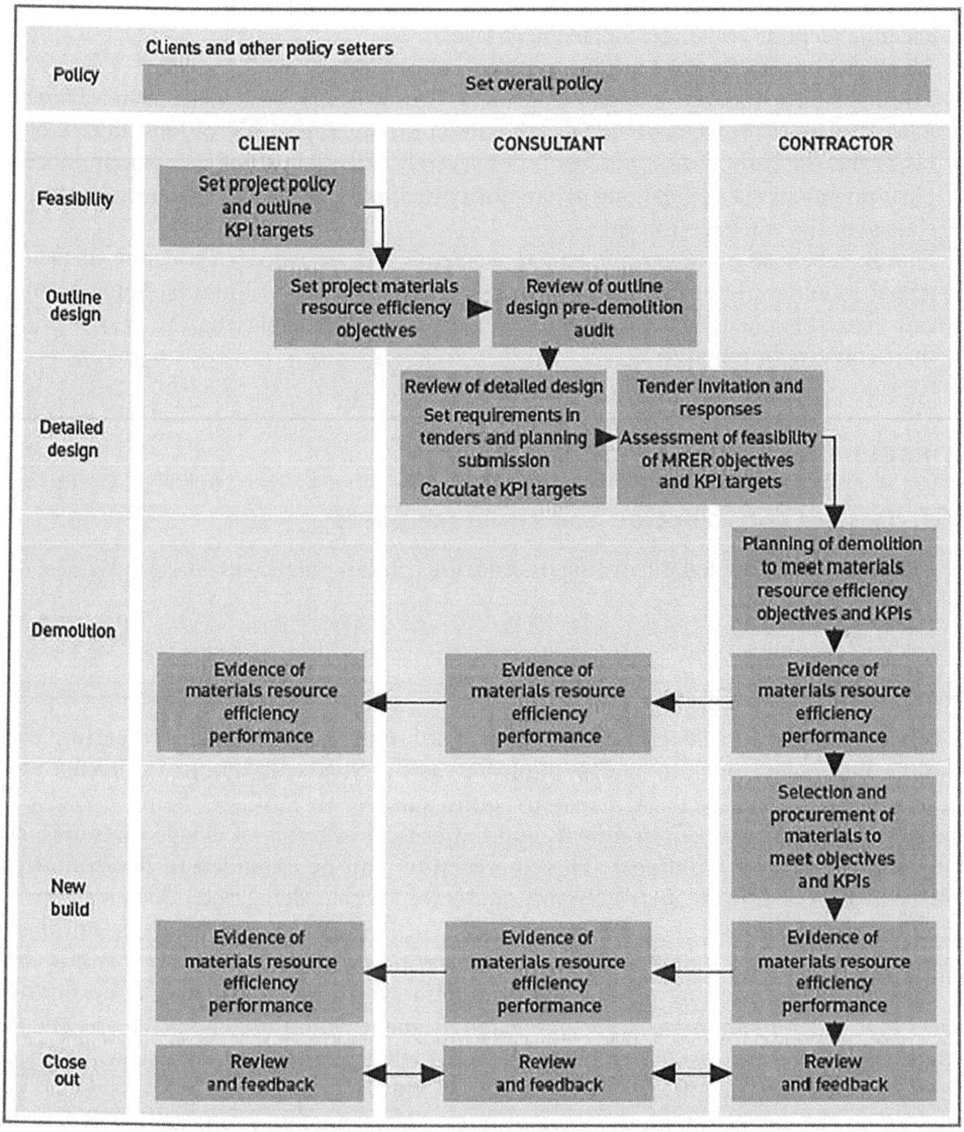

Figure 33.1 Chart providing a simplified overview of the process and actions to be taken by the various parties [Waste and Resources Action Programme (WRAP)]. KPI: Key Performance Indicators; MRER: Materials Resource Efficiency in Regeneration.

- Attention must be paid to slump retention, with a possible use of retarding agent to correct too early stiffening. This may happen when recycled FA is used, even when all aggregate fractions are fully presaturated before batching;
- Curing of horizontal surface should be performed more thoroughly, due to the higher level of potential plastic shrinkage;
- The evolution of tensile strength must be checked if this property is critical in the envisaged application;
- The elastic modulus decreases significantly (up to 30%) as the level of drying shrinkage increases (up to 100%), while the creep strains may double or more. However,

the cracking risk induced by restrained shrinkage does not seem to be markedly influenced, except at very high replacement level;

- The resistance to flexural fatigue is slightly degraded;
- The hardened material is more porous, as being richer in cement paste. Therefore, most transport properties evolve in the direction of easier flows of ions, liquid, or gas. However, the carbonation process is not severely affected, which is fortunate since this phenomenon is the critical one in a majority of applications, according to the exposure classes in the construction site;
- Finally, where an architectonic concrete is required, the aspect of concrete surface may reveal its internal constitution and the presence of some pollutants. Some architects can view this positively as designers who are keen of displaying the previous life of their construction materials.

33.4 PRELIMINARY QUESTIONS

33.4.1 Type of construction and concrete parts

Here, a review is provided regarding the various areas where the use of RAC can be considered.

33.4.1.1 Usage

The two main segments of concrete market are buildings and civil engineering works. Regarding buildings, concrete can be found in cast-in-place concrete or in precast pieces. For the former, internal vertical parts are quite suitable to RAC, as being invisible, not exposed to any aggressive environment, and not critical in terms of delayed strains (except in the case of high-rise buildings). This appreciation can be extended to horizontal parts (slabs, beams, decks), if the spans remain moderate, so that deflections do not control the section heights (otherwise, the high deformability of RAC might lead to an extra thickness of the slabs). Regarding foundations, RAC is appropriate as long as the soil is not chemically aggressive (e.g., through the presence of sulfates). Hence, RAC may contain ancient cements that are not sulfate resistant. If the new cement would match such a specification, the RAC could have a satisfactory behavior, but such a quality would need to be checked by appropriate testing methods. With regard to exposed building envelopes (walls, tiles), depending on the local climate, the issue of freeze/thaw must be examined, as RA is being rarely frost-resistant.

The case of precast concrete used in buildings can be separate in plain concrete, reinforced concrete, and prestressed elements. Plain concrete masonry blocks are an important sector where RA could be readily valorized, even if precast processes based on direct demolding require a sharp control of stiff concrete water content, which can be an issue with RA of variable porosity. The case of prestressed pieces as precast beams raises the matter of higher RAC deformability. In some countries (including France), RAC is forbidden for such applications. At least, one should account for higher prestressing losses and possible higher counterdeflections.

In civil engineering works, it is possible to consider the use of RAC in different applications. As for roads and airfields, it was shown in Chapter 11 that the higher heterogeneity and lower fatigue strength of RAC were not in favor of a high level of replacement in top layers, when the design is based upon fatigue considerations. This is the case in France, but not in other countries like USA where RAC is extensively used in all pavement layers. Base and subbase lean concrete layers are in any case good destinations for RA, as commonly made in some German highway projects. In such application, a short circuit recycling process can be proposed, where the material generated by an old pavement demolition is reused on site in a new one. Parking lots, pedestrian paths, and filling materials for trenches are other applications where RA can be chosen. Here again, a direct recycling from the original structure is easy to organize, with a clear traceability of constituents.

Reinforced concrete bridges with small span are another area of possible application of RAC, as shown in Chapter 23. The case of tunnels is also worth considering, subject to prevent sulfate attack risk, as was considered for building foundations. For cast-in-place tunnels, a limited rate of RA in concrete should not enable cracking: the case is similar to the slab-on-grade application related in Chapter 22. For tunnels, another alternative material, the excavated rock, can also be of interest, but this is out of the scope of current work.

Dams are another type of civil engineering works that may consume large quantities of concrete. Here, the main issue deals with heat of hydration and thermal-induced cracking. It was shown that recycling at limited rate does not induce a higher binder dosage. Furthermore, the highest degree of relaxation accompanying the increase of creep clearly favors the prevention of thermal cracking. This is even a rare case where RAC exhibits superior properties, as compared to a conventional mix only produced from NAs. The only question lies in the low probability of finding a close source of RA for dams that are generally built in remote areas, far from cities. However, the partial retrofit of an existing dam structure can provide an opportunity for local reuse of demolished concrete. In any case, the risk of Alkali-Silica Reaction must be managed according to local standards. The same considerations apply to other mass concrete application.

The case of marine applications deserves some specific considerations. As already stated, the replacement of a part of the NA by a recycled one always provokes an increase of transport properties, including chloride ion diffusivity. This means that the compatibility between material properties, concrete cover, and life expectancy requires a special investigation, contrary to a simple conservation of current practice applying to NA concrete of equal strength. In some cases, there may exist a safety margin that will be diminished by the partial replacement process. In other cases, the classical durability provisions of the mix-design—minimum cement dosage, maximum water–cement ratio—will need a modification (see Chapters 12 and 34) or to carry out a performance-based durability approach.

33.4.1.2 Specified strength

A common stereotype is the following: RAC can only be of low specified compressive strength. As a matter of fact, this assumption is not confirmed by experience. For instance, Dao and his colleagues were able to perform a 100% recycling of an airfield pavement slab, with the new concrete having a cylinder compressive strength of 62 MPa (Dao 2012). The case of ultrahigh performance, fiber-reinforced concrete is also of interest: here again, a full recycling process where all aggregate comes from a previously crushed UHPC is equally feasible (Sedran et al. 2009).

However, Dao (2012) has shown that, for a given replacement rate, the relative decrease of compressive strength increased with the level of original strength. Although it was possible to produce high-strength concrete with RA from an original medium-strength concrete, the cement demand was higher than in the case where both mixes had comparable strengths. Therefore, a general conclusion is that the lower the specified strength, the easier the sustainable recycling process. But in specific cases, it can be worth producing high-performance concrete with RA.

33.4.1.3 Site conditions

Production of an RAC requires some practical conditions that need to be met for successful application:

- Availability of a source of RA at a low distance from the concrete production place (e.g., less than 25 km). This source can be either permanent (as a recycling platform) or temporary (a demolition site with a mobile quarry unit). In all cases, it should be operated by competent professionals able to show their experience and their adherence to a credible quality assurance plan (see Chapter 28);
- Room and equipment for RA stocks at the concrete production place. Quite often, especially in urban locations, this aspect is a real barrier. The producer must invest in related installations, but may be reluctant if the market of RAC is not yet established;
- In terms of local climate, where concrete is feasible, RAC should also be. However, the case of moderate freezing risk is to be considered. Hence, not all existing concrete are air-entrained in such an area, so that some RAC aggregates can be frost sensitive. In dry climates, the need for careful curing is reinforced by the recycling process;
- Finally, there is also an important human factor. As long as recycling concrete into concrete is not a routine practice, such a process requires awareness, competency, and acceptance from all stakeholders, from the RA production to the construction owner, including design office, concrete manufacturer, contractor, site engineer, and quality controller. This book may hopefully help to reach such a situation.

33.4.1.4 Acceptable risk level with regard to the application

Recycling is a moral duty, but construction quality and soundness are other ones too. Standards are developed to ensure that all phases of the process of material selection, concrete production, and placement are performed with an acceptable level of risk. However, the fact of using constituents of unknown origin can only increase the likelihood of unexpected degradations. As long as the resource in RA is only a part of the global consumption of aggregate in concrete, it is therefore reasonable to exclude, or to critically examine, the use of RAC in the following cases:

- where exceptional lifespan is required for concrete pieces. For instance, some precast concrete containers produced to store nuclear wastes, designed for several centuries;
- where maintenance or repairs are difficult and very costly: high-rise buildings, long-span bridges, etc.;
- where service interruptions have huge impacts and social costs: some bridges in urban areas, highly trafficked highways, metro tunnels corresponding to very high flows of users, etc.;
- iconic constructions where concrete surfaces are essential for the owner or the city image.

Fortunately, these cases only represent a very small part of opportunities to use RAC within the construction world.

33.4.2 RA resource

To develop the use of RA in concrete, it is necessary to have a good quality resource (consistency of the characteristics, purity of the resource, etc.). It is also necessary that the resource is available near the places where it will be used (concrete production units) to minimize the cost of delivery and in a sufficient quantity that allows the concrete producer to have RA whenever he needs them.

33.4.2.1 *Availability vs. distance to concrete production*

In the scope of RECYBÉTON project, Mongeard and Dross (2016) studied sorting and recycling facilities and their distribution in France. The study was made for the 12 biggest cities in France where the quantity of CDW is considered sufficient to develop such a sector.

The distances between production units of aggregates (NA and RA) and production units of concrete (ready-mixed concrete and precast concrete units) were compared. In all cases, the comparison is in favor of the RA. However, the study also considers the waste collection sites that currently do not produce RA. It is also important to consider that, even if a production site of RA exists and is not too far from the concrete production units, in some cases, the quality of the produced aggregates will not be sufficient to be used in concrete (Figures 33.2 and 33.3).

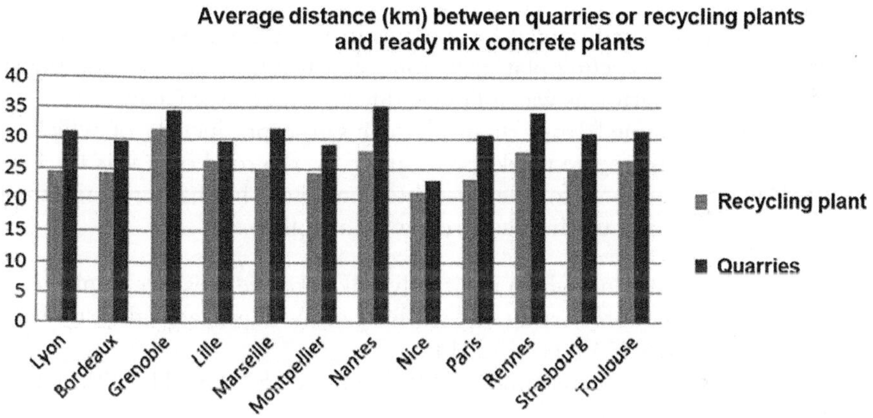

Figure 33.2 Average distances between ready-mix concrete plants and aggregate sources.

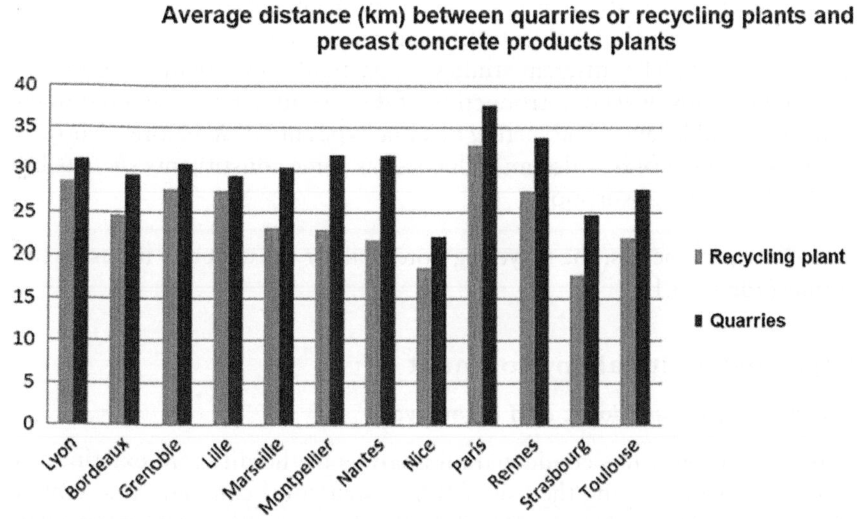

Figure 33.3 Average distances between precast concrete product plants and aggregate sources.

33.4.2.2 Type of products

The kinds of CDW that can be recycled in concrete are as follows:

- Selected demolition material (that include mostly concrete—purity should be 95% of concrete and stony materials to maximize its reuse in new concrete, 90% and 70% are also possible, but such aggregates should be limited to few concretes mostly in X0 environment).
- Return concrete. This source of concrete represents 3%–5% of delivered concretes. It is a very interesting source, as it is rather pure and not contaminated by harmful species as plaster.

The RA obtained from those two sources can be delivered separately as recycled FA and recycled CA. It can also be delivered to the concrete production unit as a premix of RA and NAs (depending on local regulations). Either this premix can be produced on a quarry or on a recycling platform (if benefits from return freight are clear).

33.4.2.3 Quality of available resource

RECYBÉTON studied the quality of RA resources in France. The variability of aggregate characteristics on sixteen recycling platforms (one-shot testing) was studied, and the production of two recycling platforms was followed during 2 years (see Chapter 29).

The quality of RA obtained from these platforms (the production of which was not intended to the concrete aggregate market but rather for the road one) was for half of the production (55%) not yet adapted for their use into concrete. The characteristics involved in most cases and preventing the use into concrete were as follows:

- The soluble sulfate content (which is directly linked to the presence of plaster and can be improved through more sorting and washing of the aggregates) determined according to §10.2 of NF EN 1744-1+A1 (2014).
- The classification of RA (determined according to NF EN 933-11 (2009)) and mostly the floating content (plastics, wood, etc.) of the aggregates. As for the previous characteristic, this one can be improved through more sorting and washing of the aggregates.
- The water absorption of RA (determined according to NF EN 1097-6 (2014)). It is the range of this characteristic that could cause problems (the control of that parameter has a direct impact on the total water/cement ratio of concrete and then on the durability of the material). The different studies, in particular on premix of natural and RA, show that with a low level of introduction of RA the influence of that parameter can be kept to a reasonable level. The sorting of some types of CDW before their introduction in the recycling facility has also an influence, as some constituents such as bricks have a high level of water absorption.

According to those parameters, the recycling potential for the concrete industry in France is 39 million tons (Table 33.1).

33.4.3 Legal and political environment

33.4.3.1 Standards, regulations, and incentives

The RECYBÉTON project has conducted an analysis of the different existing texts (standards and regulations) regarding the use of RA in structural concretes and concrete products (see Chapter 34). The report was based on the distinction between texts related to the products, the structure design, and the execution of works.

Table 33.1 Potential of recycled resources for concrete manufacturing (Mongeard and Dross 2016)

Inert wastes	*Total quantity of waste generated (million tons and %)*		*Potential resource for use in concrete (% of potential valorization and million tons)*	
Unpolluted soil and stones	140	61%		
Mixed inert wastes	31	14%	30%	9
Other materials from roadway demolition and rocky materials	25	11%	75%	19
Bituminous mixtures containing no tar	13	6%		
Concrete	18	7%	60%	11
Bricks, tiles, ceramic, and slate	2	1%		
Other inert wastes	2	1%		
Total	231			39

The normative texts included:

- Materials: specifications texts and test standards;
- Design rules on constructions with reinforced concrete or prestressed concrete;
- (Eurocodes, fib model code);
- Execution rules (European and French standards, e.g., French Unified Technical Document, etc.)

The examined regulation texts cover various legislative and regulatory French codes with a specific point on the identified obstacles. At last, some texts from local initiatives based on application of regulations, such as guides, have helped to highlight a strong use in road engineering.

This analysis of French texts has outlined the strengths and the weaknesses of the current texts regarding the use of RA. It shows that this use obeys the following general philosophy:

- A characterization of their quality in correlation with the variability in their properties;
- (classification, water absorption, resistance to freeze–thaw , etc.);
- A moderation in their use in structural concretes (low incorporation rates).

In a second time, RECYBÉTON has analyzed standards, regulations, and guidance documents on the use of RA in concrete from different countries.

In Europe, 13 countries, whose existence of provisions considering the use of RA was demonstrated, were identified. Several countries, including the Netherlands, Germany, Austria, and UK, have tested the recycling into concrete for several years. Out of Europe, seven countries have demonstrated the recycling into concrete in recent years. These countries include Japan, China, the United States, and Canada.

Other documents showing high interest in recycling into concrete were also investigated.

The analysis of these foreign texts shows that the use of RA generally obeys the following philosophy:

- A tradition in road applications for which the use of RA is well controlled.
- A characterization of their performances (value levels and variability).
- A marked difference exists when considering recycled CA and RA with, for the latter, limitations of its use in structural concrete for which risks are higher.

A marked disparity between the different countries on the rate of incorporation of recycled coarse aggregates in structural concretes was noted. Most of them establish limitations in the substitution rate. Other ones allow rates up to 100% of recycled coarse aggregates in concrete (e.g., the Netherlands, Switzerland, Japan, and China). However, such permission is allowed if the traceability of the recycled coarse aggregates is guaranteed, and if additional controls are carried out. Some countries have introduced coefficients in the calculation of concrete properties, such as elastic modulus, shrinkage, or creep coefficient, when using recycled coarse aggregates (e.g., the Netherlands, Switzerland, and Spain).

Apart from Switzerland and Norway, countries prohibit the use of RA in prestressed concrete elements.

Regarding the environmental aspects, several countries (e.g., Austria, Germany, and the Netherlands) have implemented a series of tests to control the release of dangerous substances (i.e., heavy metals).

33.4.3.2 Position of national/local authorities—green labels

Among all the labels and certifications devoted to the construction sustainability, it seems that three of them really promote the use of RA. These labels and certifications are detailed in Chapter 35.

The first one is "Verde" (Green) assigned by the Green Building Council of Spain, which specifies the use of RAs. At a regional scale, the Instituto Valenciano de la Edificación (Building Institut of Valencia) creates "Perfil de Calidad" (PdC, Profile of Quality) in which a specific criterion promoting RAC exists ("Hormigones reciclados de resistencia no superior a 40 N/mm^2", RACs with strength not higher than 40 N/mm^2). This criterion gives four points among the 40 points necessary to obtain the label PdC.

The second one is Minergie Eco introduced in Switzerland in 2006. Using RAC induces a benefit directly. When it can be used, its volume fraction must not be less than 50%. In addition, this label defines an exclusion criterion including insufficient use of RAC. Thereby, as noted by Bougrain et al. (2017), even though no statistical study has attempted to establish a link between the evolution of market shares of RA and the introduction of this label, the supply of RA has risen, as well as higher value-added use. Thus, RAC can be used in supporting elements in construction. Public owners have shown an interest in obtaining this label. The canton of Vaud played a particularly key role in the development of the market. In the agglomeration of Zurich, the proportion of RAC is steadily increasing and now accounts for around 10% of concrete requirements. In particular, the city's buildings office agreed with local consultants and companies that all of its new buildings would incorporate recycling concrete. The "Werdwies" district project in Zurich is exemplary since the recycling concrete was used for the sealed basements, the ceilings having in part large spans, and the concrete walls of the ground floor.

The third one is BREEAM (Building Research Establishment's Environmental Assessment Method) (BREEAM 2016), which promotes the use of RA by according credits for applications in two sections: Material and Wastes. In the Waste section, it encourages, first, the sustainable management of the construction site (and reuse where feasible) of construction waste. In the same Waste section, credits can be obtained as a function of the percentage levels of recycled or secondary aggregate[1] specified against set targets. In the Material section, one credit is obtained if at least 25% of the high-grade aggregate uses are provided by secondary or RA.

[1] By-products of industrial processes that can be processed to produce secondary aggregates. Secondary aggregates are subdivided into manufactured and natural, depending on their source.

For all other certifications, the recommendations regarding the use of RA in concrete are illustrated, when they exist, by the need to evaluate environmental impacts of the materials and thus try to reduce them without specific imposed thresholds.

Recycled concrete is not yet used a lot. The development of labels specifying the quantities required of RA in concrete has permitted to increase the market and to improve sectors. It is important to keep in mind that when a project is conducted with concrete made with RA the project owner plays the most important role.

33.5 DECISION MAKING

33.5.1 The questions to address

In a more and more constrained context in terms of sustainable development and circular economy, it is important in the domain of construction to have in mind the end of life of the constructed concrete. The European Directive (2008/98/CE) indicates a prioritization of waste prevention and waste management. Thus, the best to do is to prevent waste, then to reuse, then to recycle, and at last to recover, including energy. Here the focus is given on recycling. When is it interesting to put RA in concrete and at which proportion? We are going to try to answer this question (Figure 33.4).

33.5.2 Technical conditions

The contracting authority is surrounded by experts to verify the conformity of the building with the regulations and the program of operation. However, at the present stage, since the RAC sector is not sufficiently structured, many partners in the construction industry do not know the product. This implies that the design and execution studies should be larger and longer for an operation using RAC than for an operation using conventional materials.

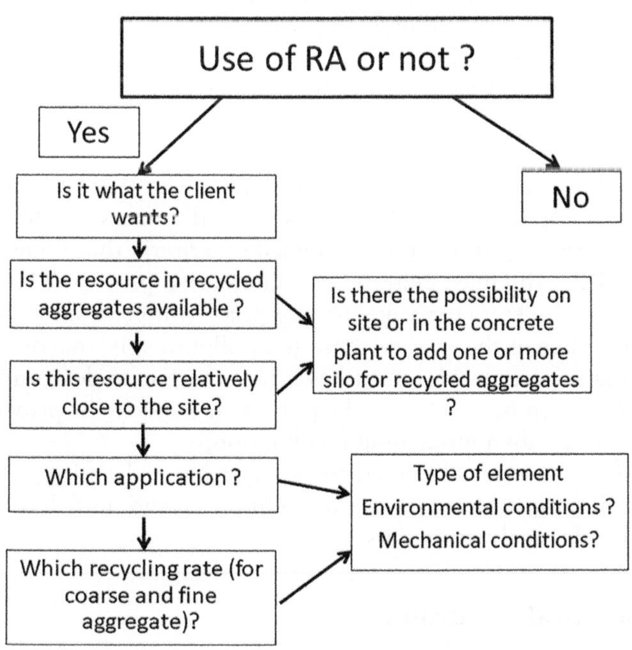

Figure 33.4 The questions to address for decision making.

The designer has to define the environmental classes (exposure class) of the building and indicate more precisely the mechanical, physical, and esthetic characteristics expected of the materials, to verify that the finished product complies with the requirements.

If the choice of using RA is done by the contractor, the supplier of aggregates must indicate the rates of incorporation of recycled materials and then prove that the process of manufacturing RAC makes it possible to guarantee the absence of pollutants (plastics, plaster/lime, asbestos, etc.) in the finished product.

The concrete supplier must realize the concrete formulations, validate them, and do the compliance tests. Above all, he must guarantee the announced physical and mechanical characteristics. This process requires a time that must be considered in the planning of the construction execution. Finally, the designer is required to prepare his performance calculation notices in line with the special features announced by the manufacturer, instead of the standard features incorporated by default in calculation engines or charts. He must guarantee an implementation in accordance with the rules of the art and carry out checks on the materials.

The technical control board must certify the conformity of the implementation of the materials with respect to the regulations and the technical opinions, and be sensitized to this new material.

The insurer must provide its guarantees on the building only if it complies with the regulations and if the products implemented comply with the technical approvals. This implies that previous interlocutors have ensured regulatory compliance and implementation.

For conventional concrete, the definition of the desired characteristics is easy to do, since they have standard and reproducible formulations to match the usual strength classes. But this is not the case for RAC where certain characteristics are lower than for a comparable conventional concrete. Therefore, the formulation depends on the type and amounts of recycled materials incorporated into the concrete composition. Thus, all the specificity of the use of RAC consists of three points: to define the characteristics of RAC so that it has the mechanical, physicochemical, and esthetic characteristics aimed and adapted to its usage; to define an adequate formulation allowing to reach these characteristics; and to prove that these characteristics have been achieved.

The success of such a project depends on four actors who guarantee the points mentioned earlier:

- the designer, who must have the will to use this innovative product and who must have the knowledge to optimize the characteristics really necessary to the concrete so as not to make overquality and not to limit itself on criteria that do not have place to be,
- the supplier of RA, who must guarantee its fabrication process. Thus, the constancy of the aggregate's parameters for a precise site (for example, a same original demolition site) and a guarantee to the quality chain (no pollutants) is unavoidable,
- the concrete manufacturer who must have the expertise to determine the formulation and explain the differences between the particular concrete he provides and the classical concrete that has an approaching formulation,
- the control office, who is accredited to certify technical and regulatory compliance. This organization must be aware of this technology and must have the knowledge to verify these nonclassical parameters.

33.5.3 Environmental conditions

It is instinctive to consider that using RA in concrete is "better" than using NAs. However, this instinctive point of view needs to be argued and even proved. In terms of environmental impacts, the tool used is the LCA method presented in Chapter 30. This method provides a way to evaluate environmental impact indicators with regard to a particular project. Various impact indicators are evaluated, such as climate change, abiotic depletion of resources, energy consumption, hazardous waste, etc. (see Chapter 30). Chapter 30 shows this method, within this study, through an experimental site (in Chaponost, near Lyon) and a parameter study related to transport distances. This study highlights the fact that the positive effect of RA use to manufacture new concrete must not be affected by an increase of cement content, because it induces an increase of environmental impact and price. The solution can be found in an equilibrium between RA and cement content vs. geographic conditions as distance between demolition, recycling, and depot sites.

A difficulty is that this method (LCA) does not show big differences between RA and NA. This is particularly because inventory data used for NA and RA have no big differences. Therefore, by construction, there are not big differences in the results while comparing the two. We can conclude that LCA is not the best method to evaluate the potential of using RAs. However, we can also try to interface this method with the way practitioners work on site. We could then add others indicators impacts such as "tons of aggregates which were not taken from the ground", as proposed in Chapter 30. It induces avoided impacts, such as the ones due to the extraction of nonrenewable materials or to landfilling waste materials, that are poorly addressed in LCA methods.

33.5.4 Economic conditions

On the scale of building construction, the structural work represents one-third of the cost of a building, and this represents a high cost of personnel and machinery compared to the cost of materials. As the technical and financial offers do not show the cost of materials, but rather the cost of the components, the difference in cost of RAC solution compared to a conventional concrete will be hardly detectable in the offer, except if it is the subject of a well-identified variant.

The use of recycled materials is based on the political or strategic buyer's will. It can be encouraged by a valorization of this approach in the environmental certifications that are put forward by public or private contractors. The state, via environmental regulations, such as classified polluting installations (ICPE), is also aware of the concern to limit the storage of waste, or find solutions to dump them, in embankments for example. A moderate additional cost would be acceptable in return for the limitation of waste production or nonreuse of materials.

33.6 FAVORING CIRCULAR ECONOMY THROUGH A GLOBAL APPROACH

The built environment is an excellent support for experimentation to deal with the subjects of circular economy through a multiactor approach and in a transverse and systemic manner.

If the word building means an isolated object, the term "built environment" implies thinking the interactions and synergies that take place both between the buildings themselves and between the buildings and their environment.

For example, in France, the "Club Métiers Deconstruction" of the association ORÉE (2015) has addressed, since the beginning of 2015, the challenges facing contractors on the subjects of the deconstructing buildings and infrastructures and of the Public Works waste (inert and finishing works) generated. The subjects that have already been explored are, amongst others, the best organization practices of the deconstruction sites, the traceability, the ecodesign of buildings in view of anticipating their deconstruction, and the economic interest of deconstruction (specifically reuse). Local authorities can use these guides to set prescriptions in call to tender for public buildings construction, considering the local conditions of demolition waste disposal and/or of natural resources availability.

On the one hand, circular economy provides ideas for solutions to reduce the impact of the built environment and thus participate in mitigating climate change. Indeed, in its principle, it aims at preventing and reducing the use of resources (therefore of the associated emissions), optimizing and closing the loops of materials and energy flows. It seeks to reproduce the principle according to which Nature produces no waste and implement it into human activities. In a global approach, circular economy deals with both direct and indirect emissions (Gray emissions).

The performance of the built structure can be anticipated according to four complementary criteria: sobriety, optimization of the building, modularity (which consists of changing the building to adapt it to the needs of users), and reversibility (which makes it possible to consider a recovery of material and soils). These complementarities enable a lesser consumption of resources and protection of environments in the short, medium, and long term. They enable a global vision of the building's life cycle, from design to use and maintenance phase right through to deconstruction. At this level, regional authorities can set orders to contractors through various prescriptions.

Circular economy and reuse within territories must be anticipated, and land should be reserved for this goal. This type of space faces an issue of acceptability: these are often storage areas, which are perceived as not being highly esthetic and a source of disturbance. However, these areas, which are somewhere between wasteland and reserve, have a great potential, under the condition that they are well designed. In circular economy logic, they can be a support for a variety of uses related to the sorting and reemployment of materials, contributing to the creation of wealth and employment on the territory.

33.6.1 Circular economy in French law

[Law No. 2015-992] of 17 August 2015 relative to energy transition for green growth (LTECV [Loi de Transition Energétique pour la Croissance Verte]) advocates the transition to a circular economy that aims at exceeding the current economic model by promoting a sober and responsible consumption of natural resources and primary raw materials. Circular economy implies the prevention of waste production by following the ranking of methods for processing waste, with special care to use methods, which range from reuse to a valorization of waste.

Decree No. 2016-288 of 10 March 2016 specified the measures relative to distributors of building materials, products and equipment for professionals taking back waste from the same types of materials, products, and equipment they sell.

Companies must be included in this process to reach the goal set by LTECV of 70% valorization of building materials between today and 2020.

Taking back of waste from the same types of building materials, products, and equipment must be organized by all the distributors of these materials whose production unit exceeds a surface area of 400 m^2 and a turnover of million euros. There are already examples of companies combining into interest groups to share processing and valorization areas on dedicated platforms.

In conclusion, there exist various levers to improve the environmental efficiency of buildings regarding climate change, thanks to circular economy approaches. In the building/demolition phase (sobriety, management of resources and environment, lifespans, etc.),

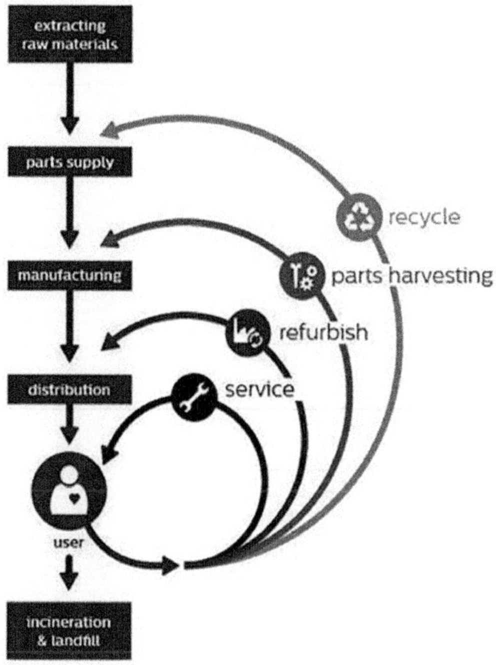

Figure 33.5 General organization of circular economy.

attention must be drawn to the need for a change of scale and the setting up of synergies and exchanges (Figure 33.5).

33.7 CONCLUSION

This section shows that the development of the recycling of concrete demolition waste to produce new concrete is a major component of sustainable construction, and beyond that of sustainable urban and territorial development.

The success of this development passes by the involvement of public authorities, foremen, contractors, construction and raw materials suppliers, and demolition and recycling companies. Places of discussion, exchange of good practices, practical guides, and tools for analysis of territorial and economic opportunities are necessary.

It is not possible to give the best solutions to reach the best play, considering the great number of actors. International and national political authorities must fix general objectives translated in directives. Local authorities have to take into consideration local opportunities to develop recycling, upper by the necessary managing of CDW and downer by the preservation of construction raw materials. They fix the frame of the dialog between local actors. The last level is that of the project. The contractor authority must give prescriptions to the general contractor. This actor must include in the project the use of RAC, after economic, territorial, and environmental analysis. This needs the participation and adhesion of building companies, materials suppliers, demolition and recycling companies, and concrete producers.

This book aims at contributing to the success of this development.

Section IX

Implementation of concrete recycling

F. de Larrard
LCR (LafargeHolcim Research Centre)

Previous sections provide the reader with a comprehensive review of what is known about concrete recycling, from an academia point of view. However, implementation and practical use goes beyond knowledge. In all countries, a number of legal and/or technical texts define what is permitted or advised in the construction sector and serve as references to all involved professionals.

Standards are first reviewed in Chapter 34. In the vast list of standards enforced in countries (published by International Standards Organisation (ISO), Comité Européen de Normalisation (CEN), or national organizations), some do not address concrete recycling but define processes which are potentially impacted by the use of demolition materials as aggregates or cement raw materials. Others allow and/or limit these practices. Depending on cases, adherence to standards is mandatory or is specified by the project owner. Even if it was not possible to provide in this book a comprehensive picture of concrete recycling standardization in all countries, the authors made their best to present the main general concepts and the most important European standards. This reading is particularly advised to all professionals who intend to go to recycling.

Besides standards, governments and local authorities publish texts that govern or promote the use of recycled materials. Voluntary systems as environmental certification can also greatly influence the degree of implementation of concrete recycling. Also, based on local markets and traditions, some countries have developed practices that can be duplicated in other places where less local experience is available. These topics are addressed in Chapter 35, with more details about the French and European contexts, closing the list of monographs that constitute this book.

Chapter 34

Standards

S. Fonteny
UNPG

W. Pillard
EGF.BTP

P. Francisco
CERIB

CONTENTS

Abstract

The definition of the characteristics of RA and of the final product they are used for are strictly described in documents as standards, regulations, guides or business references. Drafted and applied at various levels (International, European or local level), these documents are made to specify the requirements, to describe the methods, to ensure the provisions are achieved and to guarantee the good behavior of the final product. In Europe, in Asia or in America, the framework is quite the same but the specifications can vary depending on the history of use of RA and on the final destination. In Europe, works made with the EC (European Community) and the CEN (Comité Européen de Normalisation) allow countries to have a common basis with the Construction Product

Regulation and with product and test standards. Countries can draft and publish local standards to answer to national considerations. Furthermore, some other documents as guides can be drafted at a national level or business references, as private specifications, can be applied for specific works. Outside of Europe, as none organization as EC or CEN exist, each country had to develop its own normative corpus, with the exception of ISO (International Standard Organization). Product and test standards were drafted according to the ISO as European countries in CEN did. Documents can be similar but manners to consider the RA and to apprehend their uses and their behavior in the final concrete are different. Requirements are somehow different in terms of thresholds and codifications. Restrictions or authorizations of uses are also different. However, some specific topics are commonly shared as ASR (alkali-silica reaction), soluble sulfates; they can be considered as major concerns regarding the variability of RA.

34.1 INTRODUCTION

Standards are documents established in a consensual way between different actors and which aim to facilitate exchanges between them. In the construction sector, we find two important organizations: ISO for international standards and CEN for European standards. Their works are all published by national standard bodies (e.g., AFNOR in France or DIN in Germany). Obviously, they are not the only documents that can be used: national standards and other specific documents such as guides, recommendations, booklets...

With the awareness of the need to preserve natural resources in connection with the depletion of their access, the will to use stocks of RA and to develop a circular economy, some countries have drafted new national documents or updated existing ones to include the use of RA. These national requirements and provisions have sometimes served as a base for the draft or the revision of international standards. Indeed, it is clear that some countries have been proactive in integrating RA into their national application documents and/or in broader geographical application documents published by CEN or ISO.

One of the issues of the RECYBÉTON project was to determine to which extent RA are taken into account in France and, as well, at European and global levels. It is important at this point to make the difference between standards and regulations. Regulation are mainly composed of texts published by laws or decrees so that their use is mandatory in the country where they are enforced. As for standards, their use is voluntary.

34.2 DIFFERENT TYPES OF STANDARDS

In the field of construction and thus in the domain of concrete (mainly ready mixed concrete and precast concrete products), standards are more prominent as they constitute the major proportion of reference texts grouped into four categories:

- Specifications standards defining requirements for concrete, precast concrete products and for their constituents;
- Test methods to determine the characteristics of materials (in terms of value or thresholds);
- Standards for the execution of works which give the traditional way to execute structures and the associated quality-control system;
- Standards for design requirements and works design (called "Eurocodes" in Europe (EN 1992-1-1 2004)). Their purpose is to harmonize calculation methods for buildings and civil engineering works.

International standards (EN or ISO) are set by national standards bodies who can publish national standards on the same topics. These national standards should neither supplant nor contradict International standards. The interweaving of ISO, CEN and national standards and other texts can be outlined as shown in Figure 34.1.

34.3 RECYCLED AGGREGATES FOR CONCRETE

Specifications for RA are described through documents that define their constituents and performance classes in respect to the required characteristics. The characteristics measurement is made by application of the test standards. Other documents as guides/recommendations can complete the standards.

34.3.1 Specifications

34.3.1.1 In Europe

The use of recycled materials as aggregates, to produce new concrete or for a use related to concrete, is covered at European level by various standards published by CEN.

The first of these is EN 12620 "Aggregates for concrete". This standard, in particular through its version published in 2008 (EN 12620+A1 2008) (called EN 12620 in this chapter) specifies the characteristics that aggregates from recycled materials must fulfill for the production of concrete. The scope of the standard clearly states that it is relevant for the factory production control of aggregates from recycled materials. Furthermore, in clause 5.8, Table 20 of EN 12620 describes the types of constituents and the codes to use for the complete designation based upon the content of constituents (Table 34.1).

For each constituent (e.g., Rc) or authorized combinations (Rc+Ru and X+Rg), EN 12620 defines categories to use or to prescribe. E.g. category Rc_{90} means the RA is composed of at least 90% of concrete (≥90%). For Rcu_{95}, the whole content of concrete and unbound materials is not inferior to 95%. As well, this means that the other constituents don't exceed 10% for Rc_{90} and 5% for Rcu_{95}. For the reuse into new concretes, a minimal content of 70% for Rc and for Rcu (if prescribed) is obvious and even more depending of the class of new concrete produced. This is a matter treated in EN 206 and its national supplements.

(EN 13139 2002) "Aggregates for mortar" also allows to use recycled materials. However, the major difference with EN 12620 is the lack of any particular specification for RA.

The CEN TC104, Technical committee for the standardization of concrete, introduced in EN 206-1 and then in EN 206 requirements for RA within Annex E. More requirements are introduced in the National Supplement of EN 206 that European countries may publish.

The CEN TC229, Technical committee for the standardization of precast concrete products, introduced in (EN 13369 2004) requirements on RA within Annex Q.

34.3.1.2 National requirements for RA

The drafting of national requirements, complementary to the European provisions stated in the EN standards, is common. National documents are intended to provide detailed thresholds or levels, to pay attention to some characteristics or to endorse national practices not accounted by EN standards.

The additional specifications developed by European countries are logically linked to EN 12620 (Table 34.2) for the use in the concrete structure and intend to clarify requirements or to add local variations as the codification of RA linked to the constituents and their

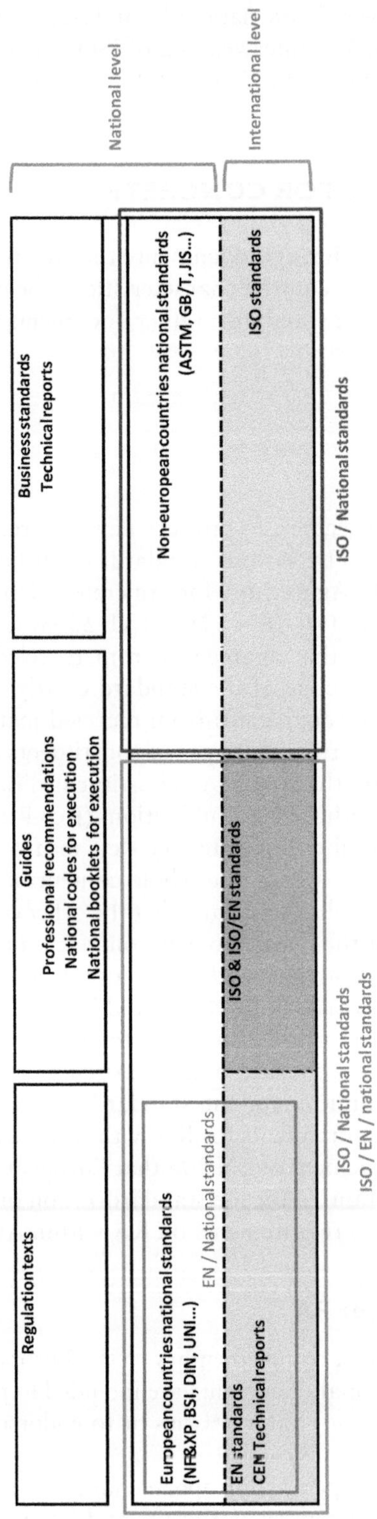

Figure 34.1 Interweaving of ISO, CEN and national texts.

Table 34.1 Types of RA and constituents

Code for constituent	*Materials*
Rc	Concrete as ready mixed concrete, precast concrete, concrete mansonry unit, mortar
Ru	Unbound materials: unbound aggregates, natural stone, hydraulically bound aggregates
Rb	Clay product as tiles and clay mansonry units, non-floating concrete
Ra	Bituminous mixtures
Rg	Glass
FL	Floating elements (indicated by volume)
X	Other materials as clay, sols, metals, wood, plastics, plaster, etc.

Table 34.2 National supplementary documents to EN 12620

Countries	*Supplementary documents to EN 12620*	*Other documents*
Austria	Standards (B3140 2013; B3131 2006) "rules of transposition for EN 12620"	Regulation texts: red directive and (Green Directive)
Belgium	None	(PTV 406 2003): RA—aggregates from concrete wastes, mixed wastes, debris, masonry and asphalt wastes—crushed stone and serious—aggregates for concrete, unbound mixtures and materials treated with hydraulic binders
France	(NF P18-545 2011): aggregates—defining elements, conformity and coding (NF EN 206/CN 2014): concrete	None
Germany	(DIN 4226-100 2002): aggregates for concrete and mortars—part 100: RA	None
Italy	UNI C88—aggregates for concrete—additional instructions for application of EN 12620—parts 1 & 2: part 1: designation and conformity criteria's & part 2: requirements	None
Netherlands	(NEN 5905+A1 2008): Dutch supplement to NEN EN 12620 "aggregates for concrete"	None
Norway	None	(NB26 2014): material recycling concrete and masonry for the production of concrete—Publication no 26
Spain	None	draft for regulation "EHE" (EHE Project 2003)
Switzerland	(SIA 2030 2010): concrete recycling	—
United Kingdom	None	Guide (PD 6682-1 2009): aggregates—part 1: aggregates for concrete—guidelines for BS EN 12620

origins (road, building…). These national specifications always aim to ensure suitability for use by the mastery of characteristics. Common specifications or requirements to test are dealing with:

- Soluble sulfates, although the measurement method may be in water or in acid,
- Constituents of RA (to avoid plaster for example),

- Water absorption (especially for fine aggregate),
- Components involved in ASR.

Usually, these specific requirements imply more frequent tests as compared to natural aggregates (NA).

34.3.1.3 Countries outside Europe

These countries do not apply EN standards for the specifications of aggregates and therefore have drafted specific corpus of documents based on their knowledge and experience. These documents contain similarities, some of them are inherited from a common history and from a shared area of influence (Commonwealth).

Due to the considered approach, the codification of the RA is very different from a country to another. It depends on the nature of the constituents or on the final uses:

- In Japan, density and water absorption are two criteria to classify RA;
- Australia has defined two classes and ten sub-classes or grades to specify RA for structural and non-structural concrete;
- In Brazil, the way to classify is close to the European way and some classes are very similar to Rc (concrete), Ru (unbound materials), X (others as metals, wood, plastics, etc.) and FL (floatings) but a new class is introduced for solvents and paints, constituents non taken into account in other countries;
- For Canada, seven classes are defined for RA: MR-1 to MR-7 according to different percentages of concrete residues, asphalt residue and NA.

Some thresholds are very different from European ones. A significant example is soluble sulfates in China where the threshold is 2.0% maximum while 0.2% or 0.8% are applied in Europe. However the methods used for the extraction of sulfates are different. Some characteristics are also monitored as in Europe, e.g., those dealing with ASR.

34.3.2 Test standards

As for product standards, the test standards can be international or national documents.

The consideration of RA in the test standards is intrinsically linked to the presence of specifications in the product standards and in other documents.

34.3.2.1 European test standards

As the four test standards drafted by the ISO Technical Committee for Concrete are too old to take RA into account, the EN standards are mainly used across Europe.

There are almost as many standards as characteristics to measure (even more if considered (EN 1744-1 2012) which integrates different methods to measure soluble sulfates).

For RA, Annex H of EN 12620 has to be taken as a reference for the dedicated characteristics to measure on RA intended to be used to produce new concretes and as well to determine which test standards are to be used.

For this purpose, standards were specifically drafted (e.g., (EN 933-11 2009)—Constituents of RA and (EN 1744-6 2006)—Determination of the influence of recycled aggregate extract on the initial setting time of cement), some have dedicated clauses (e.g., (EN 1744-1 2012)—soluble sulfates in water—clause 10.2 and (EN 1367-4 2008)—Determination of drying shrinkage—Annex A). Specific methods have to be completed to be fully applied to RA.

34.3.2.2 National, European test methods

The development of national standards is related to the lack of European standards on specific topics or the will to introduce specific methods based on national experience (Table 34.3). ASR may represent an example for which local standards are commonly present (as national specificities have to be used according to the product standards). In some cases, standards are drafted on a very specific and national topic (e.g., the French (NF P18-566 2014) standard for ombroscopy).

34.3.2.3 National standards in other countries

Three countries drafted test methods on specific characteristics or introduced RA in existing test methods (Table 34.4).

34.4 RECYCLED AGGREGATES IN CONCRETE

The use of RA, either in the concrete structure or not, cast on site or precast is made through standards, guides, specification documents and even regulations.

Table 34.3 National test methods

Austria	The water absorption should be performed on the fraction 4/32 mm
France	• (XP P18-544 2015) Determination of the alkali-soluble active lime water can be used to determine the potential of alkaline active release that may have a negative influence in concrete; • (NF P18-566 2014) "particle analysis—test using an ombroscopy device" is applicable to RA; • (NF P18-576 2013) can be applied for the measurement of sand friability coefficient
Germany	• The water absorption is measured over a period of 10 min and without the fraction less than 0.0125 mm (DIN 4226-100 2002) • Leaching tests are to be carried twice a year and many of these measures are carried out with German standards • Annexes D (determination of water absorption), E (determination of the content of acid-soluble chlorides), G (harmful substances), H (verification of resistance to freeze-thaw test for aggregates—test on concrete) & I (effects of RA on soil and water of (DIN 4226-100 2002) are to apply on RA
Spain	The draft EHE (EHE Project 2003) regulation introduces national tests applicable to RA for the Shape coefficient, Water absorption, Clay particles and Soft particles
United Kingdom	For concretes for roads and highways, a different method to (EN 933-11 2009) has been developed for the United Kingdom Department for the construction of roads and motorways—SHW)

Table 34.4 Non-European, national specific test methods

Australia	(AS 1141) methods for sampling and testing of aggregates
Quebec	(LC 21-901 2008)—determination of the composition of a recycled material containing residues of asphalt and cement concrete (similar to (EN 933-11 2009))
USA	• (ASTM C88 2013) standard "test method for soundness of aggregates by use of sodium sulfate or magnesium" is equivalent to (EN 1367-2 2010); • (ASTM C294 2012) "Descriptive Nomenclature for constituents of concrete aggregates" is similar to (EN 932-3 1996; EN 933-11 2009) • (ASTM C666/C666M 2015) for the resistance to fast freezing/thawing

Testing standards are also used to measure the characteristics of the concrete produced. These tests on concrete are made on fresh concrete and hardened concrete. So both states are taken into account in the test methods.

Test methods have been standardized for ISO as (ISO 1920-X). They can be published national standardization bodies as national documents. For the time being, RAC are considered as NAC in these standards.

34.4.1 Specifications of use for RA

34.4.1.1 Specifications of use of in the European application documents

The standards for the specification of structural concrete and precast concrete products are:

- EN 206 published by CEN. This document refers to EN 12620 and RA are treated at a national level.
- (EN 13369 2004) Common rules for precast concrete products. This standard refers to the above standard for the constituent materials of precast concrete products:
    - For structural concrete products (masonry units apart), product standards refer to (EN 13369 2004), which refers itself to the provisions of EN 206-1 for the constituent materials of concrete (RA included). National provisions valid in the place of use indicate how the informative annexes (annex E of EN 206 and annex Q of EN 13369) are used.
    - For other concrete products (including masonry units), the technical specifications (largely defined in harmonized European standards) of these concrete products are mainly translated in terms of performance requirements, verified by tests performed directly on the finished product for the production control. The characteristics of the raw materials are part of the documentation of production control.

34.4.1.2 National practices specifications (local standards, guidelines or regulations)

EN 206 permits its adaptation in CEN countries by the way of adding a national supplement. Denmark, France, Germany, Italy, Sweden and United Kingdom drafted such texts or guidance document for the application of EN 206. Regulatory texts or specific guides have also been published in some countries (Table 34.5).

34.4.1.3 Specifications of use in Asia, America and Oceania

As for the documents for specifications for RA, the concerned countries have developed specifications (Table 34.6) that clearly specifies the requirements for concrete produced with RA.

34.4.2 Test standards

34.4.2.1 Common test standards in Europe

In Europe, Specific test standards have been drafted for fresh and hardened concretes and published by CEN:

- Tests on fresh concrete: (EN 12350-X)
- Tests on hardened concrete: (EN 12390-X)

Table 34.5 National documents for the specifications of concrete produced with RA or for RA

Countries	*Standard (national supplement or adaptation of EN 206)*	*Other document*
Austria	None	Red and green guidelines to use with Ö Norm EN 206 (for the road and for works prefabricated) Green directive (for roads) is for quality certification towards the environmental compatibility of RA with three classes A+, A & B depending of the contents of elements in the eluates
Belgium	None	Specifications of use: (TRA550 2011) RA must comply with the specifications of the PTV 406 and have to be assessed by BENOR; Only crushed concrete Rcu>90% can be used.
Denmark	(DS 2426 2011): guide for EN 206-1 The overall capacity for RA as crushed concrete and crushed tiles, is established in accordance with DS/EN 12620	None
France	In (NF EN 206/CN 2014), table specifies NA.2 RA defines specific codes, linked to EN 12620, for: $Rcu_{95} \rightarrow CR_B$ $Rcu_{90} \rightarrow CR_C$ $Rcu_{70} \rightarrow CR_D$ Table NA.5, clause 5.13 specifies the maximum content by mass for coarse RA and fine aggregate depending of the exposure class of the new concrete. Furthermore for coarse RA, 3 types (1, 2 and 3) are defined	(Fascicle 65 2014) with a restriction on the origin of RA (which should originate in demolition of civil engineering structures) and with a limit for the strength class for the concrete
Germany	(DIN 4226-100 2002): national declination of EN 206-1 (DIN 206-1)	Guides for the use of recycled concrete in accordance with DIN EN 206-1 and (DIN 1045-2)
Italy	(UNI 8520-2 2005): aggregates for concrete—additional instructions for the application of EN 12620—requirements	Departmental Document of 2008
Netherlands	None	Recommendations (CUR 112 2012) (concrete made with coarse RA) & (CUR 106 2014) (concrete with fine particles of recycled as sand)
Norway	None	(NB26 2014) Guide "recycling of concrete and masonry materials for the production of concrete"
Russia	None	Recommendations published in 1987 for the use in concrete for road engineering. However, recommendations for the use of RA for all types of concrete are present
Spain	None	Spanish draft rules for the use of RA in structural concrete "(EHE Project 2003)—Spanish regulations for the use of recycled" It is interesting to note that RA are defined not by their constituents but by characteristics such as fine content, shape, water absorption, resistance to fragmentation...

(*Continued*)

Table 34.5 (Continued) National documents for the specifications of concrete produced with RA or for RA

Countries	*Standard (national supplement or adaptation of EN 206)*	*Other document*
Sweden	(SS EN 137003 2008): national complement to EN 206-1	(BA 99): Swedish Board of Housing, Building and Planning: Additional manual to the one for the design of structures
Switzerland	None	(SIA 2030 2010) technical specification defining the specifications of RA as well as the requirements for the use in concrete
United Kingdom	(BS 8500-2:2006+A1:2012) concrete. Complementary British Standard to BS EN 206-1. Specification for constituent materials and concrete: national supplement	None

Table 34.6 Specifications for the use of RA in concrete in international countries

Countries	*Standard*	*Other document*
Australia	None	Guide (HB 155 2002): guide to the use of recycled concrete and masonry materials
Brazil	None	National rules 307/02 CONAMA
Canada—Quebec	(NQ 2560-600 2002)	None
China	(JGJ/T 240 2011) technical provisions for the use of RA (GB/T 14902 2012) ready mixed concrete	None
Japan	Requirements for RA H (JIS A 5021 2011), L (JIS A 5023 2012) & M (JIS A 5022 2012) and their use in concrete: A fourth standard (JIS A 5308 2014) for the "classic" concrete allows the use of H-type aggregates under certain conditions	None
New Zealand	None	Best practice guide for the use of RA in new concrete
USA	None	Technical report (ACI 555R-01 2002) Document (ACI E1-07 2007)

In the current state of these standards, there is no specific test to concrete incorporating RA. In EN 206, RA within the specifications described in its clause 5.1.3 are deemed suitable for the production of concrete. It is thus the concrete material as formulated which is subject to testing. In EN 206-Annex A, clause A.4 indicates to consider the necessity to perform tests to determine the drying shrinkage, creep and module elasticity when the produced concrete contains RA. Furthermore, the retention time of the consistency of concrete made with RA has to be monitored as well as the influence of RA by using the existing test standards for the concrete material.

34.4.2.2 *Specific national test methods*

As for aggregates, many countries have drafted methods to test local characteristics listed in supplements or national application guides of EN 206 (e.g., resistance to internal frost in the air and in the water). Below, Table 34.7 indicates examples for France.

Table 34.7 Specific French test standards for concrete

Test standard for fresh concrete	*Test standard for hardened concrete*
workability flow test—(NF P18-452 1988)	• Dimensional changes—(NF P18-427 1996) • Test methods on reactivity to alkalies—(NF P18-594 2015) • Internal freezing—(NF P18-424 2008; NF P18-425 2008) • Scaling—(XP P18-420 2012) • Surface hardness—(P18-417 1989)

34.4.3 Calculation and design works

CEN has drafted a substantial number of documents for the design of structures named "Eurocodes". Eurocode 2 does not address RA in any of its various parts. Eurocode 6 is a reference to the applicable product standards. In fact, it can be accepted that RA are allowed as long as these products standards apply to these materials.

Some non-European countries have developed or keep national documents on the design of structures (Table 34.8). For example, China drafted a standard JGJ/T 240 "Technical specifications for the use of RA".

Table 34.8 Examples of international documents for the design of structures

Countries	*Document*	*Comment*
China	(JGJ/T 2040 2011) "technical specifications for the use of RA"	The law of variation of the elastic modulus depending on the resistance class is modified with respect to the common rules in China (GB 50010) when type II or III aggregates are used
Denmark	(DS 411 1999) standard "structural use of concrete"	The document limits the use of RA by the introduction of additional requirements in the standard (DS 2426 2011)
Spain	(EHE Project 2003)—draft of Spanish regulations for the use of recycled aggregates	Some properties are being applied correction factors: tensile strength, elastic module, shrinkage and flowing
Norway	(NS 3473 2003) standard "concrete structures—design and retailing rules"	Application of conventional design rules below some percentage of substitution of the document (NB26 2014) In addition, some adjustments are taken into account for lower densities and their impact on the resistance, withdrawal…
Netherlands	Provisions (CUR 112 2012)	Below a percentage of 50% of coarse RA substitution in volume, the rules for calculation and design remain unchanged. The coarse RA must comply with density requirement and maximum rate of substitution according to the exposure classes. The concerned strength classes are restricted (C12/15–C35/45) Beyond 50% substitution by volume, a correction factor has to be applied for the modulus of elasticity, creep and the shrinkage
Switzerland	Technical specifications (SIA 2030 2010)	In design calculations, for RC-M concretes, the values of deformation of the law for the concrete performance are changed.

34.4.4 Execution of works

Documents for the execution of works reflect the quality-control requirements defined to ensure the proper execution of the work including material specifications, independent internal and external controls on materials, products...

At a European level, (EN 13670 2009) standard on the execution of concrete structures is published, however this standard does not mention the RA.

Regarding roads, (EN 13877-1 2013) standard: Concrete pavements—Part 1: Materials requires aggregates complying with the applicable EN 12620 or with national provisions. That indirectly may allow the use of RA.

For national documents, France has different documents which allows the use of RA under specific conditions:

- The fascicles of the CCTG (General Technical Conditions of Contract) no 65 for the "Execution of civil engineering works in reinforced or prestressed concrete" revised in 2014 (Fascicle 65 2014) (and published on 2018) introduces the possibility to use RA from deconstruction of engineering works. Only RA type 1 (according to (NF EN 206/CN 2014)) resulting from the deconstruction of structures and whose traceability is ensured can be used for concretes with a strength class of less than C35/45 in class XC1, XC2, XC3, XC4 or XF1 with a maximum substitution rate of 20%.
- Other fascicles directly or indirectly indicate the possible use of RA (e.g., (Fascicle 29 2006) "Construction of concrete tanks" and fascicle 29 "Execution of roads coverings and public spaces in modular products") provided their performance is equivalent to the one of natural materials.
- (NF P 98-170 2006): Concrete Pavements of cement-Implementation and control indicating that aggregates must comply with the NF EN 12620 and (NF P18-545 2011) standard.
- Table 34.9 indicates examples of references documents drafted by public or private companies in France: SNCF (National Railways Company) and EDF (the major company for Electric services).

The first common and important point to note in these two examples is the reference to standards in order to have a basis for the specifications and test methods. This reference is somehow a guarantee to avoid specific requirements. The second common point is the restriction of performance classes for the RA used and concretes produced with these aggregates.

Table 34.9 Example of business references in France

EDF	General technical conditions of contract for structural works—no 91.C.020.04 (CCTG EDF 2010)	Aggregates must fulfill the requirements of the French standard XP P18-545 (now (NF P18-545 2011)) The aggregates fulfill, at least, the B codes for specifications (according to NF P 18-545 standard). The freeze—thaw resistance is required only for XF3 or XF4 concretes according to NF EN 206, and unless additional requirements
SNCF	(IN 0034 2011): execution of works with reinforced concrete and prestressed concrete	The crushed concrete aggregates and the aggregates recovered from the process water or from the fresh concrete, and treated in a washing/screening plant are allowed for the production of strength class concrete ≤C20/25 as long as the percentage of substitution doesn't exceed 5%

34.5 CONCLUSION

The use of RA to produce a new concrete needs to be specified in a framework composed of several types of documents.

In the first type, product and test standards for aggregates define the suitable characteristic for RA and the way to measure them for their final uses. Then other product standards are needed to specify their use in the final product (in a ready mixed concrete, in precast concrete, etc.). Then test standards are needed to measure the impact of RA on the characteristics of the final product, but most often the tests used for products incorporating NA are unchanged.

Other documents as regulations, texts for design and execution of works or local business references can be part of the framework for the use of RA in a concrete to make a civil work, a building or a road.

Some exceptions apart (e.g., test methods for RA (EN 933-11 2009) and (EN 1744-6 2006) or specifications for the use in Japan and China), most documents do not specifically address recycled aggregate concrete (RAC). Only few clauses concern RA as they are considered as another source of aggregates.

Considering all documents collected in various countries and examined, the first point to note is that RA are considered in the product and test standards for aggregates and concrete. In Europe, this is the result of the works made under the CEN and this is the basis for a common language. But, according to the history of use and national experiences, local documents are unavoidable and somehow necessary to ensure the quality of the civil works and the buildings. For countries out or Europe, the scheme is quite the same as each country developed its own normative corpus to achieve the same results and respond to the same needs.

Finally, the situation differs also from a country to another in terms of practice, barriers and incentives. These aspects are reviewed in the next chapter of this book.

Chapter 35

National practices

Regulations, barriers and incentives

S. Braymand
Université de Strasbourg

W. Pillard
EGF.BTP

R. Bodet
UNPG

P. Francisco
CERIB

CONTENTS

Abstract

The various chapters of this book highlighted that the use of recycled aggregate (RA) in concrete is a mastered technique since the sector of recycling is enough structured (in terms of RA quality and traceability of controls). Beyond these technical aspects, the degree of dissemination in the various countries and even within the same country is rather diverse, and mostly is linked with the demand in RA for road applications. So, to promote the recycling of aggregates for concrete structure requires to investigate in the statutory aspects and in the local initiatives. The regulations deal generally with the recycling through a more global approach which is the one of the waste management and circular economy. The voluntary initiatives developed in the labels of environmental certification constitute a very interesting approach because they propose complements to the simple statutory approach. However these certifications define through a system of notation, different levels of environmental performance including various criteria: the criterion "waste," remains quite general, and does not include specific targets as "use of RAC", except for the Swiss approach (Minergie-Eco), therefore, the consequence is not too much incentive. It might, however, be decisive for the acceptance of offers and this is an aspect to be strengthened. More incentive measures can be envisaged such as the landfill taxation for CDW (Construction and Demolition Waste Materials), or to provide for a clause in public contracts that values tenders including recycled aggregates.

35.1 INTRODUCTION

The planet produces between 3.4 and 4 billion tons of waste (Seghier 2009), of which 1.7–1.9 billion tons of municipal waste, 1.2–1.67 billion tons of non-hazardous industrial waste (including construction wastes) and 490 million tons of hazardous industrial waste. Of the 225–300 (BRGM 2010; SOES 2017a; ADEME 2016a) million tons of construction and demolition waste produced per year in France (225 MT in 2014), only a part of the concrete waste is recycled, mainly for road works. In 2014, 17 million tons of pure demolition concrete waste were produced and 76% of the total volume of inert waste are used primarily for road works or quarry backfill (CERC 2018; UNPG 2016); 62% of the total inert waste, including the majority of concrete is recycled for road work (Stassi 2016). The remnant stays in deposit sites. It is estimated that more than 90% of pure concrete waste is recycled as it is the "noble" and most wanted part of inert waste. On the other hand, 21.4 millions of recycled aggregates are produced by recycling plants (UNPG 2016). Although this volume (21.4 Mt) represents only 6.5% of the French national production of aggregates, it must be emphasized that this quantity does not include the part of recycled materials directly recycled on work site (71 Mt in 2014) (SOES 2017a). Furthermore, the building sector that has higher requirements than the road construction, still remains a marginal destination for RA in France while some countries have developed real networks (Puttallaz et al. 2016). High quality recycling is not developed because of the low actual demand for RA as constituents of structural concrete for buildings (Bougrain et al. 2017).

Recently, France has adopted the framework provided by Directive 2008/98/EC (European Commission 2008) on waste to promote the circular economy. The aim is to achieve a 70% (by weight) recycling target by 2020 for non-hazardous CDW while the valorisation percentage (reuse and recycling) is still around 60%. However, the use of RA for construction road doesn't encourage the operators to opt for solutions with added value like it is suggested by circular economy principles (Le Moigne 2014, 2018); especially the objectives of public command (article 79 of the energetic transition law dated August 17, 2015).

Based on these observations and steered by the European objective (available in French Energetics' transition law) as well as by environmental and economic imperatives, the operators of the construction sector are encouraged to modify this situation. To increase this rate of recycling in concrete, in addition to the technical possibilities noted in the previous sections, regulatory and economic obstacles were identified. It was also necessary to identify existing incentives for the development of concrete recycling to subsequently propose new development.

To achieve this objective, an analysis of national French, European and International texts of mandatory or optional application (technical guides and environmental quality certifications) are presented in this section in addition to Chapter 34 which deals with standards. It was necessary to propose a map of countries (at European and worldwide level) for which the use of RA into concrete has been developed.

Since the incentives or the barriers are not limited to regulation texts or voluntary approaches, the impact of various public policy measures that could be envisaged in France is finally presented. This framed this analysis in the French economic and political context.

35.2 REGULATION TEXTS

35.2.1 French legislative and regulatory codes

French legislative and regulatory codes dealing with management of construction wastes and management of recycling were studied and were classified in order to evaluate the impact of their application: an incentive or an obstacle to concrete recycling (Bodet et al. 2014) (Table 35.1).

It can be noticed that none of the French codes directly deals with RA in concrete but, more generally, these codes deal with wastes, their storage and their recycling.

35.2.2 European and International legislative and regulatory codes

Among the European and International legislative and regulatory texts, two national legislative documents deserve to be raised (Bodet et al. 2015).

- The British WRAP protocol, United Kingdom (NIEA 2013).
- The Swiss Directive OFEV (2006), Switzerland (OFEV 2006).

In United Kingdom, the suitability of RA from inert waste must be assessed in accordance with the requirements of the WRAP Quality Protocol (Waste & Resources Action Program) (NIEA 2013). This document sets the requirements for the classification of recycled materials as well as the assessment of their compliance. It sets criteria for end of waste status, within the meaning of Article 3 paragraph 1 of the Framework Directive of the EU on Waste (2008/98/EC) (European Commission 2008). It sets also criteria for the production and use of aggregates from inert waste for use in road techniques. In this way, the recycled material stops to be a waste as soon as it is demonstrated that the product complies with the WRAP Quality Protocol. Producers must demonstrate that all these criteria are met. WRAP protocol specifies the conditions for which wastes, considered as inert waste (Table 35.2), are suitable for the production of RA. This protocol considers that recoverable raw materials can be used as aggregates for concrete (used for road application) under the condition of being

Table 35.1 Legislative and regulatory codes

Text	*Article*	*Subject*	*Incentive or obstacle*
Code of environment (Code of environment 2018)	L541-14-1 R 541-41-1 à 18	Regulatory framework of departmental plans for prevention and waste management	Incentive
	L515-3	Regional quarry plan—sustainable management of mineral resources made from recycled materials	Incentive
	L 541-2 L 541-7, L 541-43	Definition of the responsibilities of the producer and the holder of the waste	Both
	L 541-2	Obligation for the contracting authority to transmit to the company any useful information (before the execution of the works) to valorize or to eliminate the waste	Incentive
	2010-1579 ordinance completing dispositions in L542-2 to 46	Demolition materials and even more deconstructed concretes coming out of the site from which they are produced get a waste status	Obstacle
	Order dated June 19, 2015 completing D541-12-14	Quality management system for waste status	Both
Code of public works (Code of public works 2006)	Article 5 completed by an ordinance dated July 23, 2015	The contracting authority must evoke the risks and must characterize the materials contained in the building, to indicate the presence of dangerous products	Obstacle
Code of public health (Code of public health 2018)	2010-1579 order	Prohibition for the manufacture to use building products materials and concrete waste contaminated or likely to be	Obstacle
Code of work (Code of work 2018)	L 4531-1	Obligation for the contracting authority to assessing the health risks related to asbestos	Obstacle
Order dated February 29 (Code of environment 2018)	Article 1 & 2	On recycling site, keeping of an admission register and a sales register of deconstruction materials	Incentive
Code of customs (Code of customs 2018)	Articles 266, 268, 285 Bulletin dated April 18, 2016	General tax on polluting activities (TGAP): the facilities for treatment, storage or transit of demolition materials are not subject to the general tax on polluting activities	Incentive
European directive (European Commission 2008)	2008/98/CE "Waste framework"	Conditions for an exit from waste status	Obstacle at time, could evolve

Table 35.2 Acceptable inert waste input materials—WRAP Quality Protocol—case of construction and demolition waste

Type and restrictions	*Waste code*
Concrete (must not include concrete sludge)	17 01 01
Bricks	17 01 02
Tiles and ceramics	17 01 03
Mixtures of concrete, bricks, tiles and ceramics other than those mentioned in 17 01 06	17 01 07
Mixed construction and demolition wastes other than those mentioned in 17 09 01, 17 09 02 and 17 09 03	17 09 04

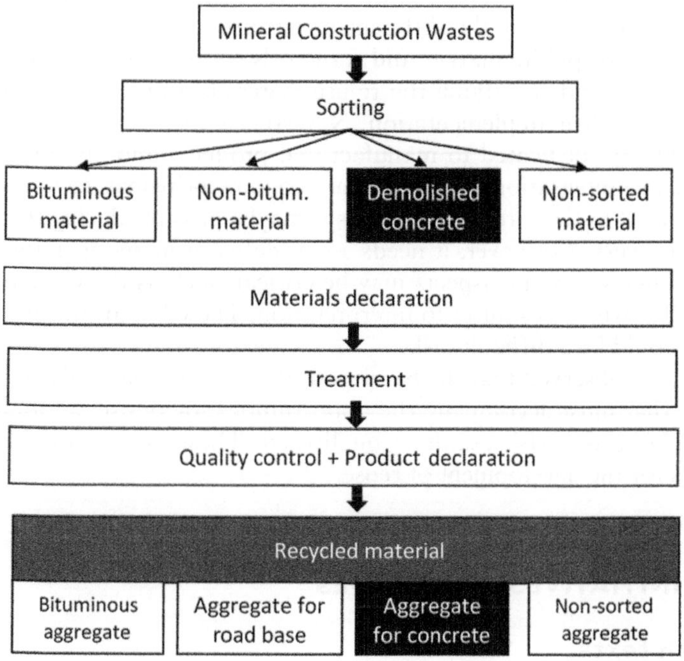

Figure 35.1 Recycling of mineral construction waste— With permission of OFEV (OFEV 2006).

approved by industry standards and Factory Production Control. Table 35.2 (NIEA 2013) lists all the input materials and their relevant "waste code"[1] or European Waste Catalogue (EWC) code considered inert and acceptable to produce RA under this Quality Protocol.

The Swiss Directive for the recovery of mineral construction wastes (OFEV) (OFEV 2006) defines the different categories of construction wastes and their range of utilization as well as the methods for recovering mineral construction wastes. Four material types are classified: bituminous aggregate, road base, aggregate for concrete, non-sorted aggregate. The Quality Protocol for their reuse acceptance is also defined (Figure 35.1).

35.2.3 Feedback on barriers and incentives on the reclamation and reuse of construction materials and products: framework documents review

In order to present feedback on the reclamation and the reuse of construction materials and products in the building and public works sector, ADEME published a report (ADEME 2016b). The levers identified to lift the constraints on reclamation were identified. Even though this study does not concern the recycling of construction materials but the reclamation by reusing construction materials and products without transformation, the identified brakes and incentives are similar to ones that concern the recycling of concrete into concrete.

It was observed that few legislative and regulatory texts explicitly deal with the reuse of building materials and products. Waste prevention and management regulations aim

[1] "Waste code" refers to the six-digit code for a type of waste in accordance with the List of Wastes (England) Regulations 2005, List of Wastes (Wales) Regulations 2005 and List of Wastes (Northern Ireland) Regulations 2005, as amended. Where it refers to hazardous wastewaste, the code includes an asterisk.

to encourage reuse initiatives. However, the difficulties of understanding the status of the material (waste or product) by the different operators, makes this encouragement rather inefficient. The code of public market and standards of the private market allow the possibility of prescribing and specifying the re-use materials and the re-use products as well as the modalities for their implementation. Nevertheless, for the building sector, most of the technical texts are dedicated to manufactured products and almost not applicable to re-use products. The integration of the circular economy principles (eco-design of the parts of the work, reuse...) in these regulatory texts is maybe possible during their revision but it remains a long process. Moreover, it needs a consensus of many operators with generally some diverging interests. Some aspects may be critical, such as the Construction Products Regulations (CPR), which is subject to interpretation. The CE marking and the declaration of performance could be a difficult task.

Eventually, it was observed that the barriers to the reuse of materials and products relate to the vigilance that must accompany the opportunities for re-use: aptitude for the use of products, control of claims risks, sanitary quality, etc. These aspects, that could be barriers, are prerequisites for the development of reuse.

35.3 LOCAL INITIATIVES AND GUIDES

35.3.1 French texts

At present, there is no French national text giving guidelines for recycling concrete into concrete. This is one of the aims of RECYBÉTON. However, some regional and national guides dealing with the use of RA for road application were considered as examples for analyzing an appropriate methodology (Bodet et al. 2014). This methodology could be transposed to the case of the recycling of concrete into concrete.

CEREMA (CEREMA 2016), proposed a guide on "Environmental acceptability of alternative materials in road engineering". This guide was elaborated on the basis of several regional guides (ADEME 2011b; Destombes 2003; OFRIR 2016; CEREMA 2014).

This guide evaluates a method following several steps:

- Production of alternative materials
  - Characterization and reception of deconstruction materials
  - Treatment of the alternative material
  - Production of the road material
- Definition of the range of use according to the level of exposure to meteoric waters.
- Definition of the limitations for the uses in accordance to the environment and to the manufacturing of the road material.
- Environmental Quality Protocol
  - Environmental conformity: pollutant content (eventually by lixiviation test) (NF EN 12 457-2 2002)
  - Test frequency
  - Set up and stock management
  - Traceability

35.3.2 European and International texts

The French Federation for Building (FFB) reported a European Benchmark on the management of construction waste from building site (FFB 2017). This benchmark feedback lead to propose improvement tracks and recommendations. These recommendations are available in several steps:

- Upstream of the site
  - Educate and train project management teams and project management
  - Improve knowledge about waste
  - Organize waste management
- On construction site
  - Improve sorting practices and site logistics
- Downstream of the site
  - Promote the recycling of inert
  - Improve sorting performance in the center
  - Consolidate the mesh of collection and treatment points
- Develop the reuse of building materials and products
- Develop the use of recycled materials
  - Gain the trust of users
  - Reinforce the prescription of recycled aggregates
- Write an operational sorting guide
- Promote a logistic adapted to sorting at source
- Train staff in sorting: training should integrate economic and environmental benefits associated with efficient waste management to motivate people

Anyway, most technical and guidelines texts consist in applications of standards (Bodet et al. 2015). They are considered in Chapter 34. However, three examples are underlined below:

- Case of Switzerland: ECO-CFC guide (Green building sheets according to the code of construction costs), 2011 (ECO-BAU 2011); Application Guide for the Use of Mineral Recycling Materials, 2016 (Puttallaz et al. 2016).—KBOB recommendations (Co-ordination Conference for Construction Services and Buildings of Public Owners), 2007. (ECO-BAU 2007)
- Case of Australia: Guide to the use of recycled concrete and masonry materials, 2002 (Standards Australia 2002).
- Case of United States: Technical report from ACI, Removal and Reuse of Hardened Concrete, 2001 (ACI 555R-01 2002).

The Swiss application guide for the use of mineral recycling materials (Puttallaz et al. 2016), (that is new and extended version of the KBOB recommendations (ECO-BAU 2007)) detailed recommendations for recycling concrete (concrete with at least 25% of RA, under this rate, concrete is considered as "normal"). Around 7% of the Switzerland concrete production (1 million of m^3) contain recycling materials. Considering properties of recycling aggregates concrete and the defined terminology according to the SIA 2030 Technical report (SIA 2010), two type of recycled concrete are proposed: RC-C (at least 25% of recycled concrete aggregates) and RC-M (at least 25% of recycled mixed aggregates). Recommendations are detailed as follow:

- The roles of different operators (client or contracting authority, project manager, authorized representative, contractor) are detailed by step of the operation. For example, the client set the used products (RC-C, RC-M...) during the project phase.
- Data sheets summarize the range of use according to the typology (according to SIA 2030) and the recycled materials characteristics.

- When the client requests the use of RAC, the particular conditions of the request for proposal documents must contain all the useful information.
- Environmental and technical data must comply with certifications.

These recommendations are reported in the Green building sheets according to the code of construction costs (Eco CFC (ECO-BAU 2011)).

Otherwise, according to the SIA 2030 Technical report, Switzerland allows adding RA to natural aggregate, and a declaration is mandatory only for replacement rates higher than 25%. Indeed, under a rate of 25% of RA, concrete is considered as a normal concrete NC. This provision greatly facilitates the re-use/recycling of a good part of construction and demolition waste in the Swiss market.

The Australian "guide to the use of recycled concrete and masonry materials" is based on Japanese experience (Vivian 2009; Standards Australia 2002). Obstacles and incentives for recycling are identified.

The main obstacles are:

- The development of infrastructure for recycling requires a significant investment at the beginning while the private sector is more concerned with short-term return.
- The regulations and texts are unsuitable for the specific use of recycled materials (no investment).
- A lack of experience feedback is noted.

Some incentives are proposed (see below), they concern public administration and enterprises. They mainly consist in general recommendations providing guidelines and focusing on the main obstacles listed above:

- Define a legal framework for recycling in structural concretes;
- Define more precisely a classification of recycled aggregates;
- Improve the organization and management of companies in the direction of recycling;
- Have better control by the administration of the quantities of old concrete on the demolition sites;
- Reduce the amount of concrete waste by increasing the cost of landfilling;
- Financial support for recycling by the government to reduce high investment costs.

The technical report from ACI on "removal and reuse of hardened concrete" summarize the state of the art and propose limits for part of usual characteristics of recycled aggregates (ACI 555R-01 2002). The maximum admissible quantities of plaster, clods and other impurities (densities $<1,950\,kg/m^3$) are limited to $10\,kg/m^3$. The maximum admissible quantities of particulates (greater than 1.2 mm size) of bitumen, plastics, wood, paper and other similar types of particles are limited to $2\,kg/m^3$. Regarding concrete manufacturing, this ACI report recommends avoiding the use of the 0–2 mm recycled fraction and recommends presaturation of recycled CA before mixing.

35.4 CONTRIBUTION OF ENVIRONMENTAL QUALITY CERTIFICATIONS

The appellation Green Building has become a trend-setting movement in the construction and real estate industries worldwide (Pillard et al. 2018). Assessing the quality of a construction by a sustainability labeling is related to environmental, ethical, economic, social and commercial considerations. The benefit for a building owner is to objectively prove that the construction has a sustainable value and that the company reaches social responsibility objectives.

Methods for environmental quality certification aim to limit the environmental impact of construction while improving the comfort and the quality of life of the occupants. They consist in the optimization of choices for materials and equipment (criteria related to ecology, energy optimization, rational management of the building, etc. ...).

Some targets (objectives) are set by impact with some prerequisites (or some minimum criteria) required for each of these targets as a basis for certification. Then, evaluation is processed by target and notes (or performance levels) are determined. The certification is obtained by an overall score for the project.

The environmental quality procedures apply during the various phases of preparation, implementation and management of operations.

35.4.1 Main certifications

Many of the certifications around the world are based on the recognition of the GBC (Green Building Council, an international organization including national members (GBC 2018)). There are currently 72 Green Building Council members around the world, each at different stages of their own green building journey (see Figure 35.2). There are three membership levels for national Green Building Councils: established, emerging or prospective. These levels depend on the different stages of their own green building journey.

Some of these certifications processes (and others not included in BGC) were studied by (Pillard et al. 2018) in order to identify levers to recycle waste construction materials into concrete. Most of European certifications were analyzed (Table 35.3).

35.4.2 Incentive certifications

At first, a state of the art of European practices highlighted the diverse nature of the socio-economic context of construction (Pillard et al. 2018).

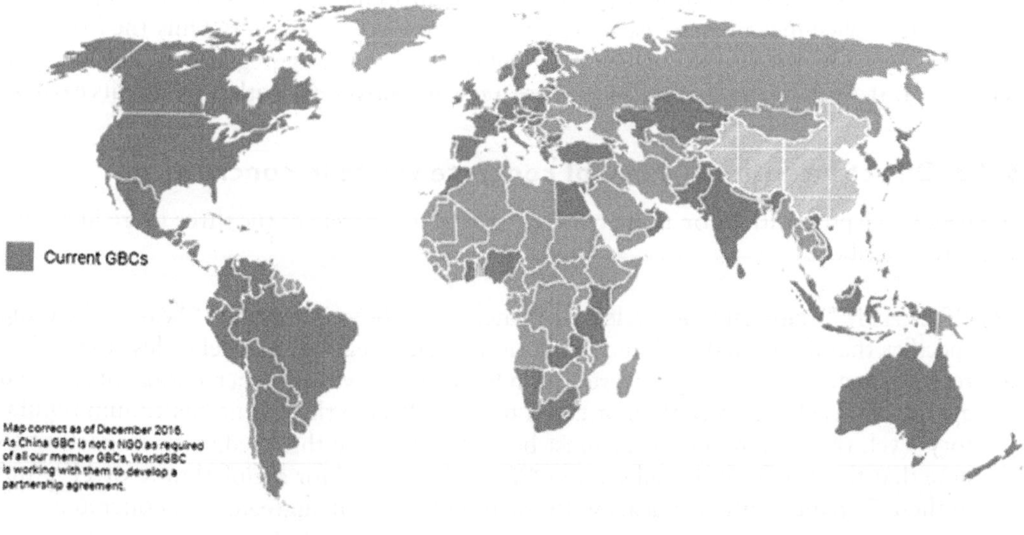

Figure 35.2 Green Building Council members around the world. (MENA: Middle East and North Africa.) With permission of World Green Building Council (GBC 2018).

Table 35.3 Analyzed environmental quality certifications

Label	*Country*	*Year*	*GBC*
NFQE (CSTB 2015)	France	1996	Yes
HQE-BD sustainable building (Certivea 2016)	France	2016	Yes
BDM & BDF sustainable building (BDF 2017; BDM 2008)	France, Mediterranean and Île de France Regions	2008/2017	No
BREEAM (BREEAM 2016)	United Kingdom+several adaptations in the world	1990	UK: partnership Other countries: yes
DGNB (DGNB 2016)	Germany	2008	Yes
VERDE (VERDE 2016)	Spain	2002	Yes
VALIDEO (VALIDEO 2001)	Belgium	2011	No
MINERGIE-ECO (MINERGIE 2016)	Switzerland	2008	No
LEED (LEED 2017)	United States	1998	Yes

Table 35.4 Incentive quality environmental certifications

Incentive on increasing sorting of wastes and valorization network (upstream)	BREEAM-LEED-NFQE-HQE BD-DGNB VALIDEO-BDM
Incentive by life cycle assessment performance Responsible materials/material reuses	BREEAM-LEED-HQE BD-DGNB VERDE-BDM & BDF
Direct and explicit incentive by use of concrete waste in concrete	BREEAM (1 subsection) MINERGIE-ECO (exclusion criteria) VERDE (low points)

A comparative analysis of the certifications showed that there is a strong will to develop the circular economy. However, most of the time its implementation consists of a performance-based and global approach (requirements such as results of LCA[2] for building or such as responsible choice for material sourcing) or in an incentive to sort waste upstream. There are no explicit incentives to use recycled aggregates (see Table 35.4). Only the Minergie Eco label explicitly proposes a minimum quantity of recycled aggregates in concrete (see Section 35.4.3). Breeam and Verde also propose a few incentives but with poor credits earned.

35.4.3 Direct incentives by use of concrete waste in concrete

Here bellow the propositions on using concrete waste in concrete of the three more incentive labels are described:

- *BREEAM*: "Aggregates Recycled in Concrete" subsection of the "Waste" heading specifies that if national building regulations permit regulatory level of less than 50% recycled aggregates, exemplary credit can be obtained when the percentage of recycled aggregates used is greater than or equal to 35%. Where there is no maximum regulatory level, the 50% requirement must be met to provide this credit. It's important to note that the weight of the subsection "Waste" is only 7 for a global note of 119, and in these 7 credits, only one deals with the use of recycled aggregates in concrete.

[2] Life Cycle Assessment.

- *VERDE*: A first interesting criterion (in "Natural resources" heading) is based on the calculation of the recycled materials cost percentage compared to the total cost of the used materials, and this whatever the recycled material is. The aim of this approach is based on the fact that the use of recycled materials must be done in a global vision, so that more the costs are high less the use is interesting. Two thresholds are used: 10% and 20%. "Recycled concrete with a strength not exceeding 40 N/mm^2" subsection presents a more incentive criterion. This qualification (40 N/mm^2) for concrete implies that recycled aggregate are not used for high performance concrete. It allows to obtain credits which can, represent a maximum of four points on a total of 40. In the best case the use of "recycled concrete with mechanical resistance less than 40 MPa" will represent 10% of the global note.
- *MINERGIE-ECO*: Minergie-Eco, established in 2006 in Switzerland, complements the Minergie certification, which did not address the issue of choice of materials and environmental impacts. Minergie-Eco fosters a low impact on the building environment and the preservation of resources, from construction to demolition. Minergie-Eco comprises an exclusion criterion for applicants who do not comply with concrete recycling provision. Local lack of natural aggregates in several cantons and the competitive price of recycled materials justify the existence of this criterion. Between all the proposed criteria there is in this label a clear incentive since the criterion "raw materials widely available and higher share of recycling material" refers unambiguously to that requirement. A set of requirements relate to the local availability of the resource. Indeed, it should be noted that for an operation to be labeled Minergie-Eco, it is imperative that 50% of the concrete parts are concrete made of recycled concrete (designed by RC-concrete), according to technical specifications SIA 2030. Figure 35.3 shows the parts of the structure in which it is possible to use RAC. On the other hand, if the distance between the concrete plant and the site exceeds 25 km or if the cost of recycled concrete is much higher than that of conventional concrete, an exception may be granted.

35.5 INCENTIVES: NATIONAL EXPERIENCES AND PROSPECTIVE STUDIES

Among the incentives identified previously (see Section 35.3), some of them has been deepened to simulate the effect of these potential measures on the volumes of RA that could be recycled in concrete. Then, in order to reinforce the validity/interest of these measures, the conclusions of ADEME study (ADEME 2016b) on the identification of the brakes and levers for the reuse of building products and materials, and the case of the political measures in Switzerland are presented.

35.5.1 Impact of potential public policy measures in the French context

Recycling concrete into concrete represents a real benefit but this network is underdeveloped today. Therefore, public policy appears as one solution to spur the demand for competitive RA (Bougrain et al. 2017).

In this framework, following the incentives detected previously, several public actions have been considered. Based on construction operations statistical data applying to France, the share of concrete construction methods has been estimated. This initial hypothesis has served as a basis for the simulations calculated subsequently, the conclusions of which are given below.

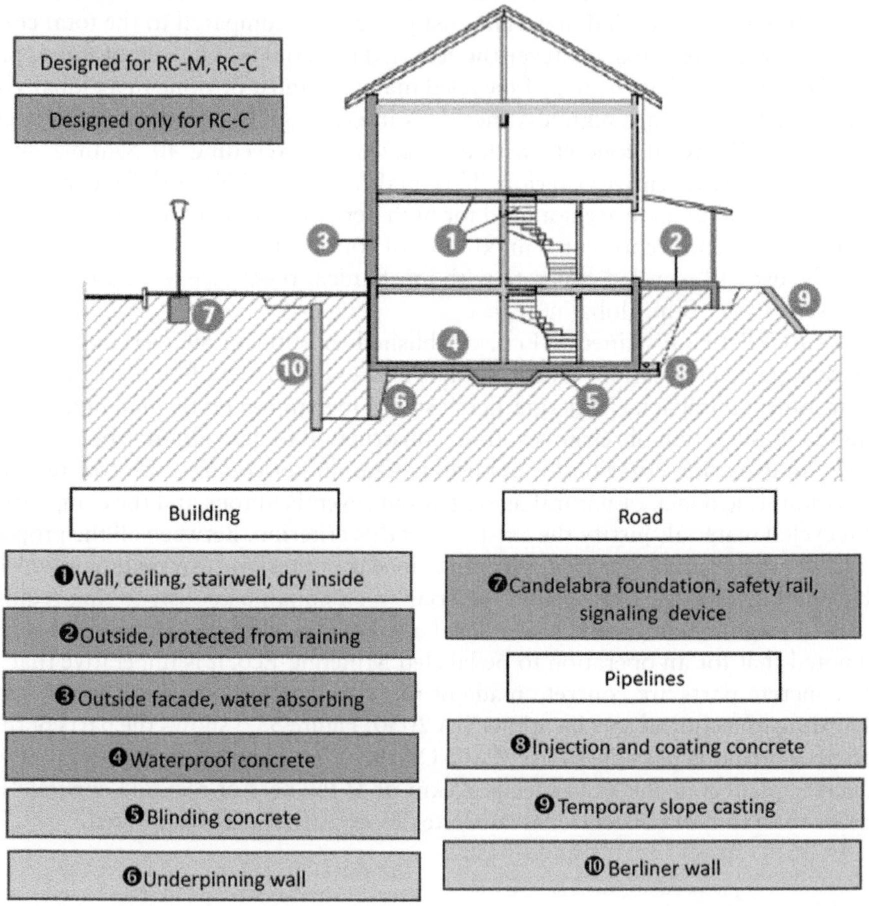

Figure 35.3 Possible use of recycled concrete— With permission of KBOB- MINERGIE-ECO (MINERGIE 2016). Where RC-C, concrete recycled concrete; RC-M, mixed recycled concrete. (RC-C (at least 25% of recycled concrete aggregates) and RC-M (at least 25% of recycled mixed aggregates).)

The proposed incentives, as part of a circular economy approach, are twofold: a supply policy and a demand policy (market launch). The two actions under a policy of supply are, first, the creation of a zero-interest loan for the recycling equipment and, secondly, the imposition of a tax on landfill. The two actions under a policy of demand are, first, the creation of a new criterion in an environmental certification (HQE) and, secondly, the implementation of a rate of RA in the public markets for building construction sector (if they are located within 25 km of a recycling platform).

35.5.1.1 Landfill taxation for CDW

The interest of the tax would be to modify the arbitration of the economic agents and to reinforce the interest of an upstream sorting and to create a better valuation of the building site wastes. In the current state of knowledge and information available, the beneficial impacts of imposing a landfill tax remain difficult to quantify. Unanticipated effects may emerge and may have against-productive effects: wild landfill, waste transport to neighboring countries. The impact on the balance between natural aggregate (NA) prices and RA prices is difficult to assess because it faces many of the strategies of local actors and the availability of the material. On the other hand, as long as the NA is generally cheaper to produce and is not subject to long-distance hauling, concrete producers are not pushed to buy RA and to produce RAC. As a general conclusion, tax modifications are classical public policy tools, and their application to the topic of concrete recycling remains to be studied more deeply.

35.5.1.2 Zero interest rate to waste management plants that invest in machinery

The introduction of a zero-rate loan for recycling equipment would have only a limited impact as constructors invest only at the time of a demand.

35.5.1.3 The integration of indicators promoting the use of RA in concrete in the French environmental assessment system (HQE)

Creating a new theme like "Integration of secondary raw materials" in the current HQE system, which already includes 28 themes, might create a demand. Even the place for such a criterion already exists, but it is not yet activated. The development of this "quality concrete" RA market would be essentially Parisian under the dominance of the Ile-de-France Region in tertiary HQE projects, but in the future more HQE projects could be expected for the whole national territory. The (Bougrain et al. 2017) study tries to assess the quantity of RA that could be used in RAC due to a new HQE provision (Table 35.5). On the one hand, the volumes are quite limited as compared to existing resource (see Chapter 29). On the other hand, the symbolic impact could be significant, and cases where a lack of available RA would impair the application of the HQE recycling provision would be scarce.

It should be noticed that these estimations were calculated on the basis of a lot of hypotheses. These hypotheses concern the part of construction made of concrete among the HQE certifications, the project locating, rate of concrete volume to building surface, and rate of RA in concrete.

35.5.1.4 The development of green public procurement requiring a percentage of recycled aggregates in their call for tenders (when transport distances do not exceed 25 km)

The implementation of a concrete recycling rate in public building markets (for projects where concrete facilities are located less than 25 km from a recycling platform) is the measure whose impact is the most certain and strongest. This policy would send a credible signal to all stakeholders of the deconstruction/reconstruction market:

- Long-term demand and long-term guarantee would encourage recycling platforms to invest in equipment allowing better treatment of aggregates.

Table 35.5 Estimative recycled concrete aggregate used by HQE incentives (Bougrain et al. 2017)

Hypothesis RA in concrete	*Coarse aggregate mass (Ton/year)*		*Fine aggregate mass (Ton/year)*	
	Low estimation	*High estimation*	*Low estimation*	*High estimation*
10% of coarse	52,471	72,504	0	0
≤30% of coarse	98,130	131,256	0	0
≤45% of coarse + 15% of fine	132,375	175,319	22,830	29,376

Table 35.6 Estimative recycled aggregate used by public procurement requiring incentives (Bougrain et al. 2017)

	Coarse aggregate mass (MTon/year)		*Fine aggregate mass (MTon/year)*	
Hypothesis RA in concrete	*Low estimation*	*High estimation*	*Low estimation*	*High estimation*
10% of coarse	0.8	0.9	0	0
≤30% of coarse	2.5	2.7	0	0
≤45% of coarse + 15% of fine	3.8	4.1	0.8	0.9

- With new prospects for recovery, deconstruction companies and project owners would be more aware of sorting at the deconstruction stage. The volume of materials brought to recycling platforms could then increase and benefit to the entire industry.

Of course, the prescriber must check the availability of a local resource in recycled aggregate before imposing a certain replacement rate (see Chapter 33). The implementation of this measure should primarily be done at the regional level since the problems of management and recovery of waste is local (Table 35.6).

35.5.1.5 Best incentive

In conclusion, the two supply policy actions are of less interest than those of a demand-side policy. A high tax or a zero-rate loan combined with an increasing demand for RA would be more efficient.

Among the two actions on demand, the use of green public procurement seems to be the most effective. Requiring a small percentage of RA in public procurement for buildings would of course have a limited impact on the market. However, it would play a major role to launch the market for RAC. This policy would incite recycling plants to invest in crushing and screening equipment and to propose RA that can be used in structural concrete.

35.5.2 Case of Switzerland

The Swiss Council of State has created a "waste and mineral resources" commission. Representatives of the concerned cantonal departments and private associations were then assigned to organize the management of mineral waste and their recycling. Three levers were identified (Puttallaz et al. 2016):

- Encouraging the use of recycled materials on the construction market.
- Assigning to a particular region, through its worksites, an exemplary role in the implementation of recycling materials.
- Raising the awareness of the construction sector actors to gradually remove the barriers related to the recycling of materials, by strengthening the dialogue between the different entities and by developing and providing pragmatic tools for the implementation of recycling materials.

35.5.3 French roadmap for incentive measures

In the context of reducing emissions into the air, the French Ministry of the Environment is currently seeking to prioritize clean projects on site waste management models with short flows. For example, a roadmap was published on April 13, 2018 detailing one of several measures aimed at ensuring that territorial actors take ownership of the air quality performance requirements as a function of the context [DICOM-DGPR]. Thus, indirectly and locally, this measure could change the management of construction waste and more particularly could encourage their reuse and/or recycling on the same deconstruction site or in its vicinity. Even though, this is a general incentive, this measure could change the local practices of construction sites.

In addition, the French government has drafted a roadmap on the circular economy to rapidly implement the energy transition law. In this context, 50 measures were published in April 2018 including targets and incentives to reduce waste, improve recycling and save raw materials. Today, the main guidelines have been set to encourage dialogue between actors and to ensure that the most appropriate measures are taken with the aim of changing the economic model. For instance:

- Anticipate the management of deconstruction (improved waste diagnostic) for a more efficient collection and recycling rate of C&D waste.
- An increased role of local observatories to manage waste and raw materials more efficiently.
- Adapt taxation to make the recovery of waste less expensive than their elimination (VAT, TGAP)

These future measures are clearly aimed at optimizing the material recovery of all mineral waste, including deconstructed concrete.

35.6 CONCLUSIONS: RECYBÉTON'S OUTPUTS

In the continuity of the RECYBÉTON project, next works must focus on the scenarios allowing to develop all the texts (technical and/or statutory) raising the subject of the recycled aggregates. Chapter 34 focused on the normative aspects by identifying the authorities and the concerned texts. Chapter 35 completes these normative aspects by identifying from a statutory point of view barriers and opportunities brought by the environmental labels, which prefigure the future regulations.

At a time when the environmental considerations become a major stake of our society (circular economy, E+C label), tracks to be envisaged concern at the French level the code of the environment and the public works contract regulations and this in relation with the European regulations (e.g., 2008/09/CE Waste framework).

For the voluntary incentives, certification standards show that the use of RA in the concrete is a criterion of allocation of points often flooded in more global criteria. As a consequence, the relative weight of recycling becomes relatively weak. It may, however, be decisive for the acceptance of offers and this is an aspect to be strengthened. The introduction of more explicit sub-criteria like that implemented in Minergie-Eco (Switzerland) is a way of improvement which would be advisable to explore for certifications such as NF HQE. Finally, the introduction of recycling specifications in public contracts is probably the most efficient way to foster the process of recycling concrete into concrete.

ACKNOWLEDGEMENTS

Authors thanks all members of the French National Project RECYBÉTON who contributed to analyze many regulation texts and certification texts.

Conclusion

F. de Larrard
LCR (LafargeHolcim R&D)

H. Colina
ATILH

After a comprehensive review of most aspects of the concrete recycling process, it appears that this practice is not new, nor technically challenging. With suitable RA, recycling at limited rate can be carried out with confidence, as acknowledged by the current state of standardization (see e.g., EN 206). As a matter of fact, the main barriers are more of economic nature. To crush demolished reinforced concrete parts is not cheaper than to use pieces of stone, so that the production cost of RA is at least as high as the one of NA. Therefore, one can state that concrete recycling will develop where natural aggregates cost are rising, either because of long-distance hauling and/or because an anticipated shortage in the availability of local natural aggregates. However, the construction industry is a large vessel who needs strong efforts to deviate from its route. Fully addressing the challenge of circular economy will often require public policies (green public procurement, taxes, incentives etc.).

RECYBÉTON demonstrated the feasibility of recycling concrete into concrete at all rates, from 0% to 100% of the concrete aggregate phase. When RA particles come with their original mortar as a coating, it was found that their incorporation requires more cement and more admixtures as soon as the rate is above 30%–40%, impairing the sustainability of the process. Fortunately, a lower percentage would suffice to absorb the whole RA stream in concrete industry. However, high-rate recycling can be sustainable in particular cases, e.g., when a large construction project is planned close to a significant demolition site. Additionally, in the future, more efficient processing of RA could reach a better separation between the original virgin aggregate from its cementitious matrix, providing to the concrete industry a renovated aggregate keeping all its original characteristics.

In terms of scientific and technical progresses, RECYBÉTON brought or confirmed a wealth of interesting results. The feasibility of incorporating fine RA in the cement process was demonstrated. A global assessment of all engineering properties of RAC was performed, showing that even if RA slightly alter the material owing to their porosity, any common set of specifications for structural concrete can be addressed with RA, even at high rate. On the practical side, it is noteworthy that the five experimental construction sites developed smoothly, with no change in usual practices and no identified drawbacks (to date). Also, the oldest reported construction, built in 2005 (see Chapter 16) did display a satisfactory behavior in spite of its close-to-100% recycling rate.

Of course, the history of concrete technology as a whole is not finished, and the same statement applies to RAC. Research needs in the various areas were already identified in many chapters of this book. According to the authors, if the aim is to secure high-rate recycling in the mid-term, the most important blocks are the following:

- Characterization of RA should be improved (some methods originally developed for NA showing their limits with RA);
- More efficient and cost-effective methods for RA processing are needed, whatever the objective (in-line detection of harmful species, sorting of alien particles, aggregate/

mortar separation) in order to allow a high-rate recycling at constant binder and admixture dosages;

- Improvement methods should be developed for fine RA, the incorporation of which remaining more detrimental than the one of coarse RA for most concrete properties;
- Accelerated carbonation of RA is currently studied as a way to improve the CO_2 balance of RAC.[1] Hence, using alternative fuels in the cement production and re-carbonating all recycled concrete could theoretically transform concrete as a carbon-neutral technology;
- In terms of RAC mechanical properties, creep and fatigue in compression require more investigations (tests and modeling);
- Structural testing of reinforced RAC members also needs further efforts, e.g., in the range of shear design, slab punching behavior etc.
- Reliability of RAC structures should be studied, maybe with a re-evaluation of safety coefficients in design codes when the rates of RA are higher than those already accepted by the standards;
- Low recycling rate concrete's durability is currently addressed through a prescriptive approach, as NA concrete. For higher dosages, it will be covered by performance-based methods,[2] currently studied for all structural concretes. However, RAC deserves specific studies with regard to the acceptable amounts of alien species. As compared to current product RA standards, higher levels of certain substances are probably acceptable, but experience is lacking to further open the door.

Finally, available results from R&D already allow updating many standards and codes, in order to foster the loop of recycling concrete into concrete.

Paris, June, 2018.

[1] This topic is covered in the French "FASTCARB" national project.
[2] Currently studied within the French "PERFDUB" national project.

Appendix

A.1 REFERENCE CONSTITUENTS

The following sections summarize the main properties of the reference constituents used to design the reference concrete mixes shared by the partners within the RECYBÉTON project. For repeatability purposes, the necessary amount of aggregate was produced in one shot at the beginning of the project, and then stored with the cements for all the partners. Admixtures, as they have a limited lifespan, and limestone filler because it was assumed to be more regular, were distributed on demand to the partners.

A.1.1 Admixtures

A polycarboxylatether (PCE) based superplasticizer and a chloride-free retarder were used when necessary in the reference mixtures (Table A.1).

A.1.2 Cements

Three different cements were used as reference materials (Table A.2):

- A classical CEM II/A-L 42.5 N CE CP2 NF available on the market, named C
- A CEM I specifically produced for the project. Recycled FA was introduced in the kiln as substitution of about 15% of the traditional raw materials (see Chapter 4), named CR1
- A blended cement named CR2, specifically produced for the project, in which about 25% of fine RA were crushed, milled, and then mixed and homogenized with clinker as another main constituent, replacing limestone. In a future version of the NF EN 197-1 standard, this type of cement could perhaps be considered as a new cement: CEM II/B-RFA where RFA stands for Recycled Fine Aggregates (Chapter 5).

The saturation amount of superplasticizer was measured according to AFREM slurry method (de Larrard et al. 1996). The packing densities were calculated on normal paste

Table A.1 Main properties of the superplasticizer and the retarder

Property	*Superplasticizer*	*Retarder*
Density (kg/L)	1.06	1.08
Dry extract (% in mass)	30	20
Recommended range of dosage (in kg/100 kg of cement)	0.2–3	2–8

Table A.2 Main properties of the three cements

Property	*C*	*CR1*	*CR2*
Components (%)			
Clinker K	87	100	73.4
Limestone L or LL	11	0	0
Fine RA	0	0	26.6
Others	2	0	0
Gypsum	3.4	6	6
Composition of the clinker (%)			
C_3S	61	61.8	53.9
C_2S	11.1	17.8	4.8
C_3A	7.9	6.1	6.8
C_4AF	12	11.6	6
Blaine specific surface (cm^2/g)	3,700	3,630	4,548
2 day class (MPa)	25.1	35	20.8
28 day class (MPa)	51.8	62.2	41.5
Specific gravity (kg/L)	3.09	3.14	2.97

Table A.3 Other properties of the cements

Property	*C*	*CR1*	*CR2*
Saturation amount of superplasticizer (in % of dry extract/mass of cement)	0.35	0.50	0.40
Packing density without superplasticizer	0.548	0.469	0.557
Packing density with a saturation amount of superplasticizer	0.598	0.498	0.638

(Sedran et al. 2007). Table A.3 shows that the compatibility between CR1 and the selected superplasticizer is probably poor, as the packing density of CR1 was not significantly improved by the superplasticizer.

A.1.3 Filler limestone

A limestone filler was used to improve the granular skeleton or as cement replacement to control the compressive strength. Its properties are summarized in (Table A.4 and Figure A.1).

The saturation amount of superplasticizer was measured according to AFREM slurry method (de Larrard et al. 1996) on a mix of 50% in mass filler and 50% of CEM II/A-LL. Knowing the saturation amount of cement, the one of filler was deduced.

Table A.4 Properties of the limestone filler

Property	
Specific gravity (kg/L)	2.7
Blaine specific surface (cm^2/g)	4,620
$CaCO_3$ (%)	98.8
Saturation amount of superplasticizer (in % of dry extract/mass of cement)	0.15
Packing density without superplasticizer	0.591
Packing density with a saturation amount of superplasticizer	0.681

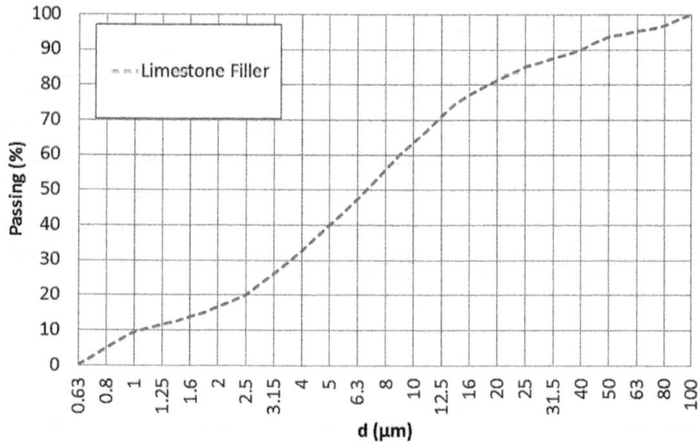

Figure A.1 Size distribution of limestone filler (values extrapolated under 1 μm).

The packing densities were measured on normal paste (Sedran et al. 2007) on two mixes: one with 90% in mass of filler and 10% of CEM II/A-LL, the second one with 80% in mass of filler and 20% of CEM II/A-LL. The packing density of the filler was linearly extrapolated based on these two values

A.1.4 Aggregates

Detailed information can be found in Chapters 3, 10 and 15. The following tables and figures summarize the main properties of natural and recycled reference aggregates (Tables A.5–A.7 and Figures A.2 and A.3).

Table A.5 Nature, origin of reference aggregates, and classification of RA according to NF EN 933-1

Material	*Nature*	*Origin*	*Type*
Fine RA-0/4	Recycled concrete	Recycling plant, Gonesse (France, dpt 95)	—
Coarse RA-4/10	Recycled concrete	Recycling plant, Gonesse (France, dpt 95)	Rcu_{98}
Coarse RA-10/20	Recycled concrete	Recycling plant, Gonesse (France, dpt 95)	Rcu_{99}
Fine NA-0/4	Semi-crushed silico calcareaous	Sandrancourt (France, dpt 78)	—
Coarse NA-4/10	Crushed limestone	Givet (France, dpt 08)	—
Coarse NA-6,3/20	Crushed limestone	Givet (France, dpt 08)	—

Table A.6 Basic properties of aggregates

Material	*Mvr (kg/L)*	*Abs (%)*	*LA (%)*	*MDE (%)*
Fine RA-0/4	2.08 (2.18)	8.9 (7.3)	—	—
Coarse RA-4/10	2.29	5.6	—	—
Coarse RA-10/20	2.26	5.8	37	23
Fine NA-0/4	2.58 (2.59)	0.8 (1)	—	—
Coarse NA-4/10	2.71	0.51	—	—
Coarse NA-6,3/20	2.71	0.46	16	17

Values in brackets measured on material retained on the 63-μm sieve. Real density Mvr and water absorption Abs according to NF EN 1097-6, Los Angeles LA according to NF EN 1097-2 and Micro Deval MDE according to NF EN 1097-1.

Table A.7 Supplementary properties of aggregates

Material	K_g	K_t	C_{exp}	*E (GPa)*	β	*H (mm)*
Fine RA-0/4	4.42	0.364	0.690	49.5	0.563	237.1
Coarse RA-4/10	5.173	0.364	0.569	49.5	0.593	267.3
Coarse RA-10/20	5.173	0.364	0.554	49.5	0.594	277.8
Fine NA-0/4	5.767	0.453	0.676	66.2	0.632	199.8
Coarse NA-4/10	5.767	0.453	0.580	66.2	0.611	221.4
Coarse NA-6,3/20	5.767	0.453	0.583	66.2	0.627	220.2

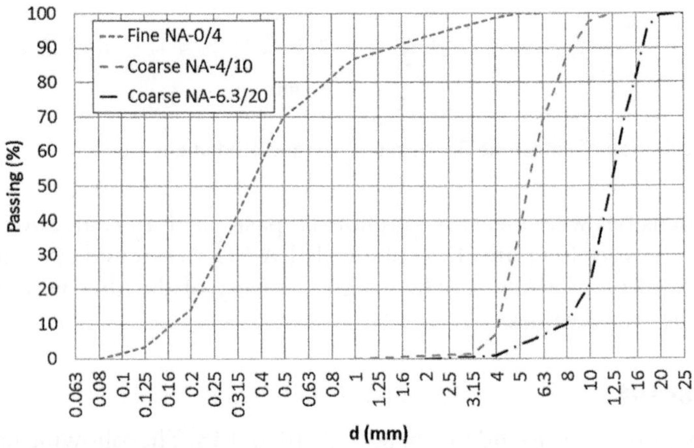

Figure A.2 Size distribution of natural aggregates NA.

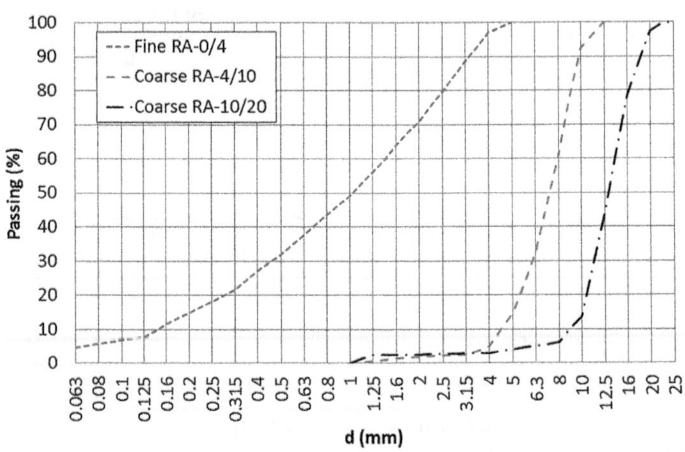

Figure A.3 Size distribution of recycled aggregates RA.

Compression coefficient K_g, tensile strength coefficient K_t and elastic modulus were fitted on concrete mixes (Chapter 10). Overall experimental packing density C_{exp} of each material and final height H were measured on a 7 kg aggregate sample compacted in a Ø160 mm cylinder (LCPC 2004; de Larrard et al. 2003). From these values, the virtual

Table A.8 Reference concrete C25/30 made with ordinary cement C; composition per cubic meter with dry aggregates

Constituent (kg/m³)	*C25/30-0S-0G*	*C25/30-0S-30G*	*C25/30-0S-100G*	*C25/30-30S-0G*	*C25/30-30S-30G*	*C25/30-100S-100G*
Added water	190	210	244	213	228	303
Cement C (Cem II/A-L 42,5)	270	276	282	276	277	326
Limestone filler	45	31	31	31	31	50
Fine NA 0/4	780	813	806	549	500	
Fine RA 0/4				235	218	673
Coarse NA 4/10	267	228		190	171	
Coarse RA 4/10			163		145	304
Coarse NA 6,3/20	820	462		829	552	
Coarse RA 10/20		296	701		167	442
Superplasticizer	1.31	1.51	1.4	1.16	1.08	1.18
Retarder	0	0		1.1	1.1	2.6
Free water (kg/m³) W_{eff}	180	185	189	185	185	199
Free water/Equ. Cement W_{eff}/C_{equ} [a]	0.65	0.65	0.65	0.65	0.65	0.59
Fine NA 0/4 (%vol)	43	45	45	30.3	27.6	0
Fine RA 0/4 (%vol)	0	0	0	16.1	14.9	49.6
Coarse NA 4/10 (%vol)	14	12	0	10	9	0
Coarse RA 4/10 (%vol)	0	0	10.25	0	9	20.4
Coarse NA 6,3/20 (%vol)	43	24.3	0	43.6	29	0
Coarse RA 10/20 (%vol)	0	18.7	44.75	0	10.5	30
Superplasticizer (dry extract % of cement mass)	0.15	0.17	0.15	0.13	0.12	0.11
Retarder (% in liquid of cement mass)	0	0	0	0.4	0.4	0.8
Mean target Rc_{28} (MPa)[b]	34	33	31	32	32	30
Mean experimental Rc_{28} (MPa)	33.9		32	33.6		28.5

[a] According to EN NF 206/CN.

[b] Some Rc_{28} expected values are higher than the 30 MPa required. This is due to the fact that XC1 class according to NF EN 206/CN impose a $W_{eff}/C_{equ} \leq 0.65$.

packing density β for each size of grain was calculated with the Compressible Packing Model (de Larrard 1999a).

A.2 REFERENCE CONCRETE MIXES

The composition of reference concrete mixes are given in Tables A.8–A.10. The method for obtaining these dosages is detailed in (Sedran 2017).

The set of requirements was as follows:

- Mean compressive strength at 28 days of 30 and 40 MPa for the C25/30 and C35/45 respectively
- Slump between 160 and 210 mm (S4 according to EN 206/CN) after 90 min
- Class of environment XC1 for the C25/30 and XF1 for C35/45 according to NF EN 206/CN (which was applied, except for the RA replacement ratio, higher than permitted for a number of mixes)

Table A.9 Reference concrete C35/45 made with classical cement C; composition per cubic meter with dry aggregates

Constituent (kg/m³)	*C35/45-0S-0G*	*C35/45-0S-100G*	*C35/45-30S-30G*	*C35/45-100S-100G*
Added water	185	238	220	284
Cement C (Cem II/A-L 42,5)	299	336	321	381
Limestone filler	58	53	44	70
Fine NA 0/4	771	782	491	
Fine RA 0/4			214	663
Coarse NA 4/10	264		168	
Coarse RA 4/10		158	142	299
Coarse NA 6,3/20	810		542	
Coarse RA 10/20		682	164	435
Superplasticizer	2.1	2.18	1.64	2.78
Retarder	0	0	1.3	3
Free water (kg/m³) W_{eff}	175	185	179	184
Free water/Equ. Cement W_{eff}/C_{equ}[a]	0.59	0.54	0.57	0.47
Fine NA 0/4 (%vol)	43	45	27.6	0
Fine RA 0/4 (%vol)	0	0	14.9	49.6
Coarse NA 4/10 (%vol)	14	0	9	0
Coarse RA 4/10 (%vol)	0	10.25	9	20.4
Coarse NA 6,3/20 (%vol)	43	0	29	0
Coarse RA 10/20 (%vol)	0	44.75	10.5	30
Superplasticizer (dry extract % of cement mass)	0.21	0.20	0.16	0.22
Retarder (% in liquid of cement mass)	0	0	0.4	0.8
Mean target Rc_{28} expected (MPa)	40	40	40	40
Mean experimental Rc_{28} (MPa)	38.7	39.3	40.1	38.5

[a] According to EN NF 206/CN.

Table A.10 Reference concrete made with special cement CR1 and CR2; composition per cubic meter with dry aggregates

Constituent (kg/m³)	*C25/30-0R-0R-CR1*	*C35/45-0R-0R-CR1*	*C25/30-0R-0R-CR2*
Added water	191	180	183.7
CR1 (CEM I)	252	250	
CR2 (CEM II/B-SBC)			293
Limestone filler	108	140	0
Fine NA 0/4 (%vol)	760	759	790
Coarse NA 4/10	260	259	270
Coarse NA 6,3/20	799	797	830
Superplasticizer	2.16	2.9	1.13
Retarder	0	0	2.93
Free water (kg/m³) W_{eff}	181.3	170.9	175.6
Free water/Equ. Cement W_{eff}/C_{equ}[a]	0.65	0.62	0.60
Fine NA 0/4 (%vol)	43	43	43

(Continued)

Table A.10 (Continued) Reference concrete made with special cement CR1 and CR2; composition per cubic meter with dry aggregates

Constituent (kg/m³)	*C25/30-0R-0R-CR1*	*C35/45-0R-0R-CR1*	*C25/30-0R-0R-CR2*
Coarse NA 4/10 (%vol)	14	14	14
Coarse NA 6,3/20 (%vol)	43	43	43
Superplasticizer (dry extract % of cement mass)	0.26	0.35	0.12
Retarder (% in liquid of cement mass)	0	0	1
Mean Rc_{28} expected (MPa)	37.5[b]	40.0	30.0
Mean experimental Rc_{28} (MPa)	38		31

[a] According to EN NF 206/CN.
[b] Rc_{28} expected value higher than the 30 MPa required. This is due to the fact that XC1 class according to EN 206/CN impose a $W_{eff}/C_{equ} \leq 0.65$.

The codification of the mixes is the following:

- C25/30-XS-YG: for a C25/30 made of cement C in which fine RA represents X% in mass of the total fines and coarse RA represent Y% in mass of coarse aggregate;
- C35/45-XS-YG: for a C35/45 in which fine RA represents X% in mass of the total fines and coarse RA represent Y% in mass of coarse aggregate;
- The term CR1 or CR2 is added when cement CR1 or CR2 is used instead of cement C

References

Original reports from the RECYBETON French national project are accessible at the following Internet direction: https://www.pnrecybeton.fr/publications/rapports-recybeton/ They are written in French with an English summary.

Abbas, A., Fathifazl, G. (2009) Durability of Recycled Aggregate Concrete Designed with Equivalent Mortar Volume Method. *Cement and Concrete Composites,* 31(8), 555–563.

Abbas, A., Fathifazl, G., Fournier, B., Isgor, O.B., Zavadil, R., Razaqpur, A.G., Foo, S. (2009) Quantification of the Residual Mortar Content in Recycled Concrete Aggregates by Image Analysis. *Materials Characterization,* 60(7), 716–728.

Abbas, A., Fathifazl, G., Isgor, O.B., Razaqpur, A.G., Fournier, B., Foo, S. (2007) Proposed Method for Determining the Residual Mortar Content of Recycled Concrete Aggregates. *Journal of ASTM International,* 5(1), 101087.

ACI Committee 224. *Control of Cracking in Concrete Structures Report ACI-224R-01.* American Concrete Institute: Detroit, MI, 2001.

ACI Committee 318. *Building Code Requirements for Structural Concrete and Commentary,* American Concrete Institute: Detroit, MI, 2011.

ACI Committee 555. (2002) Removal and Reuse of Hardened Concrete. *ACI Materials Journal,* 99(3), 300–325.

ACI Committee 555. *Removal and Reuse of Hardened Concrete.* American Concrete Institute Report. N° 555R–01, American Concrete Institute: Detroit, MI, 2002.

Adams, M.P. (2012) Alkali-Silica Reaction in concrete containing Recycled Concrete Aggregates, Thesis for the degree of Master of Science in Civil Engineering, Oregon State University, 148 p.

ADEME (December 2016) Déchets –Chiffres clés édition 2016 – ADEME.

ADEME. (2011a) Ademe, Analyse technico-economique de 39 plateformes francaises de tri/valorisation des dechets du BTP, synthèse réalisée pour le compte de l'ademe par Treize Developpement et Pöyry, 2011, 25 p.

ADEME. (2011b) ADEME Les matériaux de recyclage en Languedoc-Roussillon. Synthetis note, 2011.

ADEME. (2016a) déchets chiffres-clés. ADEME Editions-Faits et chiffres. 96p, December.

ADEME. (2016b) Indentification des freins et des leviers au réemploi de produits et matériaux de construction. Final Technical Report, France, April, 149 p.

AFGC. *Concrete Design for a Given Structure Service Life – Durability Management with Regard to Reinforcement Corrosion and Alkali-Silica Reaction – State of the Art and Guide for the Implementation of a Predictive Performance Approach Based Upon Durability Indicators.* AFGC: Paris, April 2007.

AFNOR FD P18-542. (2015) Aggregates - Criteria for qualification natural aggregates for hydraulic concrete with respect to the alkali-reaction.

AFNOR NF EN 12350-2. (2012) *Testing Fresh Concrete – Part 2*: Slump test.

AFNOR XP P18-462. (2012) Testing hardened concrete – Chloride ions migration accelerated test in non-steady-state conditions – Determining the apparent chloride ions diffusion coefficient.

AFNOR XP P18-463. (2011) Concrete – Testing gas permeability on hardened concrete.

AFNOR, NF EN 12390-13. (Février 2014) Essai pour béton durci - Partie 13 : Détermination du module sécant d'élasticité en compression.

AFNOR, XP P18-543. (2015) Aggregates – Petrographic investigation of aggregates relating to alkali-silica reaction.

AFNOR. FD P18-456. (2004) Concrete - Reactivity of a concrete formula with respect to the alkali-aggregate reaction - Criteria for interpretation of the performance test results.

AFNOR. FD P18-464. (2014) A Concrete — Provisions for the prevention of alkali silica reactions.

Aït Alaiwa, A., Lavaud, R. (2014) Validité des normes d'essais sur les granulats recyclés. Rapport RECYBÉTON N°R/14/RECY/010.

Ajdukiewicz, A., Kliszczewicz, A. (2002) Influence of Recycled Aggregates on Mechanical Properties of HS/HPC. *Cement and Concrete Composites,* 24(2), 269–279.

Ajdukiewicz, A.B., Kliszczewicz, A.T. (2007) Comparative Tests of Beams and Columns Made of Recycled Aggregate Concrete and Natural Aggregate Concrete. *Advanced Concrete Technology,* 5(2), 259–273.

Akbarnezhad, A., Ong, K.C.G. *Separation Processes to Improve the Quality of Recycled Concrete Aggregates (RCA). Handbook of Recycled Concrete and Demolition Waste.* Woodhead Publishing Limited, F. Pacheco-Torgal, V. W. Y. Tam, J. A. Labrincha, Y. Ding and J. de Brito, eds., Cambridge, UK, 2013.

Akbarnezhad, A., Ong, K.C.G., Tam, C.T., Zhang, M.H. (2013) Effects of the Parent Concrete Properties and Crushing Procedure on the Properties of Coarse Recycled Concrete Aggregates. *Journal of Materials in Civil Engineering,* 25, 1795–1802.

Akbarnezhad, A., Ong, K.C.G., Zhang, M.H., Tam, C.T. (2013b) Acid Treatment Technique for Determining the Mortar Content of Recycled Concrete Aggregates. *Journal of Testing and Evaluation,* 41, 20120026.

Akbarnezhad, A., Ong, K.C.G., Zhang, M.H., Tam, C.T., Foo, T.W.J. (2011) Microwave-Assisted Beneficiation of Recycled Concrete Aggregates. *Construction and Building Material,* 25(8), 3469–3479.

Al Bayati, H.K.A., Das, P.K., Tighe, S.L., Baaj, H. (2016) Evaluation of Various Treatment Methods for Enhancing the Physical and Morphological Properties of Coarse Recycled Concrete Aggregate. *Construction and Building Materials,* 112, 284–298.

Alhozaimy, A.M. (2009) Effect of Absorption of Limestone Aggregates on Strength and Slump Loss of Concrete. *Cement and Concrete Composites,* 31, 470–473.

Ali, A.M., Farid, B.J., AL-Janabi, A.I.M. (1990) Stress-Strain Relationship For Concrete in Compression Made of Local Materials. *Engineering Sciences Journal,* 2, 1319–1047.

Almusallam, A., Maslehuddin, M., Abdul-Waris, M., Khan, M. (1998) Effect of Mix Proportions on Plastic Shrinkage Cracking of Concrete in Hot Environments. *Construction and Building Materials,* 12, 353–358.

Ambrós, W.M., Sampaio, C.H., Cazacliu, B.G., Miltzarek, G.L., Miranda, L.R. (2017) Usage of Air Jigging for Multi-Component Separation of Construction and Demolition Waste. *Waste Management,* 60, 75–83. Available online 28 November 2016.

American Concrete Institute. (August 2007). Education Bulletin E1-07: Aggregates for concrete. USA.

Amorim, P., de Brito, J., Evangelista, L. (2012) Concrete Made with Coarse Concrete Aggregate: Influence of Curing on Durability. *ACI Materials Journal,* 109(2), 195–204.

Anand, C.K., Amor, B. (2017) Recent Developments, Future Challenges and New Research Directions in LCA of Buildings: A Critical Review. *Renewable and Sustainable Energy Reviews,* 67, 408–416.

Arezoumandi, M., Smith, A., Volz, J.S., Khayat, K.H. (2014) An Experimental Study on Shear Strength of Reinforced Concrete Beams with 100% Recycled Concrete Aggregate. *Construction and Building Materials,* 53, 612–620.

Arezoumandi, M., Smith, A., Volz, J.-S., Khayat, K.-H. (2015) An Experimental Study on Shear Strength of Reinforced Concrete Beams with 100% Recycled Concrete Aggregate. *Engineering structures,* 88, 154–162.

Arora S., Singh, S.P. (2016) Analysis of Flexural Fatigue Failure of Concrete Made with 100% Coarse Recycled Concrete Aggregates. *Construction and Building Materials,* 102, 782–791.

Arora, S., Singh, S.P. (2015) Flexural Fatigue Analysis of Concrete made with 100% Recycled Concrete Aggregates. *Journal of materials and engineering structures,* 2, 77–89.

Artoni, R., Cazacliu, B., Hamard, E., Cothenet, A., Parhanos, R.S. (2017) Resistance to Fragmentation of Recycled Concrete Aggregates. *Materials and Structures,* 50(1), 11.

Arulrajah, A., Piratheepan, J., Disfani, M.M., Bo, M.W. (2012) Geotechnical and Geoenvironmental Properties of Recycled Construction and Demolition Materials in Pavement Subbase Applications. *Journal of Materials in Civil Engineering, ASCE,* 25(8), 1077–1088,

AS 1141 Australian Standards (2012). Methods for testing and sampling aggregates. Australia.

Asakura, H., Watanabe, Y., Ono, Y., Yamada, M., Inoue, Y., Alfaro, A.M. 2010. Characteristics of Fine Processed Construction and Demolition Waste in Japan and Method to Obtain Fines Having Low Gypsum Component and Wood Contents. *Waste Management and Research,* 28, 634–646.

ASI. ÖNORM B 3131. (October 2006) Gesteinskörnungen für Beton - Regeln zur Umsetzung der ÖNORM EN 12620. Austria.

ASI. ÖNORM B 3140. (August 2013) Rezyklierte Gesteinskörnungen für das Bauwesen. Austria.

Asif, M., Muneer, T., Kelley, R. (2007) Life Cycle Assessment: A Case Study of a Dwelling Home in Scotland. *Building and Environment,* 42(3), 1391–1394.

Aslani, F., Nejadi, S. (2012) Mechanical Properties of Conventional and Self-Compacting Concrete: An Analytical Study. *Construction and Building Materials,* 36(6), 330–347.

Assié, S. (2004) Durabilité des bétons auto-plaçants, INSA-Toulouse, *Ph.D Thesis,* (in French).

ASTM C127. *Standard Test Method for Density, Relative Density (Specific Gravity), and Absorption of Coarse Aggregate,* ASTM International, West Conshohocken, PA, 2015.

ASTM C128. *Standard Test Method for Density, Relative Denstiy (Specific Gravity), and Absorption of Fine Aggregate,* ASTM International, West Conshohocken, PA, 2015.

ASTM C192. *Standard Practice for Making and Curing Concrete Test Specimens in the Laboratory,* ASTM International, West Conshohocken, PA,2007.

ASTM C294. *Standard Descriptive Nomenclature for Constituents of Concrete Aggregates.* ASTM International: West Conshohocken, PA, 2012.

ASTM C457/C457M-12. *Standard Test Method for Microscopical Determination of Parameters of the Air-Void System in Hardened Concrete.* ASTM International: West Conshohocken, PA, 2012.

ASTM C666/C666M. *Standard Test Method for Resistance of Concrete to Rapid Freezing and Thawing.* ASTM International, West Conshohocken, PA, 2015.

ASTM C88. *Standard Test Method for Soundness of Aggregates by Use of Sodium Sulfate or Magnesium Sulfate.* ASTM International, West Conshohocken, PA, 2013.

Aubert, J.E., Escadeillas, G., Leklou, N. (2009) Expansion of Five-Year-Old Mortars Attributable to DEF: Relevance of the Laboratory Studies on DEF? *Construction and Building Materials,* 23(12), 3583–3585.

Aubert, J.E., Escadeillas, G., Leklou, N. (2013) Five-Year Monitoring of Curing Solutions of Heat-Cured Mortars Affected by Delayed Ettringite Formation. *Advances in Cement Research,* 25(3), 155–163.

BA 99 Boverket. Swedish manual on the use of recycled building materials "BÅ 99", 2000. Sweden.

Baalbaki, W., Benmokrane, B., Chaallal, O., Aitcin, P.C. (1991) Influence of Coarse Aggregate on Elastic Properties of High-Performance Concrete. *ACI Materials Journal,* 88(5), 499–503.

Babu, V.S., Mullick, A.K., Jain, K.K., Singh, P.K. (2014) Strength and Durability Characteristics of High-Strength Concrete with Recycled Aggregate – Influence of Mixing Techniques. *Journal of Sustainable Cement-Based Materials,* 3(2), 88–110.

Bach, H. *Evaluation of Attrition Tests for Railway Ballast.* Graz University of Technology: Graz, 2013.

Bai, W.H., Sun, B. X. (2010) Experimental Study on Flexural Behavior of Recycled Coarse Aggregate Concrete Beam. *Applied Mechanics and Materials,* 29, 543–548.

Bairagi, N.K., Kishore, R., Pareek, V.K. (1993) Behaviour of Concrete with Different Proportions of Natural and Recycled Aggregates. *Conservation and Recycling,* 9(1–2), 109–120.

Banfill, P.F.G. (2011) Additivity Effects in the Rheology of Fresh Concrete Containing Water-Reducing Admixtures. *Construction and Building Materials,* 25(6), 2955–2960.

Barbudo, A., Agrela, F., Ayuso, J., Jiménez, J.R., Poon, C.S. (March 2012). Statistical Analysis of Recycled Aggregates Derived from Different Sources for Sub-Base Applications. *Construction and Building Materials,* 28(1), 129–138.

Barles, S. (2014) L'écologie territoriale et les enjeux de la dématérialisation des sociétés: l'apport de l'analyse des flux de matières. *Développement durable et territoires,* 5(1).

Barnes, M.S., Diamond, S., Dolch, W.L. (1978) The Contact Zone between Portland Cement Paste and Glass Aggregate Surfaces. *Cement and Concrete Research,* 8, 233–244.

Baroghel-Bouny, V., Chaussadent, T., Croquette, G., Divet, L., Gawsewitch, J., Godin, J. *Caractéristiques microstructurales et propriétés relatives à la durabilité des bétons.* Méthode d'essai N°58. LCPC: Paris, 2002.

Baron, J., Lesage, R. (1976) *La composition du béton hydraulique du laboratoire au chantier,* rapport de recherche LPC n°64. LCPC: French.

Barra, M., Vazquez, E. (1996) The influence of retained moisture in aggregates from recycling on the properties of new hardened concrete. *Waste Management,* 16(1–3), 113–117.

Bartos, P. *Fresh Concrete : Properties and Tests.* Elsevier: Amsterdam, 1992.

Bažant, Z.P., Chern, J.C. (1985) Concrete Creep at Variable Humidity: Constitutive Law and Mechanism. *Materials and Structures,* 18(103), 1–20.

Bazant, Z.P., Kaplan, M.F. *Concrete at High Temperatures: Material Properties and Mathematical Models.* Longman: London, 1996.

BDF. (2017) La démarche Bâtiment Durable Franciliens. France.

BDM. (2008) La démarche Bâtiment Durable Méditerranéens. France.

Behera, M., Bhattacharyya, S.K., Minocha, A.K., Deoliya, R., Maiti, S. (2014) Recycled Aggregate from C&D Waste & Its Use in Concrete—A Breakthrough Towards Sustainability in Construction Sector: A review. *Construction and Building Materials,* 68, 501–516.

Belén, G.F., Martinez-Abella, F., Carro Lopez, D., Seara-Paz, S. (2011) Stress-Strain Relationship in Axial Compression for Concrete Using Recycled Saturated Coarse Aggregate. *Construction and Building Materials,* 25(5), 2335–2342.

Belin, P., Habert, G., Thiery, M., Roussel, N. (2014) Cement Paste Content and Water Absorption of Recycled Concrete Coarse Aggregates. *Materials and Structures,* 47, 1451–1465.

Bello, L. (2014) Mise au point d'une méthodologie pour formuler de nouveaux bétons légers autoplaçants durables, Université de Montpellier II, Ecole Doctorale I2S Mécanique et Génie Civil.

Bello, L., Garcia-Diaz, E., Rougeau, P. (2017) An Original Test Method to Assess Water Absorption/Desorption of Lightweight Aggregates in Presence of Cement Paste. *Construction and Building Materials,* 154, 752–762.

Benboudjema, F., Torrenti, J. M. (2008) Early Age Behaviour of Concrete Nuclear Containments. *Nuclear Engineering and Design,* 238(10), 2495–2506.

Bendimerad, A.Z. (2016) Comportements Au Jeune Âge et Différé Des Bétons Recyclés: Influence de La Saturation Initiale En Eau et Du Taux de Substitution, Thèse de doctorat, Ecole Centrale de Nantes.

Bendimerad, A.Z., Roziere, E., Loukili, A. (2015) Combined Experimental Methods to Assess Absorption Rate of Natural and Recycled Aggregates. *Materials and Structures,* 48(11), 3557–3569.

Bendimerad, A.Z., Rozière, E., Loukili, A. (2016) Plastic Shrinkage and Cracking Risk of Recycled Aggregates Concrete. *Construction and Building Materials,* 121, 733–745.

Bentur, A., Alexander, M.G. (2000) A Review of the Work of the RILEM TC159-ETC: Engineering of the Interfacial Transition Zone in Cementitious Composites. *Materials and Structures,* 33, 82–87.

Benz, D.P., Garbockzi, E.J. (1991) Percolation of Phases in a Three-Dimensional Cement Paste Microstructural Model. *Cement and Concrete Research,* 21, 1368–1376.

Bernard, O., Ulm, F.J., Lemarchand, E. (2003) A Multiscale Micromechanics-Hydration Model for the Early-Age Elastic Properties of Cement-Based Materials. *Cement and Concrete Research,* 33, 1293–1309.

Bio Intelligence Service. (2015) Construction and Demolition Waste management in FRANCE, European Commission DG ENV.

Bissonnette, B., Pierre, P., Pigeon, M. (1999). Influence of Key Parameters on Drying Shrinkage of Cementitious Materials. *Cement and Concrete Research,* 29, 1655–1662.

Blengini, G.A., Garbarino, E., Šolar, S., Shields, D.J., Hámor, T., Vinai, R., Agioutantis, Z. (2012) Life Cycle Assessment Guidelines for the Sustainable Production and Recycling of Aggregates: The Sustainable Aggregates Resource Management Project (SARMa). *Journal of Cleaner Production,* 27, 177–181.

Bodet, R., Fonteny, S., Pillard, W. (2014) Détermination des freins à l'utilisation des granulats recyclés en France. RECYBETON Technical report R/14/RECY/005, September, 112 p.

Bodet, R., Fonteny, S., Pillard, W. (2015) Technical report n° R15/RECY/025- Synthèse des documents internationaux sur l'utilisation des granulats recycles dans les bétons (Synthesis of international documents on the use of recycled aggregates in concretes), RECYBETON project, in French.

Bogue, R.H. (1929) Calculation of the Compounds in Portland Cement. *Industrial & Engineering Chemistry Analytical Edition,* 1, 192–197.

Bougrain, F., Moisson, P.H., Belaïd, F. (2017). Impact de différentes mesures de politique publique sur le recyclage du béton. RECYBETON Technical report R/17/RECY/44, July, 63 p.

Boulay, C. (2007) Développement D'un Dispositif de Mesure Du Retrait Endogène D'un Béton Au Jeune Âge. In: *Actes des 8èmes journées Scientifiques du RF2B, Montréal*, pp. 48–57.

Bovea, M.D., Powell, J.C. (2016) Developments in Life Cycle Assessment Applied to Evaluate the Environmental Performance of Construction and Demolition Wastes. *Waste Management,* 50, 151–172.

Bradley, G. (2004) Design for Deconstruction.

Braga, A.M., Silvestre, J.D., de Brito, J. (2017) Compared environmental and economic impact from cradle to gate of concrete with natural and recycled coarse aggregates. *Journal of Cleaner Production,* 162, 529–543.

Braga, M., de Brito, J., Veiga, R. (2012) Incorporation of Fine Concrete Aggregates in Mortars. *Construction and Building Materials,* 36, 960–968.

Braga, M., de Brito, J., Veiga, R. (2014) Reduction of the Cement Content in Mortars Made with Fine Concrete Aggregates. *Materials and structures,* 47, 171–182.

Brand, A.S., Roesler, J.R., Salas, A. (2015) Initial Moisture and Mixing Effects on Higher Quality Recycled Coarse Aggregate Concrete. *Construction and Building Materials,* 79, 83–89.

Braunschweig, A., Kytzia, S., Bischof, S. (2007) Recycled concrete: Environmentally beneficial over virgin concrete? In: *Towards Life Cycle Sustainability Management, LCM 2011 Conference, 2011*, Berlin, Germany, p. 635.

Bravo, M., de Brito, J., Pontes, J., Evangelista, L. (2015) Durability Performance of Concrete with Recycled Aggregates from Construction and Demolition Waste Plants. *Construction and Building Materials*, 77, 357–369.

Braymand, S., Roux, S., Deodonne, K., Feugeas, F., Fond, C. (2017) Use of recycled aggregates of concrete in total replacement of natural materials: Influence on the determination of concrete formulation parameters. *HISER International Conference - Advances in Recycling and Management of Construction and Demolition Waste, Delft*, pp. 189–192.

Braymand, S., Roux, S., Fares, H., Déodonne, K., Feugeas, F. (2016) Separation and Quantification of Attached Mortar in Recycled Concrete Aggregates. *Waste Biomass Valorization,* 8(5), 1393–1407. doi: 10.1007/s12649-016-9771-2.

Braymand, S., Roux, S., Fares, H., Feugeas, F. (2017) Multi-criteria study for recycled concrete aggregate separation processes. *Proceedings of International HISER Conference on Advances in Recycling and Management of Construction and Demolition Waste, Delft, The Netherlands,* pp. 97–100.

Braymand, S., Roux, S., Kunwufine, D., Mihalcea, C., Feugeas, F., Fond, C. *Les granulats recyclés de bétons: Un matériau à fort potentiel de valorisation dans les bétons.* Matériaux. Montpellier, France, 2014.

Breccolotti, M., D'Alessandro, A., Roscini, F., Bonfigli, M.F. (2015) Investigation of Stress - Strain Behaviour of Recycled Aggregate Concrete Under Cyclic Loads. *Environmental Engineering and Management Journal,* 14, 1543–1552.

Breccolotti, M., Materazzi, A.L. (2013) Structural Reliability of Bonding between Steel Rebars and Recycled Aggregate Concrete. *Construction and Building Materials*, 47, 927–934.

BREEAM. (2016) BREEAM International new construction. Technical Manuel SD 233 1.0, United Kingdom.

BRGM. (November 2010) Projet ANR COFRAGE – Caractérisation du gisement des déchets de chantier du BTP à l'échelle du territoire français. Rapport final BRGM/RP-59115. 66p.

Bresson, J. (2003) Guide pour le recyclage des déchets de béton de l'Industrie du Béton. DDP 92, CERIB.

Briffaut, M., Benboudjema, F., Torrenti, J.M., Nahas, G. (2011) A Thermal Active Restrained Shrinkage Ring Test to Study the Early Age Concrete Behaviour of Massive Structures. *Cement and Concrete Research*, 41(1), 56–63.

Brokk, A.B. *A Handbook of Demolition with Brokk*. Brokk Company: Morphett Vale, 2000.

Bru, K., Touzé, S., Bourgeois, F., Lippiatt, N., Ménard, Y. (2014) Assessment of a Microwave-Assisted Recycling Process for the Recovery of High-Quality Aggregates from Concrete Waste. *International Journal of Mineral Processing*, 126, 90–98.

Bru, K., Touzé, S., Parvaz, D.B. (2017) Development of an innovative process for the up-cycling of concrete waste. *Proceedings of International HISER Conference on Advances in Recycling and Management of Construction and Demolition Waste, Delft, The Netherlands*, pp. 55–58.

Brunetaud X. (2005) Étude de l'influence de différents paramètres et de leurs interactions sur la cinétique de l'amplitude de la réaction sulfatique interne au béton, Thesis. Châtenay-Malabry, École Centrale de Paris.

BSI. (2009) Aggregates – Part 1: Aggregates for concrete – Guidance on the use of BS EN 12620. UK.

BSI. *Concrete. Complementary British Standard to BS EN 206-1. Specification for Constituent Materials and Concrete*. National Supplement: London, UK, 2012.

Buck, A.D. (1972) Recycled concrete. DTIC Document.

Buck, A.D. (1977) Recycled Concrete as a Source of Aggregate. *ACI Journal*, 74 (5), 212–219.

Building Research Establishment. Digest 433, *Recycled Aggregates*, IHS BRE Press, Bracknell, RG12 8FZ, UK 1998.

Butler, L., West, J.S., Tighe, S.L. (2011) The Effect of Recycled Concrete Aggregate Properties on the Bond Strength between RCA Concrete and Steel Reinforcement. *Cement and Concrete Research*, 41(10), 1037–1049.

Butler, L., West, J.S., Tighe, S.L. (2012) *Effect of Recycled Concrete Aggregate Properties on Mixture Proportions of RA Concrete Developed for Structural Applications, TRB 2012 Annual Meeting*, 18 p.

Buyle-Bodin, F., Hadjieva-Zaharieva, R. (2002) Influence of Industrially Produced Recycled Aggregates on Flow Properties of Concrete. *Materials and Structures*, 35(8), 504–509.

Cabral, A.E.B., Schalch, V., Dal Molin, D.C.C., Ribeiro, J.L.D. (2010) Mechanical Properties Modeling of Recycled Aggregate Concrete. *Construction and Building Materials*, 24, 421–430.

Calvo Pérez, B., Parra y Alfaro, J.L., Astudillo, B., Sanabria, C.M., Carretón, R. (2002) Aridos reciclados para hormigones y morteros. Caracterizatión mineralógica y quimica, LOEMCO, Escuela Técnica Superior de Ingenieros de Minas de Madrid.

Cameron, D.A., Gabr, A.R. (2012) Permanent strain testing of recycled concrete aggregate for evaluation of unbound bases. *2nd International Conference on Transportation Geotechnics (ICTG) Japan*, September 10–12, 2012.

Carreira, D.J., Chu, K.H. (1985) Stress-Strain Relationship for Plain Concrete in Compression. *ACI Materials Journal*, 82(6), 797–804.

Cassagnabère, F., Mouret, M., Escadeillas, G., Broilliard, P. (2009) Use of Flash Metakaolin in a Slip-Forming Concrete for the Precast Industry. *Magazine of Concrete Research*, 61(10), 767–778.

Cast3M. Commissariat à l'Energie Atomique CEA – DEN/DM2S/SEMT, Cast3m finite element code. http://www-cast3m.cea.fr/, accessed 2014.

Casuccio, M., Torrijos, M.C., Giaccio, G., Zerbino, R. (2008) Failure Mechanism of Recycled Aggregate Concrete. *Construction and Building Materials*, 22(7), 1500–1506.

Cazacliu, B. (2013) Solution d'innovation pour la mesure de l'humidité des granulats (SIMH), rapport final du projet pour la region Pays de la Loire, 98 p.

Cazacliu, B., Huchet, F. (juin 2016) Technologies de séparation à sec des sables et des fines de recyclage - Etude bibliographique, rapport de recherché, PN Recybeton R15RECY019, 62 p.
Cazacliu, B., Sampaio, C.H., Miltzarek, G., Petter, C., Le Guen, L., Paranhos, R., Huchet, F., Kirchheim, A.P. (2014) The Potential of Using Air Jigging to Sort Recycled Aggregates. *Journal of Cleaner Production*, 66, 46–53.
CCTG EDF. (Novembre 2010) Cahier des Clauses Techniques Générales applicables aux travaux de gros œuvre. France.
CEB-FIP. (2010) *Fib Model Code for Concrete Structures*. Ernst & Sohn: London. ISBN: 978-3-433-03061-5, 434 pages.
Cedolin, L., Dei Poli, S., Iori, I. (1983) Experimental Determination of the Fracture Process Zone in Concrete. *Cement and Concrete Research*, 13(4), 557–567.
Cellules Economiques. Régionales de la Construction = Observatoires de la filière Construction regroupant les experts du Ministère de la Transition Ecologique et Solidaire, de la CAPEB, de la FFB, de la FNTP et de l'UNICEM. www.cerc-actu.com/ accessed 2016.
CEN. *Aggregates for Concrete*, 2008.
CEN. *Aggregates for Mortar*, 2002.
CEN. *Common Rules for Precast Concrete Products*. CEN, 2004.
CEN. *Common Rules for Precast Concrete Products*. European standard, Afnor, Paris, 2013.
CEN. *Concrete Pavements*. Materials. CEN, 2013.
CEN. *Concrete, Specification, Performance, Production and Conformity*, 2013.
CEN. *Execution of Concrete Structures*. CEN, 2009.
CEN. Testing fresh concrete (various parts). ISO, 2009.
CEN. Testing hardened concrete (various parts). ISO, 2009.
CEN. *Tests for Chemical Properties of Aggregates - Part 1: Chemical Analysis*. CEN, 2012.
CEN. *Tests for Chemical Properties of Aggregates - Part 6: Determination of the Influence of Recycled Aggregate Extract on the Initial Setting Time of Cement*. CEN, 2006.
CEN. *Tests for Geometrical Properties of Aggregates - Part 11: Classification Test for the Constituents of Coarse Recycled Agggregate*. CEN, 2009.
CEN. *Tests for Thermal and Weathering Properties of Aggregates - Part 2: Magnesium Sulfate Test*, 2010.
CEN. *Tests for Thermal and Weathering Properties of Aggregates - Part 4: Determination of Drying Shrinkage*. CEN, 2008.
CERC (January 2018) Déchets et recyclage du Bâtiment et des Travaux Publics en France : les contributions du Réseau des CERC Enseignements de la consolidation des diagnostics départementaux Déchets et Recyclage du Bâtiment et des Travaux Publics des CERC. www.cerc-actu.com.
CEREMA. (2014) Graves de valorisation-Graves de déconstruction. Guide Rhône-Alpes d'utilisation en Travaux Publics, April, 24 p.
CEREMA. (2015) Technical Guide: Acceptabilité environnementale des matériaux alternatifs en construction – Les matériaux de deconstruction issus du BTP, CEREMA, Editions Reference, Sourdun, France.
CEREMA. (2016) Acceptabilité environnementale de matériaux alternatifs en technique routière. Les matériaux de déconstruction issus du BTP. Technical guide. France, January, 39 p.
Certivea. *Référentiel HQE Bâtiment Durable*. Annexe Technique: France, 2016.
CGDD SOeS-Data-Lab. (March 2017) Entreprises du BTP : 227,5 millions de tonnes de déchets en 2014.
Chakradhara Rao, M., Bhattacharyya, S.K., Barai, S.V. (2010) Influence of Field Recycled Coarse Aggregate on Properties of Concrete. *Materials and Structures*, 44(1), 205–220.
Champeau, B., Potin, J.N. (2003) Valorisation des sables. *Mines et Carrières*, 207, 41–49.
Chateau, L. *Nouveaux systèmes constructifs démontables en rénovation ou déconstruction pour valorisation et recyclage simplifiés et attractifs des produits et matériaux (DEMODULOR)*. ADEME: Juillet, 2015.
Chatterjee, T.K. (1991) Burnability and Clinkerization of Cement Raw Mixes, Mysore Cements Limited., India.

Chatterji, S. (1989) Mechanisms of alkali-silica reaction and expansion. *8th International Conference on Alkali-aggregate Reaction in Concrete, Kyoto, Japan*, pp. 101–105.

Chen, J., Brown, B. (2012) University of wisconsin system solid waste research program student project report leaching characteristics of recycled aggregate used as road base May 2012 student investigators: Jiannan Chen, Brigitte Brown Advisors: Tuncer B . Edil, James Tinjum Un, (May), 1–22.

Chen, J., Tinjum, J.M., Edil, T.B. (2013) Leaching of Alkaline Substances and Heavy Metals from Recycled Concrete Aggregate Used as Unbound Base Course, Transportation Research Record, TRR 2349, pp. 81–90.

Choi W.C., Yun, H.D. (2012) Compressive Behavior of Reinforced Concrete Columns with Recycled Aggregate Under Uniaxial Loading. *Engineering Structures*, 41, 285–293.

Christensen, N.H. (1979) Burnability of the Raw Mixes at 1,400°C, the Effect of the Fineness. *Cement Concrete Research*, 9, 9–219.

Christensen, N.H., Johansen, V. (June 1979) Role of liquid phase and mineralizers. *FLS-review in Proceedings of Cement Production and Use Conference; The engineering Foundation, New Hampshire, USA*.

Chrysochoou, M., Dermatas, D. (2007) Evaluation of Ettringite and Hydrocalumite Formation for Heavy Metal Immobilization: Literature Review and Experimental Study. *Journal of Hazardous Materials*, 136(1), 20–33.

Code of customs. (January 1, 2018) France, Consolidated version.

Code of environment. (April 15, 2018). France, Consolidated version.

Code of environment. (April 9, 2018) France, Consolidated version.

Code of public health. (April 16, 2018) France, Consolidated version.

Code of public procurement. (August 2006), Décret n° 2006-975 du 1er août 2006 portant code des marchés publics, France.

Coelho, A., de Brito, J. (2013) Economic Viability Analysis of a Construction and Demolition Waste Recycling Plant in Portugal – Part I: Location, Materials, Technology and Economic Analysis. *Journal of Cleaner Production*, 39, 338–352.

Coelho, A., De Brito, J. *Conventional Demolition Versus Deconstruction Techniques in Managing Construction and Demolition Waste (CDW), Handbook of Recycled Concrete and Demolition Waste*. Woodhead Publishing Limited, F. Pacheco-Torgal, V. W. Y. Tam, J. A. Labrincha, Y. Ding and J. de Brito, eds., Cambridge, UK, 2013.

Comité Européen de Normalisation. (2014) Produits de construction - Évaluation de l'émission de substances dangereuses - Partie 2 : Essai horizontal de lixiviation dynamique des surfaces. CEN TS 16637-2.

Comité Européen de Normalisation. (Décembre 2002) Caractérisation des déchets – Lixiviation – Essai de conformité pour lixiviation des déchets fragmentés et des boues –Partie 2 : essai en bâchée unique avec un rapport liquide/solide de 10 l/kg et une granularité inférieure à 4 mm EN 12457-2.

Commissariat Général au Développement Durable. (2010) 254 millions de tonnes de déchets produits par l'activité construction en France en 2008. *Chiffres et Statistiques*, 164.

Commissariat Général au Développement Durable. (2015) Bilan 2012 de la production de déchets en France - À partir de l'enquête sur les déchets et les déblais produits par la construction et secteur de la dépollution en 2008. *Chiffres et statiques*, 615.

Commissariat Général au Développement Durable. (Octobre 2010) 254 millions de tonnes de déchets produits par l'activité de construction en France en 2008. *Chiffres et statistiques*, 164. Also published as Déchets gérés par les établissements des travaux publics: quantités et modes de gestion en 2008 Chiffres et statistiques n 230 - July 2011.

Conseil des Communautés Européennes (CCE). (4 Avril 2011) Règlement n° 305/2011 du Parlement européen et du Conseil du 9 mars 2011 établissant des conditions harmonisées de commercialisation pour les produits de construction et abrogeant la directive 89/106/CEE du Conseil Règlement Produits de Construction. *Journal Officiel des Communautés Européennes*, 88, 5–43.

Coquillat, G. (1982) Recyclage de matériaux de démolition dans la confection de béton. CEPTB – Service d'Etude des matériaux, Unité : technologie des bétons, No 80-61-248, Saint Remy les Chevreuse.

Coronado, M., Dosal, E., Coz, A., Viguri, J.R., Andrés, A. (2011) Estimation of Construction and Demolition Waste (C&DW) Generation and Multicriteria Analysis of C&DW Management Alternatives: A Case Study in Spain. *Waste and Biomass Valorization*, 2(2), 209–225.

Cortas, R. (2012) Nouvelle approche expérimentale pour la maîtrise de la fissuration du béton jeune: Influence de la nature et de la saturation des granulats, *Ph.D thesis*, Ecole Centrale de Nantes.

Cortas, R., Rozière, E., Staquet, S., Hamami, A., Loukili, A., Delplancke-Ogletree, M.P. (2014) Effect of the Water Saturation of Aggregates on the Shrinkage Induced Cracking Risk of Concrete at Early Age. *Cement and Concrete Composites*, 50, 1–9.

CPR. (March 9, 2011) European Parliament and of the Council. Regulation (EU) No 305/2011/EC of the European Parliament and of the Council of 9 March 2011 laying down harmonised conditions for the marketing of construction products and repealing Council Directive 89/106/EEC Text with EEA relevance.

Cree, D., Green, M.F., Noumowé, A. (2013) Residual Strength of Concrete Containing Recycled Materials After Exposure to Fire: A Review. *Construction and Building Materials*, 45, 208–223.

Crumbie, A.K. (1994) Characterisation of the microstructure of concrete, *Ph.D thesis*, University of London.

CSTB. (2015) HQE. Référentiel pour la qualité environnementale des bâtiments. Certification NF HQE-TM, 176 p.

Cuenca-Moyano, G.M., Zanni, S., Bonoli, A., Valverde-Palacios, I. (2017) Development of the Life Cycle Inventory of Masonry Mortar Made of Natural and Recycled Aggregates. *Journal of Cleaner Production*, 140, 1272–1286.

Cui, Z.L., Lu, S.S., Wang, Z.S. (2004) Influence of Recycled Aggregate on Strength and Preview the Carbonation Properties of Recycled Aggregate Concrete. *Journal of Buildings and Materials*, 50, 1–9.

CUR SBRCUR. CUR-Aanbeveling 106. (2014) Beton met fijne fracties uit recyclinggranulaten als fijn toeslagmateriaal. Netherlands.

CUR SBRCUR. CUR-Aanbeveling 112. (2014) Beton met betongranulaat als grof toeslagmateriaal. Netherlands.

Cyr, M., Lawrence, P., Ringot, E., Carles-Gibergues, A. (2000) Variability of efficiency factors characterizing mineral admixtures. *RILEM - Materials and Structures*, 33(231), 466–472.

Cyr, M., Mouret, M., Cassagnabere, F., Nguyen, V.N. (2014) Evolution de la rhéologie du béton recyclé frais. Rapport RECYBÉTON N° R/14/RECY/012.

Da Costa, M. (1999) Processes of demolition of structures (in Portuguese), Civil Engineering MSc Thesis, Aveiro, Portugal, Universidade de Aveiro.

Danish Standards Association. (1999) Norm for betonkonstruktioner DS411. Brand. Denmark.

Danish Standards Association. (2011) Beton – Materialer – Regler for anvendelse af EN 206-1 i Danmark. Denmark.

Dao, D.T. (2012) (Multi)-recyclage du béton hydraulique ((Multi)-recycling of concrete), *Ph.D thesis*, doctoral school SPIGA, in French.

Dao, D.T., Sedran, T., de Larrard, F. (23-26 September 2014) Optimization of the recycling of concrete in concrete: application to an airport slab. *12th international symposium on concrete road, Prague, Czech Republic.*

De Brito, J., Alves, F.T. (2010) Concrete with Recycled Aggregates: The Portuguese Experimental Research. *Materials and Structures*, 43(S1), 35–51

De Brito, J., Saikia, N. *Recycled Aggregate in Concrete: Use of Industrial, Construction and Demolition Waste, Green Energy and Technology Collection.* Springer: London, UK, 2013.

De Brito, J.; Barra, M.; Ferreira, L. Influence of the Pre-Saturation of Recycled Coarse Concrete Aggregates on Concrete Properties. *Magazine of Concrete Research,* 2011, 63(8), 617–627.

de Juan, M.S., Gutierrez, P.A. (2009) Study on the Influence of Attached Mortar Content on the Properties of Recycled Concrete Aggregate. *Constructtion and Building Materials,* 23(2), 872–877.

de Larrard, F. (1999a) Concrete mixture-proportioning: A scientific approach, Modern Concrete Technology series No. 9, In: Bentur, A., Mindness, S. (Eds.), E & FN SPON, Routledge, 420 pages, ISBN 0-419-23500-0, London.

de Larrard, F. (1999b) Why Rheology Matters? *Concrete international,* 21(8), 79–81.

de Larrard, F., Belloc, A. (1997) The Influence of Aggregate on the Compressive Strength of Normal and High-Strength Concrete. *ACI Materials. Journal,* 94(5), 417–426.

de Larrard, F., Bosc, F., Catherine, C., De Florenne, F. (August-September 1997) The AFREM Method for the Mix Design of HPC. *Materials and Structures*, 30(201), 439–446.

de Larrard, F., Bosc, F., Catherine, C., De Florenne, F. (mars-avril 1996) La nouvelle méthode des coulis de l'AFREM pour la formulation des bétons à hautes performances, Bulletin des Laboratoires des Ponts et Chaussées n°202, pp. 61–69.

de Larrard, F. (2000) « Structures granulaires et formulation des bétons », Etudes et recherches des Laboratoires des Ponts et Chaussées, OA 34, 414 p., Avril, report n° OA 34, ed. LCPC, 414p, ISBN 2-7208-2006-8.

de Larrard, F., Dao, D.T., Mialon, D., Rogat, D. (December 2014) Recyclage du béton dans le béton – Le chantier expérimental de Chaponost (Rhône), Revue Générale des Routes et Aérodromes No. 924, pp. 72–77.

de Larrard, F., Lédée, V., Sedran, T., Frédérick Brochu, F., Ducassou, J.B. (2003) New test for measuring the compactness of granular fractions on the shock table, Bulletin des laboratoires des ponts et chaussées - 246–247 - september-october-november-december 2003 - réf. 4492, pp. 101–115.

de Larrard, F., Sedran, T. (1994) Optimization of Ultra-High-Performance Concrete by the Use of a Packing Model. *Cement and Concrete Research*, 24(6), 997–1009.

de Larrard, F., Sedran, T. (2007) The BetonlabPro software package, Version 3. *Bulletin des laboratoires des ponts et chaussées*, 270–271, LCPC, 75–85.

de Sa, C., Benboudjema, F. (May 2016) Fissuration des parties d'ouvrage liées à la dessiccation – Modèle de prédiction de la fissuration des dalles du chantier de Chaponost, Recybéton report No. R/16/RECY/031.

De Schepper, M., De Buysser, K., Van Driessche, I., De Belie, N. (2013) The Regeneration of Cement Out of Completely Recyclable Concrete: Clinker Production Evaluation. *Construction and Building Materials*, 38, 1001–1009.

de Schutter, G. (1999) Hydration and Temperature Development of Concrete Made with Blast-Furnace Slag Cement. *Cement and Concrete Research,* 29, 143–149.

Debieb, F., Courard, L., Kenai, S., Degeimbre, R. (2010) Mechanical and Durability Properties of Concrete Using Contaminated Recycled Aggregates. *Cement and Concrete Composite,* 32(6), 421–426.

Decision 2014/955/EU. (2014) Decision 2014/955/EU of December 18th 2014 amending Decision 2000/532/EC on the list of waste pursuant to Directive 2008/98/EC.

Delobel, F., Bulteel, D., Mechling, J.M., Lecomte, A., Cyr, M., Rémond, S. (2016) Application of ASR Tests to Recycled Concrete Aggregates: Influence of Water Absorption. *Construction and Building Materials,* 124, 714–721.

Dent Glasser, L.S. (1979) Osmotic Pressure and the Swelling of Gels. *Cement and Concrete Research,* 9, 515–517.

Dent Glasser, L.S., Kataoka, N. (1981) The Chemistry of alkali-aggregate reaction. *5th International Conference on Alkali-Aggregate Reaction, Cape Town, South Africa.*

Deodonne, K. (2015) Etudes des caractéristiques physico-chimiques des bétons de granulats recyclés et de leur impact environnemental. *Ph.D thesis in Civil Engineering.* Strasbourg University. Strasbourg. France.

de-Oliveira, M.B., Vazquez, E. (1996) The Influence of Retained Moisture in Aggregates from Recycling on the Properties of New Hardened Concrete. *Waste Management,* 16(1), 113–117.

Destombes, M.A. (2003) Guide technique pour l'utilisation des matériaux régionaux d'Ile de France – les bétons et produits de démolition recyclés, Technical report.

DGNB. (2016) DGNB Global Benchmark for sustainability, Germany. www.dgnb-system.de.

Dhir, R.K., Limbachiya, M.C., Leelawat, T. (1999) Suitability of Recycled Concrete Aggregate for Use in BS 5328 Designated Mixes. *Institute of Civil Engineering – Structures Buildings*, 134(3), 257–274.

Dhir, R.K., Paine, K.A. *Performance Related Approach to Use of Recycled Aggregates*. WRAP and University of Dundee, Project code: AGG0074, Eds: Waste & Resources Action Programme, 65 p., Oxon, UK, February 2007.

Dhonde, H.B., Mo, Y.L., Hsu, T.T.C., Vogel, J. (2007) Fresh and Hardened Properties of Self-Consolidating Fiber-Reinforced Concrete. *ACI Materials Journal*, 104(5), 491–500.

Diamond, S. (1989) ASR: Another look of mechanism. *8th International Conference on Alkali-Aggregate Reaction, Kyoto, Japan*, pp. 83–94.

DICOM-DGPR/PLA (Avril 2018) Feuille de route de l'économie circulaire (FREC) – 50 mesures pour une éconmie 100% circulaire.

DIN 4226-100 (Sept., 2000). Gesteinskörnungen für Beton und Mörtel – Teil 100: Rezyklierte Gesteinskörnungen.

DIN. *Gesteinskörnungen für Beton und Mörtel - Teil 100*. Rezyklierte Gesteinskörnungen: Germany, 2002.

Ding, T., Xiao, J., Tam, V.W.Y. (2016) A Closed-Loop Life Cycle Assessment of Recycled Aggregate Concrete Utilization in China. *Waste Management*, 56, 367–375.

Directive 2008/98/CE. (2008) Directive 2008/98/CE du Parlement européen et du Conseil du 19 novembre 2008 relative aux déchets et abrogeant certaines directives. Available on http://eur-lex.europa.eu/LexUriServ/LexUriServ.do?uri=OJ:L:2008:312:0003:0030:fr:PDF.

Divet, L., Randriambololona, R. (1998) Delayed Ettringite Formation: The Effect of Temperature and Basicity on the Interaction of Sulphate and C-S-H Phase. *Cement and Concrete Research*, 28(3), 357–363.

Djerbi Tegguer, T.A. (2012) Determining the Water Absorption of Recycled Aggregates Utilizing Hydrostatic Weighing Approach. *Construction and Building Materials*, 27(1), 112–116.

Domingo-Cabo, A., Lazaro, C., Lopez-Gayarre, F., Serrano-Lopez, M.A., Serna, P., Castano-Tabares, J.O. (2009) Creep and Shrinkage of Recycled Aggregate Concrete. *Construction and Building Materials*, 23(7), 2545–2553.

Dong, Z., Keru, W. (2001) Fracture Properties of High Strength Concrete. *Journal of Materials in Civil Engineering*, 13(1), 86–88.

Dreux G., Festa, J. (1995) Nouveau guide du béton : composants et propriétés, composition et dosage, fabrication, transport et mise en oeuvre, contrôle et normalisation, ed. Eyrolles.

Dron, R., Brivot, F., Chaussadent, T. (1998) Mécanisme de la réaction alcali-silice. *Bulletin de Liaison des Laboratoires des Ponts et Chaussées*, 214, 61–68.

Dugat, J., Roux, N., Bernier, G. (May 1996) Mechanical Properties of Reactive Powder Concretes. *Materials and Structures*, 29(4), 233–240.

ECO-BAU. (2007) KBOB – Béton de granulats recycles. Recommandations, Switzerland, 6p.

ECO-BAU. (2011) ECO-CFC Fiches de construction écologique selon le code des frais de construction (CFC), Switezerland, 73 p.

Ecoinvent. *Ecoinvent Report, Data v2.2*. Swiss Centre for Life Cycle Inventories: Dübendorf, Switzerland, 2011.

Eguchi, K., Teranishi, K., Akira, N., Hitoshi, K., Kimihiko, S., Masafumi, N. (2007) Application of Recycled Coarse Aggregate by Mixture to Concrete Construction. *Construction and Building Materials*, 21, 1542–1551.

EHE Project. (2003) Spanish Standardization Committee. Draft of Spanish Regulations for the sue of recycled aggregates in the production of structural concrete. Spain.

Eligehausen, R., Balazs, G.L. Bond and Detailing. *Bulletin d'Information CEB*, 217, 173–226.

Elsharief, A., Cohen, M.D., Olek, J. (2005) Influence of Lightweight Aggregate on the Microstructure and Durability of Mortar. *Cement and Concrete Research*, 35, 1368–1376.

EN 1097-1. *Tests for Mechanical and Physical Properties of Aggregates - Part 1: Determination of the Resistance to Wear (Micro-Deval)*, 2011.

EN 1097-2. *Tests for Mechanical and Physical Properties of Aggregates Part 2: Methods for the Determination of Resistance to Fragmentation*, 2010.

EN 1097-6. *Tests for Mechanical and Physical Properties of Aggregates - Part 6: Determination of Particle Density and Water Absorption*, 2014.

EN 13369. *Common Rules for Precast Concrete Products*, 2013.

EN 1744-1+A1. *Tests for Chemical Properties of Aggregates - Part 1: Chemical Analysis*, 2012.

EN 1744-5. *Tests for Chemical Properties of Aggregates - Part 5: Determination of Acid Soluble Chloride Salts*, 2006.

EN 933-3. *Tests for Geometrical Properties of Aggregates - Part 3: Determination of Particle Shape.* Flakiness index, 2012.

Engagements pour la croissance verte (May, 2018), Engagement pour la croissance verte relatif à la valorisation et au recyclage des déchets inertes du BTP, https://www.ecologique-solidaire.gouv.fr/engagements-croissance-verte.

Engelsen, C.J., Van Der Sloot, H., Wibetoe, G., Justnes, H., Lund, W., Stoltenberg-Hansson, E. (2010) Leaching Characterisation and Geochemical Modelling of Minor and Trace Elements Released from Recycled Concrete Aggregates. *Cement and Concrete Research*, 40(12), 1639–1649.

Engelsen, C.J., Van der Sloot, H., Wibetoe, G., Petkovic, G., Stoltenberg-Hansson, E., Lund, W. (2009) Release of Major Elements from Recycled Concrete Aggregates and Geochemical Modelling. *Cement and Concrete Research*, 39, 446–459.

Engelsen, C.J., Wibetoe, G., Van der Sloot, H., Lund, W., Petkovic, G. (2012) Field Site Leaching from Recycled Concrete Aggregates Applied as Sub-Base Material in Road Construction. *Science of the Total Environment*, 427–428, 86–97.

Erichsen, E., Ulvik, A., Sævik, K. (2011) Mechanical Degradation of Aggregate by the Los Angeles-, the Micro-Deval- and the Nordic Test Methods. *Rock Mechanics and Rock Engineering*, 44, 333–337. doi: 10.1007/s00603-011-0140–y.

Ermco. (2014) Annual ready- mixed concrete industry statistics.

Etxberria, M. *The Role and Influence of Recycled Aggregate. Recycled Materials in Building and Structures.* RILEM: Barcelona, Spain, 2004.

Etxeberria, M., Mari, A.R., Vazquez, E. (2007a) Recycled Aggregate Concrete as Structural Material. *Materials and Structures,* 40(5), 529–541.

Etxeberria, M., Vazquez, E., Mari, A., Barra, M. (2007b) Influence of Amount of Recycled Coarse Aggregates and Production Process on Properties of Recycled Aggregate Concrete. *Cement and Concrete Research,* 37(5), 735–742.

Eurocode 2. *Design of Concrete Structures - Part 1-1: General Rules and Rules for Buildings.* Amendment A1, 2014.

European Commission. (2008) Directive 2008/98/EC on waste (Waste Framework Directive).

European standard. Eurocode 2. *Design of Concrete Structures – Part 1-1 : General Rules and Rules for Buildings*, 2004.

Evangelista, L., de Brito, J. (2007a) Mechanical Behaviour of Concrete Made with Fine Recycled Concrete Aggregates. *Cement and Concrete Composites,* 29(5), 397–401.

Evangelista, L., de Brito, J. (2007b) Environmental life cycle assessment of concrete made with fine recycled concrete aggregates. In: *Portugal SB07 Sustainable Construction, Materials and Practices-Challenge of the Industry for the New Millenium.* Luis Bragança, Manuel Pinheiro, Said Jalali, Ricardo Mateus, Rogério Amoêda, Manuel Correia Guedes, Ed., IOS Press: Amsterdam, pp. 789–794.

Evangelista, L., de Brito, J. (2010) Durability performance of concrete made with fine recycled concrete aggregates. *Cement & Concrete Composites,* 32(1), 9–14.

Evangelista, L., Guedes, M., de Brito, J., Ferro, A.C. (2015) Physical, Chemical and Mineralogical Properties of Fine Recycled Aggregates Made from Concrete Waste. *Construction and Building Materials,* 86, 178–188.

Faessel, P., Robinson, J.R., Morisset, A. *Tables d'états limites ultimes des poteaux en béton armé.* SDTBTP: Paris, 1971.

Faleschini, F., Zanini, M.A., Pellegrino, C., Pasinato, S. (2016) Sustainable Management and Supply of Natural and Recycled Aggregates in a Medium-Size Integrated Plant. *Waste Management,* 49, 146–155.

Fan, Y., Xiao, J., Tam, V.W. (2014) Effect of Old Attached Mortar on the Creep of Recycled Aggregate Concrete. *Structural Concrete*, 15(2), 169–178.

Farah M., Grondin F., Matallah M., Loukili M., Saliba J. (2013) Multi-scales characterization of the early-age creep of concrete. *Proceedings of 9th Concreep, Boston, USA.*

Fares, H. (2009) Propriétés Mécaniques et physico-chimiques de bétons autoplaçants exposés à une température élevée. L2MGC Université de Cergy-Pontoise, *Ph.D thesis.*

Fares, H., Remond, S., Noumowe, A., Cousture, A. (2010) High Temperature Behaviour of Self-Consolidating Concrete: Microstructure and Physicochemical Properties. *Cement and Concrete Research,* 40(3), 488–496.

Fascicule 29. (2006) Bulletin Officiel du Ministère de l'Équipement. Cahier des clauses techniques générales - Fascicule 29 - Exécution des revêtements de voirie et espaces publics en produits modulaires. France.

Fascicule 65 (V 1.0 décembre 2017) Journal Officiel de la République Française - Cahier des clauses techniques générales - Fascicule 65 applicable aux marchés de génie civil- Exécution des ouvrages de génie civil en béton armé et précontraint (The Client General Specifications for public works - Execution of civil engineering reinforced and prestressed concrete structure).

Fascicule 74. (1998) Bulletin Officiel du Ministère de l'Équipement. Cahier des clauses techniques générales - Fascicule 74 - Construction de réservoirs en béton. France.

Fathifazl, G., Razaqpur, A.G., Isgor, O.B., Abbas, A., Fournier, B., Foo, S. (2009) Flexural Performance of Steel-Reinforced Recycled Concrete Beams. *ACI Structural Journal,* 106, 858–867.

Fathifazl, G., Razaqpur, A.G., Isgor, O.B., Abbas, A., Fournier, B., Foo, S. (2011b) Creep and Drying Shrinkage Characteristics of Concrete Produced with Coarse Recycled Concrete Aggregate. *Cement and Concrete Composites,* 33(10), 1026–1037.

Fathifazl, G., Razaqpur, A.G., Isgorc, O. B., Abbasd, A., Fournier, B., Foof, S. (2011a) Shear Capacity Evaluation of Steel Reinforced Recycled Concrete (RRC) Beams. *Engineering Structures,* 33(3), 1025–1033.

FD CEN/TR 15739. (2010) AFNOR, Precast concrete products - Concrete finishes – Identification.

FD P 18-456. (2004) Concrete - Reactivity of a concrete formula with respect to the alkali-aggregate reaction—Criteria for interpretation of the performance test results, in French.

FD P 18-464. (2014) Concrete - Provisions for the prevention of alkali silica reactions, in French.

FD P 18-503. (1989) AFNOR, Surfaces et parements de béton – éléments d'identification.

FD P 18-541. (2015) Aggregates - Guide for the drafting of the quarry specification for the purposes of the prevention of deleterious effects of alkali-aggregate reactions, in French.

FD P 18-542. (2017) Aggregates - Criteria for qualification natural aggregates for hydraulic concrete with respect to the alkali-reaction, in French.

Feraille-Fresnet, A. *Du matériau à l'ouvrage : quelques apports méthodologiques relatifs à l'Analyse de Cycle de Vie.* Habilitation à Diriger des Recherches: Paris Est, 2016.

Ferraris Chiara, F. *Measurement of Rheological Properties of High Performance Concrete: State-of-the-Art Report.* National Institute of Standards and Technology: Gaithersburg, p. 33, 1996.

Ferraris, C F., de Larrard, F. *Testing and Modelling of Fresh Concrete Rheology.* National Institute of Standards and Technology, Gaithersburg, Md, USA, 59, 1998.

Ferreira, L., de Brito, J., Barra, M. (2011) Influence of the Pre-Saturation of Recycled Coarse Concrete Aggregates on Concrete Properties. *Magazine of Concrete Research,* 63, 617–627.

FFB. (2017) Fédération Française du Bâtiment. Benchmark européen sur la gestion des déchets de chantier –Final Report, France, December, 249 p.

FHWA. (august 2007) Long-Life Concrete Pavements in Europe and Canada, report n°FHWA-PL-07-027, 80 p.

Fleischer, W., Rubby, M. (1999). Recycled aggregates from old concrete highway pavements. *Proc., Int. Seminar Exploiting Wastes in Concrete, Thomas Telford, London,* pp. 151–161.

Florea, M.V.A., Brouwers, H.J.H. (2013) Properties of Various Size Fractions of Crushed Concrete Related to Process Conditions and Re-use. *Cement and Concrete Research,* 52, 11–21.

FNTP-FFB-CERIB. (March 2009) Méthodologie d'application du concept de performance équivalente des bétons, Recommandations professionnelles provisoires (Concrete equivalent performance concept implementation methodology, Interim professional recommendations), FNTP-FFB-CERIB, in French.

Folino, P., Xargay, H. (2014) Recycled Aggregate Concrete – Mechanical Behavior Under Uniaxial and Triaxial Compression. *Construction and Building Materials,* 56, 21–31.

Fouré, B. (mai 1996) Empirical constitutive law for concrete in compression and extrapolation to very high strength concrete. *AFPC / AFREM Fourth International Symposium on Utilization of high strength/high performance concrete BHP 96 Presses de l'ENPC*, Paris, pp. 29–31, vol. 2, pp. 663–668.

Fouré, B. *Comportement en compression, flexion, flexion composée et flambement. Granulats et bétons légers. Bilan de dix ans de recherches (joint work).* Presses de l'ENPC: Paris, 1986, pp. 321–348, Chapter 11.

Fouré, B., de Larrard, F., Paultre, P. (mai 1996) Justifications sous sollicitations normales. Bulletin des Laboratoires des Ponts et Chaussées, special XIX, Extension des règlements BAEL / BPEL aux bétons à 80 MPa, pp. 31–41.

Fouré, B.,. Chapitre 1. *Introduction: Modélisation et règlements. Comportement mécanique du béton (joint work).* In Comportement mécanique du béton, J.M. Reynouard and G. Pijaudier-Cabot, Eds., Hermès / Lavoisier editors: Paris, 2005, pp. 21–64.

Fundal, E. *Burnability of Cement Raw Meal with Matrix Correction.* World Cement Res & Dev, April 1996, pp. 63–86.

Galbenis, C.T., Tsimas, S. (2006) Use of Construction and Cemolition Wastes as Raw Materials in Cement Clinker Production. *China Particuology,* 4(2), 83–85.

Galvín, A.P., Agrela, F., Ayuso, J., Beltrán, M.G., Barbudo, A. (2014) Leaching Assessment of Concrete Made of Recycled Coarse Aggregate: Physical and Environmental Characterisation of Aggregates and Hardened Concrete. *Waste Management* (New York, N.Y.), 34(9), 1693–1704.

Galvín, A.P., Ayuso, J., Agrela, F., Barbudo, A., & Jiménez, J.R. Analysis of leaching procedures for environmental risk assessment of recycled aggregate use in unpaved roads. Construction and Building Materials, 40, 1207–1214. 2013

Galvín, A.P., Ayuso, J., Jiménez, J.R., Agrela, F. (2012) Comparison of Batch Beaching Tests and Influence of pH on the Release of Metals from Construction and Demolition Wastes. *Waste Management* (New York, N.Y.), 32(1), 88–95.

Garboczi, E.J.; Bentz, D.P. (1996) Modelling of the Microstructure and Transport Properties of Concrete. *Construction and Building Materials,* 10, 293–330.

Gaspar, L., Stryk, J., Marchtrenker, S., Bencze, S. (2015) Recycling Reclaimed Road Material in Hydraulically Bound Layers. *Proceedings of the Institution of Civil Engineers – Transport,* 168(3), 276–287.

GB/T 14902 CNIS (China National Institute of Standardization). (2012) 预拌混凝土 – Ready Mixed Concrete. China.

GBC. (2018) World Green Bulding Council. www.bgc.org.

Geng, S., Wang, Y., Zuo, J., Zhou, Z., Du, H., Mao, G. (2017) Building Life Cycle Assessment Research: A Review by Bibliometric Analysis. *Renewable and Sustainable Energy Reviews,* 76, 176–184.

Gesoğlu, M., Özbay, E. (2007) Effects of Mineral Admixtures on Fresh and Hardened Properties of Self-Compacting Concretes: Binary, Ternary and Quaternary Systems. *Materials and Structures,* 40(9): 923–937.

Gomes, F., Brière, R., Feraille, A., Habert, G., Lasvaux, S., Tessier, C. (2013) Adaptation of Environmental Data to National and Sectorial Context: Application for Reinforcing Steel Sold on the French Market. *International Journal of Life Cycle Assessment,* 18(5), 926–938.

Gomez-Soberon, J.M. (2002) Creep of Concrete with Substitution of Normal Aggregate by Recycled Concrete Aggregate. *ACI Material Journal,* 209, 461–474.

Gómez-Soberón, J.M. (2003) Relationship between Gas Adsorption and the Shrinkage and Creep of Recycled Aggregate Concrete. *Cement Concrete and Aggregates,* 25, 42–48.

Gomez-Soberon, J.M.V. (2002) Porosity of Recycled Concrete with Substitution of Recycled Concrete Aggregate An Experimental Study. *Cement and Concrete Research,* 32(8), 1301–1311.

Gonçalves, A., Esteves, A., Vieira, M. (2004) Influence of recycled concrete aggregates on concrete durability. *International RILEM Conference on the Use of Recycled Materials in Buildings and Structures.*

González, J.G., Robles, D.R., Valdés, A.J., Morán del Pozo, J.M., Romero, M.I.G. (2013) Influence of Moisture States of Recycled Coarse Aggregates on the Slump Test. *Advanced Materials Research,* 742, 379–383.

Gonzalez-Fonteboa, B., Martinez-Abella, F. Shear Strength of Recycled Concrete Beams. *Construction and Building Materials,* 21(4), 887–893.

Granju, J.L. *Béton armé - Théorie et applications selon l'Eurocode 2.* Eyrolles : Paris, 2012.

Green Directive. *Sterreichischer Güteschutzverband Recycling-Baustoffe. Directive sur les matériaux de construction recyclés - Directive verte,* 7ème édition. Wien, Austria. 7th Directive, 2007.

Grondin, F., Bouasker, M., Mounanga, P., Khelidj, A., Perronnet, A. (2010) Physico-Chemical Deformations of Solidifying Cementitious Systems: Multiscale Modelling. *Materials and Structures,* 43(1–2), 151–165.

Grondin, F., Matallah, M. (2014) How to Consider the Interfacial Transition Zones in the Finite Element Modelling of Concrete? *Cement and Concrete Research,* 58, 67–75.

Grübl, P., Nealen, A. (1998) Construction of an Office Building Using Concrete Made From Recycled Demolition Material. *Darmstadt Concrete,* 13, 163–177.

Guerra, M., Ceia, F., de Brito, J., Júlio, E. (2014) Anchorage of Steel Rebars to Recycled Aggregates Concrete. *Construction and Building Materials,* 72, 113–123.

Guo, M., Grondin, F., Alam, S.Y., Loukili, A. (2016) Fracture process analysis of recycled aggregate concrete with combined acoustic emission and digital image correlation techniques, *Proceeding of FraMCoS-9, Berkeley, USA.*

Gustavsson, M., Karawacki, E., Gustafsson, S.E. (1994) Thermal Conductivity, Thermal Diffusivity, and Specific Heat of Thin Samples from Transient Measurements with Hot Disk Sensors. *Review of Scientific Instruments,* 65, 3858.

Haase, R., Dahms, J. (1998). Material Cycles on the Example of Concrete in the Northern Parts of Germany. *Beton,* 48(6), 350–355.

Habert, G., Castillo, E., Morel, J.C. (2010) Sustainable indicators for resources and energy in building construction. *Second International Conference on Sustainable Construction Materials and Technologies, Ancona, Italy.*

Hájek, P., Fiala, C., Kynčlová, M. (2013) Life Cycle Assessments of Concrete Structures – A Step Towards Environmental Savings. *Structural Concrete,* 12(1), 13–22.

Hamard, E., Cazacliu, B. (2014) Etude de la validité des normes d'essai mécaniques pour les granulats recycles, report R14RECY011 for the Projet National Recybeton, in French, 48 p.

Hamard, E., Cazacliu, B. (Sept 2014) Influence du type de concassage sur les différentes fractions granulaires. Rapport PN Recybeton R/14/RECY/008, 33 p.

Hammer, T.A., Fossa, K.T., Bjøntegaard, Ø. (2007) Cracking Tendency of HSC: Tensile Strength and Self Generated Stress in the Period of Setting and Early Hardening. *Materilas and Structures,* 40, 319–324.

Hanehara, S., Oyamada, T., Fujiwara, T. (2008b) Reproduction of delayed ettringite formation in concrete and its mechanism. *1st International Conference on Microstructure Related Durability of Cementitious Composites, Nanjing, China,* pp. 143–152.

Hanehara, S., Oyamada, T., Fukuda, S., Fujiwara, T. (2008a) Delayed ettringite formation and alkali aggregate reaction. *8th International Conference on Creep, Shrinkage and Durability of Concrete and Concrete Structures CONCREEP'08, 1051-1056, Ise-Shima, Japan.*

Hansen, C.T., Mattock, A.H. (1966) Influence of Size and Shape of Member on the Shrinkage and Crrep of Concrete. *Journal of American Concrete Institute,* 63(10), 267–290.

Hansen, T.C. (1986) Recycled Aggregates and Recycled Aggregate Concrete Second State-of-the-Art Report Developments 1945–1985. *Materials and Structure,* 19, 201–246.

Hansen, T.C. (1992) Recycling of demolished concrete and masonry, RILEM Report 06, E&FN Spon, London.

Hansen, T.C., Boegh, E. (1985) Elasticity and Drying Shrinkage of Recycled Aggregate Concrete. *ACI Journal Proceedings,* 82(5), 648–652.

Hansen, T.C., Narud, H. (1983) Strength of Recycled Concrete Made from Crushed Concrete Coarse Aggregate. *Concrete International,* 5(1), 79–83.

Hansen, W., Almudaiheem, J.A. (1987) Ultimate Drying Shrinkage of Concrete - Influence of Major Parameters. *ACI Materials Journal,* 84, 217–223.
Harrisson, A. (Jan 2010) Free Lime: The Fifth Phase. *International Cement Review,* 59–61.
Hayakawa, M., Itoh, Y. (1982) A new concrete mixing method for improving bond mechanism. In: Bartos, P. (Ed.), *Bond in Concrete.* Applied Science Publishers: London, pp. 24–33.
HB 155 Australia. (2002). Guide to the use of recycled concrete and masonry materials. Australia.
Hendriks, C., Xing,W. (2004) Suitable separation treatment of stony components in construction and demolition waste (CDW). In: Vázquez, E., Hendriks, C.F., Janssen, G.M.T. (Eds.), *Proceedings of the International RILEM Conference on the Use of Recycled Materials in Building and Structures. RILEM Publications SARL,* pp. 166–172.
Heng, N., Hashimoto, C., Wanatabe, T., Ueda, T. (2005) Effect of the Mixing Method with Oscillation of Mixture Inserting Vibrators on Properties of Concrete with Low Treated Recycled Aggregate. *JCA Proceedings of Cement & Concrete,* 58, 525–532.
Higuchi, Y. (1980) Coated-Sand Technique Produces High Strength Concrete. *Concrete International,* 2(4), 75–76.
Hobbs, D.W. (1993) Deleterious Alkali–Silica Reactivity in the Laboratory and Under Field Conditions. *Magazine of Concrete Research,* 45(163), 103–112.
Homand-Etienne, F., Houpert, R. (1989) Thermally Induced Microcracking in Granites: Characterization and Analysis. *International Journal of Rock Mechanics and Mining Sciences & Geomechanics Abstracts,* 26, 125–134.
Hong, S.H., Glasser, F.P. (2002) Alkali Sorption by C-S-H and C-A-S-H Gels: Part II Role of Alumina. *Cement and Concrete Research,* 32, 1101–1111.
Hu, J., Kejin, W. (2011) Effect of Coarse Aggregate Characteristics on Concrete Rheology. *Construction and Building Materials,* 25(3), 1196–1204.
Huang, W.L., Lin, D.H., Chang, N.B., Lin, K.S. (2002) Recycling of Construction and Demolition Waste Via a Mechanical Sorting Process. *Resources, Conservation and Recycling,* 37, 23–37.
Huntzinger, D.N., Eatmon, T.D. (2009) A Life-Cycle Assessment of Portland Cement Manufacturing: Comparing the Traditional Process with Alternative Technologies. *Present and Anticipated Demands for Natural Resources: Scientific, Technological, Political, Economic and Ethical Approaches for Sustainable Management,* 17(7), 668–675.
Idir, R., Feraille, A., Serres, N., Braymand, S. (September 2015) Evaluation environnementale du béton de granulats recyclés – CEREMA, ENPC, ICUBE study R/15/RECY/024.
IFSTTAR. (2011) Méthode d'essai n°78 - Essai sur granulats pour béton. Mesure de l'absorption totale d'eau par un sable concassé.
IFSTTAR. (2017) Technical guide GTI5, Recommandations pour la prévention des désordres dus à la réaction sulfatique interne (Recommendations for preventing disorders due to delayed ettringite formation), ed. Ifsttar, 70p, in French.
Ignjatovic, I.S., Marinkovic, S.B., Miskovic, Z., Savic, A.R. (2013) Flexural Behavior of Reinforced Recycled Aggregate Concrete Beams Under Short-Term Loading. *Materials and Structures,* 46(6), 1045–1059.
IN 0034 SNCF. (December 2011) Exécution des Ouvrages en Béton Armé et Béton Précontraint. France.
Ioannidou, D., Nikias, V., Brière, R., Zerbi, S., Habert, G. (2015) Land-Cover-Based Indicator to Assess the Accessibility of Resources Used in the Construction Sector. *Resources, Conservation and Recycling,* 94, 80–91.
ISO 1920-10. (2010) Testing of concrete – Part 10: determination of static modulus of elasticity in compression.
ISO 1920-X. Testing of concrete – Various parts. ISO, 2016.
ISO 6892-1. *Metallic Materials – Tensile Strength – Part 1: Method of Test at Room Temperature.* International Organization for Standardization: Genève, 2009.
Jansson, R. (2004a) Material properties related to fire spalling of concrete, Thesis work, Lund Institute of Technology.
Jansson, R. (2004b) Measurement of concrete thermal properties at high temperatures. In: *Fib task group 4.3 workshop "Fire design of concrete structures: What now? What next?", Milan.*

Jezequel, F. (2013) Etude de variabilité des caractéristiques de granulats recyclés issus de diverses sources et suivi. Rapport RECYBÉTON N° R/14/RECY/013.

Jezequel, F. (2014) Incertitude de mesure et pertinence des essais de masse volumique et absorption d'eau sur les granulats recyclés. Rapport RECYBÉTON N°R/14/RECY/014.

JGJ 52. (2006) Standard for technical requirements and test method of sand and crushed stone (or gravel) for ordinary concrete, Chinese Standard.

JGJ/T 2040. (2011) CNIS (China National Institute of Standardization). 再生骨料应用技术规程 - technical provisions for the use of RA. China.

Ji, C., Hong, T., Jeong, J., Kim, J., Lee, M., Jeong, K. (2016) Establishing Environmental Benchmarks to Determine the Environmental Performance of Elementary School Buildings Using LCA. *Energy and Buildings,* 127, 818–829.

Ji, T., Chen, C.Y., Chen, Y.Y., Zhuang, Y.Z., Chen, J.F., Lin, X. (2013) Effect of Moisture State of Recycled Fine Aggregate on the Cracking Resistibility of Concrete. *Construction and Building Materials,* 44, 726–733.

JIS A 5021. (2011) JSA (Japanese Standards Association). コンクリート用再生骨材H – H Class Recycled Aggregates for concrete. Japan.

JIS A 5022. (2012) JSA (Japanese Standards Association). 再生骨材Mを用いたコンクリート Concrete containing M Class Recycled Aggregates. Japan.

JIS A 5023. (2012) JSA (Japanese Standards Association). 再生骨材Lを用いたコンクリート - Concrete containing L Class Recycled Aggregates. Japan.

JIS A 5308. (2014) JSA (Japanese Standards Association). レディーミクストコンクリート – Ready Mixed Concrete. Japan.

Johansson, R. *Air Classification of Fine Aggregates.* Doktorsavhandlingar Vid Chalmers Tekniska Högskola, N.S., 3817. Chalmers University of Technology: Göteborg, 2014.

Jolliet, O., Saade, M., Crettaz, P., Shaked, S. *Analyse du cycle de vie Comprendre et réaliser un écobilan.* Presses polytechniques et universitaires romandes: Lausanne, 2010.

Jones, T.N. (1988) A New Interpretation of Alkali-Silica Reaction and Expansion Mechanisms in Concrete. *Chemistry and Industry,* 74, 40–44.

Josa, A., Aguado, A., Cardim, A., Byars, E. (2007) Comparative analysis of the life cycle impact assessment of available cement inventories in the EU. *Cement and Concrete Research,* 37(5), 781–788.

Josa, A., Aguado, A., Heino, A., Byars, E., Cardim, A. (2004) Comparative Analysis of Available Life Cycle Inventories of Sement in the EU. *Cement and Concrete Research,* 34(8), 1313–1320.

Joseph, M., Boehme, L., Sierens, Z., Vandewalle, L. (2015). Water Absorption Variability of Recycled Concrete Aggregates. *Magazine of Concrete Research,* 67, 592–597.

JSCE. *Standard Specifications for Concrete Structures.* Japan Society of Civil Engineers: Tokyo, 2002.

Jullien, A., Proust, C., Martaud, T., Rayssac, E., Ropert, C. (2012) Variability in the Environmental Impacts of Aggregate Production. *Resources, Conservation and Recycling,* 62, 1–13.

Kadri, E.H., Aggoun, S., De Schutter, G. (2009) Interaction between C3A, Silica Fume and Naphthalene Sulphonate Superplasticiser in High Performance Concrete. *Construction and Building Materials,* 23(10), 3124–3128.

Kaihua, L., Jiachuan, Y., Qiong, H., Yao, S., Chaoying, Z. (2016) Effects of Parent Concrete and Mixing Method on the Resistance to Freezing and Thawing of Air-Entrained Recycled Aggregate Concrete. *Construction and Building Materials,* 106, 264–273.

Kameche, Z.E., Ghomari, F., Khelidj, A., Choinska, M. (2012) La perméabilité relative comme indicateur de durabilité : Influence de l'état hydrique du béton et de la taille des éprouvettes. AUGC 2012.

Kang, T.H.K., Kim, W., Kwak, Y.K., Hong, S.G. (2014) Flexural Testing of Reinforced Concrete Beams with Recycled Concrete Aggregates. *ACI Structural Journal,* 111(3), 607–616.

Karihaloo, B.L., Abdalla, H.M., Xiao, Q.Z. (2006) Deterministic Size Effect in the Strength of Cracked Concrete Structures. *Cement and Concrete Research,* 36(1), 171–188.

Katz, A. (2003) Properties of Concrete Made with Recycled Aggregate from Partially Hydrated Old Concrete. *Cement and Concrete Research,* 33, 703–711.

Kavyrchine, M., Fouré, B., Bronsart, O. (November 1976) Flambement de poteaux en béton léger armé sous charge de courte durée - Recherche "IU", CEBTP report.

Kawabata, Y., Yamada, K. (2015) Evaluation of Alkalinity of Pore Solution Based on the Phase Composition of Cement Hydrates with Supplementary Cementitious Materials and its Relation to Suppressing ASR Expansion. *Journal of Advanced Concrete Technology,* 13, 538–553.

Kawai, K. (2011) Application of Performance-Based Environmental Design to Concrete and Concrete Structures. *Structural Concrete,* 12(1), 30–35.

Khalaf, F.M., DeVenny, A.S. (2004) Recycling of Demolished Masonry Rubble as Coarse Aggregate in Concrete: Review. *ASCE Journal of Materials in Civil Engineering,* 16, 331–340.

Khaliq, W., Kodur, V. (2011) Thermal and Mechanical Properties of Fiber Reinforced High Performance Self-Consolidating Concrete at Elevated Temperatures. *Cement and Concrete Research,* 41, 1112–1122.

Khelidj, A., Loukili, A. (1998) Etude Expdrimentale Du Couplage Hydro-Chimique Dans Les Bétons En Cours de Maturation:incidence Sur Les Retraits. *Materials and Structures,* 31, 588–594.

Khoshkenari, A.G., Shafigh, P., Moghimi M., Bin Mahmud, H. (2014) The Role of 0–2 mm Fine Recycled Concrete Aggregate on the Compressive and Splitting Tensile Strengths of Recycled Concrete Aggregate Concrete. *Materials and Design,* 64, 345–354.

Khoury, E., Cazacliu, B., Remond, S. (2017) Impact of the Initial Moisture Level and Pre-Wetting History of Recycled Concrete Aggregates on their Water Absorption. *Materials and Structures,* 50, 229.

Kikuchi, M, Mukai, T., Koizumi, H. (1988) Properties of concrete products containing recycled aggregate. *Proc. 2nd international symposium on demolition and re-use of concrete and masonry,* vol. 2, pp. 595–604.

Kim, H., Bentz, D. (2008) Internal Curing with Crushed Returned Concrete Aggregates for High Performance Concrete. In *NNRMCA Concrete Technology Forum: Focus on Sustainable Development*, pp. 1–12.

Kim, J.K., Kim, Y.Y. (1996) Experimental Study of the Fatigue Behavior of High Strength Concrete. *Cement and Concrete Research,* 26(10), 1513–1523.

Kim, S.W., Yun, H.D. Evaluation of the Bond Behavior of Steel Reinforcing Bars in Recycled Fine Aggregate Concrete. *Cement and Concrete Composites,* 46, 8–18.

Kim, S.W., Yun, H.D. Influence of Recycled Coarse Aggregates on the Bond Behavior of Deformed Bars in Concrete. *Engineering Structures,* 48, 133–143.

Kim, T.H. (2014). A Study on Carbon Emission Impact Analysis of Concrete Mixing Recycled Aggregate. *Journal of Korea Society of Waste Management,* 31, 96–104.

Kim, Y.J., Choi, Y.W. (2012) Utilization of Waste Concrete Powder as a Substitution Material for Cement. *Construction and Building Materials,* 30, 500–504.

Kjellsen, K.O., Wallevik, O.H., Fjalberg, L. (1998) Microstructure and Microchemistry of the Paste-Aggregate Interfacial Transition Zone of High-Performance Concrete. *Advances in Cement Research,* 10, 33–40.

Kleijer, A. L., Lasvaux, S., Citherlet, S., Viviani, M. (2017) Product-Specific Life Cycle Assessment of Ready Mix Concrete: Comparison between a Recycled and an Ordinary Concrete. *Resources, Conservation and Recycling,* 122, 210–218.

Klinkenberg, L.J. *The Permeability of Porous Media to liquid and Gases.* Drilling and Production Practice, American Petroleum Institute: New York, 1941.

Knaack, A.M., Kurama, Y.C. (2015) Behavior of Reinforced Concrete Beams with Recycled Concrete Coarse Aggregates. *Journal of Structural Engineering,* 141, B4014009.

Knoeri, C., Sanyé-Mengual, E., Althaus, H.J. (2013) Comparative LCA of Recycled and Conventional Concrete for Structural Applications. *The International Journal of Life Cycle Assessment,* 18(5), 909–918.

Kofoworola, O.F., Gheewala, S.H. (2008) Environmental Life Cycle Assessment of a Commercial Office Building in Thailand. *The International Journal of Life Cycle Assessment,* 13(6), 498.

Kohno, K., Okamoto, T., Isikawa, Y., Sibata, T., Mori, H. (1999) Effects of Artificial Lightweight Aggregate on Autogenous Shrinkage of Concrete. *Cement and Concrete. Research,* 29, 611–614.

Koji, S. (2010) The current state and future prospects of waste and recycling in Japan. In: *Proceedings of the First International Conference on SustainableUrbanization, Hong Kong*, pp. 837–845.

Kong, D., Lei, T., Zheng, J., Ma, C., Jiang, J., Jiang, J. (2010) Effect and Mechanism of Surface-Coating Pozzolanic Materials Around Aggregate on Properties and ITZ Microstructure of Recycled Aggregate Concrete. *Construction and Building Materials*, 24, 701–708.

Kordina, K., Blume, F. (1985) Empirische Zusammenhänge zur Ermittlung der Schubtragfähigkeit stabförmiger Stahlbetonelemente. *Deutscher Ausschuss für Stahlbeton, Heft*, 364.

Kou, S., Poon, C., Wan, H. (2012) Properties of Concrete Prepared with Low-Grade Recycled Aggregates. *Construction and Building Materials*, 36, 881–889.

Kou, S.C., Poon, C.S. (2010) Properties of Concrete Prepared with PVA-Impregnated Recycled Concrete Aggregates. *Cement and Concrete Composites*, 32(8), 649–654.

Kou, S.C., Poon, C.S. (2012) Enhancing the Durability Properties of Concrete Prepared with Coarse Recycled Aggregate. *Construction and Building Materials*, 35, 69–76.

Kou, S.C., Poon, C.S. (2013) Long Term Mechanical and Durability Properties of Recycled Aggregate Concrete Prepared with the Incorporation of Fly Ash. *Cement and Concrete Composites*, 37, 12–19.

Koulouris, A., Limbachiya, M.C., Fried, A.N., Roberts, JJ. Use of recycled aggregate in concrete application: Case studies. Sustainable waste management and recycling: construction demolition waste. In: *International Conference, Concrete and Masonry Research Group, Kingston University*, September 14–15 2004, pp. 245–257.

Kumar, R., Singh, B., Bhargava, P. (2011) Flexural Capacity Predictions of Self-Compacting Concrete Beams Using Stress–Strain Relationship in Axial Compression. *Magazin Concrete Research*, 63(1), 49–59.

Kurowa, R., Tsuji, M., Sawamoto, T., Tanaka, Y. (1999) Effect of Water Condition in Aggregate and Mixing Procedure Under Decreased Pressure on Compressive Strength of Recycled Aggregate Concrete. *JCA Proceedings of Cement & Concrete*, 53, 535–542.

Kwan, W.H., Ramli, M., Kam, K.J., Sulieman, M.Z. (2012) Influence of the Amount of Recycled Coarse Aggregate in Concrete Design and Durability Properties. *Construction and Buildings Materials*, 26, 565–573.

Laby, F. (2007) Le recyclage des déchets de démolition pourrait être amélioré par la déconstruction sélective, Actu-Environnement.com.

Laneyrie, C. (2014) Valorisation des déchets de chantiers du BTP: Comportement à haute température des bétons de granulats recyclés, Thèse de l'université de Cergy-Pontoise.

Laneyrie, C., Beaucour, A.L., Green, M., Hebert, R., Ledesert, B., Noumowé, A. (2016) Influence of Recycled Coarse Aggregates on Normal and High Performance Concrete Subjected to Elevated Temperatures. *Construction and Building Materials*, 111, 368–378.

Laneyrie, C., Beaucour, A.L., Noumowe, A. (2014) Évaluation des méthodes de caractérisation des granulats naturels appliquées aux recyclés. Rapport RECYBÉTON N°R/14/RECY/009.

Langton, C.A., Roy, D.M. (1980) Morphology and microstructure of cement paste/rock interface regions. *7th International Congress on the Chemistry of Cement, Paris, France*, pp. 127–132.

Lavaud, R. (2017) Aggregates characterization, unpublished internal report, Technodes S.A.S. HEIDELBERGCEMENT GROUP.

Law No. 2015-992. (2015) Loi n° 2015-992 du 17 août 2015 relative à la transition énergétique pour la croissance verte.

LC 21-901. (December 2008) Transports Québec. Détermination de la composition d'un matériau recyclé contenant des résidus d'enrobé et de béton de ciment. Québec.

LCPC. (1993) Essai pour déterminer les alcalins solubles dans l'eau de chaux, Méthode LPC n°37, ed. IFSTTAR.

LCPC. (2003) Technical guide "Recommandations pour la durabilité des bétons durcis soumis au gel" (Recommendations for durability of concrete submitted to freeze-thaw), ed. LCPC, in French.

LCPC. (2007) Recommandations pour la prévention des désordres dus à la réaction sulfatique interne, Guide technique, GTRSI, ed. Ifsttar, ISSN 1151-1516.

LCPC. (juillet 2004) Essais de compacité des fractions granulaires à la table à secousses. Méthode d'essai des LPC n°61, ed LCPC.

Le Guen, L. (Sept 2015) Technologie de tri sélectif des granulats béton concassé et détection en continu de la présence d'éléments indésirables dans les granulats recyclés. Rapport PN Recybeton R/15/RECY/018, 83 p.

Le Moigne, R. (January 2018) *L'économie circulaire 2ème édition. Stratégie pour un monde durable.* Dunod, Paris.

Le Moigne, R. (September 5 2014) Economie circulaire : les nouveaux business models. Futuribles.com.

Le Saoût, G., Kocaba, V., Scrivener, K. Application of the Rietveld Method to the Analysis of Anhydrous Cement. *Cement and Concrete Research,* 41, 133–148.

Le, T. (2015) Influence de l'humidité des granulats de béton recyclé sur le comportement à l'état frais et durcissant des mortiers. *Ph.D Thesis*, Ecole des Mines de Douai, France.

Le, T., Le Saout, G., Garcia-Diaz, E., Remond, S. (2017) Hardened Behavior of Mortar Based on Recycled Aggregate: Influence of Saturation State at Macro- and Microscopic Scales. *Construction and Building Materials,* 141, 479–490.

Le, T., Rémond, S., Le Saout, G., Garcia-Diaz, E. (2016) Fresh Behavior of Mortar Based on Recycled Sand – Influence of Moisture Condition. *Construction and Building Materials,* 106, 35–42.

Lea. (2003) In Hewlett, P.C. (Ed.), *Lea's Chemistry of Cement and Concrete*, 4th edition. Elsevier: Oxford.

LEED. (2017) LEED v4 for building design and construction, United State, 161 p.

Legrand, C. (1972) Contribution to the Study of the Rheology of Fresh Concrete. *Materials and Structures,* 5(5), 275–295.

Leklou, N. (2008) Contribution à la connaissance de la réaction sulfatique interne. Thesis. Université Paul Sabatier III.

Li, W., Xiao, J., Sun, Z., Kawashima, S., Shah, S.P. (2012) Interfacial Transition Zones in Recycled Aggregate Concrete with Different Mixing Approaches. *Construction and Building Materials,* 35, 1045–1055.

Liang, Y., Ye, Z., Vernerey, F., Xi, Y. (2015) Development of Processing Methods to Improve Strength of Concrete with 100% Recycled Coarse Aggregate. *Journal of Materials in Civil Engineering,* 27, 04014163.

Limbachiya, M.C. *RILEM International Symposium on Environment-Conscious Materials and Systems for Sustainable Development.* RILEM Publications SARL. N. Kashino and Y. Ohama, Paris, 2004.

Limbachiya, M.C., Dhir, R.K., Leelawat, T. (2000) Use of Recycled Concrete Aggregate in High-Strength Concrete. *Materials and Structures,* 33, 574.

Limbachiya, M.C., Marrochino, E., Koulouris, A. (2007) Chemical-Mineralogical Characterisation of Coarse Recycled Concrete Aggregate. *Waste Management,* 27, 201–208.

Lin, Y.H., Yaw-Yauan, T., Ta-Peng, C., Ching-Yun, C. (2004) An Assessment of Optimal Mixture for Concrete Made with Recycled Concrete Aggregates. *Cement and Concrete Research,* 34(8), 1373–1380.

Linß, E., Mueller, A. (2004) High-Performance Sonic Impulses—An Alternative Method for Processing of Concrete. *International Journal of Mineral Processing,* 74, S199–S208.

Lipovac, J.C., Boutonné, A. (2014) Villes durables: leviers de nouveaux modèles économiques et de développement?. *Développement durable et territoires,* 5(1).

Liu, C.L, Bai, G., Wang, L., Quan, Z. (2010) Experimental study on the compression behavior of recycled concrete columns. *2nd International Conference on Waste Engineering and Management – ICWEM.*

Liu, K., Yan, J., Hu, Q., Sun, Y., Zou, C. (2016) Effects of Parent Concrete and Mixing Method on the Resistance to Freezing and Thawing of Air-Entrained Recycled Aggregate Concrete. *Construction and Building Materials,* 106, 264–273.

Liu, Q., Xiao, J.Z., Sun, Z.H. (2011) Experimental Study on the Failure Mechanism of Recycled Concrete. *Cement and Concrete Research,* 41(10), 1050–1057.

Liu, Y., Wang, W., Chen, Y.F., Ji, H. (2016) Residual Stress-Strain Relationship for Thermal Insulation Concrete with Recycled Aggregate After High Temperature Exposure. *Construction and Building Materials,* 129, 37–47.

Lo, T., Cui, H.Z., Tang, W.C., Leung, W.M. (2008) The Effect of Aggregate Absorption on Pore Area at Interfacial Zone of Lightweight Concrete. *Construction and Building Materials,* 22, 623–628.

López Gayarre, F., González Pérez, J., López-Colina Pérez, C., Serrano López, M., López Martínez, A. (2016) Life Cycle Assessment for Concrete Kerbs Manufactured with Recycled Aggregates. *Journal of Cleaner Production,* 113, 41–53.

Lotfi, S., Deja, J., Rem, P., Mróz, R., Van Roekel, E., Van der Stelt, H. (2014) Mechanical Recycling of EOL Concrete into High-Grade Aggregates. *Resources, Conservation and Recycling,* 87, 117–125.

Lotfi, S., Eggimann, M., Wagner, E., Mroz, R., Deja, J. (2015) Performance of Recycled Aggregate Concrete Based on a New Concrete Recycling Technology. *Construction and Building Materials,* 95, 243–256.

Mahmoud, S. Classification of recycled sands and their applications as fine aggregates for concrete and bituminous mixtures, *Ph.D*, Kassel University, Deutschland.

Maki, I., Fukuda, K., Imura, T., Yoshida, H., Ito, S. (1995) Formation of Belite Clusters from Quartz Grains in Portland Cement Clinker. *CCR,* 25(4), 835–840.

Malesev, M., Radonjanin, V., Marinkovic, S. (2010) Recycled Concrete as Aggregate for Structural Concrete Production. *Sustainability,* 2, 1204–1225.

Malhotra, V.M. (1978) Use of recycled concrete as a new aggregate. *Proceedings of Symposium on energy and resource conservation in the cement and concrete industry, CANMET, Report N°76-9, Ottawa.*

Manzi, S., Mazzotti, C., Bignozzi, M.C. (2013) Short and Long-Term Behavior of Structural Concrete with Recycled Concrete Aggregate. *Cement and Concrete Composites,* 37, 312–318.

Marinkovic, S. (2013) Life cycle assessment (LCA) aspects of concrete. In: Pacheco-Torgal, F., Jalali, S., Labrincha, J., John, V.M. (Eds.), *Eco-Efficient Concrete.* Woodhead Publishing: Cambridge.

Marinković, S., Dragaš, J., Ignjatović, I., Tošić, N. (2017) Environmental Assessment of Green Concretes for Structural Use. *Journal of Cleaner Production,* 154, 633–649.

Marinkovic, S.B., Ignjatovic, I. (2013) Life-cycle assessment of concrete with recycled aggregates. In: *Handbook of Recycled Concrete and Demolition Waste.* Woodhead Publishing series in Civil and Structural Engineering, , number 47, (§ 23.2 Properties of concrete with recycled aggregates), F. Pacheco-Torgal, V. W. Y. Tam, J. A. Labrincha, Y. Ding and J. de Brito, eds., Cambridge, UK.

Marinkovic, S.B., Radonjanin, V. (2013) Life cycle assessment of concrete with recycled aggregate. In: Pacheco-Torgal, F., Tam,V. W. Y., Labrincha, J. A., Ding, Y. de Brito, J. eds., *Handbook of Recycled Concrete and Demolition Waste*, vol. 23. Woodhead Publishing and Fernando Pacheco: Cambridge, UK.

Marroccoli, M., Telesca, A., Ibris, N., Naik, T. R. (2016) Construction and demolition waste as raw materials for sustainable cements. *Fourth International Conference on Sustainable Construction Materials and Technologies.* www.claisse.info/Proceedings.htm.

Martínez-Lage, I., Martínez-Abella, F., Vázquez-Herrero, C., Pérez-Ordóñez, J. L. (2012) Properties of Plain Concrete Made with Mixed Recycled Coarse Aggregate. *Construction and Building Materials*, 37, 171–176.

Martin-Morales, M., Zamorano, M., Valverde-Palacios, G., Cuenca-Moyano, M., Sănchez-Roldăn, Z. (2013) *Quality Control of Recycled Aggregates (RAs) from Construction and Demolition Waste (CDW) in Handbook of Recycled Concrete and Demolition Waste.* Woodhead Publishing, F. Pacheco-Torgal, V. W. Y. Tam, J. A. Labrincha, Y. Ding and J. de Brito, eds., Cambridge, UK.

Maruyama, I., Sato, R. (2005) A trial of reducing autogenous shrinkage by recycled aggregate. In *Proceedings of the Fourth International Research Seminar. Report TVBM-3126, Gaithersburg, Maryland, USA*, pp. 264–270.

Materrio (Nov., 2017). www.materrio.construction: recyclage et valorisation des matériaux, www.unicem.fr/ 2017/11/27/www-materrio-construction-recyclage-et-valorisation-des-materiaux/.

Mazars, J. (1984) Application de la mécanique de l'endommagement au comportement non linéaire et à la rupture de béton de structures, Thèse de doctorat, Université Paris VI.

Mbemba-Kiele, E.P. (2010) Influence Du Vent et de La Cure Sur Le Comportement Des Bétons Au Trés Jeune Âge, Thèse de doctorat, Ecole Centrale de Nantes.

Mechling, J.M., Lecomte, A., Merriaux, K. (2003) The Water Absorption Measurement of Mineral Admixture for Concretes by Evaporometry. *Materials and Structures,* 36(255), 32–39.

MEDDE. (2012) Ministère de l'Ecologie, du Développement durable et de l'Energie. National framework for waste prevention.

MEDDE. (October 2013) Direction Générale des Risques – Service des risques technologiques "Nomenclature des installations classées – liste des activités soumises à la TGAP".

Mefteh, H., Kebaïli, O., Oucief, H., Berredjem, L., Arabi, N. (2013) Influence of Moisture Conditioning of Recycled Aggregates on the Properties of Fresh and Hardened Concrete. *Journal of Cleaner Production,* 54, 282–288.

Mehta, P.K. (1986) Hardened cement paste-microstructure and its relationships to properties. *8th International Congress on the Chemistry of Cement. Rio de Janeiro, Brazil,* pp. 113–121.

Meinel, A. (2010) Fine and Very Fine Screening. *Mineral Processing,* 51, 2–8.

Menard, Y., Bru, K., Touze, S., Lemoign, A., Poirier, J.E., Ruffie, G., Bonnaudin, F., Von Der Weid, F. (2013) Innovative Process Routes for a High-Quality Concrete Recycling. *Waste Management,* 33, 1561–1565.

MINERGIE. (2016) Label MINERGIE-ECO® s'appliquant aux bâtiments, Switzerland.

Mohamed Mohamed, A.S. (2011) Influence de la valorisation des microfibres vegetales su la formulation et la resistance aux cycles de gel-degel de BAP, *Ph.D Thesis.* University of Cergy-Pontoise-L2MGC.

Momber, A.W. (2004) Aggregate Liberation from Concrete by Flow Cavitation. *International Journal of Mineral Processing,* 74, 177–187.

Mongeard, L., Dross, A. (June 2016) La ressource en matériaux inertes recyclables dans le béton en France. RECYBETON report R/16/RECY/032, 72 p.

Monteiro, P.J.M., Maso, J.C., Olivier, J.P. (1985) The Aggregate-Mortar Interface. *Cement and Concrete Research,* 15, 953–958.

Montero, A., Tojo, Y., Matsuo, T., Matsuto, T., Yamada, M., Asakura, H., Ono, Y. (2010) Gypsum and Organic Matter Distribution in a Mixed Construction and Demolition Waste Sorting Process and Their Possible Removal from Outputs. *Journal of Hazardous Materials,* 175, 747–753.

MOP law no 85-704. (12 july 1985) Law on public project contracting and its relationship to private project management ("Loi MOP").

Moreno, J., Cazacliu, B., Artoni, R., Cothenet, A. (2016) Recycled Concrete Aggregate Friability During Mixing New Concrete. *Construction and Building Materials,* 116, 299–309.

Moreno, J., Cothenet, A., Cazacliu, B. (2015) Influence du malaxage sur la rhéologie des bétons de granulats recycles. Rapport R15RECY016 for the Projet National Recybeton, 41 p.

Mounanga, P., Baroghel-Bouny, V., Loukili, A., Khelidj, A. (2006) Autogenous Deformations of Cement Pastes: Part I. Temperature effects at early age and micro–macro correlations. *Cement and Concrete Research,* 36, 110–122.

Mousavi, M., Ventura, A., Antheaume, N. (2017) LCA Modeling of Cement Concrete Waste Management, 35e Rencontres de l'AUGC, Nantes, pp. 134–137.

MRF. (2014) Diagnostic rapide et environnemental appliqué aux matériaux recyclés issus du BTP (Projet DREAM)-Synthèse, MRF Agence DLB, Convention ASEME 1206C0068, 22 p.

Mueller, A. (2015) Recycled Aggregate Characterization Methods. *Presented at the III Progress of Recycling in the Built Environment, I. Martins, C. Ulsen and S. C. Angulo,* pp. 244–276.

Mukai, T., Kikuchi, M. (1988) Properties of Reinforced Concrete Beams Containing Recycled Coarse Aggregate. *Demolition and Reuse of Concrete and Masonry,* 2, 670–679.

Mulder, E., de Jong, T.P.R., Feenstra, L. (2007) Closed Cycle Construction: An Integrated Process for the Separation and Reuse of C&D Waste. *Waste Management,* 27, 1408–1415.

Müller, A., Wienke, L. (2004) Measurements and models for the gravity concentration of C&D waste through jigging. In: Vázquez, E., Hendriks, Ch. F., Janssen, G.M.T. (Eds.), *Proceedings of the International RILEM Conference on the Use of Recycled Materials in Building and Structures.* RILEM Publications SARL, Paris, pp. 115–122.

Müller, C., Reiners, J., Palm, S. (2015) Closing the loop: What type of concrete re-use is the most sustainable option (Technical Report No. A-2015-1860), European Cement Research Academy, p. 41.

Multon, S., Cyr, M., Sellier, A., Leklou, N., Petit, L. (2008) Coupled Effects of Aggregate Size and Alkali Content on ASR Expansion. *Cement and Concrete Research*, 38(3), 350–359.

Mulugeta, M., Engelsen, C.J., Wibetoe, G., Lund, W. (2011) Charge-Based Fractionation of Oxyanion-Forming Metals and Metalloids Leached from Recycled Concrete Aggregates of Different Degrees of Carbonation: A Comparison of Laboratory and Field Leaching Tests. *Waste Management*, 31(2), 253–258.

Murata, J., Kukokawa, H. (1992) Viscosity Equations for Fresh Concrete. *ACI Materials Journal*, 89(3), 230–237.

Nagataki, S., Gokce, A., Saeki, T., Hisada, M. (2004) Assessment of Recycling Process Induced Damage Sensitivity of Recycled Concrete Aggregates. *Cement and Concrete Research*, 34(6), 965–971.

NB26 Norsk Betongforening. (2014) Publikasjon nr.26 - Materialgjenvinning av betong og murverk for betongproduksjon. Norway. August 2003.

NEN 5905+A1. (August 2008) Nederlandse aanvulling op NEN-EN 12620 "Toeslagmaterialen voor beton". Netherlands.

NEN 7375 Netherlands standardisation institute. (2004) Leaching characteristics - Determination of the leaching of inorganic components from moulded or monolitic materials with a diffusion test - Solid earthy and stony materials. NEN 7375.

Neveu, A. (2016) Simulation numérique de la fragmentation des granulats, These de doctorat de l'Ecole doctorale SPIGA.

Neveu, A., Artoni, R., Richard, P., Descantes, Y. (2016) Fracture of Granular Materials Composed of Arbitrary Grain Shapes: A New Cohesive Interaction Model. *Journal of the Mechanics and Physics of Solids*, 95, 308–319.

Neville, A. (2004) The Confused World of Sulphate Attack on Concrete. *Cement and Concrete Research*, 34, 1275–1296.

Neville, A., Dilger, W., Brooks, J. (1983). *Creep of Plain and Structural Concrete*. Construction Press, London.

NF 033 revision. (2016) AFNOR, Référentiel de certification du béton prêt à l'emploi.

NF DTU 21 P1-2. (2017) Building works—Execution of concrete structures—Part 1-2: General criteria for the selection of materials, in French.

NF DTU 21. (2017) AFNOR, Travaux de bâtiment, Éxécution des ouvrages en béton (in French).

NF EN 1097-1. (2011) French Standard, Tests for mechanical and physical properties of aggregates. Part 1: Determination of the resistance to wear (micro-Deval). Afnor ed. Paris. France.

NF EN 1097-2. (2010) AFNOR, Tests for mechanical and physical properties of aggregates - Part 2: methods for the determination of resistance to fragmentation.

NF EN 1097-6. (2014) AFNOR, Tests for mechanical and physical properties of aggregates - Part 6: determination of particle density and water absorption.

NF EN 1097-6. *Tests for Mechanical and Physical Properties of Aggregates—Part 6: Determination of Particle Density and Water Absorption*, 2001.

NF EN 12 457-2. (2002) AFNOR Caractérisation des déchets - Lixiviation - Essai de conformité pour lixiviation des déchets fragmentés et des boues - Partie 2 – NFEN 12 457-2.

NF EN 12350-7. (2012) AFNOR, Testing fresh concrete - Part 7: air content - Pressure methods.

NF EN 12390-3 AFNOR. (2012) Testing hardened concrete - Part 3: compressive strength of test specimens.

NF EN 12390-5 AFNOR. (2012) Essais pour béton durci. Partie 5 : Résistance à la flexion sur éprouvettes.

NF EN 12390-6 AFNOR. (2012) Testing hardened concrete - Part 6: tensile splitting strength of test specimens.

NF EN 12620 + A1. (2008) AFNOR, Aggregates for concrete.

NF EN 13670/CN. (2013) AFNOR, Execution of concrete structures - National addition to NF EN 13670.

NF EN 15804. (2012) Afnor: "Sustainability of con-struction works, Environmental product declarations, Core rules for the product category of construction products" (No. NF EN 15804).

NF EN 1744-1+A1. (February 2014) AFNOR, Essais visant à déterminer les propriétés chimiques des granulats - Partie 1 : analyse chimique, NF EN 1744-12010 +A1 2013+A1.

NF EN 1744-5. (2007) AFNOR, Tests for chemical properties of aggregates - Part 5: determination of acid soluble chloride salts.

NF EN 1744-5. (January 2007) AFNOR, Essais pour déterminer les propriétés chimiques des granulats - Partie 5 : détermination des sels chlorures solubles dans l'acide, NF EN 1744-5.

NF EN 1744-6. (2007) AFNOR, Tests for chemical properties of aggregates - Part 6: determination of the influence of recycled aggregate extract on the initial setting time of cement.

NF EN 196-1. (2006) AFNOR, Methods of testing cement- Part 1: determination of strength, 2016.

NF EN 196-3. (2017) AFNOR, Methods of testing cement- Part 3: determination of setting times and soundness.

NF EN 197-1. (2012) AFNOR, Cement – Part 1: Composition, specifications and conformity criteria for common cements.

NF EN 1992-1-1. (2005) Eurocode 2 - Calcul des structures en béton - Partie 1-1 : Règles générales et règles pour les bâtiments. Association française de normalisation, La Plaine Saint-Denis.

NF EN 1992-1-1/NA. (2016) Eurocode 2: Design of concrete structures—Part 1-1: General rules and rules for buildings - National annex to NF EN 1992-1-1:2005 - General rules and rules for buildings.

NF EN 1992-2. (Mai 2006) AFNOR, Eurocode 2 - Calcul des structures en béton - Partie 2: ponts en béton - Calcul des dispositions constructives.

NF EN 206. (2014) AFNOR Concrete—Specification, performance, production and conformity.

NF EN 206/CN. (2014) AFNOR, Concrete—Specification, performance, production and conformity—National addition to the standard NF EN 206.

NF EN 933-1. (May 2012) AFNOR, Essais pour déterminer les caractéristiques géométriques des granulats - Partie 1 : détermination de la granularité - Analyse granulométrique par tamisage, NF EN 933-1.

NF EN 933-11. (2009) AFNOR, Tests for geometrical properties of aggregates - Part 11: Classification test for the constituents of coarse recycled aggregate.

NF EN 933-3. (2012) AFNOR, Tests for geometrical properties of aggregates - Part 3: determination of particle shape - Flakiness index.

NF EN 933-9. (June 2013) AFNOR, Essais pour déterminer les caractéristiques géométriques des granulats - Partie 9: qualification des fines - Essai au bleu de méthylène, NF EN 933-9 1999+A1.

NF EN ISO 14040. (2006) Afnor: "Management environnemental-Analyse du cycle de vie-Principes et cadre" (No. NF EN ISO 14040).

NF EN ISO/CEI 17025. (2015) AFNOR, General requirements for the competence of testing and calibration laboratories.

NF ISO 5725-2. (1994) Application de la statistique - Exactitude (justess et fidélité) des résultats et méthodes de mesure.

NF P18-417. (December 1989) AFNOR. Béton - Mesure de la dureté de surface par rebondissement à l'aide d'un scléromètre. France.

NF P18-424. (2008) AFNOR. NF P18-424. Concrete – Freeze test on hardened concrete – Freeze in water – Thaw in water.

NF P18-425. (2008) AFNOR. NF P18-425. Concrete – Freeze test on hardened concrete – Freeze in air – Thaw in water.

NF P18-427. (1996) AFNOR, French standard, Béton - Détermination des variations dimensionnelles entre deux faces opposées d'éprouvettes de béton durci (Concrete – Determination of dimensional variations between two opposite faces or a hardened concrete specimen), in French.

NF P18-452. (2017) AFNOR, Concrete— measuring the flow time of concretes and mortars using a workabilitymeter.

NF P18-454. (2004) AFNOR. NF P18-454. Concrete - Reactivity of a concrete formula with regard to the alkali-aggregate reaction - Performance test.

NF P18-459. (2010) AFNOR, Concrete - Testing hardened concrete - Testing porosity and density.

NF P18-459. (2010) Concrete – Testing hardened concrete – Testing porosity and density.

NF P18-542. (May 1988) AFNOR. Bétons - Mesure du temps d'écoulement des bétons et des mortiers aux maniabilimètres. France.

NF P18-544. (2015) AFNOR NF P18-544. Aggregates – Determination of active lime water-soluble alkalis.

NF P18-545. (2011) AFNOR, Aggregates - Defining elements, conformity and coding, 2011, in French.

NF P18-566. (December 2014) AFNOR. Granulats - Analyse granulométrique - Essai à l'aide d'un appareil d'ombroscopie. France.

NF P18-576. (February 2013) AFNOR. Granulats - Détermination du coefficient de friabilité du sable. France.

NF P18-594. (2015) AFNOR NF P18-594. Aggregates - Test methods on reactivity to alkalis.

NF P98-170. (April 2006) AFNOR. Chaussées en béton de ciment - Exécution et contrôle. France.

NF P98-232-4. (1994) AFNOR Test relating to pavements- Determination of the mechanical characteristics on material bound with hydraulic binder-Part 4: Flexural test.

NF P98-233-1. (1994) AFNOR Test relating to pavements- Determination of fatigue resistance on material bound with hydraulic binder-Part 1: Flexural fatigue with constant stress.

Ngo, T.T., Kadri, E.H., Bennacer, R., Cussigh, F. (2010) Use of Tribometer to Estimate Interface Friction and Concrete Boundary Layer Composition during the Fluid Concrete Pumping. *Construction and Building Materials,* 24(7), 1253–1261.

Nguyen, T.D., Le Saout, G., Devilllers, P., Garcia-Diaz, E. (2014) The effect of limestone aggregate porosity and saturation degree on the interfacial zone. *2th International Symposium on Cement-Based Materials For Nuclear Waste (NUWCEM 2014), Avignon, France.*

Nguyen, V.H., Leklou, N., Aubert, J.E., Mounanga, P. (2013) The Effect of Natural Pozzolan on Delayed Ettringite Formation of the Heat-Cured Mortars. *Construction and Building Materials,* 48, 479–484.

NIEA Nothern Ireland Environment Agency. (2013) WRAP quality protocol – Aggregates from inert waste. Waste & Resources Action Programme- Technical report, October, 24 p.

Niry, R.R. (2015) Comportement des bétons à haute température : influence de la nature du granulat. Thèse de l'université de Cergy-Pontoise.

Nixon P.J. (1978) Recycled Concrete as an Aggregate for Concrete-A Review. *Materials and Structures,* 11(65), 371–378.

Noguchi, T. (2015) Outline of AIJ Recommendation for Mix Design, Production and Construction Practice of Concrete with Recycled Concrete Aggregate. *Concrete Journal,* 53(2), 165–171.

Noguchi, T., Kitagaki, R., Tsujino, M. (2011) Minimizing Environmental Impact and Maximizing Performance in Concrete Recycling. *Structural Concrete,* 12, 36–46.

NQ 2560-600. (2002) BNQ Granulats – Matériaux recyclés fabriqués à partir de résidus de béton, d'enrobés bitumineux et de briques – classification et caractéristiques. Quebec.

NS 3473. (2003) Norsk Betongforening. Prosjektering av betongkonstruksjoner - Beregnings- og konstruksjonsregler. Norway.

Oberholster, R.E. (1983) Alkali reactivity of siliceous rock aggregates: Diagnosis of the reaction, testing of cement and aggregate and prescription of preventive measures. In: Idorn, G.M., Rostam, S. (Eds.), 6th International Conference on Alkali-Aggregate Reaction in Concrete. Danish Concrete Association, pp. 419–433.

Odler I., Chen, Y. (1995) Effect of Cement Composition on the Expansion of Heat-Cured Cement Pastes. *Cement and Concrete Research,* 25(4), 853–862.

OFEV. (2006) Directive pour la valorisation des déchets de chantier minéraux. *Environnement pratique,* 31, 36 p.

OFRIR. (2016) Observatoire français des ressources dans les infrastructures. http://ofrir2.ifsttar.fr/.

Ogawa, H., Nawa, T. (2012) Improving the Quality of Recycled Fine Aggregate by Selective Removal of Brittle Defects. *Journal of Advanced Concrete Technology,* 10(12), 395–410.

Oikonomou, N.D. (2005) Recycled Concrete Aggregates. *Cement and Concrete Research,* 27(2), 315–318.

Oksri-Nelfia, L., Mahieux, P.Y., Amiri, O., Turcry, Ph., Lux, J. (2016) Reuse of Recycled Crushed Concrete Fines as Mineral Addition in Cementitious Materials. *Materials and Structures*, 49, 3239–3251.

Oliveira, W.S. (1992) Réactivité et aptitude à la cuisson du cru et son influence sur la spécification des réfractaires pour zones de cuisson et de transfert des fours rotatifs à ciment. *Ciments, Bétons, Plâtres et Chaux*, 3/92(796), 169–176.

Olivier, J.P., Maso, J.C., Bourdette, B. (1995) Interfacial Transition Zone in Concrete. *Advances in cement Based Materials*, 2, 30–38.

Olorunsogo F.T., Padayachee, N. (2002) Performance of Recycled Aggregate Concrete Monitored by Durability Indexes. *Cement and Concrete Research*, 32, 179–185.

Omar, M., Loukili, A., Pijaudier-Cabot, G., Le Pape, Y. (2009) Creep-Damage Coupled Effects: Experimental Investigation on Bending Beams with Various sizes. *Journal of Material Civil Engineering*, 21(2), 65–72.

Omary, S. (2017) Effet de l'incorporation des granulats recycles sur le comportement et la durabilite vis-a-vis du gel-degel des betons, Thèse de l'Université de Cergy Pontoise.

Omary, S., Ghorbel, E., Wardeh, G. (2015) Influence de l'incorporation des granulats recyclés de démolition sur les propriétés physiques et mécaniques des bétons fluids. *Proceedings of 33th Rencontres Universitaires de Génie Civil, Orléans, France, In French.*

Omary, S., Ghorbel, E., Wardeh, G. (2016) Relationships between Recycled Concrete Aggregates Characteristics and Recycled Aggregates Concretes Properties. *Construction and Building Materials*, 108, 163.

ORÉE. (2015) Circular economy. Serving the preservation of resources and the climate. A flows and channels approach to a territorial ecosystem. www.oree.org/3priorites/economie-circulaire/ressources.html.

Orsetti, S. (1997) Influence des sulfates sur l'apparition et le developpement de pathologies dans les materiaux de genie civil traites ou non aux liants hydrauliques. Cas du platre dans les granulats issus de produits de demolition. Thèse de l'Université Paris 6.

Ortiz, O., Castells, F., Sonnemann, G. (2009) Sustainability in the Construction Industry: A Review of Recent Developments Based on LCA. *Construction and Building Materials*, 23(1), 28–39.

Otsuki, N., Miyazato, S., Yodsudjai, W. (2003) Influence of Recycled Aggregate on Interfacial Transition Zone, Strength, Chloride Penetration and Carbonation of Concrete. *Journal of Materials in Civil Engineering*, 15, 443–451.

Padmini, A.K., Ramamurthy, K., Mathews, M.S. (2009) Influence of Parent Concrete on the Properties of Recycled Aggregate Concrete. *Construction and Building Materials*, 23(2), 829–836.

Palmieri, R., Bonifazi, G., Serranti, S. (2014) Automatic detection and classification of EOL-concrete and resulting recovered products by hyperspectral imaging. In: VoDinh, T., Lieberman, R.A., Gauglitz, G.G. (Eds.), Advanced Environmental, Chemical, and Biological Sensing Technologies Xi. Spie-Int Soc Optical Engineering: Bellingham, p. 91060D.

Paranhos, R.S., Cazacliu, B.G., Sampaio, C.H., Petter, C.O., Neto, R.O., Huchet, F. (2016) A Sorting Method to Value Recycled Concrete. *Journal of Cleaner Production*, 112(Part 4), 2249–2258.

Park, J., Tae, S., Kim, T. (2012) Life Cycle CO2 Assessment of Concrete by Compressive Strength on Construction Site in Korea. *Renewable and Sustainable Energy Reviews*, 16(5), 2940–2946.

Pavoine, A., Brunetaud, X., Divet, L. (2012) The Impact of Cement Parameters on Delayed Ettringite Formation. *Cement and Concrete Composites*, 34(4), 521–528.

Pavoine, A., Divet, L. (2007) Réactivité d'un béton vis-à-vis d'une réaction sulfatique interne, Techniques et méthodes des Laboratoires des Ponts et Chaussées, Méthode d'essai des lpc, 66, 19 p.

PDR. (February 2000) US Départment of Housing & Urban Development - Office of Policy Development and Research, A guide to Deconstruction.

Pedro, D., de Brito, J., Evangelista, L. (2014) Influence of the Use of Recycled Concrete Aggregates from Different Sources on Structural Concrete. *Construction and Building Materials*, 71, 141–151.

Pedro, D., de Brito, J., Evangelista, L. (2015) Performance of Concrete Made with Aggregates Recycled from Precasting Industry Waste: Influence of the Crushing Process. *Materials and Structures*, 48, 3965–3978.

Pepe M. (2015) A Conceptual Model for Designing Recycled Aggregate Concrete for Structural Applications, Springer theses, ISSN 2190-5053, 167 p.

Petavratzi, E., Kingman, S.W., Lowndes, I.S. (2007) Assessment of the Dustiness and the Dust Liberation Mechanisms of Limestone Quarry Operations. *Chemical Engineering and Processing: Process Intensification*, 46, 1412–1423.

Petitpain, M., Dehaudt, S., Jacquemot, F., Rougeau, P. (2017) Annales du BTP. Economie circulaire – Utilisation des matières premières secondaires dans les bétons, vol. 68, N° 4, ESKA.

Pettingell, H. (June 2008) An Effective Dry Sand Manufacturing Process from Japan; Potential to Replace Natural Sand Entirely in Concrete. *Quarry Management Magazine*, 1–6.

Pillard, W., Bodet, R., Braymand, S. (2018) Certifications pour évaluer la qualité environnementale des bâtiments, RECYBETON Technical report, 35 p.

PN RECYBETON, thème 2. (2014) Validité des normes d'essais sur les granulats recyclés, rapport interne (rédigé par le CTG : Aït Alaiwa A., Lavaux R.), Axe 2.1, 30p.

PN RECYBETON. (2011) RECYclage Complet Des BETONs, France. www.pnrecybeton.fr/.

Poole, A.B. (1992) Alkali-silica reactivity mechanisms of gel formation and expansion. *9th International Conference on Alkali-Aggregate Reaction, London, England*, vol. 1, pp. 782–789.

Poon, C., Chan, D. (2007) The Use of Recycled Aggregate in Concrete in Hong Kong. *Resources, Conservation and Recycling*, 50(3), 293–305.

Poon, C.S., Chan, D. (2007) Effects of Contaminants on the Properties of Concrete Paving Blocks Prepared with Recycled Concrete Aggregates. *Construction and Building Materials*, 21(1), 164–175.

Poon, C.S., Shui, Z.H., Lam, L., Fok, H., Kou, S.C. (2004) Influence of Moisture States of Natural and Recycled Aggregates on the Properties of Fresh and Hardened Concrete. *Cement and Concretes Research*, 34, 31–36.

prCEN/TS 12390-12. (2010) Testing hardened concrete - Part 12: Determination of the potential carbonation resistance of concrete: Accelerated carbonation method.

prEN 12390-13. (2012) Testing hardened concrete - Part 13: Determination of secant modulus of elasticity in compression. European committee for standardization, Bruxelles.

Prezzi, M., Monteiro, J.M., Sposito, G. (1997) The Alkali-Silica Reaction, Part 1: Use of the Double Layer Theory to Explain the Behaviour of Reaction-Products Gels. *ACI Materials Journal*, 94, 10–17.

Prince, M.J.R. and Singh, B. (2014) Bond Strength of Deformed Steel Bars in High-Strength Recycled Aggregate Concrete. *Materials and Structures*, 48(12), 3913–3928.

Prince, M.J.R., Singh, B. (2013) Bond Behaviour of Deformed Steel Bars Embedded in Recycled Aggregate Concrete. *Construction and Building Materials*, 49, 852–862.

PTV 406 COPRO. (October 2003) Granulats recyclés - Granulats de débris de béton, de débris mixtes, de débris de maçonnerie et de débris asphaltiques - Pierres concassées et graves - Granulats pour béton, MTLH et GNT. Belgium.

Purnell, P., Black, L. (2012) Embodied Carbon Dioxide in Concrete: Variation with Common Mix Design Parameters. *Cement and Concrete Research*, 42(6), 874–877.

Putallaz, J.C. et al. (January 28, 2016) Guide technique d'application pour l'utilisation de matériaux minéraux de recyclage. Technical report from Canton of Valais, AVE WBV and AVGB, Mineral Resources commission, 76 p.

Quattrone, M., Cazacliu, B., Angulo, S.C., Hamard, E., Cothenet, A. (2016) Measuring the Water Absorption of Recycled Aggregates, What is the Best Practice for Concrete Production? *Construction and Building Materials*, 123, 690–703.

Radocea, A. (1994) A Model of Plastic Shrinkage. *Magazine of Concrete Research*, 46, 125–132.

Raeis, S., Reza, B., Daniotti, B., Pelosato, R., Dotelli, G. (2015) Properties of Cement–Lime Mortars vs. Cement Mortars Containing Recycled Concrete Aggregates. *Construction and Building Materials*, 84, 84–94.

Rafla, K. Empirische Formeln zur Berechnung der Schubtragfähigkeit von Stahlbetonbalken (Empirical formulas for the calculation of shear capacity of reinforced concrete beams). *Strasse, Brücke, Tunnel*, 23(12), 311–320.

Ravina, D., Shalon, R. (1968) Plastic Shrinkage Cracking. *ACI Journal Proceedings,* 65(22), 282–294.

Ravindrarajah Sri, R., Tam, C.T. (1985) Recycling Concrete as Fine Aggregate in Concrete. *International Jouranl of Cement Composites Lightweight Concrete,* 9(4), 235–241.

RECYC-QUEBEC. (1999) Guide d'information sur le recyclage des matériaux secs.

Reggad, A. (1993) Influence de la finesse de mouture d'un cru quartzeux et alcalin sur la texture du clinker et les propriétés des ciments. *Ciments, Bétons, Plâtres et Chaux,* 3/93(103), 242–244.

Repellin, D. B. *Comptabilité des flux de matières dans les régions et les départements. Guide méthodologique.* Ministère de l'Écologie, du Développement durable et de l'énergie – CGDD (coll. Repères: La Défense, 2014.

Ricaud J.M., Masson R. (2009). "Effective properties of linear viscoelastic heterogeneous media: Internal variables formulation and extension to ageing behaviours", *International Journal of Solids and Structures,* 46, 1599–1606.

Richardson, A., Coventry, K., Bacon, J. (2011) Freeze/Thaw Durability of Concrete with Recycled Demolition Aggregate Compared to Virgin Aggregate Concrete. *Journal of Cleaner Production,* 19(2–3), 272–277.

Ricotier, D. *Dimensionnement des structures en béton selon l'Eurocode 2.* Editions du moniteur: Paris, 2012.

RILEM – TC 107-CSP (1998) Standardized Test Methods for Creep and Shrinkage. *Materials and Structures,* 31, 507–512.

RILEM. (1994) Specifications for Concrete with Recycled Aggregates. *Material and Structures,* 27, 557–559.

Rilem. (August 1995) Compressive strength for service and accident conditions" RILEM Draft Recommendation 129-MHT Test Methods for Mechanical Properties of Concrete at High Temperatures. *Materials and Structures,* 28(7), 410–414.

Robinson, J.R., Modjabi, S.S. (Sept. 1968) La prévision des charges de flambement des poteaux en béton armé par la méthode de M. P. Faessel. Annales de l'ITBTP, pp. 1295–1316.

Robinson, J.R., Morisset, A. (1969) Paramètres fondamentaux de la fissuration des tirants en béton armé. Annales de l'ITBTP, N° 254, série Béton, Béton armé (102), pp. 227–246.

Rogers, C.A., Hooton, R.D. (1991) Reduction in Mortar and Concrete Expansion with Reactive Aggregates Due to Alkali Leaching. *Cement, Concrete and Aggregates,* 13(1), 42–49.

Rønning, T.F. (2001) Freeze-thaw resistance of concrete: effect of: curing conditions, moisture exchange and materials, *Ph.D thesis in Civil Engineering. Norwegian Institute of Technology.* Division of Structural Engineering. Trondheim, Norway.

Roziere, E., Cortas, R., Loukili, A. (2015) Tensile Behaviour of Early Age Concrete: New Methods of Investigation. *Cement and Concrete Composites,* 55, 153–161.

Ryu J.S. (2002) An Experimental Study on the Effect of Recycled Aggregate on Concrete Properties. *Magazine of Concrete Research,* 54(1), 7–12.

Ryu, J.S. (2002) Improvement on Strength and Impermeability of Recycled Concrete Made from Crushed Concrete Coarse Aggregate. *Journal of Materials Science Letters,* 21, 1565–1567.

Sagoe-Crentsil, K.K., Brown, T., Taylor, A.H. (2001) Performance of Concrete Made with Commercially Produced Coarse Recycled Concrete Aggregate. *Cement and Concrete Research,* 31, 707–712.

Salgues, M., Souche, J.C., Devillers, P., Garcia-Diaz, E. (2016) Influence of Initial Saturation Degree of Recycled Aggregates on Fresh Cement Paste Characteristics: Consequences on Recycled Concrete Properties. *European Journal of Environmental and Civil Engineering,* 22(9), 1146–1160. doi: 10.1080/19648189.2016.1245630.

Saliba, S., Grondin, F., Matallah, M., Loukili, A., Boussa, H. (2013) Relevance of a Mescoscopic Modeling for the Coupling between Creep and Damage in Concrete. *Mechanics of time-dependent materials,* 17(3), 481–499.

Samouh, H., Soive, A., Roziere, E., Loukili, A. (2016) Experimental and Numerical Study of Size Effect on Long-Term Drying Behavior of Concrete: Influence of Drying Depth. *Materials and Structures,* 49, 1–20.

Sampaio, C.H., Cazacliu, B.G., Miltzarek, G.L., Huchet, F., le Guen, L., Petter, C.O., Paranhos, R., Ambrós, W.M., Oliveira, M.L.S. (2016) Stratification in Air Jigs of Concrete/Brick/Gypsum Particles. *Construction and Building Materials,* 109, 63–72.

Sampaio, C.H., Tavares, L.M.M. (2005) Beneficiamento gravimétrico. Uma introdução aos processos de concentração mineral e reciclagem de materiais por densidade. Editora da Ufrgs.

Sánchez de Juan, M., Alaejos Gutierrez, P. (9-11 november 2004) Influence of attached mortar content on the properties of recycled concrete aggregate, RILEM, Actes du colloque Use of Recycled Materials in Building and Structures, Barcelona, Espana.

Sanchez de Juan, M., Aleajos Gutiérrez, P. (8-11 November 2004) Influence of recycled aggregate quality on concrete properties. *International Rilem Conference on the Use of Recycled Materials in Buildings and Structures, Barcelona, Spain*, pp. 545–553.

Sánchez-Roldán, Z., Martín-Morales, M., Valverde-Palacios, I., Valverde-Espinosa, I., Zamorano, M. (2016) Study of Potential Advantages of Pre-Soaking on the Properties of Pre-Cast Concrete Made with Recycled Coarse Aggregate. *Materiales de Construcción*, 66, e076.

Sarhat, S., Sherwood, E. (2013) Residual Mechanical Response of Recycled Aggregate Concrete after Exposure to Elevated Temperature. *Journal of Material in Civil Engineering*, 25, 1721–1730.

Sato, R., Maruyama, I., Sogabe, T., Sogo, M. Flexural Behavior of Reinforced Recycled Concrete Beams. *Journal of Advanced Concrete Technology*, 5, 43–61, 2007.

Schmidt, R., Kern, A. (2001) Quantitative XRD Phase Analysis. *World Cement*, 32, 35–42.

Schnellert, T., Kehr, K., Müller, A. (2011) Development of a separation process for gypsum-contaminated concrete aggregates. In: John, V.M., Vazquez, E., Angulo, S.C., Ulsen, C. (Eds.), Proceedings of the Second International RILEM Conference on Progress of Recycling in the Built Environment. RILEM Publications SARL, Paris, pp. 477–482.

Schoon, J., De Buysser, K., Van Driessche, I., De Belie, N. (2015)Fines Extracted from Recycled Concrete as Alternative Raw Material for Portland Cement Clinker Production. *Cement & Concrete Composites*, 58, 70–80.

Schouenborg, B, Aurstad, J., Pétursson, P. (2004) Test methods adapted to alternative aggregates. In: Vázquez, E., Hendriks, Ch.F., Janssen, G.M.T. (Eds.), International RILEM conference on the use of recycled materials in buildings and structures, Barcelona, pp. 1154–1162, published by RILEM, Paris.

Schubert, S., Hoffmann, C., Leemann, A., Moser, K., Motavalli, M. Recycled Aggregate Concrete: Experimental Shear Resistance of Slabs Without Shear Reinforcement. *Engineering Structures*, 41, 490–497.

Schwartzentruber, A., Catherine, C. (October 2000) La méthode du mortier de béton équivalent (MBE)—Un nouvel outil d'aide à la formulation des bétons adjuvantés, (Method of the concrete equivalent mortar (CEM)—A new tool to design concrete containing admixture). *Material and Structures*, 33(8), 475–482. in French.

Sciumè, G., Benboudjema, F., De Sa, C., Pesavento, F., Berthaud, Y., Schrefler, B.A. (2013) Multiphysics Model for Concrete at Early Age Applied to Repairs Problems. *Engineering Structures*, 57, pp. 374–387.

Scott, H.C. (2006) Mitigating alkali silicate reaction in recycled concrete, *Ph.D Thesis*, University of New Hampshire.

Scrivener, K.L. (1999) Characterization of the ITZ and its quantification by test methods. In: Alexander, M.G., Arliguie, G., Ballivy, G., Bentur, A., Marchand, J. (Eds.), Engineering and transport properties of the interfacial Transition Zone in Cementitious Composites, RILEM Report. RILEM Publications, Paris, 20, pp. 3–14.

Scrivener, K.L., Bentur, A., Pratt, P.L. (1988) Quantitative Characterization of the Transition Zone in High Strength Concretes. *Advances in Cement Research*, 1, 230–237.

Scrivener, K.L., Crumbie, A.K., Laugesen, P. (2004) The Interfacial Transition Zone (ITZ) between cement past and aggregate in concrete. *Interface Science*, 12, 411–421.

Scrivener, K.L., Gartner, E.M. (1987) Microstructural Gradients in Cement Paste Around Aggregate Particles. *MRS Proceedings*, 114, 77–86.

Scrivener, K.L., Pratt, P.L. (1996) In Maso, J.C. (Ed.), *Interfacial Transition Zone in Concrete*. E & FN Spon: London, pp. 3–17.

Seara-Paz, S., Gonzalez-Fonteboa, B., Eiras-López, J., Herrador, M.F. (2013) Bond Behavior between Steel Reinforcement and Recycled Concrete. *Materials and Structures*, 47(1), 323–334.

Sedran, T. (2013). Mise au Point des Formules de Béton de Référence, RECYBETON report, 37 p.

Sedran, T. (2017) Mise au point des formules de béton de reference. National Project Recybeton report n° R/13/RECY/003, in French.

Sedran, T., de Larrard, F. (1994) RENÉ-LCPC: Un Logiciel Pour Optimiser La Granularité Des Matériaux de Génie Civil. Bulletin de liaison des Laboratoires des Ponts et Chaussées, N°194, 87–93.

Sedran, T., de Larrard, F., Le Guen, L. (October-Décember 2007) Determination of the compaction of cements and mineral admixtures using the Vicat needle, Technical note in Bulletin des Laboratoires des Ponts et Chaussées, 270–271.

Sedran, T., Durand, C., de Larrard, F (2009). *An example of UHPFRC recycling, Concevoir et construire en BFUP: Etat de l'art et Perspectives*. BFUP 2009: Marseille.

Sedran, T., Durand, C., de Larrard, F. (2010) An example of UHPFRC recycling. In: Resplendino, J., Toulemonde, F. (Eds.), *Designing and Building with UHPFRC: State of the Art and Development*, Chapter 44. Wiley: Hoboken, NJ, 814 p.

Sedran, T., Le Mouel, J. (2016) Comportement du béton recyclé durci en fatigue par flexion pour utilisation routière, National Project Recybeton report n° R/16/RECY/035, in French.

Seghier, C. (June 12, 2009) La production mondiale de déchets progresse toujours. Actu environnement.com.

Serres, N., Braymand, S., Feugeas, F. (2016) Environmental Evaluation of Concrete Made from Recycled Concrete Aggregate Implementing Life Cycle Assessment. *Journal of Building Engineering*, 5, 24–33.

SETRA. *Methodological Guide: Acceptability of alternative materials in road construction – Environmental assessment*, SETRA Editions. SETRA: Sourdun, France, 2011.

SETRA-LCPC. (1997) French design Manual for pavement structures, Technical guide, ed. SETRA-LCPC, 248 p.

SETRA-LCPC. (1998) Catalogue des structures types de chaussées neuves, ed. SETRA-LCPC, in French.

Shannag, M.J. (2000) High Strength Concrete Containing Natural Pozzolan and Silica Fume. *Cement and Concrete Composites*, 22(6), 399–406.

Shehata, M.H., Christidis, C., Mikhaiel, W., Rogers, C., Lachemi, M. (2010) Reactivity of Reclaimed Concrete Aggregate Produced from Concrete Affected by Alkali-Silica Reaction. *Cement and Concrete Research*, 40, 575–582.

Shehata, M.H., Thomas, M.D.A. (2000) The Effect of Fly Ash Composition on the Expansion of Concrete Due to Alkali–Silica Reaction. *Cement and Concrete Research*, 30(7), 1063–1072.

Shen, K.L., Soong, T.T., Chang, K.C., Lai, M.L. (1995) Seismic Behaviour of Reinforced Concrete Frame with Added Viscoelastic Dampers. *Engineering Structures*, 17(5), 372–380.

Shen, L., Struble, L., Lange, D. (2009) Modeling Dynamic Segregation of Self-Consolidating Concrete. *ACI Materials Journal*, 106(4), 375–380.

Shima, H., Tateyashiki, H., Matsuhashi, R., Yoshida, Y. (2005) An Advanced Concrete Recycling Technology and its Applicability Assessment through Input-Output Analysis. *Journal of Advanced Concrete Technology*, 3, 53–67.

SIA 2030. (June 2010) Société Suisse des ingénieurs et architectes. Béton de recyclage, Switzerland.

SIA. (2010) Béton de recyclage. Cahier technique 2030, Switzerland, 6 p.

Silva, R.V., de Brito, J., Dhir, R.K. (2014) Properties and Composition of Recycled Aggregates from Construction and Demolition Waste Suitable for Concrete Production. *Construction and Building Materials*, 65, 201–217.

Silva, R.V., de Brito, J., Dhir, R.K. (2014) The Influence of the Use of Recycled Aggregates on the Compressive Strength of Concrete: A Review. *European Journal of Environmental and Civil Engineering*, 19(7), 825–849.

Silva, R.V., de Brito, J., Dhir, R.K. (2015) Tensile Strength Behaviour of Recycled Aggregate Concrete. *Construction and Building Materials*, 83, 108–118.

Silva, R.V., de Brito, J., Dhir, R.K. (2017) Availability and Processing of Recycled Aggregates within the Construction and Demolition Supply Chain: A Review. *Journal of Cleaner Production*, 143, 598–614.

Silva, R.V., de Brito, J., Evangelista, L., Dhir, R.K. (2016) Design of Reinforced Recycled Aggregate Concrete Elements in Conformity with Eurocode 2. *Construction and Building Materials,* 105, 144–156.

Silva, R.V., Neves, R., de Brito, J., Dhir, R.K. (2015) Carbonation Behaviour of Recycled Aggregate Concrete. *Cement and Concrete Composites,* 62, 22–32.

Sim, J., Park, C. (2011) Compressive Strength and Resistance to Chloride Ion Penetration and Carbonation of Recycled Aggregate Concrete with Varying Amount of Fly Ash and Fine Recycled Aggregate. *Waste Management (Oxford),* 31(11), 2352–2360.

Slowik, V., Ju, J.W. (2011) Discrete Modeling of Plastic Cement Paste Subjected to Drying. *Cement and Concrete Composites,* 33(9), 925–935.

Slowik, V., Schmidt, M., Fritzsch, R. (2008) Capillary Pressure in Fresh Cement-Based Materials and Identification of the Air Entry Value. *Cement and Concrete Composites,* 30(7), 557–565.

SNBPE. (2011) BETie, Béton et impacts Environnementaux, tool for calculation of environemental impact of concrete, available on www.snbpe.org.

Sobhan, K., Gonzalez, L., Reddy, D.V. (2016) Durability of a Pavement Foundation Made from Recycled Aggregate Concrete Subjected to Cyclic Wet–Dry Exposure and Fatigue Loading. *Materials and Structures,* 49, 2271–2284.

Sobhan, K., Krizek, R.J. (1999) Fatigue Behavior of Fiber-Reinforced Recycled Aggregate Base Course. *Journal of Material Civil Engineering*, 11(2), 124–130.

SOES/CGDD/Minister of the Environment. (2017a) Entreprises du BTP: 227,5 millions de tonnes de déchets en 2014. *Data-Lab*, mars, 4p.

SOeS. (2017b), Enquête Déchets et déblais – 2014, www.statistiques.developpement-durable.gouv.fr/logement-construction/r/entreprises-btp-enquetes-thematiques.html?tx_ttnews%5Btt_news%5D=25289&cHash=067708a56cc57823d0c2bf6d61f9647c.

Soualhi, H., Kadri, E.H., Ngo, T.T., Bouvet, A., Cussigh, F., Benabed, B. (2015) Rheology of Ordinary and Low-Impact Environmental Concretes. *Journal of Adhesion Science and Technology,* 29(20), 2160–2175.

Soualhi, H., Kadri, E.H., Ngo, T.T., Bouvet, A., Cussigh, F., Tahar, Z.E.A. (2016) Design of Portable Rheometer with New Vane Geometry to Estimate Concrete Rheological Parameters. *Journal of Civil Engineering and Management,* 3730(June), 1–9.

Souche, J.C. (2015) Etude du retrait plastique des bétons à base de granulats recyclés avec mesure de l'influence de leur degré de saturation, Thèse de doctorat, Université de Montpellier.

Souche, J.C., Devillers, P., Salgues, M., Garcia-Diaz, E. (2017) Influence of Recycled Coarse Aggregates on Permeability of Fresh Concrete. *Cement and Concrete Composite,* 83, 394–404.

SRBTP. (2014) Syndicat des Recycleurs du BTP, Guide de conception et de fonctionnement des installations de traitement des déchets du BTP, 144 p.

Sri Ravindrarajah, R., Tam, C.T. (1987) Recycling Concrete as Fine Aggregate in Concrete. *International Journal of Cement Composite and Lightweight Concrete,* 9(4), 235–241.

SS EN 137003. (2008) SIS. Betong - Användning av EN 206-1 i Sverige. Sweden.

Standards Australia. (December 2002) Guide to the use of recycled concrete and mansonry material. Technical Guide HB 155-20022, Australia.

Stanton, T.E. (1940) Expansion of Concrete Through Reaction between Cement and Aggregates. *Proceedings of The American Society of Civil Engineers,* 66(10), 1781–1812.

Stassi, F. (October 7, 2016) Le recyclage du béton bute sur plusieurs obstacles. Usine nouvelle.com.

Stefan, L., Benboudjema, F., Torrenti, J.M., Bissonnette, B. (2010) Prediction of Elastic Properties of Cement Pastes at Early Ages. *Computational Materials Science,* 47, 775–784.

Sucic, A., Lotfy, A. (2016) Effect of New Paste Volume on Performance of Structural Concrete Using Coarse and Granular Recycled Concrete Aggregate of Controlled Quality. *Construction and Buildings Materials,* 108, 119–128.

Tahar, Z.E.A., Kadri, E.H., Ngo, T.T., Bouvet, A., Kaci, A. (2016) Influence of Recycled Sand and Gravel on the Rheological and Mechanical Characteristic of Concrete. *Journal of Adhesion Science and Technology,* 30(4), 392–411.

Tahar, Z.E.A., Ngo, T.T., Kadri, E.H., Bouvet, A., Aggoun, S. (2017) Effect of Cement and Admixture on the Utilization of Recycled Aggregates in Concrete. *Construction and Building Materials,* 149, 91–102.

Takahashi, H., Ando, M. (2009). DEM simulation of crushing for concrete blocks by mobile crusher. In: Nakagawa, M., Luding, S. (Eds.), *Powders and Grains 2009.* Amer Inst Physics: Melville, pp. 843–846.

Tam, V.W., Gao, X., Tam, C., Chan, C. (2008) New Approach in Measuring Water Absorption of Recycled Aggregates. *Construction and Building Materials,* 22(3), 364–369.

Tam, V.W., Tam, C.M., Le, K.N. (2007) Removal of Cement Mortar Remains from Recycled Aggregate Using Pre-Soaking Approaches. *Resources Conservation and Recycling,* 50(1), 82–101.

Tam, V.W.Y. (2008) Economic Comparison of Concrete Recycling: A Case Study Approach. *Resources, Conservation and Recycling,* 52, 821–828.

Tam, V.W.Y. *Improving Waste Management Plans in Construction Projects), Handbook of Recycled Concrete and Demolition Waste.* Woodhead Publishing Limited, F. Pacheco-Torgal, V. W. Y. Tam, J. A. Labrincha, Y. Ding and J. de Brito, eds., Cambridge, UK, 2013.

Tam, V.W.Y., Tam, C.M. (2007) Assessment of Durability of Recycled Aggregate Concrete Produced by Two-Stage Mixing Approach. *Journal of Materials Science,* 42, 3592–3602.

Tam, V.W.Y., Tam, C.M. (2008) Diversifying Two-Stage Mixing Approach (TSMA) for Recycled Aggregate Concrete : TSMA S and TSMA Sc. *Construction and Building Materials,* 122(10), 2068–2077.

Tam, W.Y.V., Gao, X.F., Tam, C.M. (2005) Micro-Structural Analysis of Recycled Aggregate Concrete Produced from Two-Stage Mixing Approach. *Cement and Concrete Research,* 35(6), 1195–1203.

Tamimi, A.K. (1994) The Effects of a New Mixing Technique on the Properties of the Cement Paste-Aggregate Interface. *Cement and Concrete Research,* 24, 1299–1304.

Tasong, W.A., Lynsdale, C.J., Cripps, J.C. (1999) Aggregate-Cement Paste Interface. *Cement and Concrete Research,* 29, 1019–1025.

Tavakoli, M., Soroushian, P. (1996) Drying Shrinkage Behavior of Recycled Aggregate Concrete. *Concrete International,* 18(11), 58–61.

Tavakoli, M., Soroushian, P. (1996) Strengths of Recycled Aggregate Concrete Made Using Field-Demolished Concrete as Aggregate. *ACI Materials Journal,* 93, 178–181.

Taylor, H.F.W. *Cement Chemistry,* 2nd edition, T Telford: London, 1997.

Taylor, H.F.W., Famy, C., Scrivener, K.L. (2001) Delayed Ettringite Formation. *Cement and Concrete Research,* 31(5), 683–693.

Taylor, J.C., Hinczak, I., Matulis, C.E. Rietveld Full-Profile Quantification of Portland Cement Clinker: The Importance of Including a Full Crystallography of the Major Phase Polymorphs. *Powder Diffraction,* 15, 7–18.

Teh, S.H., Wiedmann, T., Castel, A., de Burgh, J. (2017) Hybrid Life Cycle Assessment of Greenhouse Gas Emissions from Cement, Concrete and Geopolymer Concrete in Australia. *Journal of Cleaner Production,* 152, 312–320.

Teramoto, Y., Tsuji, M., Kobayashi, T. (1998) Application of Mixing Technique Under Decreased Pressure on Recycled Concrete. *Cement Science,* 52, 456–461.

Thiery, M., Dangla, P., Belin, P., Habert, G., Roussel, N. (2013) Carbonation Kinetics of a Bed of Recycled Concrete Aggregates: A Laboratory Study on Model Materials. *Cement and Concrete Research,* 46, 50–65.

Thomas, C., Setién, J., Polanco, J.A., Alaejos, P., Sánchez de Juan, M. (2013) Durability of Recycled Aggregate Concrete. *Construction and Buildings Materials,* 40, 1054–1065.

Thomas, C., Sosa, I., Setién, J., Polanco, J.A., Cimentada, A.I. (2014) Evaluation of the Fatigue Behavior of Recycled Aggregate Concrete. *Journal of Cleaner Production,* 65, 397–405.

Thomas, C., Sosa, I., Setién, J., Polanco, J.A., Lombillo, I., Cimentada, A.I. (2014) Fatigue Limit of Recycled Aggregate Concrete. *Construction and Building Materials,* 52, 146–154.

Thomas, J.J., Jennings, H. Materials of Cement Science Primer TEA-21 Year 5 Final Report 98 p., http://iti.northwestern.edu/publications/utc/tea-21/FR-5-Jennings-Thomas.pdf, accessed on Dec. 12, 2018.

Thomas, M.D.A., Blackwell, B.Q., Nixon, P.J. (1996) Estimating the Alkali Contribution from Fly Ash to Expansion Due to Alkali-Aggregate Reaction in Concrete. *Magazine of Concrete Research,* 48(177), 251–264.

Tomas, J., Schreier, M., Gröger, T., Ehlers, S. (1999). Impact Crushing of Concrete for Liberation and Recycling. *Powder Technology,* 105, 39–51.

Topçu, I.B., Sengel, S. (2004) Properties of Concretes Produced with Waste Concrete Aggregate. *Cement and Concrete Research,* 34(8), 1307–1312.

Topic, J., Prosek, Z., Plachy, T. (2017) Influence of increasing amount of recycled concrete powder on mechanical properties of cement paste. *IOP Conference Series: Materials Science and Engineering,* 236, 012094. doi: 10.1088/1757-899X/236/1/012094.

Torgal, F.P. (2013) In: Pacheco-Torgal, F., Tam, V., Labrincha, J., Ding, Y., De Brito, J. *Handbook of Recycled Concrete and Demolition Waste.* Woodhead, Cambridge, UK.

Torrenti, J.M., Benboudjema, F. (2005) Mechanical Threshold of Concrete at an Early Age. *Materials and Structures,* 38(277), 299–304.

Touzé, S., Bru, K., Ménard, Y., Weh, A., Von der Weid, F. (2017) Electrical Fragmentation Applied to the Recycling of Concrete Waste – Effect on Aggregate Liberation. *International Journal of Mineral processing,* 158, 68–75.

Tovar-Rodríguez, G., Barra, M., Pialarissi, S., Aponte, D., Vázquez, E. (2013) Expansion of Mortars with Gypsum Contaminated Fine Recycled Aggregates. *Construction and Building Materials,* 38, 1211–1220.

TRA 550. (November 2011) CRIC-OCCN. Règlement d'application BENOR BETON TRA 550. Belgium.

Tsoumani, A.A., Barkoula, N.-M., Matikas, T.E. (2015) Recycled Aggregate as Structural Material. *Waste Biomass Valor,* 6, 883–890.

Turcry, P. (2004) Retrait et Fissuration Des Bétons Autoplaçans : Influence de La Formulation, Thèse de doctorat, Ecole Centrale de Nantes et Université de Nantes.

Turcry, P., Loukili, A. (2006) Evaluation of Plastic Shrinkage Cracking of Self-Consolidating Concrete. *ACI Materials Journal,* 103(4), 272–279.

Ulsen, C., Kahn, H., Hawlitschek, G., Masini, E.A., Angulo, S.C. (2013) Separability Studies of Construction and Demolition Waste Recycled Sand. *Waste Management,* 33, 656–662.

UNI 8520-1. (September 2005) Aggregati per calcestruzzo - Istruzioni complementari per l'applicazione della EN 12620 - Parte 1: Designazione e criteri di conformità, Italy.

UNI 8520-2. (September 2005) Aggregati per calcestruzzo - Istruzioni complementari per l'applicazione della EN 12620 – Requisiti. Italy.

UNPG. (2016) L'industrie Française des granulats en 2015, UNICEM-UNPG technical paper.

VALIDEO. *Construire durable aujourd'hui Certification des compétences et réalisations en construction durable.* SECO, Brussels, 2011.

Van den Heede, P., De Belie, N. (2012) Environmental Impact And Life Cycle Assessment (LCA) of Traditional and "Green" Concretes: Literature Review and Theoretical Calculations. *Cement and Concrete Composites,* 34(4), 431–442.

Vandecasteele, C., Heynen, J., Goumans, H. (2013) Materials Recycling in Construction: A Review of the Last 2 Decades Illustrated by the WASCON Conferences. *Waste and Biomass Valorization,* 4(4), 695–701.

Vázquez, E., Barra, M., Aponte, D., Jiménez, C., Valls, S. (2014) Improvement of the Durability of Concrete with Recycled Aggregates in Chloride Exposed Environment. *Construction and Building Materials,* 67, 61–67.

Vegas, I., Broos, K., Nielsen, P., Lambertz, O., Lisbona, A. (2015) Upgrading the Quality of Mixed Recycled Aggregates from Construction and Demolition Waste by Using Near-Infrared Sorting Technology. *Construction and Building Materials,* 75, 121–128.

VERDE. (2016) Certificate, Spain, www.gbce.es.

Vicat, L. (1817) Principaux résultats de diverses expériences sur les chaux de construction, les mortiers ordinaires et les bétons. Annales des Ponts et Chaussées, vol. 5, pp. 387–392, Paris.

Vieira, D.R., Calmon, J.L., Coelho, F.Z. (2016) Life Cycle Assessment (LCA) Applied to the Manufacturing of Common and Ecological Concrete: A review. *Construction and Building Materials,* 124, 656–666.

Vieira, J.P.B., Correia, J.R., de Brito, J. (2011) Post-Fire Residual Mechanical Properties of Concrete Made with Recycled Concrete Coarse Aggregates. *Cement and Concrete Research,* 41, 533–541.

Villain, G., Thiery, M., Platret, G. (2007) Measurement Methods of Carbonation Profiles in Concrete: Thermogravimetry, Chemical Analysis and Gammadensimetry. *Cement and Concrete Research,* 37, 1182–1192.

Vitale, P., Arena, N., Di Gregorio, F., Arena, U. (2017) Life Cycle Assessment of the End-of-Life Phase of a Residential Building. *Special Thematic Issue: Urban Mining and Circular Economy,* 60, 311–321.

Vivian, W.Y.T. (2009) Comparing the Implementation of Concrete Recycling in the Australian and Japanese Construction Industries. *Journal of Cleaner Production,* 17, 688–702.

Vrijders, J. (2017) Lisa Wastiels: Environmental impact of concrete with recycled aggregates an evaluation through 4 case studies. *HISER International Conference - Advances in recycling and management of construction and demolition waste, Delft,* pp. 255–259.

Wallevik, J.E. (2006) Relationship between the Bingham Parameters and Slump. *Cement and Concrete Research,* 36(7), 1214–1221.

Wallevik, O.H., Wallevik, J. E. (2011) Rheology as a Tool in Concrete Science: The Use of Rheographs and Workability Boxes. *Cement and Concrete Research,* 41(12), 1279–1288.

Wang, H., Gillott, J.E. (1991) Mechanisms of Alkali-Silica Reaction and Significance of Calcium Hydroxide. *Cement and Concrete Research,* 21, 647–654.

Wang, Z., Wang, L., Cui, Z., Zhou, M. (2011) Effect of Recycled Coarse Aggregate on Concrete Compressive Strength. *Transactions of Tianjin University,* 17, 229–234.

Wardeh, G., Fiorio, B., Ghorbel, E., Gomart, H. (2015) Adhérence béton recyclé / armatures HA. R/15/RECY/021 RECYBETON report. L2MGC, Cergy-Pontoise.

Wardeh, G., Ghorbel, E. (2013) Prediction of Fracture Parameters and Strain-Softening Behavior of Concrete: Effect of Frost Action. *Materials and Structures,* 48(1–2), 1–16.

Wardeh, G., Ghorbel, E. (2015) Analyse du coefficient de ductilité en courbure des sections en béton armé incorporant des granulats recyclés. *9ème colloque National AFPS, Marne-La-Vallée, France.*

Wardeh, G., Ghorbel, E., Gomart, H. (2015) Mix Design and Properties of Recycled Aggregate Concretes: Applicability of Eurocode 2. *International Journal of Concrete Structures and Materials,* 9(1), 1–20.

Wasserman, R., Bentur, A. (1996) Interfacial Interactions in Lightweight Aggregate Concretes and Their Influence on the Concrete Strength. *Cement and Concrete Composites,* 18, 67–76.

Waste and Resources Action Programme (WRAP). (October 2007) Aggregates Research Programme. Engineering properties of concrete containing recycled aggregates. Banbury (UK).

Wee, T., Chin, M., Mansur, M. (1996) Stress-Strain Relationship of High-Strength Concrete in Compression. *Journal of Materials in Civil Engineering,* 8(2), 70–76.

Weil, M., Jeske, U., Schebek, L. (2006) Closed-Loop Recycling of Construction and Demolition Waste in Germany in View of Stricter Environmental Threshold Values. *Waste Management & Research,* 24(3), 197–206.

Weimann, K., Giese, L.B., Mellmann, G., Simon, F.G. (2003) Building Materials from Waste. *Materials Transactions,* 44, 1255–1258.

Wijayasundara, M., Crawford, R.H., Mendis, P. (2017) Comparative Assessment of Embodied Energy of Recycled Aggregate Concrete. *Journal of Cleaner Production,* 152, 406–419.

Wil, V.S. (2015) Stochastic Service-Life Modeling of Chloride-Induced Corrosion in Recycled-Aggregate Concrete. *Cement & Concrete Composites,* 55, 103–111.

Wittmann, F.H. (1976) On the Action of Capillary Pressure in Fresh Concrete. *Cement and Concrete Research,* 6, 49–56.

WRAP. Linking demolition and new build – a step by step guide - The efficient use of materials in regeneration projects, www.wrap.org.uk/construction,The Old Academy, Oxon, UK.

Wu, K.R., Chen, B., Yao, W., Zhang, D. (2001) Effect of Coarse Aggregate Type on Mechanical Properties of High-Performance Concrete. *Cement and Concrete Research,* 31(10), 1421–1425.

Wu, P., Xia, B., Zhao, X. (2014) The Importance of Use and End-of-Life Phases to the Life Cycle Greenhouse Gas (GHG) Emissions of Concrete – A Review. *Renewable and Sustainable Energy Reviews,* 37, 360–369.

Wyrzykowski, M., Trtik, P., Munch, B., Weiss, J., Vontobel, P., Lura, P. (2015) Plastic Shrinkage of Mortars with Shrinkage Reducing Admixture and Lightweight Aggregates Studied by Neutron Tomography. *Cement and Concrete Research,* 73, 238–245.

Xia, H., Bakker, M.C.M. (2014) Reliable Classification of Moving Waste Materials with LIBS in Concrete Recycling. *Talanta,* 120, 239–247.

Xiao, J., Falkner, H. (2007) Bond Behaviour between recycled aggregate concrete and steel rebars. *Construction and Building Materials,* 21(2), 395–401.

Xiao, J., Fan, Y., Tawana, M.M. (2013) Residual Compressive and Flexural Strength of a Recycled Aggregate Concrete Following Elevated Temperatures. *Berlin – Structural Concrete,* 14(2), 168–175.

Xiao, J., Lei, B., Zhang, C. (2012) On Carbonation Behavior of Recycled Aggregate Concrete. *Science China Technological Science,* 55(9), 2609–2616.

Xiao, J., Li, H., Yang, Z. (2013) Fatigue Behavior of Recycled Aggregate Concrete Under Compression and Bending Cyclic Loadings. *Construction and Building Materials,* 38, 681–688.

Xiao, J., Lu, D., Ying, J. (2013) Durability of Recycled Aggregate Concrete: An Overview. *Journal of Advanced Concrete Technology,* 11(12), 347–359.

Xiao, J., Sun, Y., Falkner, H. (2006) Seismic Performance of Frame Structures with Recycled Aggregate Concrete. *Engineering Structures,* 28(1), 1–8.

Xiao, J.Z., Ying, J.W., Tam, V.W.Y., Gilbert, I. (2014) Test and Prediction of Chloride Diffusion in Recycled Aggregate Concrete. *Science China Technological Science,* 57(12), 2357–2370.

Xiao, J.Z., Zhang, C.Z. (2007) Fire Damage and Residual Strengths of Recycled Aggregate Concrete. *Key Engineering Materials,* 348–349, 937–940.

Xiao, J.Zh., Li, J.B., Zhang, Ch. (2006) On Relationships between the Mechanical Properties of Recycled Aggregate Concrete: An Overview. *Materials and Structures,* 39, 655–664.

Xing, W., Hendriks, C. (2006) Decontamination of Granular Wastes by Mining Separation Techniques. *Journal of Cleaner Production,* 14, 748–753.

Xing, W., Pietersen, H., Hendriks, C., Rem, P. (2002) *Improve the Quality of Construction and Demolition Waste by Separation Techniques.* In Advances in Building Technology. Elsevier: Oxford, pp. 1439–1446.

Xing, Z., Beaucour, A.L., Hebert, R., Noumowe, A., Ledesert, B. (2015) Aggregate's Influence on Thermophysical Concrete Properties at Elevated Temperature. *Construction and Building Materials,* 95, 18–28.

XP P 18-420. (2012) Scaling test for hardened concrete surfaces exposed to frost in the presence of a salt solution, in French.

XP P 18-543. (2017) Aggregates - Petrographic analysis of aggregates applied to alkali-aggregate reactions, in French.

XP P 18-544. (2015) Aggregates—Determination of active lime water-soluble alkalis, in French.

XP P01-064/CN. (2014) Afnor: Contribution des ouvrages de construction au développement durable - Déclarations environnementales sur les produits - Règles régissant les catégories de produits de construction - Complément national à la NF EN 158044 (No. XP P01-064/CN).

XP P18-420 AFNOR. (May 2012) Béton - Essai d'écaillage des surfaces de béton durci exposées au gel en présence d'une solution saline. France.

XP P18-458 AFNOR. (2008) Tests for hardened concrete - Accelerated carbonation test - Measurement of the thickness of carbonated concrete.

XP P18-544. (2015) AFNOR, Aggregates - Determination of active lime water-soluble alkalis.

Yagishita, F., Sano, M., Yamada, M. (1994) Behavior of Reinforced Concrete Beams Containing Recycled Coarse Aggregate. *Demolition and Reuse of Concrete & Masonry RILEM Proceeding,* 23, 331–342, E&FN Spon, London.

Yang, J., Qiang, D., Yiwang, B. (2011) Concrete with Recycled Concrete Aggregate and Crushed Clay Bricks. *Construction and Building Materials,* 25(4), 1935–1945.

Yang, R., Lawrence, C.D., Sharp, J.H. (1996) Delayed Ettringite Formation in 4-year Old Cement Pastes. *Cement and Concrete Research,* 26(11), 1649–1659.

Yang, Z., Brown, H., Cheney, A. (2006) Influence of Moisture Conditions on Freeze and Thaw Durability of Portland Cement Pervious Concrete. In: *Concrete Technology Forum: Focus on Pervious Concrete. Citeseer.*

Yermak, N. (2015) Comportement à hautes températures des bétons additionnés de fibres, Thèse de l'université de Cergy-pontoise.

Yermak, N., Pliya, P., Beaucour, A.L., Simon, A., Noumowe, A. (2017) Influence of Steel and/or Polypropylene Fibres on the Behaviour of Concrete at High Temperature: Spalling, Transfer and Mechanical Properties. *Construction and Building Materials,* 132, 240–250.

Yildrin, S.T., Meyer, C., Herfellner, S. (2015) Effects of Internal Curing on Strength, Drying Shrinkage and Freeze-Thaw Resistance of Concrete Containing Recycled Concrete Aggregates. *Construction and Building Material,* 91, 288–296.

Yoda, K., Harada, M., Sakuramoto, F. *Field Application and Advantage of Concrete Recycled In-Situ Recycling Systems.* Thomas Telford Services: London, pp. 437–446, 2003.

Yun, H.D., You, Y.C., Lee, D.H. Effects of Replacement Ratio of Recycled Coarse Aggregates on the Shear Performance of Reinforced Concrete Beams Without Shear Reinforcement. *LHI Journal,* 2(4), 471–477.

Zaharieva, R., Buyle-Bodin, F., Wirquin, E. (2004) Frost Resistance of Recycled Aggregate Concrete. *Cement and Concrete Research,* 34, 1927–1932.

Zega, C.J., Di Maio, A.A. (2006) Recycled Concrete Exposed to High Temperatures. *Magazine and Concrete Research,* 58(10), 675–682.

Zega, C.J., Di Maio, A.A. (2009) Recycled Concrete Made with Different Natural Coarse Aggregates Exposed to High Temperature. *Construction and Building Materials,* 23, 2047–2052.

Zega, C.J., Villagrán-Zaccardi, Y.A., Maio, A.A.D. (2009) Effect of Natural Coarse Aggregate Type on the Physical and Mechanical Properties of Recycled Coarse Aggregates. *Materials and Structures,* 43, 195–202.

Zerbi, T., Landò, R., Vinai, R., Grigoriadis, K., Soutsos, M. (2017) Indexing and sorting robot based on hyperspectral and reflectance information for CDW recycling. *Proceedings of HISER International Conference,* 21, 22 and 23 June 2017.

Zhang, H., Zhao, Y. (2016) Performance of Recycled Concrete Beams Under Sustained Loads Coupled with Chloride Ion Ingress. *Construction and Building Materials,* 128, 96–107.

Zhang, J., Shi, C., Li, Y., Pan, X., Poon, C.S., Xie, Z. (2015) Influence of Carbonated Recycled Concrete Aggregate on Properties of Cement Mortar. *Construction and Building Materials,* 98, 1–7.

Zhao, Z. (2014) Re-use of fine recycled concrete aggregates for the manufacture of mortars. *Ph.D Thesis,* Ecole des Mines de Douai, France.

Zhao, Z., Damidot, D., Rémond, S., Courard, L. (2015) Toward the quantification of the cement paste content of fine recycled concrete aggregates by salicylic acid dissolution corrected by a theoretical approach. *14th International Congress on the Chemistry of Cement, Beijing, China.*

Zhao, Z., Kwon, S.H., Shah, S.P. (2008) Effect of Specimen Size on Fracture Energy and Softening Curve of Concrete: Part I. Experiments and Fracture Energy. *Cement and Concrete Research,* 38(8–9), 1049–1060.

Zhao, Z., Rémond, S., Damidot, D., Xu, W. (2013a) Influence of Hardened Cement Paste Content on the Water Absorption of Fine Recycled Concrete Aggregates. *Journal of Sustainable Cement-Based Materials,* 2(3–4), 186–203.

Zhao, Z., Remond, S., Damidot, D., Xu, W. (2015) Influence of Fine Recycled Concrete Aggregates on the Properties of Mortars. *Construction and Building Materials,* 81, 179–86.

Zhao, Z., Wang, S., Lu, L., Gong, C. (2013b) Evaluation of Pre-Coated Recycled Aggregate for Concrete and Mortar. *Construction and Building Materials,* 43, 191–196.

Zhou, J.H., He, H.J., Meng, X.H., Huan, S. (March 14–17, 2010) Experimental study of recycled concrete columns under large eccentric compression. *Proceedings of the 12th international conference on engineering, science, construction, and operations in challenging, environments.* doi: 10.1061/ 411096(336)54.

Zhuang, X.J. (2007) Experimental Investigation on Complete Stress-Strain Curve of Recycled Concrete Under Uniaxial Loading. *Journal of Tongji University,* 35(11), 1445–1449.

Zhutovsky, S., Kovler, K., Bentur, A. (2002) Efficiency of Lightweight Aggregates for Internal Curing of High Strength Concrete to Eliminate Autogenous Shrinkage. *Materilas and Structures,* 35, 97–101.

Zimbelman, R. (1985) A Contribution to the Problem of Cement Aggregate Bond. *Cement and Concrete Research,* 15, 801–808.

Zuo, J., Zillante, G., Wilson, L., Davidson, K., Pullen, S. (2012) Sustainability Policy of Construction Contractors: A Review. *Renewable and Sustainable Energy Reviews,* 16(6), 3910–3916.

Index

R

S

W